UTAH STRATIGRAPHIC CORRELATION CHART

Compiled by Bruce Tohill

GEOLOGY OF THE CORDILLERAN HINGELINE

photo by P. J. Varney

Mount Timpanogos from American Fork Canyon. Visible portion is basinal facies Permo-Pennsylvanian Oquirrh Group thrust over shelf sediments along the Charleston-Nebo thrust.

J. GILMORE HILL, Editor

ROCKY MOUNTAIN ASSOCIATION OF GEOLOGISTS
1615 CALIFORNIA STREET
DENVER, COLORADO 80202

1976

Copyright 1977 by the Rocky Mountain Association of Geologists. All rights reserved. This book or any part thereof may not be reproduced in any form without written permission of the Rocky Mountain Association of Geologists.

INTRODUCTION

The Board of Directors and membership of the Rocky Mountain Association of Geologists are pleased to present the 1976 Symposium and Field Trip. The 1976 program objective is to reveal another frontier area for your consideration. The Cordilleran Hingeline has much to offer the explorationist in its complex structure and stratigraphy.

A symposium of guidebook, speakers, and field trip is a rare opportunity for specialists from industry, government, and education to come together and share facts and opinions. We believe ideas will be stimulated and expanded by this joint contribution.

On behalf of the Board and membership of the Association, I thank the individuals responsible for giving to us of their time and talent.

R. A. Matuszczak, President
Rocky Mountain Association of Geologists

ACKNOWLEDGMENTS

Snowbird, Utah. Most of us didn't know where or what Snowbird was until asked to help put on this field trip and symposium. We quickly found out. In some ways it is like putting on a field trip on the moon due to the distance from Denver, but we hope it will be well worth the effort. The committee chairmen and their associates have worked long hours to put this show on, and are to be congratulated for their efforts.

Since we are in Utah we are especially thankful for the help provided by the Utah Geological Association, the Utah Geological and Mineral Survey, and various professors at Utah, Utah State, Brigham Young, and Weber State Universities as well as various members of the U.S. Geological Survey who work the symposium area.

Tom Odiorne, who is in charge of the field trip, Gil Hill, who edited the guidebook, and Dudley Bolyard who put together the symposium with their committee members have borne the brunt of the technical portion of the trip and are to be commended. Andy Alpha and his wife made a special trip to make the sketches found in the guidebook. Jack Rathbone took the time to take the excellent photographs. Marshall Crouch and Gary Nydegger have made repeated contacts with the Snowbird personnel to assure you of accommodations. Olin Isham and his committee raised the necessary funds from generous advertisers seen in the guidebook. Don Hembre has kept us reasonably solvent. Bill Jirikowic has handled publicity with a bit of help from Amoco and Chevron who discovered Ryckman Creek field after the field trip theme was selected. Herb Duey has arranged for the buses, and Ed Riker is making our stay as enjoyable as possible. A special thanks, also, to my wife, Faye, who has organized the women's entertainment, and Wanda Rey of the RMAG office for helping everybody get the job done.

I hope that the efforts made by the committees will make your stay in the hingeline enjoyable and informative.

My thanks to all who have helped.

Harry K. Veal
General Chairman

ROCKY MOUNTAIN ASSOCIATION OF GEOLOGISTS

OFFICERS 1976

President	Roger A. Matuszczak	Amoco Production Co.
President-Elect	Norman H. Foster	Filon Exploration Co.
First Vice President	David B. Mackenzie	Marathon Oil Co.
Second Vice President	John D. Pruit	Petroleum, Inc.
Secretary	C. Dennis Irwin	Consultant
Treasurer	Michael S. Johnson	Consultant
Councilor	John B. Ivey	Amuedo and Ivey
Past President	Raymond G. Marvin	Gower Oil Company

SYMPOSIUM COMMITTEE

General Chairman	Harry K. Veal	Independent
Finance Chairman	Donald R. Hembre	Lewis and Clark Exploration Co.
Housing Chairman	Marshall C. Crouch III	Consultant
Publicity Chairman	William F. Jirikowic	General American Oil Co.
Registration Chairman	Gary Nydegger	Kansas-Nebraska Natural Gas Co.
Social Chairman	Edward P. Riker	Amerada-Hess Corp.
Ladies' Activities	Faye Veal	
Symposium Meeting Chairman	Dudley W. Bolyard	Bolyard Oil and Gas, Ltd.
Transportation Chairman	Herbert D. Duey	Northwest Exploration Co.

SYMPOSIUM PUBLICATION COMMITTEE

Editor	J. Gilmore Hill	Peppard-Souders & Associates
Associate Editors	Peter E. Coffin	Anadarko Production Co.
	Albert P. Geyer	Beren Corp.
	Mark A. Guinan	Trend Exploration Ltd.
	W. Jerry Koch	Filon Exploration Co.
	Allan R. Larson	Mesa Petroleum Co.
	Richard O. Louden	Tenneco Oil Co.
	D. Keith Murray	Colorado Geological Survey
	Frank Royse, Jr.	Chevron Oil Co.
	Richard C. Schneider	Amoco Production Co.
	Peter J. Varney	Exeter Drilling and Exploration Co.
Editorial Assistants	H. T. Hemborg	Tenneco Oil Co.
	Lorna A. Porter	Peppard-Souders & Associates
	Billy G. Robertson	Louisiana Land & Exploration Co.
	Robert T. Ryder	U.S. Geological Survey

Advertising Chairman .... Olin L. Isham .... Champlin Petroleum Co.
  Committee Members: Robert D. Abbott, CIG Exploration; Robert Allen, Tricentrol U.S.; John R. Barwin, Mesa Petroleum Co.; Donald W. Beardsley, Chevron Oil Co.; Jerry N. Cox, Cities Service Oil Co.; Denny P. Furst, Texas Oil & Gas Corp.; Graig E. Gunter, Wichita Industries; Marvin A. Heany, Equity Oil Co.; L. Clark Kiser, Energetics, Inc.; James W. Mallin, Depco, Inc.; Billie J. McAlpine, Consultant; William F. Oline, Mountain Fuel Supply Co.; Harold W. Price, Phillips Petroleum Co.; Alan R. Ramer, Grayrock Corp.; Fred H. Reiter, Louisiana Land and Exploration Co.; H. Glenn Richards, Dixel Resources; Gerald L. Stone, Southland Royalty Co.; William Troxel, Schlumberger Well Services; William A. Wilson, Michigan-Wisconsin Pipeline Co.; Robert Zinke, Independent.

Photography	John H. Rathbone	Consultant
Roadlog Chairman	Howard H. "Tom" Odiorne	Crystal Exploration and Production Co.
Assistant Chairman	Michael W. Boyle	Crystal Exploration and Production Co.

  Committee Members: Sidney R. Ash, Weber State College; Wilson W. Bell, Odessa Natural Corp.; John C. Kirkpatrick, Aminoil USA, Inc.; James H. Madsen, Jr., University of Utah; Richard W. Moyle, Weber State College; Robert E. Near, Energy Reserves Group.

Sketches .... Andrew G. Alpha .... Mobil Oil Corp.

TABLE OF CONTENTS

Correlation Chart and Index Map	inside front cover
Title page	1
Introduction	2
Acknowledgements	2
Rocky Mountain Association of Geologists, 1976 Officers; Symposium Committee; and Symposium Publication Committee	3
Table of Contents	4, 5
Editor's Introduction	6
List of Speakers, 1976 Symposium	7
Index to Advertisers	8, 9
Historical Sketches	26, 172, 192, 218, 324

STRUCTURAL GEOLOGY

"What is the Wasatch Line?," by Wm. Lee Stokes	11
"Tectonic Evolution in Utah's Miogeosyncline-Shelf Boundary Zone," by Gary W. Crosby	27
"Structural Evolution of Central Utah-Late Permian to Recent," by James L. Baer	37
"A Paleostructural Interpretation of the Eastern Great Basin Portion of the Basin and Range Province, Nevada and Utah," by E. L. Howard	47
"A Discussion of the Geology of the Southeastern Canadian Cordillera and its Comparison to the Idaho-Wyoming-Utah Fold and Thrust Belt," by K. H. Hayes	59

STRATIGRAPHY AND PETROLEUM POTENTIAL

"Stratigraphy of Younger Precambrian Rocks along the Cordilleran Hingeline, Utah and Southern Idaho," by Lee A. Woodward	83
"Depositional Environment of the Mineral Fork Formation (Precambrian), Wasatch Mountains, Utah," by Peter J. Varney	91
"The Cambrian Section of the Central Wasatch Mountains," by Christina Lochman-Balk	103
"Ordovician Sedimentation in the Western United States," by Reuben J. Ross, Jr.	109
"Mississippian Carbonate Shelf Margins, Western United States," by Peter R. Rose	135
"Relationships of Pennsylvanian-Permian Stratigraphy to the Late Mesozoic Thrust Belt in the Eastern Great Basin, Utah and Nevada," by John E. Welsh	153
"Stratigraphy and Petroleum Potential of the Permian Kaibab Beta Member, East-Central Utah," by L. Clark Kiser	161

"Permian Phosphoria Carbonate Banks, Idaho-Wyoming Thrust Belt,"
  by Marvin D. Brittenham 173

"Permian and Lower Triassic Reservoir Rocks of Central Utah," by C. Dennis Irwin 193

"Lower Triassic Facies in the Vicinity of the Cordilleran Hingeline: Western Wyoming,
  Southeastern Idaho, and Utah," by W. Jerry Koch 203

"Lower Mesozoic and Upper Paleozoic Petroleum Potential of the Hingeline Area,
  Central Utah, (Reprint)," by Floyd C. Moulton 219

"Petrology of Entrada Sandstone (Jurassic), Northeastern Utah," by E. P. Otto and
  M. Dane Picard 231

"Regional Stratigraphy and Depositional Environments of the Glen Canyon Group and
  Carmel Formation (San Rafael Group)," by William E. Freeman 247

"Jurassic Salts of the Hingeline Area, Southern Rocky Mountains," by Alan R. Hansen 261

"Stratigraphy, Sedimentology, and Petroleum Potential of Dakota Formation,
  Northeastern Utah," by R. L. Vaughn and M. Dane Picard 267

"Tertiary Tectonics and Sedimentary Rocks along the Transition, Basin and Range Province
  to Plateau and Thrust Belt Province, Utah," by Robert E. McDonald 281

"Oil Shows in Significant Test Wells of the Cordilleran Hingeline," by Harry K. Veal 319

SELECTED OIL FIELDS

"Reservoir Variations at Upper Valley Field, Garfield County, Utah,"
  by George C. Sharp 325

"The Geology of the Pineview Field Area, Summit County, Utah," by P. D. Maher 345

GEOTHERMAL RESOURCES

"Geothermal Energy, Cordilleran Hingeline-West," by B. Greider 351

AREAL GEOLOGY

"Stratigraphic and Structural Setting of the Cottonwood Area, Utah,"
  by Max D. Crittenden, Jr. 363

"Geology of the Coalville Anticline, Summit County, Utah," by Lyle A. Hale 381

FIELD TRIP ROAD LOG

"Hingeline Sediments of the Overthrust Belt," by Howard H. "Tom" Odiorne,
  Chairman and 1976 Road Log Committee 387

EDITOR'S COMMENTS

The Cordilleran Hingeline can be defined as the zone of westward downwarping or flexure of the basement rocks required to accommodate the thick, late Precambrian to Mesozoic-age stratigraphic section of the Cordilleran geosyncline. The Hingeline forms a transition zone from the geosyncline onto the western cratonic shelf, thus it is an essential element of the eastern margin of the Cordilleran geosyncline and one of the fundamental tectonic features of western North America.

To my knowledge, "Cordilleran Hingeline" has never been formally proposed or defined in the geologic literature. The term apparently originated with petroleum geologists and initially was applied to a geographically limited portion of the "Hingeline" in central Utah extending from about Nephi southward to Richfield. A geographic bias for applying the term Cordilleran Hingeline primarily in Utah still exists, thus most of the papers in this guidebook concern Utah, and to some extent, immediately adjacent states.

The Cordilleran Hingeline, since its apparent inception in late Precambrian time, has had a persistent although varied effect throughout the subsequent geologic history of the region. It has served as a locus for important facies changes, erosional truncations, and depositional wedge-edges. In addition, at least until tectonic destruction of the Cordilleran geosyncline during Mesozoic time, it probably served as a focus for updip migration of hydrocarbons from source rocks within the geosyncline. The Mesozoic and Cenozoic tectonism of the area has also been controlled, at least partially, by the position of the Cordilleran Hingeline.

The mineral wealth along the Hingeline has been long recognized and exploited, primarily by production of metallic and non-metallic minerals, energy minerals, and coal. The area's potential for significant hydrocarbon reserves also has been long recognized; but geologic complexity and high exploration costs have until recently greatly inhibited hydrocarbon exploration. The discovery of Pineview field in the Utah overthrust belt precipitated the most extensive exploration effort in the history of the Cordilleran Hingeline with significant activity extending from Montana to central Utah. The subsequent discoveries of Rykman Creek and Yellow Creek fields in Wyoming have confirmed the hydrocarbon potential of the area and the Cordilleran Hingeline and the associated overthrust belt promise to be one of the most active and potentially rewarding exploration frontiers in the United States.

The purpose of the symposium is to bring together in a single source a collection of papers on the structural and stratigraphic geology and hydrocarbon potential along and adjacent to the Cordilleran Hingeline. Additional papers discuss the geology of specific areas, existing significant oil fields, and the geothermal potential of the area. These papers both synthesize much of the existing data and present new ideas and interpretations.

As will be seen from the various papers, many geologic problems still remain to be adequately defined and solved. Additionally, alternative interpretations of existing data are possible, and in many cases no "correct" or final interpretation has yet evolved. Such an area may be overwhelmingly intimidating to the faint-of-heart, but for the imaginative and courageous explorationist it offers the opportunity for new and innovative interpretations that can contribute to the discovery of significant hydrocarbon reserves.

The value of any guidebook lies almost solely in the quality of papers it contains. On behalf of the R.M.A.G., I express my highest appreciation for the many fine papers submitted. Many papers were prepared at the cost of considerable personal time and expense to the authors and as geologists we all have a debt of gratitude for their willingness to share their research and ideas. I also wish to acknowledge our appreciation to those companies which allowed the release of valuable proprietary data in the preparation of the papers.

Most of the credit for preparing this symposium belongs to the Associate Editors and Editorial Assistants. Without their willingness to contribute a maximum effort whenever needed, it would have been impossible to compile this guidebook. However, I as editor, am solely responsible for any omissions, inconsistencies, and errors that may occur. We are also indebted to Amoco Production Company and Phillips Petroleum Company for graciously providing draftsmen to prepare some of the illustrations.

Last, but far from least, I wish to thank Peppard-Souders and Associates for their encouragement, support, and willingness to allow me to serve as editor. Without their contribution, it would have been impossible for me to have devoted the necessary time to this guidebook.

J. Gilmore Hill
Editor

LIST OF SPEAKERS AND TITLES, 1976 SYMPOSIUM

1. Gregg, C. Clare (American Quasar Petroleum Company): "Overthrust Geology From the Air"
2. Oriel, Steven S. (U.S. Geological Survey): "Oil in the Idaho-Wyoming Thrust Belt"
3. Crittenden, Max D., Jr. (U.S. Geological Survey): "Progress and Problems in Understanding Thrusts in Northern Utah"
4. Royse, F., Jr., Warner, M. A., and Reese, D. L. (Chevron Oil Company): "Thrust Belt Structural Geometry and Related Stratigraphic Problems, Wyoming-Idaho-Northern Utah"
5. Clement, James H. (Shell Oil Company): "Geological-Geophysical Illustrations of Structural Interpretations in Rocky Mountain Overthrust Belts and Basement 'Block' Faulted Terranes"
6. Sandberg, Charles A. (U.S. Geological Survey): "Petroleum Geology of Devonian and Mississippian Rocks of Cordilleran Miogeosyncline, Western United States"
7. Rose, Peter R. (Energy Reserves Group): "Mississippian Carbonate Shelf Margins, Western United States"
8. Marcantel, Jonathan B. (Shell Oil Company): "Permo-Pennsylvanian of Eastern Nevada"
9. Hansen, Alan R. (Consultant): "Jurassic Salts of the Hingeline Area, Southern Rocky Mountains"
10. Furer, Lloyd C. (Amoco Production Company): "Petrology and Stratigraphy of the Non-Marine Upper Jurassic-Lower Cretaceous Rocks of Western Wyoming and Southeastern Idaho"
11. Maher, Patrick D. (Energetics, Inc.): "Geology of the Pineview Field Area, Summit County, Utah"
12. Dunnewald, John B. (Belco Petroleum Corporation) and Gorton, Kenneth A. (Consultant): "Nugget Oil Accumulations at Dry Piney, Tip Top, and Hogsback Fields, Sublette County, Wyoming"
13. Newman, Gary W. (American Quasar Petroleum Company): "Eocene Structuring in the Eastern Basin and Range"

INDEX TO ADVERTISERS

Ackman, Edward J.	420
Allison Drilling Company	428
American Mud Company	415
American Stratigraphic Company	423
Amoco Production Company	418
Anderson Drilling Company	425
Arapahoe Petroleum, Inc.	416
Argonaut Oil & Gas Company	427
Ashland Exploration Company	416
Bass, Perry R., and Bass Enterprises Production Company	425
Bayles Laboratories	427
Beren Corporation	427
Birdwell Division, Seismograph Service Corporation	416
Boshard, J. R.	415
Brinkerhoff Drilling Company, Inc.	428
Brown Palace Hotel	430
Buckley Powder Company	422
Compagnie Generale De Geophysique	411
Chambers, Jerry — Oil Producer	419
Chaparral Resources, Inc.	420
Chevron Oil Company	429
CIG Exploration, Inc.	419
Cities Service Company	432
Continental Laboratories, Inc.	425
Core Laboratories, Inc.	423
Davis Oil Company	429
Depco, Inc.	415
Diamond Shamrock Oil & Gas Company	415
Dresser Atlas Division, Dresser Industries, Inc.	424
Dunlap, William H.	426
Edcon Exploration Data Consultants, Inc.	423
Edwards Oil Properties, Inc.	412
Energetics, Inc.	415
Energy Consulting Associates	426
Energy Reserves Group, Inc.	415
Equity Oil Company	422
Exeter Exploration Company	429
Filon Exploration Corporation	422
Geophoto Services, Inc.	419
Gear Drilling Company	416
Graham, A. Thomas	426
Griffith, Earl G.	418
Grayrock Corporation	423
Haskins-Pfeiffer-Owings, Inc.	425
Helton Engineering & Geological Services, Inc.	428
Industrial Gas Services, Inc.	418

Jensen-Mark Corporation	410
Kimbark Operating Company	418
Ladd Petroleum Corporation	414
Long Company Technical Services	410
Louisiana Land and Exploration Company	429
McAlpine, Billie J.	414
Milchem, Inc.	419
Monaco Engineering, Inc.	418
Mountain Fuel Supply Company	412
Mountain Petroleum Ltd.	424
North American Exploration Company, Inc.	414
Northwest Exploration Company	430
Oil Well Perforators, Inc.	417
Owens, Willard Associates, Inc.	410
Pacific West Exploration Company	425
Peerless Printing Company	431
Pennzoil Company	424
Peppard-Souders & Associates	409
Petroleum Geophysical Company	414
Petroleum, Inc.	420
Petroleum Information Corporation	421
Phillips Petroleum Company	430
Petroleum Supervision and Management, Inc.	424
Rainbow Resources, Inc.	412
Ross, Richard B.	424
Seismograph Service Corporation	413
Shell Oil Company	427
Signal Drilling Company, Inc.	412
Skelly Oil Company	428
Snowbird	427
Southland Royalty Company	428
Speedy Copy Service, Inc.	414
Terra Resources, Inc.	426
Toltek Drilling Company	418
Tooke Engineering	412
TransOcean Oil, Inc.	424
Trollinger Geological Associates, Inc.	420
Veal, Harry (Contribution)	
Veezay Geoservice Inc.	410
Walter Duncan Oil Properties	414
W. A. Wahler & Associates	412
Webb Resources, Inc.	422
Western Well Logging, Inc.	430
Woods Petroleum Corporation	426
Zinke, Robert	427

WHAT IS THE WASATCH LINE?

by

Wm. Lee Stokes[1]

INTRODUCTION

In his presidential address to the Geological Society of America delivered in 1950 on the subject, "Tectonic Theory Viewed from the Basin Ranges", Chester A. Longwell said: "We must be reconciled not to know the basic answers, while we devote our interest and our energies to finding out exactly what it is to be explained." (Longwell, p. 432). Much of what he had to say has a bearing on the Wasatch Line, even though that feature had not yet been named. The Wasatch Line was so designated in print by Marshall Kay in the following words: "The flexure that defined the miogeosyncline in the Cambrian (Wasatch Line) continued as the site of differential movement through the Paleozoic and earlier Mesozoic. Systems from the Ordovician through the Jurassic are generally more fully represented and thicker in the same areas in which the Lower Cambrian is present and the whole Cambrian thicker." (Kay, 1951, p. 14).

To reinforce and clarify his definition, Kay presented a diagrammatic map (his plate 3, reproduced here as Fig. 1) on which the 2,000 ft isopach for Cambrian rocks is designated parenthetically as the Wasatch Line. If this definition is to be taken as binding, the Wasatch Line extends from well into northwestern Mexico, across the entire western United States, and into Canada as far north as latitude 56°N. This same line is repeated on several other maps in the same publication, thus Kay's plate 3 shows the Wasatch Line as being the zero edge of Lower Cambrian (left map), and as roughly dividing Cambrian more than 5,000 ft thick on the west from that of lesser thickness to the east. Also, according to Kay, the Wasatch Line runs near the eastern edge of the Lower Ordovician (left map, Kay's plate 4), parallel to, but some distance from the eastern edge of the combined Upper Ordovician and Silurian (middle map, Kay's plate 4) and roughly divides Devonian sediments 2,500 to 5,000 ft thick on the west from lesser thicknesses to the east (right map, Kay's plate 4). No attempt is made to show map patterns of subsequent systems but cross-sections of the Mississippian, Pennsylvanian, and Permian are included in Kay's figure 2, (Restored Sections of Paleozoic in Miogeosynclines, Nevada to Wyoming, p. 13).

In dealing with definitions such as this, the concept of the person who originally proposed the term must bear considerable weight. It is a credit to Marshall Kay's genius that he recog-

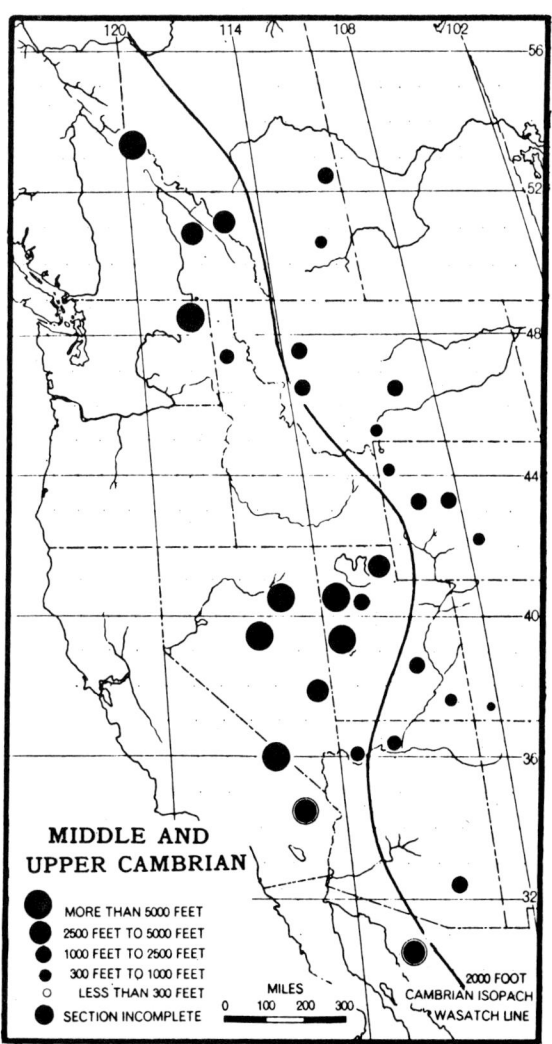

Fig. 1 — Wasatch Line according to original definition by Marshall Kay.

nized the Wasatch Line as being a major feature of North American geology. In his original definition he stressed not only its antiquity, but also its persistent effects over an immense time period. But if we were reading only Kay's definition we might conclude that its influence came to an end in the Jurassic, and that it had no effects in Nevada, for instance. These limitations are difficult to accept. Call it what we will, the geologic effects along the Wasatch Line became more intense and far-reaching with the passage of time. In fact, late Mesozoic and Cenozoic tectonism certainly has greater variety and importance than that of the Paleozoic. The

---

[1] Dept. of Geology and Geophysics, University of Utah. Salt Lake City, Utah.

present-day evidences of discontinuity along the Line are possibly greater than at any previous time and no geologist would hesitate to affirm a fundamental connection of some sort between past and present tectonics.

It is the purpose of this paper to discuss some of the facts and factors that might be taken into account in filling Longwell's admonition to find out "exactly what is to be explained."

THE LAS VEGAS LINE

A second and important step forward in understanding the Wasatch Line came through studies of John E. Welsh that are contained chiefly in an unpublished report to Shell Oil Company, "Paleozoic Hinge Line of Southern Nevada", submitted in 1955, and also in Welch's doctoral dissertation "Biostratigraphy of the Pennsylvanian and Permian Systems in Southern Nevada", submitted to the Department of Geology, University of Utah in 1959, and on file in the Department. The essence of Welsh's conclusions as summarized in his dissertation are:

"The evidence from this study indicates that the Wasatch Tectonic hinge line and the Las Vegas Tectonic hinge line form a continuous fault system. Furthermore, this structural trend connects directly with the Garlock fault system of southern California.

The implications of this discovery by the writer . . . that the Garlock lateral fault trend may be traced by stratigraphic means across southern Nevada and connected with the Wasatch Line, are that the stresses that produced these major tectonic features were operative as a single mechanical system.

The stratigraphic evidence and tectonic patterns of the Paleozoic systems suggest that some of the crustal blocks of the western United States were active as early as the Cambrian time. Movement along these inherited structural lineaments recurred throughout geologic history." (Welsh, 1959, p. 88). See Figure 2.

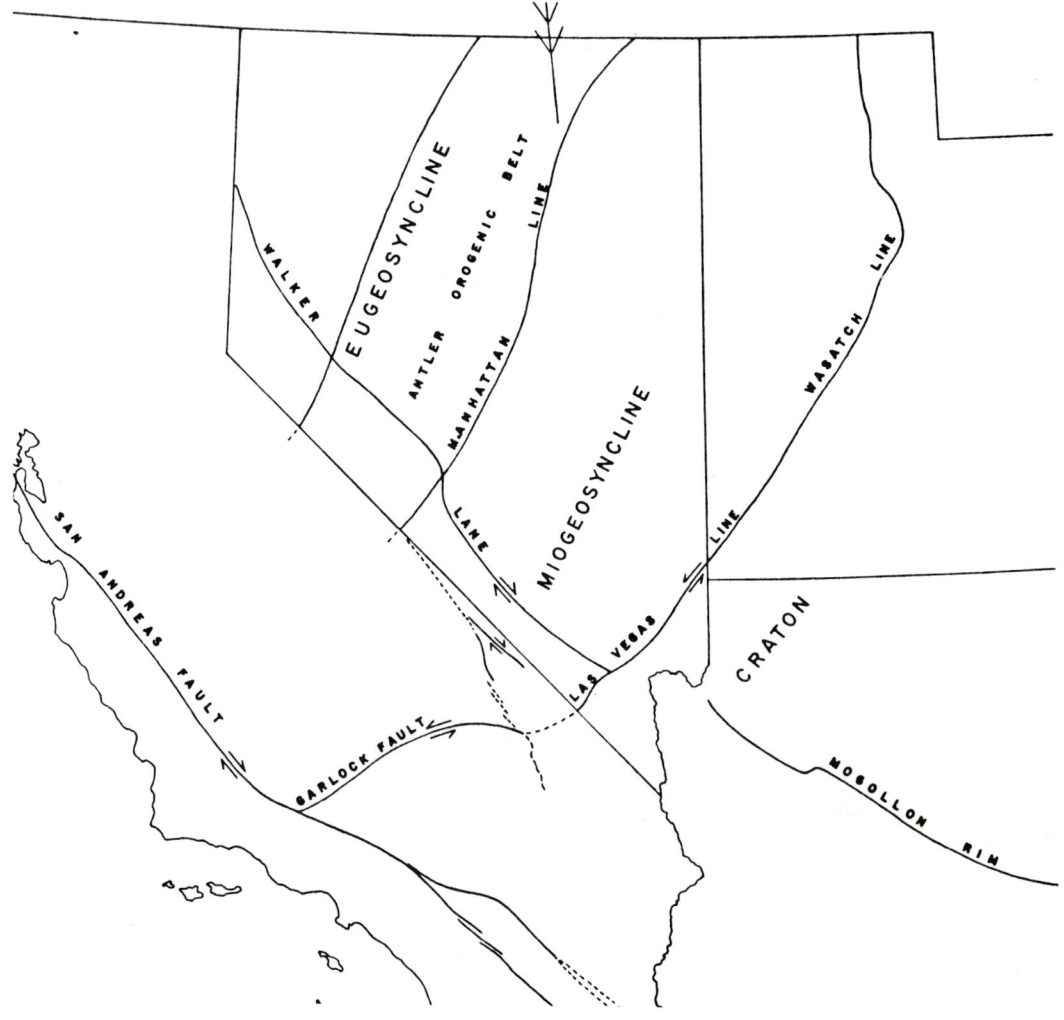

Fig. 2 — Las Vegas Line and associated tectonic features, from John E. Welsh, with permission.

OTHER RELATED FEATURES

It is difficult to discuss the Wasatch Line, or any other tectonic feature for that matter, without reference to contiguous elements (Fig. 3). Unfortunately, the naming of tectonic and structural features is a very haphazard process and only through the very uncertain route of popular usage do the bounds and meanings of such features become established.

The Wasatch Line may be thought of in terms of its original definition by Kay as marking the eastern boundary of the Cordilleran geosyncline. There is no disagreement about the name of this great primary geosyncline of North America, but its inner and outer parts have received different designations. Kay consistently referred to the inner, or miogeosyncline, as the Millard Belt. To Eardly, this is the Rocky Mountain geosyncline, and to Gilluly it is the Eastern Assemblage. The great interior east of the Wasatch Line is referred to generally as a shelf. It may be the Utah-Wyoming shelf if the Uinta Range is ignored. If the Uinta Range is regarded as a valid boundary, the area to the south of it is usually called the Colorado Plateau, that to the north, the Wyoming shelf.

When branches of the Wasatch Line southward across the Grand Canyon or southwestward into the Mojave Desert are considered, new problems of terminology arise. In the thinking of some, a vaguely defined area of southern California and much of southwestern Nevada is Mojavia, an ancient somewhat stable landmass that is the southwest termination of the Transcontinental arch. This may be a good name, but the boundaries are vague. It is suggested here that the Las Vegas-Garlock lineation be considered a northern or northwestern boundary for this tectonic unit.

The Transcontinental Arch

Numerous geologic elements enter into the definition of the Transcontinental arch or "continental backbone" that extends northeasterly entirely across North America. Little is found in geologic writings regarding the significance of this ancient feature in relation to the Cordilleran geosyncline, or the Wasatch Line, or any of the younger orogenies that have shaped the western United States. Yet its influence has been profound. In the first place, its origin as an orogenic belt resulted in a local thickening of the crust so that after about 1,500 m.y. ago, it proved to be more resistant to all types of deformation than belts on either side. All reconstructions of Precambrian North America show this southwestward projection of older rocks and generally refer to it as part of the Churchill Province (Condie, 1969).

It will be observed that the Las Vegas Line coincides fairly well with the northwest border of the Transcontinental arch across southwest Utah and Nevada. This is expectable, as ancient cratons tend to become shelves or shields with geosynclines or basin deposits on their flanks. Through a great deal of later time, the same ancient feature continued to react in a positive way while territory to the northwest was distinctly negative. Greatest thicknesses of late Precambrian and Cambrian rocks occur in south-central Nevada, and the basins containing them clearly conform in trend to the nearby Las Vegas Line. Although the evidence is in part of a negative nature, there is little justification for aligning the continental margin or successive depocenters southward across the Basin and Range section of Arizona into Mexico, parallel to the southern stretch of the Wasatch Line as drawn by Kay. Features that were determined by the southern reaches of the Transcontinental arch are therefore the Paleozoic-early Mesozoic shore-lines of shelf-basin transitions, or Las Vegas Line, the Southern Great Basin volcanic field, and the southwestern branch of the Intermountain seismic belt. Features that cut into or across the Transcontinental arch are the belt of great normal faults typified by the Hurricane fault, the Mogollon Rim or highland, the Arizona sag which was not apparent until the Devonian, and the south branch of the Intermountain seismic belt.

The Mogollon Rim

A somewhat more definite tectonic feature, in many ways resembling the Wasatch Line, is the Mogollon Rim (or highland) which forms a northwest-trending backbone of Arizona. It is merely a quirk of nomenclature that the Utah feature came to be called a line, while the Arizona counterpart became a rim or highland. Many of the same contrasting geologic characteristics that are listed for opposite sides of the Wasatch Line in Utah could be repeated for the Mogollon Rim in Arizona. Any thinking about one might, with profit, be applied to the other, of course, with due attention paid to both comparisons and constrasts.

The Mogollon transition zone, and the Wasatch transition zone, if we may put them on a comparative basis by using a unifying designation, trend toward each other, but do not join or mutually affect each other in an obvious way. The triangular area between them, through which the Colorado River flows, has been termed the Grand Canyon bight in allusion to the fact that it was an indentation or bend in the Cretaceous and Jurassic coast lines. It had other significant features that will become apparent later.

Igneous Geology

Igneous geology had nothing to do with deliniating the original Wasatch Line, and igneous rocks are not directly associated with it except in a few areas. The association of igneous effects with the Line may be merely coincidental or it may have tectonic significance. Evidence either way is not yet conclusive.

One of the greatest and most diverse igneous piles of the western United States centers near Marysvale, Piute County,

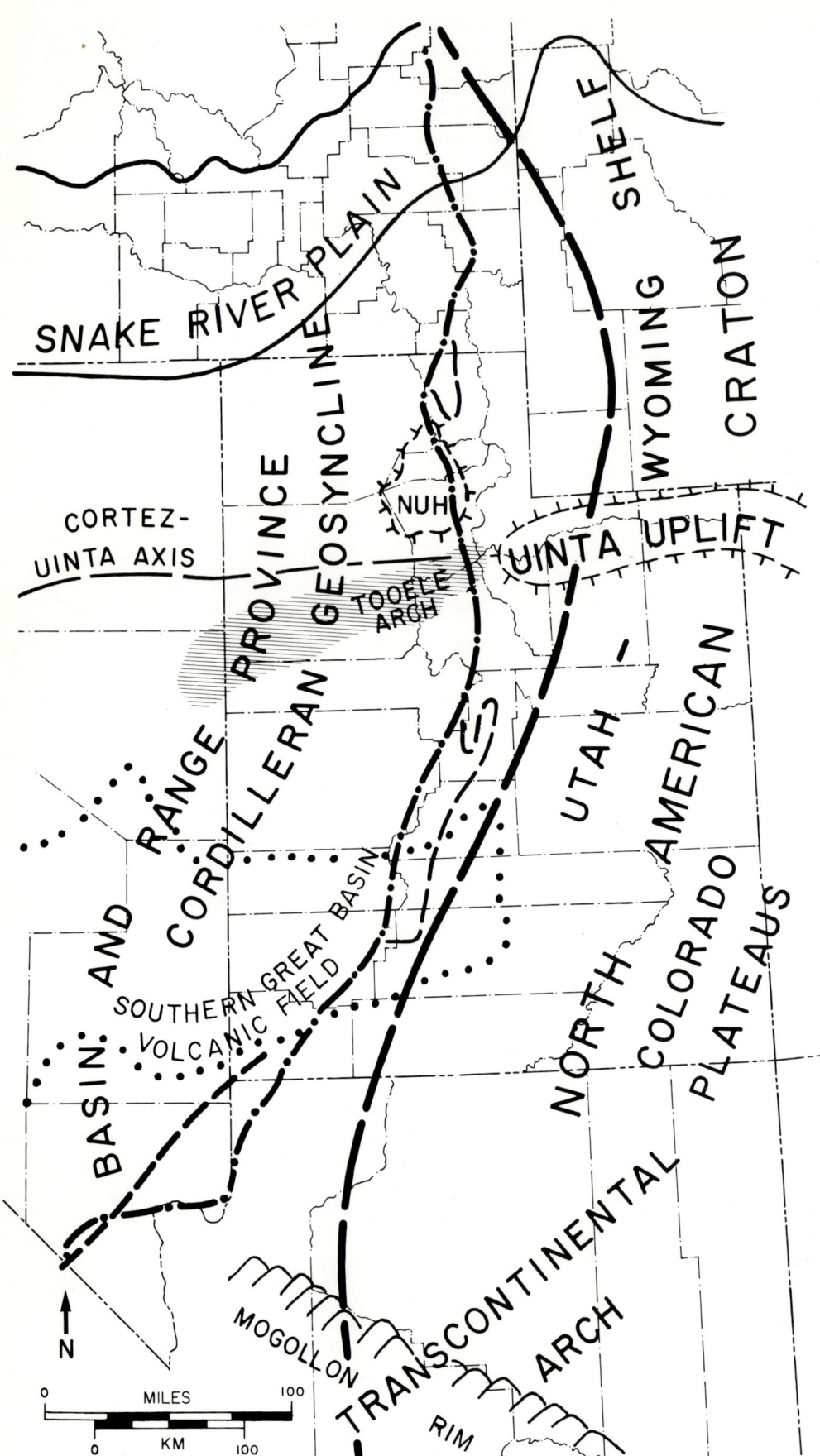

Fig. 3 — Tectonic lines and other features of the Great Basin-Colorado Plateaus transition. Heavy long dash, Wasatch Line of Kay; short dash, Las Vegas Line of Welsh; dot-dash, eastern boundary Great Basin; dots, Southern Great Basin volcanic field, NUH, Northern Utah highland of Eardley. All other features are labeled. Cortez-Uinta axis from Roberts, et al., 1965, p 1928; Tooele arch from various authors, Utah Geological and Mineralogical Survey Bull. 54a.

Utah. This was a center of acid and intermediate volcanic action chiefly between 26 m.y. and 20 m.y. ago. According to Rowly, et al, (1975), volcanic products inundated an area nearly 200 km in diameter to a thickness of at least 3000 m. The Marysvale pile may or may not be localized by the Wasatch Line. For one thing, it is the eastern end of the so-called Wah Wah-Tushar mineral belt which trends N 82° E. The mineral deposits along this line cannot be correlated with any confidence with either the Las Vegas Line or the Wasatch Line. The localization of the Marysvale volcanic center at the point where the Wasatch Line branches probably has tectonic significance. One hesitates to call this a triple junction, for it certainly is not one in the plate-tectonics sense. Nevertheless, three important lines of disturbance converge here. The northward branch is the main Wasatch Line, the southwestward branch is the Las Vegas Line, the southward branch is Kay's Wasatch Line. As a tentative hypothesis, it may be that the weakening of the crust by distension or pulling apart here may have permitted the exit of magmatic products at certain favorable times.

Farther north along the Wasatch Line, there seems to be no particularly strong concentration of igneous rock. The extrusive-intrusive field of the Tintic district lies near the line, but again, it is clearly a part of the well-marked Deep Creek-Tintic mineral belt, trending N 82° W. Still farther north, near Salt Lake City, the Uinta axis crosses the Wasatch Line with unmistakable east-west igneous effects, best illustrated by the alignment of the Bingham-Last Chance-Big Cottonwood-Clayton Peak-Park City stocks. This is the expression of the Oquirrh-Uinta mineral belt, trending N 73° E.

Still farther north, the Wasatch Line intersects the eastern edge of the Snake River plain, the most obvious cross-cutting igneous feature of North America. It approaches the Wasatch Line at an acute angle along the Idaho-Wyoming border. If any valid generalizations can be made about the Wastch Line and the igneous belts, it is that the igneous belts end near, or at the Line, and where they terminate, the effects are usually greatly intensified. Thus the Marysvale, Tintic, Cottonwood-Park City, and Absaroka-Yellowstone areas must be counted as major volcanic manifestatons by any standard.

A better case can be made for some sort of causal relationship between the Las Vegas Line (or branch) and the localization of Tertiary volcanic activity. The great Southern Great Basin volcanic or ignimbrite field clearly lies upon, or at least near, and parallel with the Las Vegas Line across southwestern Utah and southern Nevada. The boundaries and shape of this great field are drawn in different ways by different workers, but there is general agreement as to the overall northeasterly trend and curving crescent shape. The field is noted for massive extrusions of tuffaceous material, much of which is ignimbrite. Because the various units have spread widely, it has been difficult to localize the vents or fissures from which they emerged. Speculation tends to place the vents in the area of thicker accumulations, that is generally southward rather than northward, and hence, close to where a projection of the Las Vegas branch might pass.

PERTURBATIONS OF THE WASATCH LINE

It is easy to draw the Wasatch Line in one sweeping arc from north to south and to visualize it as having similar characteristics from one end to the other. This tends to obscure the very great differences that appear along various stretches at the present time. These differences may tell more about the history and meaning of the Line than the similarities. Several of the most obvious geologic features that seem to have influenced, or been influenced by the Wasatch Line, are the Uinta axis or line, the Utah-Wyoming thrust belt, and the Marysvale bifurcation. This by no means exhausts the possible topics.

The Uinta Line or Axis

No matter on what basis the Wasatch Line is drawn across northern Utah, it cannot be a smooth curve where it crosses the Uinta Range or its westward projection. The Uinta Range represents an aulocogen filled with sediments mostly over 1000 m.y. old belonging to a tectonic configuration that predates the earliest manifestation of the Wasatch Line. Apparently, an east-west rift cutting deeply into the North American continent received the thick deposit of clastic sediments that was to become the Uinta Range less than 60 m.y. ago. During about 1000 m.y., the rift and its contained sediments were in a state of comparative quiescence, but constituted a zone of relative weakness along which there were periods of slight subsidence or uplift. Until the final strong compression that formed the present range, reactions were mild, and the site was either an embayment that received somewhat thicker sediment, or a low shoal or peninsula with correspondingly thinner deposits (see isopach maps in Hansen, 1965).

Another peculiar structure of north-central Utah is the Northern Utah highland recognized and mapped by Eardley on the basis of studies of the Precambrian rocks of the region (Eardley, 1939). The most extensive remnants of the Northern Utah highland are in the Farmington Mountain horst and Antelope Island. These well-known features are north-trending uplifts cut from the older positive area of uncertain shape and size.

One thing is certain, the Northern Utah highland does not lie on a projection of the Uinta aulocogen, and although it may belong to the same tectonic family, it is difficult to relate the two in subsequent reactions. They did combine as a unit in one historical episode, and that was in their reactions to the thrust faulting of the Sevier orogeny. Traces of the thrust swing westward under Great Salt Lake in a great arc to avoid the combined Uinta projection and the Northern Utah highland (Crittenden, 1976). Whereas the thrust sheets traveled well onto the margins of the Uinta basin to the south and the Green River basin to the north, they were evidently held back or forced upward so as to be destroyed by erosion in the central Wasatch Range. A buttressing effect is plainly evident.

By contrast to the reactions to thrust faulting, the same area seems to have had no resistance to the later episode of normal faulting. The Wasatch fault zone transects the projection of the Uinta Range and in the central Wasatch cuts indiscriminately and cleanly through rocks of almost every type and age, including the 18,000 ft thick, predominantly quartzite, Cottonwood Canyon Formation and great thicknesses of the gneissic Farmington Complex.

The configuration of this section during the early Paleozoic may never be known from lack of evidence. The Cambrian is thin from both non-deposition and erosion, Ordovician and Silurian are lacking, and Devonian is patchy in the central Wasatch. What is found on Antelope Island leads to the possibility that the summit of the Northern Utah highland may have received no sediment until Early Mississippian time (Larsen, 1954). This is certainly true of parts of the western Uinta Range where Madison Limestone rests directly on Precambrian quartzite (Williams, 1953). Apparently, any isopach lines drawn on the basis of Paleozoic rocks would have to swing westward around the Uinta extension along somewhat the same paths as the later thrust faults. Such a bend can be drawn on the basis of criteria set up by Kay who showed no compelling data for drawing the Line as straight or as far east as he does.

Nevada-Utah Thrust Belt

A number of major thrust faults cut the surface along a 200 mile-wide belt that coincides with the Wasatch Line from southern Nevada into Alberta. Much work by many geologists, generally in extremely difficult terrane, has been done merely to map the traces of the component faults of this great system, and no comprehensive tectonic analysis of it has been offered. What is briefly mentioned here deals mainly with the part of the system crossing Utah and contiguous portions of Nevada and Wyoming. The most comprehensive survey of this section available is that of R. L. Armstrong (1968).

Armstrong divides the part of the thrust belt in Nevada and Utah into four sectors: southern Nevada-southwest Utah sector, Wah Wah-Canyon Range sector, Nebo-Charleston sector, and northern Utah sector. An additional division, the Great Salt Lake sector between the Nebo-Charleston and northern Utah sectors would seem to be justified. The number of major thrust-fault traces that cut the surface along the system as mapped by Armstrong (his Fig. 3) ranges from one to five, for most of the distance there are two or three.

Although details vary and are complex, this fault system can be related without too much difficulty to previous tectonic configurations. All cross sections show the faults rising upward from the top of the crystalline basement, or deep in the quartzite section through an eastward-thinning transition zone to cut the surface at or near the margin of the shelf. That the fault planes have followed certain favorable stratigraphic horizons that converge and originally were rising eastward seems clear enough. It is in connection with the Uinta axis that the greatest disruptions of the fault belt occur. Were it not for the peculiar westward detour that the faults make in passing around the combined Uinta-Northern Utah highland structural knot in north-central Utah, it seems likely that the entire system would have a smooth arc-like sweep from the west-central Wyoming border to central Utah. As it is, the faults behave as though they were controlled largely by the configuration of the Precambrian basement — they pressed eastward into the depressions of the Uinta and Green River basins and were retarded by the Teton and Uinta-Northern Utah highland positive areas. To the south, the trend is practically straight and the fault planes emerge as though they were being directed upward along the front of a block with a fairly straight trend and a fairly level threshold or rim.

No matter what the mechanism of thrusting may have been, the map relations of the thrusts show that the belt has a trend very near that of the Wasatch Line as this would be modified by the westward projecting Uinta axis, and the Las Vegas branch. In other words, the thrust belt follows the Paleozoic pattern, rather than the present one.

The Marysvale Bifurcation

A definite branching or splitting of the Wasatch Line takes place at the Marysvale volcanic center. One branch, which is the Las Vegas Line of Welsh, curves southwestward and leaves Utah slightly north of the southwest corner. The other continues southward into eastern Arizona about 50 mi east of the corner. Certain characteristics that combine and coincide in the Wasatch Line north of the Marysvale volcanic center no longer occur together on either of the branches. Certain effects are more pronounced on one branch, others on the other branch. Only a few are shared. The great normal faults continue on a northerly course, there are none on the Las Vegas Line. By contast, there are no thrusts on the southerly branch, but many along the Las Vegas Line. There are many erosional and depositional edges as well as abrupt facies change in the Paleozoic rocks across the Las Vegas Line, few or none across the southerly branch.

Seismic effects are shared between the two branches, in fact, seismologists show an inverted Y-shaped bifurcation of the Intermountain seismic belt in this region (Keller, Smith, and Braile, 1975). The density of epicenters is much stronger along the Las Vegas Line, but strangely, this term does not occur in any geophysically oriented papers I have seen. The southerly branch has fewer epicenters, and the belt in which they occur seems to turn in the vicinity of the Grand Canyon to follow the Mogollon Rim.

THE WASATCH LINE THROUGH TIME

The Precambrian Beginnings

The earliest geologic features that coincide in location and trend with the Wasatch Line are apparent in the late Precambrian. This has been pointed out by J. H. Stewart, who noted many features of Precambrian western North America that seem to set a pattern for later developments. Stewart (1972) believes that there was a fragmentation of crustal material about 850 m.y. ago during which a large slice of western North America rifted away and disappeared. The relatively straight, slightly sinuous zone of rifting became the new continental margin, and the space to the west vacated by the massive removal of the bordering land mass became the site of the Cordilleran geosyncline. The initial deposits of the geosyncline are of Windemere age and they show a high degree of north-south continuity not apparent in the older Belt-Purcell series (Crittenden and Peterman, 1975; Stewart, 1972).

Notable among deposits that help delineate the ancestral Wasatch Line are the diamictite (tillite?) formations. As shown by Stewart's map, these stretch from Alaska to California, and their easternmost occurrences follow roughly the Wasatch Line. There are significant variations, however. On the basis of the California occurrences and the lack of known diamictites in Arizona or Mexico, a much better correlation is made along the Las Vegas branch of the Wasatch Line than along the Plateau branch.

The Paleozoic Cordilleran Geosyncline

The Cordilleran geosyncline is one of the great repositories of late Precambrian and Paleozoic sediments of the world. Because this is also true of the Appalachian geosyncline, a certain geologic symmetry of the North American continent was noted rather early and continued to be a cornerstone in such later contributions as Sloss' recognition of sequences as possible improvements on the time-honored systems. Basic to the idea of symmetry is the observation that the central area of the continent is a thinly veneered cratonic or shelf area of the continent with geosynclinal belts on either side. The Appalachian Mountains and Rocky Mountains contain the disturbed contents of these two flanking geosynclines. To complete the mirror-image picture, were two impressive borderlands—Appalachia to the east, and Cascadia to the west.

The hard facts soon marred a beautiful theory. The ancient central positive craton apparently trends diagonally across the continent, and does not divide it neatly into two parts. The histories of the flanking geosynclines differ greatly, especially after the Early Triassic. The Rockies were created a full three periods later than the Appalachians, and by a totally different set of forces. Appalachia became Euro-Africa, and Cascadia seems never to have existed.

The destruction of symmetry was not complete, what remains is useful and fundamentally informative. There was symmetry for a time. During the Late Cambrian and Early Ordovician, the borderlands of North America had a greater degree of symmetry than they were ever to have later. It is during this period that the conditions which Kay visualized as being typical of the early stages of geosynclinal growth prevailed in western North America. This was the time when one might say there clearly was a eugeosyncline and a miogeosyncline and a Wasatch Line also.

Open seas lay to the west and lapped far beyond the continental border into the interior of the continent. There was a rather abrupt shoaling along the Wasatch Line from deeper water on the west to shallower to the east. West of the Line, submergence was almost total and uninterrupted—east of the Line, there were frequent emergences with consequent unconformities and lapses in the record.

During middle and late Paleozoic time, there were more or less drastic effects which broke up the tectonic pattern of the earlier periods. The Antler orogeny crumpled the continental edge and destroyed the eugeosyncline. Clastic sedimentation prevailed where carbonates had dominated before. The Wasatch Line is evident in many maps that might be drawn of Devonian and Mississippian rock units, but after this time, another stress-strain system came into being that temporarily broke across and obliterated it. This was the Ancestral Rockies revolution. The trend is definitely west-northwest from Oklahoma to western Utah. The significant point is that, although the Wasatch Line suffered a tectonic eclipse during the late Paleozoic, it reasserted itself in the Early Triassic.

Early Mesozoic — A Page Missing?

Between the Late Permian and Late Jurassic, the record of what was happening along the Wasatch Line leaves much to be desired. The Late Permian Park City-Phosphoria Groups are found fairly well represented on both sides of the Line, and the paleogeography can be reconstructed in a satisfactory way. Not so with succeeding Triassic and Jurassic rocks, which are thick and varied east of the Line, but almost totally absent west of it. The termination of formations is abrupt and complete along the Wasatch Front, and although there are remnants west of corresponding faults in central and southern Utah, these are relatively small and incomplete.

Many formations thicken westward and are abruptly cut off by the Wasatch Front. The Woodside, Thaynes, Ankareh, Gartra, Chinle, Nugget, Twin Creek, and Stump formations with an aggregate total thickness of over 8,000 ft reach the front of the Wasatch Range near Salt Lake City where they terminate against the Wasatch fault. This is not in itself too remarkable, but no trace of any of these units except for small remnants of Lower Triassic have been found to the west.

On first thought, one would conclude that since the Wasatch fault is a normal fault, the early Mesozoic section has been down-dropped to the west and is present in the trench under Salt Lake Valley. But it should also be exposed somewhere in the succession of uplifted blocks to the west. There is no evidence that the missing Mesozoic section was over this region when normal faulting took place.

Following this line of thought leads to the conclusion that the Mesozoic section was removed sometime between the late Middle Jurassic (time of Stump deposition) and the middle Tertiary. This is a long time interval and there is plenty of indirect evidence that erosion was going on in the Great Basin to supply raw material for the Late Jurassic, Cretaceous, and early Tertiary formations of the former shelf area to the east. Formations with a definite western origin are the Curtis, Morrison, Cedar Mountain, Dakota, Mancos (broad sense), Mesa Verde (broad sense), and North Horn. Taken together these aggregate many thousands of cu mi in volume and their bulk must be equated roughly at least with what was removed or could have been removed in the source area. This problem of source area has baffled almost everyone who has thought about it. There is one unknown factor, and that is the amount of early Mesozoic sedimentary material that may have occupied the eastern part of the miogeosyncline parallel to the Wasatch Line. The Triassic and Jurassic formations are thickening rapidly where they are cut off in the central Wasatch. They may have been thousands of ft thicker not far to the west. Many of these formations are soft sandstone, siltstone, and shale, not durable rocks that would be expected to leave large identifiable clasts. How voluminous the early Mesozoic strata may have been along the Wasatch Line, and what the behavior of the miogeosyncline was during the corresponding time interval, are among the items of information needed before the full history of the hinge area can be reconstructed.

The Mesozoic Orogenies and the Wasatch Line

The Sevier orogeny has grown steadily in importance in the geologic literature since it was first described by Harris in 1959. Harris presented maps showing the orogeny to be confined to a belt of south-central Utah (Harris, 1959). What caused the Sevier orogeny to assume increasing importance was the discovery that other areas all along the Cordillera had been affected by disturbances at the same time as southwest-central Utah. Many large tracts are still not dated precisely, but it seems likely that the Sevier orogeny will be found to have had effects well into the northern Rocky Mountains.

The most notable effect of the Sevier orogeny was eastward thrusting that caused piling up and duplication of miogeosynclinal rocks, especially along the convex central section of the Wasatch Line in Utah-Wyoming. A great deal of speculation has appeared in print as to the nature of this thrusting (Armstrong, 1972). Chief ideas are, a) gravitational sliding, b) extension in the hinterland with regional décollement to the west, and c) compression with crustal shortening to the west.

The problem of how Late Cretaceous and early Tertiary thrusting relates to the Wasatch Line has many aspects and only a few general statements are possible in this brief review: a) there is a concentration of fault traces along the Wasatch Line, b) the movement of fault blocks has been from west to east, and c) movement was concentrated in the late Late Cretaceous.

No matter what forces were acting to initiate and propogate eastward movement on the thrust plates, there was something in the pre-existing arrangement of stratigraphic and structural elements that caused the fault planes to slice upward in the vicinity of the Line. Two possible controls come to mind. The initial dip of successive Paleozoic strata was to the west and this had been accentuated over a lengthy period of relative down-sinking. Fault planes following any particular favorable lithologic horizons might tend to rise eastward and break surface where such planes ended or became too thin to exercise significant effects. This same concept would govern any movement along surfaces between the solid crystalline basement and overlying sediments. That the surface of the basement rises eastward is also a consequence of the downsinking which caused the formation of the Cordilleran geosyncline. Any restoration of the Paleozoic geosyncline as shown by east-west cross-sections must invariably show an eastward rise of the top of the crystalline basement. This configuration alone would seem sufficient to create a concentration of thrust faults along the same flexure that had formerly separated the geosyncline from the craton or shelf.

These mechanical reconstructions presuppose the application of a force to the west which drove the thrust plates eastward. The relief of pressure under this arrangement was possible only through the breaking away of slices along planes that were forced to rise upward by pre-existing structural and stratigraphic conditions. A hinge-like configuration of the basement would supply the necessary arrangement.

At least one more possibility is apparent. An extensive area west of the line may have been vertically elevated so as to have been bodily many thousands of ft higher than the belt to the east. The Wasatch Line under this interpretation may have become a zone of actual fracture and dislocation — upthrown rocks to the west, downthrown rocks to the east. Gravity would then provide the necessary energy to initiate and propogate the thrust faults. Successive slices would move downhill only as far as energy was available, and would be expected to pile up not far away from the eastern edge of the uplifted belt.

All cross-sectional arrangements that can be visualized to explain what is seen in the thrust belt involve an abrupt change in the configuration of the basement along the Wasatch Line. Under some arrangements, the slope that existed at the close of miogeosynclinal deposition may remain fairly unchanged with pressure being applied to the west sufficient to drive thrust plates along suitable planes toward the surface. At the opposite extreme, the slope would be reversed so as to tilt eastward rather than westward giving to gravity the role of prime mover in creating the thrust blocks. How the slope on the basement is drawn depends a great deal on how one conceives the deeper crust to behave. Does it bend over a short radius, or does it break along steep faults? In view of the almost total lack of thermal effects along most of the Wasatch Line, and hence, probable less plasticity or fluidity, the idea that the deep-seated rocks failed by fracturing seems more likely. Also, to continue speculation on this particular problem, I believe it is probable that vertical uplift, with eastward tilting of the former eastern miogeosyncline is required to account for what happened along and east of the Line. That such movements are possible is not, however, mere speculation — the present Wasatch fault, for most of its distance, has a displacement on the order of 15,000 ft. Rocks of every description have been neatly sliced off with no extrusions of volcanic material, and surprisingly little thermal effects of other kinds.

The Wasatch Line in the Neogene

A great reversal of tectonic elements took place after the Laramide orogeny. This might be called the second great reversal, the first having been accomplished in the Triassic, when the Mesocordilleran high began to rise. The younger reversal has been referred to as the decline of the Great Basin and the Basin-Range orogeny. The essential action was a subsidence of the Great Basin in relation to the shelf or cratonic area to the east. The subsidence was accomplished by the sinking of scores of elongated blocks to produce the distinctive horst-and-graben structure that is unusual, if not unique, in global geology. The disturbance is usually referred to as having affected the area between the west-facing scarp of the Wasatch Range and the east-facing scarp of the Sierra Nevada. The parallelism and nearly coincident trends of the easternmost great normal faults with the Wasatch Line as drawn by Kay must have deep-seated significance. As a matter of fact, in the thinking of most geologists, the Wasatch fault and Wasatch Front (Fig. 4) are synonymous with the Wasatch Line. Kay's Wasatch Line seems to be 30-40 km east of the Wasatch and Sevier faults, and it is evident that Kay was not greatly influenced by present-day structural elements (Fig. 5) in drawing the Line.

That tensional or extensional tectonics was operative in bringing about the collapse of the Great Basin is obvious. Furthermore, the coincidence of the beginning of collapse and the change in direction of motion of the Pacific plate would seem to account for the tension or at least the cessation of compression that had prevailed in the previous orogenic episodes.

Some would have the collapse begin along the central axis of the Great Basin with progressive breaking outward toward the margins. Others see evidence of early faulting at least along the Wasatch margin. The evidence from geomorphology is that the trench system of the Wasatch Line was deep enough to divert drainage westward out of the Uinta Range across the Wasatch Line and into the Great Basin, probably in the Miocene.

Geophysical investigations along the Wasatch Front have resulted in discoveries that have been described as follows:

> "The gravity data indicate that in the valley areas between this fault block (Oquirrh Mountains, Boulder Ridge, northern East Tintic) and the Wasatch fault block, an intermont trough (designated by us as the Wasatch structural trough) more than 100 mi in length comprises a great belt of grabens and smaller fault blocks whose dislocations are varied and more complex than previously realized. Several large block fragments lying just west of the Wasatch block have apparently dropped deeper than some other fragments, as if slipping into a great crevass. From north to south, the major grabens are the Farmington, Jordan Valley, Utah Valley, and Juab Valley grabens." (Cook and Berg, 1961)

Other workers have given specific estimates of the depth of fill in the Wasatch structural trough. Arnow and Mattick (1968) estimate the fill to range from 600 to 4,800 ft, and Mattick (1970) gives the maximum thickness in the Jordan Valley as about 5,000 ft. Mikulich and Smith (1974) state that the thickness of Tertiary and Quaternary sediments under the Great Salt Lake ranges from about 1600 ft in the north to more than 10,000 ft in the south. Hintze (1972) gives a range of 8,000 to 10,000 ft for the deeper fill of Utah Valley. Further south near Cedar City, along the Hurricane fault which has about the same relation to the Wasatch Line as the Wasatch fault, Cook and Hardman (1967) have detected a similar system of grabens. Thickness of unconsolidated fill under Cedar Valley a few miles west of the fault is nearly 4,000 ft.

Without knowing the age of the deposits that lie at the bottoms of the depressions of the Wasatch structural trough, it is difficult to judge the age of the associated faulting. The 1,000 ft core, described by Eardley et al., (1973), taken near Burmester, Tooele Valley, is considered to have bottomed in sediments about 3.4 m.y. old. Extrapolating the rate of sedimentation downward to 10,000 ft would give a rough age of 34 m.y. for the first sediments on the bedrock. This is almost at the

Fig. 4 — Wasatch Front between Spanish Fork Canyon, right, and Hobble Creek Canyon, left, Utah County, Utah. Shows faceted spurs, Bonneville lake terrace, and several fault scarps. Most recent movement appears to have been on scarp that crosses the converging roads a short distance before they join to enter the canyon. This section of the Wasatch Range, called Maple Mountain, is entirely Pennsylvanian Oquirrh Formation. *Photo – George E. Brogan, Courtesy Woodward-Clyde Consultants.*

Eocene-Oligocene boundary. Using a more conservative thickness for unconsolidated fill of 5,000 ft gives a date of 17 m.y., somewhere in the middle Miocene.

Although the Wasatch fault is active along this stretch and the maximum displacement of 15,000 ft could have been reached in Quaternary time alone, the evidence is that the depression of grabens that are not necessarily bounded by the Wasatch fault, but are closely related to it, commenced in middle Tertiary time. How this relates to the decline of the rest of the Great Basin and to the theory that failure began in the center and is spreading outward will have to be decided on better evidence. The present-day fault configuration associated with the Wasatch Line is shown on Figure 5.

The Wasatch Line Today

The most prominent feature of Utah geography is the curving, elevated belt of mountains and plateaus that divides the state into distinctly different western and eastern provinces. It is surely no mere coincidence that this "backbone" coincides for the most part, with the Wasatch Line, and in the minds of many, it *is* the Wasatch Line. A brief tabulation of geologic and geographic features east and west of the line as it is typically developed in Utah probably tells more about its origin and nature than any reconstruction of past effects.

The foregoing comparisons call attention to differences across the Wasatch Line. Several geologic features clustered along the Line and relatively rare on either side should also be noted.

Earthquake epicenters. — Any map of earthquake epicenters shows a curving band extending along the Garlock fault, Las Vegas Line and Wasatch Line from south-central California to the Canadian border. This is the Intermountain seismic belt named by Smith and Sbar (1974). According to these workers this zone is the next most active to that of California and Nevada seismic zones so far as the western United States is concerned.

Active faults. — Many fresh-appearing scarps occur along the Wasatch fault and its branches. None of the displacements have occured in historic time but the physiographic evidences

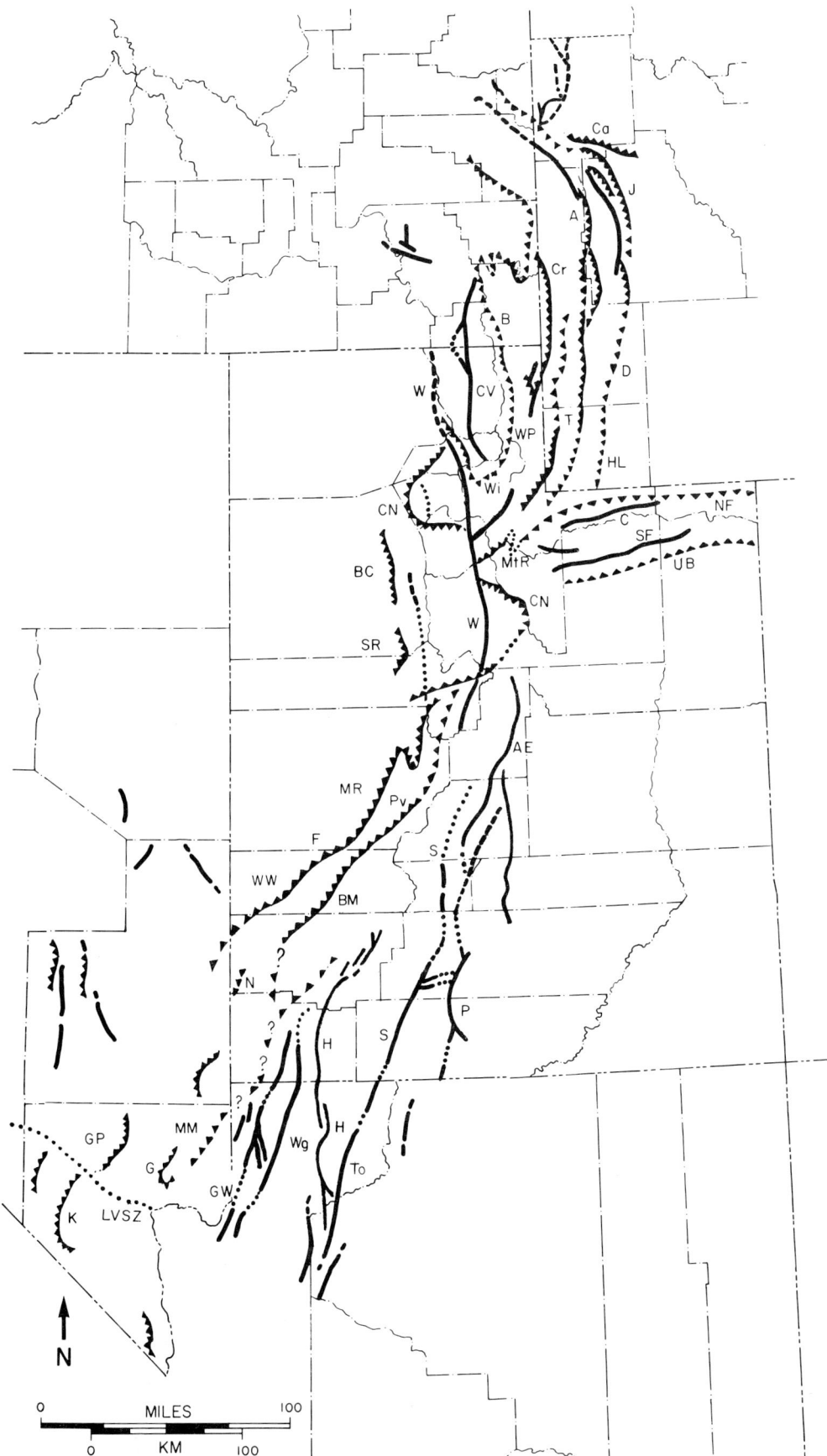

Fig. 5 — Major normal and thrust faults associated with the Wasatch Line. Thrust faults are barbed, normal faults are plain lines. A, Absaroka; AE, Ancient Ephraim; B, Bannock; BC, Broad Canyon; BM, Blue Mountain; C, Crest; Ca, Cache; CM, Cedar Mountain; CN, Charleston-Nebo; Cr, Crawford; CV, Cache Valley; D, Darby; F, Frisco; G, Glendale; GP, Gass Peak; GW, Grand Wash; H, Hurricane; HL, Hogsback-Labarge; J, Jackson; K, Keystone; LVSZ, Las Vegas shear zone; MR, Mineral Range; MM, Muddy Mountains; N, Needles; NF, North Flank; P, Paunsagunt; Pv, Pavant; S, Sevier; SF, South Flank; T, Teton; To, Toroweap; UB, Uinta boundary; W, Wasatch; Wi, Willard; WP, Woodruff-Paris; WW, Wah Wah; Wg, Washington.

	West	**East**
Drainage	Internal drainage, run-off does not reach the ocean. Great Salt Lake receives most of the drainage but there are also small closed basins.	External drainage, runoff reaches the Pacific Ocean by way of the Colorado River and its tributaries. Some crosses the Wasatch Line into the Great Basin.
Chief Geologic Activity	Deposition dominates, material from the mountains is deposited on their flanks or in nearby lake beds. Some is carried away by wind.	Erosion dominates, the Colorado River and its tributaries work at high gradients and energy levels.
Landscape	Sawtooth mountains (sierras), sloping alluvial fans, and wide flat-bottomed valleys	Angular mesas and buttes, stripped surfaces that follow hard flat-lying beds, and deep canyons with step-like walls.
Structure	Complex structure with many faults, both normal and reverse. Horst-and-graben dominate.	Simple structure, the beds are mainly nearly flat-lying as originally deposited. There are a few broad uplifts bordered by long, curving deep-seated faults.
Rock Types	Exposed rocks are mainly limestone and dolomite, practically all of marine origin. Very little sandstone.	Most exposed formations are sandstone and shale mostly of non-marine origin. Very little limestone in exposed sections.
Age of Sediments	Mostly Paleozoic and late Cenozoic. There are thick and extensive formations of Cambrian, Ordovician, Silurian, Devonian, Mississippian, Pennsylvanian, and Permian age making up the exposed ranges. These are more or less buried in soft deposits of Miocene, Pliocene, and Pleistocene age.	Mostly Mesozoic and early Cenozoic. Formations are chiefly of Triassic, Jurassic, Cretaceous, Paleocene, and Eocene age. Only in in the deeper canyons and major uplifts are formations older than Triassic exposed. The Uinta Basin is chiefly Eocene.
Geologic History	Western Utah was a geosyncline during the Paleozoic, uplifted in the Mesozoic, and collapsed and sank in the mid-Cenozoic.	Eastern Utah was a shelf area under shallow water or exposed to erosion during much of the early Paleozoic, there were local uplifts in the Pennsylvanian Period, shallow seas in the early Mesozoic, and deeper seas in the late Mesozoic. Lakes abounded in the early Cenozoic and erosion dominated in the late Cenozoic.
Igneous Rocks and Events	Many extrusive and intrusive igneous rocks. Extrusive rocks are both silicic and basic, the former being of mid-Tertiary age, the latter of later Tertiary-Quaternary age. Tuffaceous rocks including ignimbrites are prominent to the south. Intrusions of quartz monzonite are common.	Few extrusive rocks except those spilling over from the Wasatch Line to cover the High Plateaus. Four mountain groups with cores of intrusive material: Henry Mountains, LaSal Mountains, Abajo Mountains, and Navajo Mountain.
Mineral Resources	Chiefly metallic minerals in ore-deposits of hydrothermal origin. Deposits of copper, gold, silver, lead, zinc, iron, molybdenum, tungsten, uranium, mercury, antimony, and less common metals are widely dispersed. Mostly these are associated with the igneous intrusions.	Chiefly hydrocarbons such as coal, oil, natural gas, oil sands, oil shale, gilsonite, and less common solid organic materials. Also most of the uranium deposits are here associated with minor vanadium and copper.
Heat Flow	Relatively high heat flow that amounts to 2 or more HFU.	Relatively low heat flow amounting to less than 2 HFU.
Underlying Crust	Crust is relatively thin, about 30 km (18.6 mi).	Crust is relatively thick, about 40 km (24.8 mi).
Earthquake Waves	Pn waves have velocities of 7.4 to 7.6 km/sec (4.6 to 4.7 mi/sec).	Pn waves have velocities of 7.8 to 7.9 km/sec (4.8 to 4.9 mi/sec).
Magnetic Variations	The Curie point is at relatively shallower depths.	The Curie point is at relatively greater depths.
Gravity Variations	Higher anomaly values of the gravity field, —150 to —175 milligals.	Lower anomaly values for the gravity field than for the Basin and Range, —200 to —280 milligals.
Electrical Variations	Relatively higher electrical conductivity of the upper mantle.	Relatively lower electrical conductivity of the upper mantle.

place many of them within a few thousand years at most. Fresh scarps are especially notable from near Levan to Brigham City, Utah. Farther south from central Utah to the Grand Canyon many dated lava flows are cut and displaced by faults in ways that prove ongoing movements over a period of several million years.

Hot springs. — Many springs occur along the Wasatch Line simply because of the greater supply of water that falls on the associated mountains and plateaus. Thermal springs follow the same pattern and may do so partly because of more active groundwater circulation. However, the presence of hot springs indicates a heating mechanism at depth that must be related to other activities along the Wasatch Line.

IS THE WASATCH LINE MIGRATING?

Shuey *et al.* (1973) present evidence that the transition between the Colorado Plateaus and Basin Ranges as drawn on geophysical criteria is considerably east of the boundary as drawn by physiographers. Their map (Fig. 5, p. 109) illustrates this, and emphasizes particularly the fact that block faulting is found well eastward of the physiographic boundary in the Colorado Plateaus province (Fig. 6). It might be added that the easternmost grabens and half-grabens are apparently very young—post-glacial movement can be proven in some faults of the Wasatch Plateau, for example.

An eastward migration of volcanic effects in the Grand Canyon area has been well documented. Hawaiite basalt (Fig. 7)

Fig. 6 — Air view looking north along the axis of Joes Valley graben, summit of Wasatch Plateau, Emery County, Utah. Central block has dropped between curving parallel faults. Area is now occupied by Joes Valley Reservoir. *Photo – U.S. Bureau of Reclamation.*

Fig 7 — Cinder cone and basaltic lava flow, Diamond (Damron) Valley, Washington County, Utah. Locality is about 5 mi east of Grand Wash fault; extensive basalt flows lie to the north and east. *Photo–Wm. Lee Stokes.*

appeared in the vicinity of the Grand Wash fault 6.5 m.y. ago; in the central belt of the Shivwits Plateau, 5 to 15 mi to the east, about 3 m.y. ago; and along the Hurricane and Toroweap faults, 40 mi or so east, at 1 m.y. ago or less.

Seismic data may tell the same story. Smith (1974) has shown that many earthquakes occur east of the Wasatch fault and suggests that the zone of seismic activity may be migrating eastward. This suggested migration may be the explanation of a puzzling quiet condition along the segment of the Wasatch fault in the central Wasatch Mountains.

THE WASATCH LINE AND GLOBAL TECTONICS

The Wasatch Line by any definition is a major feature of North America, and as such, should be explainable in terms of the new global tectonics. To what plates, subplates, subduction zones, suture lines, triple junctions, transcurrent or transform faults, or rift zones can it be related, and to what episodes of drift, collision, or fragmentation has it responded in the past? These are difficult questions, but they do call attention to the fact that the Wasatch Line is part of a system of geological discontinuities and structural weakness that involves much of western North America.

It cannot be mere coincidence that five great structural elements, Wasatch Line, Las Vegas Line, Garlock fault, Transverse Ranges, and Murry fracture zone form a continuous configuration (Fig. 8). Part of such a chain might be coincidental, but not all. If there is a break, it might be expectable at the transition from continent to ocean. In this case, a meaningful connection with the Murray fracture zone does not exist.

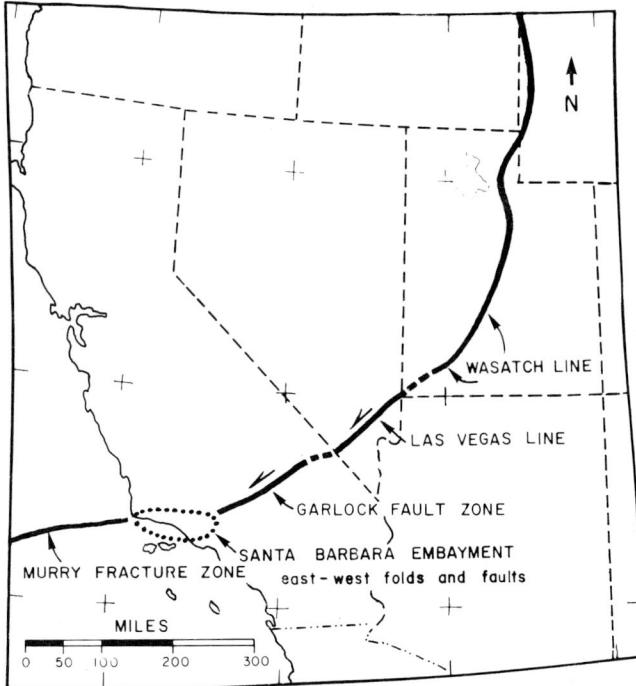

Fig. 8 — Some major dislocations of the southwestern United States.

This is not the place to discuss the nature of the discontinuities that exist across the Las Vegas Line, across the Garlock fault zone, or in the Santa Barbara embayment. Aside from being discontinuities, these land features also posses the common characteristic of a similar direction of dislocation. The Las Vegas Line, Garlock fault, and Santa Ynez fault all currently have left-lateral shifts and the same may be true of the Wasatch Line south of the Uinta projection.

Another broad generalization seems to be called for. The area enclosed by the arc of related structures displays a set of unusual, if not unique, geological features that seem to be related to large-scale accretionary processes. Thus, there are thousands of cu km of both clastic and carbonate sediments, similar amounts of lava and fragmental extrusives, plus massive intrusions such as the Sierra Nevada and Idaho batholiths. Taken together, these have added at least 600,000 sq mi of territory to western North America since late Precambrian time.

A third fact seems notable—there are a great many features of different sorts such as shorelines, edges of igneous provinces, seismic zones, and mineral belts that face northwesterly, that is, have northeasterly or easterly trends. Looked at in the broadest possible way, these lines seem to be parallel with the Las Vegas-Garlock configuration on one hand, and the southeastern margin of the Columbia Plateau in the other.

Many of these northeast-trending features seem to be extensional in nature and suggest motions toward the northwest. This is in fact, the direction that this segment of the crust seems to be moving at the present time. What has been going on in this area for at least a half-billion years might be described as a series of events, motions, and processes that have had the effect of filling a great concavity in the western side of North America between the southwestward projection of the Transcontinental arch and the Canadian Shield. This process seems to have gone on independently of movements of the large lithospheric plates. The westward motion of North America had its own effects and these are manifest chiefly in north-trending features such as the Basin-Range faults, the Cascade volcanoes, and the Sierra Nevada.

SUMMARY

Considering the total known history of the Wasatch Line, the following tentative sequence of events is suggested:

(1) A disruptive event took place in the late Precambrian, 800 to 1200 m.y. ago that removed a sizeable section of western North America. This was a tensional episode, and the zone of rifting was controlled to a large extend by the strength of the upper crust. Rifting followed the northwest and west edge of the Transcontinental arch into southern Utah and then swung northward along and around the Wyoming prong of the shield area. This ancestral rift zone became the Wasatch Line.

The process of rifting created fragments of unknown size and of unequal thickness. Some of these may have drifted away to unknown destinations, but a few remained in the vicinity of the continent to become locked into subsequent sedimentary and tectonic arrangements. Perhaps such anomalous splinters as the Northern Utah highland, Snake Range, and Raft River Range are fundamentally such Precambrian islands or microcontinents. What sort of sub-marine surface this disruption would create or leave behind is not clear, but it seems not to have been greatly different from oceanic crust. The rifted margin was probably not one gigantic scrap, but was nevertheless abrupt—no different from other well authenticated rifted continental margins today.

(2) Westward draining streams, plus an oceanic current system with clockwise motion proceeded to fill the concavity left by earlier disruptions. This soon created a more normal shelf-slope configuration with an actual continental edge established in central Nevada. This was the time of growth of the Cordilleran geosyncline. Thermal manifestations and interplate

reactions seem to have been minimal.

The Antler and Ancestral Rockies orogenic episodes, neither of which follow the classic textbook pattern, affected western North America, but did little to alter conditions along the Wasatch Line.

(3) The break-up of Pangaea and the opening of the Atlantic Ocean were attended by westward drift of North America and its collision with the Pacific plates. Details are complex and there is as yet no clear-cut ruling theory of what went on in the Great Basin. One thing seems sure. Compressional tectonics held sway throughout the Mesozoic. Uplift certainly affected the former geosynclinal sediments and during the Sevier orogeny both thrust faulting and folding took place on the Wasatch Line. The concept that the former front of the Precambrian continental block served as a buttress against which softer rocks rose and locally moved eastward seems basically sound. This would seem not to have been the closing of an ancient rift with opposing sides with similar structure and lithology. One wall of the rift had disappeared and it was a reaction of a continental shelf-slope sedimentary section against a somewhat abrupt cratonic edge. This edge should not be visualized as having its original squared-off profile, it had certainly been considerably beveled during late Precambrian erosion.

(4) With the change of plate motion from compressional to strike-slip in the mid-Tertiary the Wasatch Line again manifested its inherent weakness. Old faults were reactivated with downward movement on the west. Rifts were again opened along much the same patterns as the original disruptions of the late Precambrian. There was one great difference. The original rifting had removed the western block perhaps in rather large fragments. Some of these may remain in a matrix of later sedimentary and volcanic deposits, but the Tertiary disruption is probably much more complex than its predecessor.

CONCLUSION

In concluding this paper, there is a feeling of having perhaps written a memorial, or even an obituary to the Wasatch Line. For what must be good reasons, the name designating the subject of this guidebook and associated field trip is the Cordilleran Hingeline. I think this may be a good term, a hinge is certainly more descriptive of what we are dealing with than is a line. Cordilleran Hinge is a good companion term to Cordilleran geosyncline and Mesocordilleran high. But Wasatch Line is in the literature, terminology, and thinking, at least to a moderate extent, and has priority. Perhaps the various papers of the guidebook will swing the balance one way or the other. The editor may exercise the power of his pen to aid in a decision. When the present surge of investigation is over, will the question of what is the Wasatch Line, or the Cordilleran Hinge, be answered? Time will tell.

REFERENCES CITED

Arnow, Ted, and Mattick, R. E., 1968, Thickness of valley fill in the Jordan Valley, east of the Great Salt Lake, Utah: U.S. Geol. Survey Prof. Paper 600-B, p. B79-B82.

Armstrong, R L., 1968, Sevier orogenic belt in Nevada and Utah: Geol. Soc. America Bull., v. 79. p. 429-458.

_____, 1972, Low-angle (denudation) faults, hinterland of the Sevier belt, eastern Nevada and western Utah: Geol. Soc. America Bull., v. 83, p. 1729-1754.

Condie, K. C., 1969, Geologic evolution of the Precambrian rocks in northern Utah and adjacent areas: Utah Geol. and Mineral Survey Bull. 82, p. 71-95.

Cook, K. L., and Berg, J. W. Jr., 1961, Regional gravity survey along the central and southern Wasatch Front, Utah: U.S. Geol. Survey Prof. Paper 316-E, p. 75-89.

_____, and Hardman, Elwood, 1967, Regional gravity survey of the Hurricane fault area and Iron Springs district, Utah: Geol. Soc. America Bull., v. 78, p. 1067-1076.

Crittenden, M. D. Jr., 1976, Map to accompany field trip to Northern Wasatch Range: Geol. Soc. America annual meeting, 1976.

_____, and Peterman, Z. E., 1976, Provisional Rb/Sr age of the Precambrian Uinta Mountain Group, northeastern Utah: Utah Geology, v. 2, p. 75-77.

Eardley, A. J., 1939, Structure of the Wasatch-Great Basin region: Geol. Soc. America Bull., v. 50, p. 1277-1310.

_____, Shuey, R. T. Nash, W. P., Gvosdetsky, V., Picard, M. D., Grey, D. C., and Kukla, J., 1973, Lake cycles in the Bonneville Basin, Utah: Geol. Soc. America Bull., v. 84, p. 211-216.

Harris, H. D., 1959. A late Mesozoic positive area in western Utah-Nevada: Am. Assoc. Petroleum Geologists Bull. v. 43, p. 2636-2652.

Hansen, W. R., 1965, Geology of the Flaming Gorge area Utah-Colorado-Wyoming: U.S. Geol. Survey Prof. Paper 490, 196 p.

Hintz, L. F., 1972, Wasatch fault zone east of Provo, Utah: in Environmental Geology of the Wasatch Front, 1971, Utah Geol. Assoc. Pub. 1, p. F1-F9.

Kay, Marshall, 1951, North American geosynclines: Geol. Soc. America Memoir 48, 143 p.

Keller, G. R., Smith, R. B., and Braile, L. W., 1975, Crustal structure along the Great Basin-Colorado Plateau transition from seismic refraction studies: Jour. Geophysical Research, v. 80, p. 1093-1098.

Larsen, W. M., 1954, Petrology and structure of Antelope Island, Davis County, Utah: unpub. Ph.D. dissertation, University of Utah, 185 p.

Longwell, C. R., 1950, Tectonic theory viewed from the Basin Ranges: Geol. Soc. America Bull., v. 61, p. 413-434.

Mattick, R. E., 1970, Thickness of unconsolidated to semiconsolidated sediments in Jordan Valley, Utah: U.S. Geol. Survey Prof. Paper 700-C, p. C119-C124.

Mikulich, M. J., and Smith, R. B., 1974, Seismic reflection and aeromagnetic surveys of the Great Salt Lake, Utah: Geol. Soc. America Bull., v. 85, p. 998.

Roberts, R. J., Crittenden, M. D., Jr., Tooker, E. W., Morris, H. T., Hose, R. K., and Cheney, T. M., 1965, Pennsylvanian and Permian basins in northwestern Utah, northeastern Nevada, and southcentral Idaho: Am. Assoc. Petroleum Geologists Bull., v. 49, no. 11, p. 1926-1956.

Rowley, P. D., Anderson, J. J., and Williams, P. L., 1975, A summary of Tertiary volcanic stratigraphy of the southwestern High Plateaus and adjacent Great Basin, Utah: U.S. Geol. Survey Bull. 1405-B, 20 p.

Smith, R. B., 1974, Seismicity and earthquake hazards of the Wasatch Front, Utah: Earthquake Information Bulletin, v. 4, p. 12-17.

_____, and Sbar, M., 1974, Contemporary tectonics and seismicity of the western United States, with emphasis on the Intermountain seismic belt: Geol. Soc. America Bull., v. 85, p. 1205-1218.

Shuey, R. T., Schellinger, D. K., Johnson, E. H., and Alley, L. B., 1973, Aeromagnetics and the transition between the Colorado Plateau and Basin Range Provinces: Geology, vol. 1. p. 107-110.

Stewart, J. H. 1970, Upper Precambrian and Lower Cambrian strata in the southern Great Basin, California and Nevada: U.S. Geol. Survey Prof. Paper 620, 206 p.

_____, 1972, Initial deposits in the Cordilleran geosyncline: evidence of a Late Precambrian (<850 m.y.) continental separation: Geol. Soc. America Bull., v. 83, p. 1345-1360.

Welsh, John E., 1959, Biostratigraphy of the Pennsylvanian and Permian Systems in southern Nevada: Unpub. Ph.D. dissertation, University of Utah.

Williams, Norman C., 1953, Late Precambrian and early Paleozoic geology of the western Uinta Mountains, Utah: Am. Assoc. Petroleum Geologists Bull. v. 37, P. 2734-2742.

JOHN F. STEWARD

Mention John Steward's name to a group of geologists today and you will probably be rewarded with perplexed expressions.

COLORADO magazine in 1969 carried a picture of John Steward taken 98 years earlier. His name is misspelled and the caption identifies him as a "guide and trapper."

John F. Steward was actually Assistant Geologist with Major John Wesley Powell's Colorado River Expedition of 1871. The fact that he has remained in obscurity for more than a century may be attributed in part to the nature of Powell's report on the **Exploration of the Colorado River of the West.** The report was published in 1875. For some strange reason, perhaps to simplify its writing and to heighten its dramatic impact, Major Powell incorporated events that occurred on both the first expedition of 1869 and the second expedition of 1871. The cast he chose for his report was that of the 1869 expedition, and the members of the 1871 expedition were written out of the script.

Although he never mentioned the matter publicly, it was a decision that Major Powell later regretted. Thirty years had passed when Fred Dellenbaugh, who had been the artist for the 1871 expedition at the tender age of 17, began work on his book, **The Romance of the Colorado River.** (Dellenbaugh's career at this time included extensive Arctic and South American exploration, recognition as an artist and lecturer, being one of the founders of the Explorers Club, and marriage to the actress, Harriet Otis.) Major Powell had retired in 1894, but when he heard about Dellenbaugh's projected book, he wrote to insist, "I hope that you will put on record the second trip and the gentlemen who were members of that expedition."

John Steward and Major Powell met during the Civil War at Vicksburg, where both were collecting fossils from the same limestone outcrop. Steward was born in 1841 in Plano, Illinois, and served in Company F, 127th Illinois Volunteer Infantry. He was wounded following the Vicksburg campaign. The wound and a back injury troubled him later during the river exploration. After the Civil War, Steward was employed by the Marsh Harvester Company of Plano, Illinois, where an invention of his was rewarded with a substantial sum of money. At the time of his going on the 1871 Colorado River expedition he was already financially independent.

Steward is described as "a handsome fellow, just under six feet in height, of dark complexion and manly bearing. He was independent, forceful, and energetic. His likes were many, his friendships sincere; his dislikes were intense, his enmities outspoken. Dellenbaugh characterized him as 'of the most independent sort imaginable.' " Steward was capable of unexpected

John F. Steward, Assistant Geologist, Colorado River Expedition of 1871, as photographed by E. O. Beaman in Glen Canyon. This photograph was also widely distributed as a stereoscopic pair for use in hand viewers. — U.S. Geological Survey photo

gestures of generosity and affection. The members of the 1871 expedition who remained to map the Grand Canyon region under the direction of the topographer A. H. Thompson were surprised to receive a huge box of candy from Steward on his return to Chicago.

Steward afterward was occupied with business concerns, but he maintained an interest in geology for the remainder of his life. He was a member of the Chicago Academy of Sciences. He was also a member of two historical societies. He died in 1915.

In common with many educated people of his era, Steward had a command of shorthand. During the river trip he kept a journal in shorthand which he later transcribed. The Utah State Historical Society has done geologists everywhere a great kindness by publishing his journal as a part of their large 1948-1949 **Quarterly.**

— W. Lyle Dockery

TECTONIC EVOLUTION IN UTAH'S MIOGEOSYNCLINE-SHELF BOUNDARY ZONE

by

Gary W. Crosby[1]

ABSTRACT

Geologic features of Utah's miogeosyncline-shelf boundary zone demonstrate tectonic principles that appear to have worldwide application. Observations there favor gravitational gliding of the miogeosynclinal sedimentary strata off a broad regional uplift in the Sevier orogenic belt and pile up structures at the miogeosyncline-shelf boundary zone. Relative vertical movement during this tectonic interval was approximately 19 km. Pull-apart structures formed rearward of the thrust sheets leaving thin, rotated slice blocks and occasionally exposing rocks older than those above décollement zone(s). The primary uplift plus isostatically induced uplift resulting from erosion and partial tectonic denudation produced extension that caused minor normal faulting in the basement. Magnetic data do not reflect major basement displacements, and drill data reveal surprisingly old bedrock beneath intermontane basins.

INTRODUCTION

The zone containing the geosyncline-shelf boundary in Utah is exceptionally well exposed due to aridity and high relief. Therein are exhibited typical miogeosyncline and shelf facies, intense deformation in pile-up structures at the edge of the fold and thrust belt, and great volumes of interfingering, coarse, syntectonic, fluviatile and marine sediments in a foredeep basin. These features have been broken by further tectonic development subsequent to overthrusting, dominated in the main by vertical movements; these, however, serve more to expose and clarify earlier events than to obscure them. Utah is, therefore, not only an exceptional open air laboratory where depositional and structural development of a miogeosyncline and the adjoining shelf may be studied at the provincial level, but also is an area in which to formulate and test principles that may be more generally applicable to geosynclinal development worldwide and temporally.

TECTONIC MODEL

According to the thickness data of Hintze (1973) the miogeosyncline subsided deeply and received locally more than 14 km of Paleozoic-Mesozoic sediments in western Utah while the shelf remained relatively positive (Fig. 1).

Beginning in Late Mesozoic, the miogeosyncline uplifted in excess of 14 km locally exposing pre-miogeosyncline basement in the Sevier orogenic belt. Simultaneously, the shelf edge subsided up to 6 km and received thick, syntectonic, marine and transitional sediments in a foredeep basin such that the relative vertical movement between the Sevier belt and foredeep basin was approximately 19 km in central Utah. Principally during this time the Sevier orogenic belt was partially denuded by erosion and the gliding of sedimentary strata off the uplift and into the foredeep basin to form the pile-up zone at the miogeosyncline-shelf boundary.

The intermontane basins are, in the main, pull-apart structures. Range front "faults" are edges of residual blocks that are mildly tilted, but in most ranges not much deformed internally. Some basement faulting of minor displacement has occurred as a result of expansion in uplift accentuated by unloading. Major horizontal separation in the sedimentary section above the décollement(s) is primarily a result of gliding, whereas the horizontal expansion in the basement due to uplift is minor. Drilling in intermontane basins may encounter either subdecollement rocks or residual toreva blocks of sedimentary strata, but statistically the rocks encountered will be older than those exposed in adjacent ranges.

The foredeep basin, which deeply subsided earlier, has been uplifted recently, locally in excess of three km. This uplift lowered or even reversed the dips existing on the décollement surface(s) when the folds and thrusts in the pile-up zone developed.

DISCUSSION

The long duration subsidence and filling of the miogeosyncline in western Utah has been demonstrated by the cumulative investigations of many workers (see Hintze, 1973). The shelf to the east has been clearly defined through reconstructed latest Jurassic, Cretaceous and Paleocene depositional histories. The restored section (Fig. 2), which strikes about east-west a little south of Milford, Utah, typically illustrates the relatively negative and positive aspects of the miogeosyncline and shelf respectively during Paleozoic-Mesozoic time.

---

[1] Phillips Petroleum Company, Del Mar, California.

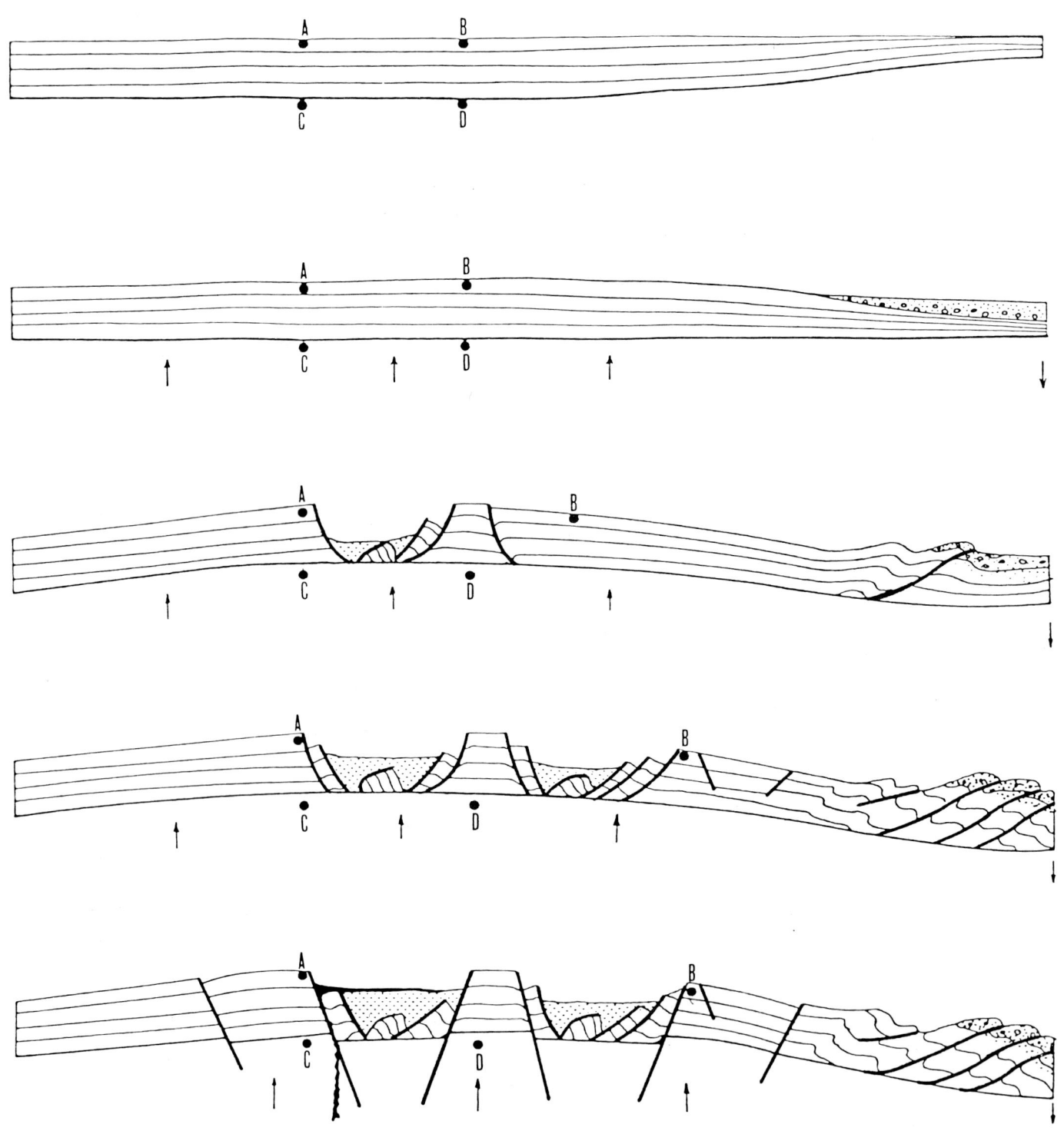

Fig. 1 — Stages in the development of the Sevier orogenic belt out of the miogeosyncline. Extensional structures developed over the uplift are genetically related to the folds and thrusts in the pile-up zone at the shelf edge. Points A and B above the detachment zone have extended 150 percent while points C and D have extended only one percent (Modified from Crosby, 1972, Fig. 1).

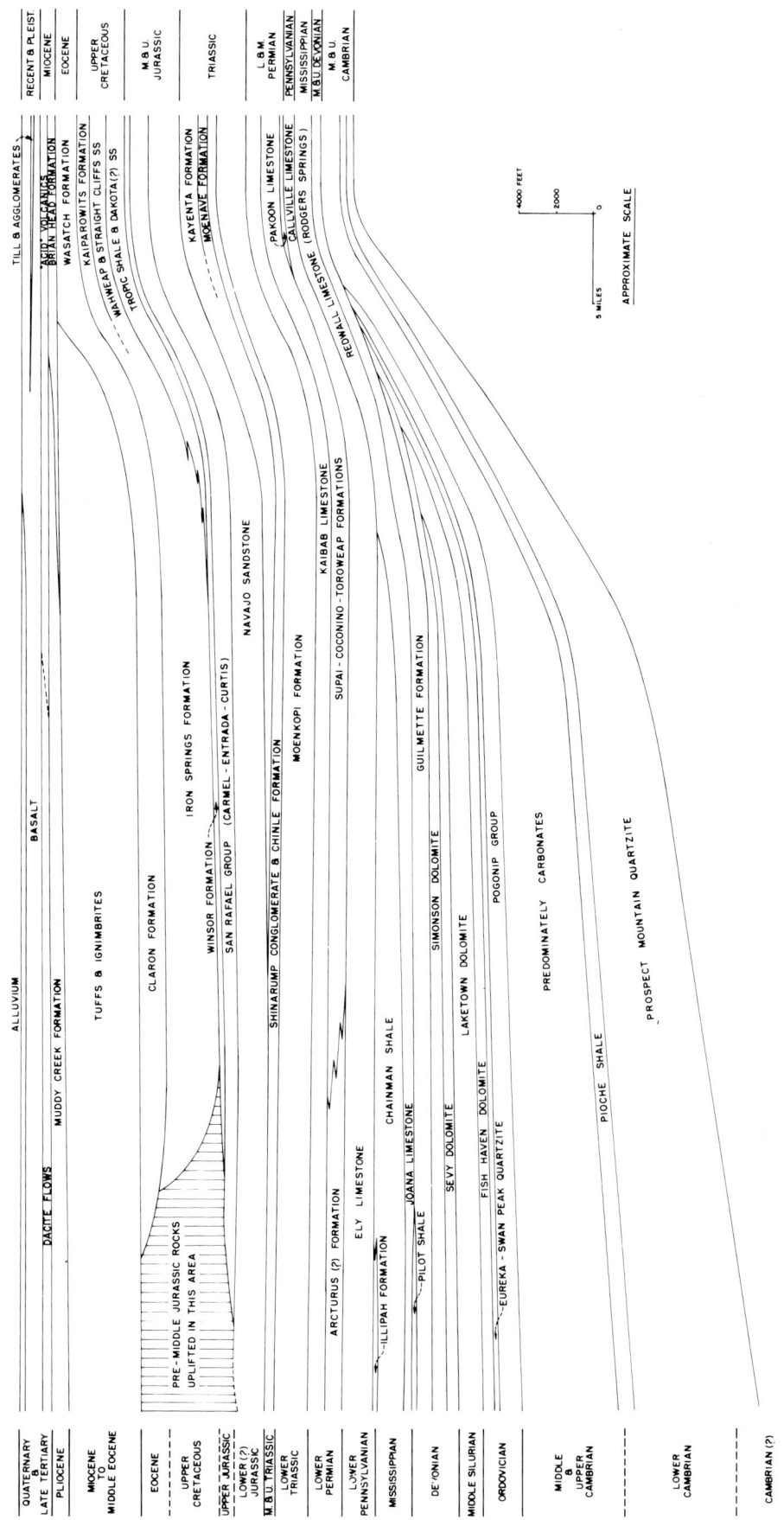

Fig. 2 — Restored section trending approximately east-west through Beaver and Iron Counties, Utah.

Thickness of Paleozoic and Triassic-Jurassic sediments in western Utah at the beginning of the Cretaceous averaged 11,600 m, but locally reached thicknesses up to 14,000 m (Crosby, 1970, p. 3508). In the uplift that followed, Eocambrian quartzites were exposed and characteristic quartz clasts were shed into the foredeep basin that was developing simultaneously at the shelf edge (Armstrong, 1968, p. 448). If the Lower Cambrian quartzites at the bottom of the pile were exposed through dominantly vertical movements, then uplift must have exceeded the thickness of the Paleozoic and Triassic-Jurassic sediments, in excess of 14,000 m locally in central Utah.

The approximate top of the Cretaceous (Upper Mesaverde equivalents) is marine, and there is approximately 4800 m of Cretaceous sediments below these marine sandstones in central Utah; hence, the relative vertical movement of the Precambrian surface during Sevier orogenic events was in the order of 19 km. There is another 3800 to 4000 m of sediment between the base of the Cretaceous and base of the Cambrian beneath the axis of the foredeep basin in central Utah; therefore, the relief on the top of the Precambrian, assuming dominantly vertical movements, was approximately 23 km. The axis of the uplift on the west was, at most, 200 km from the axis of the foredeep basin, allowing for reasonable palinspastic restoration of thrust sheets. If the relief on this surface had been represented by a plane of constant slope, its dip was 6.5°. Without palinspastic reconstruction, not required in the model presented here, the distances are shorter and the dip would be significantly greater.

Effects of Uplift. — One possible effect of the creation of such relief on a mechanically significant datum surface is gravitational gliding. The mechanism has been considered in many investigations (see Roberts and Crittenden, 1973). Attenuation rearward from the leading edge of a thrust sheet is the expected response over a broad regional uplift, unless there is fundamental crustal shortening; for example, as in consumption of a portion of the hinterland by subduction, or mobilization to form a zone of infrastructure. Lacking these, the attenuation may be expressed as thinning of the allochthonous mass by trailing out thin, rotated slices at the following edge, or by pull-apart structures that expose the rocks below the detachment.

Tectonic denudation of any form, together with erosion of the uplift, will trigger further isostatically induced uplift. Unless it can be shown that crustal blocks on each side of the uplift move closer together, elongation of any datum surface must occur. Extension develops during uplift from two causes: (1) local increase in radius of the earth, and (2) local increase in convexity of the surface in excess of the curvature of the reference spheroid (Fig. 3). Inasmuch as the geosyncline and subsequent uplift are elongate structures, the problem may be treated as two-dimensional.

The length of a line segment, S, is

$$S = \int_{x=m}^{n} \sqrt{1 + (dy/dx)^2}\, dx \qquad (1)$$

For a circle the solution of the integral is simply $a\Theta$, where a is the radius and Θ is the internal angle of the segment in radians. The change in segment length is

$$\Delta S_1 = S_u - S_o = a_u\Theta - a_o\Theta = \Theta(a_u - a_o) \qquad (2)$$

where subscripts u and o designate uplifted and original segment lengths.

As discussed above, the pre-miogeosynclinal surface was uplifted in the order of 15 km from its maximum depression. With the Sevier orogenic belt described by Armstrong (1968) serving as our model, the parameters are taken to be

$$S_o = 140 \text{ km},\; a_o = 6370 \text{ km},\; a_u = 6385 \text{ km},$$

$$\text{and } \Theta = .022 \text{ rad.};$$

therefore

$$\Delta S_1 = .022(15) = .333 \text{ km}$$

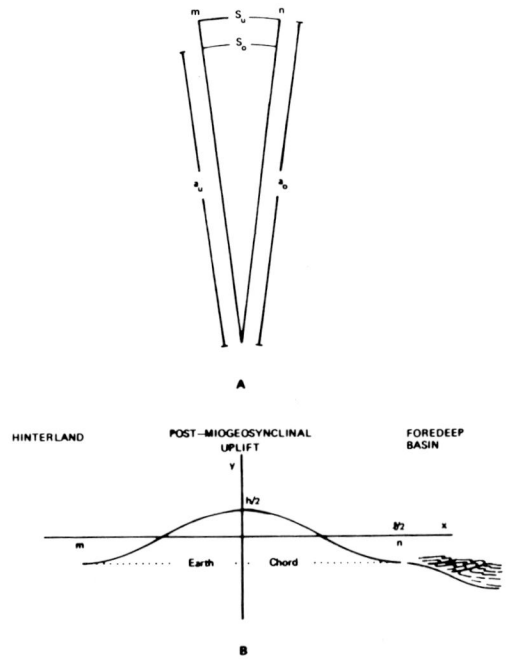

Fig. 3 — Diagrammatic representation of extension due to A — local increase in the radius of the earth, and B — local increase in the convexity of the earth.

The arching accompanying local uplift grossly takes a sinusoidal form. There is no precise solution of the integral, equation (1), for a sine curve, but its segment length can be approximated with the desired degree of precision with a Newtonian series (Dallmus, 1958, p. 911).

$$S_2 = \ell + \frac{\pi^2 h^2}{2\ell} - \frac{3\pi^4 h^4}{32\ell^3} - \frac{5\pi^6 h^6}{128\ell^5} - \colon \ldots \quad (3)$$

where ℓ is the wavelength (width of the uplift) and h is trough to peak amplitude (height of the uplift). Actually, any surface being uplifted starts to experience extension the instant it passes through the chord of the segment. For the 140 km average width of the Sevier orogenic belt the chord is at maximum depth 0.4 km below a surface concentric with the reference spheroid, and the uplift of sinusoidal form, producing extension in the top of the pre-miogeosynclinal rocks, is of the order of 5 km. This is a rough estimate, at best, principally because of the uncertainty of the position of the pre-miogeosynclinal surface west of the Sevier uplift. Substituting the appropriate values into equation (3), the new length, S_2, is 140.877 km.

Total expansion, therefore is

$$\triangle S_1 + S_2 = 0.333 + 0.877 = 1.21 \text{ km}$$

or one and a quarter km in significant figures for the basement rock surface.

It can be shown (Crosby, 1971, p. 2700) that lateral elastic expansion due to unloading is minor, only about 12 m in this case, and is insignificant to this discussion.

Normal faulting develops as a result of the changing stress state during uplift. The average dip of the faults might be expected to be about 60° (Price, 1966, p. 61). With this dip value and 1.21 km of lateral elongation, the total fault displacement, d, is

$$d = 1.21/\cos 60° = 2.42 \text{ km}$$

if the total displacement is accomplished along a single fault. The average displacement is 1210 m for two faults, 605 m for four faults, and so on.

While these are order of magnitude estimates it does appear that a mechanism is available to produce normal faults in the basement below the detachment surface(s) during gravitational gliding. Displacement likely would be distributed on many basement faults with displacement on any individual fault being relatively minor. It must be stressed, however, that these are the faults that cut the pre-miogeosynclinal rocks and displace their upper surface by the above amounts. This distinction is critical inasmuch as the sedimentary section above the basement will show a significantly greater amount of extension where pull-apart structures form as a consequence of gliding, which is likewise a form of normal faulting. Moreover, extension above the décollement zone(s) is contemporaneous with compressional deformation in the pile-up zone at the boundary between miogeosyncline and shelf, whereas faulting of the basement rocks mostly will follow folding and thrusting in the episode of further uplift triggered by tectonic denudation and erosion.

Once the basement uplift begins to be segmented by faults, the tendency for isobaric isostatic adjustments is significantly reduced. Uplift of the surface favors the horsts since, relatively, these blocks go up and the grabens subside. With respect to sea level the grabens may be going up, remaining stationary, or going down absolutely. Assuming the terrain consists of half horsts and half grabens, then incremental isostatic uplift of the base of the abnormal upper mantle, where compensation occurs during mountain building, is expressed by uplift of only half of the surface. The horst blocks are raised twice as high as the basement surface would be if it were being elevated as a unit, when the graben blocks remain stationary with respect to sea level. Hence, fault displacements calculated above for the Sevier orogenic belt could be increased by a factor of about two. Maximum cumulative fault displacement of the top of the basement is, then, in the order of 4.8 km.

This is approximately the stage of development of the Sevier orogenic belt in Utah. The foredeep basin at the shelf edge, however, has recently experienced uplift in excess of three km (Crosby, 1970, p. 3510).

Magnetic Test. — The total intensity aeromagnetic map of Utah (Shuey, 1975) generally does not show the expected expression of a high angle fault at the boundary between ranges and intermontane valleys in western Utah. The bedrock floors beneath the basins, being covered, are immune to erosion while the highest and generally the youngest portions of the stratigraphic section in the mountains is removed progressively by erosion. Unless there is some further structural complication in the tectonic evolution of the eastern Great Basin, it might be expected that statistically the age of the bedrock floors of the intermontane valleys would be younger than the strata found on the tops of adjacent ranges.

On this assumption, stratigraphic displacement across many boundary fault(s) can be inferred to be thousands of meters. Do these high angle faults cut and displace the crystalline basement? The metamorphic rocks of the basement are relatively magnetic and such major displacements should have magnetic expression, particularly where the basement complex

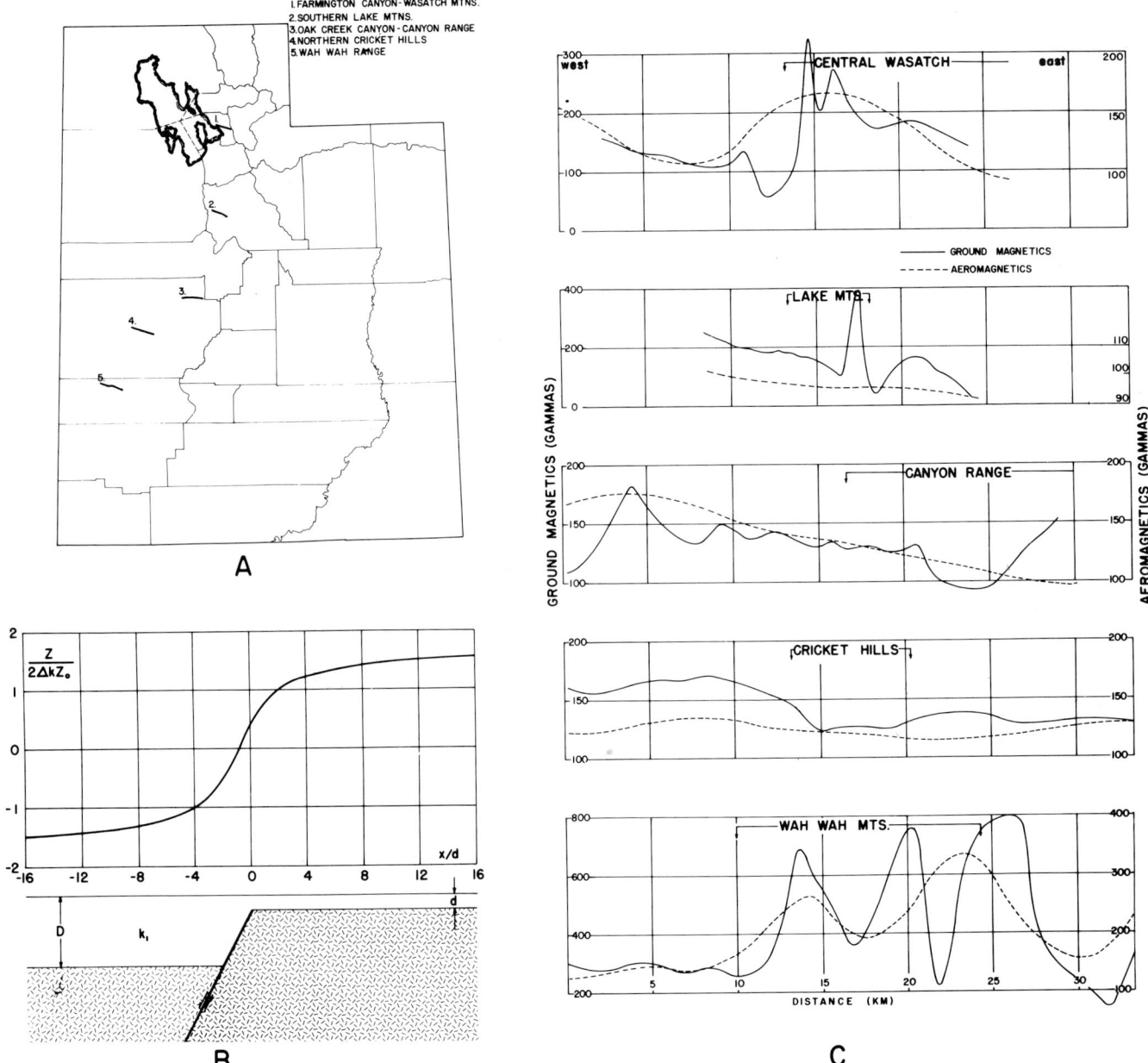

Fig. 4 — A — location of magnetic profiles across range fronts in western Utah. B — Calculated magnetic profile over a high-angle, north striking normal fault that offsets the magnetic basement. C — Ground level total intensity and aeromagnetic profiles across the five range fronts. Datums arbitrary.

is Farmington Canyon or equivalent rocks, which are assumed to underlie wide areas of Utah (Stokes, 1969, p. 38).

The state aeromagnetic map, however, is generalized. The survey was flown at high altitudes and wide flight line spacing, 2745 to 3660 m and 2 to 4 km, respectively. In a further effort to learn whether or not a magnetic expression of the faults exists, detailed total intensity ground magnetic profiles were measured across boundaries between ranges and basins in five localities (Fig. 4A): the central Wah Wah Mountains, northern Crickett Hills, central Canyon Range (Oak Creek), southern Lake Mountains and central Wasatch Range (Farmington Canyon). Traces of major overthrusts at the miogeosyncline-shelf boundary zone are east of these localities. Precambrian Farmington Canyon Complex basement rocks are exposed at the range front between Salt Lake City and Ogden, where the Uinta Mountains strike into the central Wasatch Range. This profile was selected specifically to include one line where magnetic rocks can be observed to be offset along a fault. All other lines cross ranges where strata varying in age between Precambrian Sheeprock Series and Mississippian Humbug Formation are exposed. The Wah Wah Mountains profile extends over three Tertiary stocks.

The ground magnetics are shown in Figure 4C, together with the aeromagnetics along the same line. There does not appear to be any expression of high-angle faults at or near the range fronts in the four southern profiles, in either the ground or aeromagnetics. The profiles may be compared with the calculated profile over a north-striking fault displacing the magnetic basement (Fig. 4B). The scales are dimensionless; the absolute amplitude being controlled by susceptibility contrast, displacement, and depth to basement, also the width of the anomaly is controlled by depth.

The expected expression of a fault obtains in the aeromagnetic profile running through Farmington Canyon, except that the measured profile is modified somewhat by the fact that the top of the basement dips east on both the upthrown and downthrown blocks. The ground magnetics, however, only poorly reflect the ideal fault form at the boundary between basin and range. The data, however, are suspect in this area which is extensively developed culturally. There is nowhere the magnetometer can be placed without some interference from buildings, vehicles, fences, highways, railroads, pipelines, power lines, or metal debris. The gross form of the ground magnetic profile does express the presence of a high angle fault that displaces an eastward dipping magnetic body.

The western end of the Wah Wah Mountains profile superficially resembles the ideal fault expression; however, the west range front is two to three km west of the position suggested by the profile. The measured curve is in response to a Tertiary local intrusive.

Alternatively, large scale basement offsets may be so deep or the susceptibility contrast may be so low that there is no detectable magnetic expression. It can be stated, nevertheless, that this simple test has failed to demonstrate offsets of the crystalline basement that anywhere approaches the displacements that can logically be inferred for the Phanerozoic rocks. This lack of expression is in accord with the earlier conclusion that offsets on basement faults are minor, at best.

Drill Data. — One further test can be applied. Crosby (1972) tabulated deep wells that spudded in intermontane basins and penetrated bedrock beneath thick valley fill. The age of the floor rocks was compared with the age of the nearest outcropping bedrock. Two more wells now can be added to the list; all such wells for which reliable information is available are shown in Table 1. The conclusion stated previously (1972) still holds, namely, that most wells, of the few that have drilled to bedrock, have encountered rocks older than those in the nearest exposures in the adjacent ranges. The score is six of eight, and the age discrepancy would be even greater if the comparison were made with strata on the tops of the ranges instead of the nearest outcrop.

This test must await further drilling before it can be considered statistically sound; nevertheless, results to date favor tectonic denudation.

Ancillary Evidence. — 1. Trends of horsts and grabens in the Sevier belt in Utah closely parallel the trend of the pile-up structures at and near the shelf edge. This is apparently true for both the major normal faults coincident with residual block edges in pull-apart structures, and those bounding the horst and graben blocks that demonstrably cut the basement.

2. Extension expressed by normal faulting within the sedimentary rocks will usually be in excess of that possible due solely to expansion by reason of uplift (Fig. 1). The value may be near 15 km for the Sevier orogenic belt (Stewart, 1971, p. 1036). The demonstrable amount of horizontal extension by normal faulting in rocks beneath the lowest décollement zone is something less, perhaps in the order of a km or less. There is at present no accurate basis for determining the actual extension in the Sevier belt inasmuch as outcrops of crystalline basement are few. Known exposures occur in several places in the Wasatch Mountains, Antelope and Carrington Islands in Great Salt Lake, and a possible limited exposure in the Mineral Range (Condie, 1969, p. 79).

3. Dips in sedimentary strata within the extension zone may be locally steep and fairly close folding may be observed locally. This is to be expected where the extensional faulting above the décollement zone(s) involves rotation and reverse drag, particularly where the smaller blocks take on the form of toreva blocks. Marked dips on the basement surface have not been demonstrated at the present time.

Table 1. Western Utah wells that have encountered bedrock in intermontane basins.

Well	County	Location	Floor Strata	Bdr Depth	Nearest Outcropping pre-Tertiary Strata
Utah So. Bar-B #1	Box Elder	17 T10N R7W	Miss Brazer	3580	Penn-Perm Oquirrh
Utah So. #2 Fed	Box Elder	6 T14N R9W	Miss Und	3080	Penn-Perm Oquirrh
Utah So. #1 Govt	Box Elder	14 T14N R10W	Penn Und	3880	Penn-Perm Oquirrh
Utah So. #1 Fed	Box Elder	22 T14N R10W	Penn Und	440	Penn-Perm Oquirrh
Shell #1 Govt	Millard	19 T20S R19W	€ Prospect Mt.	4180	Dev Guilmette
Whitlock #1 M. Salt	Salt Lake	24 T1N R3W	p€ Xln	3656	Penn-Perm Oquirrh
Eskdale Fed #1-23	Millard	23 T20S R19W	Sil-Ord Dol	4100	Dev Guilmette
Bridger Fed #1	Beaver	15 T26S R17W	€ Marjum	4928	Ord Eureka-Swan Pk

4. The stratigraphically deep exposure in the hinterland reflects the former uplift with subsequent erosion and tectonic denudation. In the Sevier orogenic belt, however, the average elevation is about 600 m lower than the average elevation of the cratonward edge of the fold and thrust belt (Fig. 5) which has been recently uplifted and does not show the effects of deep erosion (Crosby, 1970, p. 3509). This seeming paradox is reconciled in this hypothesis.

5. The earliest faulting that cuts the basement should be in the vicinity of the apex of the uplift where curvature is likely the greatest, and migrate outward in time. This ideal scheme may be reflected in the fact that in Utah, Idaho and Wyoming the most seismically active faults are at the east side of the Sevier orogenic belt (Cook and Smith, 1969, p. 698), and the easternmost faults, with relatively less displacement, involve little or no valley-fill deposits.

6. The elevations of Lake Bonneville shorelines in Utah apparently indicate an absolute uplift, with respect to sea level, of horsts (Bissell, 1959, p. 1710; Osmond, 1960, p. 41), and also absolute subsidence of the grabens (Crittenden, 1963, p. 28). These opposed movements are possible within the framework of this hypothesis.

7. Some range fronts exhibit linear scarps and faceted spurs which strongly express their boundary faults. In others, boundary faults can be referred to a polygonal system of faults. In still others, the range fronts are so ragged that through-going boundary faults cannot be inferred. These latter range fronts may represent pull-apart structures that are not sharpened by basement cutting faults.

8. Basement faults occur at the faulted edges of horst and graben blocks in the sedimentary section that resulted from earlier extension rearward on the gravity glide sheets. The rare observation of basement involvement in normal faulting does not necessarily mean that all faults cut basement when these rocks are not observed.

9. The rare cases of "thrusts" that place younger rocks over older in orogenic belts may be interpreted as normal faults that flatten with depth, as described by Moore (1960), and with rotation in addition to displacement at the boundaries of blocks residual in the extensional zone at the back of thrust sheets. Billings (1933, p. 162) has pointed out the difficulty in distinguishing normal faults from thrusts that place younger rocks over older.

10. The gross form of the former uplift is apparent in a profile drawn on the top of the basement in the major horsts, or the major grabens.

Problems in Observation. — Normal faulting in the hinterland may be divided into two stages: (1) that which occurs over the uplift, as a result of gliding in the flanks and is contemporaneous with it, and affects only sedimentary strata above the décollement zone(s); and (2) that which cuts

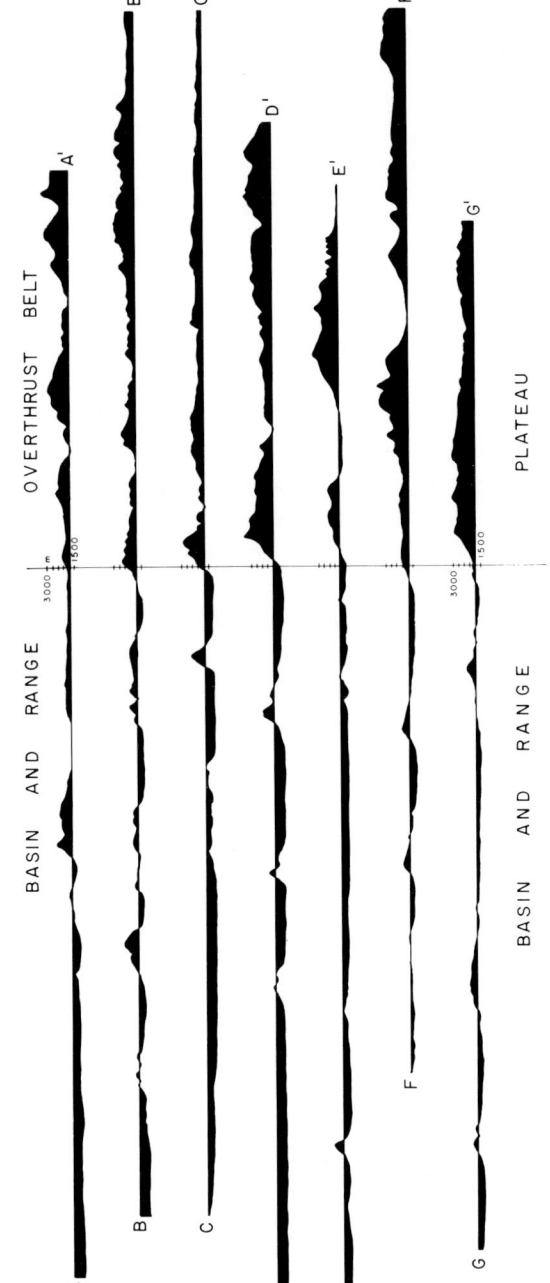
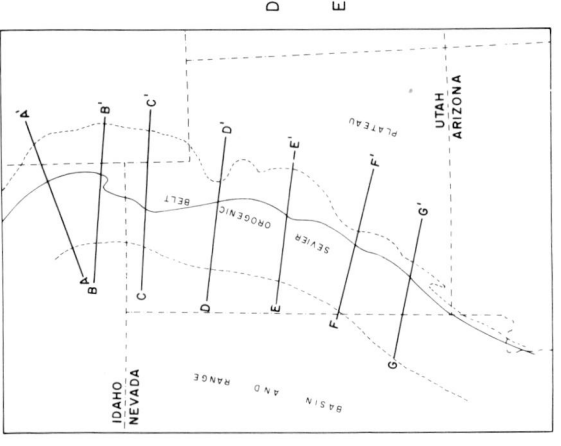

Fig. 5 — Topographic profiles across the Sevier orogenic belt (Basin and Range Province) and adjacent portions of the shelf (Plateau Province). Datum is 1800 m elevation. Solid line on location map is boundary between basin-range and plateau provinces, the Sevier orogenic belt occupies the area between the dashed lines.

and displaces the top of the basement and generally follows folding and thrusting, though its locale is the same as that of the earlier, more superficial faulting.

Separating the effects of these two faulting events may be difficult in the field. First, they possibly overlap in time. Longwell (1950, p. 427) and Nolan (1935) present evidence for Late Cretaceous normal faulting that cuts and displaces thrust sheets in Nevada and Utah, and is apparently contemporaneous with thrusting. Similar evidence for normal faulting within the thrust mass, contemporaneous with the emplacement of the Helvetischen decken in eastern Switzerland, has been illustrated by Trümpy (1969). Deep-seated normal faulting began in lastest Eocene in the Basin and Range province (Hamilton and Meyers, 1966, p. 527). Effusion of ignimbrites over a vast area in the Sevier belt began soon after thrusting in late Eocene (Mackin, 1960, p. 101), and may reflect the onset of the more fundamental faulting that cut the basement and provided fissures for the ascent of magma.

Secondly, the faults develop in the same region; namely, they are concentrated where curvature of the uplift is greatest. Thirdly, their traces will tend to coincide inasmuch as the greatest concentration of shearing strain energy in the shallow basement is at the boundaries of the faulted blocks in the sedimentary rocks above (Nadai, 1931, p. 246). Finally, faulting in the basement will also cut and displace all superjacent sedimentary rocks.

The depositional evidence for faulting, that is, the valley-fill deposits in the grabens, will tend to show a greater association with the latter faulting that cuts the basement. This is so because the entire basement surface uplifts by reason of horst and graben development in the sediments during tectonic denudation, whereas the basement surface beneath the grabens is lowered relative to the horsts during later extensional faulting in the basement.

CONCLUSIONS

The hypothesis advanced here holds that there is a genetic relationship among the foredeep basin at the shelf edge, the compressional fold and thrust belt structures, and the zone of extensional horsts and grabens in its hinterland. The surprisingly old bedrock encountered beneath valley fill sediments in intermontane basins and the lack of a magnetic expression of basement faults are not accounted for in existing tectonic hypotheses for this area. The problem of a rearward root zone for the overthrust faults, which are so dramatically displayed in the pile-up zone at the shelf edge, disappears in the tectonic scheme envisioned here.

This tentative model reconciles many observations in Utah's miogeosyncline-shelf boundary zone, but requires much additional confirming information before it can be considered demonstrated. It does have serious ramifications, however, for those considering a drilling venture in the eastern Great Basin. To the extent the model is applicable the potential for hydrocarbon accumulations in intermontane basins is extremely limited, and certainly reduced in the ranges.

REFERENCES CITED

Armstrong, R. L., 1968, Sevier orogenic belt in Nevada and Utah: Geol. Soc. America Bull., v. 79, p. 429-458.

Billings, M. P., 1933, Thrusting younger rocks over older: Am. Jour. Sci., v. 25, p. 140-165.

Bissell, H. J., 1959, Wasatch fault in central Utah (abs.): Geol. Soc. America Bull., v. 70, p. 1710.

Condie, K. C., 1969, Geologic evolution of the Precambrian rocks in northern Utah and adjacent areas, in Jensen, M. L., ed., Guidebook of northern Utah: Utah Geol. and Mineral Survey Bull. 82, p. 71-95.

Cook, K. L., and Smith, R. B., 1967, Seismicity in Utah, 1850 through June 1965: Seismol. Soc. America Bull., v. 57, p. 689-718.

Crittenden, M. D., Jr., 1963, New data on the isostatic deformation of Lake Bonneville: U.S. Geol. Survey Prof. Paper 454-E, 31 p.

Crosby, G. W., 1970, Radial movements in the western Wyoming salient of the Cordilleran overthrust belt: Reply: Geol. Soc. America Bull., v. 81, p. 3507-3512.

_____, 1971, Gravity and mechanical study of the great bend in the Mexia-Talco fault zone, Texas: Jour. Geophys. Res., v. 76, p. 2690-2705.

_____, 1972, Dual origin of Basin and Range faults, in Baer, J. L. and E. Callaghan, eds., Plateau-Basin and Range transition zone, central Utah: Utah Geol. Assoc. Publ. 2, p. 67-73.

Dallmus, K. F., 1958, Mechanics of basin evolution and its relation to the habitat of oil in the basin, in Weeks, L. G., ed., Habitat of Oil: Am. Assoc. Petroleum Geologists, Tulsa, Oklahoma, p. 883-941.

Hamilton, Warren, and Myers, W. B., 1966, Cenozoic tectonics of the western United States: Rev. Geophys., v. 5, p. 509-549.

Hintze, L. F., 1973, Geologic history of Utah: Brigham Young Univ. Geol. Studies, v. 20, pt. 3, 181 p.

Longwell, C. R., 1950, Tectonic theory viewed from the Basin Ranges: Geol. Soc. America Bull., v. 61, p. 413-434.

Mackin, J. H., 1960, Structural significance of Tertiary volcanic rocks in southwestern Utah: Am. Jour. Sci., v. 258, p. 81-131.

Moore, J. G., 1960, Curvature of normal faults in the Basin and Range province of the western United States, in Geological Survey Research, 1960, U.S. Geol. Survey Prof. Paper 400-B, p. B409-B411.

Nadai, A., 1931, Plasticity: New York, McGraw-Hill Book Co., 349 p.

Osmond, J. C., 1960, Tectonic history of the Basin and Range province in Utah and Nevada: AIME Trans., v. 217, p. 31-45.

Price, N. J., 1966, Fault and joint development in brittle and semi-brittle rock: Oxford, Pergamon Press, 176 p.

Roberts, R. J. and Crittenden, M. D., Jr., 1973, Orogenic mechanisms, Sevier orogenic belt, Nevada and Utah, in DeJong, K. A. and Scholten, R., Gravity and Tectonics: New York, John Wiley & Sons, p. 409-428.

Shuey, R. T., 1975, Aeromagnetic map of Utah: Utah Geol. and Min. Survey.

Stewart, J. H., 1971, Basin and Range structure—a system of horst and grabens produced by deep seated extensions: Geol. Soc. America Bull., v. 82, p. 1019-1044.

Stokes, W. L., 1969, Stratigraphy of the Salt Lake region, in Mead, L. J., Guidebook of northern Utah: Utah Geol. and Mineral Survey Bull. 82, p. 37-49.

Trümpy, Rudolf, 1969, Die helvetischen Decken der Ostschweiz: Eclogae Geol. Helvetiae, v. 62, p. 105-142.

photo by Peter J. Varney

View of northwest side of Big Cottonwood Canyon. The conspicuously folded and bedded rocks are part of the Upper Precambrian Big Cottonwood Formation.

STRUCTURAL EVOLUTION OF CENTRAL UTAH — LATE PERMIAN TO RECENT

by

James L. Baer[1]

INTRODUCTION

The central Utah area (Fig. 1) has been the object of a number of recent seismic surveys. These surveys have produced sufficient encouraging data to warrant the investment in five exploratory wells drilled since 1971, with another major drilling endeavor currently underway southeast of Moroni, Utah (Hanson-Moroni, Sec. 14, T15S, R3E, Sanpete County, Utah). While seismic surveys and drilling operations have been relatively active in the past five to six years, geologic investigations have gone on almost uninterrupted since the 1870's. The major work has been done by geology students and faculty of the Ohio State University, largely under the indefatigable supervision and inspiring tutelage of Professor Edmund M. Spieker. Under his direction literally hundreds of formal and informal geologic investigations have been conducted in the area. This mass of data is too numerous to acknowledge separately and certainly too important and fundamental to ignore or to be regarded lightly. However, in light of the new data both from seismic surveys and drilling operations it is relevant that this area should be reexamined and a more refined hypothesis of structural genesis offered.

Therefore it is the purpose of this paper to propose a tectonic model for the study area from Late Permian to the Recent utilizing this new data base as well as personal field observations made periodically in the area since the summer of 1956 when I first came to Sanpete Valley driving a field vehicle behind a car driven by Professor Ed Spieker.

GENERAL GEOLOGIC SETTING

The study area is within the transition zone between the block-faulted Basin and Range province and the uplifted plateaus of the Colorado Plateau province. As such it sits astride the Cordillearean Hingeline and much of its geologic activity resulted from tectonic events focused along this zone of weaker crust.

Sedimentation patterns and structural development in the area are responses to the uplift and the compressional phases of the Sevier orogeny (Late Triassic-Late Cretaceous), the uplift and milder compressional phases of the Laramide orogeny (latest Cretaceous-Oligocene), and the Basin and Range deformation (post-Oligocene through the Recent). Volcanic events

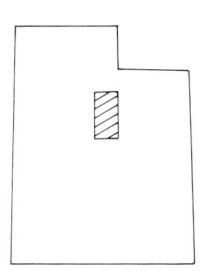

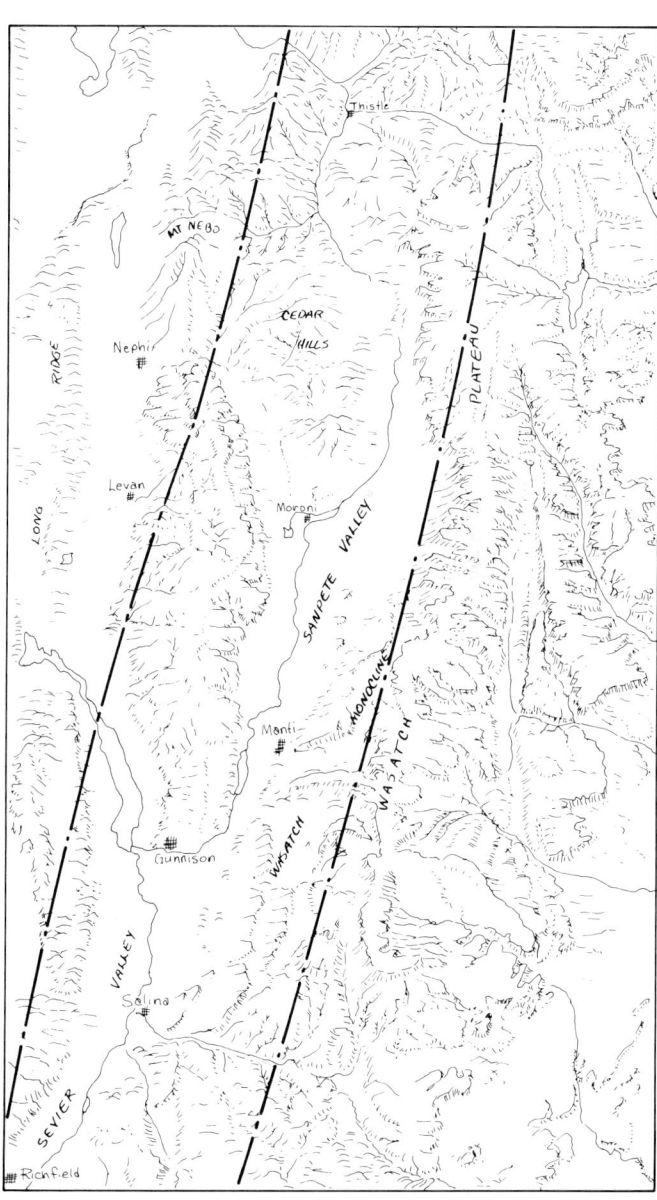

Fig. 1 — Index map, Sanpete-Sevier valleys area, central Utah.

[1] Department of Geology, Brigham Young University, Provo, Utah.

are associated with the latter two orogenic episodes. Superimposed upon these broad regional tectonic events is a dramatic sequence of evaporite diapirism that uplifted, contorted, and collapsed adjacent sedimentary units. This evaporite diapirism has been referred to by Gilliland (1963), Hardy (1952), Moulton (1975), and Runyon (In Press). Periodic diapirism occurred within the area from Middle Cretaceous through the Pleistocene. It is thought this diapirism is the principal cause for most of the unusually complex structural features such as strip-thrusting, "double angular unconformities", and dual monocline development described by Spieker (1949).

Bedrock in the study area is largely Cretaceous varying from marine sedimentary units on the east to synorogenic clastics on the west. Tertiary units, indicative of a fluvial and lacustrine setting, make up most of the low-gradient topography. These units were uplifted and eventually were overlain by indigenous as well as transported, intermediate to silicic volcanics; principally during Oligocene-Miocene time. The oldest rocks exposed in the area are Jurassic and are found largely as diapirs.

STRATIGRAPHY

It is not the purpose here to give an exhaustive treatise of the stratigraphy involved. Good general discussions on the Permian through Tertiary rocks can be found in guidebooks published by the Utah Geological Society (1946); Brigham Young University Geology Studies, number 9, part 1 (1962); and the Utah Geological Association (1972). The purpose is to examine the general lithologies, distribution, correlation, and tectonic significance of the Late Permian to late Tertiary rock units and use this information as a base for a structural genetic model.

Information on the Permian and Triassic rocks of the area is sketchy, with rocks of these ages found only in the subsurface. However, Jurassic through Tertiary rocks are found in fair to excellent outcrops throughout the area. Identification of most of these units can be done with high confidence but some outcrops, critical to the structural interpretation, defy confident identification because they lack critical fossils or distinctive lithologies. Early investigators, for instance, identified steeply dipping red beds as one Jurassic unit or another primarily on the basis of their structural attitude. However, this criterion is not valid because palynologic samples taken from some of these outcrops previously designated as Jurassic yielded diagnostic early Tertiary palynomorphs. Conglomerate units like the Jurassic Morrison Formation, Cretaceous Indianola Formation and Price River Formation, and the Cretaceous-Tertiary North Horn Formation can be confused. Recent work done by James Stolle on conglomerates near the Canyon Range that were previously mapped as Indianola yielded Paleocene microfossils and he considers these conglomerate units as Flagstaff Limestone equivalents (Personal Communication, 1974).

Permian

There are no Permian rocks exposed at the surface that are indigenous to the area. Permian rocks identified as Oquirrh Formation, Kirkman Limestone, Diamond Creek Sandstone, and Park City Formation outcrop in the Loafer Mountain area but are part of the allocthonous block of the Nebo thrust plate (Metter, 1955). All the above named units are marine in origin and are part of a shelf sequence of rocks that have been tectonically transported an estimated 5 to 30 mi by thrusting during the Sevier orogeny.

Phillips Petroleum Company wells No. 1 United States "D" (Sec. 20, T22S, R3E) and No. 1 United States "E" (Sec. 27, T19S, R3E) both encountered Permian rocks in the subsurface. In the "D" well 325 ft of rock are identified as Kaibab Formation, White Rim (?) Sandstone, and Toroweap Formation. This sequence is a series of light gray, cherty dolomites alternating with light gray sandstones and siltstones with occasional thin beds of anhydrite. Some 600 ft of similar but more thick-bedded lithologies were penetrated in the "E" well and were designated as Kaibab, Toroweap, and Elephant Canyon formations. The Permian rocks in these two wells are interpreted to represent a restricted marine environment, perhaps a lagoon within a fringing reef, or some other protected shallow-water sediment trap.

Triassic

Triassic rocks also are found at the surface only in the allocthonous plate. These rocks are representative of a marine shelf sequence of limestones, siltstones, and shales. In the "D" well 2,120 ft of sandstone, siltstone, shale, and limestone, indicative of marginal marine, tidal flat, and delta environments, was penetrated. In the "E" well, 1,780 ft of Triassic rock was encountered. The lithologies were similar to those in the "D" well, but there is a preponderance of shallow marine limestone and dolomite at the base of the section.

Extensive subaerial erosion took place after the deposition of Wordian age rocks and Triassic rocks were deposited over a significant unconformable surface throughout western Utah and eastern Nevada (Collinson, et al., 1976). Collinson describes a westward migration of Triassic sediments upon an east-facing paleoslope. The development of this disconformity coincides with the timing of the Sonoma orogeny which was most evident in western Nevada. Speculation as to the cause of this orogeny and its affect in central Utah will be presented later.

Triassic-Jurassic

The Navajo Sandstone is one of the major potential oil and gas reservoirs in the Rocky Mountain states. Discussions as to its origin and environment of deposition have been lengthy and sometimes heated. Without attempting to resolve the problems and lines of evidence as to whether the Navajo Sandstone is marine or eolian in origin, it should be stated that this unit appears to have facies suggestive of both origins. Observations from both surface outcrop and well cuttings and logs indicate that the Navajo Sandstone has both marine and nonmarine affinities depending upon place and time of deposition. In central Utah, the Navajo Sandstone appears to be dominantly nonmarine both in outcrop and where penetrated by wells.

Thicknesses of 910 ft and 740 ft of Navajo were encountered in the "D" and "E" wells respectively. Some 1400 ft of Navajo was penetrated in the Standard of California No. 1 Levan Unit (Sec. 17, T15S, R1E) and their No. 1 Sigurd Unit (Sec. 32, T22S, R1W) bottomed approximately 650 ft into sandstone that was apparently eolian in origin and is thought to be Navajo. The Navajo Sandstone outcrop at Thistle Junction is 1450 ft thick and appears to be totally eolian in origin.

Jurassic

The Arapien Shale, Twist Gulch Formation, and Morrison Formation make up the bulk of a thick sequence of evaporites, siltstones, shales, limestones, and conglomerates that makes central Utah geology what it is.

Basically the sequence changes with time from the underlying Navajo Sandstone to a thick evaporite sequence with interbeds of shale and siltstone, then to carbonates and finally to coarse clastics. Total thickness of the Jurassic is unknown. Hintze (1973) suggests a thickness in excess of 6700 ft with the majority made up of Arapian Shale. Some investigators indicate that the Arapien may exceed 10,000 ft in thickness in the subsurface. These estimates are based on seismic records that are not available for private investigation. Regardless of total thickness determinations, the Arapien with its considerable volume of evaporites indicates a significant departure of the sedimentary pattern in central Utah.

The Arapien was penetrated in all the previously mentioned wells and in addition was encountered in the Phillips Petroleum wells No. 1 Price "N" (Sec. 29, T29S, R3E) and No. 1 Wasatch Plateau Unit (Sec. 24, T21S, R1E). Arapien outcrops are readily apparent in the southern half of Sanpete Valley and dominate the Sevier Valley (Fig. 2). Wherever they are exposed the Arapien and the Twist Gulch show strong indications of diapiric activity (Figs. 3, 4).

The depositional setting of the Arapien is enhanced by correlation with adjacent Jurassic units to the east and west of the study area. East of the Wasatch monocline the evaporites quickly changes facies. Only 250 ft of evaporites was encountered in the Phillips "D" well and scattered evaporites with one salt unit with a maximum thickness of 40 ft was found in the "E" well. The record to the west is less evident but apparently the evaporites change to limestones. These changes both in thickness and lithology indicate that the central Utah area was the site of a long trough, extending over 100 mi north-south and being 20 to 25 mi wide, (Fig. 1). The axis of the trough trends N 15° E and a high-angle normal fault makes up the east boundary. Moulton (1975) refers to this fault as the

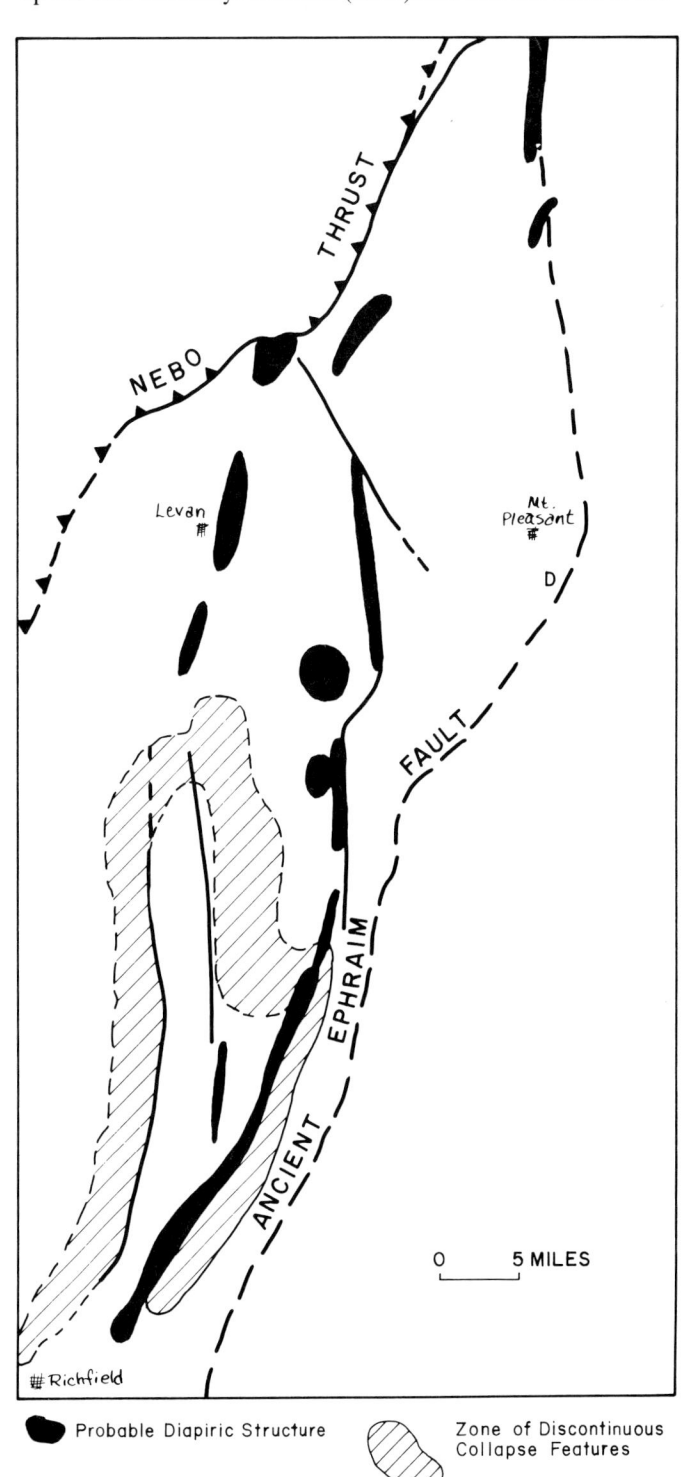

Fig. 2 — Areas of probable diapiric structure and associated collapse zones, central Utah.

ancient Ephraim fault and to the trough as the Sanpete-Sevier rift.

The trough apparently is asymmetrical with the dominant movement on the east fault allowing the thickest sediments to accumulate along the eastern margin. This half-graben tectonic activity apparently occurred throughout Jurassic and into Cretaceous time. In the Cretaceous, continued rifting produced a linear trough that received over 18,000 ft of clastics that were shed from the slowly eastward-migrating Sevier orogenic front. Burchfiel and Hickcox (1972) show this relationship along with another similar basin located in northern Utah and southwestern Wyoming (Fig. 5). The significance of these basins will be discussed later.

Cretaceous

Rocks of Cretaceous age in central Utah are of two types. A thick sequence of paludal and restricted marine rocks are the dominant facies with a thick synorogenic coarse-clastic facies in the western part of the study area.

Outcrops and subcrops of Cretaceous are numerous and the character of these rocks is fairly well described and will not be

Fig. 3 — Gypsum diapir at mouth of Salt Creek Canyon, east of Nephi, Utah.

Fig. 4 — Diapiric fold in Salt Creek Canyon, east of Nephi, Utah.

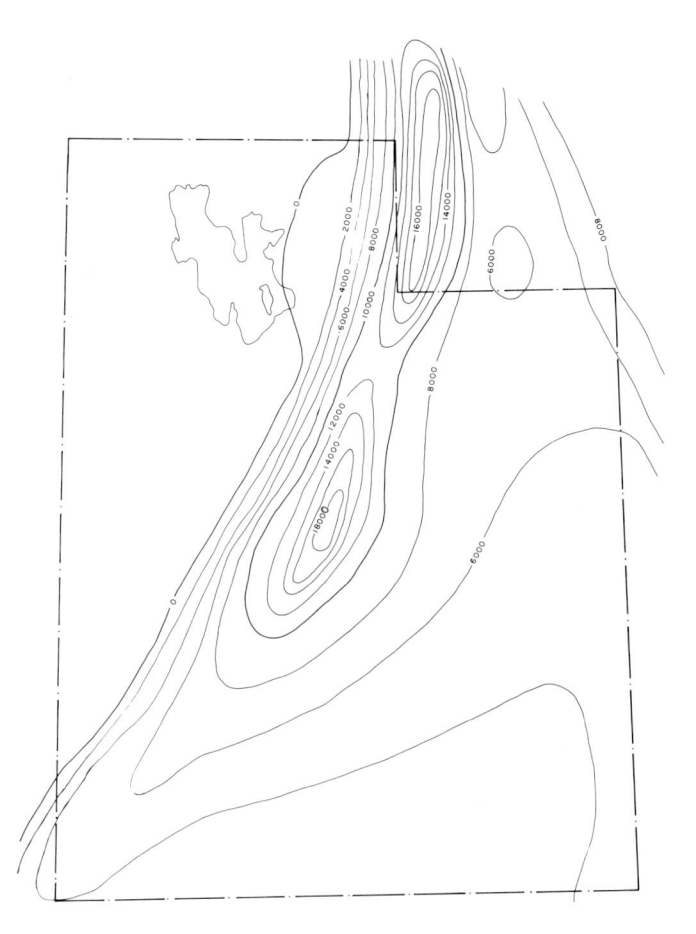

Fig. 5 — Isopach map, total Cretaceous.

repeated here. The reader is referred to Hale (1972) for a more detailed account. The Cretaceous in central Utah was a time of transition from a coastal plain-type of environment to a fluvial-lacustrine environment developed near or at sea level with slowly eroding mountains making up the western horizon and low hills beginning to develop to the east.

Tertiary

One of the world's finest examples of lake-fluvial deposition occurs in the subject area. Rocks that make up portions of the North Horn Formation and all of the Flagstaff Limestone, the Colton Formation, the Green River Formation, and the Crazy Hollow Formation are representative of lakes, deltas, and rivers that formed in a subtropical environment. The presence of fossil remains of crocodiles, hippopotami, and even camels along with a fossil flora indicative of a near sea level freshwater environment all attest that central Utah during Tertiary time was much like the Okfenokee Swamp of present-day northern Florida and southern Georgia.

This environment was fairly stable throughout the early Tertiary. Periodic volcanic eruptions, probably from volcanoes concentrated along opening rift faults to the west, spewed volcanic dust into the lakes that later became the volcanic tuffs found throughout the early Tertiary formations. The volcanic activity was attendant to broad, epeirogenic uplift that began to lift the area differentially. This uplift episode became more evident at the close of Green River deposition. The Green River Formation thins eastward onto the present Wasatch monocline and southward onto the developing Marysvale volcanic complex. This indicates that the monocline development had its inception by the end of the Eocene.

Another group of structural features became more apparent at about this time. Sedimentary loading of the evaporite beds reached the critical limits for diapirism to begin sometime during the Cretaceous. Hints of upturned Cretaceous marine shales can be seen in selected spots below the alluvial apron near the front of the Gunnison Plateau north and west of Wales, Utah. Tertiary units from the North Horn through the Green River Formation were contorted, uplifted, and in some cases collapsed as the diapirism became more active. In the Redmond, Utah area, Tertiary units correlative to the Green River and Colton formations, as well as some younger units (possibly correlative to the Crazy Hollow Formation), can be found in small fault blocks and in contorted steeply dipping rock units. The "double angular unconformity" near Christianburg is interpreted to be the result of recurrent diapirism along the western edge of the Sanpete Valley.

Diapirs are common in the area. The Sevier Valley diapir has yielded commercial salt and gypsum for over a century. Recently the gypsum diapir at the mouth of Salt Creek Canyon east of Nephi has been reopened for gypsum quarrying (Fig. 4).

Interpretation of seismic lines indicates that much of Sanpete Valley contains diapiric structures at depth. At least six wells drilled in or near the valley have encountered significant thicknesses of salt and/or anhydrite. In one well, the Phillips Petroleum No. 1 Price "N" (Sec. 29, T15S, R3E), over 2000 ft of evaporites were drilled. Figure 2 is a map of the known and suspected salt diapirs and related structures. The larger of these will be discussed in the structural genesis portion of this paper.

Of interest to the tectonic picture is the presence of Oligocene-Miocene, with some possible latest Eocene, igneous intrusions and extrusive rocks located in the Levan area and in the western portion of the Cedar Hills. John (1972) and Schoff (1951) report a series of silicic to intermediate dikes, sills, and small (the largest is slightly over one sq mi in area) plutons that intruded rocks as young as the Green River Formation. Extrusive rocks are principally water-lain agglomerates and tuffs with some ashflow tuffs (Fig. 6). The source of these extrusives is suggested by John (1972) to be the Levan area intrusive bodies. He cites as evidence the close proximity and the similarities of mineral and chemical composition. The intrusive bodies, mainly the plutons and dikes, follow a general trend of N 20-30° E.

Fig. 6 — Miocene waterlain volcanics, upper Salt Creek Canyon, Utah.

STRUCTURAL EVOLUTION OF CENTRAL UTAH

Central Utah structural geology is complicated mainly by the presence of the diapirs. Upturned sediments dating from Middle Cretaceous through Eocene age can be found throughout Sanpete and Sevier valleys. Gilliland (1963) mentions most of the angular discordance found within this area. He further mentions apparent structural relief in the order of 15,000 to 20,000 ft. His estimate based upon surface data is fairly well substantiated by recent well and geophysical data.

If the diapirism were not present, the tectonic history of central Utah would be less spectacular. From Permian to the present it would be characterized by a series of relatively mild episodes of uplift and subsidence. The implication of these broader regional episodes is related to the plate interactions along the western margin of the North American plate. These interactions are only beginning to be interpreted and are subject to considerable differences of opinion. The structural evolution of central Utah will be reviewed in respect to a postulated plate tectonic model.

Table 1 is a compilation of the central Utah stratigraphy, interpreted environments of deposition, inferred tectonics, and implied regional plate tectonic event. With this figure as a base the discussion follows.

Sequence of Structural Events

Permian rocks autochthonous to the area are thin, restricted-marine carbonates with some evaporites. These rocks are interpreted to have been deposited in shallow water depths, probably in a restricted-shelf environment. After the deposition of the Permian rocks the area was uplifted, more properly slightly tilted to the east. The Uncompahgre uplift to the east was still a positive feature but was being gradually overridden by sediments. The presence of a major erosional surface separating Triassic from Permian rocks indicates that much of the western United States was experiencing subaerial erosion for millions of years. The interpreted plate tectonic episode is that of gradual uplift as the western North American plate was either experiencing collision, subduction, or possibly both. The central Utah area was, in all probability, experiencing a slight compressional

Table 1. Inferred Sequence of Geologic Events — Central Utah

Time	Dominant Sedimentation	Inferred Environment	Inferred Tectonics	Comments
Tertiary	Volcanics Clastics Shale, Limestone Limestone Shale, Siltstone Coarse clastics	Volcanics Fluvial Lacustrine Fluvial	General uplift, volcanism more prevalent, diapirism creates many local structures, block faulting persists through present	Local diapirism causes much of present day structure. Differential uplift probably due to isostatic adjustments as a result of lower crust-upper mantle underplating
	---- Erosional Phase ----			
Cretaceous	Coarse clastics Tuffs Shale, Siltstone	Fluvial Deltaic Restricted marine	General uplift to west-orogenic front migrating eastward, some diapirism	Thick Cretaceous clastics fill in tectonic downwarp, shallow marine conditions prevail, some periodic volcanism
	---- Erosional Phase ----			
Jurassic	Coarse clastics Siltstone Limestone Fine clastics Evaporites	Fluvial Delta Marinal Marine Restricted marine	Uplift to the west with accompanying fault-basin development to the east	Rifting caused asymmetrical basin to appear in central Utah. Basin was site for thick evaporites
Triassic	Sandstone Siltstone, Shale Limestone Siltstone, Shale	Eolian Deltaic Nearshore marine Some continental deposits	Differential uplift perhaps along faults, then graben development, then general uplift	Migration of sedimentation from east to west. Both marine and nonmarine sedimentation patterns may indicate isolated seas resulting from plate collision
	---- Erosional Phase ----			
Late Permian	Cherty limestone with evaporites	Restricted marine	Epeirogenic uplift, reactivation of movement along previously existing faults suspected	Uncompahgre uplift minor influence on sedimentation — Sonoman orogeny progressing from west creating east-facing paleoslope — perhaps initial phase of plate collision to the west

strain along with the uplift.

The autochthonous Triassic rocks in the area were formed in marginal marine and deltaic environments. Directional studies indicate a general northwesterly movement of sediments across Utah from the Uncompahgre and related uplifts that were still mildly positive. Discontinuous marine basins apparently developed in Nevada during the Triassic. An island arc in the westernmost part of Nevada was the dominant tectonic feature. This feature may have begun to subduct after a brief collision phase with the interior portion of the North American plate. The result of this interaction in central Utah was gradual uplift and development of a desert that was marginal to shallow inland seas. This was probably the beginning of the Sevier orogeny that was to be more strongly felt in the Jurassic and Cretaceous.

The most divergent tectonic feature in the Jurassic of central Utah was the development of the fault basin in which the Arapien evaporites developed. This basin was in all probability asymmetrical east-west, being deeper along the more active eastern boundary fault. It was also asymmetrical in a north-south direction with the deeper, most downdropped end, being to the south. The basin was also broadly arcuate, being concave toward the west. This was not a rift basin in the sense of an active upwelling region but rather a tensional feature that may have been caused by the major uplift that developed to the west as the Sevier orogeny continued to grow. Differential rotation of intraplate subdivisions may have contributed to its development, but the dominant causal factor is thought to be the western uplift. The present-day Uinta block must have influenced the positioning of this type of basin because recent drilling north of the Uinta Mountains along the Idaho-Wyoming border indicates the presence of a similar basin in that area (Fig. 5).

As the Sevier orogeny moved slowly eastward, a linear foreland basin developed into which the synorogenic clastics were deposited. The present-day Cretaceous thicks coincide with the approximate location of the Jurassic fault basins. Thrusts that are the dynamic portion of the orogeny never reached the central Utah area except in the Nebo Thrust near Nephi. Compressional forces associated with the thrusting deformed the Cretaceous rocks to a degree but it was salt diapirism that was the principal cause of deformation. The major angular unconformity that can be seen in Salina Canyon, Spanish Fork Canyon, and Lake Fork Canyon is probably the result of both the compressional phase and the diapirism. There is no major unconformity on the west side of the evaporite basin because the salt was thinner and the unconformity here was milder and later partially obliterated by late diapirism and faulting (Fig. 7).

Near the close of the Cretaceous the area began to gradually uplift as evidenced by the transition from marine sedimentation to continental sedimentation. North Horn, Flagstaff and Green River sediments filled lakes near sea level. Diapirism reactivated throughout the early Tertiary caused islands to appear in these lakes. Jurassic and Cretaceous rocks were exposed periodically and were eroded.

During Oligocene-Miocene the area continued to uplift but more differentially with the Wasatch monocline becoming a prominent feature. The major reason for the uplift in the western United States was a reactivation of the Pacific plate moving under the North American plate. This movement is not a simple collision; strike-slip motion and volcanism also played a role. It is thought that strike-slip movement along some previously existing faults and fractures caused some strike-slip movement as far east as central Utah. Thin portions of the crust became centers for volcanism and in central Utah it was along the N 20-30° E fractures that volcanism was centered.

Uplift continued throughout the Miocene and is probably continuing today. The rate of uplift since the Eocene may be on the order of 5 to 20 m per million yrs. The ultimate cause of uplift in central Utah is suspected to be subcrustal underplating. Material, moving at or near the crustal-mantle interface, became more stable as it cooled and then welded to the crust. This additional material caused the area to adjust isostatically. Uplift for this reason acted differentially in central Utah. The Wasatch monocline uplift is partially due to recurrent movement on the ancient Ephraim fault of Moulton (1975). Later movements caused by the isostatic adjustments to crustal underplating may have caused the diapirism to reactivate. Diapiric ridges developed along faults along the east side of the Gunnison Plateau and near Indianola, Utah (Runyon, In Press).

Summary and Conclusions

Central Utah has experienced, over the last 230 m.y., gradual uplifts and depressions. The cause of these movements is gradual but persistent plate motion combined with the flow of material in the subcrust. Compressional structrual features are almost totally absent in central Utah. The principal causes of structures are tension and diapirism. The evaporite basin developed during the Jurassic is thought to be caused by continental plate rotation and/or tension caused by uplift. After the Cretaceous the dominant force was epeirogenic uplift believed to be caused by crustal underplating.

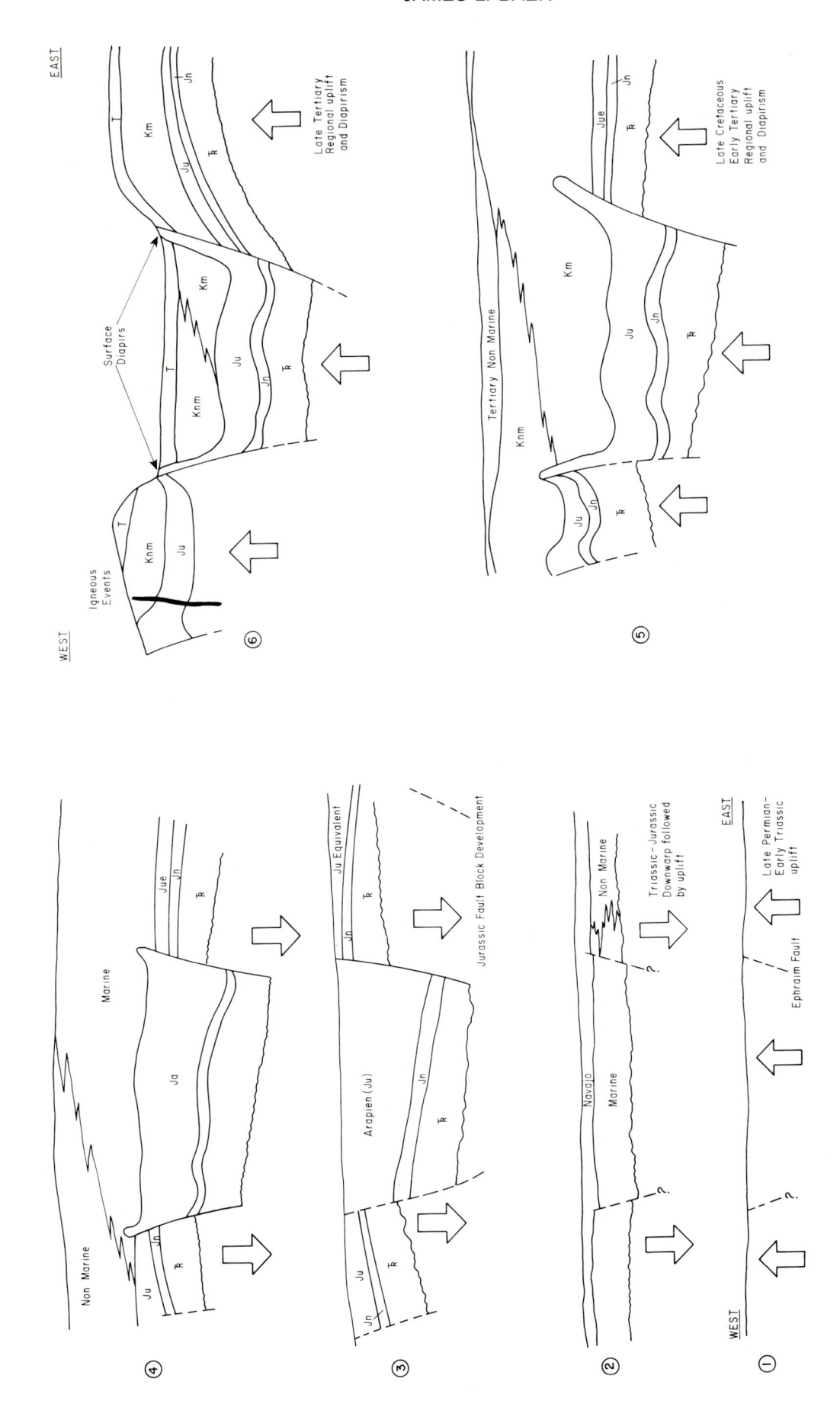

Fig. 7 — Inferred sequence of tectonic development, central Utah. Jn — Navajo Ss., Ju — Arapien Sh., Jue — Arapien equivalent, Knm — non-marine Cretaceous, Km — marine Cretaceous.

REFERENCES

Burchfiel, B. C., and Hickcox, C. W., Structural development of central Utah; Utah Geol. Assoc. Publ. 2, pp. 55-66.

Collison, J. W., Kendall, C. G. S. C., and Marcantel, J. B., 1976, Permian-Triassic boundary in eastern Nevada and west-central Utah: Geol. Soc. America Bull., v. 87, pp. 821-824.

Gilliland, W. N., 1963, Sanpete-Sevier Valley anticline of central Utah: Geol. Soc. America Bull., v. 74, pp. 115-124.

Hale, L. A., 1972, Depositional history of the Ferron Formation, central Utah: Utah Geol. Assoc. Publ. 2, pp. 29-40.

Hardy, C. T., 1952, Eastern Sevier valley, Sevier and Sanpete Counties, Utah, Utah Geol. and Mineral Survey Bull. 43, 98 p.

Hintze, L. F., 1973, Geologic history of Utah: Brigham Young Univ. Geology Studies, v. 20, Pt. 3, 181 p.

John, E. C., 1972, Petrology and petrography of the intrusive igneous rocks of the Levan area, Juab County, Utah: Utah Geol. Assoc. Publ. 2, pp. 97-107.

Metter, R. E., 1955, Geology of a part of the southern Wasatch Mountains, Utah, unpub. Ph.D. dissertation, Ohio State Univ.

Moulton, Floyd C., 1975, Lower Mesozoic and Upper Paleozoic petroleum potential of the Hingeline area, central Utah, Rocky Mtn. Assoc. of Geologists 1975 Symposium on deep drilling frontiers of the central Rocky Mountains, pp. 87-97.

Schoft, S. L., 1951, Geology of the Cedar Hills, Utah: Geol. Soc. America Bull., v. 62, pp. 619-642.

Spieker, E. M., 1949, The transition between the Colorado Plateaus and the Great Basin in central Utah: Utah Geol. Soc. Guidebook to the geology of Utah, no. 4.

Runyon, D. M., In Press, Structure, stratigraphy and tectonic history of the Indianola Quadrangle, Central Utah.

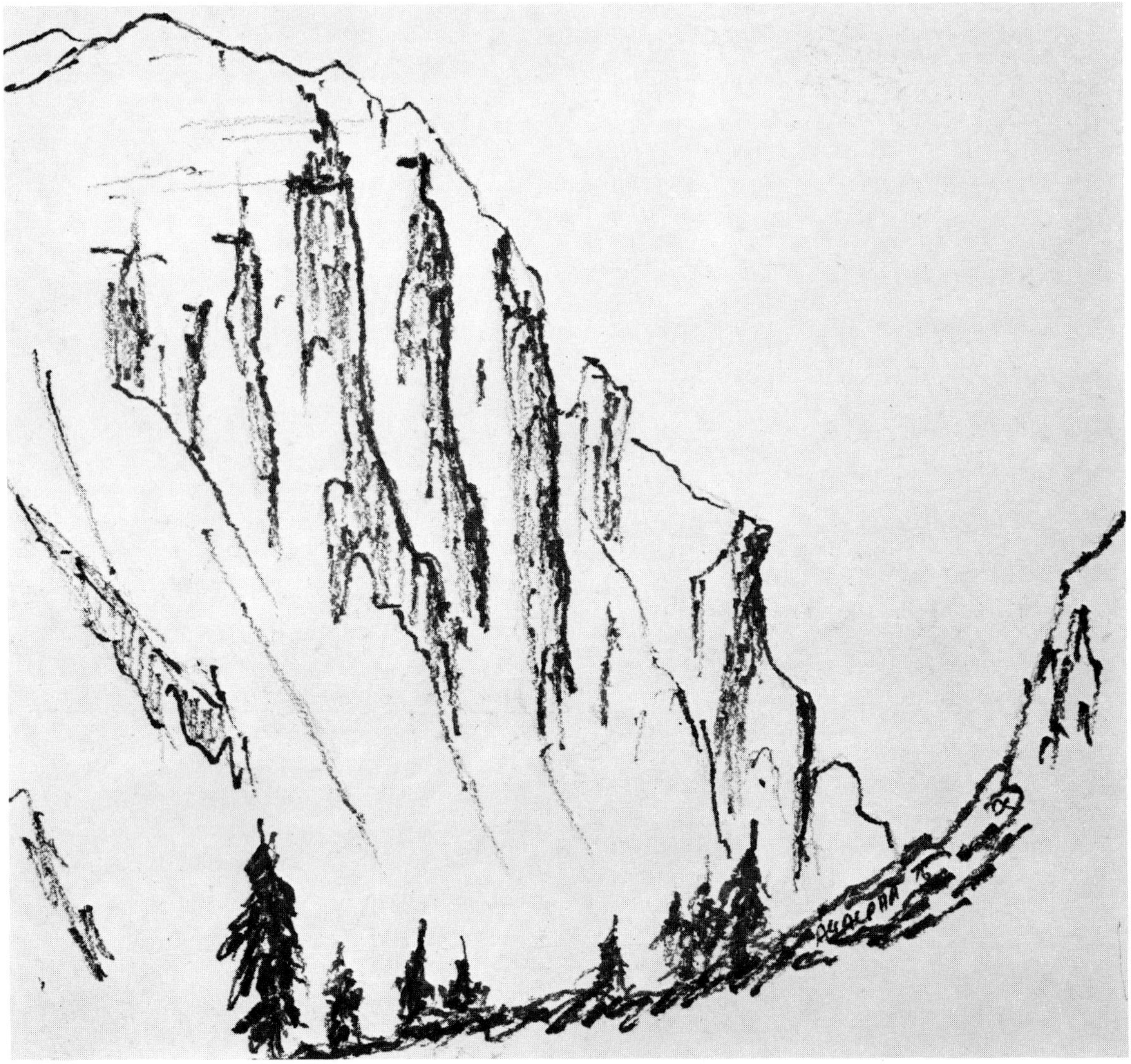

A PALEOSTRUCTURAL INTERPRETATION OF THE EASTERN GREAT BASIN PORTION OF THE BASIN AND RANGE PROVINCE NEVADA AND UTAH

by

E. L. Howard[1]

INTRODUCTION

The entire expanse of the physiographic Great Basin in the Basin and Range Province has long been subjected to many differing concepts as to the origin and timing of its complex paleostructural history. It is now generally accepted that almost all of the complexly involved fault tectonics that affected the rocks of the eastern Great Basin (eastern Nevada and western Utah) occurred since the middle of the Mesozoic Era. At this point the divergence of ideas on the sequence of ensuing structural events understandably widens. There is virtually no record of the sediments which may have been deposited between Middle Jurassic and early Tertiary time in the eastern Great Basin. As a result, the structural geologist is pardoxically confronted with an abundance of structural evidence, and yet, still lacks sufficient data to accurately piece it together in chronologic order.

This exercise is, then, a further attempt to resolve this "problem." It is the result of some eleven years of field work by the author in eastern Nevada and western Utah and is presented not as a final solution, but as a theory of practicality as applied to the perspective provided by many great geologic minds who have gone before and in the hope it will give cause for thought to those who will most certainly follow.

THE GREAT BASIN STRUCTURAL "PROBLEM"

For the most part, the theories that attempt to explain the complex structural history of the Great Basin can be broken down into two main schools of thought:

1. A majority of past and present authors believe that during the Nevadan and Laramide orogenies the entire eastern Great Basin was subjected to a series of major décollement-type (regional plane of detachment, or some variance thereof), east to west underthrusts that reached inferred displacements of a few miles to as many as 100 mi or more (Misch, 1960; Misch, Hazzard, and Turner, 1957).

2. A smaller group of mostly unpublished authors who have worked the Great Basin for private industry would agree that there was undoubtedly some "thrust" faulting associated with the Nevadan and Laramide orogenies, but that there is not enough evidence to prove that these faults were regional in scope. They also feel that many of these apparent "décollement" have been confused with gravity sliding or even normal faults that have been reoriented into an apparent thrust position by mid-Cenozoic block-fault mechanisms.

This author basically agrees with the latter group. There is undeniable evidence that mid-Mesozoic overthrusts occurred along the frontal edge of the Sevier arch as described by Harris (1959), and that the same type of faulting affected the rocks of southern Nevada at about the same time. (Longwell, 1922). In these localities and in canyon exposures along the western edge of the Wasatch Plateau, older rocks have been unquestionably thrust over younger sediments. In the central part of the Great Basin, however, the thrusts that are described as décollement are almost entirely younger rock over older rock relationships (Misch, 1960).

The given explanations for these kind of major, down-section thrusts are plausible, but it does require an unusual set of conditions to persist over an entire region in order to explain such a consistent occurrence of this type of faulting.

REGIONAL STRUCTURAL FRAMEWORK
Paleozoic

Although the pre-Mesozoic tectonics that affected the internal portion of the Great Basin are considered to have been mild in nature, they are not without importance when one considers that the genetic structural picture that evolved upward through the Paleozoic had a definite bearing on what happened structurally during the ensuing Mesozoic and Cenozoic Eras.

Beginning with Fortieth Parallel Surveys in the 1870's and into the decades to follow, many individual studies were carried out in various mining districts of Nevada and Utah.

---

[1] Reserve Oil and Gas Company, Denver, Colorado

Although many ideas of regional significance arose from these studies, there is little evidence that any real efforts were made toward the construction of a regional structural or stratigraphic synthesis.

Charles Schuchert (1923) placed on a map for the first time (Fig. 1) his concepts of the evolution of the North American continent. Drawing from the geosynclinal theories of Hall and Dana, Schuchert described the first outline of Paleozoic patterns significant to the Great Basin. He viewed the Great Basin in Paleozoic time as lying in the heart of the Cordilleran geosyncline, which was a subsiding trough between the neutral areas of Siouxia on the east, and the mobile borderland of Cascadia on the west. At this time there was little evidence to differentiate between the accumulating sedimentary facies within this broad belt.

Some five years later, T. B. Nolan (1928) presented evidence that the Cordilleran geosyncline was not a simple, uncomplicated trough filled with deep water sediments. He recognized that there was a difference between the eastern and western facies of the basin, and equated that the basin was bisected by an epeirogenic uplift that was later named by Eardly (1947); the "Manhattan geanticline". At about the same time, Marshall Kay (1947) had modified and adopted a geologic concept developed by the German geologist Hans Stille, who had recognized the Great Basin area as lying in an orthogeosyncline which was described as a highly mobile subsiding belt of sediments between broad, stable regions called "cratons". He further subdivided the orthogeosyncline into a western facies belt (eugeosyncline), which was highly unstable and receiving an abundant mixture of deep-water sediments and volcanics, and an eastern facies belt (miogeosyncline) that was receiving shallower-water type sediments and was more stable and less associated with volcanism. These subdivisions of the orthogeosyncline were later named respectively the "Frazier Belt" and the "Millard Belt" by Marshall Kay (1947) as illustrated in Figure 2.

Eardley's (1951) interpretation of the Mississippian shows an orogenic belt along the western edge of the miogeosyncline that, for the first time, was clearly dividing the facies of the two basins (Fig. 3). Later work by Roberts (1951) and Kay (1952) substantiated and further defined this orogeny, named the "Antler orogeny" by Roberts, as a massive overthrust of western facies onto eastern facies rocks.

Thus the earlier insipient Manhattan geanticline that appeared as early as the Late Ordovician had continued to slowly rise through the lower Paleozoic to eventually become a major orogenic belt that was to remain active throughout the rest of the Paleozoic Era. The effects of such a massive orogenic belt adjacent to the miogeosyncline to the east had an obvious influence on the amount and type of sediments that were

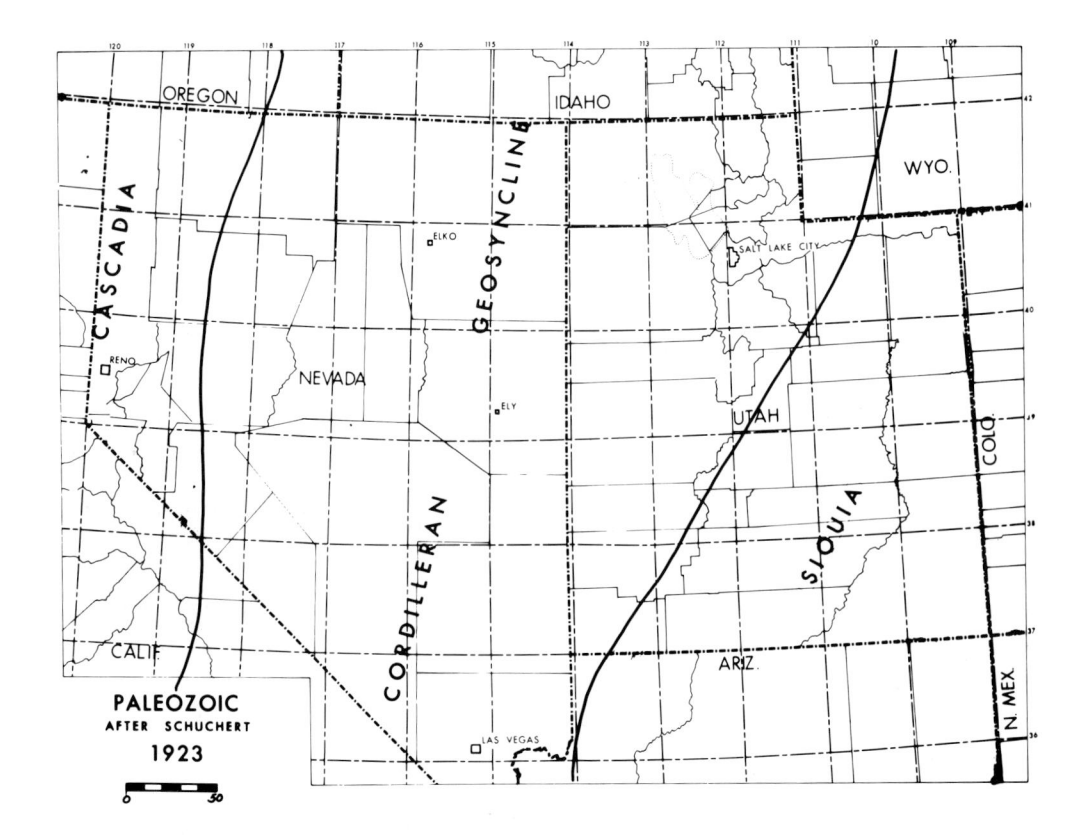

Fig. 1 — Cordilleran geosyncline and associated cratonic areas in the Great Basin as defined by Schuchert.

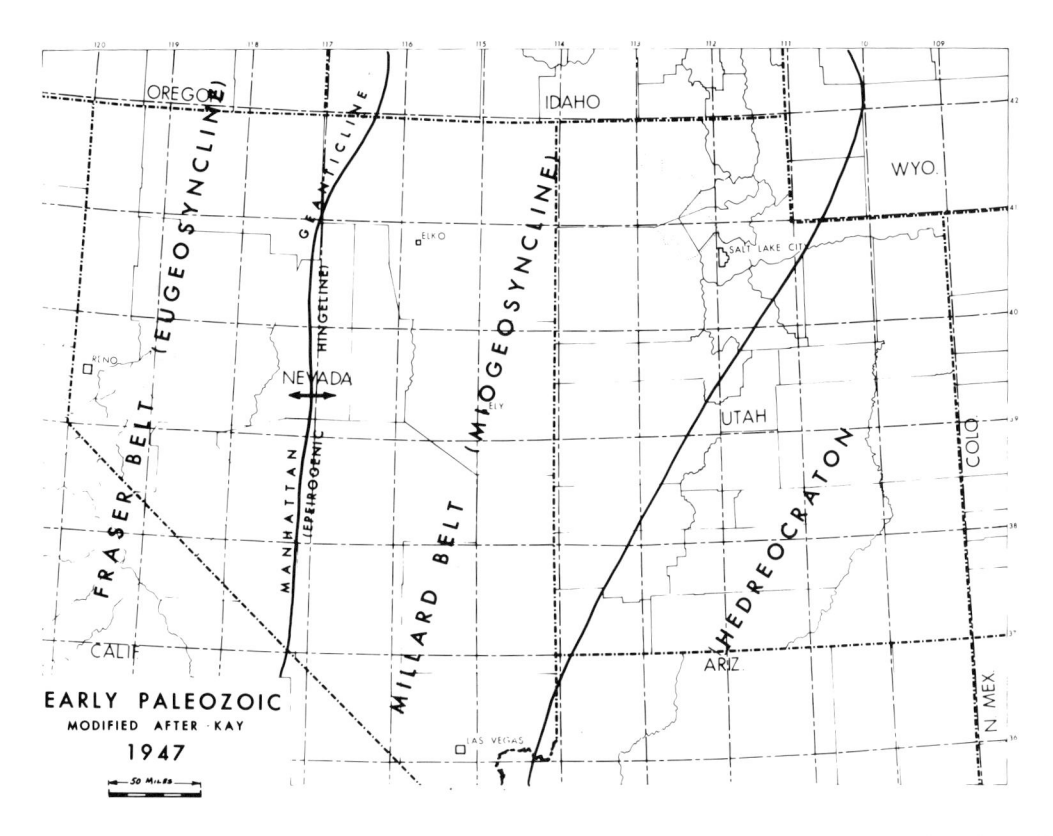

Fig. 2 — Subdivisions of the Cordilleran geosyncline in the Great Basin as defined by Kay.

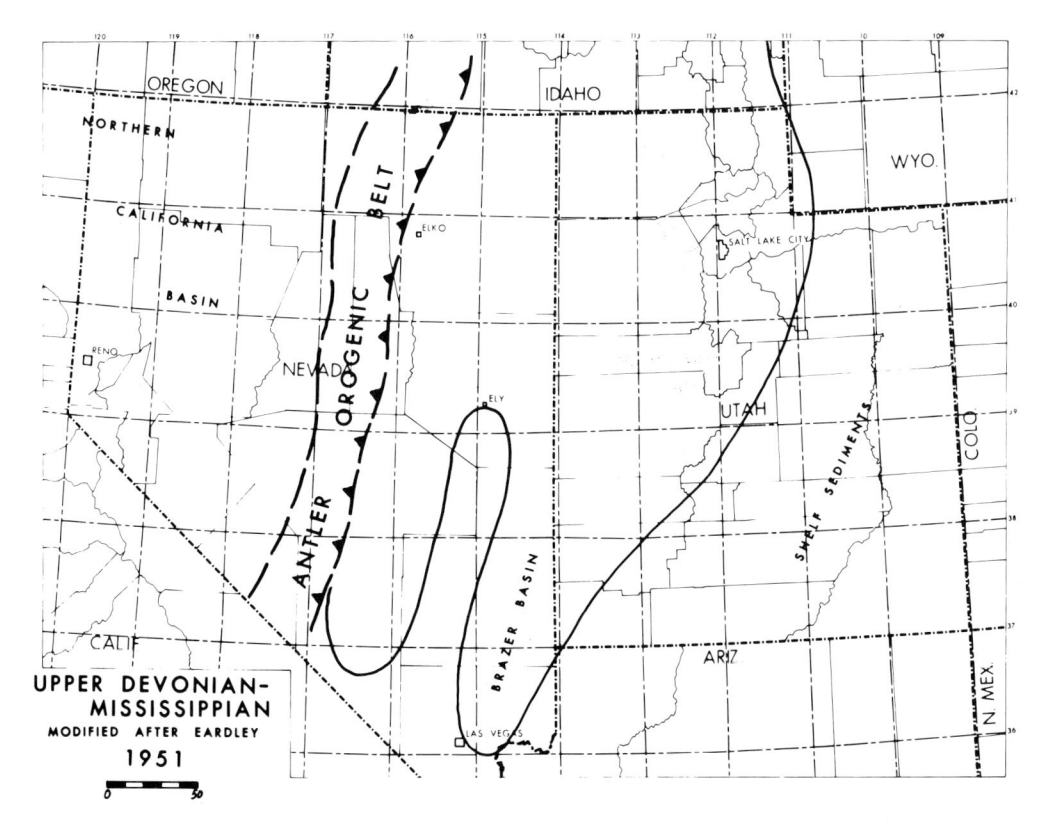

Fig. 3 — Cordilleran geosyncline in the Great Basin during Upper Devonian-Mississippian time as defined by Eardley.

deposited in the shallow trough, but there are other indications from these sediments that, with the beginning of structural activity along the Antler belt, the internal structural patterns of the miogeosyncline were also changing.

The deposition of the Joana Limestone over a broad stable shelf in Early Mississippian generally coincides with the initial eastward epirogenic surges of the Antler orogeny and could be the first predication that at least some of the forces brought to bear on the Antler belt were transcending into the miogeosyncline and subtly bringing about structural changes.

Steele (1960) describes the eastern Great Basin structural framework during the Pennsylvanian as an 'unstable shelf with subsidiary intracratonic basins and positives'. (Fig. 4). He describes a broad, subaerially exposed highland that has developed in western Utah and an intricate system of intra-basinal troughs that formed between the changing shorelines of various positive areas. By Middle Permian these channels became so restricted that evaporite basins were formed along the Butte-Deep Creek trough.

Mesozoic

At the end of Lower Triassic, the seas withdrew from the Great Basin, and with the exception of a few scattered outcrops of Cretaceous age, there is virtually no record of sediments that might have been deposited in the basin from Middle Jurassic to Eocene. When Nolan (1943) first described the Manhattan geanticline, he stated also that a second geaanticline, named the 'Mid-Cordilleran geanticline' by Schuchert (1923) became emergent during Permian and persisted into early Mesozoic time, and that its axial position was about 100 mi east of the Manhattan geanticline. Stokes (1960) recognized that the stature of this uplift appeared to be more than a mere enlargement of the older Antler orogenic belt and had more deep-seated significance than any previous uplift in the eastern Great Basin. He named the uplift the "Mesocordilleran highland" (Fig. 5). Although he was skeptical about its relative height, he felt that the highland was extensive in size, was deeply eroded, and the products of this erosion were carried eastward to provide the sediments for the Mesozoic formations in the Colorado Plateau and adjacent areas.

It is not too unreasonable to assume that this mid-Mesozoic uplift was the logical culmination of the compressive forces that actually began with the formation of the Manhattan geanticline in early Paleozoic and ended with the final collapse of the basin in mid-Cenozoic time.

As mentioned earlier, it is theoretically during the time of this uplift that the décollement-type thrust faulting was also supposed to have taken place (Misch, 1960). Even in its eroded state, if we were able to observe the surface of this

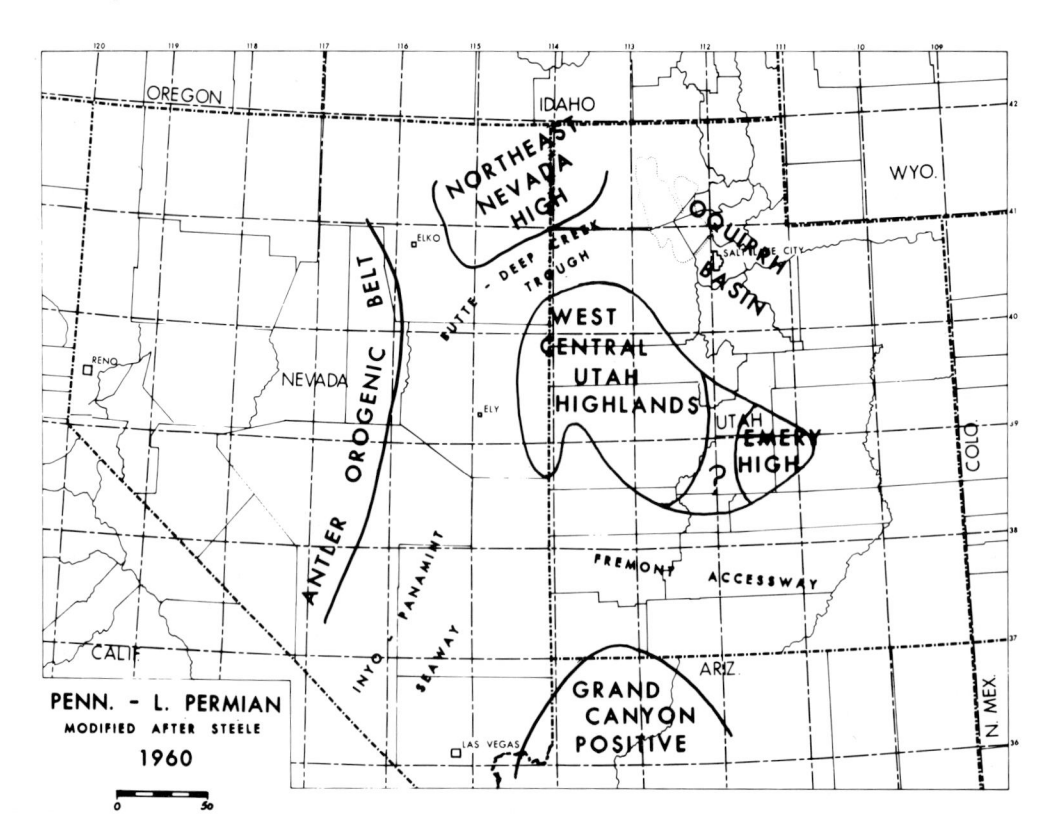

Fig. 4 — Intracratonic basins and positive areas in the eastern Great Basin during Pennsylvanian-Lower Permian time as defined by Steele.

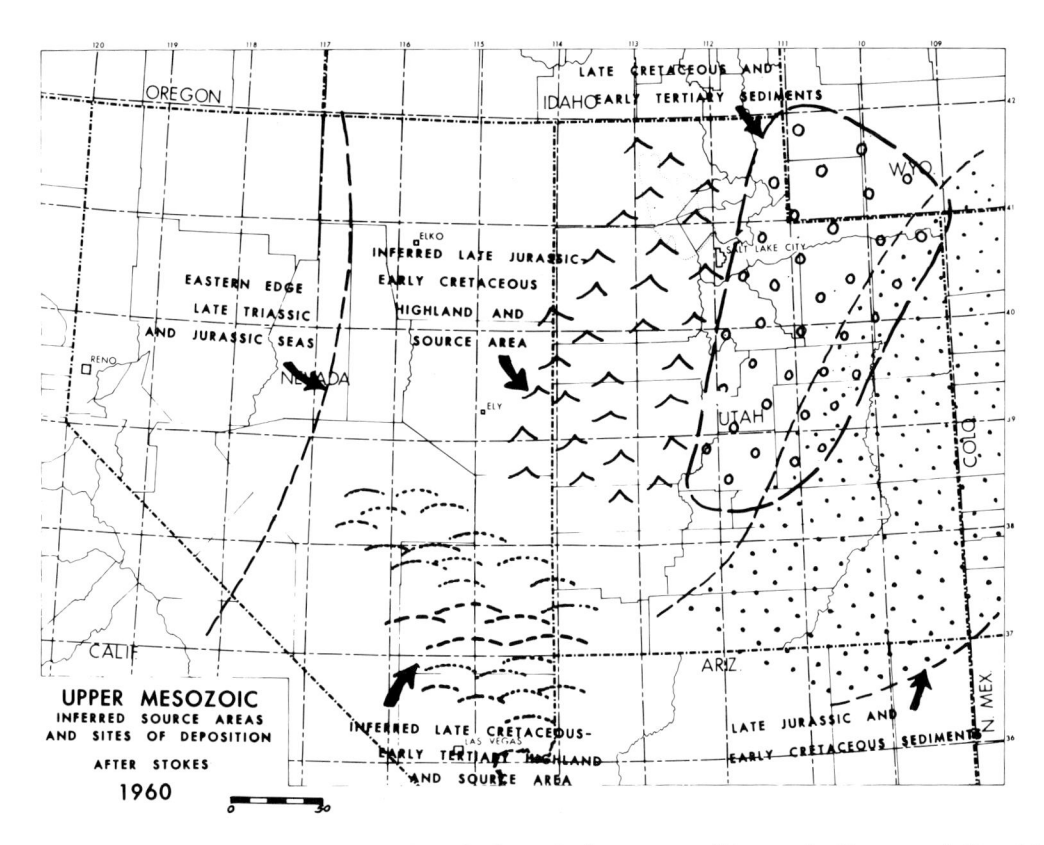

Fig. 5 — Inferred source areas and depositional sites during upper Mesozoic time as defined by Stokes.

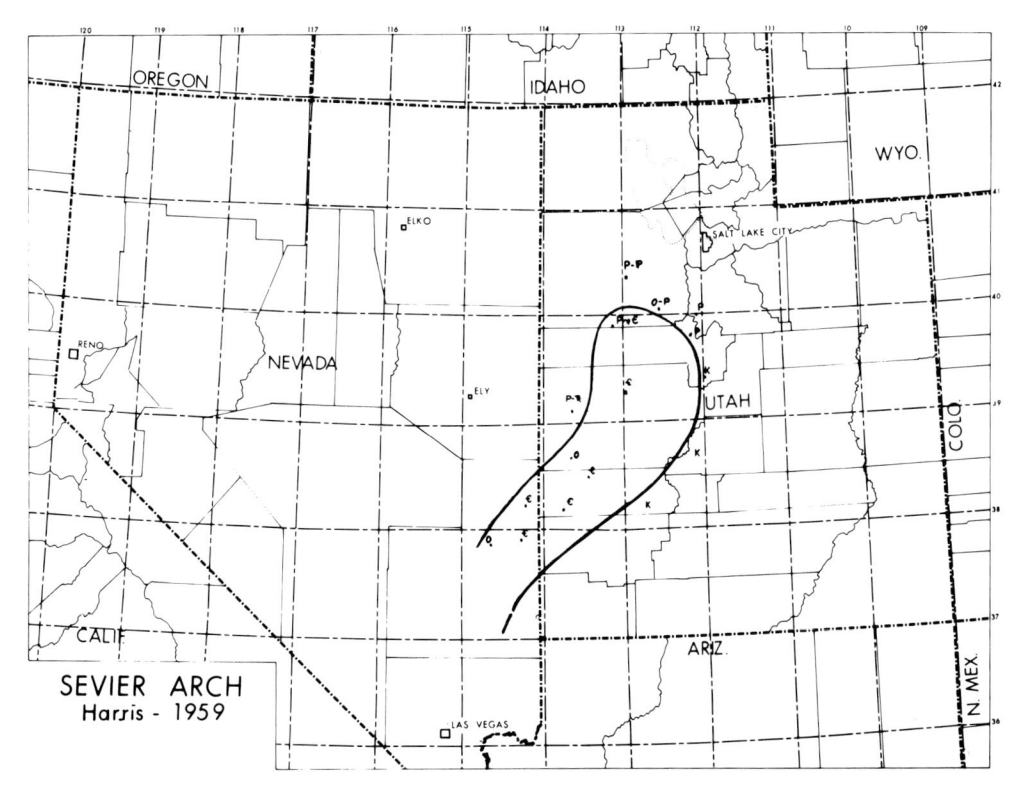

Fig. 6 — Sevier arch, as defined by Harris, showing the age of units immediately beneath the Oligocene and younger ignimbrites.

highland at some time after its peak of development, we would be able to verify such faulting and any other type of deformation that might have taken place in the eastern Great Basin during the Nevadan and Laramide orogenies.

Harris (1959) described a particular late Mesozoic positive area in western Utah and southeastern Nevada which he named the "Sevier arch" (Fig. 6). This arch was defined by investigation of older rocks that lie beneath a sequence of Oligocene and younger ignimbrite volcanics. In reality, he had exhumed the pre-Oligocene surface and revealed its physiographic setting just prior to the deposition of the welded tuffs. He was looking at a collective structural picture of what had happened in the Sevier arch area during late Mesozoic and early Cenozoic times. The method used was simple but effective.

The volcanics used by Harris to delineate the Sevier arch are not unique to his study area. Mackin (1960) recognized the great usefulness of these ignimbrites in structural and stratigraphic studies and Cook (1960) described what he called the "Ignimbrite Province" of Nevada and Utah (Fig. 7) where these rocks cover an area of over 60,000 sq mi and were laid down during a relatively short period of geologic time.

When the author entered the Great Basin in 1960 this relationship between ignimbrite volcanics and older rocks was studied whenever possible. After gathering data for some six years, it became apparent that with the exception of some local areas that exhibited obviously intense pre-ignimbrite folding, the volcanics were often in dip accord, or at slightly lesser but complimentary angles to the underlying sediments. In general, this relationship was also true where the contact between the lacustrine sedimentary rocks of the Eocene Sheep Pass sequence and older rocks could be observed.

As the data on volcanic contacts with older rocks was accumulated and placed on a map, it was apparent that significant patterns were beginning to develop. With all the contacts noted, control points tied together, and volcanics in essence removed, the geology of the Great Basin took on a new look as illustrated on Plate I[1] (In pocket). For reference, the map area of Plate I is indexed over the regional map in Figure 7.

The geologic outcrop patterns in Plate I were primarily placed by using volcanic rock contacts with older sediments. In some areas where volcanics were not present, due to erosion or other causes, and there was no major faulting indicated, it was assumed that at least the youngest rocks that still outcrop in the area were the surface rocks exposed just prior to volcanic deposition. Additionally, there are some younger volcanics in the northern part of this discussion area (mostly in northern White Pine County, Nevada) that are not included on Cook's ignimbrite map. These volcanics are not ignimbrites, but are mostly flow-rock volcanics and are usually much younger (Miocene) than the ignimbrites. However, their contact relationships with older rocks were found to have the same significance to the overall patterns as that of the older volcanics. The following are a number of interesting features that came to light as a further study of this map was made:

1. We now see a more detailed outline of the Sevier uplift and associated thrusting as described by Harris (1959) in Figure 6.

2. The verification of pre-Oligocene overthrusting in the area around and north of Las Vegas, Nevada.

3. Proceeding northward across the 38th parallel, the thrusting gives way to a pattern of folding that is evident over the entire center of the eastern Great Basin. When one considers that actual mountain outcrops in this region only represent roughly 40 percent of the surface and 60 percent is covered with late Tertiary valley fill, (Fig. 8) there is very likely a great deal more folding present. The one distinct and most important thing about this part of the map is that within this entire area of folding there were only one or two localities where the Oligocene volcanics were in actual contact with any faulting that could be construed as "thrusts".

4. Most of the younger sediments are preserved in the central part of the Basin, generally around the axis of the ancient Butte-Deep Creek trough of Steele (1960) (Fig. 4). This brings out the fact that in spite of all the "thrusting" that was supposed to have occurred during the Nevadan and Laramide orogenies the old structural pattern, developed as early as Mississippian, was essentially still intact.

5. There is evidence that the greatest observable relative movement in the basin was along lateral tear faults such as the Current fault in the Grant Range, the Shingle Pass fault in the Egan Range, many smaller faults in the Golden Gate-Seaman Range areas, and a long and significant left-lateral fault that the author has named the 'Ely-Black Rock lineament'.[2] This fault system possibly involves a broader and more complex area than is here indicated. It was placed in this position to generally align a series of structural and stratigraphic phenomena that appeared to have a regional interrelationship. The most prominent feature of this lineament is the apparent left-lateral displacement of

---

[1] The original field data was gathered for a private company over ten years ago. The present contact map is a reconstruction of that work through the use of the Nevada and Utah State Geologic Maps, some unpublished material, and personal field notes. The author apologizes for not providing more factual contact control in this paper, but any data on this map can be cross checked with the available state geologic maps through the use of the longitude and latitude lines provided on Plate I.

[2] The name 'Ely-Black Rock Lineament' was introduced during the presentation of this paper at the AAPG-SEPM Rocky Mountain Section Meeting in Billings, Montana, March 30, 1976.

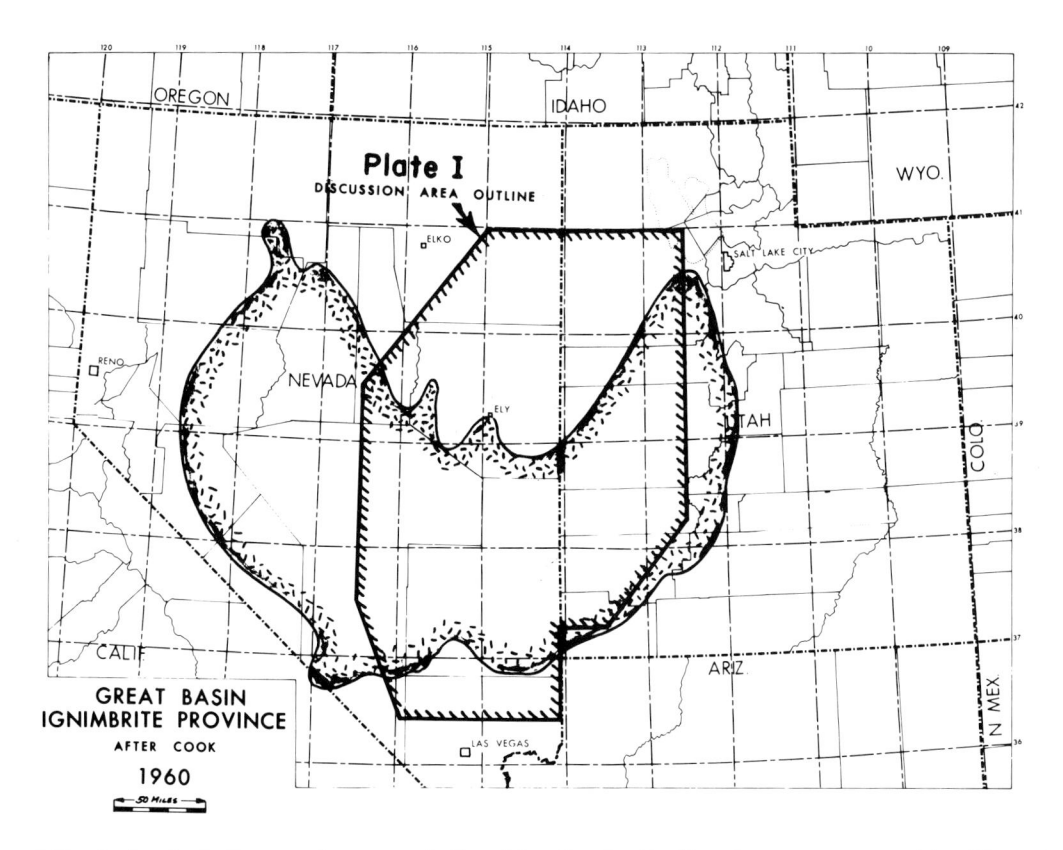

Fig. 7 — Great Basin ignimbrite province as defined by Cook. Outline is area covered by Plate 1, in pocket.

the ancient Butte-Deep Creek trough axis by some 10 mi in the southern Butte Range (Plate I and Fig. 8). A similar offset can be seen along this trend in Utah between the geology on the south end of the Conger Range and the north end of the Burbank Hills. Another feature is an indicated constriction or narrowing of the Sevier arch just west of Black Rock, Utah. The intervening area was suspected by Misch (1960, p. 34) as an aligment of some relative importance when he stated that "in central-east Nevada most of the known or suspected early Tertiary sediments appear to fall within a roughly transverse belt (Sacramento Pass, Connors Pass, Egan Range south of Ely, Currant Area)." There are other early Tertiary sedimentary accumulations along this lineament at the north end of Jakes Valley, in the Illipah Creek area and on the Emigrant Springs structure (west of the Antelope Mountains). A westward extension of this lineament would also traverse through the only known locality of preserved Cretaceous sediments in this part of the basin.

6. One speculative point about this map is the evidence that during the last half of the Mesozoic and early Tertiary there was some type of regional crustal shortening that affected the whole of the eastern Great Basin. The presence of abundant folding, tear faults, and frontal thrusts are strong indicators that such a shortening did take place.

If we consider the type of thrusts that are present in the Las Vegas areas as any criteria, the movement was probably from the west. In consideration of the amount of eastward movement necessary to account for the fault "overthrusts" along the Sevier arch, it is possible that if a palinspastic reconstruction were to be made it could straighten out the "S" curve position of the ancient Butte-Deep Creek axis to a position of from 25 to as much as 100 mi (?) back to the west (Fig. 8). The south end of this axis terminated against the Sevier arch. (Plate I). It is entirely possible that a large part of the southeast segment of the old miogeosynclinal trough was thrust-up, over the Sevier arch, during the Mesozoic orogenies and erosionally removed.

Mid-Cenozoic

It is generally accepted that there was considerable relief on the Mesocordilleran uplift near the end of Laramide time and that by the middle of the Cenozoic (mid-Miocene) a general tensional readjustment or collapse had begun over the entire basin. If we are to assume that the pre-volcanic structure map on Plate I is valid, we are faced with the problem of explaining the structural complexities that exist in the basin today by mechanisms inherent to block faulting. In any judgement we cannot assume that the "Basin and Range" block faulting has

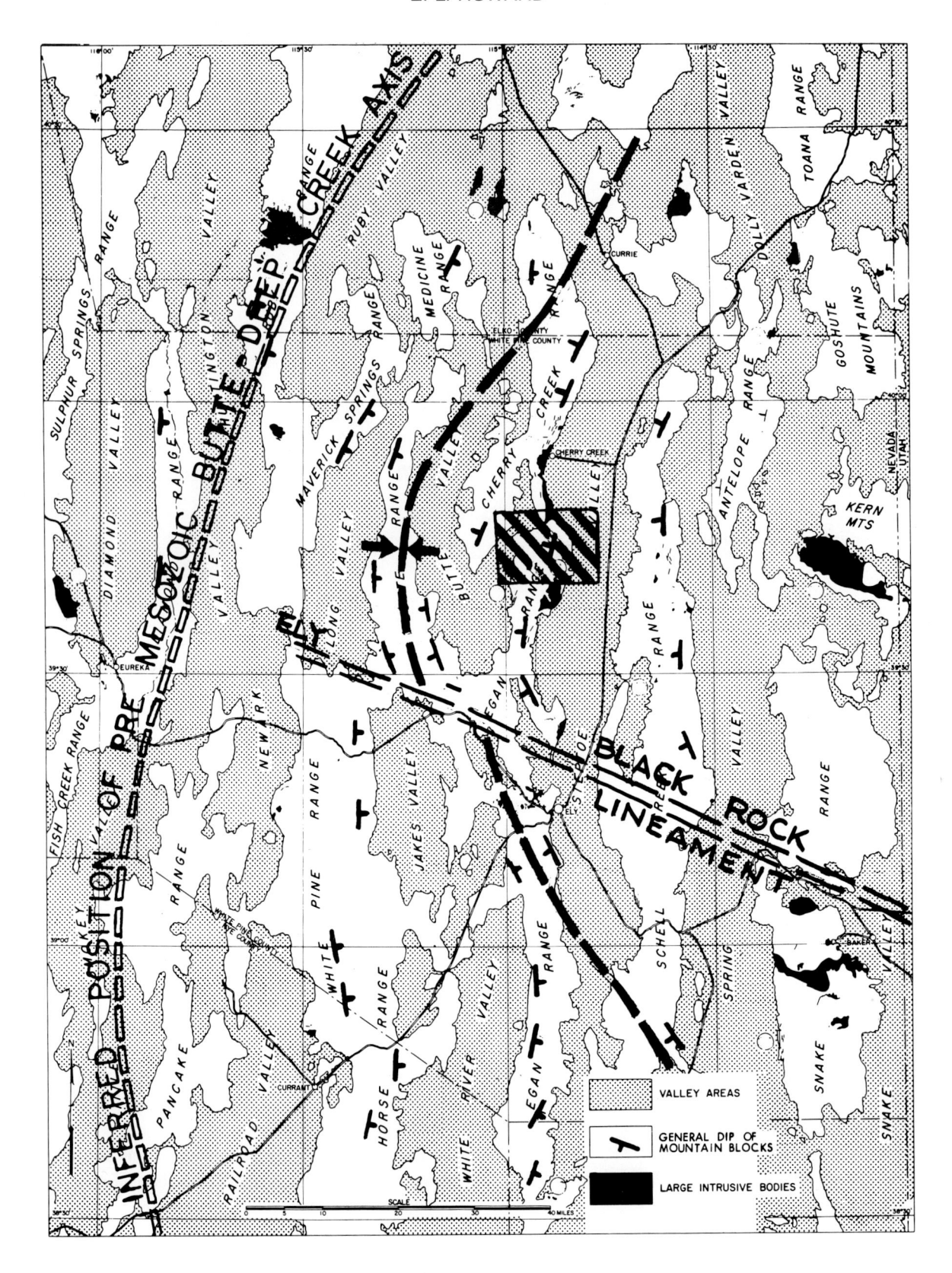

Fig. 8 — Index map showing present relationship of ranges to the Butte-Deep Creek trough and Ely-Black Rock lineament in east-central Nevada. Hatchured area is location of Figure 9.

to be necessarily uniform in its expression throughout the basin. For an entire region to be subjected to compressional forces from at least Cambrian to the mid-Cenozoic, uplifted, overthrust, and differentially eroded, it seems logical that there would be a wide variance in the type of isostatic forces present from one area to the next. The resulting amount of deformation that took place after the compressional forces were released is a direct function of how much isostatic inequilibrium had been created within any given area. The interaction could be mild, extreme, or any combination thereof. Within these parameters, especially when the inequilibrium is extreme, an explanation could evolve to account for some of the so called "thrusts".

Without going into a deep discussion about individual faults or fault systems some generalizations can be made in the light of the above assumptions:

1. The mountain ranges that lie along the axis of the old Butte-Deep Creek trough (north Cherry Creek, central Butte, central Egan, south-central Schell Creek, and Fortification ranges) appear to have had the least relative movement of any ranges in the basin, either up or down. When they are involved with adjoining blocks through faulting, the faults are usually of relatively low magnitude of displacement. They have experienced very little block "tilting" and the only real deformation is related to Nevadan and Laramide folding (Fig. 8).

2. The ranges immediately adjacent to both sides of the old axis do show more movement relative to adjoining valley blocks and exhibit a great deal more tilting, which is usually toward the axis of the ancient Butte-Deep Creek trough (Fig. 8). Depending on the conditions, the ranges can be associated with one, two, or more cycles of tilting or block rotation. In Figures 9, 10a, b, and c an actual situation involving this type of rotation is shown. A segment of the northern Egan Range mapped by Fritz (1960) is illustrated on Figure 9 (regionally located as the hatchured area on Figure 8). Cross section A-A' in Figure 10a pictures a schematic setting of this range at about the beginning of the mid-Cenozoic faulting. With the tensional release, usually associated with crustal spreading, the blocks begin to settle and rotate along faults X, Y, and Z. In Figure 10b the settling and rotation continues with more movement along the fault zones until no more rotation slippage is possible along the X, Y, and Z fault planes because of the low angle. At this point a new set of fault zones are formed at M and N. The settling and rotation continues until the fault planes X and Y are elevated to their present position in Figure 10c, where they appear to be of décollement origin.

Close examination of the map in Figure 9 will disclose there are more "thrust" faults but that every one of these faults has younger rocks thrust down on older beds.

3. When the block faulting occurred in those areas of extreme isostatic inequilibrium the tilting and rotation could be compounded by large displacements along fault systems bordering the ranges. It is not uncommon to see relative vertical displacements of over 25,000 ft and such displacements undoubtably would create enough instability to cause major segments of these ranges to break away and slide into valley depressions or over a lower segment of the same range. Such sliding could also create the "tectonic banded marbles...and recrystallized carbonate mylonites" along the décollement fault planes of Misch (1960). The key, again, is that most of these faults "thrust" younger onto older rocks. This type of faulting mostly occurs in the ranges that are farthest away from the Butte-Deep Creek axis and where the isostatic inequalibrium seems to have been the greatest. To the east, this type of extreme structural complexity is present in the Snake Range, Deep Creek Mountains, and Kern Mountains. On the west these conditions can be seen in the Ruby Range, Diamond Range, southern White Pine Range, and the Grant Range.

SUMMARY AND CONCLUSIONS

As a practical geologist, the writer clings stubbornly to the belief that all complexities are composed of many simple principles that are basic in nature. If we are to assume, as most authors do, that the basin as we see it today is the product of an extremely complex set of regional thrusts that have been further compounded by a superimposed destructive sequence of basin and range-type faults, and we try to explain it all through some very complex rhetoric, we have not served our purpose as geologists.

The evidence is clearly visible to substantiate that a simple tectonic framework was developed during late Paleozoic in the eastern Great Basin. Through the use of the Oligocene ignimbrites it can be seen that there was relatively minor Nevadan or Laramide deformation and that essentially the older Paleozoic tectonic pattern was still intact at that time. There is evidence also that many of the so-called décollement thrusts can be explained through the mechanisms that are inherent to the tensional basin and range-type faults. It is possible that these answers are too simple and all-inclusive for the amount of details to be studied, but as in so many problems, a final solution has to involve a combination of factors from seemingly incompatable theories and perhaps from things yet unknown.

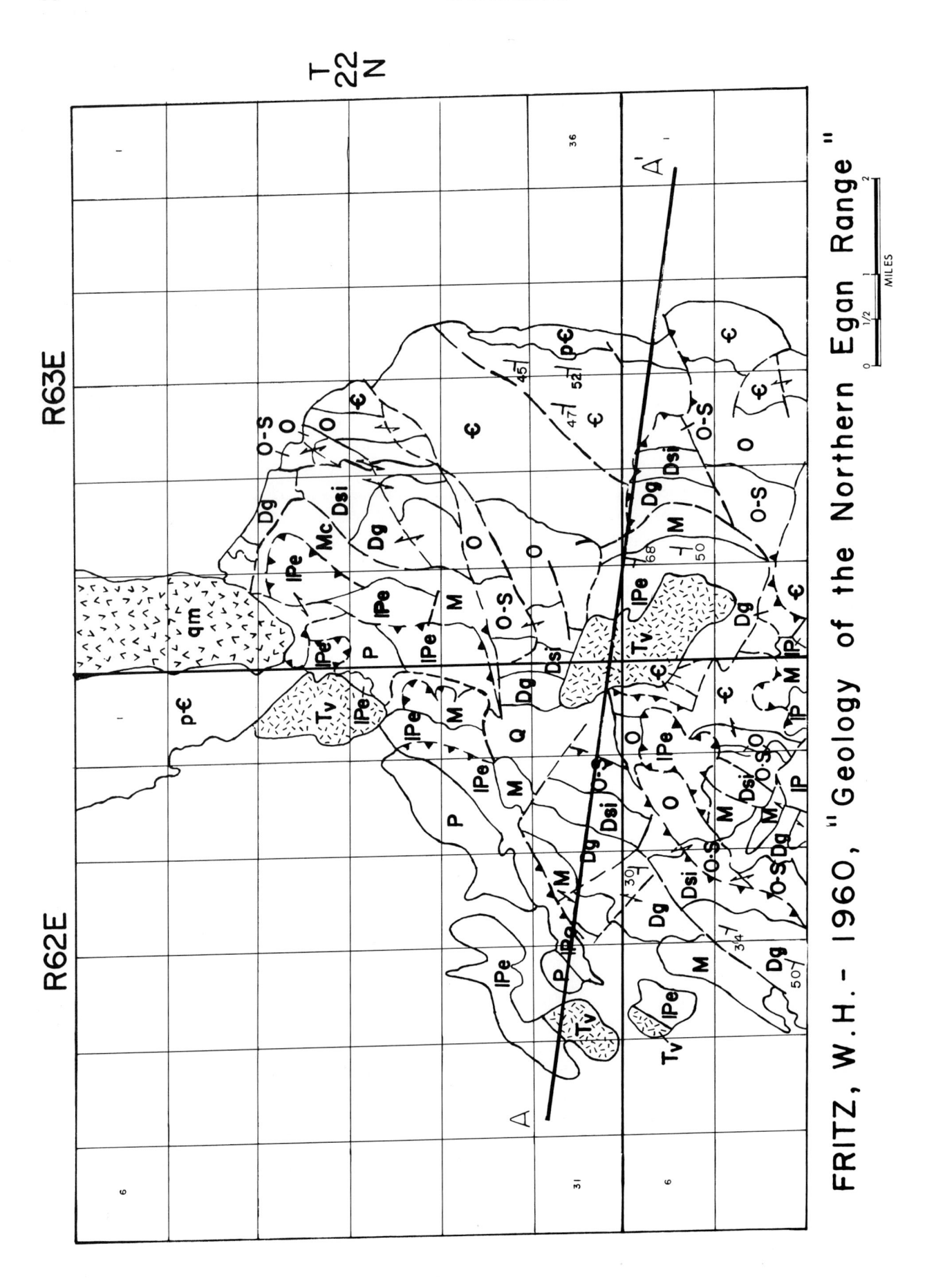

Fig. 9 — Geologic map of the northern Egan Range.

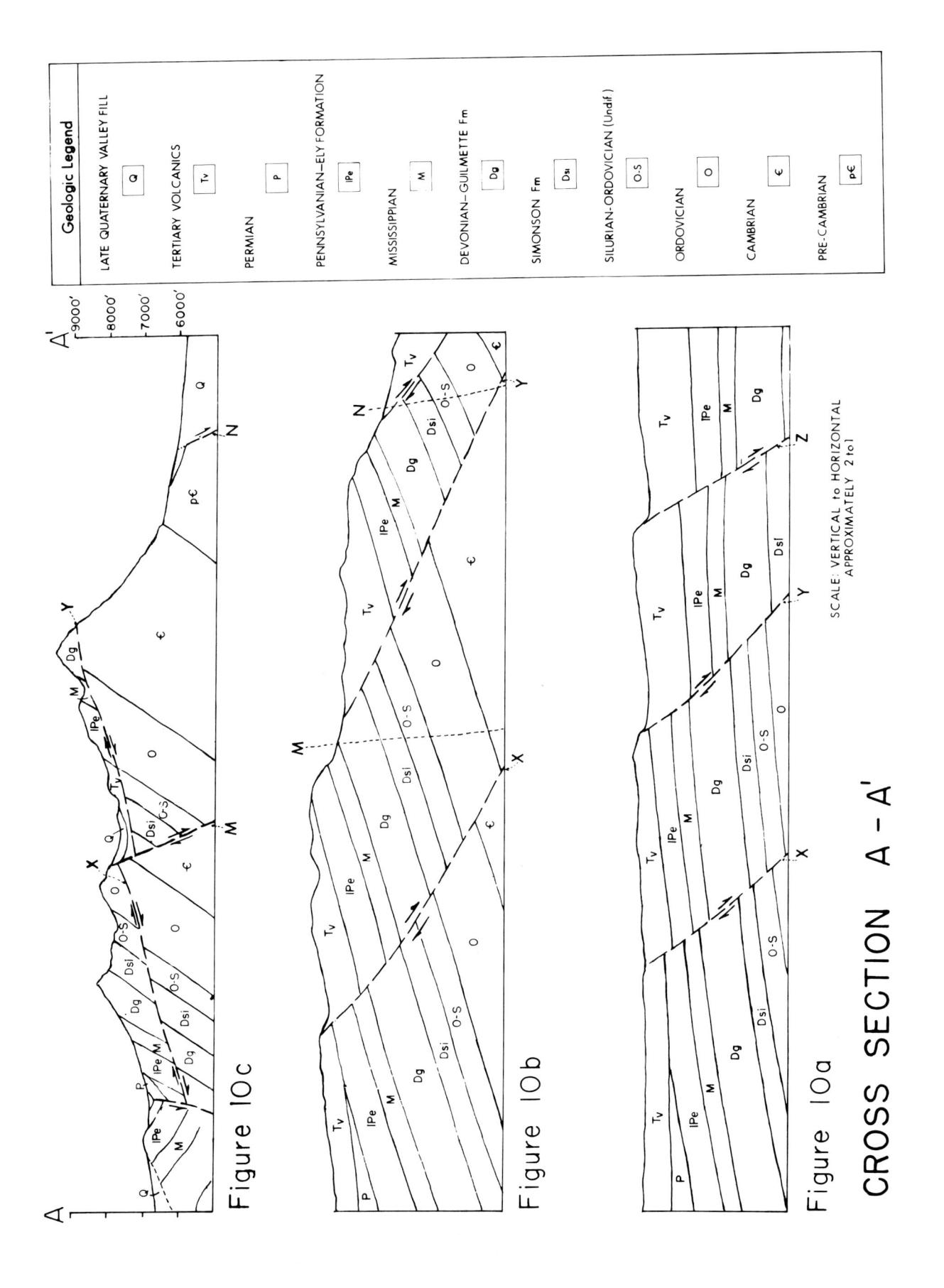

Fig. 10 — Structural evolution of the northern Egan Range during Cenozoic time.

REFERENCES CITED

Cook, Earl F., 1960, Great Basin ignimbrites: Intermtn. Assoc. Petroleum Geologists and eastern Nevada Geol. Soc., Guidebook to the geology of east-central Nevada, pp. 134-140.

Eardly, A. J., 1947, Paleozoic Cordilleran geosyncline and related orogeny: Jour. Geol., v. 55, no. 4, p. 309-342.

_____, 1951, Structural geology of North America: Harper & Brothers, New York.

Harris, H. D., 1959, Late Mesozoic positive area in western Utah: Am. Assoc. Petroleum Geologists Bull., v. 43, no. 11, p. 2636-2652.

Kay, Marshall, 1947, Geosynclinal nomenclature and the craton: Am. Assoc. Petroleum Geologists Bull., v. 31, no. 7, p. 1289-1293.

_____, 1952, Late Paleozoic orogeny in central Nevada, (Abstr.): Geol. Soc. America Bull., v. 63, p. 1269-1270.

Longwell, C. R., 1921, Geology of the Muddy Mountains, Nevada: Am. Jour. Sci., 5th Ser., v. 1, p. 39-62.

_____, 1922, The Muddy Mountains thrust in southern Nevada: Jour. Geol., v. 30, p. 63-72.

Mackin, J. H., 1960, Structural significance of Tertiary volcanic rocks in southwestern Utah: Am. Jour. Sci., v. 258, p. 81-131.

Misch, P., 1960, Regional structural reconnaissance in central-northeast Nevada and some adjacent areas: observations and interpretations: Intermtn. Assoc. Petroleum Geologists and Eastern Nevada Geol. Soc., Guidebook to the geology of east-central Nevada, p. 17-42.

_____, and Easton, W. H., 1954, Large overthrusts near Conners Pass in the southern Schell Creek Range, White Pine County, eastern Nevada (Abstr.): Geol. Soc. America Bull., v. 65, p. 1347.

_____, Hazzard, J. C., and Turner, F. E., 1957, Precambrian tillitic schists in the southern Deep Creek Range, western Utah and Precambrian units of western Utah and eastern Nevada, (Abstr.): Geol. Soc. America Bull., v. 68, p. 1837.

Nolan, T. B., 1928, A late Paleozoic positive area in Nevada: Am. Jour. Sci., 5th series, v. 41, no. 92.

_____, 1943, The Basin and Range Province, Utah, Nevada, and California: U.S. Geol. Survey, Prof. Paper 197-D.

Roberts, R. J., 1951, Antler Peak Quadrangle, Nevada: U.S. Geol. Survey Geologic Quadrangle Map GQ-10.

Sadlick, Walter, 1960, Some preliminary aspects of Chainman stratigraphy: Intermtn. Assoc. Petroleum Geologists and Eastern Nevada Geol. Soc., Guidebook to the geology of east-central Nevada, p. 81-90.

Schuchert, C., 1923, Sites and nature of the North American geosynclines: Geol. Soc. America Bull., v. 34, pp. 151-230.

Spieker, E. M., 1946, Late Mesozoic and early Cenozoic history of central Utah: U.S. Geol. Survey Prof. Paper 205-D, pp. 117-161.

Steele, Grant, 1960, Pennsylvanian — Permian stratigraphy of east-central Nevada and adjacent Utah: Intermtn. Assoc. Petroleum Geologists and Eastern Nevada Geol. Soc., Guidebook to the geology of east-central Nevada, p. 91-113.

Stokes, W. L., 1960, Inferred Mesozoic history of east-central Nevada and vicinity: Intermtn. Assoc. Petroleum Geologists and Eastern Nevada Geol. Soc., Guidebook to the geology of east-central Nevada pp. 117-121.

A DISCUSSION OF THE GEOLOGY OF THE SOUTHEASTERN CANADIAN CORDILLERA AND ITS COMPARISON TO THE IDAHO-WYOMING-UTAH FOLD AND THRUST BELT

by

K. H. Hayes[1]

ABSTRACT

Recent significant hydrocarbon discoveries in western Wyoming and northern Utah have initiated a major petroleum industry exploration effort in the eastern Cordilleran fold and thrust belt of the western United States. The Canadian disturbed belt is perhaps the best known productive analogue and experience gained from this province can be usefully applied to exploring the Idaho-Wyoming-Utah fold and thrust belt.

In both the southeastern Canadian Cordillera and the Idaho-Wyoming-Utah fold and thrust belt, the hingeline stratigraphy consists of an older miogeocline-platform assemblage, and a younger syntectonic clastic wedge sequence. These wedge-like sequences have been subsequently deformed, largely by means of thrust faults and concentric folds. Changes in the style of deformation can be related to both lateral and vertical changes in stratigraphic thickness and facies.

In Canada, the eastern elements of this system have been extensively explored, resulting in discovery of 18 TCF of natural gas, significant volumes of natural gas liquids, sulphur, and over one billion bbls of oil. Two of the major fields, Turner Valley at the eastern edge of deformation and Waterton at the foot of the Front Ranges, are discussed.

Concepts of the tectonic evolution of the southeastern Canadian Cordillera and the structural style of the Canadian Front Ranges and Foothills are reviewed. The principles of structural interpretation and exploration concepts evolving from the geological, geophysical, and geochemical exploration of the southeastern Canadian Cordillera are discussed. This more explored and geologically better known region is then compared with the Idaho-Wyoming-Utah fold and thrust belt as an analogy for geologic interpretation and exploration potential.

INTRODUCTION

The Cordilleran hingeline and overthrust region of the western United States is one of the few remaining frontiers in North America where significant potential remains for the exploration for major hydrocarbon reserves. Recent hydrocarbon discoveries at the Pineview field, Summit County, Utah, and the Ryckman Creek field, Uinta County, Wyoming, have triggered a major surge of petroleum industry activity in the Idaho-Wyoming-Utah fold and thrust belt.

Experience gained from well documented histories of hydrocarbon exploration in analogous structural provinces, such as the southern Canadian Rocky Mountain fold and thrust belt, can and should be applied to exploration in the Idaho-Wyoming-Utah fold and thrust belt. The purpose of this paper is to: 1) discuss the Canadian setting, 2) provide regional and local stratigraphic information, 3) outline possible theories of structural genesis, and 4) describe variations of structural style and accompanying geometrical constraints. Some detail on commonly referred-to-hydrocarbon-bearing thrust structures is included. Regional hydrocarbon variations and thermal maturation of hydrocarbon source-rocks pertinent to the Canadian disturbed belt are also discussed. Finally, a brief comparison is made between the southern Canadian Rocky Mountain fold and thrust belt and the Idaho-Wyoming-Utah fold and thrust belt.

---

[1] Exploration Geologist, Filon Exploration Corporation, Denver, Colorado.

The Cordillera of western North America extends from Alaska through Mexico. It is a complex entity consisting of: 1) extremely thick sedimentary wedges which have been extensively deformed, 2) batholithic intrusions with associated regional metamorphism, 3) volcanic eruptions, and 4) regional uplift. The easternmost elements of the Cordillera are the fold and thrust belts (Fig. 1).

The southeastern Canadian Cordillera consists of the southern Canadian Rocky Mountains and adjacent Foothills located along the foreland margin northeast of the Rocky Mountain Trench. It forms a distinctive narrow and linear physiographic province of relatively rugged topography rising above the high interior plains (Fig. 1).

HISTORY OF HYDROCARBON EXPLORATION IN THE SOUTHERN CANADIAN CORDILLERA

Early efforts to develop commercial hydrocarbons in the region concentrated on surface structural features in close proximity to surface manifestations of oil and gas. These early efforts resulted in modest economic ventures. They did, however, serve the purpose of documenting the presence of hydrocarbons in the system and their relationship to surface geology.

The first major discoveries were the Turner Valley gas field in 1924 and the Turner Valley oil field in 1936; thus the boom was on. The complex thrusted nature of many surface structures led some frustrated geologists, upon having drilled the crestal position of various precisely mapped surface structures, to glibly conclude that the "basement complex" of the thrust belt most certainly consisted of Upper Cretaceous shale.

With the advent of geophysical exploration methods, there was another surge in exploration activity and several major finds resulted. More recently, with advanced geophysical methods and more logical geometrical and genetic constraints on structural interpretations, the industry has rejuvenated activity in complex areas, and has continued to find significant accumulations.

Economics have improved considerably in recent times. This improvement has provided the necessary impetus for the industry to pursue more and more expensive and difficult logistical and technical problems. To date, more than 18 TCF of natural gas, significant volumes of natural gas liquids and sulphur, as well as approximately one billion bbl of oil have been discovered in the Rocky Mountain fold and thrust belt. The location and discovery year of the fields which account for these originally-in-place reserves are shown in Fig. 2.

The structural geology of this region is presented in comprehensive and authoritative works by: Link (1954); Scott (1954); Fox (1959); Bally, *et al.*, (1966); Keeting (1966); Dahlstrom (1969, 1970); Douglas, *et al.*, (1970); Gretner (1971); Halladay and Mathewson, ed. (1971); Jones, (1971); and Evers and Thorpe, ed., (1975).

STRATIGRAPHIC FRAMEWORK OF THE SOUTHERN CANADIAN CORDILLERA

Regional Stratigraphy

The supracrustal rocks that form the southern Canadian Rocky Mountains and adjacent regions record 1,250 million years of tectonic and stratigraphic evolution. Price (1971) divides these rocks into two tectono-stratigraphic assemblages: 1) a miogeocline-platform assemblage, and 2) a syntectonic clastic wedge assemblage.

Miogeocline-platform assemblage. — A northeasterly tapering wedge-shaped sequence (greater than 45,000 ft in thickness) records the evolution of a continental margin from Middle Proterozoic to Late Jurassic time (Fig. 3). It consists predominantly of carbonate, shale, and mature sandstone units. These rocks are underlain by Hudsonian crystalline basement of Early Proterozoic age (Burwash, *et al.*, 1964). The provenance of the terrigenous detritus is largely the craton to the northeast. The enormous amount of overall subsidence which accommodated this assemblage was interrupted by epirogenic events identified by major unconformities to the east. Westward from the stable cratonic platform, these unconformities diminish and the stratigraphic sequences become more complete (Bally, *et al.*, 1966).

Syntectonic clastic wedge assemblage. — This assemblage is a younger sequence consisting of clastic deposits up to 20,000 ft in thickness, which also thins to the northeast. In contrast with the older sequence, detritus was derived from within the cordillera rather than the craton. Successive pulses of detritus began in uppermost Late Jurassic time and continued throughout the Cretaceous and early Tertiary (Fig. 3). The deposition of these clastic wedges signaled the onset of large-scale orogeny within the cordillera as a whole.

Several discrete unconformity-bounded clastic wedges are recognized along the foreland margin. Variations in thickness, grain size, and facies patterns indicate that these successive pulses of detritus were shed progressively northeastward over older shelf sequences. The non-marine deposits prograding in such a manner interfinger with, and eventually replace, marine deposits (Price and Mountjoy, 1970). Recognition of these relationships led to the concept of a migrating foredeep in advance of northeastward-spreading deformation (Bally, *et al.*, 1966). Subsidence of the foredeep, at least initially, appears to be an isostatic adjustment to the tectonic loading during supracrustal shortening (folding and thrusting) (Price, 1973). Continued subsidence was probably some function of sediment loading.

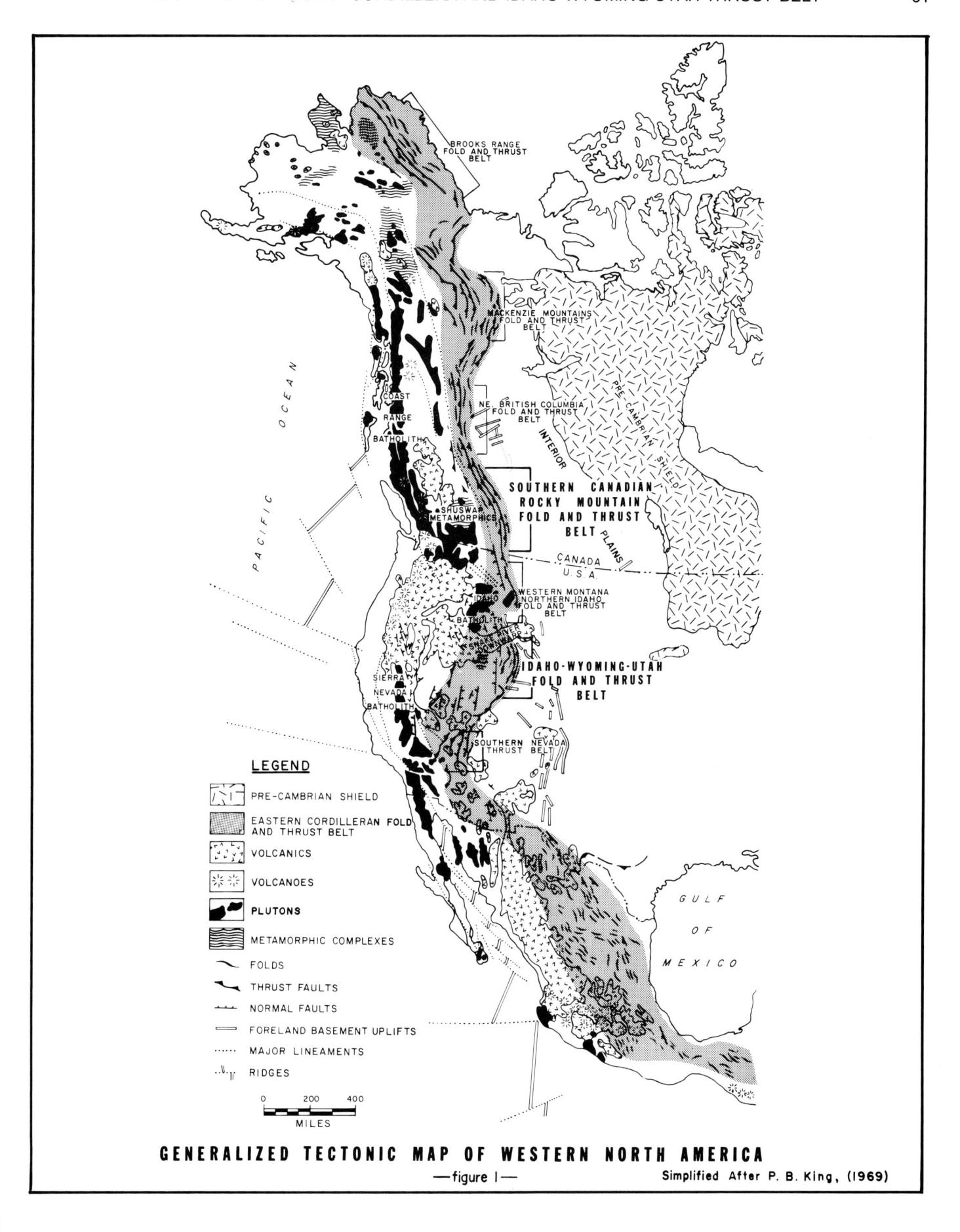

SOUTHEAST CANADIAN CORDILLERA AND IDAHO-WYOMING-UTAH THRUST BELT

GENERALIZED TECTONIC MAP OF WESTERN NORTH AMERICA

—figure I—

Simplified After P. B. King, (1969)

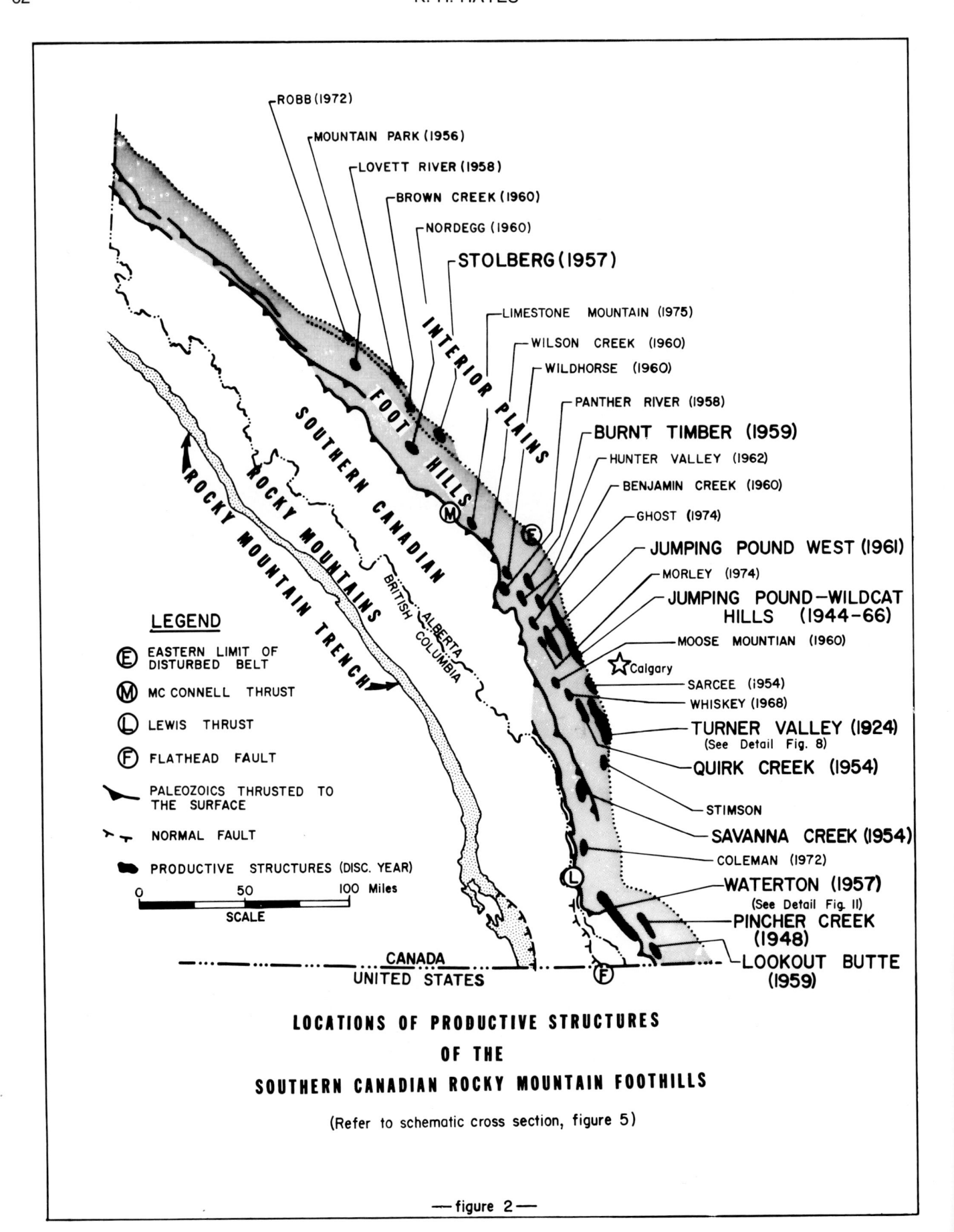

LOCATIONS OF PRODUCTIVE STRUCTURES OF THE SOUTHERN CANADIAN ROCKY MOUNTAIN FOOTHILLS

(Refer to schematic cross section, figure 5)

— figure 2 —

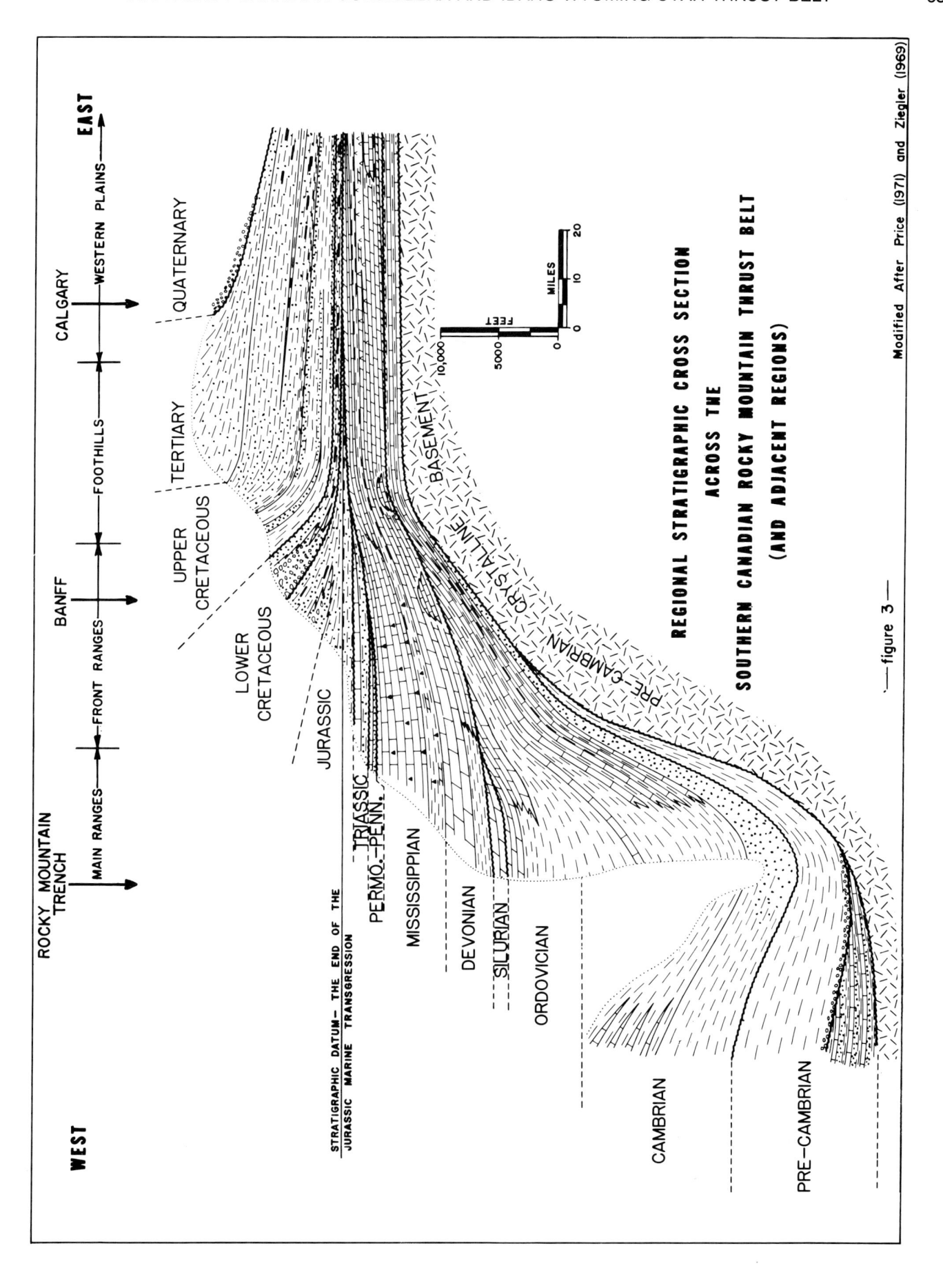

SOUTHEAST CANADIAN CORDILLERA AND IDAHO-WYOMING-UTAH THRUST BELT

REGIONAL STRATIGRAPHIC CROSS SECTION ACROSS THE SOUTHERN CANADIAN ROCKY MOUNTAIN THRUST BELT (AND ADJACENT REGIONS)

— figure 3 —

Modified After Price (1971) and Ziegler (1969)

Important Factors of Foothills and Front Range Stratigraphy

The straigraphy most pertinent to hydrocarbon exploration is shown in Fig. 4. Although generalized, it serves the purpose of illustrating the stratigraphic position of important reservoirs, "probable" hydrocarbon source-rocks, "preferred" detachment positions (décollement zones), as well as providing some seismic velocity data.

By far the most prolific reservoirs in all of the southern Canadian thrust belt are porous Mississippian carbonates of the Rundle Group, located at or near the overlying pre-Jurassic unconformity (90 percent of the in-place reserves). Accordingly, most exploration since the discovery of the Paleozoic reservoir at the Turner Valley field has been directed toward these objectives. Unfortunately, exploration drilling properly located for Mississippian reservoirs in this structural regime often has not been optimumly located to test potentially prospective horizons above or below the primary objective.

The average velocities for various stratigraphic intervals are presented in Fig. 4. The most noteworthy seismic reflections commonly used in subsurface interpretations are the Basal Mesozoic or "Near Mississippian Event", and the Cambrian or "Near Basement Event".

STRUCTURAL FRAMEWORK OF THE SOUTHERN CANADIAN CORDILLERA

The stratigraphic record preserved in the clastic wedge sequences suggests that orogenic deformation within the southern Canadian Cordillera was the result of a continuum of events dating from Late Jurassic through early Tertiary time. From Late Jurassic to earliest Upper Cretaceous time, parts of the Cordilleran geosyncline underwent major phases of deformation, regional metamorphism, granitic intrusion, and regional uplift (the Columbian orogeny). Continued and/or rejuvenated deformation during the Laramide orogeny produced the northeasternmost elements of the Cordilleran region; the linear Rocky Mountain thrust belt. The Laramide deformation of this region is considered to have taken place during and immediately subsequent to the deposition of the late Upper Cretaceous and early Tertiary non-marine clastic sequences (Douglas, et al., 1970).

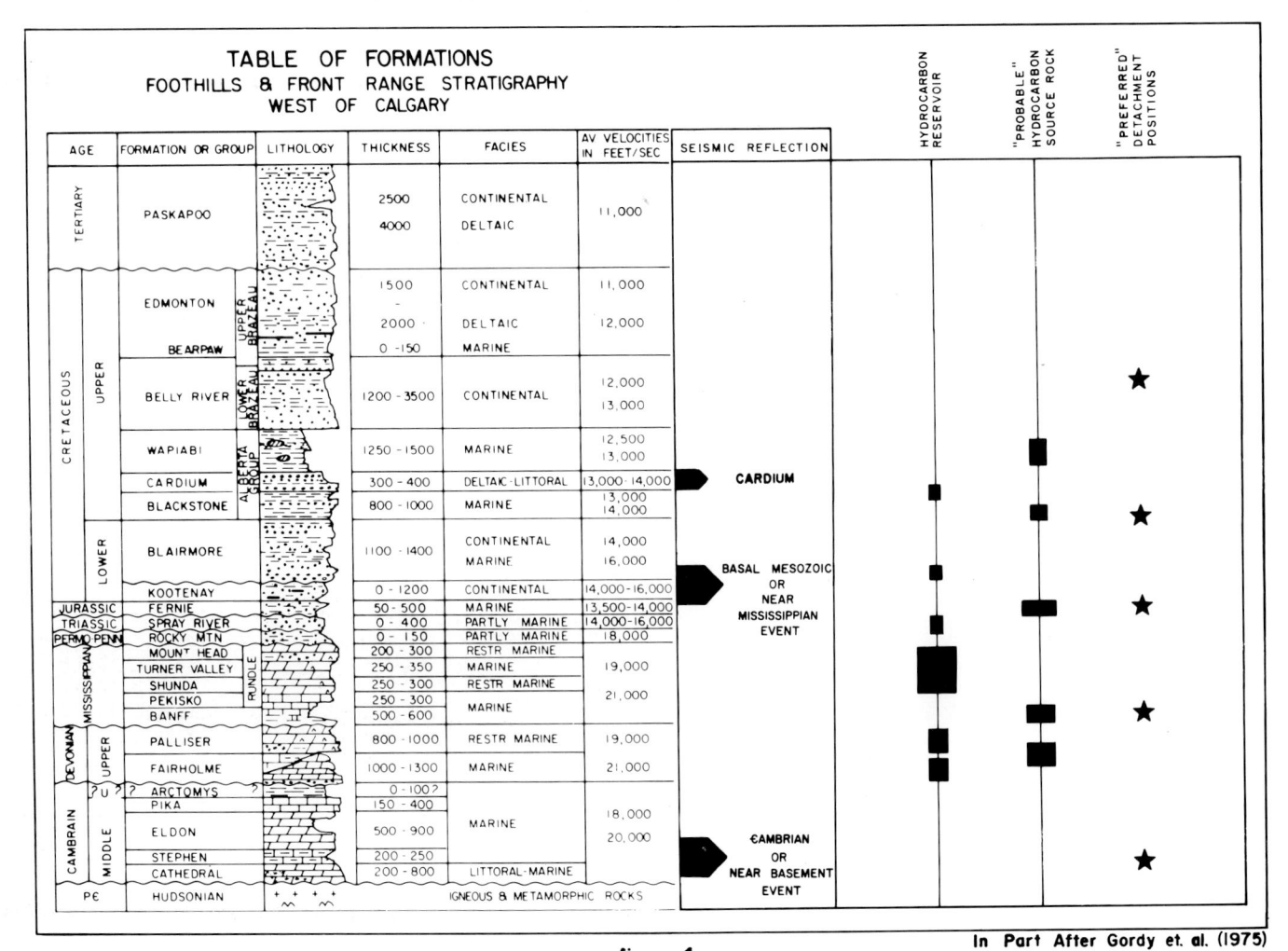

— figure 4 —

The regional structure of the southern Canadian Rocky Mountains and adjacent Foothills is dominated by several major southwest-dipping thrust sheets. Each thrust sheet has been displaced northeastward from several miles to tens of miles and stratigraphically from a few thousand to tens of thousands of feet. The basement is considered to be passive and to dip westward undisturbed beneath a major portion of the eastern cordillera.

Regional Subprovinces

Changes in structural style, which reflect both 1) the progressive southwesterly change in character and thickness of the supracrustal sedimentary sequence, and 2) changes in the structural level that is exposed, are the basis of recognizing a series of subparrallel structural subprovinces in the southern Canadian Rocky Mountains. These subprovinces as recnized by Price and Mountjoy (1970) are: 1) the Foothills, 2) Front Ranges, 3) Main Ranges, and 4) the Rocky Mountain Trench.

Foothills. — This subprovince forms the northeastern margin of the Rocky Mountains. The northeastern edge of disturbed Cretaceous marks its outer limits. Its southwestern boundary corresponds to the first thrust which carries predominantly Paleozoic rocks to the surface. This boundary is commonly an individual fault over long distances, but is an en-echelon arrangement of major thrusts in places. The structures are generally associated with flat-lying thrust faults which imbricate as they cut up-section along their leading edges. The faults, by their very nature, are often folded by younger movements which originate beneath the earlier-formed fault planes. The magnitude and intensity of deformation generally decreases eastward. The topography of the eastern Foothills is subdued because the level of exposure is predominantly erosionally recessive shales of the Upper Cretaceous. The western Foothills on the other hand, have more rugged topography due to the outcrops of erosionally resistant sandstones of the Lower Cretaceous.

Front Ranges. — The dominant level of exposure is that of Devonian to Jurassic age units, with all younger rocks commonly eroded away. The structure of this subprovince is characterized by a series of six or less major thrust slices that dip moderately to steeply westward. These form characteristically rugged linear ranges of erosionally resistant Paleozoic carbonate rocks separated by valleys carved into easily eroded Mesozoic clastics.

Main Ranges. — The dominant level of exposure of the Main Ranges is that of the thicker and more complete pre-Devonian rock units. The eastern Main Ranges consist of thick, competent sequences of distinctly bedded, relatively flat-lying carbonates which have been carved by glaciation into castellated peaks. The western Main Ranges, on the other hand, consist of thick, relatively homogeneous, less competent, shaley, and mildly metamorphosed pelitic facies which exhibit widespread complex penetrative deformation and well-developed cleavage.

Rocky Mountain Trench. — The Rocky Mountain Trench is a linear, glaciated trough, controlled in part by late normal faults along its margins.

Theories on Structural Evolution

Gravitational model. — Price and Mountjoy (1970) suggest that the structural evolution of the region involves a continuing process of bouyant upwelling and lateral spreading of a hot mobile core zone (Shuswap metamorphic complex) beneath a sedimentary cover. The rising northeasterly flow of such a core zone and equivalent progressive northeasterly growth of the foreland fold and thrust belt, is thought to have formed the structure of the southern Canadian Rocky Mountains. The lateral spreading is implied to have been a gravitational phenomenon involving the entire mass. The driving mechanism by which thrust plates have moved up the passive basement slope is believed to be one of dynamic gravitational equilibrium, such as that of a plastic solid which flows in response to its own weight. In such a manner, the northeasterly topographic slope of the surface of a mass of supracrustal rocks should yield a gravitational potential sufficient to produce low-angle west-dipping thrusts on a regional scale. The analogy is to the peripheral zone of a large ice sheet. The foreland thrust belt formed in this manner is thought to have experienced as much as 125 mi of supracrustal shortening.

Underthrusting model. — Campbell (1973), working mainly in the core zone and adjacent regions of the Main Ranges, suggests that the basement beneath the Main Ranges is involved in "thick-skinned" tectonics as opposed to the "thin-skinned" tectonics concept advanced by Price and Mountjoy. As a consequence, the total shortening of the Rocky Mountains is reduced by 60 percent and occurs almost totally within the Main Ranges. Campbell believes that the relatively uniform level of the Lower Cambrian sequences in the core zone and Main Ranges does not support the concept of a gravitational mechanism. In other words, a denuded zone required by the gravitational mechanism is apparently missing.

Campbell suggests that the flanking zones of outward-directed thrusting possibly resulted from underthrusting of the craton. The continued westward movement of cratonic crust under elevated basement massifs could lead to a detachment of the cover rocks which produced the structure of the Foothills and Front Ranges.

Discussion of Structural Style of the Foothills and Front Ranges

The southern Canadian fold and thrust belt is characterized by a suite of relatively simple structural types referred to as a "family of structures" — which have been discussed at length by Bally, *et al.*, (1966), Dahlstrom (1970), and Gretner (1972). These include: 1) low-angle (commonly folded) thrust faults which cut up through the sedimentary section in the direction of tectonic transport; 2) concentric folds and their attendant detachment zones; 3) tear faults, generally transverse, which terminate within individual plates, and locally allow differential movements of the plate; and 4) late normal faults, commonly rotating to a low angle with depth, joining previous planes of weakness such as former thrust faults (i.e., listric-normal faults).

The fact that thrust faults, in a broad sense, are concave upwards requires that the overriding plate should become folded to accommodate itself to the configuration of the fault plane. Also in this manner, older, shallower structures are necessarily deformed by younger, deeper structures while being carried "piggyback" along the underlying younger fault planes. Major thrust plates can, by means of a "bulldozer effect," cause imbricate stacks beneath themselves, which in turn deform the overriding plate. Such "pile-ups" beneath major thrust plates contain significant reserves in fields such as Waterton. It is commonly observed that complicated imbricate thrusting within the Mesozoic section originates from less complicated thrusted structures in the Paleozoic skeleton. In many cases, numerous imbrications will connect with fewer but larger thrust faults in the more competent portions of the stratigraphic section.

There are preferential positions of detachment in incompetent stratigraphic units where fault planes run parallel or subparallel to bedding. In more competent units, the faults tend to cut more oblique to bedding. The common zones of detachment, determined by statistical analysis of fault occurrences observed in the field and subsurface, are illustrated in Fig. 4.

It is interesting that these zones of "preferred" detachment positions commonly, but not always, coincide with "probable" hydrocarbon source-rock zones. The criteria often suggested to be necessary for determining "preferred" detachment positions is their relative incompetency and/or susceptibility to increased pore pressure. Mature and generating hydrocarbon source-rock intervals are, in many cases, associated with high pore pressure and thus should readily yield to horizontally-directed compression (Meissner, personal communication, 1976). Interpreted in this manner, reservoir units could directly overlie, in thrust contact, younger hydrocarbon generating source-rocks.

The overall shortening of the sedimentary cover is accomplished predominantly by means of a pattern of en-echelon thrusts. Along the gentle lateral plunges of a thrust plate displacements have been relayed or transferred to an adjacent sub-parallel thrust, which probably has a common root in a zone of regional detachment. Thrusts often die out laterally into folds which also take up some of the overall shortening. Decreasing fault throw may also translate into increased length of the vertical limb of a drag fold. Differential movements within an individual thrust plate can develop along transverse tear faults.

It appears that the dominant mode of deformation (i.e., concentric folding or thrust faulting) is a function of lithology, vertical and lateral facies changes, and the thickness of individual competent or incompetent strata within an anisotropic sequence. Major facies changes are often reflected by attendent changes in structural style.

In order to unravel the structural complexities of the southern Canadian thrust belt, certain procedures are commonly followed. First, structural cross sections should be constructed parallel to the direction of tectonic transport. Second, the regionally undisturbed (autochthonous) plate is projected by means of deep plains subsurface well control and/or seismic data beneath the disturbed belt. This defines a lower limit of deformation, above which the structure must be accounted for in the most reasonable manner. Last, certain geometric constraints are dictated by the known association of structures for the given lithologies and bed thicknesses.

A method of constructing structural cross sections in a structural province of post-depositional concentric deformation, such as that of thrust belts, is discussed at length by Dahlstrom (1969). Dahlstrom points out that there is no significant change in rock volume in this type of province, thus, bed thickness must also remain constant. Accordingly, the surface area of a bed and its length in cross-sectional plane must remain constant. Given such conditions, the geometric validity of a cross section can be simply tested by measuring the bed lengths at several marker horizons between reference positions. The bed lengths must be the same unless a zone of detachment intervenes. Overall shortening of the sedimentary package, whether by faulting and/or folding should also remain constant along strike. In addition attention must be given to matching footwall cutoffs to hanging wall cutoffs.

Culminations on surface geologic maps in thrusted provinces imply duplication of section. When thrust fault planes occur in the plane of bedding, the thrust itself is not readily detected. However, in "toe" regions of thrust plates, stratigraphic displacement and resulting duplication of section allows relatively simple detection. In this region, considerable information may be obtained regarding the stratigraphic and horizontal displacements, preferential zones of detachment, as well as the nature of the footwall deformation. Down-structure viewing of geologic maps (Mackin, 1950) coordinated with well and seismic control facilitates realistic interpretations of footwall deformation.

FOOTHILLS HYDROCARBON ACCUMULATIONS OF THE SOUTHERN CANADIAN CORDILLERA

Foothills hydrocarbon accumulations are controlled by present day structure which was formed essentially during early Tertiary thrusting. Assuming that a certain amount of generation, primary migration, and accumulation preceeded thrusting, the original traps may have been stratigraphic, localized by regional dip and porosity subcrops (Bokman, 1963). If this is the case, the Foothills accumulations may be secondary, produced at least in party by redistribution of earlier accumulations, as well as by continued generation, migration, and entrapment. Fracturing is believed to play a major role in the enhancement of reservoir permeability, and, in the more intensely deformed settings, contributes significantly to reservoir volume. With one major exception, the Turner Valley oil and gas field, the Foothills accumulations are gas fields with varying amounts of hydrogen sulphide and condensate.

The relative positions of the productive structures within the disturbed belt are illustrated in a schematic structural cross section (Fig. 5, in pocket). Field names, discovery year, producing horizons, and in-place reserves are listed below their appropriate trap type and relative position within the thrust belt. The two largest fields, Turner Valley and Waterton, are discussed in some detail in a later section.

Hydrocarbon Variations, Source-Rocks, and Organic Metamorphism

In a regional sense, the character of the accumulated hydrocarbons is determined by 1) the nature of the hydrocarbon source-rocks and their maturation history, 2) the migration and differential entrapment history, and 3) the secondary physical, thermal, and/or biological alteration of the migrating or accumulated hydrocarbons (Philippi, 1965; LaPlante, 1974; Burgess, 1974; Connan, 1974; Bailey, et al., 1974; Hood, et al., 1975; and Gussow, 1967). A brief discussion of the above generally accepted principles for the western Canadian sedimentary basin is appropriate in order to place the deformed basin margin into proper perspective.

The sedimentary section of the basin increases in thickness southwestward to a maximum of 20,000 ft. in the vicinity of the disturbed belt (Fig. 6a). Generalized hydrocarbon variations for the basin as a whole have been outlined by Hitchon (1964) and are as follows: 1) transitions from sour heavy oils and tars in the east to sweet lighter crudes to gas/gas condensate and finally to gas in the west, 2) *hydrogen sulphide content tends to increase westward (up to 86 percent in Devonian reservoirs in the Panther River structure), 3) total hydrocarbon, methane, and nitrogen content by volume decreases toward the west, and 4) carbon dioxide content by volume tends to increase to the west.

---
*Sour gas accumulations are common in carbonate provinces in which there is little iron present to react with the hydrogen sulphide to form pyrite (Bailey, et al., 1974).

Published data regarding the richness and maturity of hydrocarbon source-rocks in western Canada is limited; however, "probable" candidates for source-rock are Middle and Upper Devonian organic rich carbonates and shales, Mississippian Exshaw, Jurassic Fernie, and Cretaceous Blackstone and Wapiabi shales (Fig. 4). Source-rocks are believed to contain concentrations of plant and animal debris that have generally accumulated in association with fine-grained inorganic matter. Production and preservation of the organic debris are controlled by the physico-chemical conditions of the depositional environment. Some of the most common facies associations of source-rocks with evaporites, phosphorites, or regional transgressive sequences can simply be related to water circulation and nutrient supply (Koch, personal communication, 1976).

Coal ranks have recently been used for comparing stages of organic metamorphism in hydrocarbon source-rocks (Hood, et al., 1975). In western Canada, there is a progressive increase in the rank of coals from the plains region toward the Rocky Mountains (Fig. 6B). This is not simply a function of geologic age and level of exposure, since within individual formations the rank increases toward the west. The rank of coals is believed to be a time-temperature related metamorphic process, which is, in part, a function of depth of burial, geothermal gradient, and geologic age. Apparently, very little tectonic effect on the rank of coals has been observed even in rather intensely deformed settings. A close relationship with the pre-orogenic depth of burial is indicated (Hacquebard and Donaldson, 1974). This would appear to indicate that the initiation of hydrocarbon generation was prior to structural deformation, for the older and/or more deeply buried units.

Some scales of organic metamorphism and their relationship to thermal hydrocarbon generation from organic matter in source-rocks are presented in Fig. 7a. Applying these scales to the known coal rank distributions of western Canada is an instructive exercise. Ghosting maturation levels down to lesser known levels in the lower Mesozoic and Paleozoic implies that there are western limits at least of oil occurrence, and possibly liquid hydrocarbons in general.

Evans and Staplin (1971) mapped oil-occurrence data, reservoired-gas composition, and drill-cuttings gas composition in Paleozoic horizons of west-central Alberta (Fig. 7b). They concluded that the western limit of oil occurrence coincided with categorized "kerogen" color changes in the associated rocks, which in turn reflect thermal maturation levels of organic matter (Staplin, 1969). The western limits of oil occurrence in selected Paleozoic reservoirs are illustrated in Figure 7b. This conclusion has been borne out by the pattern of oil, gas condensate, and gas production established during nearly three-quarters of a century of exploration.

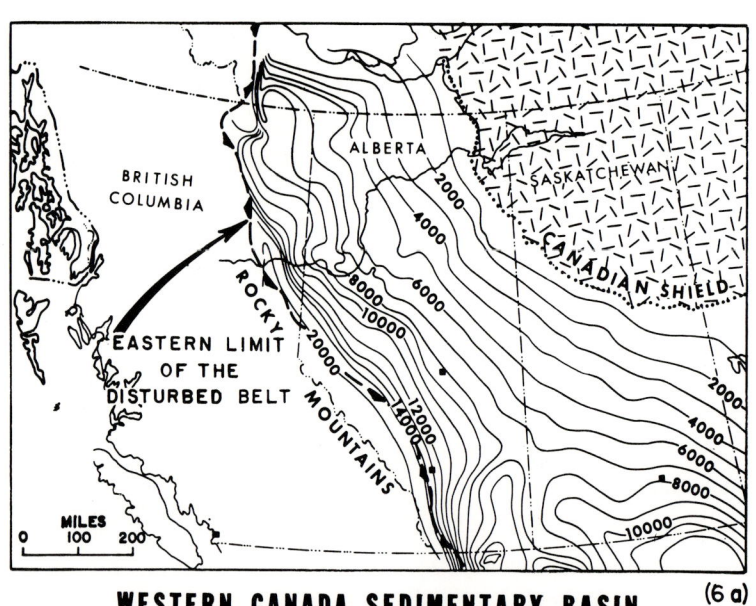

WESTERN CANADA SEDIMENTARY BASIN

CONTOURS ON P€ BASEMENT

After Bailey Et. Al. (1974)

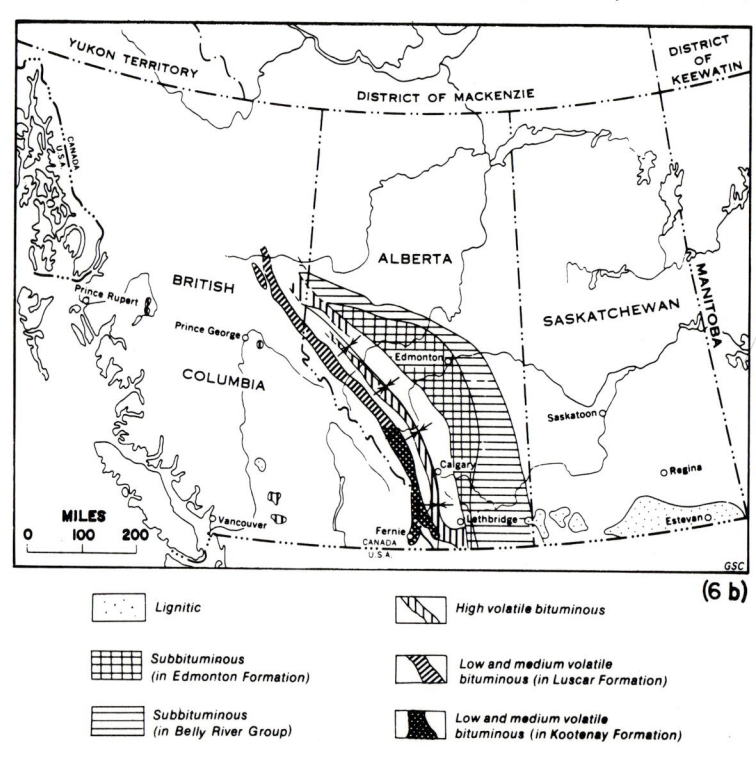

GENERAL AREAL DISTRIBUTION OF COALS BY RANK IN WESTERN CANADA

After Latour & Christmas 1970 and
Hacquebard & Donaldson 1974

—figure 6—

SOME SCALES OF ORGANIC METAMORPHISM AND THEIR RELATIONSHIP TO THERMAL HYDROCARBON GENERATION FROM ORGANIC MATTER IN SOURCE ROCKS

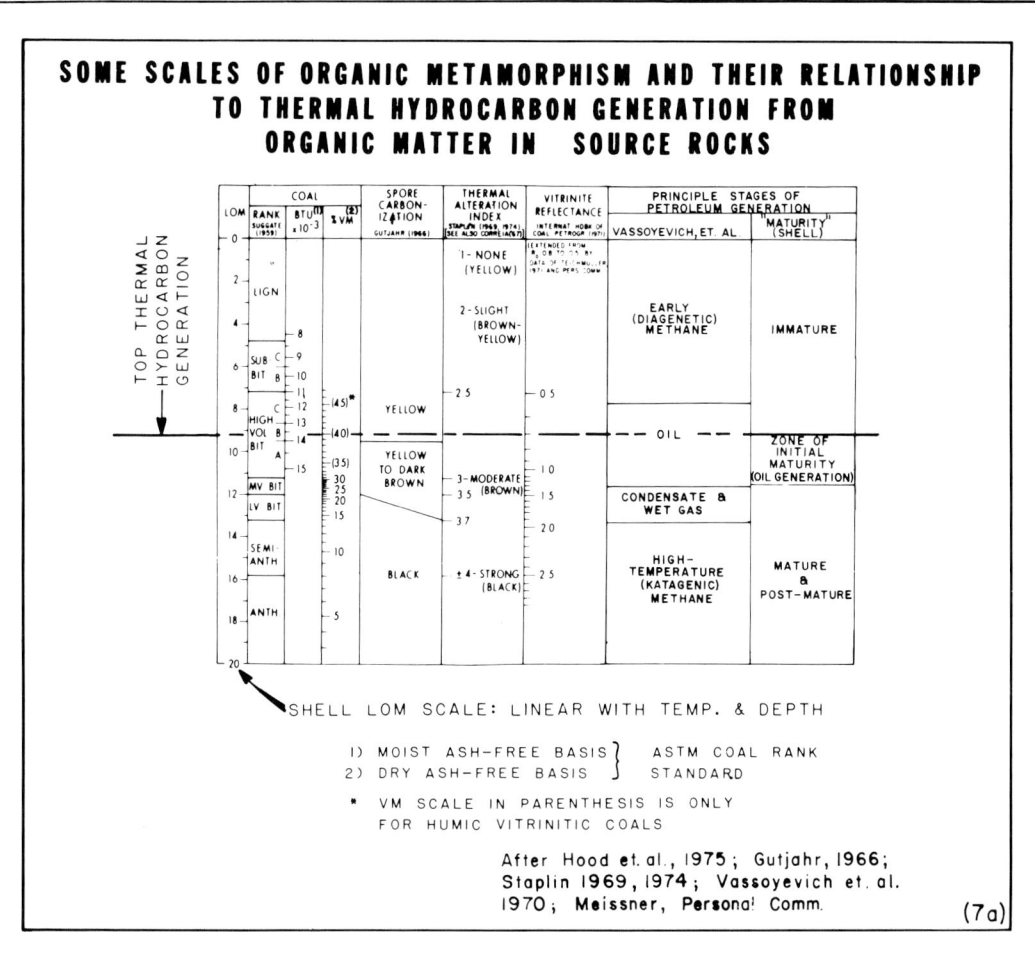

After Hood et. al., 1975; Gutjahr, 1966; Staplin 1969, 1974; Vassoyevich et. al. 1970; Meissner, Personal Comm.

(7a)

THERMAL METAMORPHISM OF PALEOZOIC UNITS IN PLAN (A) AND SECTION (B) IN THE WESTERN CANADA SEDIMENTARY BASIN

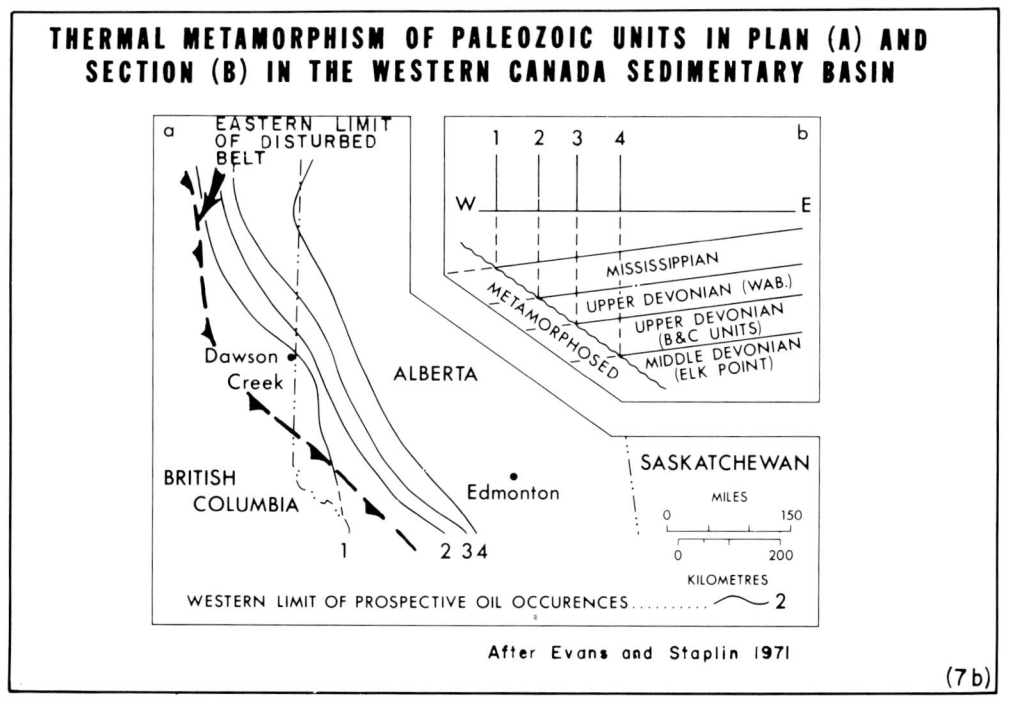

After Evans and Staplin 1971

(7b)

— figure 7 —

Some caution is necessary in applying such an exercise in the thrusted sequences of the Front Ranges and Foothills. Palinspastic structural reconstructions imply that within the deformed basin margin there will be non-representative maturalevels in thrusted plates that have been displaced horizontally from west to east a few to tens of miles. Different and often lower maturation levels will exist beneath such thrusted sequences.

The ultimate composition of the hydrocarbons is strongly influenced by processes which alter them after they have been pooled, such as continued thermal maturation of entrapped hydrocarbons, deasphalting, the action of surface-derived formation waters, and the effects of gas flushing. These processes are discussed more fully by Evans and Staplin, 1971; Bailey, et al., 1974; and Gussow, 1967.

EXAMPLES OF HYDROCARBON ACCUMULATIONS IN THE SOUTHERN CANADIAN CORDILLERA

The Turner Valley and Waterton fields are examples of major hydrocarbon accumulations which have been discovered in the southern Canadian Foothills. Industry and government agencies have provided sufficient information to enable examination of various aspects of these fields. The following descriptions summarize the surface and subsurface geology, geophysical expression, reservoir parameters, in-place reserves, and hydrocarbon characteristics of these accumulations.

Turner Valley Field

The Turner Valley oil and gas field is located 25 mi southwest of Calgary, at the eastern, outer edge of the disturbed belt (Fig. 2). Gas seepages along the Sheep River and a surface fold led to the discovery of natural gas in the Lower Cretaceous in 1913. The thrusted nature of the structure led to a great progression of interpretations. It was not until 1924 that the gas-cap portion of the Paleozoic reservoir was discovered. After twelve years of sporadic development of this portion of the reservoir, the Turner Valley Royalties No. 1 well discovered the major oil column on the west limb. Rapid development took place during World War II (Gallup, 1954; Fox, 1959). The Turner Valley field was discovered and developed without the benefit of sophisticated logging or geophysical methods. It has been referred to as a classic thrust-faulted trap for decades.

Approximately 400 wells have been drilled to date on this feature. The more than 6,000 ft of productive hydrocarbon column had original in-place reserves of 1,000,000,000 barrels of oil and 2.9 TCF of associated and solution gas (E.R.C.B., 1974).

Surface geology. — This feature is expressed at the surface as a breached, thrusted anticline or drag fold in Upper Cretaceous clastics (Fig. 8). Paleozoic rocks are not exposed at any position along strike in this feature. A simple, southwesterly plunge exists; however, the northwesterly closure is more complex, as it is interrupted by imbricate thrusts. An east-dipping thrust fault is present on the east side of the feature and is interpreted to be a remnant of a folded thrust which was in place prior to the development of younger, lower structures (Jones, 1971).

Subsurface geology. — The entire field is underlain by a major sole thrust which has been penetrated in 11 wells. It is known to have about 5,000 ft of stratigraphic displacement and at least two miles of horizontal displacement. Whether or not this fault surfaces on the east limb has been the subject of much discussion (Gallup, 1954; Fox, 1959; Dahlstrom, 1970; Jones, 1971). Both stratigraphic and horizontal displacement on the sole fault decrease along strike. Although the north end of the feature is imbricate even at the Paleozoic reservoir level, the subsurface structure is much less complicated than that of the overlying Mesozoic section (Fig. 8 and 9). Fault closure is necessary to account for the 6,000 ft of hydrocarbon column. The axial plane of the feature dips 77° to the west and the west limb dips at approximately 30°.

Geophysical expression. — Subsequent to the discovery and development of this field, it has been subjected to a great deal of geophysical investigation in order to determine its expression for analog purposes. Bally, et al., (1966) published a seismic line across this feature (Fig. 10, in pocket). The strong Cambrian event beneath the field has a velocity pull-up interpreted to be caused by the presence of the (duplicated) high-velocity Paleozoic thrust plate. This profile clearly illustrates the detached thrusted nature of the Turner Valley structure, with regionally undisturbed Paleozoic and basement rocks extending beneath the thrust belt. Another major thrust feature appears to the west of the Turner Valley anomaly on the same line. This feature, the Highwood uplift, has not proven to be productive. The Quirk Creek field, however, has subsequently been developed in the upper plates of a complex series of interrelated thrusts in front of and beneath the Highwood uplift, north of this line of section (Moore and Walidarsky, personal communication, 1976). Gravity surveys indicate an asymmetrical gravity maximum (steep limb to the east), which is thought to reflect the Turner Valley structure (Oilweek, April 1971).

Reservoirs. — The reservoir consists of carbonate units of the Rundle Group (Mississippian). The main pay zone, the Turner Valley Member, is a fossiliferous limestone, having a predominance of crinoidal material with lesser amounts of bryozoans, corals, algae, and brachiopods, suggesting a biostromal nature (Gallup, 1954).

Table 1 —Reservoir Parameters, Reserves, and Hydrocarbon Characteristics, Turner Valley Field

Reservoir Parameters (Parsons and Mulder, 1968)

Avg. Net Pay = 140 ft
Avg. ϕ = 8%
*Avg. K = 6.84 md
Avg. S_W = 10%
Gas/Water Contact −2200 ft subsea

Maximum Hydrocarbon Column > 6000 ft
Gas Productive Area: 8100 Acres

Initial Pressure 2260 psi @ −1051 ft subsea
BHT 113°F @ −1051 ft subsea

Oil/Water Contact −4300 ft subsea
(−4500 ft subsea @ north end)

Oil Productive Area: 17,000 Acres

In-Place Reserves (E.R.C.B., 1974)

		Recovery Factor**
Gas Assoc.	1.49 TCF	90%
Gas Soln.	1.36 TCF	55%
Oil	1,000,000,000 Bbls.	12%

Gas Characteristics

	Percent		Percent
Nitrogen	—	Isobutane	0.58
Carbon Dioxide	1.68	N-butane	1.27
Hydrogen Sulphide	1.43	Isopentane	0.28
Methane	80.80	N-pentane	0.32
Ethane	9.39	Hexanes	0.36
Propane	3.89	Rare Gases	—

Oil Characteristics

API Gravity: 40°
Color: Light Green
Weight % Sulphur: 0.25

---

*Fracturing is believed to be important to the enhancement of reservoir permeability.

**Large volumes of gas were flared prior to conservation regulations and gas sales. The field is not regarded to have an active water drive. Loss of effective primary gas cap/solution gas drive has resulted in a very low primary oil recovery factor.

— figure 8 —

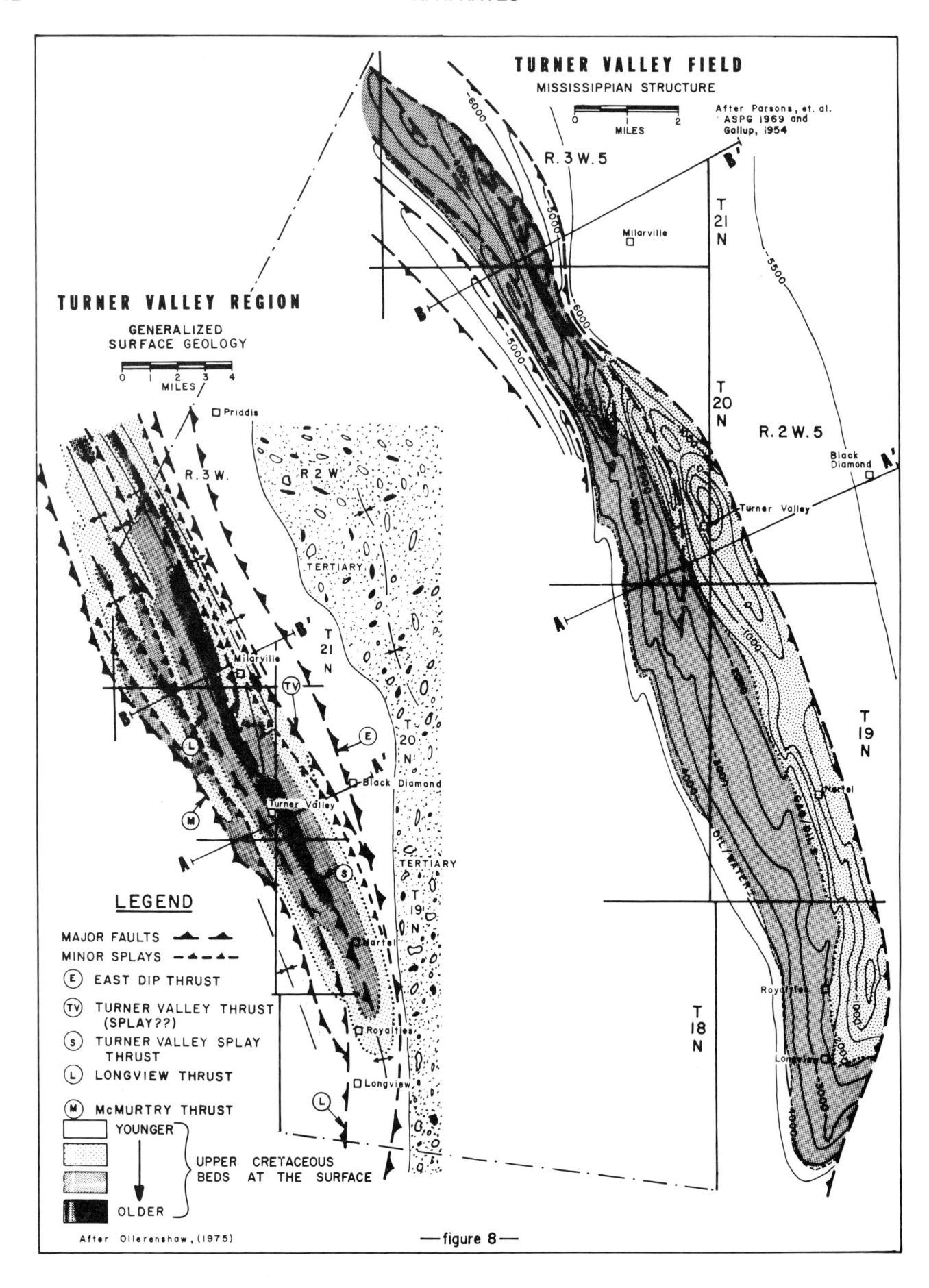

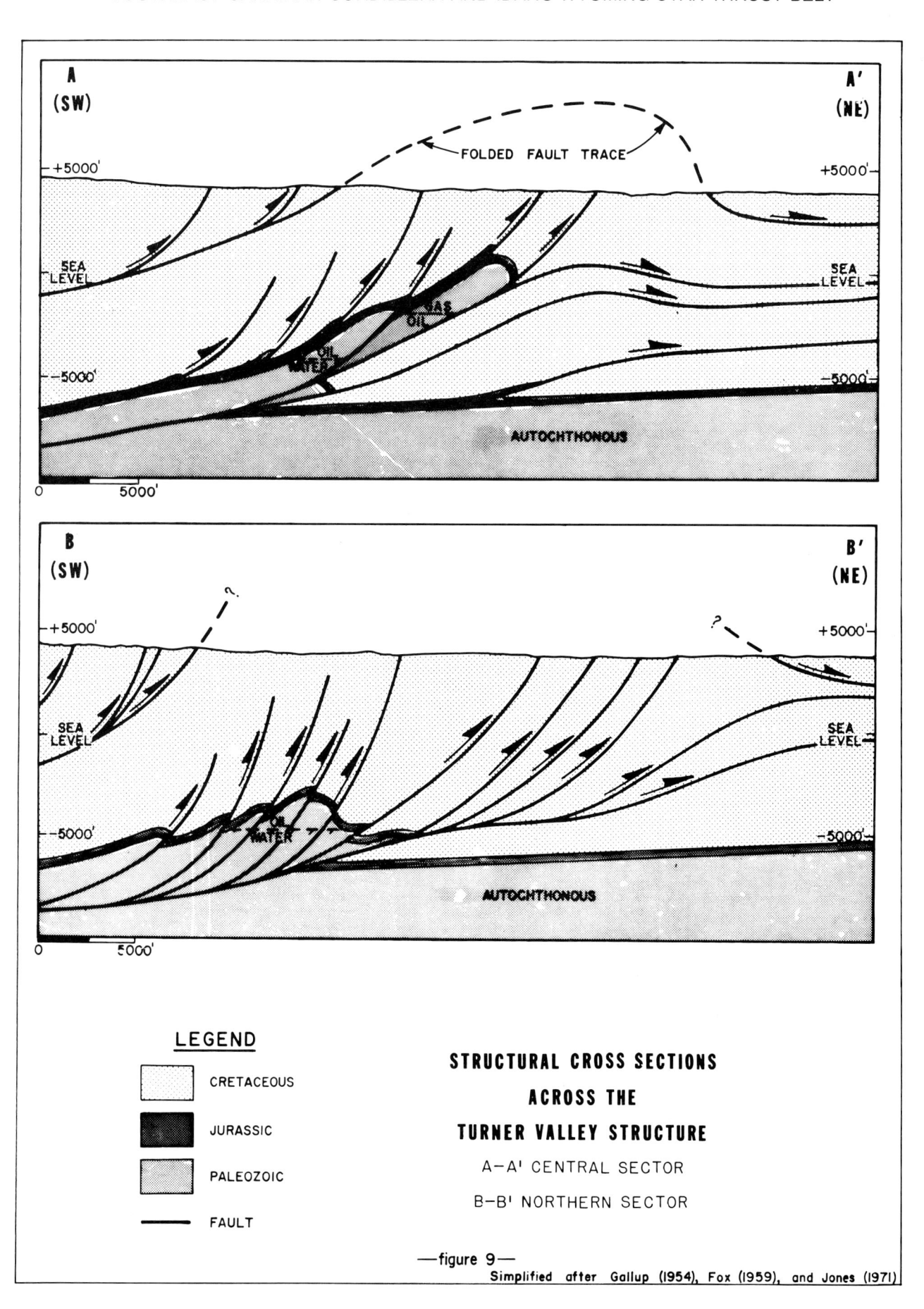

STRUCTURAL CROSS SECTIONS ACROSS THE TURNER VALLEY STRUCTURE

A-A' CENTRAL SECTOR
B-B' NORTHERN SECTOR

— figure 9 —

Simplified after Gallup (1954), Fox (1959), and Jones (1971)

Waterton Gas Field

The Waterton gas field is located along the eastern edge of the Front Ranges, 135 mi southwest of Calgary and 20 mi north of the International Boundary (Fig. 2). The field was discovered in 1957 by Shell Canada Ltd., after conducting extensive refraction and reflection seismic surveys. It is a complex thrust structure at approximately 10,000 ft depth, and its known productive extent contains in-place gas reserves of more than 4 TCF, plus significant natural gas liquids and sulphur. Waterton is a giant gas field discovered by integrated exploration techniques in a complex geological setting.

Surface geology. — The earliest efforts of oil exploration in Alberta took place a few miles south of the Waterton field in what is now Waterton National Park. At the turn of the last century, several wells were drilled in the vicinity of oil seeps along the Cameron Brook, where a minor amount of oil was discovered in Cretaceous sands beneath the Lewis thrust (Link, 1932; Fox, 1959).

The regional backdrop to the Waterton area is the relatively abrupt topographic rise of 1500 to 2500 ft from the rather subdued Foothills to the Front Ranges. Resistant Precambrian-age Belt sedimentary rocks rest on nearly flat-lying, erosionally recessive Cretaceous clastics as a result of the Lewis thrust. This thrust has experienced tens of miles of horizontal displacement and more than 20,000 ft of stratigraphic displacement. The remaining surface geology of the Waterton field consists of a highly disturbed Cretaceous clastic section (Fig. 11). To the northeast of the field, a sequence of highly imbricate thrust slices implies that the Paleozoics could be disturbed to the southwest. In general, the trap is masked by the surface geology and subsurface resolution of the structure must rely on seismic and well control.

Subsurface geology. — At the Paleozoic level, the structure has been found to be less complex than that within the overlying Mesozoic units. The traps are formed by folded leading edges of west-dipping thrust sheets which carry the Mississippian and Devonian carbonate units eastward across Jurassic and Cretaceous rocks (Fig. 11 and 12). In individual thrust sheets, the average width between the faulted leading edges and the hydrocarbon/water contact is two to three mi. Closure is completed by gentle plunges along the strike. The maximum hydrocarbon column exceeds 4000 ft.

Geophysical expression. — The internal structure is by no means well known to date; however, the major framework has been recognized by reflection and refraction seismic data. The basement is interpreted to be dipping westward, undisturbed beneath the field, at 200 ft/mi. Major leading edges are perhaps best indicated by velocity pull-up of the strong Cambrian seismic event beneath the structure (Bally, et al., 1966) (Fig. 13, in pocket). The velocity phenomenon is interpreted to be the result of thrusting, which has repeated high-velocity carbonates (20,000 ft/sec).

Reservoirs. — The main gas accumulations are found in the Mississippian Rundle Group carbonates with secondary objectives in the Devonian Wabamun (Palliser) carbonates. The Livingstone Formation is the principle target in the Rundle Group, and consists of a thick sequence of crinoidal limestone with a dolomitized mud matrix, as well as massive dolomite intervals.

Table 2. Reservoir Parameters, Reserves, and Hydrocarbon Characteristics, Waterton Field.

Reservoir Parameters (Hall, 1968)

Avg. ϕ = 3-5% (matrix)
*Avg. K = <.5 md
Gas/Water Contact −7200 ft subsea
Initial Pressure 4518 psi @ −4000 ft subsea
Maximum Hydrocarbon Column >4000 ft
Productive Area—A significant portion of this field lies undeveloped in Waterton National Park.

In-Place Reserves (E.R.C.B., 1974)
In-place Gas Reserves = 4047 BCF
Recovery Factor 85%

Deliverabilities
Average AOF = 98 MMCF/day/well
Maximum AOF = > BCF/day

Gas Characteristics (Non-Associated)

	Percent		Percent
Nitrogen	1.05	Isobutane	0.23
Carbon Dioxide	4.00	N-butane	0.50
Hydrogen Sulphide	14.95	Isopentane	0.24
Methane	72.73	N-pentane	0.22
Ethane	4.04	Hexane	0.25
Propane	1.31	Rare Gases	0.41

Gross Btu/cu ft @ 60°F and 14.65 psi = 1030
Specific Gravity = 0.765

*Fracturing is an important factor for enhanced permeability, as well as significantly increasing reservoir volume. Certain zones of lithology are preferentially fractured; however, fracturing is simply described as intense.

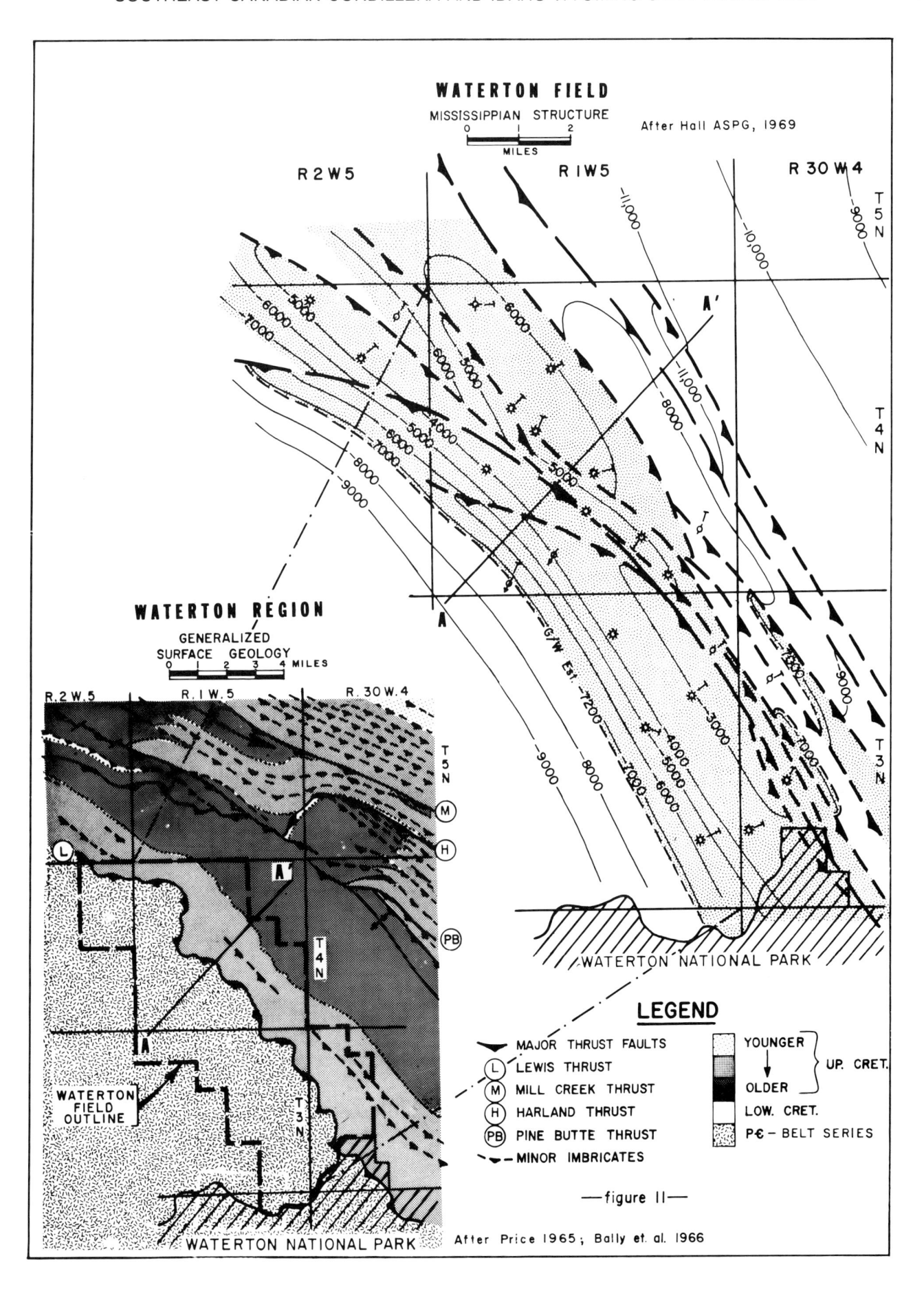

SOUTHEAST CANADIAN CORDILLERA AND IDAHO-WYOMING-UTAH THRUST BELT 75

—figure 11—

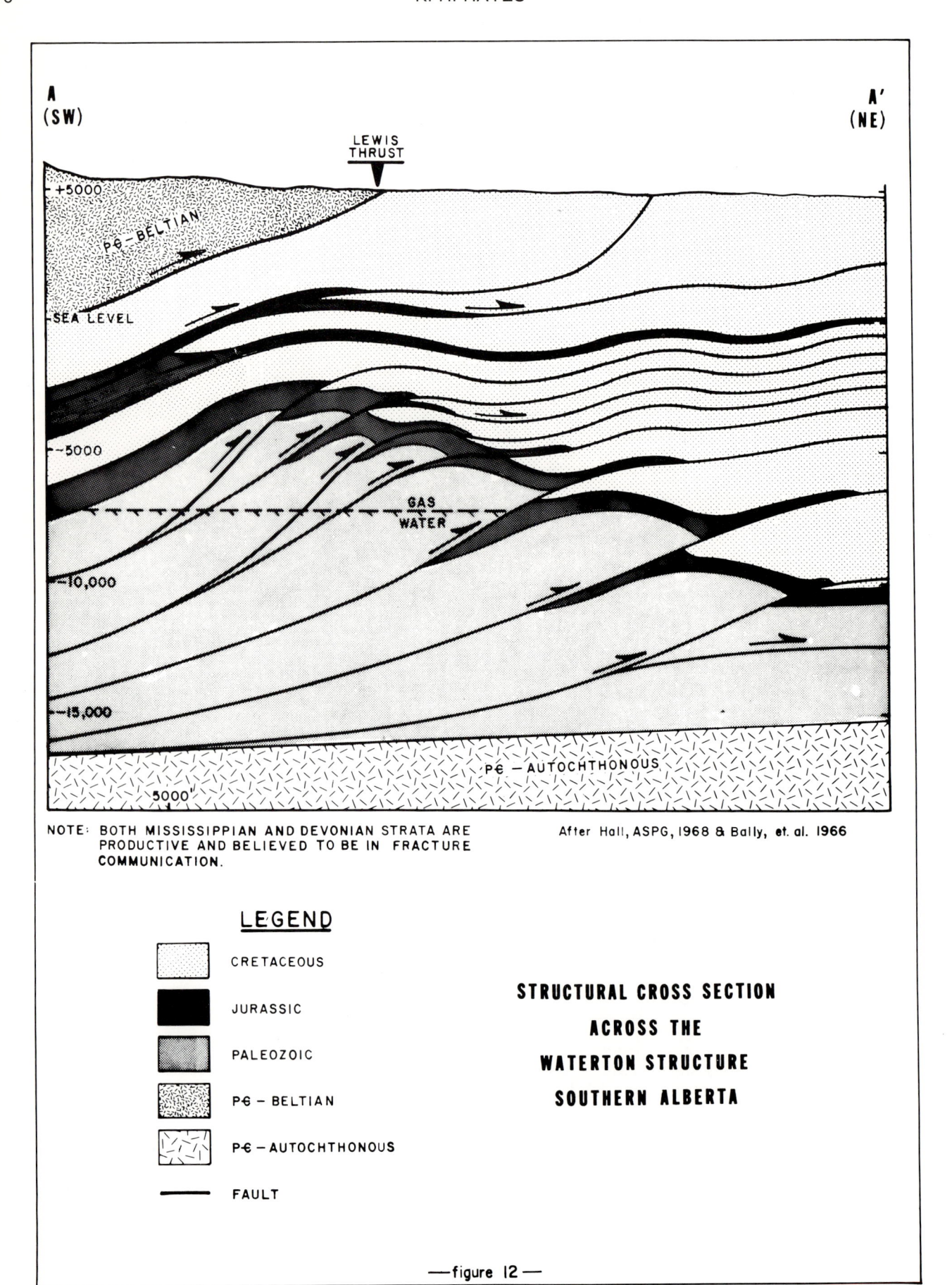

STRUCTURAL CROSS SECTION
ACROSS THE
WATERTON STRUCTURE
SOUTHERN ALBERTA

—figure 12—

IDAHO-WYOMING-UTAH THRUST BELT—A BRIEF COMPARISON

The regional stratigraphy and structural setting of the Idaho-Wyoming-Utah fold and thrust belt are comparable in many aspects to that of the southern Canadian Rocky Mountain fold and thrust belts. Major differences lie in the stratigraphy and the manner in which it controls structural expression.

Stratigraphic Framework

In the Idaho-Wyoming-Utah region, the entire stratigraphic section thickens dramatically from east to west across the Cordilleran hingeline in a manner analogous to the southern Canadian Cordillera. Many, if not all, units in this unconformity-bounded sequence exhibit major facies changes across the hingeline.

Similar to the southern Canadian setting, there is an older miogeoclinal-platform assemblage which accumulated from the Proterozoic to the end of the Jurassic marine transgression (Fig. 14). Terrigenous detritus was derived from mature sediment-source areas to the east and at times to the west. Within this assemblage, the most notable differences between the two regions are the greater thickness of the Jurassic, Triassic, Permian, and Pennsylvanian successions present in western Wyoming and adjacent regions. These successions, within the overall anisotropic sequence, play a more important role in the subsequent structural deformation than do time-equivalent strata in the southern Canadian setting.

Another readily apparent difference between the older assemblages of both regions is the paucity of major lateral facies changes of carbonates to shale in the lower Paleozoics in the Idaho-Wyoming-Utah region. Relatively abrupt facies changes occurring during this time subsequently controlled changes in structural style within the Main Ranges of western Canada. The most notable similarity of the older sequences of the two regions is the presence of upper Paleozoic age shelf carbonates, which in the Canadian disturbed belt have proven themselves to be very prolific hydrocarbon reservoirs.

A post-Jurassic sequence of composite clastic wedges is common to both areas, and their presence indicates the onset of large-scale orogenic activity. Clastic detritus was provided by newly uplifted regions within the orogenic welt to the west (Fig. 14). The concept of a migrating foredeep in advance of deformation appears to hold true for both regions. Such frontal deeps accepted locally as much as 10,000 ft of Lower Cretaceous sediments in front of the Paris-Willard thrust system and as much as 16,000 ft of Upper Cretaceous sediments in a subsiding trough in advance of the Meade thrust system (Reese, personal communication, 1976). The uplifted leading edges of successively younger and easterly thrust plates were the sediment source for these wedges. Lithoclasts present in the syntectonic clastic wedges indicate that progressively older stratigraphy was being exposed by erosion of the local source areas. The pre-Tertiary unconformity and the greater thickness of the Tertiary sedimentary packages are more significant in Wyoming and adjacent regions than anywhere in southern Alberta.

Structural Framework

A core zone comparable to the Shushwap metamorphic complex of the southern Canadian Cordillera has not been documented in the Idaho-Wyoming-Utah structural province. A denuded zone implicitly required by a gravitational mechanism also appears to be absent. The deformation of both the Wyoming-Idaho-Utah and Southern Canadian regions has proceeded in a gross sense from west to east and in an episodic manner. In light of the above, perhaps an underthrusting mechanism for the structural genesis has a certain amount of appeal. Supracrustal shortening by means of folding and thrusting has amounted to approximately 65 mi in the Idaho-Wyoming-Utah region. As in the southern Canadian setting, the crystalline basement generally does not appear to be involved in the thrusting.

The associations of structures in both regions are basically the same. These structures are: 1) thrust faults which are commonly folded, 2) concentric folds with zones of detachment, 3) tear faults, and 4) late normal faults which are often listric.

In the southern Canadian fold and thrust belt, the obvious division between the Foothills and the Front Ranges is the first exposures of the structurally competent and erosionally resistant Paleozoic carbonates. West of this boundary, the structure is predominated by interleaving stacked thrust plates consisting of Mesozoic and Paleozoic units. However, in the Idaho-Wyoming-Utah fold and thrust belt, major folds appear to be the common structure where Paleozoic units are brought to the surface by low-angle thrust faults. This is perhaps a function of the thicker sequences of incompetent strata within the Cretaceous, Jurassic, and Triassic sections of the Wyoming-Idaho-Utah province.

Late normal faults are more prevalent in the southern structural province than in provinces to the north. There is reason to believe that a significant number of the late normal faults within the thrust belt, which are subparallel to the thrust fault traces, are listric. These normal faults are interpreted to rotate back into former zones of weakness, such as pre-existing thrust fault planes, and do not involve the basement (Royse, *et al.*, (1975).

For an in-depth treatment of the tectonic development and thrust belt geometry of western Wyoming and adjacent regions, the reader is referred to papers by Armstrong and Oriel (1965) and Royse, *et al.*, (1975).

— figure 14 —

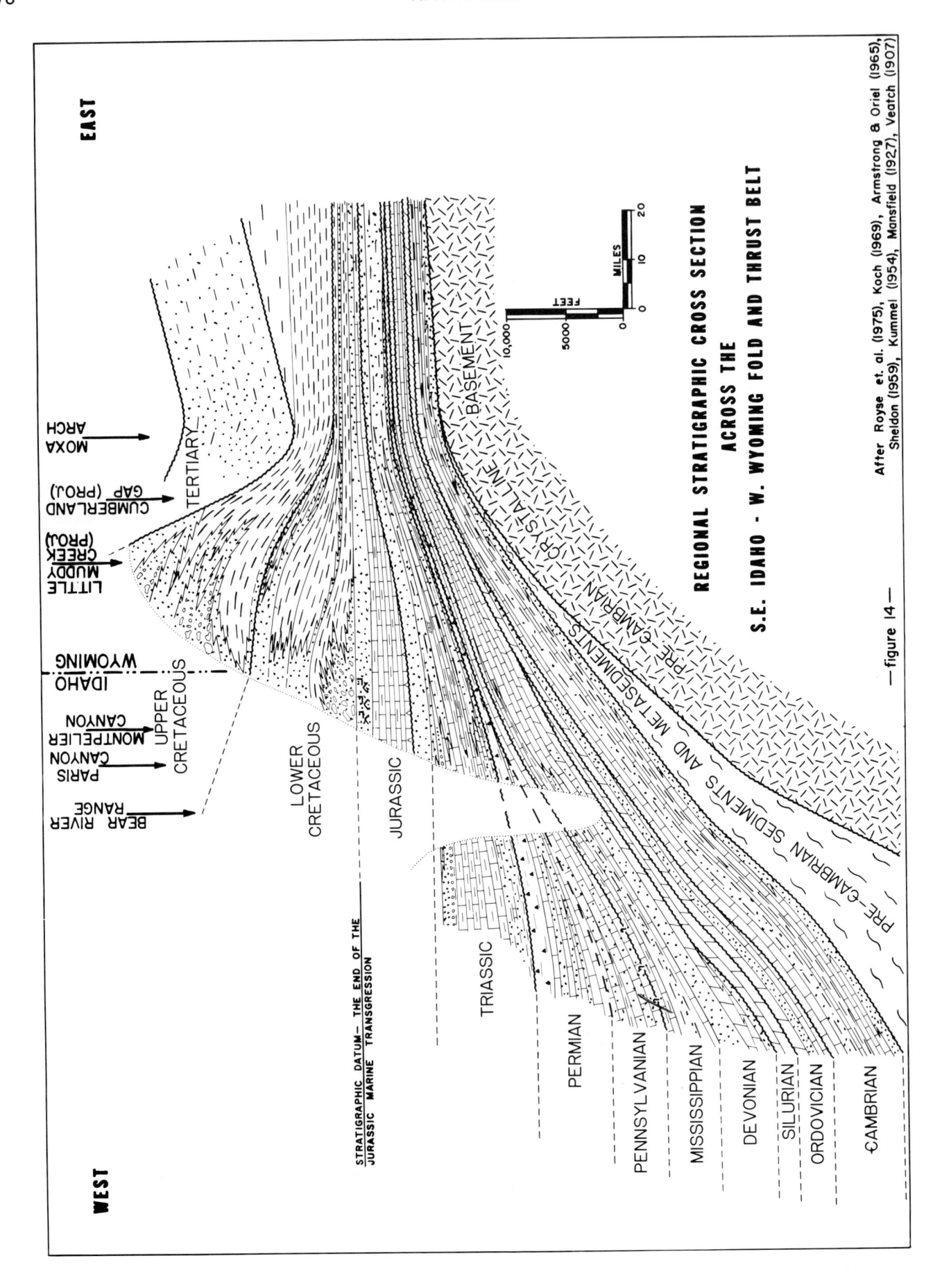

REGIONAL STRATIGRAPHIC CROSS SECTION ACROSS THE S.E. IDAHO - W. WYOMING FOLD AND THRUST BELT

After Royse et. al. (1975), Koch (1969), Armstrong & Oriel (1965), Sheldon (1959), Kummel (1954), Mansfield (1927), Veatch (1907)

Hydrocarbon Accumulations and Recent Exploration Activity

Until recently, there has been limited exploration success in the Idaho-Wyoming-Utah fold and thrust belt. Early efforts, however, did document the presence of hydrocarbons. Oil seepages from the Cretaceous clastics have long been observed along the leading edge of the Absaroka thrust plate (Veatch, 1907) and thus led to the exploitation efforts in the Spring Valley and Sulphur Creek fields. These accumulations, although located within the disturbed belt, are actually stratigraphic traps related to truncation of Cretaceous clastic units by pre-Tertiary erosion and sealed by the onlap of Tertiary sediments. It is interesting to speculate that minor hydrocarbon occurrences such as these are merely indicators of a leaking system which contains major accumulations.

Accumulations adjacent to the fold and thrust belt along the Moxa Arch in the western portions of the greater Green River basin, document the close proximity of significant reservoirs. The Dakota or equivalent strata are productive in the Bridger Lake and Church Butte structures. Significant volumes of hydrocarbons have been produced from the Tertiary Almy and Cretaceous Frontier sections in the LaBarge field. Western margins of the LaBarge complex are thrusted and to some extent thrusting influences the uppermost accumulations in the Cretaceous and Tertiary. The Jurassic Nugget produces from the high crestal areas of the Hogback, Dry Piney, and Tip Top structures beneath the eastern edge of the overthrust belt in the LaBarge complex (Monley, 1971).

The first significant accumulations discovered to date within the fold and thrust belt are the Pineview field and probably the Ryckman Creek field. The Pineview field was discovered in June, 1974, by the drilling of American Quasar *et al.*, Newton Sheep Co. No. 1, in the NW SE Sec. 4, T. 2 N., R. 7 E., Summit County, Utah (Loucks, 1975; Gregg, 1976). The producing structure is controlled by the folded leading edge of the third major thrust sheet from the east, known as the Evanston (Tunp) thrust plate. The reservoirs are the Nugget Sandstone and the overlying fractured Twin Creek Limestone, both of Jurassic age. The fault system is believed to have horizontal displacement of a few tens of miles and stratigraphic displacement in the order of 10,000 ft. Hydrocarbon columns of these two reservoirs are known to exceed 450 ft, and preliminary reserve estimates range from one hundred to several hundred million bbl of 48° API gravity oil. The subthrust structure is less known to date; however, it may have significant Cretaceous gas potential.

More recently, in February, 1976, the Ryckman Creek field was discovered by Amoco Production Co. and others by the drilling of the Champlin 224 Amoco-A well in NW NW Sec. 19, T 17 N., R 118 W., Uinta County, Wyoming. The discovery well has a column of 497 ft of natural gas, gas condensate, and oil pay in the Jurassic Nugget Sandstone. The trap at Jurassic level is the folded leading edge of the Absaroka thrust plate, the second major thrust fault system from the east. The surface expression is masked by younger underformed Tertiary cover; however, the pre-Tertiary expression is believed to be that of a breached anticline eroded to at least a Lower Cretaceous level. The discovery well drilled through a major sole fault which placed Triassic rocks over the Cretaceous Frontier or Bear River formations (Reese, personal communication, 1976).

Significant hydrocarbon shows from current activity have been keeping interest in the region at an extremely high level. Total crew-months of seismic operations are perhaps at the highest levels ever for this frontier area. As a result, numerous wildcats will be located and should spud during the forthcoming drilling season.

The stratigraphic intervals which have proven to be significant reservoirs or have potential for becoming significant reservoirs in the Idaho-Wyoming-Utah fold and thrust belt play are cited in Fig. 15. In the same illustration, "probable" hydrocarbon source-rocks and "preferred" detachment positions are highlighted. A basic amount of seismic velocity data are also included.

CONCLUSIONS

Generalizations which can be drawn from the relatively well known setting of the southern Canadian Rocky Mountain fold and thrust belt and applied with certain modifications to the Idaho-Wyoming-Utah fold and thrust belt are as follows:

1. The structural style is some function of bed thickness and lithology of the sedimentary sequence. Whether folding or faulting dominate the structural style is dependent upon the distribution and relative thickness of the competent (e.g., massive carbonates) and incompetent (e.g., shales and evaporites) stratigraphic units. Folding is more prevalent in sequences with thick incompetent members.

2. Overall shortening of the supracrustal rocks is relatively constant along the length of the thrust belt and commonly ranges between 50 and 125 mi. The net effect is to double the stratigraphic thickness by repetition of section in a thrusted sequence.

3. The deformation advanced, in general, from within the Cordillera toward its outer margins. Major thrust fault progression is generally from the west to the east. The overall chronology of movement is such that the earlier thrusts are the uppermost in the west, whereas the later ones are the the underlying thrusts. The overall magnitude and intensity of deformation decreases toward the east.

—figure 15—

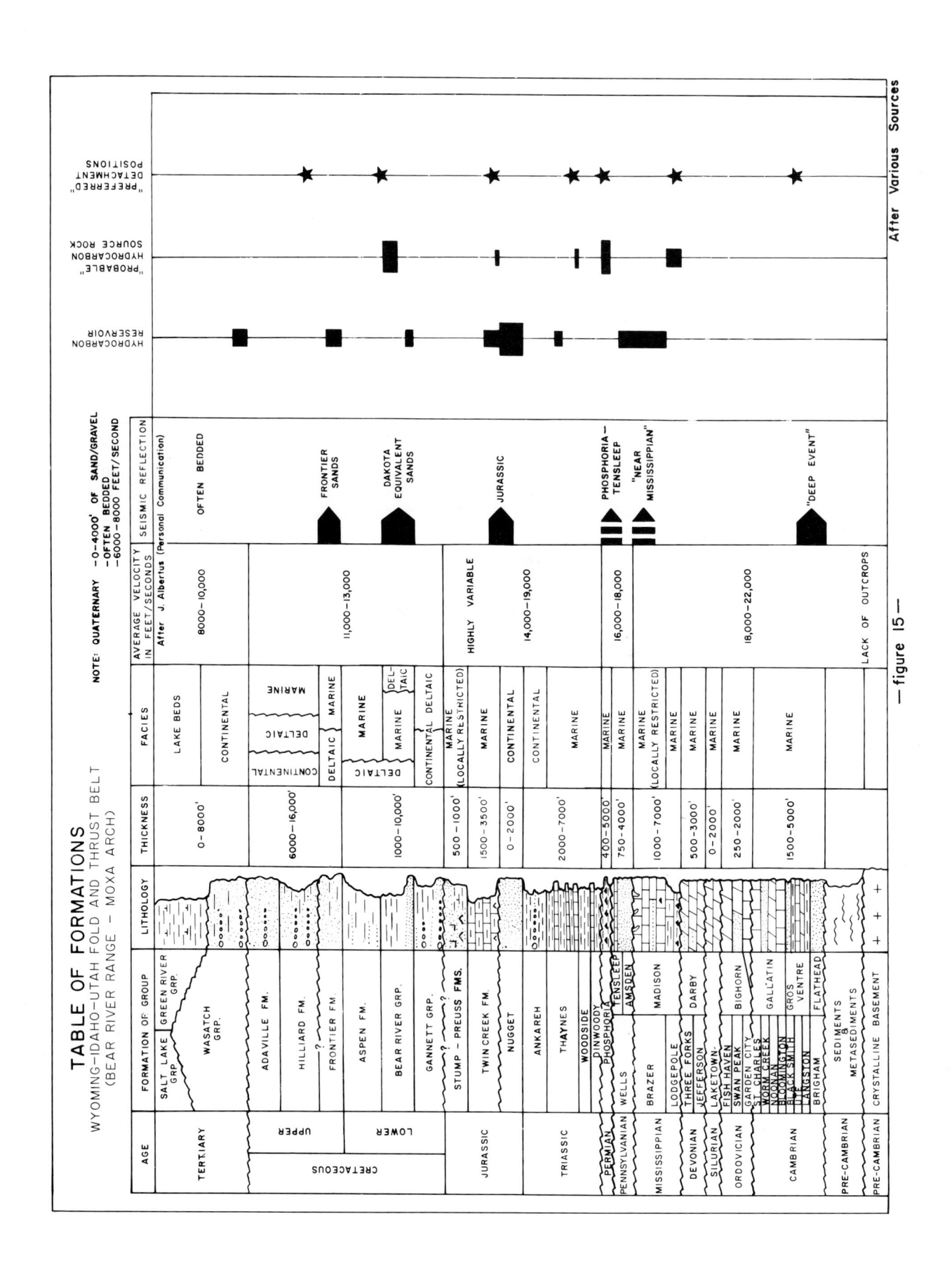

4. Thrust faults climb in the stratigraphic section in the direction of tectonic transport, toward the least overburden resistance. Thrust plates are commonly folded during emplacement as well as by younger movements on underlying thrusts.

5. Fracture-induced porosity and permeability enhancement is typical of most productive hydrocarbon-bearing thrust structures. In many of the Canadian fields, individual well productivity is thought to be inversely proportional to the distance from the leading edge of the thrust reservoir horizon.

6. Fault closure can provide a significant trapping mechanism if the appropriate reservoir and sealing facies are juxtaposed.

7. Reservoirs and hydrocarbon source-rocks are in common association in the hingeline region, where rapid thickness and stratigraphic facies changes often occur. In addition, the thrust-fault origin of many structures can place older reservoir rocks in fault contact with younger hydrocarbon source-rocks.

8. The types of hydrocarbon generated are determined by the nature of the hydrocarbon source-rocks and their maturation history. The thermal maturation levels of the hydrocarbon source-rocks in the older stratigraphic sequences were probably sufficient to have generated significant volumes of hydrocarbons prior to their structural involvement. Early accumulations were likely stratigraphic traps localized by regional dip, hingeline facies changes, and erosional truncation. Subsequently, tectonically thickened sequences (supracrustal shortening), burial by syntectonic deposits, and differential erosion altered the orderly pattern of thermal maturation and distribution of hydrocarbon types. Redistribution of hydrocarbons from primary traps into younger structural configurations combined with additional generation of hydrocarbons probably produced the thrust belt accumulations.

Hydrocarbon exploration in fold and thrust belts is complicated; however, major potential remains, and risk capital can be justified to explore these regions. In the Idaho-Wyoming-Utah fold and thrust belt, much of the early exploration was not optimumly located. Genetic and geometric constraints developed in comparable geologic settings, combined with advanced geophysical exploration methods, should greatly improve the exploration success ratio.

REFERENCES CITED AND DATA SOURCES

Armstrong, R. C. and Oriel, S. S., 1965, Tectonic development of Idaho-Wyoming thrust belt: Am. Assoc. Petroleum Geologist Bull., v. 49, no. 11, p. 1847-1866.

Bailey, N. J. L., Evans, C. R., and Milner, C. W. D., 1974, Applying petroleum geochemistry to search of oil in examples from western Canada Basin: Am. Assoc. Petroleum Geologists Bull., v. 58, no. 11, p. 2284-2294.

Bally, A. W., Gordy, P. L., and Stewart, G. A., 1966, Structure, seismic data, and orogenic evolution of southern Canadian Rocky Mountains: Bull. Can. Petroleum Geology, v. 14, no. 3, p. 337-381.

Bokman, J., 1963, Post Mississippian unconformity in western Canada Basin; in Backbone of the Americas: Am. Assoc. Petroleum Geologists Mem. 2., p. 252-263.

Burgess, J. D., 1974, Microscopic examination of kerogen (dispersed organic matter) in petroleum exploration: Geol. Soc. Alberta Sp. Paper 153, p. 19-30.

Burwash, R. A., Baadsgaard, H., Peterman, Z. E., Hunt, G. H., 1964, Geological history of western Canada; R. G. McCrossan and R. P. Glaister ed., Alberta Soc. Petroleum Geologists, p. 14-19.

Campbell, R. B., 1973, Structural cross-section and tectonic model of the southeastern Canadian Cordillera: Can. Jour. Earth Sci., v. 10, p. 1606-1620.

Coney, P. J., 1972, Cordilleran tectonics and North American plate motion: Am. Jour. Sci., v. 272, p. 603-628.

Dahlstrom, C. D. A., 1969, Balanced cross sections: Can. Jour. Earth Sci., v. 6, p. 743-757.

_____, 1970, Structural geology in the eastern margin of the Canadian Rocky Mountains: Bull. Can. Petroleum Geology, v. 18, no. 3, p. 332-406.

Dewey, J. F., and Bird, J. M., 1970, Mountain belts and the new global tectonics: Jour. Geophys. Research, v. 75, no. 14, p. 2626-2647.

Douglas, R. J. W., ed, 1970, Geology and economic minerals of Canada; Geol. Surv. of Canada Economic Geol. Rpt. No. 1, Dept. of Energy, Mines, and Resources, Canada, p. 367-546.

Eisbacher, G. H., Carrigy, M. A., and Campbell, R. B., 1974, Paleodrainage pattern and late orogenic basins of the Canadian Cordillera, in Tectonics and Sedimentation: Soc. Econ. Paleontologists and Mineralogists Spec. Pub. No. 22.

Energy Resources and Conservation Board, (E.R.C.B.), 1974, Reserves of Crude Oil, Gas, Natural Gas Liquids, and Sulphur, Province of Alberta.

Evans, C. R., and Staplin, F. L., 1971, Regional facies of organic metamorphism, geochemical exploration: Can. Inst. Mining and Metallurgy, Sp. Vol. 11, p. 517-520.

Evers, H. J., and Thorpe, J. E. ed., 1975, A guidebook to the structural geology of the Foothills, between Savanna Creek and Panther River, S.W. Alberta, Canada: Can. Soc. Petroleum Geology—Can. Soc. Exploration Geophysics.

Gallup, W. B., 1954, Geology of Turner Valley oil and gas field, Alberta, Canada, in Western Canada sedimentary basin, a symposium, Ralph L. Rutherford memorial volume, Am. Assoc. Petroleum Geologists, p. 397-414.

Gordy, P. L., Frey, F. R., and Ollerenshaw, N. C., 1975, Road log Calgary-Turner Valley-Jumping Pound-Seebe, Guidebook for structural geology of the Foothills between Savanna Creek and Panther River, S.W. Alberta, Canada: Can. Soc. Petroleum Geology—Can. Soc. Exploration Geophysics.

Gregg, C. C., 1976, Pineview field opens new Rockies oil province: World Oil, June, 1976.

Gretner, P. E., 1972, Thoughts on overthrust faulting in a layered sequence, Bull. Can. Petroleum Geology, v. 20, no. 3, p. 583-607.

Gussow, W. C., 1967, Migration of reservoir fluids: Jour. Pet. Tech., SPE 1870.

Gutjahr, C. C. M., 1966, Carbonization of pollen grains and spores and their application: Leidse Geol. Meded., v. 38, p. 1-30.

Hacquebard, P. A., and Donaldson, J. R., 1974, Rank studies of coals in the Rocky Mountains and inner Foothills belt, Canada, carbonaceous materials as indicators of metamorphism, in Geol. Soc. America Sp. Paper 153.

Halladay, I. A. R., and Mathewson, D. H., ed., 1971, A guide to the geology of the eastern Cordillera along the Trans-Canada Highway between Calgary, Alberta and Revelstoke, British Columbia: Alberta Soc. Petroleum Geologists.

Hall, N. G., 1968, Waterton field, in Gas fields of Alberta; Alberta Soc. Petroleum Geologists, p. 337-338.

Hitchon, B., 1964, Formation fluids, in Geological history of western Canada: Alberta Soc. Petroleum Geologists, p. 201-216.

Hood, A., Gutjahr, C. C. M., and Heacock, R. L., 1975, Organic metamorphism and the generation of petroleum: Am. Assoc. Petroleum Geologists Bull., v. 59, no. 6., p. 986-996

Jones, P. B., 1971, Folded faults and sequence of thrusting in Alberta Foothills: Am. Assoc. Petroleum Geologists Bull., v. 55, no. 2, p. 292-306.

Keeting, L. F., 1966, Exploration in the Canadian Rockies and Foothills: Can. Jour. Earth Sci., v. 3, no. 5, p. 713-723.

King, P. B., 1969, Tectonic Map of North America; 1:500,000, U.S. Geol. Survey.

Koch, W. J., 1969, Lower Triassic lithofacies of the Cordilleran Miogeosyncline, western U.S.: Ph.D. Dissertation, Harvard University.

LaPlante, R. E., 1974, Hydrocarbon generation in Gulf Coast Tertiary sediments: Am. Assoc. Petroleum Geologists Bull., v. 58, no. 7, p. 1281-1289.

Latour, B. A., and Christmas, L. P., 1970, Preliminary estimate of measured coal resources including reassessment of indicated and inferred resources in western Canada: Canada Geol. Survey. Paper 70-58.

Link, T. A., 1932, Oil seepages in Belt Series of Rocky Mountains near International Boundary; Am. Assoc. Petroleum Geologists Bull., v. 16, no. 3, p. 786-796.

Loucks, G. G., 1975, The search for Pineview field, Summit County, Utah: Rocky Mtn. Assoc. Geologists Symposium on deep drilling frontiers in the central Rocky Mountains, p. 255-264.

Mackin, J. H., 1950, The down-structure method of viewing geologic maps: Jour. Geology, v. 58, no. 1, p. 55-72.

Mansfield, G. R., 1927, Geography, geology, and mineral resources of part of southeastern Idaho: U.S. Geol. Survey Prof. Paper 152.

McCrossan, R. G., and Glaister, R. P., eds, 1964, Geological History of Western Canada; Alberta Soc. Petroleum Geologists.

_____, ed., 1973, The future petroleum provinces of Canada — their geology and potential: Can. Soc. Petroleum Geologists Mem. 1.

Monley, L. E., 1971, Petroleum potential of Idaho-Wyoming overthrust belt, in Future petroleum provinces of the United States — their geology and potential: Am. Assoc. Petroleum Geologists Mem. 15, v. 1, p. 509-529.

Norris, D. K., and Bailey, A. W., 1972, Coal, oil, gas, and industrial mineral deposits of the Interior Plains, Foothills, and Rocky Mountains of Alberta and British Columbia: XXIV International Geol. Congress, Guidebook for Excursion A25-C25.

Oilweek, April 1971, Gravity aids oil hunt in Alberta Foothills.

Ollerenshaw, N. C., 1975, Geological map of Rocky Mountain Foothills and Front Ranges, A compilation, in Guidebook for structural geology of the Foothills between Savanna Creek and Panther River, S. W. Alberta, Canada: Can. Soc. Petroleum Geologists—Can Soc. Exploration Geophysics.

Parsons, K. R., and Mulder, F. R., 1969, Turner Valley field, in Gas fields of Alberta: Alberta Soc. Petroleum Geologists.

Philippi, G. T., 1965, On the depth, time and mechanism of petroleum generation: Geochim et Cosmochim. Acta, v. 29, p. 1021-1049.

Price, R. A., 1965, Flathead map area, British Columbia and Alberta: Canada Geol. Survey Mem. 336.

_____, and Mountjoy, E. W., 1970, Geologic structure of the Canadian Rocky Mountains between Bow and Athabasca Rivers — a progress report: Geol. Assoc. Canada Sp. Paper No. 6.

_____, 1971, A section through the eastern Cordillera at the latitude of Kicking Horse Pass, in Guidebook for Canadian exploration frontiers symposium: Alberta Soc. Petroleum Geologists.

_____, and Douglas, R. J. W., eds., 1972, Variations of tectonic styles in Canada: Geol. Assoc. Canada Sp. Paper No. 11.

_____, Balkwill, H. R., Charlesworth, H. A. K., Cook, D. G., Simony, P. S., 1972, The Canadian Rockies and tectonic evolution of the southeastern Canadian Cordillera, in XXIV International Geol. Congress, Guidebook for Excursion AC15.

_____, 1973, Large scale gravitational flow of supracrustal rocks, southern Canadian Rockies, in Gravity and Tectonics: Wiley-Interscience.

Royse, F., Warner, M. A., and Reese, D. L., 1975, Thrust belt structural geometry and related stratigraphic problems, Wyoming-Idaho-northern Utah, in Rocky Mtn. Assoc. Geologists Symposium on deep drilling frontiers in the central Rocky Mountains, p. 41-54.

Scott, J. C., 1954, Folded faults in Rocky Mountain Foothills of Alberta, Canada, in Western Canada sedimentary basin; a symposium, Ralph L. Rutherford memorial volume, Am. Assoc. Petroleum Geologists.

Shaw, E. W., 1963, Canadian Rockies — orientation in time and space, in Backbone of the Americas: O. E. Childs and B. W. Beebe, eds, Am. Assoc. Petroleum Geologists Mem. 2, p. 231-242.

Sheldon, R. P., 1963, Physical stratigraphy and minerals resources of Permian Rocks in western Wyoming: U.S. Geol. Survey Prof. Paper 313-B.

Staplin, F., 1969, Sedimentary organic matter, organic metamorphism and oil and gas occurrence, in Can. Petroleum Geology: v. 17, no. 1, p. 46-66.

_____, Bailey, N. J. L., Pocock, S. A. J., Evans, C. R., 1974, a preprint, Diagenesis and metamorphism of sedimentary organic matter.

Vassoyevich, N. B., Korchagina, Yu. I., Lopatin, N. V., and Cheruyshev, V. V., 1970, Principle phase of oil formation: Moskov. Univ. Vestnik, No. 6, p. 3-27 (in Russian); English transl., Internat. Geol. Rev., Vol. 12, p. 1276-1296.

Veatch, A. C., 1907, Geography and geology of a portion of southwestern Wyoming: U.S. Geol. Survey Prof. Paper 56.

Wheeler, J. O., Campbell, R. B., Reesor, J. E., Mountjoy, E. W., 1972, Structural style of the southern Canadian Cordillera: XXIV Internat. Geol. Congress, Guidebook for Excursion A-01-X-01.

Ziegler, P. A., 1969, The development of sedimentary basins in western and Arctic Canada: Alberta Soc. Petroleum Geologists.

STRATIGRAPHY OF YOUNGER PRECAMBRIAN ROCKS ALONG CORDILLERAN HINGELINE, UTAH AND SOUTHERN IDAHO

by

Lee A. Woodward[1]

ABSTRACT

There are two distinct sequences of stratified Precambrian rocks in Utah. The oldest sequence is the Uinta Mountain Group-Big Cottonwood Formation of Precambrian Y age. The uppermost unit of the Uinta Mountain Group has been radiometrically dated and is about 950 m.y. old. The Big Cottonwood Formation and the Uinta Mountain Group are considered to be correlative. This sequence consists of clastic rocks that were derived from the northeast and fill an east-west-trending basin along the Uinta axis. Maximum thickness is at least 20,600 feet and may be as much as 24,000 feet for the Uinta Mountain Group and the Big Cottonwood Formation is as much as 16,000 feet thick. The Big Cottonwood is unconformably overlain by the upper sequence of Precambrian strata and Uinta Mountain Group is overlain by the Tintic Quartzite that is probably of Middle Cambrian age in the Uinta Mountains.

The upper sequence of rocks, of Precambrian Z age, marks the initial development of the Cordilleran geosyncline, with a thick basinal facies west of the hingeline and a thin shelf facies to the east. Distinctive diamictite of probable glacial origin occurs at the base of this sequence. Clastic rocks predominate in this sequence although minor limestone occurs in the basinal facies. Maximum thickness of the basinal facies is at least 20,000 feet. The Cambrian Tintic Quartzite and correlative Prospect Mountain Quartzite overlie the Precambrian Z strata.

INTRODUCTION

This paper deals with the stratified Precambrian rocks of Utah and southern Idaho. These strata comprise the Uinta Mountain Group, the Big Cottonwood Formation, and the younger formations that are overlain by the Cambrian Tintic Quartzite and the correlative Prospect Mountain Quartzite. Crystalline Precambrian rocks are not discussed.

The U.S. Geological Survey has recently divided the Precambrian into four intervals — in decreasing age — W, X, Y, and Z. Precambrian rocks discussed in this report are of Precambrian Y and Z age; Z is from the base of the Cambrian to 800 million years (m.y.) and Y is 800 to 1,600 m.y. old (James, 1972).

The objectives of this report are to summarize the current concepts concerning Precambrian stratigraphy in this region and to provide a list of references for those who may want to consult the original work.

---

[1] Department of Geology, University of New Mexico, Albuquerque, New Mexico.

STRUCTURAL SETTING

A zone of overthrust faults that formed during the Sevier orogeny trends southwesterly across Utah (Fig. 1). Miogeosynclinal rocks have moved eastward over shelf-facies rocks for distances estimated by various workers to range from as little as 12 miles (Hintze, 1960) to as much as 75 miles (Eardley, 1951, p. 330). Crittenden (1961) has summarized the evidence in northern Utah using juxtaposed thicknesses and lithologies for units ranging in age from Precambrian to Permian; his data indicate about 40 miles of movement. Miller (1966) suggested about 40-60 miles of thrust movement in southwestern Utah. Precambrian Z strata are represented by a thick basinal facies in the allochthon west of the zone of overthrusts and by a thin shelf facies east of the overthrusts.

Late Cenozoic block faulting has offset the thrust faults and has resulted in thick basin fill in the down dropped blocks covering much of the area. One of the most spectacular of the younger normal faults bounds the west side of the Wasatch uplift.

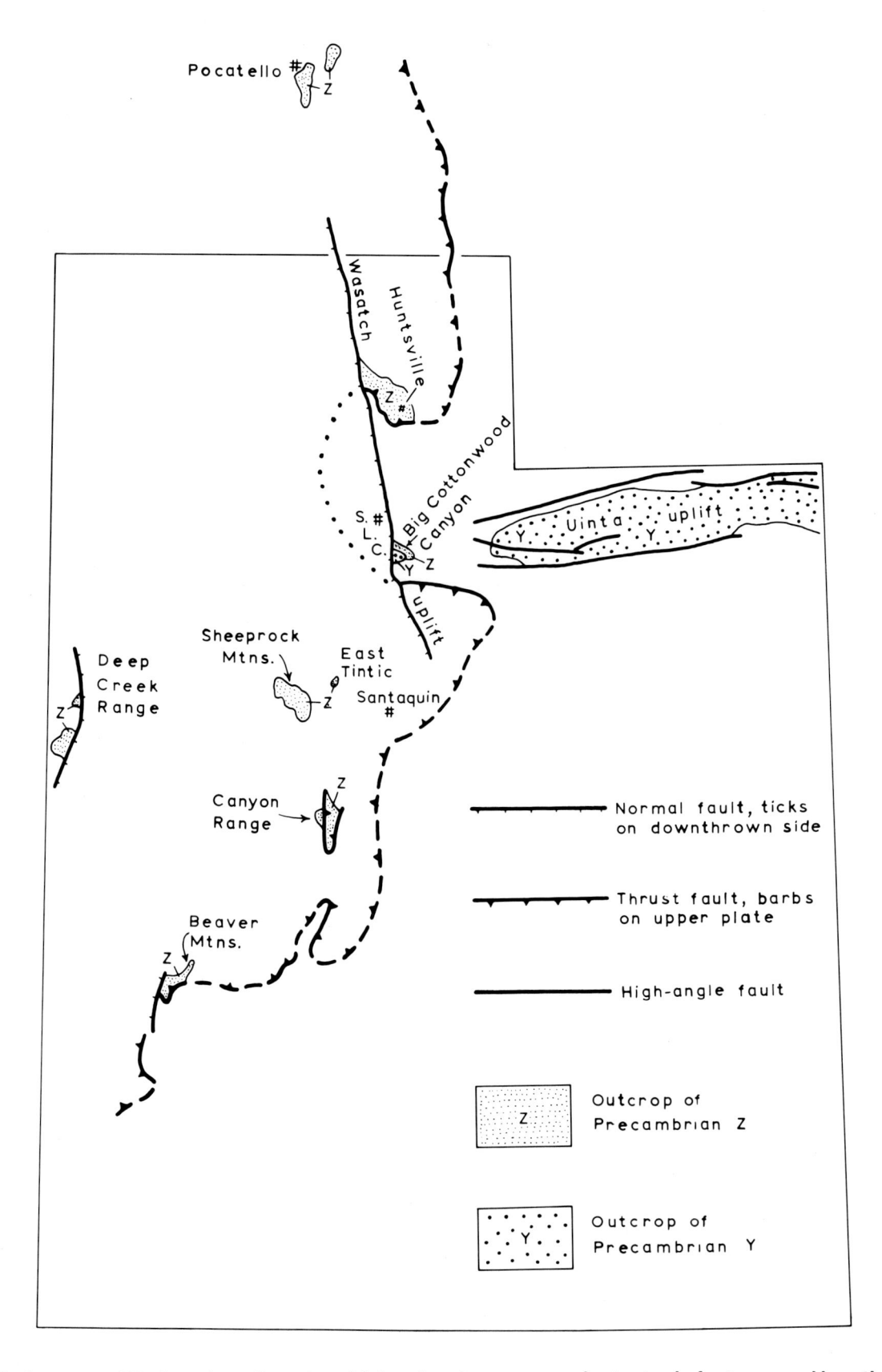

Fig. 1 — Index map of Utah and southeastern Idaho showing some major tectonic features and locations of major younger Precambrian exposures. Precambrian rocks west of major thrust fault are allochthonous. Modified from King and Beikman (1974).

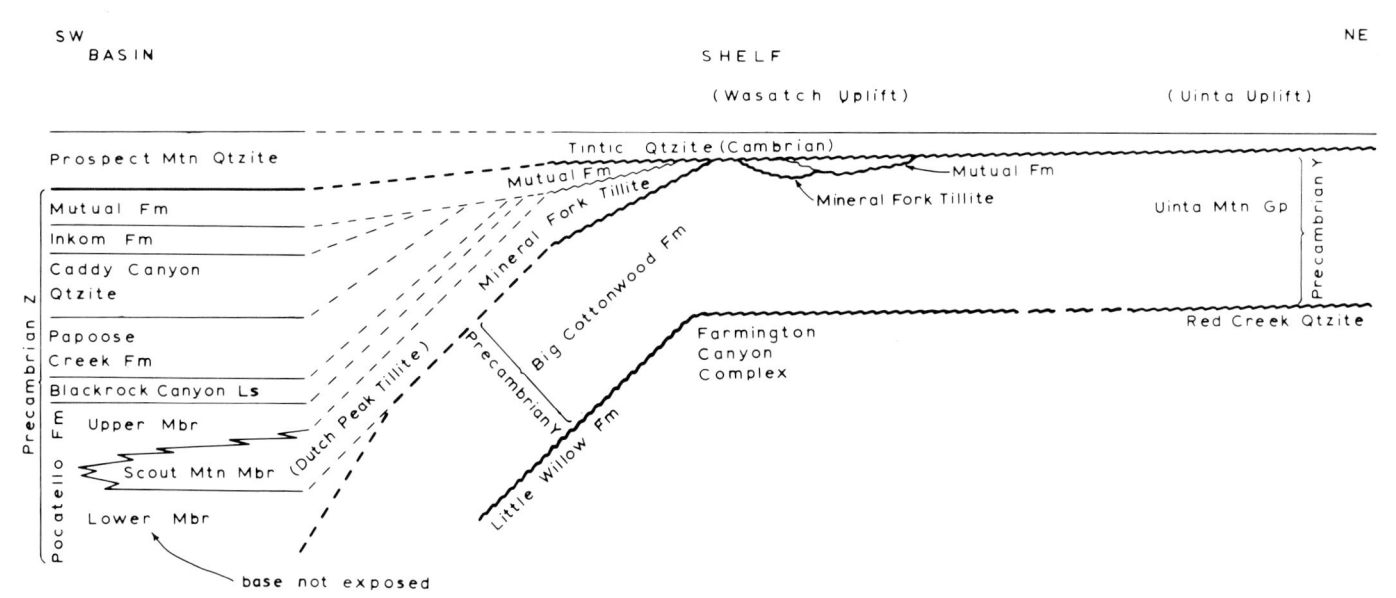

Fig. 2 — Diagrammatic northeast-southwest stratigraphic section through central Utah showing relationships of younger Precambrian strata.

STRATIGRAPHY

Stratified Precambrian rocks of Utah and southeastern Idaho are comprised of two distinct sequences: the Uinta Mountain Group and Big Cottonwood Formation of Precambrian Y age, and the units of Precambrian Z age (Fig. 2). Precambrian Y includes rocks that are 1,600 to 800 m.y. old and Precambrian Z is 800 m.y. to the beginning of the Cambrian (James, 1972).

In view of the sparsity of radiometric dates for these rocks in Utah it is necessary to assign them to Precambrian Y or Z on the basis of stratigraphic relations. In practice, the base of a distinctive and widespread diamictite unit of presumed glacial origin (Mineral Fork Tillite and correlative units) is taken as the base of Precambrian Z, although the precise age of the diamictite is not known.

The Precambrian-Cambrian boundary cannot be located precisely in the allochthon, but may occur within the Tintic Quartzite and correlative Prospect Mountain Quartzite, units commonly considered to include the basal part of the Cambrian System. East of the zone of thrusts the base of the Tintic Quartzite probably is the base of the Cambrian System.

Precambrian Y

Rocks assigned to this interval include the Uinta Mountain Group and the Big Cottonwood Formation. These units are considered to be correlative.

The Big Cottonwood Formation-Uinta Mountain Group interval does not belong to the sedimentary sequence that composes the basal units of the Cordilleran geosyncline, as the Big Cottonwood and Uinta Mountain Group appear to have been deposited in an east-west-trending basin that is transected by the northeasterly depositional strike of the rocks of Precambrian Z age (Stewart, 1972, p. 1354). The basin in which the Uinta Mountain Group and the Big Cottonwood Formation were deposited appears to have extended into the craton.

Uinta Mountain Group. — In the western part of the Uinta uplift the Uinta Mountain Group was described by Williams (1953) and by Wallace and Crittenden (1969), Wallace (1972), and Crittenden and Wallace (1973). The following descriptions are taken largely from their work.

The oldest strata exposed consist of about 2,300 feet of brick-red pebbly arkose and shale on the north slope of the range. This unit grades southward into two discrete quartzite units that are grayish-pink and grayish-white and into a red and green shale unit. The two quartzite units in turn grade westward into interbedded pebbly arkose and quartzite with lenses of olive-drab and blackish-green shale.

Above these strata and partially intertonguing with them is a brick-red quartzite ranging from about 6,000 to 11,000 feet in thickness. This unit is mostly conglomeratic and arkosic on the north flank of the Uinta Mountains and is medium-grained quartzite on the south flank.

The uppermost part of the Uinta Mountain Group was named the Red Pine Shale by Williams (1953, p. 2738). This formation is up to 5,500 feet thick and consists mostly of mottled brown, greenish-brown, and gray, thin-bedded, micaceous shale. It is gradational with the underlying strata on the north flank of the Uinta uplift but is sharply conformable on the south flank. It is unconformably overlain by the Tintic Quartzite of Cambrian age. A whole rock Rb-Sr isochron of about 950 m.y. was reported for the Red Pine Shale by Crittenden (1975).

The base of the Uinta Mountain Group is not exposed in

the western part of the uplift, but near the eastern end it is seen to be unconformable on the Red Creek Quartzite (Hansen, 1965). Hansen (1965, p. 31) reports a Rb-Sr date of 2,320 m.y. and a K-Ar date of 1,550 ± 20 m.y. for muscovite from schist in the Red Creek Quartzite. Thus, the Uinta Mountain Group is probably between about 1,500 and 950 m.y. old.

Total thickness of the Uinta Mountain Group in the western Uinta uplift is estimated to be 12,000-15,000 feet by Cohenour (1959a) and about 26,000 feet by Crittenden and Wallace (1973) on the basis of geophysical evidence, although they note that only about 13,000 feet is exposed. In the eastern Uinta uplift Hansen (1965) indicates at least 20,600 feet is present and the total may be 24,000 feet.

Big Cottonwood Formation. — The Precambrian rocks in the Wasatch Mountains near Salt Lake City have been discussed by numerous early workers (Walcott, 1886; Blackwelder, 1910, 1932; Hintze, 1913; Eardley and Hatch, 1940) and more recently by Crittenden et al. (1952). Geologic maps by Crittenden (1965a, 1965b) and by Baker and Crittenden (1961) show the distribution in the Wasatch Mountains of the upper Precambrian units described below. The descriptions that follow are mostly from the work of Crittenden and his associates with modifications by the writer.

The Big Cottonwood Formation rests unconformably on metamorphic rocks of the Farmington Canyon Complex (Eardley and Hatch, 1940) and on the Little Willow Formation (Crittenden et al., 1952; Crittenden, 1965a). Rocks of the Farmington Canyon Complex near Salt Lake City were metamorphosed about 1,600 to 1,700 m.y. ago (Crittenden and Wallace, 1973). Inasmuch as the Big Cottonwood has not undergone regional, medium-grade metamorphism, it seems likely that the Big Cottonwood is less than 1,600 m.y. old. Its minimum age is not known, but it must be older than the Mineral Fork Tillite, a unit tentatively considered to be about 750-850 m.y. old (Crittenden et al., 1972).

The Big Cottonwood Formation consists of about 16,000 feet of rusty-weathering, whitish, greenish, and gray quartzite with intercalated shale and siltstone that ranges from blue and purple in the lower part of the unit to brown and greenish-gray in the middle part and grayish-red and olive-green in the upper part of the formation. Clasts of older rocks are found in the basal part of the Big Cottonwood Formation. The type section of the formation is magnificently displayed in the lower part of Big Cottonwood Canyon, where it is unconformably overlain by the Mineral Fork Tillite and by the Mutual Formation of Precambrian Z age. Earlier usage of the term "Big Cottonwood Series" included all the quartzite, slate, and argillite above the Precambrian crystalline rocks and below the Cambrian Tintic Quartzite. Current usage generally follows the practice of Crittenden et al. (1952) in restricting the Big Cottonwood Formation to the quartzite and slate beneath the unconformity marking the base of the Mineral Fork Tillite or, where the Mineral Fork Tillite is absent, the unconformity at the base of the Mutual Formation.

The Big Cottonwood Formation is shown to crop out at Dry Mountain east of Santaquin on the Geologic Map of Northwestern Utah (Stokes, 1963, compiler); however, the description by Eardley and Hatch (1940, p. 822-823) suggests that the strata there are younger than the Big Cottonwood Formation and are probably correlative with the Mutual Formation. They describe 500 to 1,000 feet of reddish to purple quartzite that rests unconformably on the Farmington Canyon Complex and is unconformably overlain by the Tintic Quartzite. Demars (1956) has since done a more detailed study here, and he reports 863 feet of strata that are lithologically similar to the Mutual Formation.

At Long Ridge near Santaquin, Muessig (1951) reported 2,836 feet of strata which he called "Cottonwood slates" that are overlain by the Tintic Quartzite. His description of these beds (Muessig, 1951, p. 193-194) suggests that only the lower 139 feet is correlative with the Big Cottonwood Formation of current usage. These lower beds at Long Ridge consist of brown weathering, white to pink, medium-grained quartzite. The overlying 2,597 feet of pre-Tintic strata appear to be equivalent to the Mutual Formation and are discussed in more detail later.

Precambrian Z

There are profound changes in the stratigraphy of Precambrian Z rocks across the hingeline transition in Utah. Within the region considered in this report (Fig. 1) there are two contrasting facies in this interval, a thin shelf section represented by the strata in the Wasatch Mountains southeast of Salt Lake City, and a thick basinal section seen near Huntsville north of Salt Lake City, and in the Canyon Range, Beaver Mountains, and Sheeprock Mountains. The thick basinal facies is allochthonous, having been thrust eastward at least several miles (Hintze, 1960) and perhaps as much as 40 miles (Crittenden, 1961). Lack of information concerning the exact amount of displacement along the thrust faults precludes construction of precise palinspastic maps showing the original sites of deposition of the thick allochthonous basinal facies.

Lateral continuity between the thinner shelf section on the east and the thicker basinal facies to the west in central Utah is not seen because of a cover of younger rocks and tectonic disruption across thrust faults. However, two principal stratigraphic units that were deposited on the shelf, across the hingeline, and in the basin serve to tie the sections together for lithologic correlation (Fig. 2). The youngest of these extensive units is the Mutual Formation, consisting of purple and reddish quartzite and subordinate pelitic rocks, which extends from the Wasatch Mountains as far west as the Dugway Range. Another distinctive and widespread horizon is composed of diamictite: a nonsorted rock containing pebbles, cobbles, and boulders in a fine-grained matrix. The diamictite has been called the Mineral Fork Tillite in the Wasatch Mountains (Crittenden and others, 1952) and the Dutch Peak Tillite (Cohenour, 1959b) and Scout Mountain Member of the Pocatello Formation (Crittenden et al., 1971) in the basinal sections.

A remarkable north-south continuity of lithologic units and stratigraphic sequences in the basinal facies, from the Beaver Mountains, Utah to Pocatello, Idaho has been demonstrated by Crittenden, et al. (1971). Prior to this work there had been little attempt to correlate upper Precambrian formations north-south in the basinal facies of central Utah. One major exception was recognition of the uppermost Precambrian unit, the Mutual Formation, at many localities. Previously, all the upper Precambrian rocks had been assigned to either the Sheeprock Series, a name given by Cohenour (1959b) to beds in the Sheeprock Mountains, or to the Big Cottonwood Series (see Geologic Map of Northwestern Utah, Stokes, 1963) which has its type section in the Wasatch Range. It is now possible, however, to recognize the units described by Crittenden et al. (1971) at many localities in Utah.

In view of the differences between the shelf and basinal sections they are described separately.

Shelf Section

Mineral Fork Tillite. — The Mineral Fork Tillite rests unconformably on the Big Cottonwood Formation and consists dominantly of diamictite with subordinate dark gray to black quartzite, laminated pelitic rocks, and channels filled with boulder conglomerate. The diamictite is composed of pebbles, cobbles, and boulders of metamorphic rocks, granitic rocks, quartzite, and algal carbonates in a black sandy matrix. This formation is only locally present, as it occupies large, broad basins gouged out of the upper part of the Big Cottonwood Formation (Crittenden et al., 1952). This unit is up to 3,000 feet in thickness near the type section in the upper part of Mineral Fork in the Wasatch Mountains. Near Provo, Rhodes (1955) has reported that the Mineral Fork Tillite is up to 139 feet thick and is locally absent.

Hintze (1913) was the first to point out in print the glacial origin of these rocks, although he credited F. J. Pack with the original idea. Later work by Blackwelder (1932) and Crittenden et al. (1952) has confirmed Hintze's interpretation. Condie (1967) has questioned the glacial origin of these rocks; however, Condie's map of the Big Cottonwood area (1967, p. 1320) shows that he failed to recognize a very profound unconformity at the base of the Mineral Fork Tillite (cf. map by Crittenden, 1965c), and this may have led him to the wrong interpretation. The presence of striated cobbles and boulders and the general lithology of the tillite amply demonstrate its derivation from glaciers.

Mutual Formation — The Mutual Formation consists of red and purple, medium-grained to pebbly quartzite with subordinate red or less commonly green shale. This unit is unconformable on the Mineral Fork Tillite and on the Big Cottonwood Formation in the Wasatch Mountains southeast of Salt Lake City where Crittenden et al. (1952) report boulder conglomerates filling channels cut into the underlying tillite. South of Big Cottonwood Canyon the Mutual Formation is locally absent and the Cambrian Tintic Quartzite rests unconformably on the Mineral Fork Tillite and on the Big Cottonwood Formation (Crittenden, 1965a, 1965b; Baker and Crittenden, 1961). The Mutual Formation is up to 1,200 feet thick in the exposures northwest of the type section along Big Cottonwood Canyon (Crittenden et al., 1952).

South of Santaquin, Smith (1956) has described 133 feet of reddish and purple quartzite and conglomerate which he assigned to the Big Cottonwood Formation. These strata are unconformably overlain by the Tintic Quartzite (Smith, 1956, p. 7). The description given by Smith suggests that the upper Precambrian beds here are correlative with the upper part of the Mutual Formation.

The upper 2,597 feet of Precambrian strata described by Muessig (1951, p. 193-194) from Long Ridge consists dominantly of purple slate, argillite, quartzite, and grit, suggesting that these beds are equivalent to the Mutual Formation.

The Mutual Formation is widespread throughout central Utah although it may be locally missing because of erosion prior to deposition of the Tintic Quartzite.

Basinal Section

Upper Precambrian units that were deposited in a basin extending from the Beaver Mountains, Utah to Pocatello, Idaho have been described by Crittenden, et al. (1971). Many of the units they described have their type sections near Pocatello which appears to be close to the axis of the depositional basin. The available evidence suggests that the axis of the basin trended northeast, extending from the southern Deep Creek Range, Utah toward Pocatello, Idaho (Woodward, 1972). The Precambrian sequence thins southward from Pocatello, but lithologically correlative strata have been described in the Huntsville area, Sheeprock Mountains (Cohenour, 1959b), Canyon Range (Christiansen, 1952), and Beaver Mountains (Woodward, 1968).

Pocatello Formation. — The oldest rocks exposed in the basinal sequence belong to the Pocatello Formation, which in the type area near Pocatello consists of four members, a lower argillite member, a diamictite (Scout Mountain Member), a volcanic unit (Bannock Volcanic Member) that intertongues with the diamictite, and an upper argillite and quartzite member. The basal contact of the Pocatello Formation is not exposed at any reported locality, and it is not known whether the formation rests directly on older crystalline rocks or on stratified units older than the Pocatello.

In central Utah, the lower argillite member is exposed only in the Sheeprock Mountains where it is represented by about 4,160 feet of strata that have been described by Cohenour (1959b). The lower 2,690 feet consists of "black-banded phyllite" and the upper 1,470 feet is composed of quartzite, conglomerate, and dark slate. The base of this sequence is not exposed. These strata were called the Sheeprock Series (lower part) by Cohenour (1959b) and the Sheeprock Group by Harris (1958).

The next member of the Pocatello Formation is the Scout Mountain Member which consists of several thousand feet of diamictitic rocks with interbedded argillite, quartzite, and conglomerate in the Pocatello area. Lithologically similar rocks that occur at the same stratigraphic horizon in the Sheeprock Mountains were called the Dutch Peak Tillite by Cohenour (1959b). These rocks consist of pebbles, cobbles, and boulders of quartzite, metamorphic rocks, granitic rocks, and carbonates in a dark green, fine-grained matrix. This unit ranges from 0 to 4,044 feet in thickness and locally is in questionable unconformable contact with the underlying beds (Cohenour, 1959b, p. 21). The upper contact is conformable and the tillite intertongues with normal marine sedimentary rocks, as it does also near Pocatello. Near Huntsville, Utah the correlative of the upper part of the Scout Mountain Member may be the Maple Canyon Formation (Crittenden et al., 1971), which consists in ascending order of about 500 feet of olive drab argillite, 500 to 1,000 feet of green arkose, and 60 to 500 feet of conglomerate with minor argillite in the middle.

In the Pocatello area there is a thick wedge of volcanic rocks, the Bannock Volcanic Member, that intertongues with the Scout Mountain diamictites. The volcanic unit has not been observed in central Utah and presumably dies out southward from Pocatello.

The upper member of the Pocatello Formation consists of about 2,000 feet of black, thinly laminated slaty and phyllitic beds with intercalated quartzite in the type area. Correlative strata appear to be present in the Sheeprock, Canyon, and Beaver Mountains. Most earlier stratigraphers had referred to the Precambrian strata in these latter areas by numbered units. Units 1, 2, and 3 (1,480 feet thick) described by Christiansen (1952) in the Canyon Range and unit 1 (970 feet thick) in the Beaver Mountains (Woodward, 1968) are correlative with the upper part of the upper member of the Pocatello Formation. Several hundred feet of lithologically similar strata also occur in the Sheeprock Mountains (Cohenour, 1959b).

Blackrock Canyon Limestone. — This unit overlies the Pocatello Formation and in the type area near Pocatello, consists of about 800 feet of gray limestone with quartzite and minor argillite interbeds. This unit thins southward and is represented by about 610 feet of strata in the Canyon Range (units 4, 5, 6, and 7 of Christiansen, 1952, p. 721) and several hundred feet of slate, quartzite, argillite, and minor carbonate in the Beaver Mountains (lower part of unit 2 of Woodward, 1968, p. 1282). The carbonate beds in the Canyon Range and Beaver Mountains appear to represent tongues of the Blackrock Canyon Limestone that interfinger with the dominantly fine-grained clastic rocks toward the south. There are no limestone beds reported from the Precambrian of the Sheeprock Mountains. Distal tongues of this formation may be represented by about 10 feet of laminated dolomite in the basal Kelley Canyon Formation in the vicinity of Huntsville.

Papoose Creek Formation. — Conformably overlying the Blackrock Canyon Limestone is the Papoose Creek Formation which consists of about 1,800 feet of mottled gray and brown, irregularly bedded siltstone and very fine-grained quartzite near Pocatello. This formation is represented by about 335 feet of strata of unit 8 in the Canyon Range, several hundred feet of the upper part of unit 2 of the Beaver Mountains, and 300 to 700 feet of strata in the Sheeprock Mountains. This unit appears to be equivalent to the upper part of the Kelley Canyon Formation near Huntsville (Crittenden et al., 1971).

Caddy Canyon Quartzite. — The Caddy Canyon Quartzite conformably overlies the Papoose Creek Formation and is composed of several thousand feet of light colored, vitreous orthoquartzite with subordinate interbeds of argillite, siltstone, and, rarely, carbonate in the Pocatello area. This unit thins southward, being represented by 1,115 to 1,675 feet of strata in the Sheeprock Mountains, 1,350 feet in the Canyon Range (unit 9 of Christiansen, 1952), and 260 feet in the Beaver Mountains (unit 3 of Woodward, 1968). In the region near Huntsville it is 1,500 to 2,500 feet thick.

Inkom Formation. — This formation consists of greenish slate, argillite, siltstone, and very fine-grained sandstone. Locally, coarser grained sandstone or, rarely, conglomerate may be present. The Inkom is as much as 2,300 feet thick near Pocatello, but thins to about 1,200 feet in the Sheeprock Mountains, 475 feet in the Canyon Range, and 530 feet in the Beaver Mountains. Its tentative correlative near Huntsville is 360 to 450 feet thick.

In the East Tintic Mountains, Morris and Lovering (1961) described 1,673 feet of gray-green phyllitic shale and greenish-brown quartzite which they assigned to the Big Cottonwood Formation. Their description, however, suggests that these rocks are probably correlative with the Inkom Formation.

Mutual Formation. — The Mutual Formation of the basinal sequence is essentially the same as that which is present in the shelf sequence to the east. It consists of reddish and purple quartzite and conglomeratic quartzite with subordinate argillite and slate. The formation ranges in thickness from nearly 3,000 feet near Pocatello to 2,085 feet in the Beaver Mountains (units 5, 6, and 7), 1,755 feet in the Canyon Range (units 12, 13, and

14), and 900 feet in the Sheeprock Mountains. Near Huntsville there is both basalt and rhyolitic tuff interbedded with the quartzite of the Mutual.

The Mutual Formation is overlain by the Cambrian Tintic Quartzite or its western lithologic correlative the Prospect Mountain Quartzite at most localities except near Huntsville where the Mutual is overlain by the Browns Hole Formation.

Browns Hole Formation. — The Browns Hole Formation (Crittenden *et al.*, 1971) has been recognized only in the region near Huntsville where it consists of two members. The lower member consists of 180 to 460 feet of volcanics, mainly basalt, that are partly reworked to form volcaniclastic rocks, and interbedded shale and quartzite.

The upper member is terra cotta colored, fine- to medium-grained quartzite that is 100 to 150 feet thick. This unit is overlain by the basal Cambrian quartzite.

DEPOSITIONAL SETTING

Strata of the Big Cottonwood Formation-Uinta Mountain Group (Precambrian Y) were deposited in a basin that appears to be unrelated to the younger Cordilleran geosyncline. The source of the clastic sediments comprising these rocks was probably the craton to the northeast (Crittenden and Wallace, 1973). The strand line trended west-northwest through the area of the Uinta uplift, since terrestrial rocks are seen on the north flank of the uplift and marine rocks are present to the south (Crittenden and Wallace, 1973). The southern margin of the depositional basin is unknown. The rocks of the Uinta Mountain Group thus become more immature both texturally and mineralogically toward the northeast. The Big Cottonwood Formation consists of more mature sediments than the Uinta Mountain Group, as one would expect because the Big Cottonwood appears to be farther from the source area.

Initial development of the Cordilleran geosyncline began during Precambrian Z time with deposition of the Pocatello Formation and correlative rocks in a basin having an axis trending northeasterly through the Deep Creek Range toward Pocatello, Idaho (Woodward, 1972). The shoreline probably was near the site of the present Cordilleran hingeline. The Precambrian rocks of the Beaver Mountains and the Canyon Range were deposited near the southeastern edge of the basin, since the individual stratigraphic units, as well as the entire upper Precambrian sequence, thin markedly from the axis of the basin toward the south and east. The available evidence suggests that the hingeline lay between Dry Mountain where the Mutual Formation rests directly on crystalline rocks and the Canyon Range or possibly the East Tintic Mountains where strata of the basinal facies may occur. Deposition in the basin was probably continuous throughout late Precambrian time, as there is no evidence of any break in sedimentation during this interval.

The Mineral Fork Tillite was probably deposited directly from glaciers that had scooped out troughs in the underlying Big Cottonwood Formation in the Wasatch Mountains. These glaciers periodically extended westward from the shelf area into the basin and deposited material both directly from the ice and by ice-rafting (Varney, 1976). Slumping of loose glacial sediment and reworking by turbidity currents probably also occurred. Thus, the Mineral Fork Tillite, the Scout Mountain Member of the Pocatello Formation, and the Dutch Peak Tillite appear to be broadly correlative, representing a time of glaciation. Interfingering of the basinal diamictites with normal sedimentary rocks indicates contemporaneous deposition of glacial and marine sediments.

Following deposition of diamictites of the Scout Mountain Member, there was again deposition under normal marine conditions in the basin. Most of the sediments are clastics, but the presence of limestone in the section suggests a marine environment. This period of deposition is represented by the upper member of the Pocatello Formation, the Blackrock Canyon Limestone, the Papoose Creek Formation, the Caddy Canyon Quartzite, and the Inkom Formation. Meanwhile, the shelf area was probably exposed to erosion, as suggested by the nature of the profound unconformity that separates the Big Cottonwood Formation and the Mineral Fork Tillite from the overlying Mutual Formation.

The Mutual Formation was deposited as a widespread sheet of sand and pebbly sand over the entire region, including both the basin and shelf areas. Following deposition of the Mutual Formation there was erosion of the shelf area, as evidenced by a slight angular unconformity at the base of the overlying Cambrian Tintic Quartzite. In the basinal section there is concordance between the Mutual Formation and Tintic Quartzite at most localities.

REFERENCES CITED

Baker, A. A., and Crittenden, M. D., Jr., 1961, Geologic map of the Timpanogos Cave quadrangle, Utah: U.S. Geol. Survey Geologic Map GQ-132.

Blackwelder, Eliot, 1910, New light on the geology of the Wasatch Mountains, Utah: Geol. Soc. America Bull., v. 21, p. 517-542.

_____, 1932, An ancient glacial formation in Utah: Jour. Geology, v. 40, p. 289-304.

Christiansen, F. W., 1952, Structure and stratigraphy of the Canyon Range, central Utah: Geol. Soc. America Bull., v. 63, p. 717-740.

Cohenour, R. E., 1959a, Precambrian rocks of the Uinta-Wasatch Mountain junction and part of central Utah: Intermtn. Assoc. Petroleum Geologists Guidebook to the geology of the Wasatch and Uinta Mountains transition area, p. 34-39.

_____, 1959b, Sheeprock Mountains, Tooele and Juab Counties: Utah Geol. Mineralogic Survey, Bull. 63, 201 p.

Condie, K. C., 1967, Petrology of the late Precambrian tillite (?) association in northern Utah: Geol. Soc. America Bull., v. 78, p. 1317-1344.

Crittenden, M. D., Jr., 1961, Magnitude of thrust faulting in northern Utah: U.S. Geol. Survey Prof. Paper 424-D, p. D 128-D 131.

_____, 1965a, Geology of the Draper quadrangle, Utah: U.S. Geol. Survey Geologic Map GQ-377.

_____, 1965b, Geology of the Dromedary Peak quadrangle, Utah: U.S. Geol. Survey Geologic Map GQ-378.

_____, 1965c, Geology of the Mount Aire quadrangle, Salt Lake County, Utah: U.S. Geol. Survey Geologic Map GQ-379.

_____, 1975, Preliminary isotopic dating of Uinta Mountain Group: Geol. Soc. America Abstracts with Program, v. 7, p. 600.

_____, Sharp, B. J., and Calkins, F. C., 1952, Geology of the Wasatch Mountains east of Salt Lake City, Parleys Canyon to Traverse Range: Utah Geol. Soc. Guidebook no. 8, Geology of the central Wasatch Mountains, Utah, p. 1-37.

_____, Schaeffer, F. E., Trimble, D. E., and Woodward, L. A., 1971, Nomenclature and correlation of some upper Precambrian and basal Cambrian sequences in western Utah and southeastern Idaho: Geol. Soc. America Bull., v. 82, p. 581-602.

_____, Stewart, J. H., and Wallace, C. A., 1972, Regional correlation of upper Precambrian strata in western North America: 24th Internat. Geol. Cong., Montreal, sec. 1, p. 334-341.

_____, and Wallace, C. A., 1973, Possible equivalents of the Belt Supergroup in Utah: Idaho Bur. Mines and Geology, Belt Symposium v. 1, p. 116-138.

Demars, L. C., 1956, Geology of the northern part of Dry Mountain, southern Wasatch Mountains, Utah: Brig. Young Univ. Res. Studies, Geol. Ser., v. 3, no. 2, 49 p.

Eardley, A. J., 1951, Structural geology of North America: New York, Harper and Bros., 624 p.

_____, and Hatch, R. A., 1940, Proterozoic (?) rocks in Utah: Geol. Soc. America Bull., v. 51, p. 795-843.

Hansen, W. R., 1965, Geology of the Flaming Gorge area, Utah-Colorado-Wyoming: U.S. Geol. Survey Prof. Paper 490, 196 p.

Harris, DeVerle, 1958, The geology of the Dutch Peak area, Sheeprock Range, Tooele County, Utah: Brig. Young Univ. Res. Studies, Geol. Ser., v. 5, no. 1, 82 p.

Hintze, F. F., Jr., 1913, A contribution to the geology of the Wasatch Mountains: N. Y. Acad. Sci., v. 23, p. 85-143.

Hintze, L. F., 1960, Thrust faulting limits in western Utah (abs.): Geol. Soc. America Bull., v. 71, p. 2062.

James, H. L., 1972, Subdivision of Precambrian: an interim scheme to be used by U.S. Geological Survey: Am. Assoc. Petroleum Geologists Bull., v. 56, p. 1128-1133.

King, P. B., and Beikman, H. M., 1974, Geologic map of the United States: U.S. Geol. Survey, scale 1:2,500,000.

Miller, G. M., 1966, Structure and stratigraphy of southern part of Wah Wah Mountains, southwest Utah: Am. Assoc. Petroleum Geologists Bull., v. 50, p. 858-900.

Morris, H. T., and Lovering, T. S., 1961, Stratigraphy of the East Tintic Mountains, Utah: U.S. Geol. Survey Prof. Paper 361, 145 p.

Muessig, S. J., 1951, Geology of a part of Long Ridge, Utah, unpub. Ph.D. dissertation: Ohio State Univ., 212 p.

Rhodes, J. A., 1955, Stratigraphy and structural geology of the Buckley Mountain area, south-central Wasatch Mountains, Utah: Brig. Young Univ. Res. Studies, Geol. Ser., v. 2, no. 4, 57 p.

Smith, C. V., 1956, Geology of the North Canyon area, southern Wasatch Mountains, Utah: Brig. Young Univ. Res. Studies, Geol. Ser., v. 3, no. 7, 32 p.

Stewart, J. H., 1972, Initial deposits in the Cordilleran geosyncline: evidence of a late Precambrian (<850 m.y.) continental separation: Geol. Soc. America Bull., v. 83, p. 1345-1360.

Stokes, W. L., 1963 (compiler), Geologic map of northwestern Utah: Utah Geol. Mineralogic Survey.

Varney, P. J., 1976, Depositional environment of the Mineral Fork Formation (Precambrian), Wasatch Mountains, Utah: Rocky Mtn. Assoc. Geologists Symposium on Geology of the Cordilleran Hingeline.

Walcott, C. D., 1886, Second contribution to the studies on Cambrian faunas of North America: U.S. Geol. Survey Bull. 30.

Wallace, C. A., 1972, A basin analysis of the upper Precambrian Uinta Mountain Group, western Uinta Mountains, Utah, unpub. Ph.D. dissertation; Univ. California, Santa Barbara.

_____, and Crittenden, M. D., Jr., 1969, The stratigraphy, depositional environment and correlation of the Precambrian Uinta Mountain Group, western Uinta Mountains, Utah: Intermtn. Assoc. Geol., Geologic Guidebook of the Uinta Mountains, p. 127-141.

Williams, N. C., 1953, Late Pre-Cambrian and early Paleozoic geology of western Uinta Mountains, Utah: Am. Assoc. Petroleum Geologists Bull., v. 37, p. 2734-2742.

Woodward, L. A., 1968, Lower Cambrian and upper Precambrian strata of Beaver Mountains, Utah: Am. Assoc. Petroleum Geologists Bull., v. 52, p. 1279-1290.

_____, 1972, Upper Precambrian strata of the eastern Great Basin as potential host rocks for mineralization: Econ. Geology, v. 67, p. 677-681.

DEPOSITIONAL ENVIRONMENT OF THE MINERAL FORK FORMATION (PRECAMBRIAN), WASATCH MOUNTAINS, UTAH

by

Peter J. Varney[1]

ABSTRACT

The Mineral Fork Formation is comprised of five distinct rock types being mostly tillite, but with mappable amounts of fluvial conglomerate, fluvial sandstone, thin-bedded lacustrine (?) siltstone, and non-bedded siltstone. The spatial arrangement and internal characteristics of these lithologies suggests that they were deposited in glacial and near-glacial environments.

The Mineral Fork Formation unconformably overlies the shallow water marine Big Cottonwood Formation, and is in turn unconformably overlain by the continental and near-shore marine Mutual Formation. This suggests that the Mineral Fork lithologies were deposited from glaciers that advanced onto a shallow marine shelf during the late Precambrian. Following retreat of the glaciers, the Mineral Fork sediments were truncated and redeposited in part as the latest Precambrian Mutual Formation.

INTRODUCTION

The Mineral Fork Formation is the conspicuously bedded, dark colored unit exposed above Hellgate Spring on the north side of Little Cottonwood Canyon (Fig. 1). The outcrop can be seen clearly from the Symposium Headquarters at Snowbird.

In the type area, the Mineral Fork of Big Cottonwood Canyon (next canyon north of Little Cottonwood Canyon), the Mineral Fork Formation consists of as much as 3200 ft of mostly dark colored rocks of highly variable grain size, sorting and bedding characteristics. Because the unit is markedly different in appearance from the enclosing formations, it has attracted the attention of many workers (Coleman, 1926; Marsel, 1964; Condie, 1967 and 1969; Crittenden, *et al.*, 1952; and Woodward, 1972).

Hintze (1913) was apparently the first to suggest that the Mineral Fork lithologies were a result of Precambrian continental glaciation. This view was accepted for many years until challenged by Condie (1967) and modified by Dott (1961).

Clearly, the depositional environment of the Mineral Fork Formation, as a whole, is uncertain. The problem centers on two main points. First, the "tillite" apparently does not display characteristics that unequivocally indicate glacial activity, and second, there are lithologies present in the type section that probably were not deposited directly from glacier ice. This paper suggests solutions to these problems, and attempts to reconstruct a possible depositional environment and depositional history for the Mineral Fork and related units.

Fig. 1 — The conspicuously bedded dark colored unit is the Mineral Fork Formation as exposed above Hellgate Spring, Little Cottonwood Canyon. The light colored unit on the left is the Upper Precambrian Big Cottonwood Formation. The light colored rocks on the right are Cambrian Tintic Quartzite. At this location all formations have been intensely deformed along the Alta thrust.

---
[1] Exeter Exploration Company, Denver, Colorado.

Crittenden, et al. (1952) suggested that this unit be named the "Mineral Fork Tillite" because tillite-like rocks comprise the greatest percentage of the rocks exposed. Field work by the author in the summer of 1971 found that while tillite-like rocks are the most common, other rock types are present as distinctly mappable units of related but separate origin. Therefore, in this paper the rock unit as a whole is referred to as the Mineral Fork Formation.

(Author's Note: Throughout this paper, tillite-like rocks of indeterminate origin are called "tilloid" after the original descriptive manner of Blackwelder (1932) and in the sense of Harland et al. (1966, p. 251). It is not used in the genetic sense of Pettijohn (1957). An alternative term, "diamictite", was proposed by Flint (1960) for such rocks, but in the author's opinion, the term has been misused to the extent that its value is questionable.)

STRATIGRAPHICALLY ADJACENT FORMATIONS

The Mineral Fork Formation is the middle stratigraphic unit in the upper Precambrian sequence exposed in the central Wasatch Mountains. It is underlain by the Big Cottonwood Formation, and overlain by the Mutual Formation. In some places, the Mutual Formation is not present and the Cambrian Tintic Quartzite rests directly on the Mineral Fork Formation.

Big Cottonwood Formation

The Big Cottonwood Formation crops out in the western half of Big Cottonwood Canyon where it unconformably overlies the older Precambrian, highly metamorphosed Little Willow Series. In that area, it is about 16,000 ft thick (Stokes, 1969, p. 38) and consists mostly of reddish brown weathering, light grey and light greenish grey to grey, very slightly metamorphosed, silica-cemented quartzarenite. Contained within the quartzarenite are reddish brown to purplish grey weathering beds of medium grey to red and greenish grey shaly siltstone.

Quoting Crittenden, Sharp, and Calkins (1952, p. 3):

> "Ripple marks, cross-bedding, mud-cracks, and shale-flake conglomerates which are present throughout the series give evidence that it was deposited in shallow water. Individual shale and quartzite members have been mapped separately in a large part of the area of outcrop to aid in determining the actual thickness, and to clarify the relations with the adjoining units. The fact that shale units only 50 to 200 ft thick can be traced over distances of several miles suggests deposition in a broad, gradually subsiding embayment rather than in a fluviatile or flood-plain environment."

Contact with the overlying Mineral Fork Formation is sharp but unconformable to paraconformable at the locations visited by the author. However, Condie (1967, p. 1321) reported that the contact is gradational east of Lake Blanch, Mill B South Fork, Big Cottonwood Canyon (just west of the Mineral Fork).

Mutual Formation

The Mineral Fork Formation is unconformably overlain by the Upper Precambrian Mutual Formation. The Mutual consists of up to 1200 ft of greyish purple to reddish purple, medium- to coarse-grained quartzarenite with included reddish to greenish grey shaly siltstone. Near the south end of Mineral Fork Valley, the Mutual contains a basal conglomerate "containing blocks of tillite and limestone several feet across ... deposited in gullies 20 to 30 feet deep ..." (Crittenden, Sharp, and Calkins, 1952, p. 6). Some parts of the Mutual were apparently deposited in a shallow water marine environment like that of the Big Cottonwood Formation. Other parts are clearly continental.

The color differences between the underlying Mineral Fork Formation, Mutual Formation, and overlying Tintic Quartzite are quite noticeable in the field. With regard to the Mutual, Stokes (1969, p. 38) feels that "the purple color is distinctive and may be a valid feature for correlation over wide areas of the Rocky Mountains". If it were not for the color differences, the Mutual Formation would be virtually indistinguishable from the Big Cottonwood Formation upon which it rests unconformably in those places where the Mineral Fork Formation is absent.

Tintic Quartzite

The basal Cambrian Tintic Quartzite unconformably overlies both the Mutual Formation and the Mineral Fork Formation. Its light yellowish grey color is distinctive and easily seen throughout the area.

Readers desiring a more complete discussion of the Upper Precambrian stratigraphy of the central Wasatch Mountains are referred to the comprehensive paper by Lee Woodward in this volume.

MINERAL FORK FORMATION

The Upper Precambrian Mineral Fork Formation lies stratigraphically between the Big Cottonwood Formation and the Mutual Formation and Tintic Quartzite. The best exposures of the formation are in the Mineral Fork of Big Cottonwood Canyon (Fig. 2).

The formation was studied by the author in the "type" area during 1971. Formal field work consisted of three measured sections and four traverses. These were supplemented by informal visits to many points of interest within the outcrop area. The descriptive information in this section summarizes only the environmentally significant characteristics of the formation. Readers wishing additional details are referred to Varney (1972).

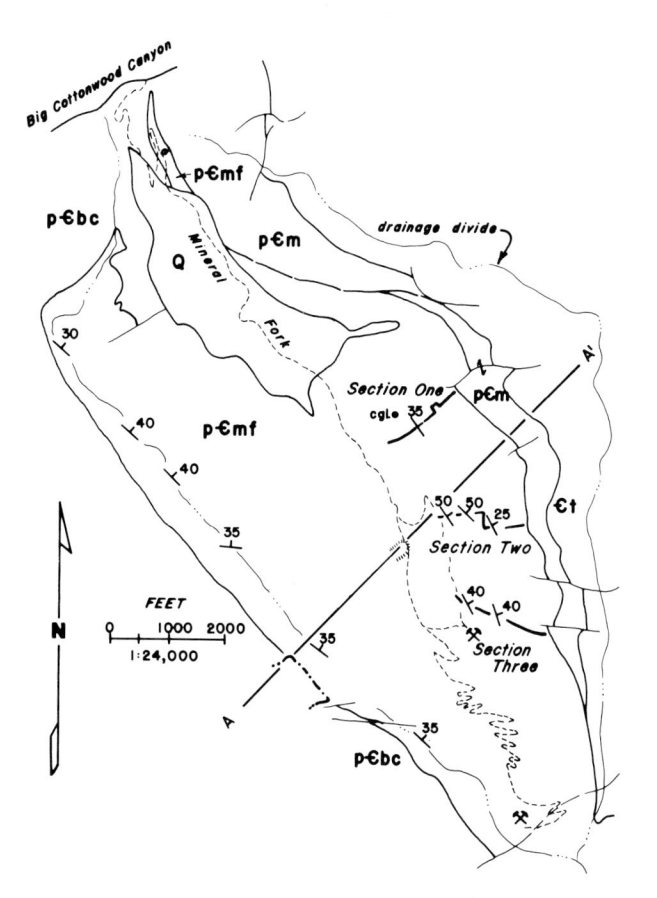

Fig. 2 — Geologic index map of the Mineral Fork of Big Cottonwood Canyon showing the locations of measured stratigraphic sections illustrated in Fig. 3. Formations shown are: Big Cottonwood Series (pЄbc), Mineral Fork Formation (pЄmf), Mutual Formation (pЄm), and Tintic Quartzite (Єt). Line A-A' is location of cross-section shown in Fig. 4. Areal geology based on U.S. Geological Survey Map GQ-378.

Lithologies

Three stratigraphic sections measured by the author (Fig. 3) illustrate that there are five distinct lithologies in the formation: tilloid, about 70%; thin bedded siltstone, about 22%; non-bedded siltstone, less than 1%; sandstone, about 7%; and conglomerate, about 2%. The lower contact of the formation is estimated on the basis of geometric reconstruction to be about 1700 ft below the base of Section 2 (Fig. 4). This yields an estimated total thickness of about 3200 ft for the formation.

Table 1 summarizes the characteristics of the five Mineral Fork Formation lithologies.

Vertical Sequence

The three stratigraphic sections shown in Fig. 3 do not demonstrate any tendency within the Mineral Fork Formation towards fining upward (or downward) sequences. The only interval that can even be correlated across all three sections with any confidence is a sandstone interval represented by units 6 to 11, Section 1; units 15 to 17, Section 2; and Unit 1, Section 3.

Sedimentary Structures

Except for horizontal, continuous bedding in the thin-bedded siltstones, primary sedimentary structure is rare in the Mineral Fork Formation, even in the lenticular sandstones. Nevertheless, some cross strata and ripple marks were found.

Seven dip and strike measurements were recorded for crossbeds in the lenticular sandstones (in about 4,000 ft of measured section). After correcting for dip of the formation (averaging 45° east), they showed an average transport direction to the northwest (Fig. 5).

The transport direction was computed using the assumption that the only deformation that the Mineral Fork Formation has undergone is simple rotation on an axis parallel to strike. There is no reason to believe that the deformation has been nearly that simple. Therefore, the suggested transport direction may not carry much meaning.

The only other sedimentary structures with directional significance are ripple marks in the sandstones and siltstones. Ripple marks are rare, and, where developed, they are generally confined to one or two bedding surfaces. The most common type of ripple-stratification is for the upper surface of a sand stringer in the thin-bedded siltstone to show some rippling. This type of structure is well developed in the Little Cottonwood Canyon exposures of the siltstone. Such structures are rarely repeated in successive layers, and may indicate that currents strong enough to form ripples were rare in the depositional area. Three ripple sets are included in Fig. 5. They show a current direction to the southwest and are responsible for the secondary mode in that direction. Obviously, such a small amount of data is not statistically significant. However, all measurements contain a dominantly westward component.

DEPOSITIONAL ENVIRONMENTS OF THE MINERAL FORK FORMATION

Most workers have suggested that the Mineral Fork Formation is a tillite, that is, that it was deposited directly from glacier ice. Recently, this idea has been challenged, particularly by Condie (1967) and Dott (1961). Condie suggests that the Mineral Fork Formation has many of the characteristics of a turbidite. Dott suggests that it may be glaciomarine. The problem is that while all lithologies within the formation have distinctive characteristics, none has characteristics that unequivocally proves its mode of origin.

From the lithologic descriptions in Table 1, it should be

Fig. 3 — Stratigraphic relationships within the Mineral Fork Formation, Mineral Fork of Big Cottonwood Canyon. The three sections represent about 1500 ft in the upper one half of the formation. The lower contact is estimated to be about 1700 ft below the base of Section 2. Numbers next to the sections are keyed to descriptions of the individual units in Appendix B, Varney, 1972.

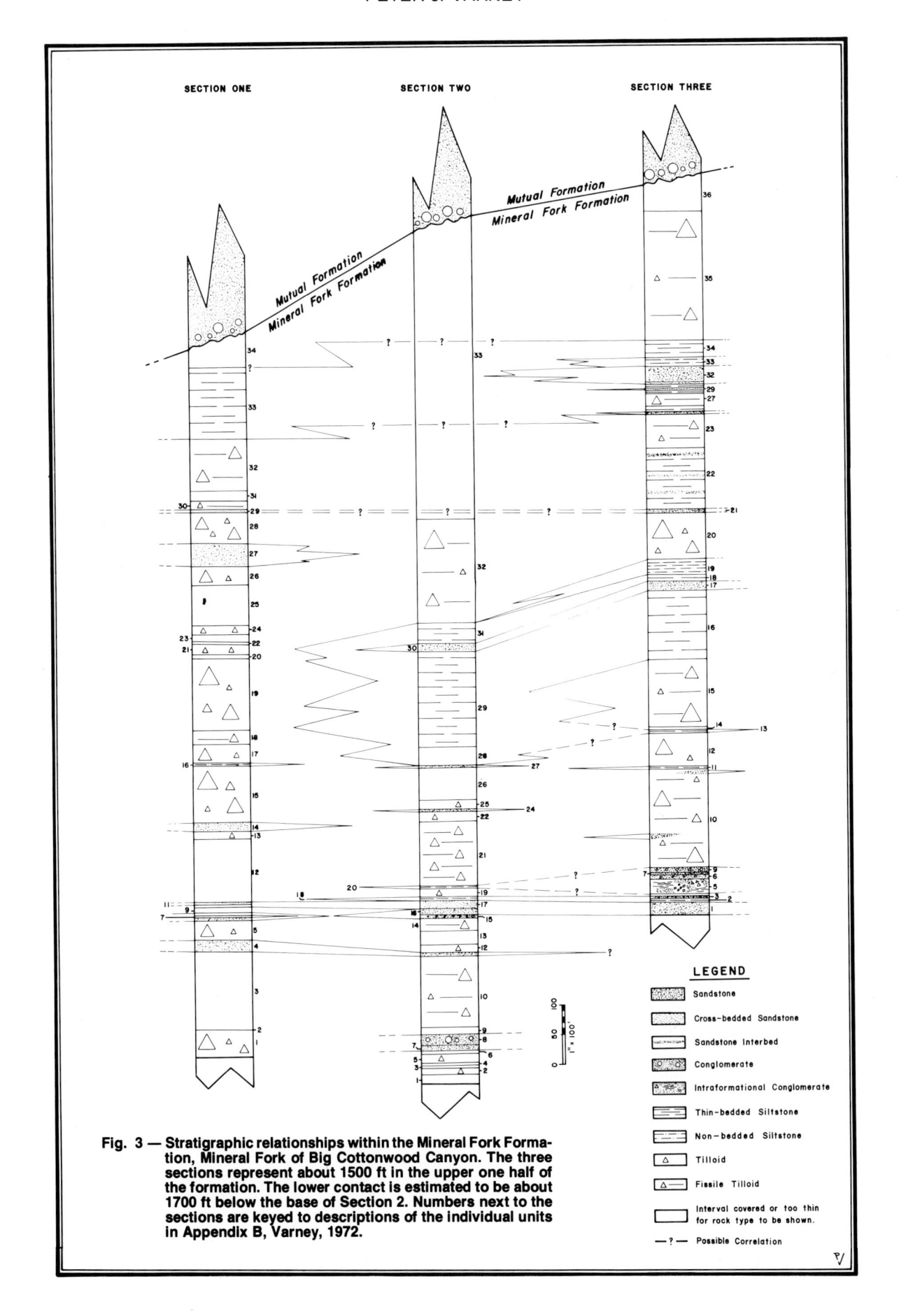

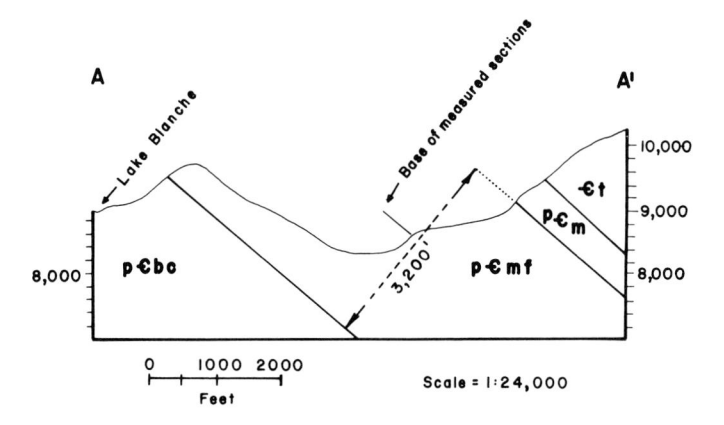

Fig. 4 — Section across the Mineral Fork along line A-A' shown on Fig. 2. The section was constructed using an average dip of 40°E. At this dip, the thickness of the formation is 3,200 ft. Formations shown are: Big Cottonwood Series (pЄbc), Mineral Fork Formation (pЄmf), Mutual Formation (pЄm), and Tintic Quartzite (Єt). (No vertical exaggeration.)

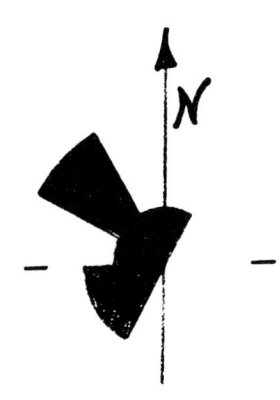

Fig. 5 — Paleocurrent directions in the Mineral Fork Formation. All measurements made in sandstone. Standard Deviation equals 1.06. All directions corrected for dip.

clear that the formation as a whole is probably not the result of a single process, but more likely is polygenetic. The following discussion constructs possible depositional environments that can explain all lithologies and associations observed. Specific environments are discussed in terms of genetically related rock types.

Conglomerate and Lenticular Sandstone

The mineralogic immaturity, lenticular shape, variable grain sizes, cross bedding, and lateral and vertical gradations observed in the conglomerates and lenticular sandstones strongly suggests that they are fluvial in origin. This is supported by thin section studies (Varney, 1972, p. 50) in which it was found that lenticular sandstone bodies have a high percentage of matrix, a characteristic that according to Picard (1971, p. 21) is found in fluvial sandstones.

Clearly, these rocks contain lithologies reworked from other parts of the formation. The complex associations observed in the intraformational conglomerates shown in Section 3 (unit 5) strongly suggest rapid channel migration and lateral cutting with associated bank failure, perhaps during flooding conditions. Furthermore, this apparently occurred prior to lithification of the included rock types.

The appearance of small, very lithic, lenticular sandstones developed in tilloid suggests that a network of channels was developed concurrent with, or between intervals of, tilloid deposition. The material in the small lenses appears very close to the composition of tilloid matrix and is observed to be gradational with tilloid in some places, suggesting fluvial reworking prior to tilloid lithification.

In one location, individual lenticular sandstones apparently are continuous along strike (Fig. 3). Tracing the lithologies along the apparent continuity shows that the horizon is one characterized by changing stream facies. This may indicate a general interval of fluvial activity between periods of deposition of the other lithologies.

Channel sandstones and conglomerates of the type described have been observed in continental environments and in tidal-flat glaciomarine environments (Miller, 1953, p. 27; Gustavson, 1974, p. 384). Because most of the sandstones and conglomerates are erosional on the underlying beds and contain evidence in some places of high energy deposition in rapidly migrating channels, they are most likely continental stream deposits.

Thin-Bedded Siltstone and Sandstone

When first examined, thin-bedded siltstone and sandstone units in the Mineral Fork Formation look very much like varves. However, Condie (1967, p. 1338) suggests that these units in the Mineral Fork and other Utah "tillites" may have originated through turbidity currents and subaqueous mudflows. His observations of the lithologic characteristics of the rocks are supported by the present study. However, the limited lateral and vertical extent of the thin-bedded deposits, presence of micaceous partings, rare ripples and drying cracks, and extremely even bedding suggests that they formed in a restricted lacustrine environment. Because they are wholly contained in tilloid — on a broad scale — they may have formed in lakes on a till plain. Because apparent ice-rafted pebble deposits are

	COLOR	VOLUME OF FORMATION	SORTING AND GRAIN SIZE	COMPOSITION	BEDDING AND SEDIMENTARY STRUCTURES
Sandstone (Figure 9)	Light yellowish to medium grey.	7%	Poorly to well sorted, fine to very coarse grained sand. Frosted quartz grains, 1 mm to 2 mm in diameter appear suspended in a matrix of finer material in the more lithic zones.	Litharenite to sublitharenite except in the thick, laterally continuous layers where it is a quartzarenite.	Commonly shows only weak, discontinuous horizontal stratification. A few lenticular sandstones are present and have well developed cross stratification.
Non-bedded Siltstone	Dark Grey	Less than 1%	Poorly to very well sorted, may generally be considered a "sandy siltstone". Rarely contains 1 or 2% clasts greater than 2 mm.	Primarily quartz with large percentage of fine mica and a small percentage of other minerals, notably hornblende.	None
Thin-bedded Siltstone (Figure 8)	Medium Dark Grey	21%	Individual laminae are size graded from fine to very fine sandstone at the base upward to silt. May be considered "Sandy Siltstone".	Primarily quartz with major amount of fine mica and "matrix".	Laterally continuous 1/4 to 1/8 in. horizontal, parallel laminations some of which show very low-amplitude undulations that may be ripple marks.
Conglomerate (Figure 7)	Light Grey to Medium Grey	2%	Poorly sorted. Clasts range from sand sized to greater than 24 in. in diameter in some places. Can be considered to sandy conglomerate.	Metamorphic clasts = 65% Sedimentary clasts = 26% Igneous clasts = 6% Carbonate clasts = 3% 100% Matrix is commonly sublitharenitic.	Only rudimentary bedding in some places. At one location, crude cross stratification was observed. Most conglomerates in the section would be considered non-bedded.
Tilloid (Figure 6)	Medium Grey to Dark Grey	70%	Greater than 2 mm portion non-sorted. Less than 2 mm portion poorly sorted. Clasts range from silt sized to boulders.	Clasts greater than 2 mm are mostly metamorphic and sedimentary. Less than 2 mm material is mostly quartz with abundant fine mica, clay and altered feldspar.	None

not particularly common in the Mineral Fork Formation, this suggests a receding glacier that constantly lost contact with down-valley lakes.

A problem does exist in this interpretation. Most of the sandstone and siltstone layers as observed in the field do not have gradational relationships. According to Bradley (1929) the graded-nongraded nature of most of the thin bedded siltstone and sandstone deposits is not characteristic of most varved sequences. Information in the literature suggests that it is not characteristic of turbidites either. However, studies of thin sections cut perpendicular to bedding in a graded portion of a thin-bedded siltstone sequence show characteristics that according to Pettijohn (1957, p. 172) are typical of glacial varves.

In Little Cottonwood Canyon, west of Alta, and immediately above Hellgate Spring, tilloidal material is interbedded with thin bedded siltstone and sandstone, with the result that the formation has a distinctly bedded aspect in that location (Fig. 1). It is entirely possible that the even layered aspect of the formation noted at this location is the result of mudflow redeposition of till into a lake that already contained a well developed sequence of thin-bedded sediment. Mudflow is an active agent in all till deposition areas, and if a flow enters a lake, it can produce features like those observed (Hartschorn, 1958, p. 481; Kuenen and Migliorini, 1950, p. 124). In a sense, this may indicate that the formation is a fresh water turbidite in part.

Rarely, markings similar to shrinkage cracks are found on surfaces in the thin-bedded sequences. These appear to indicate a period of drying, perhaps related to occasional lake drain-

Table 1. Summary of characteristics of the Mineral Fork Formation Lithologies

INTERBEDS	WEATHERING CHARACTERISTIC	CONTACTS	ALTERATION	THICKNESS
None	Weathers to medium light grey and various shades of yellow. Very resistant to erosion; well jointed breaking into slabs and blocks. Form conspicuous breaks in slope.	Grain sizes in smaller lenses grade downward particularly when underlying unit is tilloid. Cleaner, thicker horizontally bedded and lenticular sands rest erosionally on underlying units.	Secondary quartz overgrowths are abundant, as is "sericite - illite" type alteration of the feldspars and micas.	Rarely greater than 50 ft. Commonly 5 to 10 ft.
Commonly contains sublithanerite to litharenite lenses averaging 1½ by 3 feet. Lower contacts of lenses appear gradational by reworking.	Weathers to various shades of orange and red apparently in relation to iron content. Not durable, forms concave slope littered with platy fragments.	Commonly sharp, except gradational over a few inches where adjacent beds are tilloids.	Crudely developed, horizontal shaly fissility in some places.	Most units are 5 to 10 ft. thick and are broadly lenticular in shape.
Commonly contains ½ in. to 12 in. sublitharenite to quartzarenite layers that are laterally very continuous and of uniform thickness. The sandstones are not gradational with the siltstones. 2) Contains rare conglomerate interbeds that distort the underlying siltstone laminae.	Weathers medium yellowish to reddish grey. Forms straight to concave slopes littered with thin, angular plates.	Sharp with all adjacent units.	In a few places contains secondary quartz overgrowths. .. Abundant "matrix" material may be due to low grade metamorphic alteration.	May be greater than 150 ft. thick, appear to be interbeds in tilloid.
Contains cross-stratified sandstone lenses in some places.	Weathers to light grey, iron stained in a few places. Most resistant units to erosion in whole formation. Form conspicuous breaks in slope.	Commonly sharp on underlying units. Upper contacts are commonly gradational to sandstone.	Appears fresh and unaltered on outcrop.	In most places, less than 10 ft. In one location, greater than 60 ft.
At one location or another every rock type in the formation is found as an interbed in the tilloid. Sandstone lenses are the most common interbeds.	Weathers to medium yellowish orange. In most places is rounded and soft. Commonly has a platy, shaly fissility. Forms everything from concave slopes to resistant breaks in slope.	Lower contacts are sharp and appear erosional in some locations. Upper contacts are sharp but uneven except in a few places where they appear reworked.	Thermal alteration has effected the tilloid to different extents in different places. In the type area, the tilloid is only slightly metamorphosed.	Tilloid units are commonly greater than 75 ft. thick.

ing. This interpretation is supported by the truncation of thin-bedded deposits by fluvial sand and conglomerate channels in some places and by the presence of thin-bedded intraformational clasts in some conglomerate. Apparently, migrating stream channels occasionally intersected a lake or pond that subsequently drained rapidly, allowed channel migration and associated bank failure, and led to the lithologic associations as observed today.

Harrison (1975, p. 738) described Pleistocene sediments from a location in Pennsylvania that show characteristics almost identical to those of the thin-bedded Mineral Fork deposits. His interpretation is that outwash streams entering glacial lakes deposited "delta-kames" and set up conditions for rhythmic deposition of fine grained sediments in size-graded, thin layers. This suggests that thin-bedded deposits like those in the Mineral Fork Formation may be random sequences of true varves and rhythmites. More importantly it shows that the thin-bedded deposits may not be the result of a single type of process, but more likely have a fairly complex depositional history.

The thin-bedded sandstones and siltstones, then, appear to have been deposited in lakes and ponds, perhaps near an active glacier.

Tilloid and Non-Bedded Siltstone

Two conclusions can be drawn immediately from the descriptions of the tilloid given in Table 1. First, the abundant tilloid makes the Mineral Fork Formation markedly different in character than the overlying and underlying units, and second, tilloid contains most other lithologies in the formation. These general characteristics imply that the origin of the

Fig. 6 — Tilloid outcrop above Hellgate Spring, Little Cottonwood Canyon. Note the presence of abundant rounded pebbles and cobbles and general lack of sorting. The angular weathering characteristic is common here as well as in the Mineral Fork of Big Cottonwood Canyon. (Stereopair)

Mineral Fork Formation is significantly different from that of the enclosing formations, but that within the Mineral Fork, all deposits are broadly related to the same general type of activity. The answer to the question of what was the activity that led to the formation of the tilloid will probably explain all the associations observed in the formation.

There appear to be three basic processes that could have formed the observed tilloid: deposition from glaciers, mud and debris flows, or turbidity currents. A combination of all three is, of course, possible.

Condie (1967) indicated that the thin-bedded deposits in the formation could be distal turbidites. Such features, however, generally have greater areal extent, and are contained in typical marine rocks. The tilloid itself could be considered a proximal turbidite. However, the presence of large numbers of erratic stones, little identifiable primary clay in the matrix, total lack of bedding and slump structures, and complete lack of resemblance to the enclosing formations renders such an interpretation suspect.

Another possibility is that the Mineral Fork Formation may have formed through the activity of mud and debris flows.

According to Crandell (1971, p. 3-10) mud and debris flows have the following characteristics:

1. Vertical size gradations in the greater-than-two-millimeter fraction with a decrease in sizes from the base of a given unit upward. No size gradation is observed in the less-than-two-millimeter fraction.

2. Much void space is present within the matrix due to the inclusion of air bubbles at the time of flow.

3. Most mud and debris flows produce only thin veneers of material when they are flowing in restricted areas.

4. Lower contacts are most commonly non-erosional (an observation that is supported by Blackwelder, 1928).

5. Only rarely do mud and debris flows contain faceted stones.

In addition to these, Allen (1971, p. 231) indicated that mudflows should contain a high percentage of mineralogic clay.

Only one of the above features is present in the tilloid in the type section of the Mineral Fork Formation: in some places, the tilloid has a crudely developed size gradation in the greater-than-two-millimeter portion such that the clasts decrease in size upward. The tilloid is different in terms of every other char-

Fig. 7 — Conglomerate pod in very coarse grained sandstone. Photograph made on west side of Mineral Fork Valley in an apparent fluvial interval that correlates with unit 1, Section 3. Compass case is about four inches square.

Fig. 8 — Thin-bedded siltstone of unit 29, Section 2. Note the presence of numerous sandstone interbeds. Pencil is about 5 1/2 inches long.

Fig. 9 — Fluvial sandstone of unit 7, Section 2. Note cross-strata in left center of picture. Axes trend N 60° W with foresets inclined to the west. Pencil is about 5 1/2 inches long.

acteristic listed, therefore, it most likely does not have a debris flow or mudflow origin; at least in the vicinity of the type section.

In Little Cottonwood Canyon, tilloid rests non-erosionally on thin-bedded siltstone deposits in a few locations. Such a relationship can be produced by mudflow redeposition of till into a lake (Hartschorn, 1958, p. 481). Mudflow redeposition of till is a common process in active glacial areas. Therefore, while the main mass of the tilloid probably does not have a debris or mudflow origin, there is no reason to believe that parts of it could not have been reworked by these processes.

Deposition from glaciers may be the best explanation for the origin of the tilloid. Such an interpretation is not without problems, as features commonly associated with glacial activity are not present (such as striated bedrock, boulder pavements, and other features listed in Flint, 1971, p. 182).

The lack of "common glacial characteristics" can be explained, if the Mineral Fork lithologies were deposited from a "wet-base" glacier in a grounded marine shelf zone (Carey and Ahmad, 1961, p. 883). Under such conditions, ordinary glacial

Table 2. Environmental significance of 18 characteristics of the tilloid portion of the Mineral Fork Formation. Material in this table is based on information in Allen (1971), Crandell (1971), Pettijohn (1957), Picard (1971), Reading (1966) and Schwab (1976).

KEY
a = almost always
c = commonly
r = rarely
v = very rarely

	Turbidite (Proximal) 1.	Turbidite (Distal) 2.	Mudflow 3.	Glacier Complex 4.	Evidence Favors
1. Poorly sorted throughout.	a	c	a	a	1, 3, 4
2. No internal bedding.	a	r	c	a	1, 4
3. Contains no sedimentary structures.	r	r	c	c	3, 4
4. Contains a large percentage of erratic clasts larger than two millimeters in diameter.	v	r	c	c	3, 4
5. Upper size limit of clasts about five meters, but limited only by size of materials available in the source area.	r	v	c	a	4
6. Distinctly erosional lower contact.	c	r	v	a	1, 4
7. Contains lenticular interbeds of typical fluvial sandstone and conglomerate.	v	v	v	a	4
8. Contains interbedded thin-bedded siltstone that contains ice rafted stones in some places.	v	r	v	c	4
9. Contains a small percent of "faceted" and striated stones.	r	r	r	c	4
10. Greater than 3,000 feet thick in some places.	r	v	r	c	4
11. Little primary mineralogic clay in the less-than-two-millimeter size fraction.	r	r	r	c	4
12. Lower contacts are sharp, some upper contacts are gradational because of reworking.	c	r	c	a	4
13. Contains very large percentage of unaltered mineral and rock types n all size ranges.	r	r	r	a	4
14. Contains greater than 50 percent angular fragments in the less-than-two-millimeter size fraction.	r	r	c	c	3, 4
15. Contains fair percentage angular clasts in the greater-than-two-millimeter size fraction.	r	r	c	c	3, 4
16. Fine fraction modes are in the 1/4 to 1/8th-millimeter range.	—	—	—	c	—
17. Does not have an overall large- or small-clast size gradation.	r	v	r	c	4
18. Lithologically distinct from the overlying and underlying formations.	r	r	c	a	4

features would not have formed. This might show a relationship between withdrawing Late Precambrian seas and glaciation. If Eardley and Hatch are correct (1940), their Late Precambrian paleogeography would support this idea. This may show that the tilloid, and therefore the Mineral Fork Formation, is marine in part.

No single characteristic can prove or disprove a glacial origin for the formation, but the association of all observed features strongly suggests that glacial activity is responsible for the deposition of the tilloid, and ultimately, of the entire formation.

Table 2 presents the 18 most characteristic features of the tilloid portion of the formation. The likelihood of finding similar features in other environments is indicated. However, the only environment that best explains all observed features is the glacial environment. The analysis overwhelmingly supports the hypothesis of glacial origin for the tilloid, even though some "common glacial characteristics" are missing.

The occurrence of other tillite-like rocks to the west (Deep Creek Range), southwest (Sheeprock Range), south (Rock Canyon, east of Provo), north (Little Mountain, west of Ogden), and northwest (islands of the Great Salt Lake) may indicate a complex interrelationship of Late Precambrian glaciers with Late Precambrian seas. Discussion of this idea is beyond the scope of this paper, but the idea does warrant further study (see Woodward, this volume).

Non-bedded siltstone, as discussed in this report, very closely resembles the tilloid matrix, and is always associated with tilloid. Furthermore, in some places (see units 3-10 and 3-11) non-bedded siltstone grades into tilloid. The lithologic similarity and gradational relationships suggest strongly that this material either is reworked till, or is till derived from sources lacking large clasts.

CONCLUSIONS

In view of the preceding discussions, use of the genetic term "tillite" appears to be justified for the tilloidal portions of the Mineral Fork Formation. However, because the formation contains about 30 percent lacustine and fluvial lithologies, it is improper to use the term in the genetic (or unacceptable descriptive) sense for the entire unit.

The best arguments for the glacial origin seem to be:

1. The tillite is the main rock unit in the formation, and it encloses all other lithologies. If similar materials were found as, for example, lenses in a fluvial complex, the glacial origin would be questionable.

2. The Mineral Fork Formation is markedly different from the overlying and underlying units. This suggests that it was probably transported to its present site; possibly from a great distance.

Tillites in the Mineral Fork Formation were deposited from Late Precambrian glaciers into basins eroded into the Big Cottonwood Series. The great percentage of rounded pebbles, cobbles, and boulders in the tillite may indicate that glaciers followed older stream valleys and transported primarily pre-existing fluvial sands and gravels.

The location of the source area is unknown. However, Eardley and Hatch (1940, p. 841) propose that mountains were present southeast of the study area in Late Precambrian times. Limited data in the Mineral Fork Formation suggest a northwesterly transport direction, so it is possible that the Mineral Fork glaciers headed in mountains to the southeast.

Most of the till probably was deposited by a receding glacier, probably best described as a "Peidmont-type" (according to Ahlman's Classification of 1948) that originated in igneous and metamorphic highlands, and flowed out over lowlands, perhaps marginal to a Late Precambrian sea. An extensive till plain was formed, and was quickly covered by a well developed, at times all-encompassing, network of outwash streams and lakes. The streams probably were braided, rapidly shifting their channels and source areas. Small lakes were present along the ice margin and between the braided outwash stream channels, sometimes draining rapidly due either to loss of ice support, or cutting by streams. Varve and rhythmite deposition was frequently interrupted by surges of sand-sized material from the outwash streams.

Thus, the lithologies seen in the formation are the result of a complex of glacial activities as commonly observed on the outwash plains of modern glaciers.

Apparently, the Mineral Fork Formation is a remnant of a once more extensive and thicker unit. This is inferred from the unconformable upper surface. What is left represents basin fill that was protected from erosion by its relatively lower stratigraphic position. It may turn out that the extent of the original deposits, and the direction from which they came, can be determined only after extensive correlation with similar formations in nearby areas.

The preceding analysis of the general geology of the Mineral Fork Formation suggests that the formation represents a depositional phase in the following Late Precambrian sequence of events:

1. Uplift of the Big Cottonwood Series induced by uplift to the southeast and associated with withdrawal of Late Precambrian seas in a westerly or northwesterly direction.

2. Advance of glaciers from metamorphic highlands to the east or southeast onto fluvial lowlands that were formerly marginal to marine portions of the Big Cottonwood Series.

3. Retreat of the glaciers and the formation of ablation moraine.

4. Growth of proglacial streams and lakes on the previously deposited morainal materials.

5. Reduction of outwash activity and the formation of a series of lakes that occasionally drained either due to loss of ice support, or due to bed cutting by outwash streams.

6. Fluvial erosion of most of the glacial material and deposition of the Mutual Formation. (The Mutual probably was derived both from reworking of the glacial deposits and erosion of slightly re-elevated low highlands previously reduced by glaciers.)

7. Uplift and subsequent erosion of the Mutual and Mineral Fork Formations and deposition of the Tintic Quartzite as the first marine sand of the Lower Paleozoic, marking readvance of the seas.

This sequence presupposes that the Mutual Formation is largely continental and not marine. Further, it indicates that the onset of glaciation accompanied mountain building. This is not unreasonable, but there is an admitted lack of information on the supposed mountains. A similar sequence of events was postulated for the Sheeprock Mountains by Cohenour (1959, p. 94). There is a possibility of correlation between the two areas, but examination of this possibility is beyond the scope of this paper.

REFERENCES CITED

Ahlman, H. W., 1948, Glaciological research on the North Atlantic coasts: Royal Geog. Soc. (London), Research Ser. No. 1, 87 p.

Allen, J. R. L., 1971, Physical processes of sedimentation: New York, American Elsevier Publishing Company, 248 p.

Blackwelder, E., 1928, Mudflow as a geologic agent in semiarid Mountains: Geol. Soc. America Bull., v. 39, p. 465-483.

_____, 1932, An ancient glacial formation in Utah: Jour. Geology, v. 40, no. 4, p. 289-304.

Bradley, W. H., 1929, The varves and climate of the Green River Epoch: U.S. Geol. Survey Prof. Paper 158-E, p. 87-110.

Carey, S. W., and Ahmad, N., 1961, Glacial marine sedimentation, in G. O. Raasch (ed.), The geology of the Arctic: Toronto, Ontario, Canada, University of Toronto Press, v. 2, p. 865-894.

Cohenour, R. E., 1959, Sheeprock Mountains, Tooele and Juab Counties, Utah: Utah Geol. and Mineralog. Survey Bull. 63, 201 p.

Coleman, A. P., 1926, Ice ages, recent and ancient: New York, MacMillan and Company, 296 p.

Condie, K. C., 1967, Petrology of the Late Precambrian tillite (?) association in northern Utah: Geol. Soc. America Bull., v. 78, p. 1317-1344.

_____, 1969, Geologic evolution of the Precambrian rocks in northern Utah and adjacent areas: Utah Geol. and Mineralog. Survey Bull. 82, p. 71-95.

Crandell, D. R., 1971, Postglacial lahars from Mount Rainier volcano, Washington: U.S. Geol. Survey Prof. Paper 677, 75 p.

Crittenden, M. C., Jr., Sharp, B. J., and Calkins, F. C., 1952, Geology of the Wasatch Mountains east of Salt Lake City: Utah Geol. Soc., Guidebook to the Geology of Utah, no. 8, p. 1-37.

Dott, R. H., 1961, Squantum "tillite", Massachusetts — evidence of glaciation or subaqueous mass movements?: Geol. Soc. America Bull., v. 72, p. 1289-1306.

Eardley, A. J., and Hatch, R. A., 1940, Proterozoic (?) rocks in Utah: Geol. Soc. America Bull., v. 51, p. 795-844.

Flint, R. F., 1971, Glacial and quaternary geology: New York, John Wiley and Sons, Inc., 892 p.

─────────, Sanders, J. E., and Rodgers, J., 1960, Diamictite, a substitute term for symmicitite: Geol. Soc. America Bull., v. 71, p. 1809.

Gustavson, T. D., 1974, Sedimentation on gravel outwash fans, Malaspina Glacier foreland, Alaska: Jour. Sed. Petrology, v. 44, p. 374-389.

Harland, W. B., et al., 1966, The definition and identification of tills and tillites: Earth Sci. Rev., v. 2, p. 225-256.

Harrison, S. S., 1975, Turbidite origin of glaciolacustrine sediments, Woodcock Lake, Pennsylvania: Jour. Sed. Petrology, v. 45, p. 738-744.

Hartschorn, J. H., 1958, Flowtill in southeastern Massachusetts: Geol. Soc. America Bull., v. 69, p. 477-482.

Hintze, F. F., 1913, A contribution to the geology of the Wasatch Mountains, Utah: New York Acad. Sci. Annals, v. 23, p. 85-143.

Kuenen, Ph. H., and Migliorini, C. I., 1950, Turbidity currents as a cause of graded bedding: Jour. Geology, v. 58, p. 91-127.

Marsell, R. E., 1974, Glaciation, Mineral Fork Tillite: Utah Geol. and Mineralog. Survey Bull. 69, p. 67-68.

Miller, D. J., 1953, Late Cenozoic marine glacial sediments and marine terraces of Middleton Island, Alaska: Jour. Geology, v. 61, p. 17-40.

Pettijohn, F. J., 1957, Sedimentary rocks: New York, Harper and Row, 717 p.

Picard, M. D., 1971, Classification of fine grained sedimentary rocks: Jour. Sed. Petrology, v. 41, p. 179-195.

Reading, H. Gq, and Walker, R. G., 1966, Sedimentation of Eocambrian tillites and associated sediments in Finnmark, northern Norway: Palaeogeog. Palaeoclimtology, Palaeoecology, v. 2, p. 177-212.

Schwab, F. L., 1976, Depositional environments, provenance, and tectonic framework: upper part of the Late Precambrian Mount Rodgers Formation, Blue Ridge Province, southwestern Virginia: Jour. Sed. Petrology, v. 46, p. 3-13.

Stokes, W. L., 1969, Stratigraphy of the Salt Lake region: Utah Geol. and Mineralog. Survey Bull. 82, p. 36-49.

Varney, P. J., 1972, Depositional environment of the Mineral Fork Formation (Precambrian), Wasatch Mountains, Utah: University of Utah, unpublished M.S. thesis, 134 p.

Woodward, L. A., 1972, Upper Precambrian stratigraphy of central Utah: Utah Geological Association Pub. 2, p. 1-6.

THE CAMBRIAN SECTION OF THE CENTRAL WASATCH MOUNTAINS

by
Christina Lochman-Balk[1]

ABSTRACT

The Cambrian section of the central Wasatch Mountains is now a deeply eroded remnant of a sequence of a basal transgressive sandstone (the Tintic Quartzite), and a thick carbonate unit (the Maxfield Limestone) deposited in supratidal to shallow subtidal sites during several regressive-transgressive cycles. The Tintic Quartzite was deposited during latest Lower Cambrian. The two earliest Middle Cambrian faunal zones are not known in the section. The reworked muds and sands of the Ophir Shale were deposited during *Glossopleura* Zone time. The widespread carbonate-accumulating shoals and tidal flats had appeared by the beginning of the *Bathyuriscus-Elrathina* Zone, and these conditions persisted during the rest of Middle Cambrian time and most of Upper Cambrian and Lower Ordovician. However, the central Wasatch area lies across the roughly East-West axis of an early Paleozoic (probably Precambrian) positive element. Sporadic uplift of this element during early Middle Ordovician, during the Silurian, and during most of the Devonian led to the deep erosion and removal of most of the Lower Ordovician and Cambrian section. Now middle Middle Cambrian carbonates are overlain unconformably by the very Late Devonian to Kinderhookian Fitchville Dolomite.

INTRODUCTION

The Cambrian of the central Wasatch Mountains was deeply eroded during the early Paleozoic. This region is presently located along the western side of the Uinta Uplift, a structure which dates from the Late Precambrian, at least, and has been intermittently positive during much of the Phanerozoic. Cambrian sections around the Uintas indicate that this area became a cluster of islands during the Upper Cambrian and may have been completely inundated by the Early Ordovician. In the northern Wasatch Mountains relatively thick sections of Cambrian and Lower Ordovician are present. In the southern Wasatch Mountains the Cambrian sections are somewhat thinner and the Upper Cambrian is overlain only locally by a remnant of the Lower Ordovician Opohonga Limestone. However, the occurrence of Middle and Upper Cambrian and Lower Ordovician deposits in the Bighorn Mountains of Wyoming and the Black Hills of South Dakota suggests that the central Wasatch region originally was covered by carbonate units representing deposition during the entire Middle Cambrian, most of the Upper Cambrian, and the early part of the Lower Ordovician. These deposits were ultimately removed by subaerial erosion first occurring in the early Middle Ordovician and later during the Late Silurian and most of the Devonian. The varying thicknesses of the remnant Middle Cambrian carbonate unit suggest that the axis of the early Paleozoic positive element trended roughly east-west about 25 mi south of Salt Lake City near American Fork.

LITHOSTRATIGRAPHY

The formation names of the Cambrian lithic units recognized in the central Wasatch Mountains are derived, with one exception, from the much thicker, more complete, and earlier described sections of the important mining districts 30 mi to the west — the Ophir District, Oquirrh Mountains (Gilluly, 1932) and the Tintic Mining District (Lindgren and Loughlin, 1919). The units of the Mount Nebo region (Fig. 2, section 6) tie in well with those of the East Tintic Mountains only 20 mi to the northwest. In the central Wasatch Mountains the transgressive sandstone of the Tintic Quartzite and the overlying shales of the Ophir Formation are readily recognizable. The sequence of Middle Cambrian carbonate units cannot be recognized in the thin carbonate section now remaining, and it is treated as a single lithic unit — the Maxfield Formation — with the type area in the Cottonwood-American Fork Mining District (Calkins and Butler, 1943). These three lithic units were recognized as far north as the Ogden area (Fig. 2, section 1) by Eardley (1944) who also called attention to the presence within the Maxfield Limestone in this area of several lithic types which are characteristic of carbonate units recognized in the Ophir District — the Lynch dolomite type and the Hartman type — and in the Tintic District — the Bluebird type. However, these lithic types repeat through the carbonate section of the Ogden area, and clearly cannot be used in a

---

[1] New Mexico Institute of Mining and Technology, Socorro, New Mexico.

lithic correlation of the Maxfield with the succession of carbonate units recognized in the western districts. In the northern Wasatch Mountains and Bear River Range the Cambrian lithic units and their succession are distinctly different and a different terminology is used.

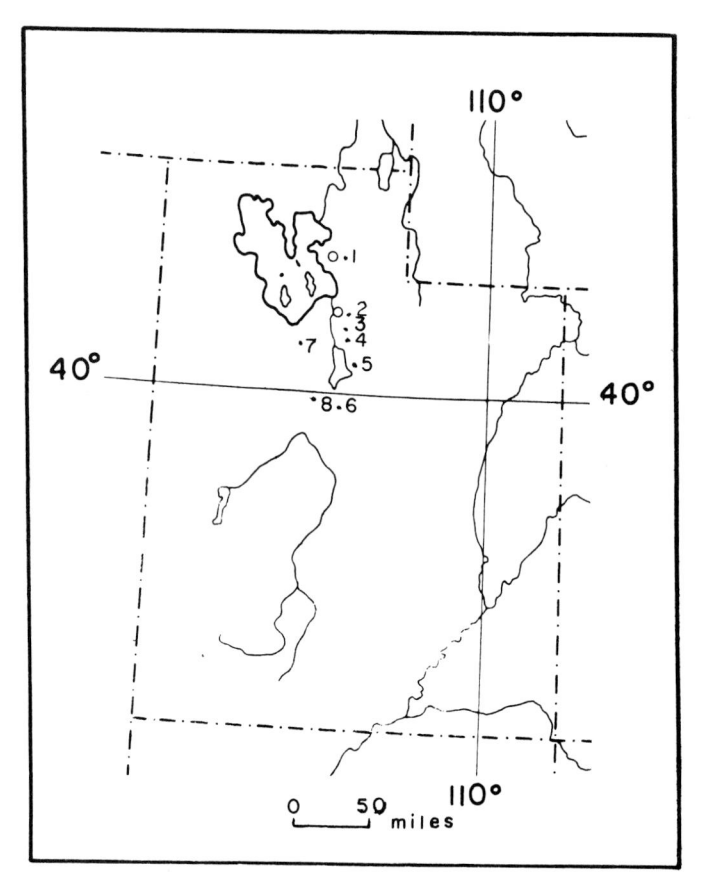

Fig. 1 — Location of Stratigraphic Sections

Section 1 — Durst Mountain, 15 miles ESE of Ogden (Eardley, 1944)

Section 2 — Parleys Canyon, 3-4 miles E of Salt Lake City

Section 3 — Cottonwood Canyon (Calkins & Butler, 1943)

Section 4 — Mouth of American Fork (Baker, 1947; Hintze, 1973, p. 7)

Section 5 — Rock Canyon (Baker, 1947; Hintze, 1969, map)

Section 6 — Mt. Nebo Area, North Canyon (C. V. Smith, 1956; Hintze, 1962; Hintze, 1973, p. 11)

Section 7 — Ophir District, Oquirrh Mountains, 32 miles SW of Salt Lake City (Gilluly, 1932, p. 7-20)

Section 8 — East Tintic District, 20 miles NW of Mt. Nebo (Hintze, 1969; Hintze, 1973, p. 58)

Tintic Quartzite

The base of the Tintic Quartzite, wherever it is exposed, rests with regional angular unconformity upon the Precambrian. It crops out as medium- to thick-bedded ledges in light colored cliffs. It is primarily a white, pink, or light gray; fine- to medium-grained; locally coarse-grained rock of clean, subrounded to rounded quartz grains cemented by silica. Locally buff, red-brown, or dark purple coloration may occur from the weathering of limonite cement. Cross-bedding is common and can occur throughout the unit. Local conglomerate lenses are also common. Clasts usually range in diameter from one to three in., are well rounded, and consist of vein quartz, jasper, or light colored Precambrian quartzites, but local basal conglomerates may contain boulders of gneiss, schist, quartzites, and jasper up to 18 in in diameter. The basal conglomerate may reach 100 ft in thickness where it occurs, but the conglomerate lenses throughout the unit are much thinner. Although accurate thicknesses are hard to measure because of faulting along the Wasatch front, 1,000 ft appears to be a maximum thickness, with many of the sections having a common average of 500 to 800 ft. The unit is much thinner in the central Wasatch Mountains than in the type area in the Tintic District where 2,300 to 3,200 ft have been reported. The contact with the overlying Ophir Formation appears lithically gradational as shale lenses appear in the upper beds of the Tintic Quartzite. Hintze (1962) states that the contact "is usually taken where light colors typical of Tintic Quartzite give way to dark browns and greens of the Ophir phyllitic shales and sandstones."

Ophir Formation

The Ophir Formation rests paraconformably upon the Tintic Quartzite. The unit consists of dark olive green, phyllitic, micaceous, fissile shales with interbeds of olive green sandstones and quartzites especially common in the lower part and interbeds of dark blue to black banded argillaceous limestones especially common near the middle. Calkins and Butler (1943) recognize three lithic members. The lower member is characterized by the numerous interbeds of quartzite and sandstone, with the shales becoming distinctly fissile as the sand interbeds thin and disappear. The weathered surfaces usually show a reddish brown coating of iron oxide. The bedding surfaces are distinctive, spangled with mica flecks, wavy and lumpy, and carrying ripple marks, raindrop impressions, trilobite exuviae, linguloid shells, and numerous trace fossils referred to as worm trails, burrows, or fucoids. These trace fossils have not been studied in detail. The middle limestone member may range up to 80 ft in thickness and carries several ore bodies in the Cottonwood Canyon-American Fork district. The base of the middle limestone member is placed at the appearance of siliceous limestone nodules in the shales, which pass upward into a blue-gray, nonmagnesian limestone with the bedding

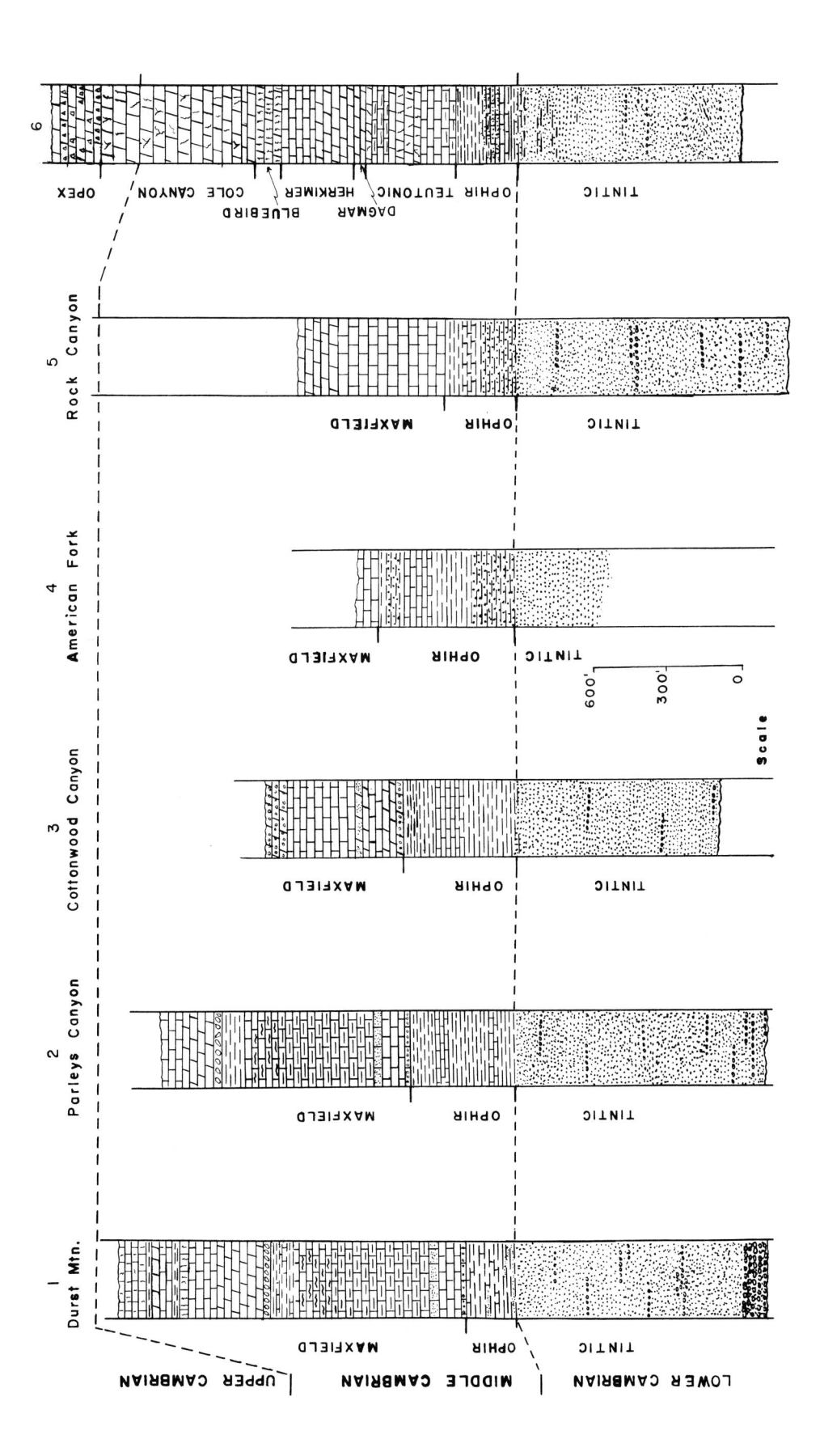

Fig. 2 — Cambrian Sections in the Central and Southern Wasatch Mountains

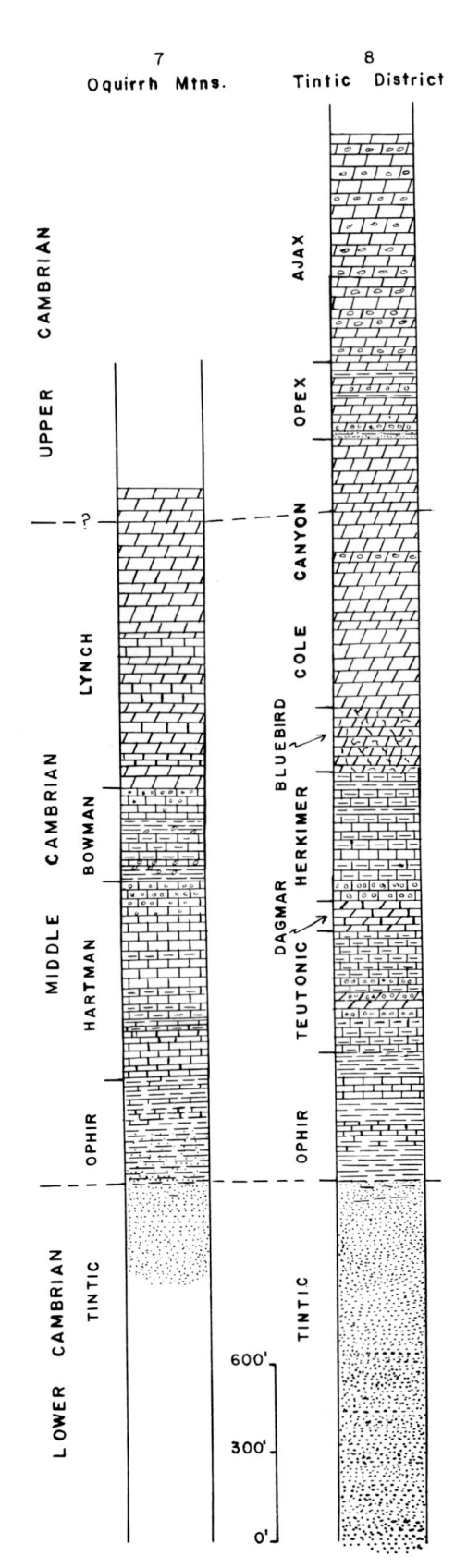

Fig. 3 — Cambrian Sections in Oquirrh Mountains and East Tintic District. Scale same as in Fig. 2.

marked by thin, wavy, siltstone laminae. Near the middle of the member the limestone becomes fairly pure, but near the top laminae of siliceous siltstone reappear abruptly and the limestone layers tend to pinch out abruptly into nodules and pods. The upper fissile shale member is markedly calcareous and weathers yellowish-brown on the exposures. However, the bedding surfaces are similar to those of the lower member; conspicuously wavy and lumpy with trilobite and linguloid body fossils and numerous trace fossils. The reported thicknesses of the formation are highly variable in the central Wasatch region, ranging from 200 ft at Durst Mountain up to 450 to 500 ft in the Cottonwood Canyon-American Fork area, to 250 ft in Rock Canyon, and 230 ft in North Canyon. It is possible that the older sections (Fig. 2, sections 2, 3, 4) are too thick because repetition of shale beds due to the extensive faulting was undetected. To the west the Ophir Formation is 320 ft thick in the type area of the Ophir District, Oquirrh Mountains and 430 ft in the Tintic District.

Maxfield Limestone

The Maxfield Limestone overlies the Ophir Formation conformably at a sharp lithic contact. The unit consists of dolomite, dolomitic limestone, and limestone beds with a variety of sedimentary textures and structures that are similar to those observed in the carbonate sequences of the Ophir and Tintic Districts. Colors are white, light to dark gray, blue-gray, and dark blue-gray to nearly black. Calkins and Butler (1943) recognized three members in the Cottonwood-American Fork area. The lower member has a basal 75 ft of very dark gray to black dolomite or limestone with abundant ooliths and pisolites. The overlying 90 ft is predominately banded, mottled, dolomitic limestones with a few oolithic beds near the base and beds with "Bluebird structures" in the upper half. In the overlying 20 ft of banded dolomite, almost black bands filled with calcite twiggy bodies alternate with thinner, light gray, unmottled bands. The abundance of small, branched and unbranched, twig-like calcite structures in some of the dolomites is a distinctive feature of the Middle Cambrian carbonate section. The bodies were referred to as vermicular, worm-like, or twig-like, and it was postulated that the bodies had an organic origin, possibly algal. The structure is well developed in and diagnostic of the Bluebird Dolomite of the Tintic District, and is now known as "the Bluebird structures". These structures are common in the Middle Cambrian carbonates throughout the eastern Great Basin (Utah and eastern Nevada) and have been studied in detail by Kepper (1974) who concludes that the Bluebird structures are inorganic in origin and "were produced through calcitization processes which involved the centripetal and centrifugal replacement of dolomite."

The uppermost 30 ft of the lower member consists of a very fine-grained, light gray to white dolomite. On some weathered surfaces thin, wavy or cross-bedded layers of silt stand out in relief. This light dolomite is the "Lynch type" as it is the characteristic lithology of the Lynch Dolomite of the Ophir District. Within a particular local area the white dolomite may be a useful horizon-marker for mapping (Calkins and Butler, 1943, p. 16), but this lithology is the carbonate sedimentation peculiar to one particular environment of deposition and as such cannot be used as a time marker even within the central Wasatch region.

The middle member of the Maxfield Limestone is predominately calcitic, mottled limestones cyclically interbedded and intercalated with shales. The dark gray limestones are mottled with light gray and buff patches. The bedding is usually thin to flaggy, but thicker-bedded dolomitic limestones with a few pisolitic and twiggy layers are present near the base. Edgewise and flat pebble limestone conglomerates are conspicuous in the upper third of the member. Fossils obtained from some of the shales were originally identified as Upper Cambrian, but restudy has placed them in the Middle Cambrian *Bathyuriscus-Elrathina* Zone.

The upper member of the Maxfield Limestone in the Cottonwood-American Fork area has a maximum thickness of 100 ft and is clearly an erosional remnant of a much thicker unit. The member consists of black to light gray banded dolomites with ooliths common and the white twiggy bodies prominent in the dark dolomite bands.

In the lower part of the Durst Mountain section (Fig. 2, section 1) of Eardley (1944) the succession of major lithic types appears similar to that in the Cottonwood-American Fork area, but each unit is much thicker. At the base is 100 ft of dark blue-gray, banded, mottled limestone with abundant ooliths and pisolites. This unit is termed the "Hartman type" by Eardley as the lithology is distinctive of the Hartman Formation in the Ophir District. An overlying 425 ft of shale and platey limestone separates the basal unit from a reappearance of the Hartman type in a 150 ft thick unit of dark-gray to black limestone with irregular mudstone bands and light tan patches. The succeeding unit is 100 ft of olive-gray shales interbedded with thin, blue-gray mottled limestones and topped by 30 ft of edgewise pebble limestone conglomerates and thin shale beds. This unit is overlain abruptly by 175 ft of banded limestones with "Bluebird structures". From this unit upward, the Durst Mountain section shows a repetition of vermicular limestones — the Bluebird type; white, laminated dolomites — the Lynch type; and tan, thin-bedded shales. It is tempting to compare the lower and middle members of the Maxfield Limestone in the Cottonwood-American Fork area with units 1 through 5 of Eardley's Durst Mountain section.

BIOSTRATIGRAPHY

Although no fossils have been reported from the Tintic Quartzite in the central Wasatch exposures, cephala and other fragments identifiable as *Olenellus* sp. have been found in upper beds at Long Ridge in the Mt. Nebo area. Specimens of *Olenellus* sp. have also been collected from the upper shaley beds of the Bingham Quartzite in the northern Wasatch Mountains. Hintze (1962) reports the presence of a sheet of amygdaloidal basalt in the Tintic Quartzite throughout the southern Wasatch-Long Ridge-East Tintic area, and observed that in the Wasatch-Long Ridge areas the flow lies from 90 to 165 ft above the base of the Tintic Quartzite but in the East Tintic Mountains the flow lies 980 ft above the base. Hintze observes that "if the flow is used as a time horizon it would indicate that the lowest part of the Tintic Quartzite in the East Tintic Mountains is not represented by any deposition in the Wasatch Mountains." This observation suggests that the Tintic of the central Wasatch area was deposited during latest Lower Cambrian *Olenellus* Zone time.

No fossils of the *Plagiura-Poliella* Zone (basal Middle Cambrian) or the overlying *Albertella* Zone have been reported from any of the central or southern Wasatch sections. In the northern Wasatch area a large *Albertella* Zone assemblage occurs in the Langston Limestone immediately overlying the Bingham Quartzite. R. A. Robison (1975, Fig. 2) shows the upper Tintic Quartzite as the time equivalent of the *Plagiura* and the *Albertella* Zones of the early Middle Cambrian, but without fossil evidence.

A *Glossopleura* Zone assemblage has been reported from the Ophir Formation at many localities in the Tintic District and the southern Wasatch area. This assemblage ranges from the base of the Ophir up to the middle limestones. From the upper half of the Ophir Formation, fossils of the *Ehmaniella* subzone of the *Bathyuriscus-Elrathina* Zone have been obtained (Robison, 1975, Fig. 2). In Fig. 2 Robison shows the deep erosion of the central Wasatch Mountains Maxfield Limestone; suggesting that only the Teutonic equivalent is present. It is possible that the fossils reported by Calkins and Butler (1943) from the middle member of the Maxfield Limestone may belong to the *Glyphaspis* fauna of the *Bathyuriscus-Elrathina* Zone.

PALEOGEOGRAPHY

The Wasatch Mountains are located just west of the position of the roughly north-south trending western boundary of the craton during the Cambrian. The continental craton, extending eastward through the central and east-central United States, was an area inundated by marine waters only at times of maximum transgression. From the craton boundary the continental shelf extended westward to central Nevada, the position of the continental slope (Cook and Taylor, 1975). The continental shelf is characterized by long time spans (a Sequence of Sloss) of slow downwarping — either continuous or intermittent — the amount of negative movement increasing from the boundary of the craton to the top of the continental slope. During the Cambrian there were distributed over the shelf numerous local tectonic elements which gave rise to more rapidly downwarping local basins and adjoining much less negative or even slightly positive linear ridges. The northern Wasatch Mountains occupied such a local basin, trending ESE-WNW. The central Wasatch region lay upon the adjoining less negative ridge to the south. The relative movements of these elements influenced both the thickness and the lithology of the Cambrian sections.

A distinctive feature of the Middle and Upper Cambrian and Lower Ordovician shelf deposits was the dominance of stromatolitic algae. These algae were able to maintain widespread carbonate reefs where the water energy was sufficient to prevent settling of clay-size particles. With optimum water and light conditions, the rapid growth of the stromatolites maintained their carbonate substrate in favorable shallow water positions — subtidal, intertidal, or even supratidal — and carbonate accumulation balanced or locally may even have exceeded the amount of shelf downwarping. During this time the shelf was a vast carbonate platform extending from the craton boundary on the east to the top of the continental slope. Large irregular stromatolite reefs were crossed by tidal channels; were flanked in interreef areas by barrier bars; and sheltered large and small, subtidal, quiet water lagoons. Low, wind and wave built carbonate islands dotted the surface of the shallow sea. Below the water-sea bottom interface carbonates accumulated; producing the characteristic limestone and dolomite sections which increase steadily in thickness from the inner edge of the shelf (the site of the central Wasatch Mountains) to the outer western edge of the shelf at the top of the continental slope.

During the Lower Cambrian the continental shelf was transgressed from the west by marine waters until the shore lay along the western border of the craton in late Lower Cambrian time. The appearance of thin shaley lenses in the typical light colored sands near the top of the Tintic is considered evidence in this region of a late Lower Cambrian marine regression. There is both lithic and faunal evidence in many Cordilleran sections of a marine regression across the continental shelf at the end of the Lower Cambrian. I interpret the contact between the Tintic Quartzite and the Ophir Formation as a paraconformity, and believe that the absence of the two early Middle Cambrian faunal assemblages in the central Wasatch area means that the region was above sea level during this time. The transgressing Middle Cambrian sea reached the Wasatch region again in *Glossopleura* Zone time and the initial deposits of interbeds of olive green and brown shales and sands represent the reworking and deposition of the weathered sands formed on the surface of the Tintic during early Middle Cambrian subaerial exposure. Hintze notes (1962) that the contact between the Tintic Quartzite and the Ophir Formation can be placed where the light colors of the Tintic are replaced by the dark browns and greens of the Ophir, although the lithology is similar on both sides of the contact. It would appear that the positive ridge of the Uintas formed a low but persistent peninsula in this area, to the south of which quiet water sites developed. The middle limestones and upper shales of the Ophir accumulated here during early *Bathyuriscus-Elrathina* Zone time. To the north the transgression continued eastward into central Wyoming and Montana. However, in the sheltered embayment to the south the transgression did not continue eastward and it appears that there were no major drainages crossing the low coastal plain to bring clastics to the shore. The sea floor remained shallow, the water turbulent, and numerous ooliths and pisolites were washed into bars. Carbonate tidal flats and shoals dominated by stromatolites developed and persisted in the area throughout the rest of the Middle Cambrian. The repetition of the various carbonate lithologies in the section demonstrates the fluctuation of various carbonate environments over the region during this time. Kepper (1972) has described in detail the different paleoenvironments which each lithofacies represents.

REFERENCES

Baker, A. A., 1947, Stratigraphy of the Wasatch Mountains in the vicinity of Provo, Utah: U.S. Geol. Survey Oil & Gas Invest. Prelim. Chart 30.

Calkins, F. C. and Butler, B. S., 1943, Geology and ore deposits of the Cottonwood-American Fork area, Utah: U.S. Geol. Survey Prof. Paper 201, p. 10-19.

Cook, H. E. and Taylor, M. E., 1975, Early Paleozoic continental margin sedimentation, trilobite biofacies, and the thermocline, western United States: Geology, v. 3, n. 10, p. 559-562.

Eardley, A. J., 1944, Geology of north-central Wasatch Mountains, Utah: Geol. Soc. America Bull., v. 55, p. 819-894.

Gilluly, James, 1932, Geology and ore deposits of the Stockton and Fairfield quadrangles, Utah: U.S. Geol. Survey Prof. Paper 173, p. 7-17.

Hintze, L. F., 1962, Precambrian and Lower Paleozoic rocks of north-central Utah: Brigham Young Univ. Geol. Studies, v. 9, pt. 1, p. 8-16.

_____, 1969, Preliminary geologic map and cross sections of the Y Mountain area, east of Provo, Utah: supplement to Brigham Young Univ. Geol. Studies, v. 16, n. 2.

_____, 1973, Geologic road logs of western Utah and eastern Nevada: Brigham Young Univ. Geol. Studies, v. 20, pt. 2, 66 p.

_____ and Robison, R. A., 1975, Middle Cambrian stratigraphy of the House, Wah Wah and adjacent ranges in western Utah: Geol. Soc. America Bull., v. 86, p. 881-891.

Kepper, J. C., 1972, Paleoenvironmental Patterns in Middle to Lower Upper Cambrian interval in eastern Great Basin: Am. Assoc. Petroleum Geologists Bull., v. 56, p. 503-527.

_____, 1974, Origin of massive banded carbonates with "Bluebird Structures" in the Cambrian of the eastern Great Basin: Jour. Sed. Petrology, v. 44, p. 1251-1261.

Lindgren, W. and Loughlin, G. F., 1919, Geology and ore deposits of the Tintic Mining District, Utah: U.S. Geol. Survey Prof. Paper 107, p. 23-32.

Lochman-Balk, Christina, 1971, The Cambrian of the Craton of the United States, *in* Holland, C. H., ed., Cambrian of the New World: London, Wiley-Interscience, p. 79-167.

_____, 1972, Cambrian System — Geologic Atlas of the Rocky Mountain Region: Rocky Mtn. Assoc. Geologists, p. 60-75.

Morris, H. T. and Lovering, T. S., 1961, Stratigraphy of the East Tintic Mountains, Utah: U.S. Geol. Survey Prof. Paper 361, p. 13-53.

Rigby, J. K. and Hintze, L. F., 1968, Guide to the geology of the Wasatch Mountain front, between Provo Canyon and Y Mountain, northeast of Provo, Utah: Brigham Young Univ. Geol. Studies, v. 15, pt. 2, 29 p.

Robison, R. A., 1975, Central and western Utah: Cambrian Paleontology and Depositional Environments: Geol. Soc. America Field Trip 9.

Smith, C. V., 1956, Geology of the North Canyon area, southern Wasatch Mountains, Utah: Brigham Young Univ. Research Studies, Geol. Ser., v. 3, n. 7.

ORDOVICIAN SEDIMENTATION IN THE WESTERN UNITED STATES

by

Reuben J. Ross, Jr.[1]

ABSTRACT

Ordovician sedimentation in the western United States was influenced mainly by the position of the Transcontinental arch relative to the earth's latitudinal belts and to sea level. Carbonates of Tremadoc and Arenig age south of the arch were mostly dolomite, formed under shallow restricted water in the horse latitudes. A gastropod-nautiloid fauna of low diversity inhabited this environment. North of the arch, abundantly fossiliferous limestone formed in well circulated equatorial waters.

Late Arenig (late Canadian) lowering of sea level exposed crystalline and older sedimentary rocks of the Canadian Shield to extreme tropical weathering. Quartz sand was released and swept westward south of the Transcontinental arch by the southern trade winds to form the Everton Formation and parts of the Simpson Group; as sea level rose, the transgressive St. Peter Sandstone and sands of the upper Simpson Group were deposited.

North of the arch, the sand (Winnipeg, Swan Peak and Eureka Quatzite) was swept westward by northern trade winds to overwhelm the carbonate deposition on the platform and shelf beneath a prograding blanket. By middle Caradoc to Ashgill (Cincinnatian) time, the source area of the sand on the Canadian Shield was inundated.

As transgression reached its maximum, evaporation of this enormous expanse of sea caused centripetal flow of normal marine surface water and centrifugal flow of hypersaline bottom water. The latter converted lime muds to dolomite. A. gastropod (*Maclurites*)-nautiloid fauna characterized the resulting facies. Sessile bottom-dwelling organisms could survive only where surface water washed carbonate banks. Dissolved silica in the under-flowing brines was deposited as chert upon dilution with normal sea water to form the peripheral facies, bearing graptolites and radiolarians.

INTRODUCTION

Ordovician sedimentation in the western United States took place in two main environments: one was on the craton and inner shelf; the other was on the outer shelf and included sediments that are generally considered to have been eugeosynclinal. The deposition on the craton and inner shelf falls into four patterns or combinations of patterns according to age — Canadian (Tremadoc-Arenig), Whiterockian (Arenig?, Llanvirn, Llandeilo?), post-Whiterockian and pre-Cincinnatian (Llandeilo — early Caradoc) and Cincinnatian (middle Caradoc — Ashgill). The type of sedimentation was influenced strongly by the Transcontinental arch and the position of the North American continent relative to latitudinal belts.

The Transcontinental arch was a linear positive feature of uncertain configuration that extended from New Mexico into central Minnesota. Many Ordovician formations lap out against the flanks of this great barrier but others in Late Ordovician time seem to have buried it. Subsequent erosion in the Devonian bevelled previously deposited strata to varying extents and so the evidence for the precise outline of the arch no longer exists. The arch probably never stood as a high mountain range, but it surely interfered with the circulation of inland seas and may from time to time have been wide enough to alter the flow of atmospheric circulation.

The latitudinal position of North America in late Caradoc (Edenian of Sweet and Bergstrom, 1974) time (Fig. 1b) has been computed by McElhinny and Opdyke (1973, pp. 3704-5, table A) based on samples from the Denley Limestone and lower third of the Steuben Limestone at Trenton Falls, New York (McElhinny and Opdyke 1973, p. 3699). That stratigraphical interval is correlated by Sweet and Bergstrom (1974) with much of the Kope Formation of Ohio and Kentucky, with the Prosser Member of the Galena Dolomite of Minnesota, with much of the Kimmswick Limestone of Missouri and with the upper Viola Limestone of south-central Oklahoma.

---

[1] U.S. Geological Survey, Denver, Colorado.

Reprinted by permission from: Bassett, M. G. (ed.), 1975, The Ordovician System: proceedings of a Paleontological Association symposium, Birmingham, Sept. 1974, University of Wales Press and National Museum of Wales, Cardiff, Wales, Great Britain.

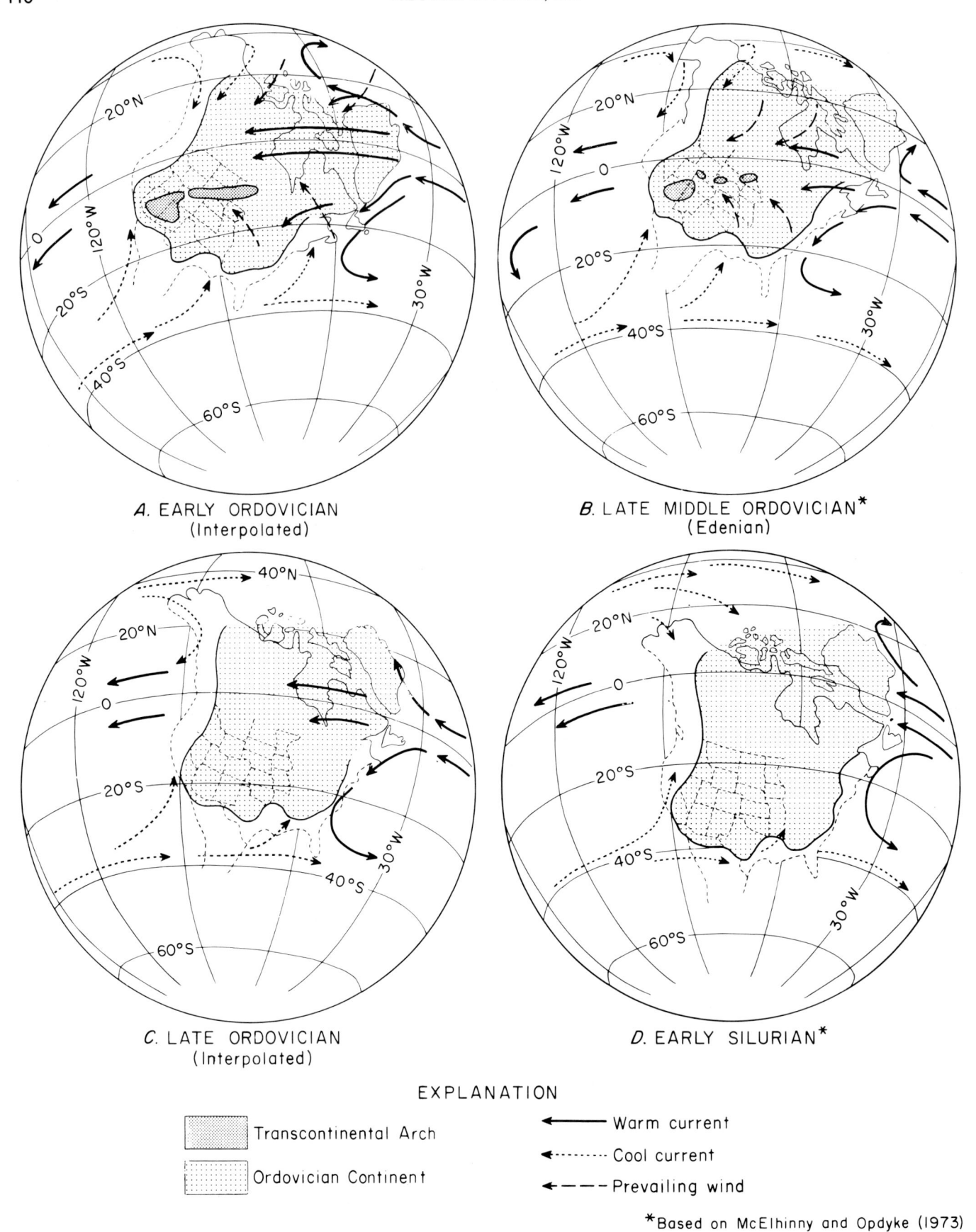

Fig. 1 — Probable positions of North America in Ordovician and Silurian time, and possible effects on ocean currents.

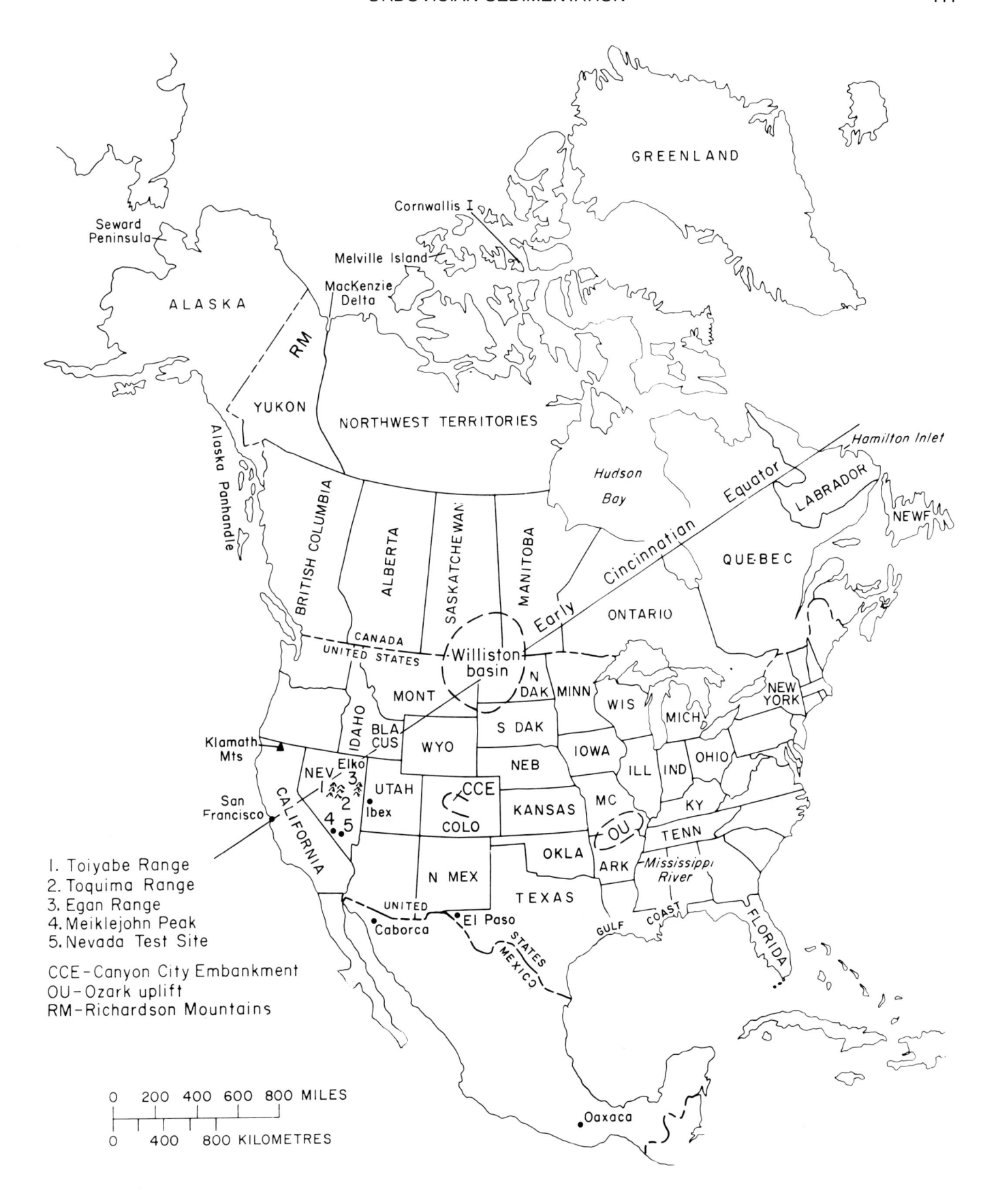

Fig. 2 — Map of North America showing locations mentioned in text.

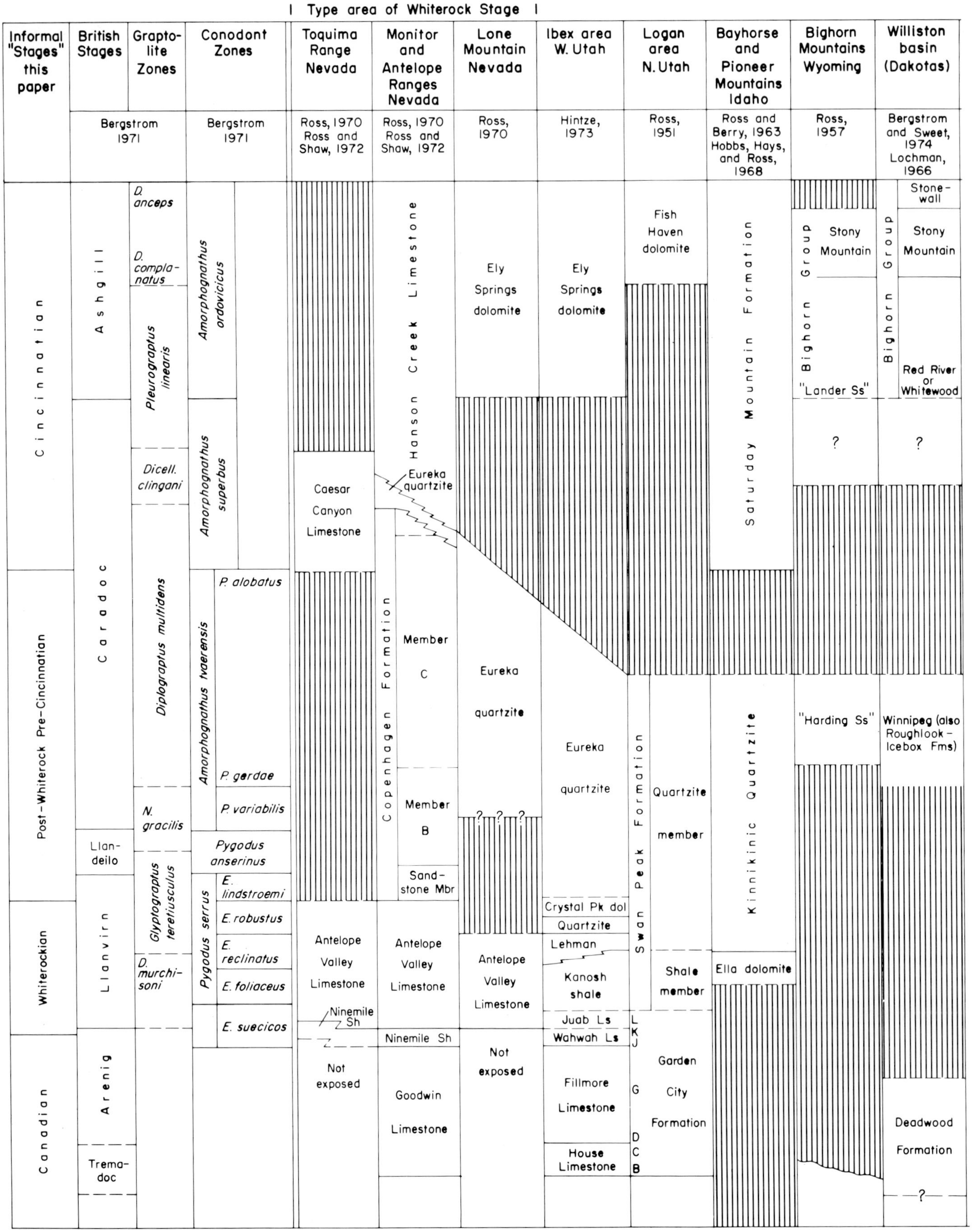

Fig. 3 — Tentative correlations of western formations with those of midwest

ORDOVICIAN SEDIMENTATION

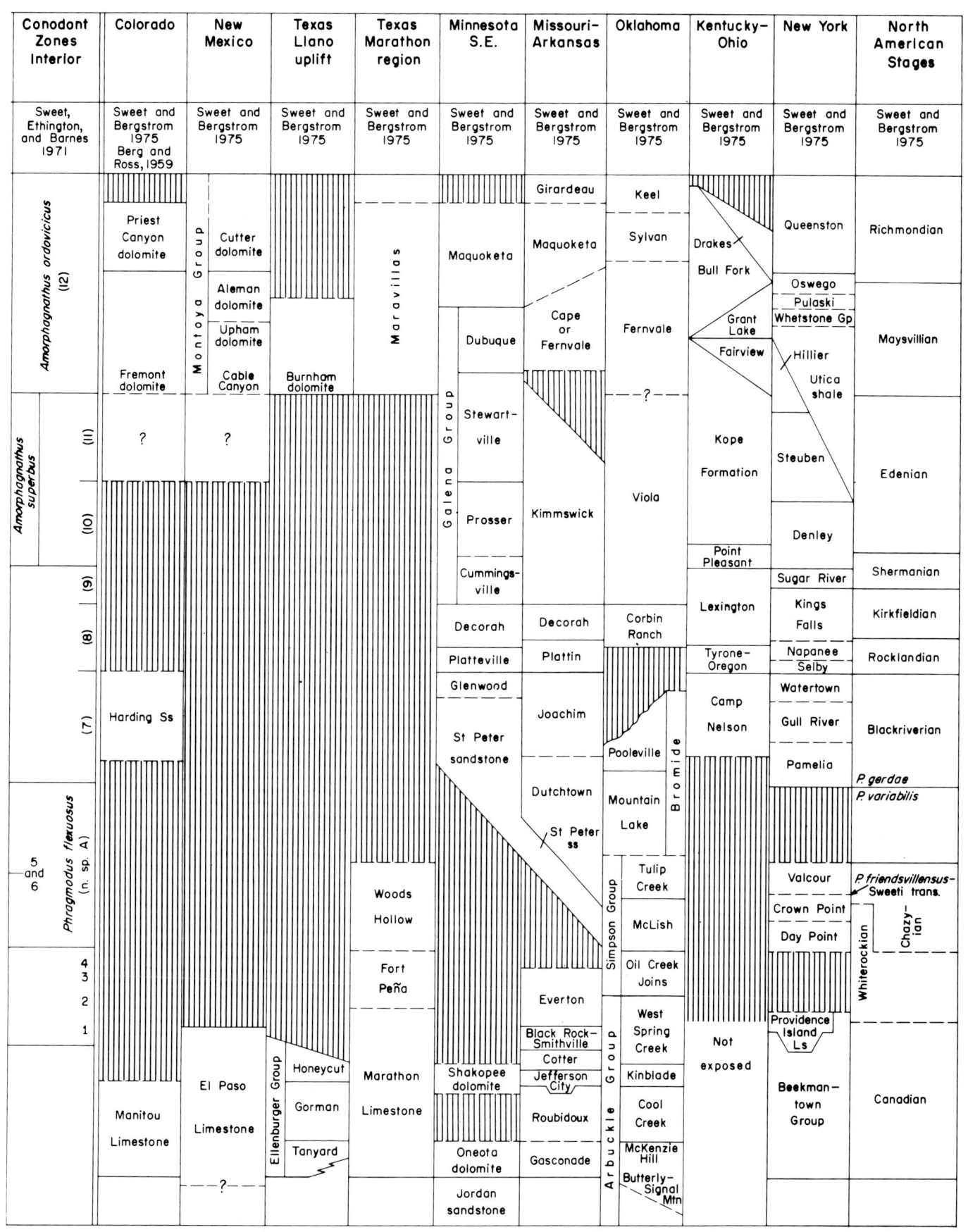

Conodont Zones Interior	Colorado	New Mexico	Texas Llano uplift	Texas Marathon region	Minnesota S.E.	Missouri-Arkansas	Oklahoma	Kentucky-Ohio	New York	North American Stages
Sweet, Ethington, and Barnes 1971	Sweet and Bergstrom 1975; Berg and Ross, 1959	Sweet and Bergstrom 1975	Sweet and Bergstrom 1975	Sweet and Bergstrom 1975	Sweet and Bergstrom 1975	Sweet and Bergstrom 1975	Sweet and Bergstrom 1975	Sweet and Bergstrom 1975	Sweet and Bergstrom 1975	Sweet and Bergstrom 1975
Amorphagnathus ordovicicus (12)	Priest Canyon dolomite; Fremont dolomite	Montoya Group: Cutter dolomite, Aleman dolomite, Upham dolomite, Cable Canyon	Burnham dolomite	Maravillas	Maquoketa	Girardeau; Maquoketa; Cape or Fernvale	Keel; Sylvan; Fernvale	Drakes; Bull Fork; Grant Lake; Fairview	Queenston; Oswego; Pulaski Whetstone Gp; Hillier; Utica shale	Richmondian; Maysvillian
Amorphagnathus superbus (11) (10)	?	?			Stewartville; Prosser	Dubuque; Kimmswick	Viola	Kope Formation; Point Pleasant	Steuben; Denley	Edenian; Shermanian
(9)(8)					Cummingsville; Decorah; Platteville; Glenwood	Decorah; Plattin; Joachim	Corbin Ranch; Tyrone-Oregon; Camp Nelson	Lexington	Sugar River; Kings Falls; Napanee Selby; Watertown; Gull River; Pamelia	Kirkfieldian; Rocklandian; Blackriverian
Phragmodus flexuosus (n. sp. A) (7)	Harding Ss			Woods Hollow	St Peter sandstone; Dutchtown; St Peter ss	Joachim; Mountain Lake; Tulip Creek; McLish; Oil Creek Joins; Everton	Pooleville (Bromide); Simpson Group			*P. gerdae*; *P. variabilis*
5 and 6; 4;3;2;1	Manitou Limestone	El Paso Limestone	Ellenburger Group: Honeycut, Gorman, Tanyard	Fort Peña; Marathon Limestone	Shakopee dolomite; Oneota dolomite; Jordan sandstone	Black Rock-Smithville; Cotter; Jefferson City; Roubidoux; Gasconade	West Spring Creek; Kinblade; Cool Creek; McKenzie Hill; Butterly-Signal Mtn (Arbuckle Group)	Not exposed	Valcour; Crown Point; Day Point; Providence Island Ls; Beekmantown Group	*P. friendsvillensus*-Sweeti trans.; Whiterockian; Chazyian; Canadian

states based largely on Bergstrom and Sweet's conodont zones.

Although the latitudinal position is established for the early Cincinnatian (Fig. 1b), estimating the North American position in the Early Ordovician (Canadian) requires that the Cambrian latitude be known. Available estimates of the Cambrian position are obviously uncertain (McElhinny, et al., 1974). However, in all estimates of Late Cambrian to Middle Ordovician latitudes the equator passes very close to the site of San Francisco; only the rotation of the North American continent varied as if pivoted at the site of San Francisco.

The position for the Canadian (Tremadoc-Arenig) is suggested by the distribution of carbonate rocks in the western United States.

Because the Ellenburger Dolomite in Texas shows evidence of hypersaline deposition (Folk in Barnes, et al., 1959, pp. 127-8, fig. 15), it was probably deposited in the southern horse latitudes; if North America is rotated to place West Texas at 20°S, the Northwest Territories will be at 20°N.

Dr. James Derby (pers. comm. March 1974) called my attention to Lower Ordovician dolomites of Ellenburger facies in the Northwest Territories of Canada. These Ellenburger-type sediments (referred to the "Cherty unit" by Macqueen, 1970, p. 229) include conodonts of the basal Ordovician *Mississquoia* zone at one locality and specimens of the uppermost Canadian gastropod *Ceratopea* cf. *C. unquis* Yochelson and Bridge at another; therefore a complete cratonic North American Lower Ordovician section is present. Furthermore the occurrence of *Ceratopea* in northern Greenland (Yochelson, 1964) would also be at latitude 20°N.

The Early Silurian position (Fig. 1d) as given by McElhinny and Opdyke (1973, Table 4) places the continent 15°-20° farther south than in the earliest Cincinnatian, and I have estimated a late Late Ordovician (Maysvillian?) position (Fig. 1c).

I have followed Bergstrom (1971, 1973) and Bergstrom, et al. (1973) in considering that most of the Whiterockian Stage based on the Antelope Valley Limestone of Nevada is a time unit, to which many of the Chazyan strata of New York and all the Marmorian rocks of Tennessee are equivalent. Few midcontinent units are currently viewed as correlative with this basic Whiterockian. Future revisions in midcontinent correlations may require adjustments to that point of view. Correlation of western stratigraphical units is compared in Fig. 3 with midcontinent and eastern units as interpreted by Sweet and Bergstrom elsewhere in this volume. [Paleontological Association Symposium, editor's note.]

CRATONIC AND INNER SHELF DEPOSITS

Canadian (Tremadoc-Arenig) Sediments (Figs. 4, 5).

Lithological and faunal contrasts. — Early Ordovician sedimentation was very different on opposite sides of the Transcontinental arch. On the shelf to the north limestone, much of it presumably of shallow-water origin, was deposited from southern California to the Dakotas (Hintze, 1951, 1953, 1973; Lochman, 1966; Lockman-Balk and Wilson, 1967; Ross, 1951, 1957, 1964a, 1970). Macrofossils include brachiopods, pelmatozoans, gastropods, cephalopods, and trilobites of great diversity. Bathyurid trilobites were abundant in the assemblage on the carbonate shelf. Westward in central Nevada olenid trilobites became increasingly abundant (Ross, 1975).

In contrast, lithologies south of the arch were predominantly dolomite, cherty dolomite and dolomitic limestone. This facies is represented in Texas by formations of the Ellenburger Group, in Kansas and much of Oklahoma by dolomite of the Arbuckle Group, in Missouri by the Gasconade, Roubidoux, Jefferson City, and Powell formations and in Minnesota by the Oneota and Shakopee dolomites. In eastern Texas the dolomite is interbedded with, and grades into, limestone but in western Texas evaporites are present in the dolomite (Barnes, et al., 1959, p. 57; Folk in Barnes, et al., 1959, p. 127) (Fig. 4). This enormous expanse of Lower Ordovician dolomite was formed subtidally in exceedingly shallow water, according to the theory of Harris (1973).

O. A. Wise, E. L. Yochelson and B. I. Clardy (unpub. data, 1974) state that the Black Rock Limestone is a lateral equivalent and member of the Smithville Formation in northern Arkansas. The Smithville is composed mostly of dolomite; among the fossils several genera of gastropods, including *Ceratopea*, trilobites, and problematical sclerosponges predominate. This is the lithological-fossil association of a restricted environment. The limestone of the Black Rock, on the other hand, is characterized by abundant brachiopods and by gastropods (*Ceratopea* is rare or absent), an assemblage requiring normal marine bottom conditions.

Webers (1972) showed that the Lower Ordovician Oneota and Shakopee dolomites of Minnesota were probably deposited under hypersaline conditions. Filter-feeding organisms made up less than 15% of the Oneota fauna and none of the Shakopee fauna; gastropods and cephalopods comprise most of the fossils.

The dolomitic strata of the Jefferson City, Cotter, and Powell formations around the Ozark uplift (Fig. 2) (Cullison, 1944) have yielded a composite fauna consisting at the generic level of gastropods (including *Ceratopea*) 27%, cephalopods 30%, trilobites 24%, brachiopods 12% and sponges 6%. That is 57% of this fauna is molluscan and only 18% is composed of sedentary or attached genera. Three of the four brachiopod genera occur in the uppermost Powell Formation and may be related to an influx of normal marine water like that producing the Black Rock Limestone in Arkansas. The fauna of the Ellenburger Group is dominated similarly by a variety of gastropods (especially *Ceratopea*) and cephalopods (Cloud and Barnes, 1948, pp. 113-7).

Hite (1970, 1973), modifying a theory of Fisher and Rodda (1968), has also related the formation of late Paleozoic and middle Mesozoic dolomite, chert and phosphorite to evaporites and refluxing brines in basinal environments. However, he believes that the chert and phosphorite precipitated where warm highly concentrated brine that was refluxing along the bottom met cool normal seawater. Essentially penecontemporaneous dolomitization of limy bottom sediments would have taken place closer to the sill of the basin.

In reviewing Cambrian and Early Ordovician sedimentation of the United States south-east of the Transcontinental arch, Harris (1973) applied Hite's theory to an enormously wide, shallow continental shelf and demonstrated that the widespread dolomites of the Early Ordovician could have resulted from refluxing of magnesium-rich brine in subtidal waters over a carbonate bottom.

Hanshaw, et al. (1971, p. 721) pointed out that "all modern dolomites are found in a dynamic environment" and that "given enough time and sufficient magnesium ion supply (i.e., an external reservoir and a hydrodynamic system), virtually any calcium carbonate terrane may be dolomitized." It is their belief that a more than ample supply of magnesium ions is available in seawater and that only a mechanism for keeping the water moving is essential for dolomitization. Hite's (1970) theory of refluxing brines, as modified by Harris (1973), satisfies such a requirement.

Effects of weather. — The contrast in sedimentation on opposite sides of the Transcontinental arch may be readily explained by a difference in climate. Because the shelf north of the arch was in the Ordovician tropical belt one may assume that warm marine waters retained normal salinity in an area of relatively heavy rainfall and low evaporation. The shelf south of the arch was in the Ordovician horse latitudes in an area of high evaporation and low rainfall, resulting in hypersaline water necessary to the theories of Hite and Harris.

Prevailing winds from the Ordovician south-east should have driven normal marine water over the wide southern shelf to replenish the surface water while the arch obstructed mixing with tropical waters. Similarly, prevailing winds from the Ordovician north-east should have moved normal marine water over a much narrower shelf in Montana, Wyoming, Utah and Nevada. Cool marine water in counterclockwise gyres should have washed the Ordovician south coast (Fig. 1b) while clockwise gyres of cool waters impinged on the Ordovician north and west coasts.

These considerations are basic and probably reasonable. How then can one explain the occurrence of Arbuckle Limestone bearing a normal varied fauna in southern Oklahoma well within the bounds of the restricted dolomitic tract?

Southern Oklahoma aulacogen. — In a study of basement rocks in southern Oklahoma, Ham, Denison and Merritt (1964, particularly pp. 142-61, pls. 4, 5) discovered that an elongated fault-bound trough, which they named the southern Oklahoma geosyncline, had come into being by Middle Cambrian time. In this trough the Arbuckle Limestone was deposited to a thickness slightly exceeding 5,000 ft (1540 m) (Ham, 1959), more than twice the thickness of the correlative strata of the Ouachita facies. Hoffman, et al. (1974) determined that this feature is not a geosyncline but a true aulacogen, which they rechristened the southern Oklahoma aulacogen. Although the aulacogen seems to fade out in the continental platform in south-eastern Colorado (King, 1969), in Early Ordovician time a sympathetic structure surely extended into Utah through west-central Colorado (Figs. 4, 5).

Isopachs of Ordovician formations in Colorado, as illustrated by Berg (1960), suggest either deposition in a narrow parallel-sided trough directly in line with the Oklahoma aulacogen, or preservation in a pre-Devonian structural trough. Within the trough a mixing of faunas is evident. R. L. Langenheim (Yochelson and Bridge, 1957, p. 291) collected near Pitkin, Colorado, a molluscan assemblage including the westernmost known specimens of *Ceratopea*. In Colorado's Manitou Limestone Berg and Ross (1959, Fig. 2) reported trilobites that characterize Lower Ordovician units in Utah, Nevada and British Columbia, but not units south of the Transcontinental arch.

In the southern Oklahoma aulacogen (Figs. 4, 5) the Arbuckle Limestone was deposited within the area of dolomitic sedimentation. Its area was at the same latitude and presumably in the same highly evaporative climate as the hypersaline shelf of western Texas. Unlike the restricted molluscan fauna of the dolomitic shelf, a diverse fauna occupied the limestone terrain of the aulacogen throughout the Early Ordovician. At least the east end, and possibly at times the entire length of the aulacogen was the site of an elongate embayment deep enough to be constantly supplied with normal marine water; such a trough could have been the passageway of normal water to the large evaporative areas of the shallower shelf.

Depth of water seems to have been at least as important as climate in determining type of carbonate sedimentation. The contrast between the deposits of the Cordilleran miogeosyncline, which extended from southern California to western Montana (Fig. 2), and the shelf south of the Transcontinental arch can be attributed to somewhat deeper water as well as to the tropical climate in the Cordilleran area.

We should note that Cooper (1956a) described a late Canadian brachiopod fauna from a borehole drilled in south-western Tennessee which included the genera *Archaeorhis*, *Diparelasma*, *Tritoechia*, *Hesperonomia*, *Orthambonites*, *Orthidium*, *Pomatotrema*, *Camerella* and *Leptella*. These brachiopods

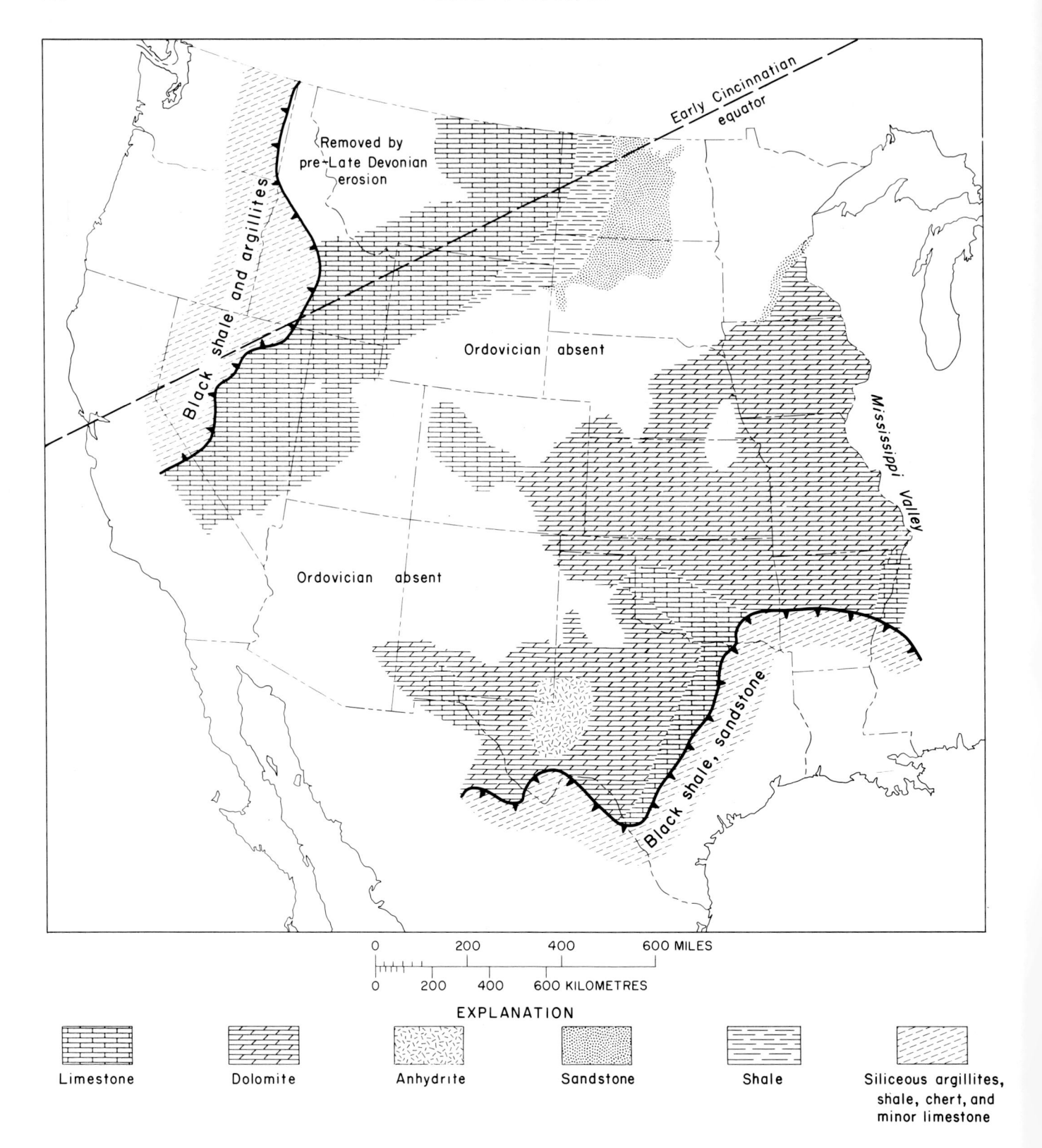

Fig. 4 — Map of Early Ordovician lithofacies of western United States.

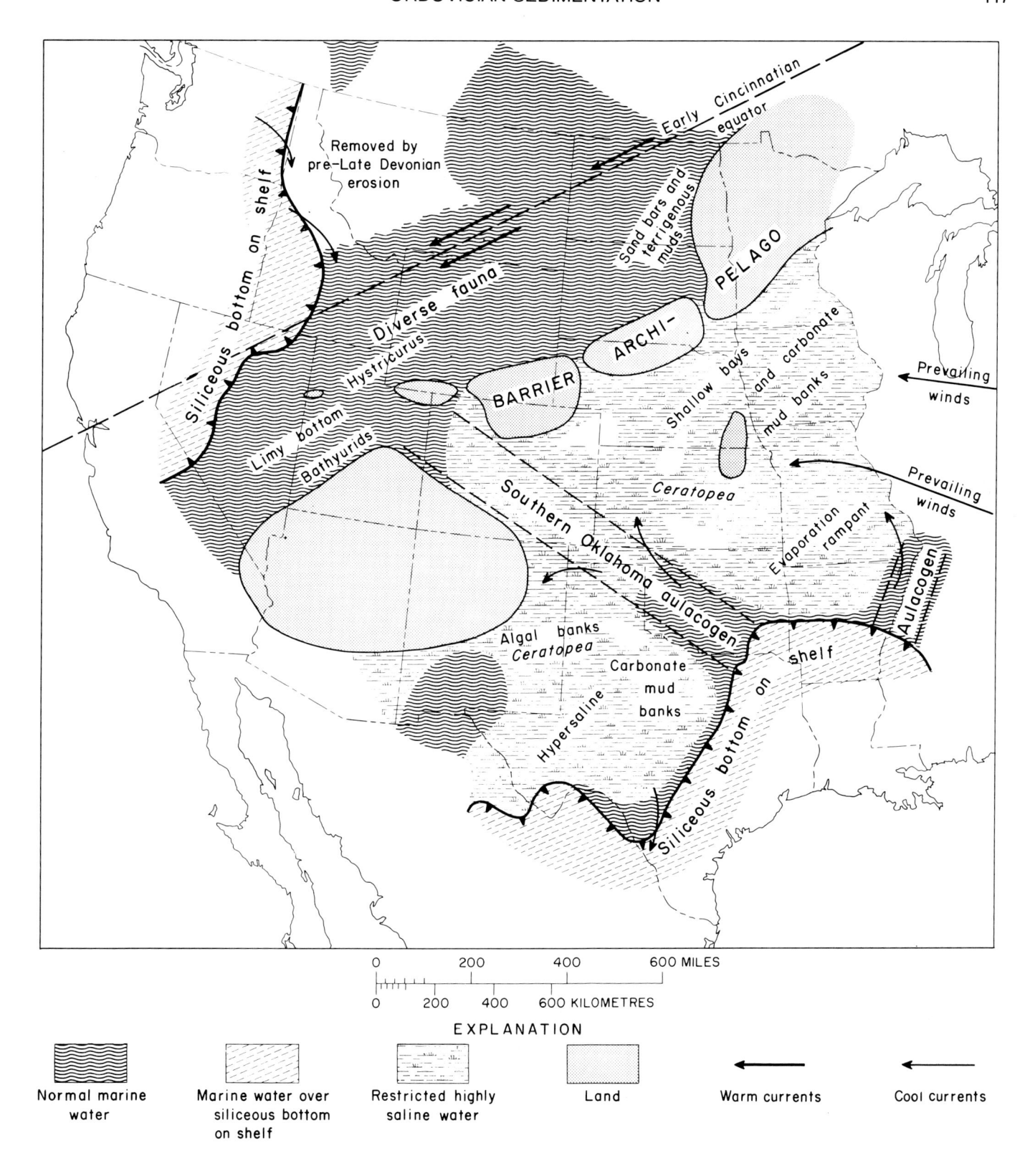

Fig. 5 — Interpretation of Early Ordovician palaeoenvironments, western United States.

characterize shelly limestone in the Arbuckle facies of Oklahoma and in the Pogonip Group of Utah and Nevada. A deep Late Cambrian and Early Ordovician basin existed in this part of the Mississippi River Valley (Bond, et al., 1971, fig. 17). This too may have been related to an early Palaeozoic aulacogen as yet undefined (Figs. 4, 5).

Middle Ordovician Quartzites (Figs. 6, 8).

The Middle Ordovician of the central and western United States is characterized by well-rounded, well-sorted quartz sands of great purity. Some of these sandstones and quartzites are diachronous; others are thought to be of different ages on the basis of regional biostratigraphy. The oldest are in the Whiterockian Oil Creek Formation of Oklahoma (Ham, 1955, pp. 28-30) and the Swan Peak Formation of Utah (Ross, 1951). Others are thought to be post-Whiterockian. Although some adjustments may be made eventually between units that are presumed to be post-Whiterockian and those that are Whiterockian, I have followed the conventional interpretation here. Despite differing ages, the origins of these quartz sands are related and merit separate note.

In the area of Michigan, southern Wisconsin, Iowa, Indiana, Illinois and Missouri (Dapples, 1955, Figs. 1, 7, 9) and Arkansas the sandstone unit is the St. Peter Sandstone. In Arkansas the Everton Formation is equivalent to slightly older units in Oklahoma where the sands are interbedded with limestones and shales as part of the Simpson Group. In the Williston basin of the Dakotas (Fig. 8) they are the sandstones of the Winnipeg Formation (Porter and Fuller, 1959, pp. 142-6; 1964, pp. 36-7; Fuller, 1961, pp. 1339-43; Ross, 1957, pp. 448-9). In the Bighorn Mountains of Wyoming (Ross, 1957, pp. 461-3) and the Front Range of Colorado the Harding Sandstone is a probable lithological equivalent. Farther west the Swan Peak Formation and the Eureka Quartzite of Utah and Nevada (Ross, 1951, 1964a, 1964b, 1967, 1970; Webb, 1956, 1958; Ketner, 1968) are the equivalents and, like the St. Peter, are diachronous. In central Idaho the same rocks are represented by the Kinnikinic Quartzite as restricted by Hobbs, et al. (1968). Norford (1969, pp. 25-7) has described the Tipperary and Mt. Wilson quartzites in the Rocky Mountains of Alberta and British Columbia.

Studies on direction of transport of the St. Peter Sandstone over most of its outcrop indicate movement from a northeasterly (easterly in Ordovician) source in a southwesterly (westerly in Ordovician) direction (Dapples, 1955, Fig. 9; Dott and Roshardt, 1972, p. 2593). However, Dott and Roshardt concluded that in southern Wisconsin net transport must have been almost due west (north-west in Ordovician), and, further, that the St. Peter was deposited as large sandwaves in about 40 m of marine water. Although such a conclusion may be considered, virtually nothing is known about the internal stratification of submarine sand bodies and so they cannot yet be meaningfully compared with aeolian dunes, which are being intensively studied (McKee, 1966).

Cross-laminations in the Mt. Wilson Quartzite in Alberta strongly suggest transport from the northeast (Norford, 1969, p. 26). Analysis of the Ordovician quartzites of the western Cordillera (Ketner, 1968) led to the conclusion that their source lay to the north in northern Alberta and that all could have been derived from erosion of Cambrian sandstones.

Agreement is general, therefore, that the Middle Ordovician sands, whether aeolian or marine, were derived from the direction of the Canadian Shield.

The areas of deposition in the early Middle Ordovician (Figs. 1a, 7, 9) would have been swept by trade winds blowing towards the north-west; based on today's orientation of North America these would be toward the south-west. In Cincinnatian time a different continental orientation would have resulted in the trade winds crossing the sites of St. Peter deposition in a direction that today would vary from southwest to due west. Deposition of sands in the Cordilleran geosyncline on the north side of the equator would have been effected by trade winds blowing towards the south-west; relative to today's orientation this would have been south and south south-westerly. Shallow-water currents capable of moving sand in an eperiric sea would have been driven by prevailing winds. Both the westerly movement of St. Peter sands in southern Wisconsin deducted by Dott and Roshardt (1972) and the southerly transport of Cordilleran sands inferred by Ketner (1968) are satisfied by these conditions.

The appearance of western United States in Middle Ordovician time may be portrayed by Shinn's (1973, particularly Figure 2) description of sedimentation along the lee coast of the Qatar Peninsula, eastern Arabia. Under the influence of the prevailing winds from the north-west barchan dunes of quartz sand migrate across the sabkha to the shore of the Persian Gulf. From the shore some of the sand is carried downslope and some is moved laterally by longshore currents.

Ketner (1968) believes that the Middle Ordovician sands were derived from pre-existing disaggregated Cambrian sandstones. The great expanses of terrain composed of crystalline rock of the Canadian Shield could have yielded great quantities of quartz if deeply weathered. A present-day example of such conditions is the exceptionally pure glass sand deposited near the coast of Liberia and derived from upcountry granitic and gneissic rocks (Rosenblum and Srivastava, 1970). This coast is between 6° and 7°N latitude and at present has one of the wettest climates on earth.

The Canadian Shield from the west shore of Hudson Bay to the Mackenzie delta (Fig. 2) would have been in the tropics in Early Ordovician time; even greater expanses of the shield and northern United States were within tropical latitudes by Cincinnatian time. However, in the Early Ordovician (Fig. 1a), no obvious source of quartz sand would have been available to the then-existing trade winds along the Appalachian coast. Only a change in orientation of the continent (Fig. 1b) brought a combination of wind direction and source area that could account for spread of the Middle Ordovician sands. Probably, much of the detrital quartz was recycled several times from late Precambrian time until Middle Ordovician deposition, accounting for the sands found almost throughout the Cambrian and Lower Ordovician.

Whiterockian (Late Arenig-Earliest Llandeilo) Sediments (Figs. 6, 7).

The cratonic shelf areas south of the Transcontinental arch show evidence of erosion of carbonate terrain toward the close of and after Canadian time. In contrast, north of the arch deposition was essentially continuous after Canadian time in miogeosynclinal areas of Utah and Nevada. Stratigraphical sections in the resulting Antelope Valley Limestone of Nevada are the bases for the Whiterockian Stage of Cooper (1956b) and of this paper.

South of the arch evidence for a drop in sea level and erosion of Lower Ordovician dolomite formations has been demonstrated in New Mexico (Flower, 1969, Fig. 2), in northern Arkansas (Young, et al., 1972, p. 70; Suhm, 1974, pp. 686-90, Fig. 7), and in Texas (Barnes, et al., 1959, pp. 37-42). Whether this erosion took place pre-Whiterockian or prior to St. Peter deposition, or both times, is not everywhere easily determined. Nor was erosion simultaneous over the entire area. Evidence for erosion beneath the St. Peter Sandstone in southern Minnesota is not compelling (Austin, 1972, pp. 467-8), and yet deep channeling is present in Arkansas (Suhm, 1974, Fig. 5). Most of the cratonic shelf area south of the Transcontinental arch was subjected to erosion at least once and possibly continuously during Whiterockian time.

To the north of the Transcontinental arch in northern Utah the carbonate deposition of the Garden City Formation continued into Middle Ordovician time (Ross, 1951, pp. 31-3). In the Ibex area of western Utah the limestone of the Early Ordovician Wahwah Formation is succeeded by the Juab Limestone of earliest Middle Ordovician age (Hintze, 1953, p. 19; Jensen, 1967, pp. 72-4); these two units are distinguished as easily on faunal as on lithological grounds. In Nevada, except for deposition of the calcareous Ninemile Shale in the late Lower Ordovician, the formation of limestone continued in Whiterockian time as the Antelope Valley Limestone. This Limestone seems to have been deposited into Porterfieldian (Llandeilo) time in the Egan Range (Ross, 1970, pp. 37-41), at Meiklejohn Peak (Ross, 1970, p. 43) and in the Ranger Mountains, Nevada Test Site (Ross, 1970, pp. 43-4).

In the more north-easterly sections of the Cordilleran miogeosyncline carbonate sedimentation was temporarily masked by an incursion of mud (lower Swan Peak Formation in central-northern Utah and Kanosh Shale in central-western Utah) and then smothered by the first incursions of quartz sand (Kinnikinnic, Swan Peak and lower Eureka quartzites).

Within the great area south of the Transcontinental arch which was subjected to post-Canadian erosion, the southern Oklahoma aulacogen persisted as a tectonically negative area. Normal marine water was inhabited by an abundant and varied fauna of the Simpson Group, virtually identical to that in the Utah-Nevada area. As noted by Derby (1973, p. 26) deposition in the aulacogen may have been interrupted at the close of the Canadian only by the span of the *Orthidiella* Zone at the beginning of the Whiterockian Stage. The limestone of the various formations of the Simpson Group within the Oklahoma aulacogen are much like those of the Antelope Valley Limestone of Nevada and the correlative formations of the Ibex area of Utah. However, the quartzites in Oklahoma intertongue with the calcareous units (Ham, 1955, Fig. 10) and did not terminate carbonate deposition prior to the Cincinnatian. The higher percentage of terrigenous components in the Oklahoma carbonates than in Utah-Nevada formations is obvious in Fig. 6.

The interplay between shape and orientation of the embayment of the aulacogen, direction of prevailing winds, position of the early Ozark uplift, and resulting longshore currents may have caused a more complex intermixing of terrigenous and calcareous sediments south of the arch than in the Cordilleran area. Indeed, wind and currents may have transported sands well offshore to form deposits like the Blakely Sandstone of the Ouachita facies.

While deposition continued in the Cordilleran miogeosyncline and in the Oklahoma aulacogen, and on the southern shelf in Arkansas, and Texas, the Lower Ordovician carbonate sediments that composed the surface of most of the interior area were subjected to weathering and erosion. Some researchers may conclude that through the action of groundwater limestone was converted to dolomite during this time of Whiterockian weathering. Although some such conversion may have taken place it could not have produced the kind of faunal restriction in effect during original deposition.

The question of the length of time during which the interior terrain was subjected to weathering and erosion is not settled. Biostratigraphical work on conodonts (Bergstrom, 1971, 1973; Bergstrom, et al., 1973; Carnes and Bergstrom, 1973; Ethington and Schumacher, 1969; Sweet and Bergstrom, 1971, 1973; Sweet, et al., 1971) indicates that these fossils, like trilobites, show differing biogeographical assemblages in the central interior and near continental margins. Some conodont faunas, like some brachiopods from strata representative of presumably

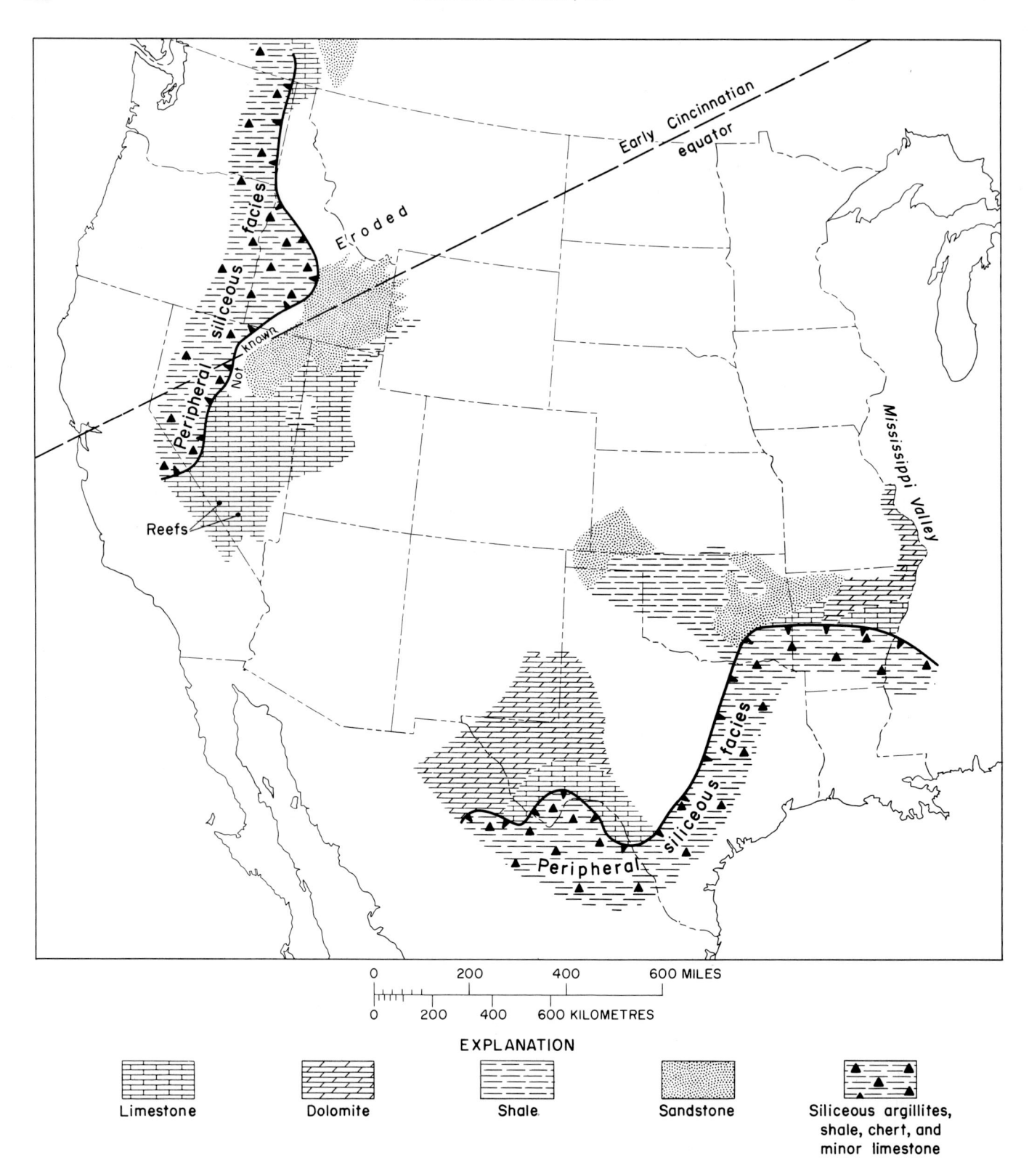

Fig. 6 — Map of early Middle Ordovician (Whiterockian) lithofacies, western United States.

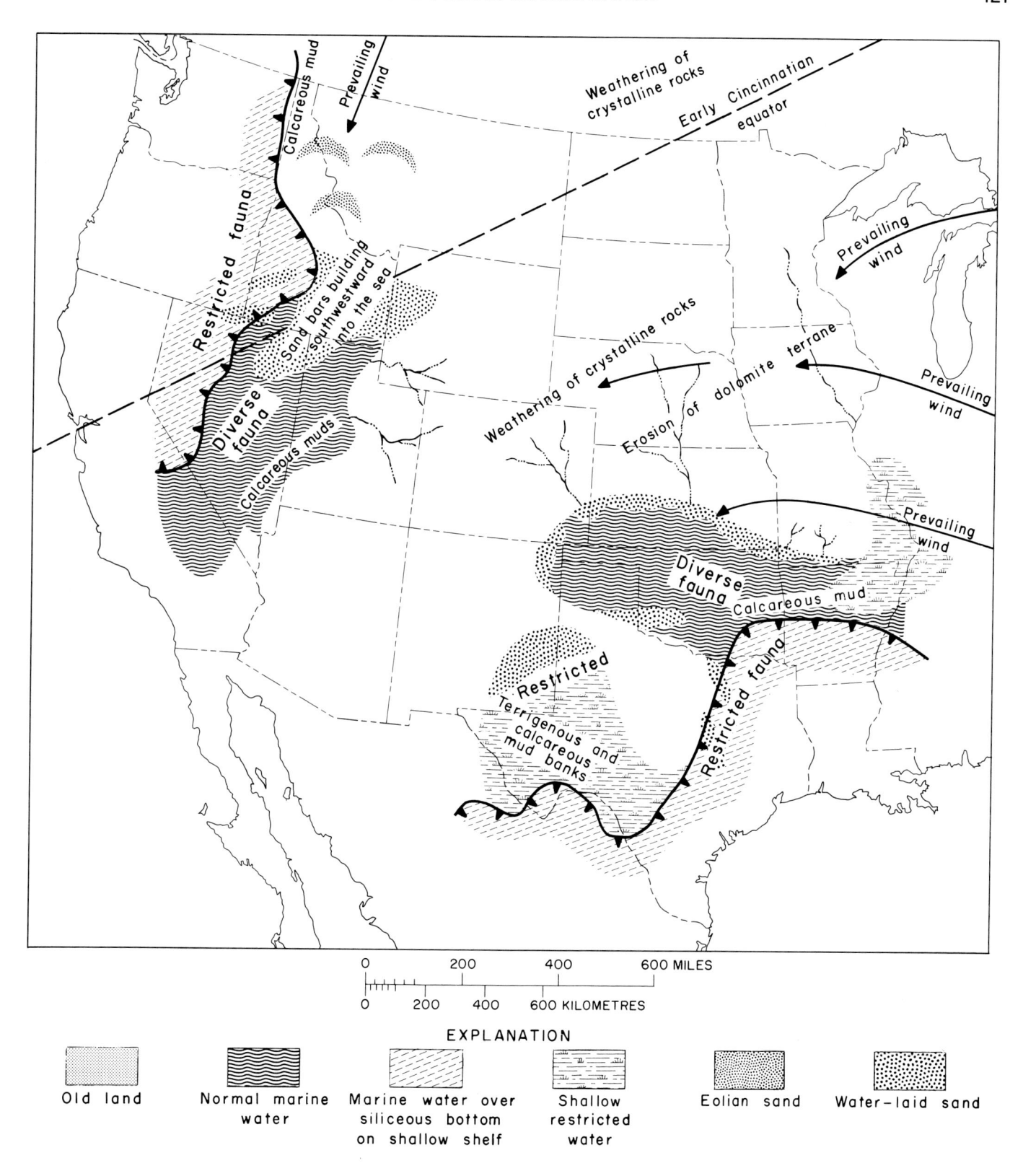

Fig. 7 — Interpretation of early Middle Ordovician palaeoenvironments, western United States.

different stages, are in fact wholly or partly correlative; examples are Whiterockian with Marmorian, late Whiterockian with early Chazyan, and Ashbyan with Porterfieldian. Lithostratigraphical units of the eastern and central interior that are now considered younger than Whiterockian may also prove to be correlative.

Post-Whiterockian Pre-Cincinnatian (Llandeilo-Early Caradoc) Sediments (Figs. 8, 9).

Transgressive sedimentation, initiated by the St. Peter Sandstone, indicates a rise in sea level south of the Transcontinental arch in the continental interior during post-Whiterockian time. With rising sea level sedimentation north of the arch was progradational.

Because the relief and the gradient on the eroded surface of the Lower Ordovician dolomite were low, a slight rise in sea level resulted in widespread and rapid invasion of the platform south of the arch. Reworking of windblown sand by the advancing sea resulted in deposition of the St. Peter Sandstone. Deposition of shaley limestone and calcareous muds in very shallow water followed, as shown by Young, *et al.* (1972) for Arkansas and by Austin (1972) for southern Minnesota. The faunas of the Glenwood-Platteville and Decorah Formations, overlying the St. Peter, are varied and include abundant brachiopods and bryozoans (Webers, 1972), indicators of well-circulated normal marine waters.

North of the Transcontinental arch the south-westward spread of quartz sand that had reached southern Idaho and northern Utah by the end of Whiterockian time continued relentlessly. Formation of carbonate banks bearing an abundant and varied fauna continued as sea level rose along the miogeosyncline from southern California to British Columbia. But as quartz sands of the Kinnikinnic, Swan Peak and Eureka formations built south-westward, more and more carbonate sedimentation was smothered. As long as the Canadian Shield was exposed to the sweep of the prevailing winds from the northeast the sand seems to have been abundantly furnished. Only the inundation of the Shield in Cincinnatian time stopped this supply of wind-blown sand.

The Copenhagen Formation in the Antelope and Monitor Ranges of central Nevada (Ross and Shaw, 1972) is the youngest stratigraphical unit beneath the regressing sandstones. It covers an age span from Llandeilo to late Caradoc. At Lone Mountain, north-east of the Monitor Range, Ethington and Schumacher (1969, pp. 445-8) found that calcareous sandy beds beneath the Eureka Quartzite proper contain conodonts typical of the lower half of the Copenhagen; presumably sand buried the Lone Mountain area while the upper Copenhagen beds were deposited in the Monitor and Antelope Ranges. Seemingly the sand failed to reach the Toquima Range, where no quartzite overlies the Caesar Canyon beds. But sand did reach the Toiyabe Range where it occurs on the south side of Wall Canyon.

Cincinnatian (Middle Caradoc to Ashgill) Sediments (Figs. 1b, 1c, 10, 11).

In the three preceding time intervals there were marked contrasts between sedimentation on opposite sides of the Transcontinental arch. In late Middle and Late Ordovician time, however, the pattern changed to one of bilateral symmetry with the arch providing the axis.

Dolomite is widespread on both sides of the arch (Fig. 10). Similar deposits formed across much of the Canadian Shield. Roehl (1976, p. 2020) has shown that "intertidal and infratidal deposits across a shallow submerged mudbank" account for the Stony Mountain Formation of the Dakotas and north-eastern Montana. Roehl (1967, p. 1985-8) attributed the difference between western and eastern faunas more to latitudinal control than to barriers and suggested that carbonate banks restricted the circulation of the extremely shallow water. He further noted that transgressive facies seem to have spread very abruptly across low topography.

Probably Roehl's conclusions are applicable to the Whitewood (South Dakota), Bighorn (Wyoming and Montana), Fremont (Colorado), Montoya (New Mexico), Fish Haven (Utah, Idaho) and Ely Springs (Nevada) formations although almost no detailed lithological analyses of a similar nature have been published on them. Despite the multiplicity of names these formations are essentially the same as the Red River and Stony Mountain formations of the Williston Basin.

The occurrence of the Fremont Dolomite in west-central Colorado (Sweet, 1954; Berg, 1960) as well as in diatremes in southern Wyoming (Chronic, *et al.*, 1969) suggests that the dolomite units covered the site of the Transcontinental arch completely in the greatest inundation in North American history.

Fuller (1961, p. 1347) noted that the lower Red River Formation includes limy beds containing a varied fossil fauna. In contrast to this indication of well-circulated marine water at the beginning of the inundation, cyclic evaporites in the upper Red River and Stony Mountain formations (Fuller, 1961, pp. 1347-51) show that there was restricted circulation of a high rate of evaporation influencing carbonate deposition. The occurrence of *Receptaculites* (probably an alga) and the large gastropod *Maclurites* as probable benthos, together with ostracodes, some trilobites and cephalopods as surface-water nekton presents a pattern almost the same as the lithological and fossil association for the Lower Ordovician Oneota and Shakopee dolomites (Webers, 1972). Corals and brachiopods may have had wide geographical but limited stratigraphical distributions and probably indicate times when the abrupt transgressions of normal marine water postulated by Roehl (1967) permitted the spread of attached bottom-dwelling forms. Such animals could also have lived high on the sides of carbonate banks above the reach of refluxing brines.

The occurrence of evaporites on the equator is seemingly anomalous. Almost certainly the Red River Formation is Cincinnatian. Assuming that the position of the equator determined by using the palaeomagnetic findings of McElhinny and Opdyke (1973) for the Denley and Steuben limestone is correct, I conclude that because of the low gradient, the sea bottom was submerged over an enormous area to such a shallow depth that tropical evaporation, possibly assisted by restriction of circulation during cyclical drops in sea level related to Gondwana glaciation, could produce evaporites.

The Canadian Shield, like the continental interior to south, was inundated; over the area of British Columbia and Alberta deposits were both limestone and dolomite. But in much of eastern Canada and in the present area of the Mississippi Valley (Fig. 10) limestone prevailed. Along the Ordovician equator dolomitic carbonates graded northward to limestone in the Hanson Creek Formation in Nevada and in some facies of the Saturday Mountain beds of Idaho, possibly because of increased depth of water. If the equator lay as indicated in Fig. 10 the area of the southern Oklahoma aulacogen and the lower Mississippi River valley should have been in the horse latitudes and subject to high rates of evaporation. By maintaining a greater water depth in these two areas deposition of limestone could have continued. Even if allowance is made for a possible southward drift of the continent during Cincinnatian time accompanied by change in latitudinal belts, depth of water may have been just as important as climate in determining the kind of carbonate deposition.

The possible effect of climate must not be minimized. The possibility must also be considered that the palaeomagnetic data (McElhinny and Opdyke, 1973) pertain to Edenian rocks which are somewhat older than the Red River western dolomite beds. The continent was considerably farther south in Early Silurian time (Fig. 1d), and so the zone of greatest evaporation should have been very close to the United States-Canada boundary; that position agrees very well with occurrences of Silurian salt deposits. Assuming that the movement of the continent was regular, rather than spasmodic, Fig. 1c may represent the position of North America in much of Cincinnatian time. If the assumption is correct we should expect more limestone than dolomite in western Canada and some indication of increased evaporation and greater restriction of marine waters in the upper Mississippi Valley toward the close of Ordovician sedimentation. Faunal evidence for southeastern Minnesota shows a drop in numbers of species after Edenian (Trenton) time (Webers, 1972, figs. VI-28, VI-29) but a slight increase in numbers within the Maquoketa Shale, particularly in that the fossils of the uppermost "Cornulites zone" (Ladd, 1929, pp. 391-5) of Iowa indicate a brief return to normal marine conditions. A similar, possibly contemporaneous marine invasion nurtured the varied fauna of the Stony Mountain shale from Manitoba to northern Wyoming (Ross, 1957), and perhaps of the Aleman Member of the Montoya Formation of New Mexico.

Let us assume that possible brief marine invasions and deepening water could not have taken place, that the Early Ordovician (Canadian) consumed at least one-third and probably nearly one-half of Ordovician (60 million years) time and that the Cincinnatian accounted for half the remainder; then we must conclude that North America moved through 20° of latitude in 15 million years. This is a rate of 2,200 mi (3540 km) in 15 million years or 23.6 cm/yr — twice the generally accepted rate of plate movement in the Cretaceous to Holocene. Therefore, 1) our estimate of the length of Cincinnatian-Early Silurian time may be too short, or 2) dating of the Steuben and Denley limestones should be considerably greater than the Red River-Stony Mountain beds of the Williston Basin, or 3) the palaeomagnetic data are in error. Obviously the palaeotectonic interpretation of Late Ordovician time in North America requires more research along several lines.

PERIPHERAL SILICEOUS FACIES

Geographical distribution. — The Ordovician rocks of the western United States form the south-west quadrant of a whole that is centred on the Precambrian North American craton. Close to the centre the rocks are mainly dolomite and limestone but peripherally they are black shale, siliceous argillites, bedded chert, minor limestone and a few greenstones. Many of the shales contain graptolites.

In Arkansas and south-eastern Oklahoma the peripheral rocks are part of the Ouachita facies (Flawn, et al., 1961; Ham, 1959). Westward, that facies is continuous into the Marathon district of western Texas (Wilson, 1954; McBride, 1970; King, 1937).

On the west the corresponding facies is mapped as the Vinini Formation or the Valmy Formation in north-eastern Nevada or the Zanzibar Limestone and Toquima Formation in the Toquima and Toiyabe Ranges (Ferguson, 1924), the Ledbetter Slate of eastern Washington and the Phi Kappa Formation of central Idaho (Ross, 1961). In addition, a more westerly belt of oceanic Ordovician rocks may have passed through the present site of the Klamath Mountains (Boucot, et al., 1973, pp. 14, 15; Berry, et al., 1973); these rocks, however, may have been added to North America at a much later date (Condie and Snansieng, 1971). The westerly belt is continuous with similar deposits in south-east Alaska. The rocks of the more easterly belt are continuous with the Glenogle Formation of British Columbia.

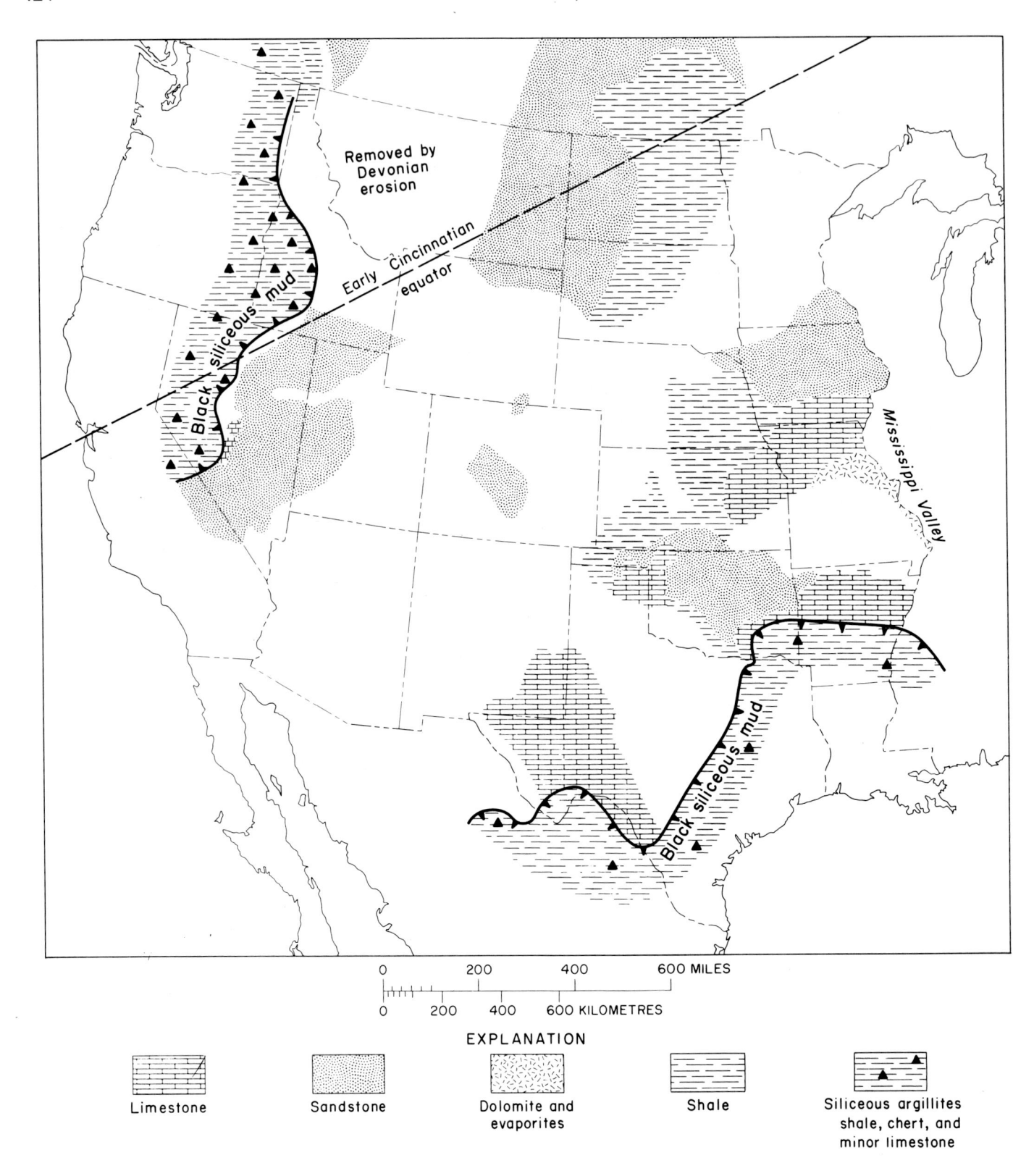

Fig. 8 — Map of post-Whiterockian and pre-Cincinnatian lithofacies, western United States.

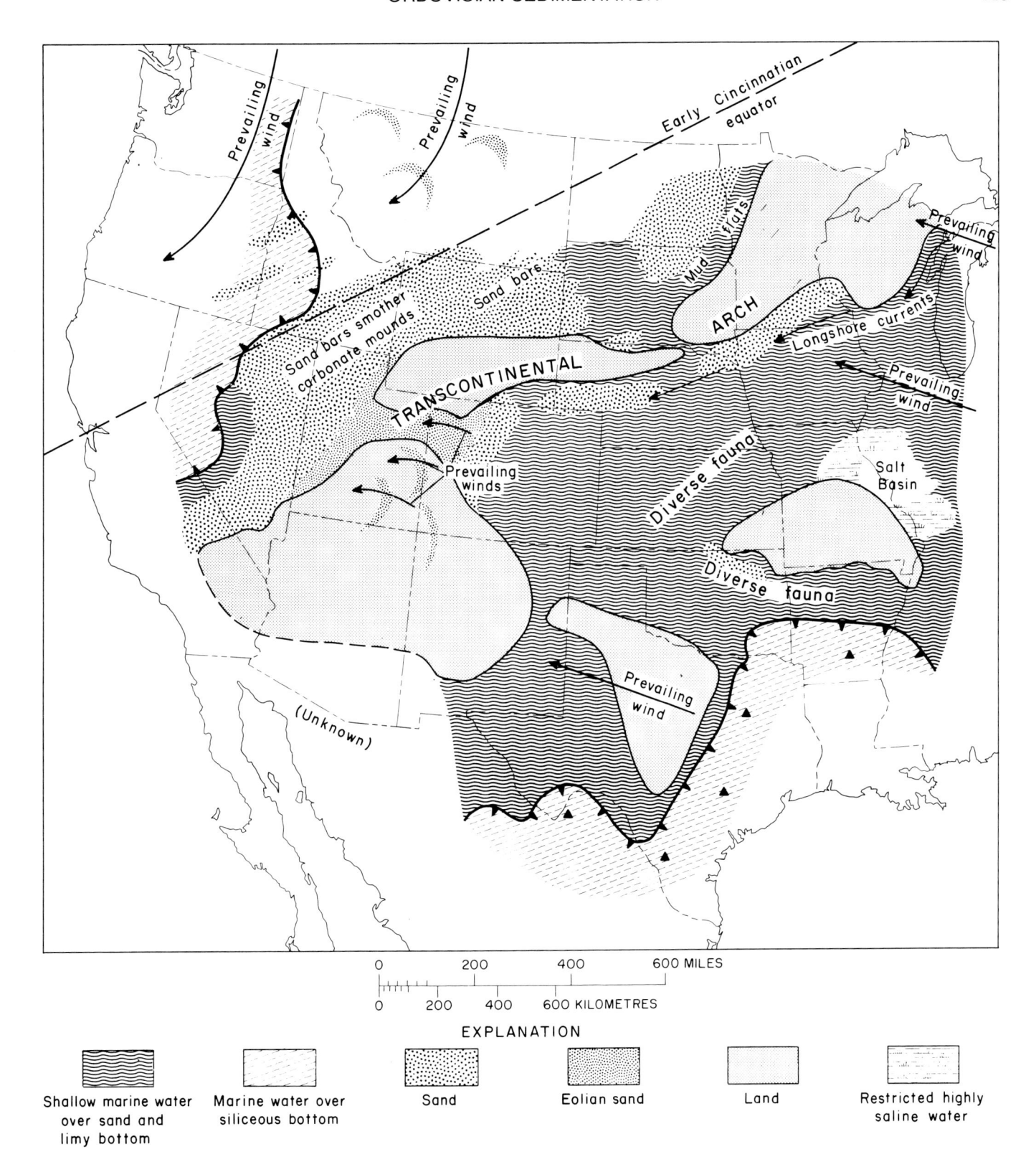

Fig. 9 — Interpretation of Middle Ordovician (post-Whiterockian and pre-Cincinnatian) palaeoenvironment, western United States.

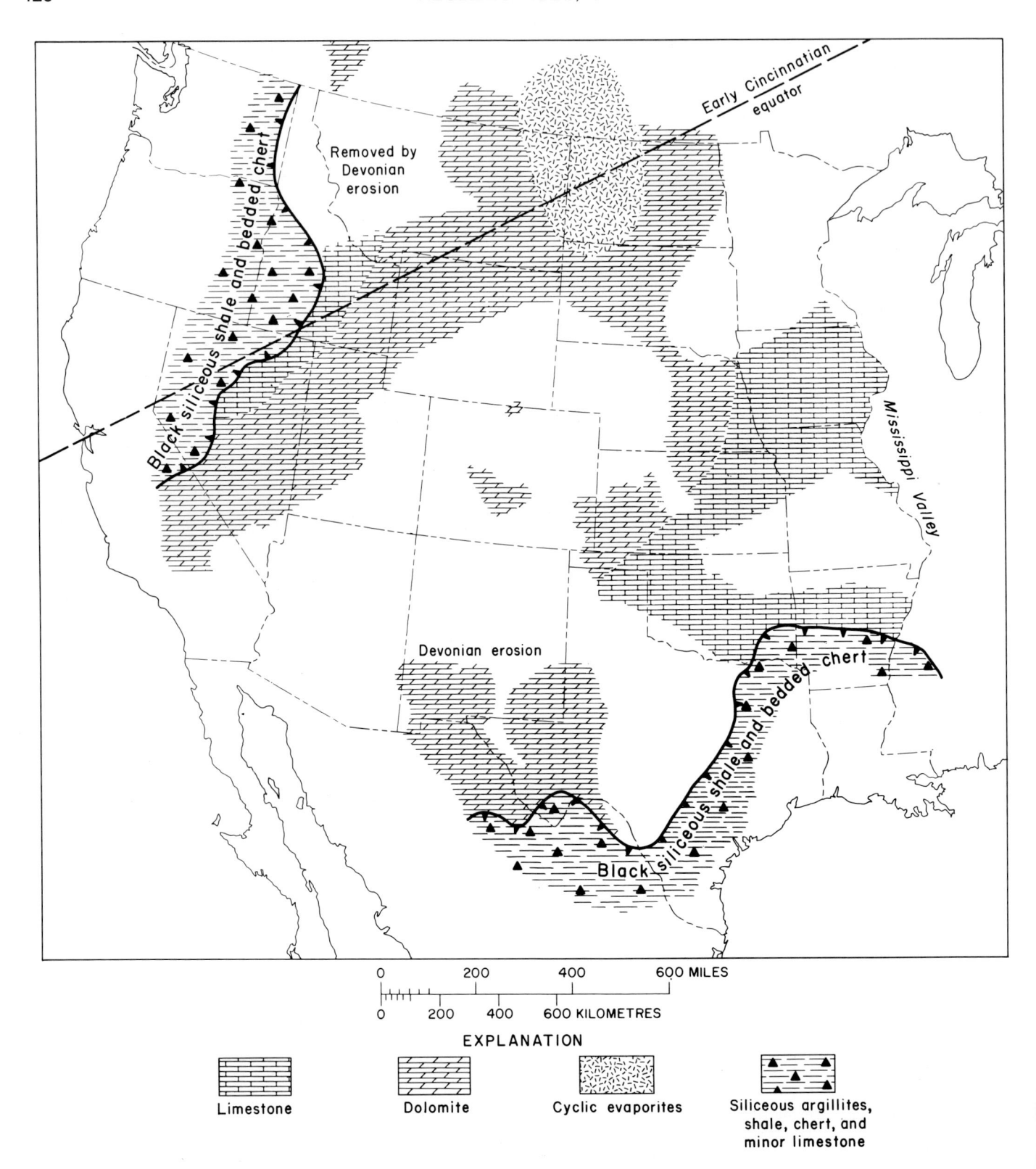

Fig. 10 — Map of Cincinnatian lithofacies, western United States.

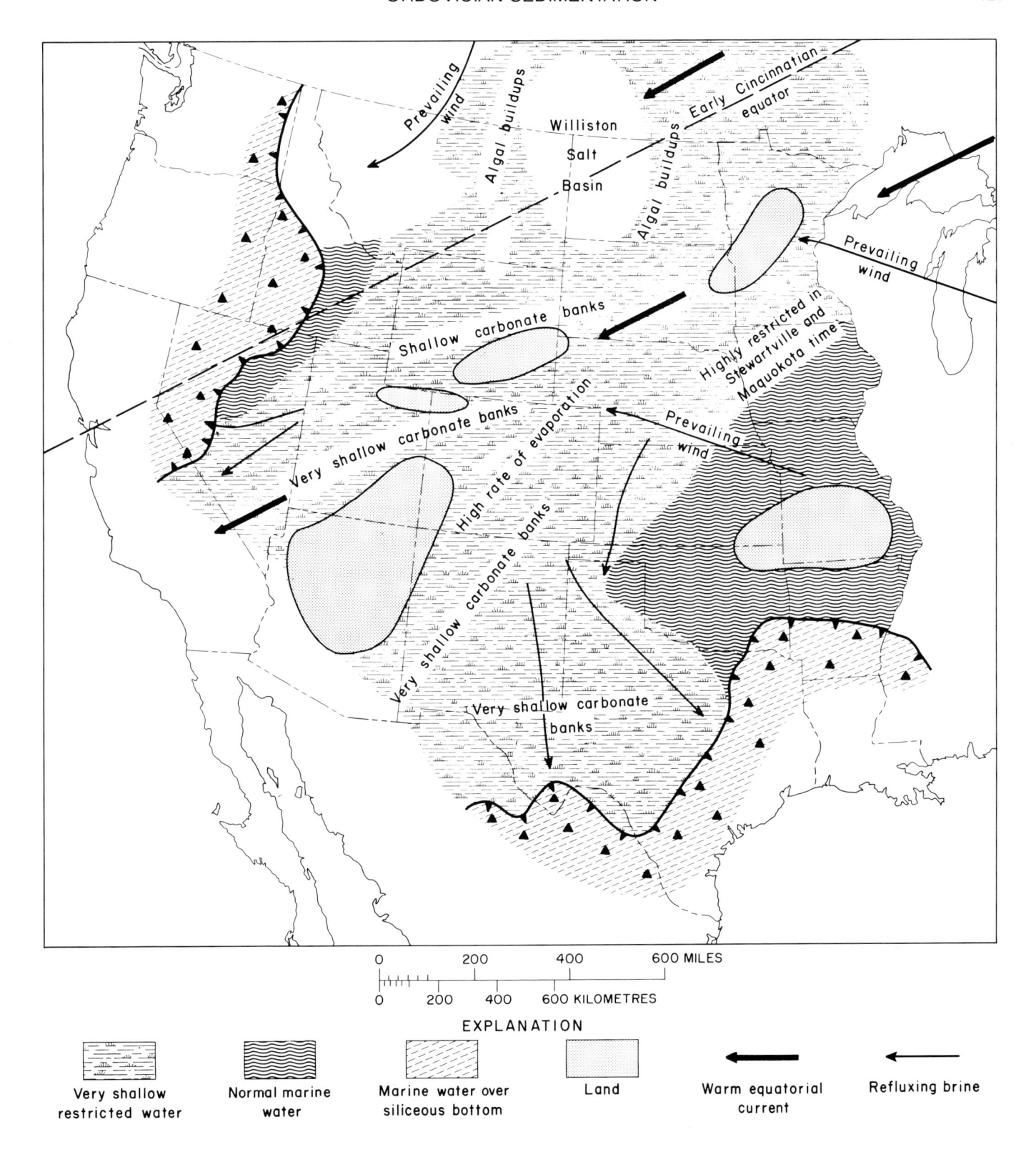

Fig. 11 — Interpretation of Cincinnatian palaeoenvironments, western United States.

Most palaeogeographical maps of the Ordovician of North America are vague about Mexico. Furthermore, the Cordilleran geosyncline is commonly shown trending south-westward into unknown reaches of the Pacific Ocean. Poole and Hayes (1971) proposed that the early Palaeozoic geosyncline of Nevada and California was cut by the San Andreas fault zone and displaced, but that the continuation is present in north-west Sonora near Caborca. Flawn (pp. 99-106 in Flawn, et al., 1961) traced the Ouachita facies south-westward from Texas into Mexico for considerable distances. It seems likely that the peripheral siliceous Ordovician facies passes through central Mexico connecting the Ouachita and Vinini-Valmy facies.

The Lower Ordovician of Oaxaca, Mexico (Robison and Pantoja-Alor, 1968) may prove to be an outer shelf deposit of the North American miogeosyncline although it may, like Florida, be part of Gondwana attached during a later collision and left behind. If the Oaxaca Ordovician was indeed part of the North American outer shelf as the faunas suggest (Yochelson, 1968; Flower, 1968c), the graptolitic shales of the siliceous facies should have passed around them on the west, south and east.

Stratigraphical distribution. — The distribution of chert in Ordovician sediments of the west differs markedly with time. In the states east of the Transcontinental arch chert seems to be very common in Lower Ordovician dolomite on the shelf and platform; the peripheral siliceous facies is composed mainly of shale and limestone in West Texas (King, 1937, pp. 28-32; Wilson, 1954, p. 2459) and of shale and sandstone in the southern Arkansas — south-eastern Oklahoma area (Ham, 1959).

West of the Transcontinental arch in Nevada, Utah, southern Idaho and in the Williston basin (eastern Montana, western North Dakota and western South Dakota) the Lower Ordovician is limestone. Chert is common in the very low part of the Goodwin Limestone (Ross, 1964, pl. 1; 1967, pl. 11; 1970, pls. 20, 21) in Nevada and in the House Limestone (Hintze, 1973, chart 29, p. 146) in westernmost Utah. In north-central Utah (Ross, 1951) black chert is abundant in a thin zone at the top of the Lower Ordovician Garden City Limestone. Despite the great expanses of the peripheral siliceous facies exposed in Nevada only about one-tenth of the strata can be identified as Lower Ordovician; these strata are mainly shale and argillite.

Evidently therefore, deposition of chert was restricted to the carbonate shelf in the Early Ordovician where it was associated with dolomite south of the Transcontinental arch and with limestone on the north.

By middle Caradoc times bedded chert and siliceous argillite constituted a very large proportion of the peripheral facies. Maximum transgression in the Cincinnatian seems to correlate with the deposition of siliceous sediments on the outer shelf. Paradoxically, the greatest marine transgression, the greatest development of dolomite in a restricted environment, and the greatest production of chert are intimately related. The temporal relationship is obvious; not so obvious, if the Hite-Harris theory of infratidal refluxing of brines is correct, is the manner in which the silica-bearing brines bypass the belt of limestone in order to reach the peripheral facies without altering the limestone.

Thickness. — The dimensions and origins of the peripheral siliceous facies are controversial. That the deposits in the Ouachita facies are comparable with those of Nevada seems obvious and yet the differences of interpretation are mostly regional and resemble those involved in the Taconic controversy.

The Ordovician formations of the Ouachita facies in Arkansas are (Ham, 1959, Fig. 1, p. 77-82):

Polk Creek Shale	100 ft	31 m
Bigfork Chert	600-700 ft	213 m
Womble Shale	600 ft	183 m
Blakely Sandstone	400 ft	122 m
Mazarn Shale	1,000 ft	305 m
Crystal Mountain Sandstone	300 ft	91 m
Collier Formation	300 ft	91 m
Lukfata Sandstone	150 ft	46 m
Total	3,550 ft	1,082 m

The total thickness is a probable maximum for the Ordovician of the Ouachita facies. These formations are confidently correlated with a carbonate sequence in the southern Oklahoma aulacogen to the west; the carbonate sequence reaches a total thickness of almost 9,000 ft (2743 m) — far greater than that of the supposedly eugeosynclinal units.

In the Marathon district, western Texas, King (1937, pp. 25-47, pls. 23, 24) described and mapped five formations aggregating about 2,000 ft (600 m):

Maravillas Chert	240-390 ft	(73-119 m)
Woods Hollow Shale	470 ft	(143 m)
Fort Peña Formation (limestone and chert)	212 ft	(65 m)
Alsate Shale (shale and to south limestone; to north saccharoidal quartz sandstone)	100-145 ft	(31-44 m)
Marathon Limestone (limestone, shale, dolomite)	815 ft	(248 m)

These formations are overlain by the Devonian Caballos novaculite of a facies similar to the Maravillas Chert. The Ordovician of the Marathon district is part of the Ouachita facies (King, 1951, pp. 148-57).

The source of the clastic components of the Ouachita

facies, such as the Blakely Sandstone, which thins northward in Arkansas, seems to have been somewhere off the present-day Gulf Coast. King (1937, pp. 44-5) and Flawn (*in* Flawn, *et al.*, 1961, p. 51) noted that limestone decreases and shale and sandstone increase in abundance towards the south-east in the Marathon district, and they inferred a land source area in that direction. Ham (1959, p. 84) also discussed the northward and westward thinning of Ordovician sandstone and concluded that the source for such lithologies must have been to the south-east of the Oklahoma-Arkansas area.

However, Wilson (1954, p. 2469) noted that boulders in a conglomerate within the Maravillas Chert are composed of calcareous rocks found only to the north of the Marathon-Ouachita facies. Nonetheless, he also concluded (pp. 2471-2) that the increase in sand toward the south in the Marathon-Solitario region of Texas, like the southward increase within the Ouachita sequence, must indicate a southerly source. However, as suggested earlier, reworking of aeolian sand into banks and bars in shallow water may account more readily for such sand bodies in the peripheral facies.

The distribution of graptolite-bearing eugeosynclinal rocks of the western Cordillera was reviewed by Ross (1961), with emphasis on the graptolites rather than on lithology. Like the siliceous rocks of the Marathon and Ouachita regions of Texas and Arkansas, the "western assemblage" of Nevada and adjoining states is composed of black shales, siliceous argillites, bedded cherts, quartz sandstone and minor amounts of limestone. Greenstone and andesitic volcanic rocks are also present (Roberts, *et al.*, 1958, p. 2832). The thickness of these Ordovician rocks has been variously estimated but never measured, owing to structural complexity. Estimates range from 2,000 ft (610 m) to 25,000 ft (7620 m) (Gilluly and Gates, 1965, p. 23; Roberts, *et al.*, 1958, pp. 2832-3).

Recent field work by F. G. Poole in the Toiyabe Range, Nevada, in an area originally mapped by Ferguson and Cathcart (1954) indicates that the Zanzibar and Toquima formations are probably about 3,000 ft thick, certainly not more than 6,000 ft (1829 m). This range of thickness agrees reasonably well with the observations of Ferguson (1924, pp. 21-5) in the Manhattan district in the Toiyabe Range. In that area a reasonable succession can be pieced together, because the entire Ordovician rock sequence is not so disrupted by faulting as in the quadrangle in which a thickness of 25,000 ft (7620 m) was estimated. In comparable strata in the Canadian Arctic on Melville Island Tozer (1956) found Ordovician and Silurian black shale and radiolarian chert totalling 3,500 ft (1067 m).

Apparently, the normal expected thickness of the siliceous Ordovician sediments peripheral to North America is 3,000-6,000 ft (914-1829 m) and seemingly greater thickness may be attributable to unusual structural circumstances. Churkin (1974, p. 177) reached a similar conclusion.

In comparing the peripheral siliceous facies with the carbonate shelf rocks Berry (1962, Table 2) demonstrated that the carbonate strata are much thicker than the siliceous facies deposited during the same span of time. Concerning the siliceous facies he concluded (1962, p. 1710) that "water may have been shallow with muds accumulating slowly over a broad shelf-like region".

Origin. — The origin of the peripheral siliceous facies, particularly of the chert, is enigmatic. Goldstein and Hendricks (1953, pp. 436-41) believed that the silica became available through the submarine weathering of volcanic ash. They further noted the wide lateral distribution of the facies and indications of shallow water deposition. They found stumps of the primitive tree *Callixylon* in the overlying but very similar Arkansas Novaculite; these stumps were in growth position and could hardly have grown in deep water.

Ketner's (1969) important examination of the Ordovician cherts, argillites, and shales in Nevada and Idaho demonstrated that all three of these lithologies are abnormally high in silica (SiO_2) compared to any modern oceanic sediment. Furthermore, the Ordovician argillites and shales differ in the same manner from normal marine shales of Palaeozoic and Mesozoic age and, obviously, from volcanic rocks.

Ketner (1969, pp. B30-B31), like McKee (1938, p. 89), further showed that the amount of silica being delivered by rivers and hot springs can easily account for the ocean's present content. Ketner (p. B33) concluded that radiolaria assisted the precipitation of silica by piling up on the bottom and supplying an excess of easily dissolved opaline silica to the bottom water. Because the water was undersaturated for opaline silica when it was supersaturated for quartz, Ketner proposed that layers of finely divided quartz crystals precipitated to form the main constituent of the eugeosynclinal sediments, the solution of one and precipitation of the other being simultaneous. He further concluded that dilution of the precipitated quartz by detrital sediments produced argillite and shale, while lack of detrital material resulted in chert.

Of special interest is Ketner's observation (p. B25) that graptolites are absent from or scarce in the chert but common in argillite and shale. P_2O_5 is three times as abundant in argillite and shale as in the chert (Ketner, 1969, Tables 2, 3, 4). The differences may be directly related.

Ketner's conclusion calls for a constant rate of siliceous precipitation punctuated by periods of detrital and graptolitic contamination. It is equally likely that chert layers reflect more rapid accumulation of SiO_2 than do the argillite and shale. Rapid accumulation would account for the absence of otherwise ubiquitous graptolites in the chert as well as the absence of inorganic detritus. Folk (1973, p. 706) proposed a similar rapid chemical origin for the Caballos Novaculite.

Hite's (1970, 1973) evidence suggests that silica is soluble not only in highly alkaline NaCO$_3$ solutions (Eugster, 1967, 1969) but also in more acid chloride brines derived from seawater. Such brines are most likely to form in very shallow marine waters over enormously wide carbonate mudflats. The farther from the open ocean supply of normal marine water, the deeper the restricted hypersaline water might be. The closer to the horse latitudes, the greater the evaporation rate. The farther from the open ocean, the less effective the tides become and the greater the dependence on prevailing oceanic currents and prevailing wind drift to ensure a supply of normal marine water.

At no time in geological history was North America covered by such widespread shallow marine water as in the late Middle and Late Ordovician. Only in the Early Ordovician was the inundation nearly as great. Evidence of dolomite and cyclical evaporite (Fuller, 1961, pp. 1347-9) in the Williston basin in upper Middle to Upper Ordovician rocks indicates that chloride brines existed inland. As indicated by various cherts, like those of the Bigfork Chert and the Maravillas Formation of the Ouachita sequence in Arkansas, and the late Caradoc and Ashgill cherts in Nevada, high concentrations of silica were reaching the continental periphery, probably by reflux of heavy highly concentrated brines.

Hite (1973) cited the lateral gradation of chert with phosphorite, carbonate and evaporites in the Permian and Jurassic of western North America. He proposed that chert, dolomite and apatite can be precipitated from warm brines when the brines are mixed with cold normal seawater. Hite (1970, pp. 57-64) had already noted the interbedding and close association of algal buildups with phosphorite and chert in the Paradox area of Utah, hardly an association that can be attributed to deep water.

Stanley and Chamberlain (1974) concluded that the Ordovician Vinini beds of Nevada were pelagic (no reason given) but were mixed with quartzites and limestones; they proposed that these two lithologies must have been redeposited from shallower depths by turbidity currents. They also noted that turbidite sequences seemed to be incomplete or reworked by marine currents. Their conclusion that deposition was on the upper continental slope is supported by no evidence that could not be equally well interpreted as indicating relatively shallow water.

Stewart (1974) described Middle and Upper Ordovician shale and chert from south-western Nevada, in which cross-laminae in silty limestone interbeds show that currents flowed from the east-north-east. Translated to Middle and Late Ordovician continental positions such currents would have been flowing nearly due west almost at right angles to the coast. That direction would have coincided with the wind drift from the convergence of trade winds close to the equator. It therefore seems likely that the cross-laminae were formed in shallow water, an interpretation that agrees with the presence of the limestones themselves.

Those who believe that the presence of radiolaria in cherts indicates deep, even abyssal, depths of deposition should certainly read the warnings of Campbell (1954, pp. D16-D18). Radiolaria are marine forms, living at all depths and in all climatic zones. In the future, detailed study may result in the recognition of indicators of temperature and depth. The mere presence of radiolaria does not indicate deep-water sedimentation.

Although Churkin (1974, pp. 179-81) insists that graptolitic beds with which the cherts are associated must have been deposited in water deeper than 3500 m, I must repeat my previous conclusion (Ross, 1961, pp. 337-9) that no evidence demands such excessive depths and must express the belief that bedded cherts may have formed in surprisingly shallow water. The preservation of phosphatic fossils and absence of calcareous shells may reflect, more than water depth, the corrosive nature of the brines in which the silica was carried to the area of its precipitation.

Heavy brines may flow down the continental shelf and slope to great depths as relatively coherent masses. For example, warm water of high salinity leaves the Mediterranean Sea via the Straits of Gibraltar flowing along the bottom (Kuenen, 1950, pp. 42-3, Figs. 21, 22). Because it is heavier than upper waters of the Atlantic but lighter than waters at depths greater than 1200 m the Mediterranean effluent spreads out within the normal ocean waters at an average depth of 1000 m as a great tongue covering several thousand sq km. The refluxed Ordovician brines were probably of higher salinities and therefore could have spread at depths greater than the modern Mediterranean effluent. Therefore there is no obvious reason why cherts could not have formed at very great depths. Criteria for determining depths of deposition of chert are not universally applicable.

ALASKA

The Ordovician history of Alaska relative to the rest of North America is poorly understood. The "panhandle" is believed by Hones, *et al.* (1972) to be a displaced piece of oceanic crust originally continuous with eugeosynclinal rocks exposed in the Klamath Mountains of northern California (Boucot, *et al.*, 1973). Monger and Ross (1971) speculated that this terrain was part of a volcanic island arc separated from the west coast of the continent by a spreading centre and not physically attached to the coast until after Permian time. They (pp. 268-70) also noted that faunal distributions closely paralleled those found in the late Palaeozoic of Japan and that much of the fauna was of Russian aspect; possibly the Alaskan panhandle was a remnant left from a previous collision with Asia. Ordovician strata in the panhandle are graptolitic shales, cherts and volcanics (Buddington and Chapin, 1929; Ross, 1961; Eberlein and Churkin, 1970; Churkin, 1973).

Whether Ordovician Alaska was as closely connected with Ordovician Yukon as it is now is uncertain. We may be tempted to separate the two along the belt of graptolite-bearing shales which extends from the Glenogle Shale of the Rocky Mountains (along the Alberta-British Columbia boundary) through the Richardson Mountains (Ross, 1961, Fig. 1) and to extrapolate this belt north-eastward through Cornwallis Island into the Franklinian geosyncline. However, Lenz (1972, pp. 332-8, Figs. 5-7) has shown that the presumably eugeosynclinal graptolitic shales are not continuous along that route. Furthermore, the tectonic map of North America (King, 1969) indicates that no simple tectonic relationship between Alaska and adjacent Canada is likely. At present we must consider that much of Alaska was a part of North America in Ordovician time; Richards' (1974) evidence suggests that Alaska is composed of several tectonic plates or parts thereof. Churkin (1974), on the other hand, believes that Alaska was intimately linked to North America throughout the Ordovician.

In central and northern Alaska, Brosge and Dutro (1973) reported that Ordovician rocks are mainly graptolitic shales, volcanics and very few carbonates. A small trilobite assemblage (Ross, 1965) from the Seward Peninsula has Baltic affinities; a brachiopod fauna from Jones Ridge near the Yukon-Alaska border (Ross and Dutro, 1966) is similar to a fauna from Perce, Quebec and to genera reported from the Klamath Mountains (Boucot, et al., 1973, pp. 14, 15). Flower (1968a, 1968b) described two small suites of Ordovician nautiloids from the Seward Peninsula; the Early Ordovician forms have no distribution outside Alaska but the Middle Ordovician genera occur in the Chazy Group of New York.

During the preparation of this paper, J. R. Derby and T. A. Hendricks provided essential advice and sources of information on midcontinent and Appalachian Ordovician stratigraphy. R. J. Hite gave welcome guidance on geochemical questions pertaining to formation of dolomite and evaporites. R. L. Ethington furnished unpublished information on Lower Ordovician formations of the Ouachita facies. A preliminary draft was criticized by Keith Ketner; critical reading by Derby, W. C. Sweet, Stig Bergstrom, W. B. N. Berry and B. S. Norford led to the final version. I could not have assembled this manuscript without the criticism and advice of these colleagues although I did not always take their advice.

DISCUSSION

C. R. Scotese. Several of your maps were sub-labelled "interpolated"; how were these interpolations made, and if the question is applicable was it linear or non-linear?
R. J. Ross, Jr. The interpolated positions of North America were not derived mathematically but by inspection.
C. J. Stubblefield asked whether phosphatized graptolites mentioned in the abstract were preserved in phosphatic nodules.
R. J. Ross, Jr. Although I know of no specific examples of graptolites preserved in phosphatic nodules, such nodules are present in many of the graptolite-bearing beds. Work by Ketner indicates that the P_2O_5 content of these beds is respectable.

REFERENCES

Austin, G. S., 1972. Paleozoic lithostratigraphy of south-eastern Minnesota, pp. 459-473 *in* Sims, P. K., and Morey, G. B. (eds.), Geology of Minnesota: a centennial volume, xvi + 632 pp., Minnesota Geological Survey.

Barnes, V. E., Cloud, P. E., Jr., Dixon, L. P., Folk, R. L., Jonas, E. C., Palmer, A. R., and Tynan, E. J., 1959. Stratigraphy of the pre-Simpson Paleozoic subsurface rocks of Texas and south-east New Mexico, Univ. Tex. Publs., 5924 (2 vols.), 1-836.

Berg, R. R., 1960. Cambrian and Ordovician history of Colorado, pp. 10-17 *in* Weimer, R. J., and Haun, J. D. (eds.), Guide to the geology of Colorado, 303 pp., Geological Society of America, Rocky Mountains Association of Geologists and Colorado Science society.

_____, and Ross, R. J., Jr., 1959. Trilobites from the Peerless and Manitou formations, Colorado. J. Paleont., 33, 106-119.

Bergstrom, S. M., 1971. Correlation of the North Atlantic Middle and Upper Ordovician conodont zonation with the graptolite succession, Mm. Bur. Rech. gol. minr., 73, Colloque Ordovicien-Silurien, Brest, Septembre 1971, 177-187.

_____, 1973. Biostratigraphy and facies relations in the lower Middle Ordovician of easternmost Tennessee, Am. J. Sci., 273A, 261-293.

_____, Ethington, R. L., and Jaanusson, V., 1973. On the stage subdivision of the North American lower Middle Ordovician: age of strata at the top of Whiterock reference sequences in Nevada, Abstr. with Progr. Geol. Soc. Am., 5, 299.

Berry, W. B. N., 1962. Comparison of some Ordovician limestones, Bull. Am. Ass. Petrol. Geol., 46, 1701-1720.

_____, Lindsley-Griffin, N., Potter, A. W., and Rohr, D. M., 1973. Early Middle Ordovician graptolites from the eastern Klamath Mountains, Siskiyou County, California, Abstr. with Progr. Geol. Soc. Am., 5, 11.

Bond, D. C., Atherton, E., Bristol, H. M., Buschbach, T. C., Stevenson, D. L., Becker, L. E., Dawson, T. A., Farnalld, E. C., Schwalb, H., Wilson, E. M., Statler, A. T., Stearns, R. G., and Buehner, J. H., 1971. Possible future petroleum potential of Region I — Illinois Basin, Cincinnati Arch, and northern Mississippi embayment, Mem. Am. Ass. Petrol. Geol., 15 (s), 1165-1218.

Boucot, A. J., Dean, W. T., Martinsson, A., Potter, A., Rexroad, C., Rohr, D., Savage, N. M., and Wright, A. J., 1973a. Biogeographic relations of the pre-late Devonian of the eastern Klamath belt, northern California, Abstr. with Progr. Geol. Soc. Am., 5, 14.

_____, 1973 b. Pre-late Middle Devonian biostratigraphy of the eastern Klamath belt, northern California, Ibid., 5, 15.

Brosge, W. P., and Dutro, J. T., Jr., 1973. Paleozoic rocks of northern and central Alaska, Mem. Am. Ass. Petrol. Geol., 19, 371-375.

Buddington, A. F., and Chapin, T., 1929. Geology and mineral deposits of south-eastern Alaska, Bull. U.S. Geol. Surv., 800, 1-398, pls. 1-22.

Campbell, A. S., 1954. Radiolaria. pp. D11-D163 *in* Moore, R. C. (ed.). Treatise on invertebrate paleontology, Part D, Protista 3, Protozoa (chiefly Radiolaria and Tintinnina). xii + 195, Geological Society of America and University of Kansas Press.

Carnes, J. B., and Bergstrom, S. M., 1973. On the stage subdivision of the North American lower Middle Ordovician: biostratigraphy of the Ashby Stage, Abstr. with Progr. Geol. Soc. Am., 5, 307.

Chronic, J., McCallum, M. E., Ferris, C. S., Jr., and Eggler, D. H., 1969. Lower Paleozoic rocks in diatremes, southern Wyoming and northern Colorado, Bull. Geol. Soc. Am., 80, 149-156.

Churkin, M., Jr., 1973. Paleozoic and PreCambrian rocks of Alaska and their role in its structural evolution, Prof. Pap. U.S. Geol. Surv., 740, 1-64.

_____, 1974. Paleozoic marginal ocean basin-volcanic arc systems in the Cordilleran foldbelt *in* Dott, R. H., Jr., and Shaver, R. H. (eds). Modern and ancient geosynclinal sedimentation, Spec. Publs. Soc. Econ. Paleont. Miner. Tulsa, 19, 174-192.

Cloud, P. E., Jr., and Barnes, V. E., 1948. The Ellenburger Group of central Texas, Univ. Tex. Publs., 4621, 1-473, pls. 1-44.

_____, 1957. Early Ordovician sea in central Texas *in* Ladd, H. S. (ed.). Treatise on marine ecology, part 2, Mem. Geol. Soc. Am., 67 (2), 163-214.

Condie, K. C., and Snansieng, S., 1971. Petrology and geochemistry of the Duzel (Ordovician) and Gazelle (Silurian) formations, northern California, J. Sedim. Petrol., 41, 741-751.

Cooper, G. A., 1956a. A new Upper Canadian fauna from a deep well in Tennessee, J. Paleont., 30, 29-34, pl. 5.

_____, 1956b. Chazyan and related brachiopods, Smithson. Misc. Collns., 127; Pt. 1, i-xvi, 1-1024; Pt. 2, 1025-1245, pls. 1-269.

Cullison, J. S., 1944. The stratigraphy of some Lower Ordovician formations of the Ozark Uplift, Bull. Mo. Sch. Mines Tech. Ser., 15 (2), 1-112, pls. 1-35.

Dapples, E. C., 1955. General lithofacies relationship of St. Peter sandstone and Simpson Group, Bull. Am. Ass. Petrol. Geol., 39, 444-467.

Derby, J. R., 1973, Lower Ordovician-Middle Ordovician boundary in western Arbuckle Mountains, Oklahoma, pp. 24-26 in Rowland, R. T., Regional geology of the Arbuckle Mountains, Guidebook for field trip no. 5, Geological Society of America Annual Meetings, 1973.

Dott, R. H., Jr., and Roshardt, M. A., 1972. Analysis of cross-stratification orientation in St. Peter sandstone in south-west Wisconsin, Bull. Geol. Soc. Am., 83, 2589-2596.

Eberlein, G. D., and Churkin, M., Jr., 1970. Paleozoic stratigraphy in the north-west coastal area of Prince of Wales Island, south-eastern Alaska, Bull. U.S. Geol. Surv., 1284, 1-67.

Ethington, R. L., and Schumacher, D., 1969. Conodonts of the Copenhagen Formation (Middle Ordovician) in central Nevada, J. Paleont., 43, 440-484, pls. 67-69.

Eugster, H. P., 1967. Hydrous sodium silicates from Lake Magadi, Kenya: precursors of bedded chert, Science, N. Y., 157, 1177-1180.

——————, 1969, Inorganic bedded chert from the Magadi area, Kenya, Contr. Miner. Petrol., 22, 1-31.

Ferguson, H. G., 1924. Geology of ore deposits of the Manhattan district, Nevada, Bull. U.S. Geol. Surv., 723, 1-163.

——————, and Cathcart, S. H., 1954. Geology of the Round Mountain quadrangle, Nevada, U.S. Geol. Surv., Geological Quadrangle Map series GQ-40.

Fisher, W. L., and Rodda, P. E., 1968. Stratigraphy and genesis of dolomite, Edwards Formation (Lower Cretaceous) of Texas in Angino, E. E., and Hardy, R. G. (eds.). A symposium on industrial mineral exploration and development, Spec. Publs. Kansas Geol. Surv., 34, 52-73.

Flawn, P. T., Goldstein, A., Jr., King, P. B., and Weaver, C. E., 1961. The Ouachita System, Publs. Bur. Econ. Geol. Univ. Tex., 6120, 1-401, pls. 1-15.

Flower, R. H., 1968a. Endoceroids from the Canadian of Alaska, Mem. Inst. Min. Technol. New Mex., 21, 13-17, pls. 3, 4.

——————, 1968b. A Chazyan cephalopod fauna from Alaska. Ibid., 21, 21-23, pl. 5.

——————, 1968c. Cephalopods from the Tinu Formation, Oaxaca State, Mexico, J. Paleont., 42, 804-810, pl. 105.

——————, 1969. Early Paleozoic of New Mexico and the El Paso region, pp. 31-101 in The Ordovician symposium, 3rd annual field trip, 162 pp., El Paso Geological Society and Society of Economic Paleontologists and Mineralogists, Permian Basin Section.

Folk, R. L., 1973. Evidence for peritidal deposition of Devonian Caballos novaculite, Marathon Basin, Texas, Bull. Am. Ass. Petrol. Geol., 57, 702-725.

Fuller, J. G. C. M., 1961. Ordovician and contiguous formations in north Dakota, south Dakota, Montana, and adjoining areas of Canada and United States. Ibid., 45, 1334-1363.

Gilluly, J., and Gates, O., 1965. Tectonic and igneous geology of the northern Shoshone Range, Nevada, Prof. Pap. U.S. Geol. Surv., 465, 1-153.

Goldstein, A., Jr., and Hendricks, T. A., 1953. Siliceous sediments of Ouachita facies in Oklahoma, Bull Geol. Soc. Am., 64, 421-441.

Ham, W. E., 1955. Regional stratigraphy and structure of the Arbuckle Mountain region, Guide Bk. Okla. Geol. Surv., 3 (2), 28-35.

——————, 1959. Correlation of pre-Stanley strata in the Arbuckle-Ouachita Mountain regions, pp. 71-86, in Cline, L. M., Hilseweck, W. J., and Feray, D. E. (eds.), The geology of the Ouachita Mountains, a symposium, 208 pp., Dallas Geological Society and Ardmore Geological Society.

——————, Denison, R. E., and Merritt, C. A., 1964. Basement rocks and structural evolution, southern Oklahoma, Bull. Okla. Geol. Surv., 95, 1-302, pls. 1-16.

Hanshaw, B. B., Back, W., and Deike, R. G., 1971. A geochemical hypothesis for dolomitization by ground water, Econ. Geol., 66, 710-724.

Harris, L. D., 1973. Dolomitization model for Upper Cambrian and Lower Ordovician carbonate rocks in the eastern United States, Jl. Res. U.S. Geol. Surv., 1, 63-78.

Hendricks, T. A., 1959. Structure of the frontal belt of the Ouachita Mountains, pp. 44-56 in Cline, L. M., Hilsewek, W. J., and Feray, D. E. (eds.), The geology of the Ouachita Mountains, a symposium, 208 pp., Dallas Geological Society and Ardmore Geological Society.

Hintze, L. F., 1951. Lower Ordovician detailed stratigraphic sections for western Utah, Bull. Utah Geol. Miner. Surv., 39, 1-100.

——————, 1953. Lower Ordovician trilobites from western Utah and eastern Nevada. Ibid., 48 [for 1952], i-vi, 1-249, pls. 1-28.

——————, 1973. Geologic history of Utah, Geology Stud. Brigham Young Univ., 20 (3), 1-181.

Hite, R. J., 1970. Shelf carbonate sedimentation controlled by salinity in the Paradox Basin, south-east Utah, in Rau, J. L., and Dellwig, L. F. (eds.), Third symposium on salt, Cleveland, Ohio, 48-66.

——————, The role of the evaporite basin in the origin of marine phosphorite deposits — a new theory, Abstr. fourth int. symposium on salt, Houston, Texas, 15.

Hobbs, S. W., Hays, W. H., and Ross, R. J., Jr., 1968. The Kinnikinnic Quartzite of central Idaho — redefinition and subdivision, Bull. U.S. Geol. Surv., 1254-J, J1-J22.

Hoffman, P., Dewey, J. F., and Burke, K., 1974. Aulacogens and their genetic relation to geosynclines, with Proterozoic example from Great Slave Lake, Canada, Spec. Publs. Soc. Econ. Paleont. Miner. Tulsa, 19, 38-55.

Jensen, R. G., 1967. Ordovician brachiopods from the Pogonip Group of Millard County, western Utah, Geology Stud. Brigham Young Univ., 14, 67-100, pls. 1-6.

Jones, D. L., Irwin, W. P., and Ovenshine, A. T., 1972. South-eastern Alaska — a displaced continental fragment? Prof. Pap. U.S. Geol. Surv., 800-B, B211-B217.

Ketner, K. B., 1968. Origin of Ordovician quartzite in the Cordilleran miogeosyncline. Ibid., 600-B, B169-B177.

——————, 1969. Ordovician bedded chert, argillite, and shale of the Cordilleran eugeosyncline in Nevada and Idaho. Ibid., 650-B, B23-B34.

King, P. B., 1937. Geology of the Marathon region, Texas. Ibid., 187, 1-148.

——————, 1951. The tectonics of middle North America, 203 pp., Princeton University Press.

——————, 1969. Tectonic map of North America, U.S. Geological Survey.

Kuenen, Ph. H., 1950. Marine Geology, x + 568 pp., Wiley, New York.

Ladd, H. S., 1929. The stratigraphy and paleontology of the Maquoketa shale of Iowa, part 1, Rep. Iowa Geol. Surv., 34, 309-448, pls. 4-12.

Lenz, A. C., 1972. Ordovician to Devonian history of northern Yukon and adjacent district of MacKenzie, Bull. Can. Petrol. Geol., 20, 321-361.

Lochman, C., 1966. Lower Ordovician (Arenig) faunas from the Williston Basin, Montana and north Dakota, J. Paleont., 40, 512-548, pls. 61-65.

Lochman-Balk, C., and Wilson, J. L., 1967. Stratigraphy of Upper Cambrian-Lower Ordovician subsurface sequence in Williston Basin, Bull. Am. Ass. Petrol. Geol., 51, 883-917.

McBride, E. F., 1970. Stratigraphy and origin of Maravillas Formation (Upper Ordovician), west Texas. Ibid., 54, 1719-1745.

——————, and Thomson, A., 1970a. The Caballos novaculite, Marathon region, Texas, Spec. Pap. Geol. Soc. Am., 122, 1-129.

——————, and Thomson, A., 1970b. Stratigraphy and origin of the Caballos novaculite, pp. 58-65, in McBride, E. F. (ed.), Guidebook to the stratigraphy, sedimentary structures, and origin of the flysch and pre-flysch rocks of the Marathon Basin, Texas, American Association of Petroleum Geologists — Society of Economic Paleontologists and Mineralogists, Annual Meeting, Dallas, Texas, 1969, 104 pp., Dallas Geological Society.

McElhinny, M. W., Giddings, J. W., and Embleton, B. J., 1974. Paleomagnetic results and the late Precambrian glaciations, Nature, Lond., 248, 557-561.

——————, and Opdyke, N. D., 1973. Remagnetization hypothesis discounted: a paleomagnetic study of the Trenton Limestone, New York State, Bull. Geol. Soc. Am., 84, 3697-3708.

McKee, E. D., 1938. The environment and history of the Toroweap and Kaibab formations of northern Arizona and southern Utah, Publs. Carnegie Instn., 492, 1-268, pls. 1-48.

——————, 1966. Structures of dunes at White Sands National Monument, New Mexico (and a comparison with structures of dunes from other selected areas). Sedimentology, 7 (special issue), 3-69, pls. 1-8.

MacQueen, R. W., 1970. Lower Paleozoic stratigraphy and sedimentology; eastern MacKenzie Mountains, northern Franklin Mountains (96 C, D, E, F; 106 G, H), Geol. Surv. Pap. Can., 70-1, part A, 225-230.

Monger, J. W. H., and Ross, C. A., 1971. Distribution of fusulinaceans in the western Canadian Cordillera. Can. J. Earth Sci., 8, 259-278.

Norford, B. S., 1969. Ordovician and Silurian stratigraphy of the southern Rocky Mountains, Bull. Geol. Surv. Can., 176, [i-xiv] 1-90, pls. 1-19.

Poole, F. G., and Hayes, P. T., 1971. Depositional framework of some Paleozoic strata in north-western Mexico and south-western United States, Abstr. with Progr. Geol. Soc. Am., 3, 179.

Porter, J. W., and Fuller, J. G. C. M., 1959. Lower Paleozoic rocks of northern Williston Basin and adjacent areas, Bull. Am. Ass. Petrol. Geol., 43, 124-189.

——————, 1964. Ordovician-Silurian, part I — Plains, pp. 34-42 in Geologic history of western Canada, Alberta Society of Petroleum Geologists, Calgary.

Richards, H. G., 1974. Tectonic evolution of Alaska, Bull. Am. Ass. Petrol. Geol., 58, 79-105.

Roberts, R. J., Hotz, P. E., Gilluly, J., and Ferguson, H. G., 1958. Paleozoic rocks of north-central Nevada. Ibid., 42, 2813-2857.

Robison, R. A., and Pantoja-Alor, J., 1968. Tremadocian trilobites from the Nochixtlan region, Oaxaca, Mexico, J. Paleont., 42, 767-800, pls. 97-104.

Roehl, P. O., 1967. Stony Mountain (Ordovician) and Interlake (Silurian) facies analogs of Recent low-energy marine and subaerial carbonates, Bahamas, Bull. Am. Ass. Petrol. Geol., 51, 1979-2023.

Rosenblum, S., and Srivastava, S. P., 1970. Silica sand deposits in the Monrovia area, Liberia, Mem. Rep. Liberan Geol. Surv., Bur. Nat. Resour. Surv., 47, (USGS IR-LI-41), 12 pp.

Ross, R. J., Jr., 1951. Stratigraphy of the Garden City formation in northeastern Utah and its trilobite faunas, Bull. Peabody Mus. Nat. Hist., 6, 1-161, pls. 1-36.

―――――, 1957. Ordovician fossils from wells in the Williston Basin, eastern Montana, Bull. U.S. Geol. Surv., 1021-M, 439-510, pls. 37-44.

―――――, 1961. Distribution of Ordovician graptolites in eugeosynclinal facies in western North America and its paleogeographic implications, Bull. Am. Ass. Petrol. Geol., 45, 330-341.

―――――, 1964a. Middle and Lower Ordovician formations in southernmost Nevada and adjacent California, Bull. U.S. Geol. Surv., 1180-C, 1-101, pl. 1.

―――――, 1964b. Relations of Middle Ordovician time and rock units in Basin Ranges, western United States, Bull. Am. Ass. Petrol. Geol., 48, 1526-1554.

―――――, 1965. Early Ordovician trilobites from Seward Peninsula, Alaska, J. Paleont., 39, 17-20, pl. 8.

―――――, 1967. Some Middle Ordovician brachiopods and trilobites from the Basin Ranges, western United States, Prof. Pap. U.S. Geol. Surv., 523-D, D1-D43, pls. 1-11.

―――――, 1970. Ordovician brachiopods, trilobites, and stratigraphy in eastern and central Nevada. Ibid., 639, 1-103, pls. 1-22.

―――――, 1975. Early Paleozoic trilobites, sedimentary facies, lithospheric plates and ocean currents, Fossils and Strata, 4.

―――――, and Dutro, J. T., Jr., 1966. Silicified Ordovician brachiopods from east central Alaska, Smithson. Misc. Collns., 149 (7), 1-22, pls. 1-3.

―――――, and Shaw, F. C., 1972. Distribution of the Middle Ordovician Copenhagen Formation and its trilobites in Nevada, Prof. Pap. U.S. Geol. Surv., 749, i-iii, 1-33, pls. 1-8.

Shinn, E. A., 1973. Sedimentary accretion along the leeward, S.E. coast of Quatar Peninsula, Persian Gulf, pp. 199-209 *in* Purser, B. H. (ed.), The Persian Gulf, 471 pp., Springer-Verlag, New York.

Stanley, K. O., and Chamberlain, C. K., 1974. Bathymetry and origin of eugeosynclinal quartzites and limestones of the Vinini Formation (Ordovician), north-central Nevada, Abstr. with Progr. Geol. Soc. Am., 6, 260-261.

Stewart, J. H., 1974. Depositional environments of Paleozoic eugeosynclinal rocks, western Great Basin — two examples. Ibid., 6, 261.

Suhm, R. W., 1974. Stratigraphy of the Everton Formation (early medial Ordovician), northern Arkansas, Bull. Am. Ass. Petrol. Geol., 58, 685-707.

Sweet, W. C., 1954. Harding and Fremont formations, Colorado. Ibid., 38, 284-305.

―――――, and Bergstrom, S. M., 1971. The American Upper Ordovician Standard XIII: a revised time-stratigraphic classification of North American upper Middle and Upper Ordovician rocks, Bull. Geol. Soc. Am., 82, 613-628.

―――――, 1973. Biostratigraphic potential of the Arbuckle Mountains sequence as a reference standard for the Midcontinent Middle and Upper Ordovician, Abstr. with Progr. Geol. Soc. Am., 5, 355.

―――――, 1975. Conodont biostratigraphy of the Middle and Upper Ordovician of the United States, *in* Bassett, M. G. (ed.), The Ordovician System: proceedings of a Palaeontological Association symposium, Birmingham, September 1974, University of Wales Press and National Museum of Wales, Cardiff.

―――――, Ethington, R. L., and Barnes, C. R., 1971. North American Middle and Upper Ordovician conodont faunas *in* Sweet, W. C., and Bergstrom, S. M. (eds.), Symposium on conodont biostratigraphy, Mem. Geol. Soc. Am., 127, 163-193.

Tozer, E. T., 1956. Geological reconnaissance Prince Patrick, Eglington and Western Melville islands, Arctic Archipelago, Northwest Territories, Geol. Surv. Pap. Can., 55-5, 1-32.

Webb, G. W., 1956. Middle Ordovician detailed stratigraphic sections for western Utah and eastern Nevada, Bull. Utah Geol. Miner. Surv., 57, 1-77.

―――――, 1958. Middle Ordovician stratigraphy in eastern Nevada and western Utah, Bull. Am. Ass. Petrol. Geol., 42, 2335-2377.

Webers, G. F., 1972. Paleoecology of the Cambrian and Ordovician strata of Minnesota, pp. 474-484, *in* Sims, P. K., and Morey, G. B. (eds.), Geology of Minnesota: a centennial volume, xvi + 632 pp., Minnesota Geological Survey.

Wilson, J. L., 1954. Ordovician stratigraphy in Marathon folded belt, west Texas, Bull. Am. Ass. Petrol. Geol., 38, 2455-2475.

Yochelson, E. L., 1964. The early Ordovician gastropod *Ceratopea* from east Greenland, Meddr. Gronland, 164 (7), 1-10, pl. 1.

―――――, 1968. Tremadocian mollusks from the Nochextlan region, Oaxaca, Mexico, J. Paleont., 42, 801-803.

―――――, and Bridge, J., 1957. The Lower Ordovician gastropod *Ceratopea*, Prof. Pap. U.S. Geol. Surv., 294-H, 281-304, pls. 35-38.

Young, L. M., Fiddler, L. C., and Jones, R. W., 1972. Carbonate facies in Ordovician of northern Arkansas, Bull. Am. Ass. Petrol. Geol., 56, 68-80.

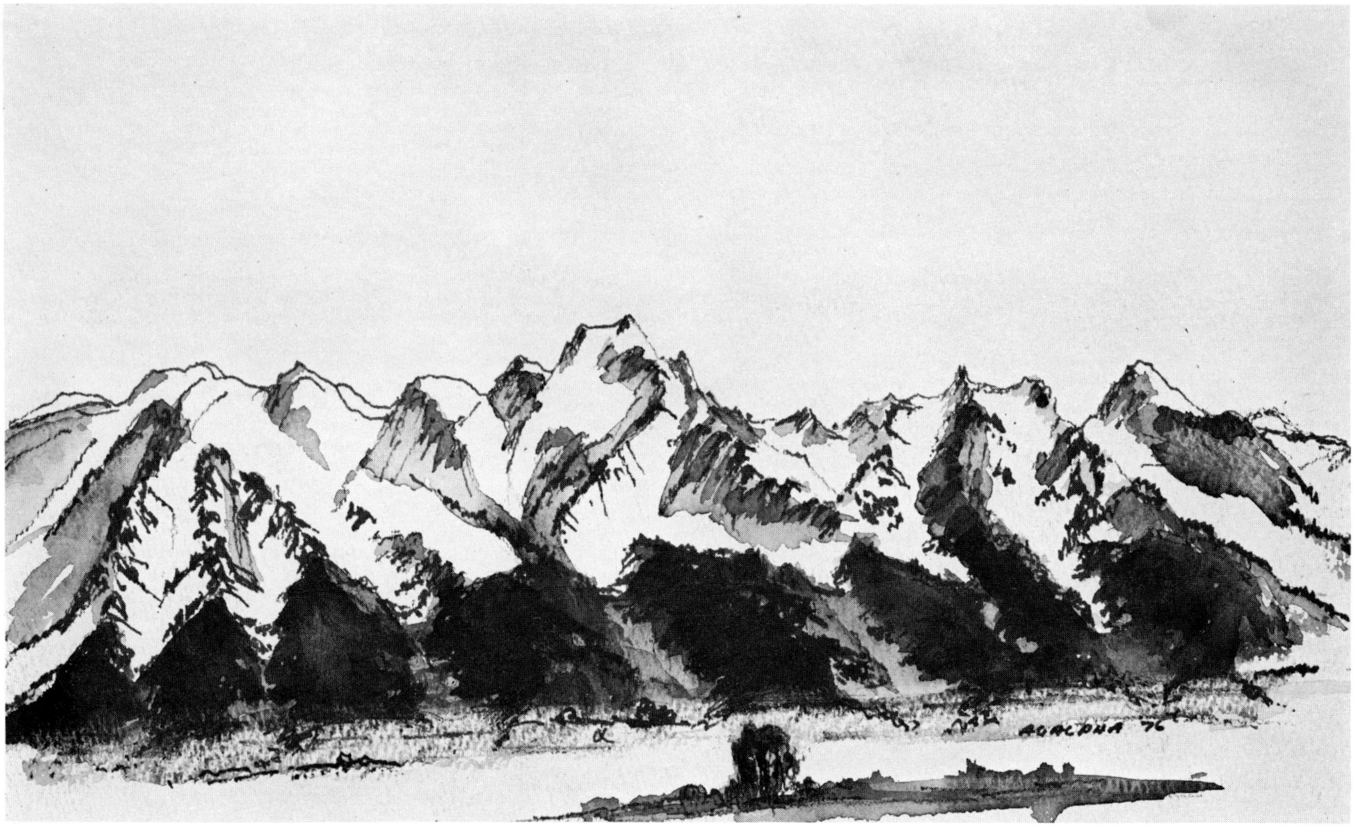

MISSISSIPPIAN CARBONATE SHELF MARGINS, WESTERN UNITED STATES

by

Peter R. Rose[1]

ABSTRACT

Regional linear carbonate shelf margins, or stratigraphic reefs, are postulated to have developed during Mississippian time along the eastern flank of the Cordilleran miogeosyncline in the Western United States. These shelf margins are analogous to well-documented ancient and modern geologic counterparts, such as the Guadalupian reef of the West Texas Permian basin, the Lower Cretaceous Edwards shelf margin of the Gulf Coast, and the modern Florida reef tract-Straits of Florida province. Two Mississippian shelf margins are believed to have existed: the lower one developed as an integral part of a widespread carbonate depositional complex of Kinderhookian through early Meramecian age; the upper shelf margin developed west of the earlier stratigraphic reef as part of a regional carbonate depositional complex of middle Meramecian to late Chesterian age. Evidence for both shelf margins consists of (1) a linear physical barrier, (2) restricted sediments in the shelf interior, (3) abrupt basinward thinning of sediments and basin-starvation just seaward of the shelf edge, (4) profound facies changes coincident with the basinward thinning — from light-colored, skeletal, shelf carbonate rocks to dark, fine-grained, silty, basinal carbonate rocks, and (5) the consistent regional occurrence of the first four patterns. Seaward topographic relief along the front edge of the lower shelf margin was probably about 200-400 m, and maximum relief along the central sector of the upper shelf margin may have approached 1,000 m.

INTRODUCTION

In the western United States a widespread sheet of shallow-marine limestone and dolomite in the lower part of the Mississippian is present through much of the Rocky Mountain province and extends westward into the Cordilleran miogeosyncline. This succession comprises carbonate rocks of Kinderhookian, Osagean, and early Meramecian ages, and has many provincial rock-stratigraphic names, including Allan Mountain Limestone and Castle Reef Dolomite, Lodgepole and Mission Canyon Limestones, Madison Limestone or Group, and Gardison, Deseret, Leadville, Redwall, and Monte Cristo Limestones. For simplicity these rocks will be called the lower depositional complex throughout the report. Regionally the lower depositional complex thickens westward from the Transcontinental arch of Eardley (1962), a persistent ancestral cratonic arch that stretches southwest across eastern South Dakota, Nebraska, and Colorado. West of the Wasatch line (Kay, 1951), in the Basin and Range province, the carbonate part of the lower depositional complex thins markedly, and, farther west, the identity of the entire complex is lost in a flysch sequence.

Carbonate rocks of Late Mississippian age are also present in the Rocky Mountain and Great Basin regions, but they occupy a much narrower arcuate belt that stretches southeast across eastern Idaho and turns southwest across western Utah. These rocks, generally of middle Meramecian through Chesterian age, are variously named the Great Blue, Ochre Mountain, Monroe Canyon, Brazer, and White Knob Limestones and the Scott Peak, South Creek, and Surrett Canyon formations. In this report these Upper Mississippian rocks are included in the upper depositional complex, most of which lies west of the area where the carbonate part of the lower depositional complex is recognizable. Like the lower depositional complex, the upper depositional complex thickens westward into the Cordilleran geosyncline, and the carbonate component ultimately loses its identity in central Idaho and westernmost Utah.

I have made no attempt to adjust the maps to compensate for thrusting in the Sevier orogenic belt, except to indicate

---

[1] U.S. Geological Survey, Denver, Colorado 80225.

Peter Rose current affiliation Energy Reserves Group, Inc., Houston, Texas.

This interpretation has developed over the past several years, partly through constructive discussions with colleagues such as C. A. Sandberg, F. G. Poole, and others who have been working on the Mississippian of the Western United States. Although the concepts presented here could not have evolved without their stimulation, the overall interpretation, and the resulting shelf-margin concept devolve solely upon me.

The existence of a starved basin in Utah and Idaho was suspected as early as 1969 by C. A. Sandberg on the basis of conodont samples from Laketown Canyon, Utah, that he processed and determined for A. E. Roberts (written commus. to A. E. Roberts and W. J. Sando, Mar. 26, 1969). Postulation of this starved basin, substantiated by later fieldwork, has been communicated by Sandberg in recent years through informal discussions and prepared talks.

the probable prethrust position of the Antler orogenic belt and flysch trough in central Idaho. The original facies belts in western Wyoming, eastern Idaho, central Utah, and eastern Nevada, therefore, may have been wider than they are today.

Other thick and widespread carbonate sheets, such as the Guadalupian of West Texas and New Mexico or the Lower Cretaceous of northern Mexico and the Texas Gulf Coast, are distinguished by prominent and abrupt linear shelf margins — the stratigraphic reefs of Dunham (1970), which are sometimes called "barrier reefs" or simply "reefs." These regional linear features represent the seaward limit of shallow-water carbonate accumulation, and the slope break just seaward represents the beginning of the fore-reef decline into a starved basin (Adams et al., 1951) deprived of substantial carbonate sediments. Such ancient shelf margins have their modern analogs among the reef tracts of Florida and the Bahamas, the Great Barrier Reef of Australia, and many other reef tracts throughout the world.

Recognition and delineation of ancient shelf margins are important because they define depositional strike along basin flanks and strongly influence the character and orientation of adjacent petroleum-reservoir rocks and seals. Identification of carbonate-shelf margins allows interpretation of the geometric arrangement and depositional mode of both shelf and basin sediments, as well as reliable estimation of ancient bathymetry and basin configuration. Thus, the accurately identified shelf margin may serve as a geologic keystone upon which the successful stratigraphic interpretation of the entire basin may rest.

Purpose and Scope

This paper is a regional synthesis based on published surface information and available commercial subsurface logs. It considers the effect of two previously unrecognized Mississippian shelf margins on stratigraphic nomenclature and on the geologic history of the Cordilleran miogeosyncline, with the intent of ultimately simplifying regional Mississippian stratigraphic nomenclature. Finally, it outlines future research necessary to confirm the existence of these postulated shelf margins and locate them accurately.

Previous Work

The Madison and its equivalents have widespread distribution, prominent topographic expression, and economic significance as aquifers, petroleum reservoirs, and sources of industrial stone. However, these units have not been considered previously as a complete regional carbonate depositional complex extending throughout the Western United States. Parts of the lower carbonate depositional complex have been analyzed in subregional studies by Andrichuk (1955), Sando and Dutro (1960), Sando (1967a, b, 1972, 1974), Craig (1972), Hintze (1973), and Gutschick (1976). Regional syntheses (Sando, 1976; Sando et al., 1975, 1976) for the Northern Rocky Mountains present concepts similar to those presented herein, except for the postulation of substantial seaward topographic relief along the fronts of two carbonate shelf margins or "reefs." These reports were not known to me until after my report was submitted for publication. The broad similarity of independently developed concepts lends support to the validity of the interpretations.

Sources of Data

This interpretation is based upon my review and synthesis of Mississippian stratigraphic columns presented in published reports of more than 100 outcrop areas, and in commercial sample logs for more than 120 wells throughout the Rocky Mountain and Cordilleran region. The references for each locality or well and the appropriate published references are available in a separate report (Rose, 1976).

CONCEPTUAL MODEL OF CARBONATE DEPOSITION

During the past 20 years, recognition of similar patterns of sedimentation among the more widely known modern depositional complexes and their ancient geologic counterparts has led to a conceptual model useful in interpreting other carbonate successions. As the natural processes that control different sedimentary features have become better understood, stratigraphic responses to factors such as climate, geologic setting, and subsidence rate can often be predicted.

The literature pertaining to this carbonate depositional model is voluminous and will not be reviewed here, except to cite a few outstanding contributions. Rich (1951) defined three dominant environments of deposition — undaform, clinoform, and fondoform — which correspond geometrically to topset, forset, and bottomset environments of deltaic sedimentation, and he recognized the general applicability of his model to most regional terrigenous and carbonate depositional systems. Work by King (1948), Adams, et al. (1951), Van Siclen (1958), Jackson (1964), and Meissner (1972) confirmed the applicability of Rich's concept in Pennsylvanian and Permian carbonate rocks of West Texas. Ginsburg (1956) and Imbrie and Purdy (1962) demonstrated that variations in hydrography, submarine topography, and geography control seawater properties and the distribution of marine organisms, and thus determine sediment types in coastal South Florida and the Bahamas. Illing, et al. (1965) and Shinn, et al. (1965) described occurrences of evaporite and dolomite in modern tidal-flat sediments of the Persian Gulf and the Bahamas, respectively, which compared well with ancient examples of evaporites and dolomites reported by King (1948), Laporte (1967), and Fisher and Rodda (1969); thus variations in mineralogy also may be related to the carbonate depositional model. Individual versions of this carbonate depositional model are present explicitly or implicitly in the work of many current carbonate specialists, and authors such as Irwin (1965) and Coogan (1969) have devoted entire papers to the subject.

At present, the most common carbonate depositional model consists of a shallow-marine sedimentary prism deposited on the flanks of a hypothetical cratonic mass, continental margin, or other positive depositional or tectonic feature. Nearly all carbonate production takes place in very shallow marine water (Stockman, et al. 1967; Neumann and Land, 1975), and little sediment accumulation can take place above high-tide level. In areas of shallow water where sediment production exceeds subsidence rate, carbonate sediments accummulate up to approximate sea level. Any additional carbonate detritus is thereafter distributed laterally by currents, tides, and storms, eventually building a broad, very shallow submarine plain. Accordingly, the three-dimensional carbonate body, its growth limited upward by sea level and rate of subsidence, thickens away from the positive element, accreting across its subsiding flanks toward open, deeper marine waters. At the point in this hypothetical traverse across the shelf where production and accumulation of shallow-marine carbonate sediments is exceeded by subsidence, the sediment package ceases to thicken (Fig. 1) and thereafter thins progressively toward deeper water to some area where sedimentation is slow and relatlively uniform in a starved basin. It is important to recognize that most starved-basin sediments have been transported from the shelf where they were manufactured. In most examples, it is at or near the shelf margin that most constructional marine communities, such as coral-algal reefs, as well as many coarse-grained bioclastic sediments are located (Wilson, 1974). Moreover, a constructional ridge at the shelf margin may serve to restrict ingress and egress of sea water onto and off the shelf, thus promoting evaporative, highly saline conditions in the shelf interior.

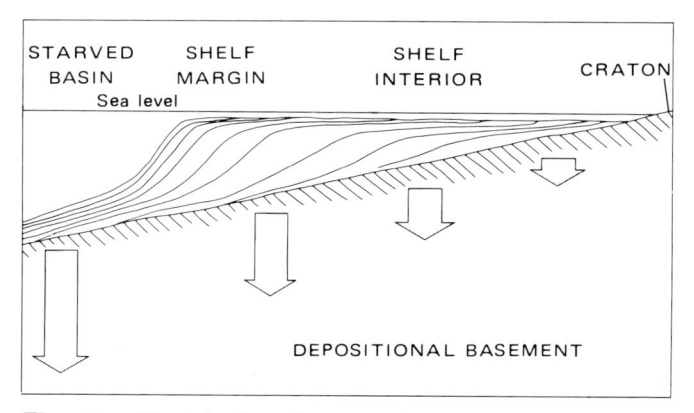

Fig. 1 — Model of carbonate deposition. Lengths of arrows are proportional to rate of subsidence.

The end product of this carbonate depositional model is a wide, flat, very shallow depositional plain with evaporites and muddy dolomitic sediments deposited in the shelf interior and light-colored, coarser grained, skeletal, calcium carbonate sediments deposited on the shelf margin. This plain terminates seaward at an abrupt break in slope where the depositional surface declines into the deeper water of a starved basin in which sediment accumulation has been much slower and subsidence has been greater. These basinal sediments tend to be dark in color and fine grained. Through time a three-dimensional prism is established, with vertical relief from the shelf edge to the basin floor commonly measured in hundreds of metres and, in at least one example (Enos, 1974), relief of as much as 1,000 m. Slope of the fore-reef surface may be as much as 35° (Wilson, 1974) but commonly is much more gentle, perhaps 1°-2°. Basinal carbonate sediments may be one-half to one-tenth as thick as their shelf equivalents.

What makes shelf margins appear so spectacular on geologic cross sections maps is the excessive vertical exaggeration of cross sections or the small contour intervals chosen for mapping, and the abrupt appearance of basinward thinning on the slope, in contrast to a regular pattern of thickening across the shelf toward the shelf margin.

THE CARBONATE DEPOSITIONAL MODEL THROUGH TIME

As long as subsidence is gradual and carbonate production and accumulation keep pace with or exceed subsidence, the shelf margin generally remains stationary or accretes seaward. But what happens if sea level drops, if subsidence ceases, or if uplift takes place on the craton? The former carbonate shelf is then flooded by terrigenous detritus derived from the craton. If the net sea-level change is not excessive, most of the terrigenous detritus is transported across the then-exposed carbonate terrane and dumped into the basin seaward of the old shelf margin (Fig. 2). Continuation of these submergent-emergent alternations can thus produce carbonate and terrigenous successions whose geometric relationship is reciprocal (Wilson, 1967; Meissner, 1972) but whose apparent stratigraphic relationship is equivalent. In some examples, the transport of sand and silt across the old carbonate shelf may be so efficient that the terrigenous phase on the shelf is locally nonexistent and is frequently unrecognized.

The often substantial depositional topography involved in shelf margins, as well as the reciprocal carbonate-terrigenous depositional cycles described above, makes the application of traditional stratigraphic methods difficult. Such methods tend to be dominated by concepts of original horizontality and constant basinward thickening of sedimentary units, and have frequently been unsuccessful in dealing with carbonate sequences in the shelf-to-basin zone. As a result, many carbonate shelf margins have gone unrecognized or, at best, misunderstood. P. B. King's (1959) engrossing and sympathetic account of the difficulties experienced by early West Texas geologists trying to

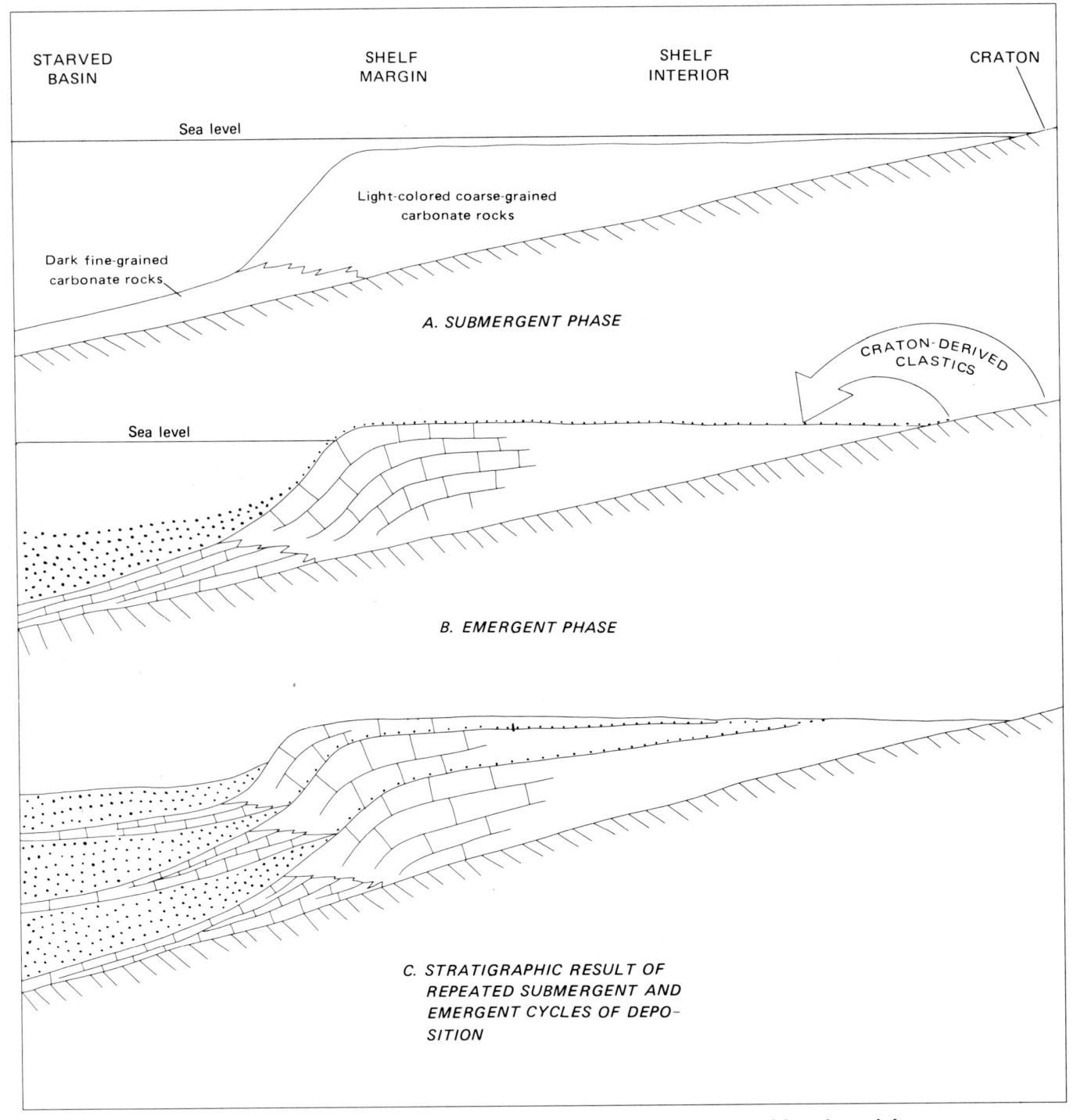

Fig. 2 — The result of cratonic uplift on the carbonate depositional model.

understand the Permian Central Basin platform illustrates the problem.

Geological literature on the Mississippian of the Western United States generally contains two kinds of misinterpretations. In some examples, shelf and shelf-margin carbonate lithosomes having reciprocal thickness relationships with immediately younger basinal lithosomes have been misinterpreted as lateral facies equivalents (Fig. 3A). In contrast, incomplete paleontological sampling has, in other examples, further confused the picture by inferring a hiatus between the thin, slowly deposited starved-basin sediments and the overlying, younger, thick, rapidly deposited, terrigeneous clastic basin-fill sediments. This inference has led to postulation of regional uplift and erosion (Fig. 3C). These two misinterpretations can be compared with the proposed shelf-margin interpretation (Fig. 3B).

LOWER DEPOSITIONAL COMPLEX

Strata in the lower depositional complex range in age from early Kinderhookian to early Meramecian. In eastern Utah and west-central Wyoming my detailed examination of the Madison Limestone (or Group) suggests that the lower part is transgressive and open marine and that in contrast, the upper part is regressive and contains evidence of shallower and more restricted depositional conditions increasing upward. Because the lower part of the Madison passes laterally into a shallow shelf facies, this distinction fades out eastward toward the Transcontinental arch where the Madison is mapped as an undivided lithic unit. Regional eastward thinning in Colorado, New Mexico and eastern Wyoming is in part related to late Paleozoic erosion of the top of the Leadville, Redwall, and Madison Limestones and their equivalents. To the west and north, however, the lower transgressive and upper regressive phases (Lodgepole and Mission Canyon Limestones, respectively) become more evident in a belt that is relatively narrow, trending northeast across central Utah and north through western Wyoming, and then widens to encompass most of Montana, including the Williston basin. Sando (1967b) called this belt the Montana province and the cratonic area to the east and south, the Wyoming province (Fig. 4). Sando called the broad miogeosynclinal area to the west, where the upper regressive phase is absent and a thick Upper Mississippian limestone succession is present, the Idaho province.

In the Montana province the lower transgressive phase is termed Lodgepole Limestone (Fig. 5); in Utah it has been called upper part of the Firchville Formation and Gardison Limestone, and in southern Nevada, Dawn and Anchor Limestone Members of the Monte Cristo Limestone (Langenheim *et al.*, 1962). These rocks are typically dark-gray, thin-bedded, cherty, fossiliferous, carbonaceous, silty limestones. The Lodgepole contains scattered bioherms. Comparison with modern carbonate sediments suggests deposition in a marine setting in water perhaps 30-100 m deep.

The upper regressive phase in Sando's Montana province is termed the Mission Canyon Limestone; it has been called Brazer Dolomite in north-eastern Utah, Deseret Limestone in central Utah, and Bullion Dolomite and Yellowpine Limestone Members of the Monte Cristo Limestone in southern Nevada. This facies includes a diverse assemblage of light-colored, thick- to massive-bedded, relatively pure dolomites and limestones containing scattered anhydrite or gypsum, and solution breccias that represent former evaporite deposits (Fig. 6). The mineralogy, textures, and sedimentary structures indicate deposition on a very shallow carbonate shelf, part of which was probably in intertidal and even supratidal settings.

In Sando's Idaho province the lower depositional complex comprises a lower, Lodgepole-type, dark limestone succession (called the Joana Limestone in western Utah and eastern

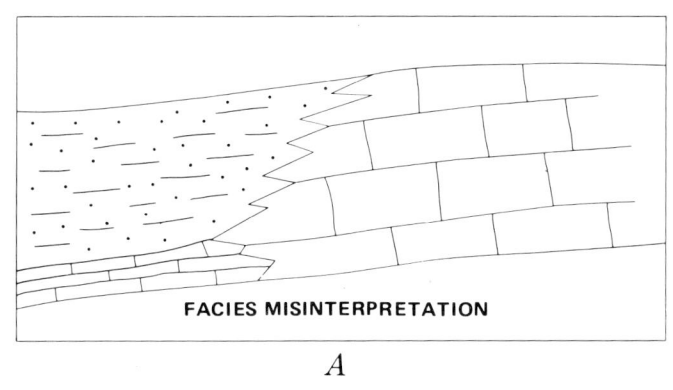

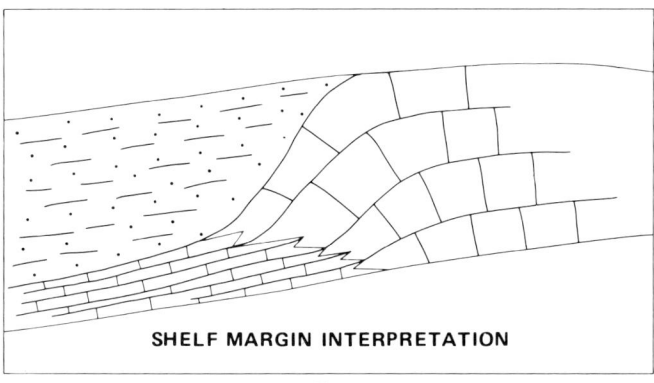

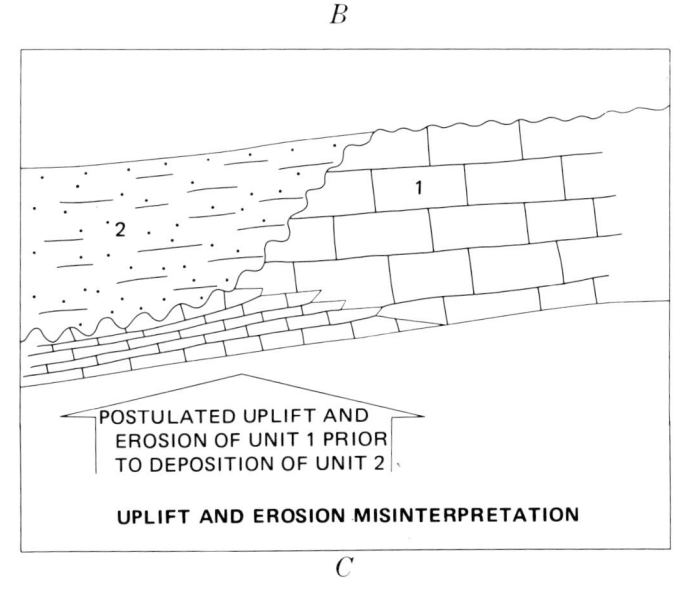

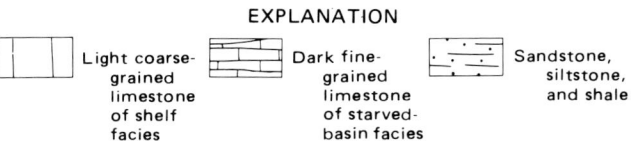

Fig. 3 — Comparison of interpretations of stratigraphic relationships of lower depositional complex. A, Facies misinterpretation. B, Proposed interpretation of shelf margin with original depositional topography and later infilling of starved basin. C, Uplift and erosion misinterpretation.

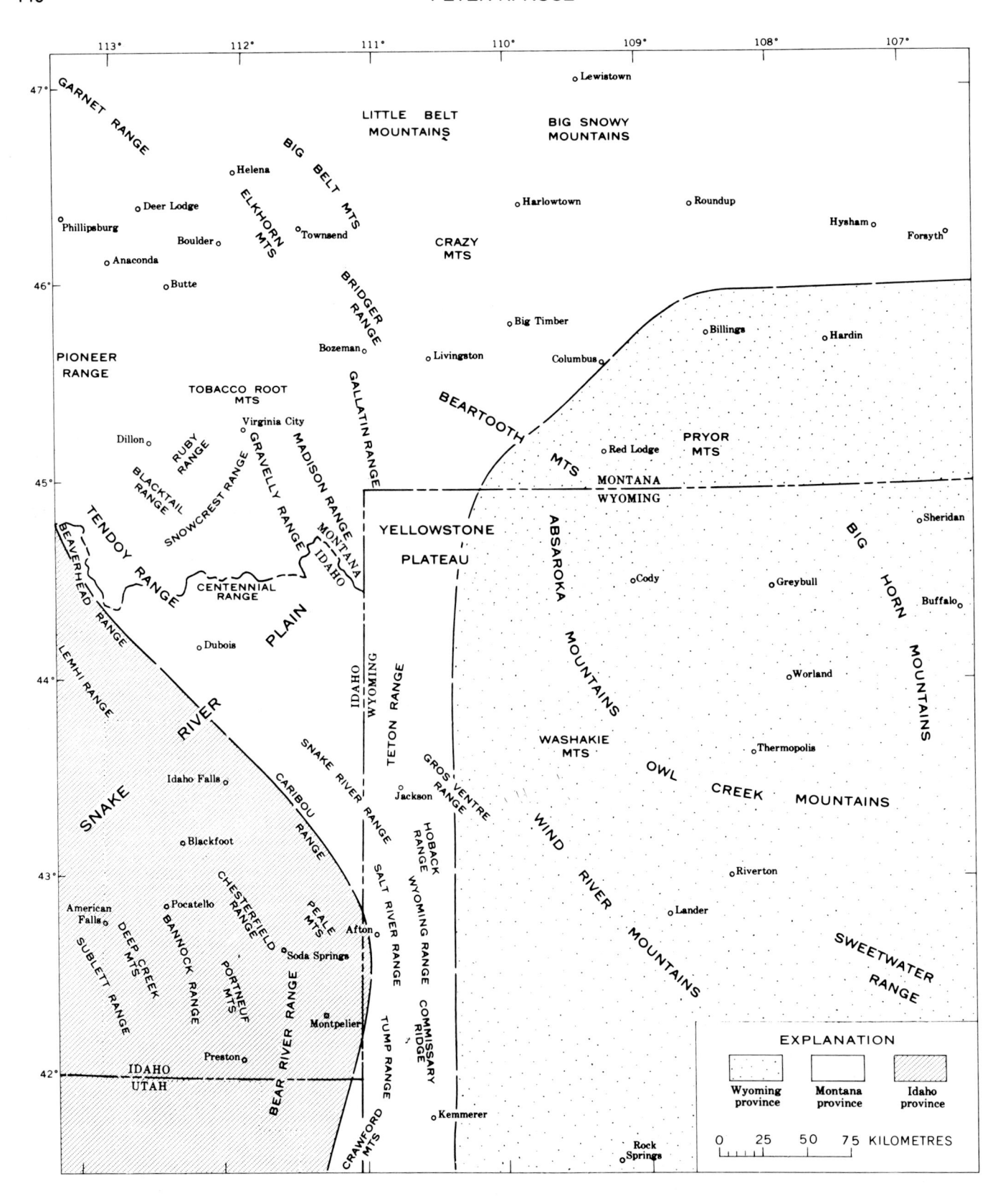

Fig. 4 — Map of the northern Cordilleran region, showing location of principal mountain ranges and Mississippian depositional provinces (modified from Sando, 1967b).

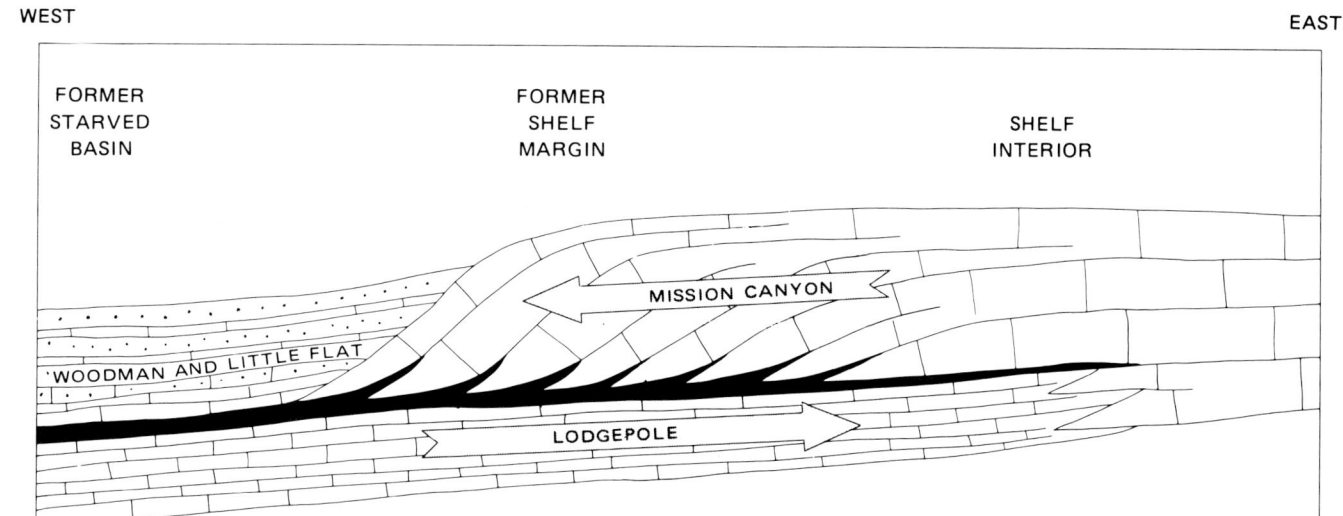

Fig. 5 — Schematic stratigraphic cross section showing the transgressive-regressive (Lodgepole and Mission Canyon Limestones) couplet of the lower depositional complex, the stratigraphic relationship of the fore-reef phosphatic zone (shown in black), and the basin-filling terrigenous clastic deposits (Woodman, Little Flat, and equivalent formations). In the shelf interior Lodgepole and Mission Canyon may be undivided, as Madison Limestone.

Nevada) and the lower part of an overlying thin-bedded, dark-colored, calcareous siltstone or silty limestone succession called Little Flat Formation in southeastern Idaho and northeastern Utah and the Upper member of the McGowan Creek Formation (Sandberg, 1975) in east-central Idaho. Elsewhere equivalents of this dark calcareous siltstone or silty limestone succession have been called Woodman Formation in western Utah and Peers Spring Formation (as used by Langenheim, 1963) in southeastern Nevada. It should be emphasized that only the lower part of this silty succession, about 100 m thick, is equivalent to the upper regressive carbonate phase (=Mission Canyon). The upper part of this silty succession is genetically related to the upper depositional complex. The true Osagean age of the lower part of the Little Flat and Woodman formations has been recently established by C. A. Sandberg (oral commun., 1975) on the basis of conodonts collected by Sandberg and R. C. Gutschick. The age of the lower part of the Peers Spring Formation has not been demonstrated paleontologically. Although the formation has been generally regarded (Langenheim, 1963) as entirely Late Mississippian, lithologic correlation with the Woodman suggests that the age of the basal part of the Peers Spring may likewise be Osagean.

Farther west, approaching the flysch deposits derived from the Antler orogen (Poole, 1974), equivalents of the Woodman and and Little Flat calcareous siltstones grade into dark silty shales of the Chainman Shale.

Of particular significance are the areal distribution and stratigraphic position of a thin phosphatic zone with respect to the transgressive-regressive lower carbonate couplet (Lodgepole and Mission Canyon) and to the siltstones (Little Flat and Woodman) that fill the formerly starved basin (Fig. 5). The phosphatic zone is best developed either in the area of the lower shelf margin, where it occurs invariably at the boundary between the deeper water transgressive facies and the overlying regressive shelf facies, or in the starved basin west of the shelf margin, where it occurs at the base of the silty Woodman and Little Flat (Gutschick, 1976, Fig. 4b; Sando and others, 1976). If one assumes that this phosphatic zone represents very slow accumulation in an upwelling marine setting, its maximum development would be expected at the toe of the westward-accreting Mission Canyon stratigraphic reef, in the starved basin where sedimentation is known to have been slow. C. A. Sandberg and R. C. Gutschick are currently pursuing the details of this economically significant stratigraphic unit.

EVIDENCE FOR LOWER SHELF MARGIN

Five lines of evidence support the existence of a prominent, accretionary, constructional carbonate shelf margin, or stratigraphic reef (Dunham, 1970) having substantial seaward topographic relief that developed during the latter phases of deposition of the lower carbonate complex:

1. Isopach mapping (Fig. 7) indicates a curvilinear, presumably constructional, ridge trending northeast across Utah from southern Nevada, then curving north and northwest across eastern Idaho, and finally crossing into westernmost Montana.

2. The widespread occurrence of dolomites and evaporites in the shelf regions of central Utah and Wyoming (Fig. 6) suggests the presence of a physical barrier that restricted circulation of normal marine waters onto the shelf.

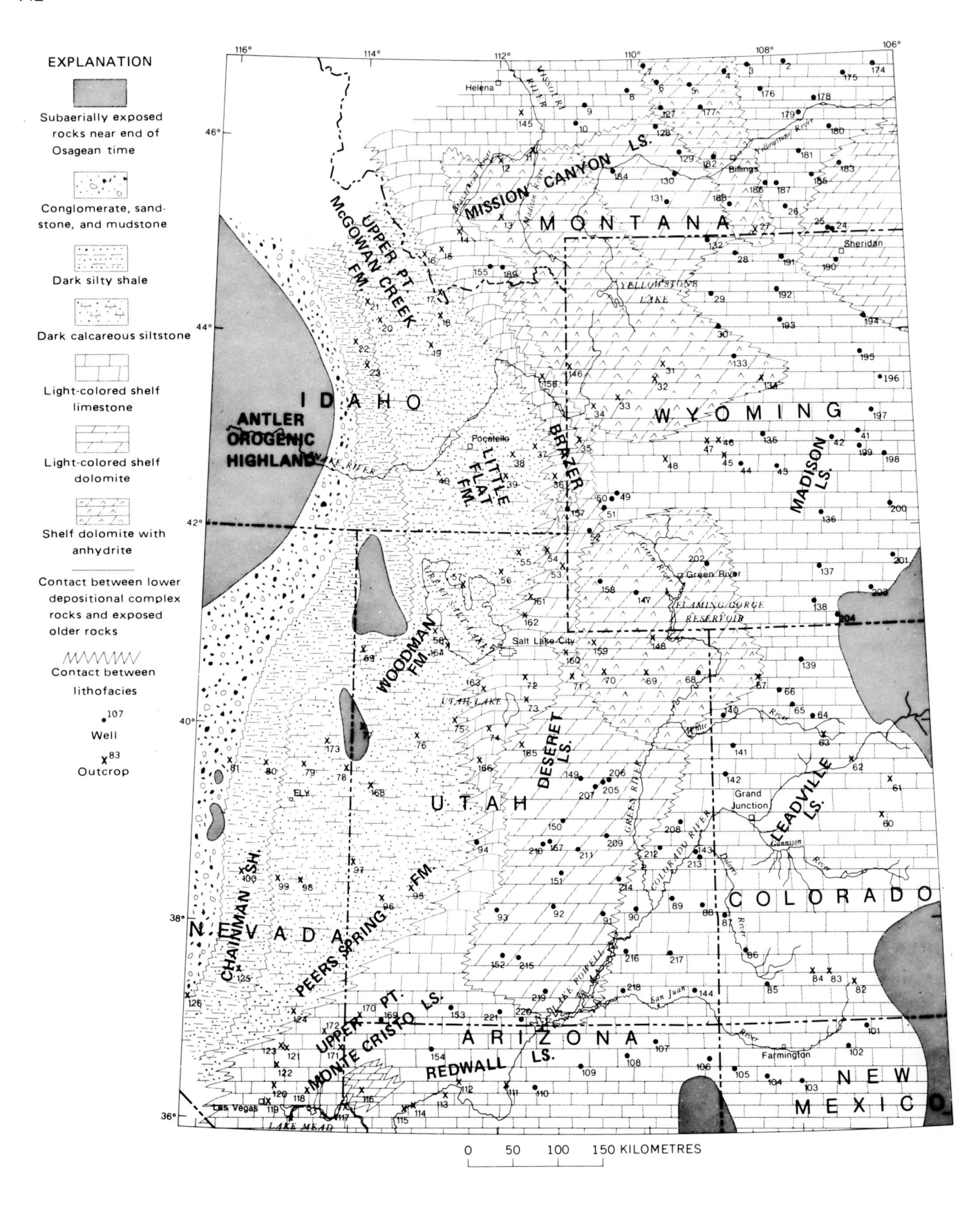

Fig. 6 — Lithofacies map of the lower depositional complex near the end of Osagean time showing provincial formation names.

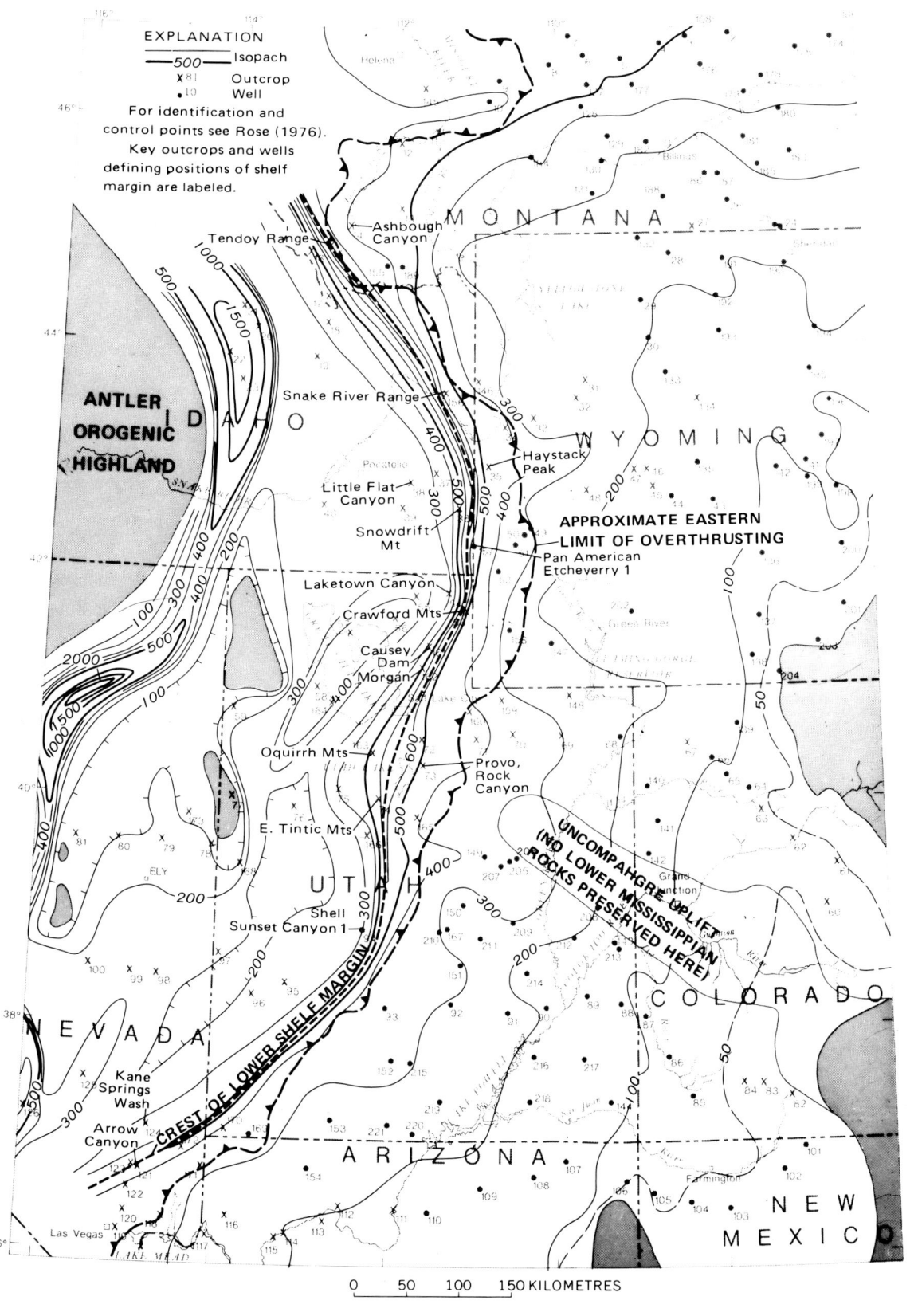

Fig. 7 — Isopach map of the lower depositional complex. Flysch trough east of Antler orogenic highland modified from Poole (1974). Eastern edge of Antler orogenic highland in Idaho restored to compensate for eastern overthrusting. Patterned areas indicate nondeposition or erosion of lower depositional complex. Hachures indicate areas of thinning. Isopach interval 50 (dashed), 100, and 500 m.

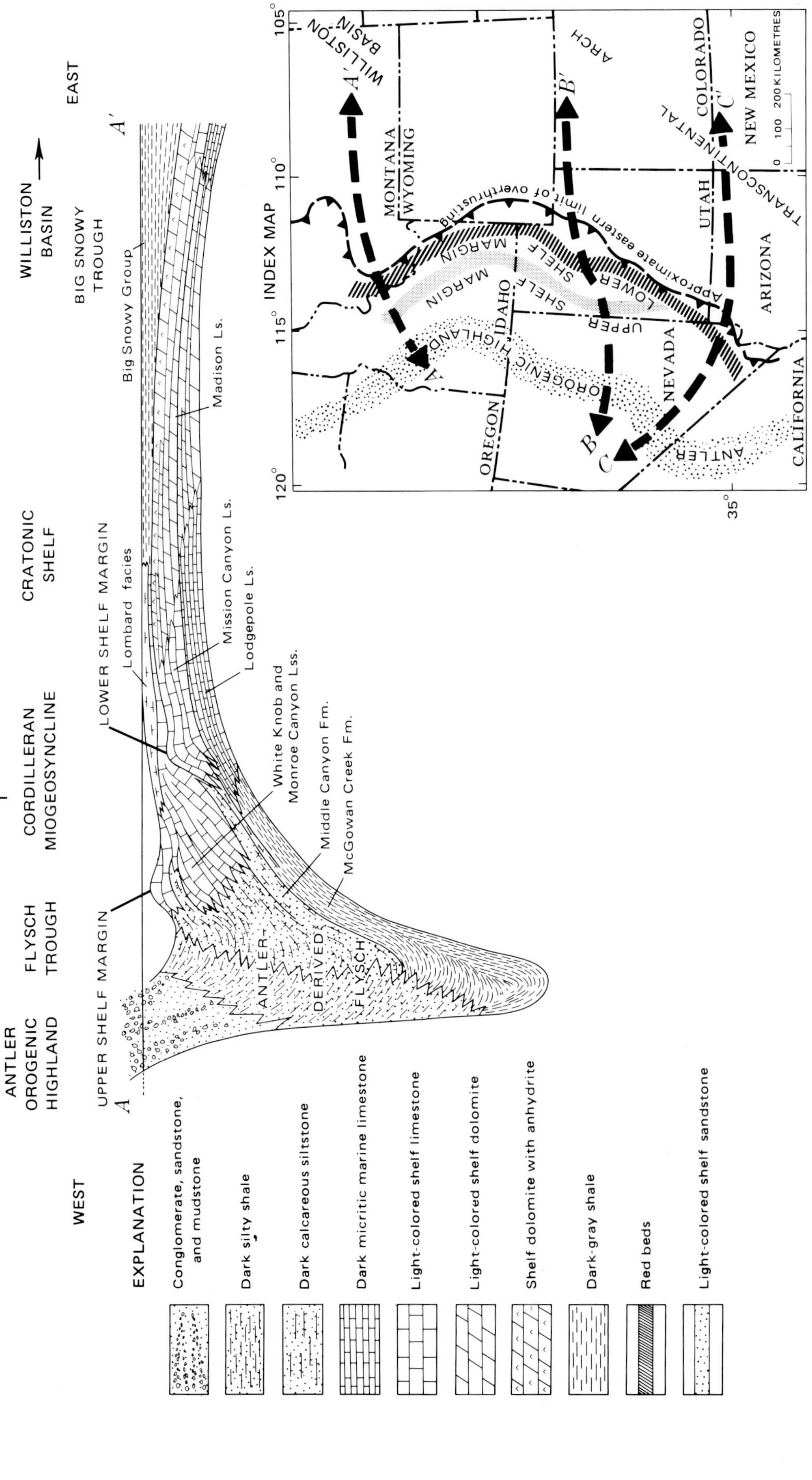
EXPLANATION
Conglomerate, sandstone, and mudstone
Dark silty shale
Dark calcareous siltstone
Dark micritic marine limestone
Light-colored shelf limestone
Light-colored shelf dolomite
Shelf dolomite with anhydrite
Dark-gray shale
Red beds
Light-colored shelf sandstone
WEST
EAST
A
A′
ANTLER OROGENIC HIGHLAND
FLYSCH TROUGH
IDAHO
MONTANA
CORDILLERAN MIOGEOSYNCLINE
CRATONIC SHELF
WILLISTON BASIN
BIG SNOWY TROUGH
UPPER SHELF MARGIN
LOWER SHELF MARGIN
Lombard facies
Big Snowy Group
Madison Ls.
Mission Canyon Ls.
Lodgepole Ls.
White Knob and Monroe Canyon Lss.
Middle Canyon Fm.
McGowan Creek Fm.
ANTLER-DERIVED FLYSCH
INDEX MAP
105°
110°
115°
120°
35°
WILLISTON BASIN
TRANSCONTINENTAL ARCH
MONTANA
WYOMING
IDAHO
OREGON
NEVADA
UTAH
COLORADO
ARIZONA
NEW MEXICO
CALIFORNIA
Approximate eastern limit of overthrusting
LOWER SHELF MARGIN
UPPER SHELF MARGIN
ANTLER OROGENIC HIGHLAND
A
A′
B
B′
C
C′
0 100 200 KILOMETRES

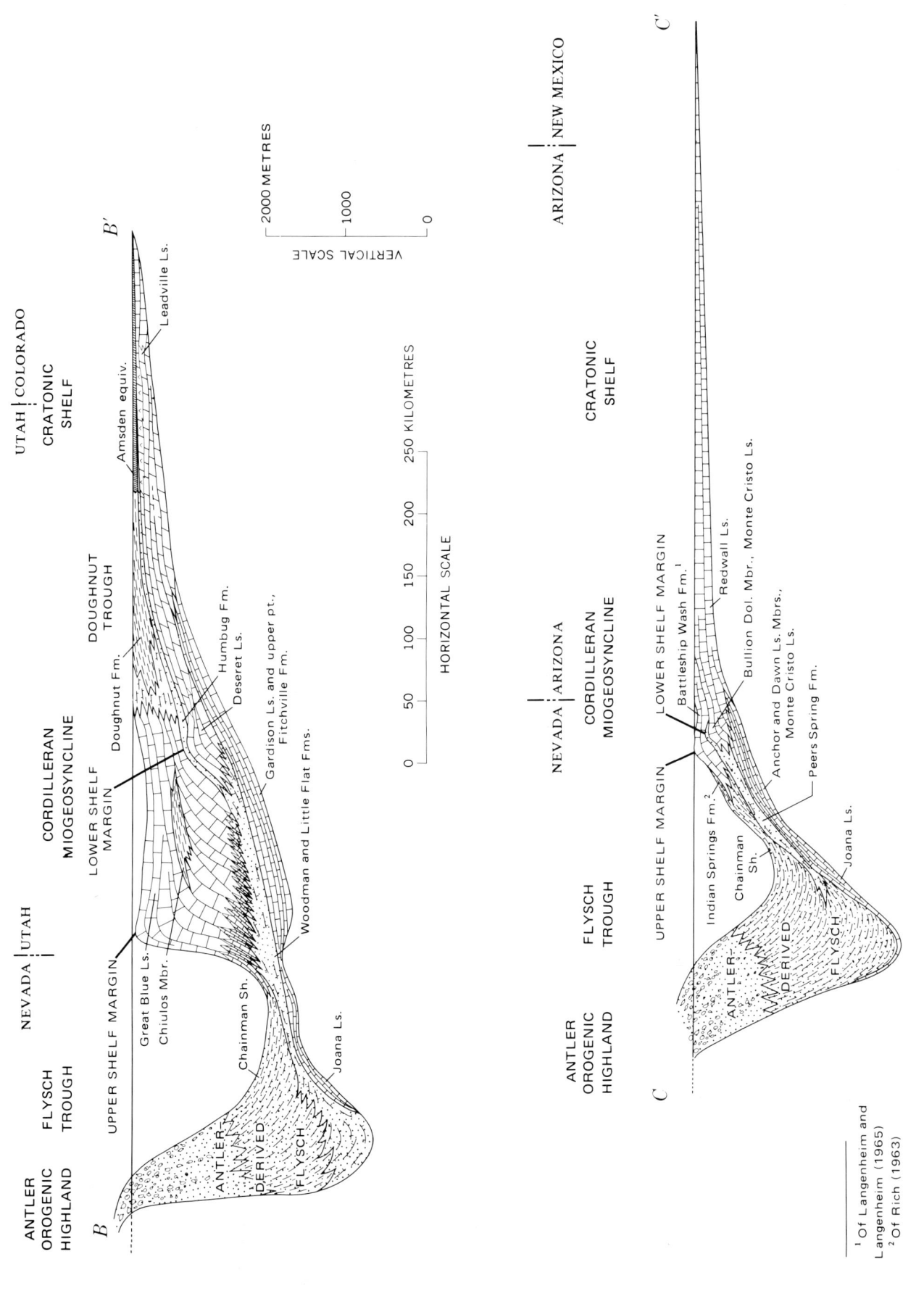

Fig. 8 — Generalized regional Mississippian stratigraphic cross sections showing local names of units and relative thicknesses. Stratigraphic datum is hypothetical late Chesterian sea level. Vertical exaggeration X65.

3. Abrupt basinward thinning occurs directly west of the curvilinear carbonate ridge. This thinning is gentlest and least apparent in southwestern Utah and southern Nevada, where the lower depositional complex is about 500 m thick along the crest of the linear ridge and a little less than 250 m thick about 30 km basinward to the northwest. In central Utah, the lower depositional complex thins from nearly 600 m to 200 m in a distance of less than 20 km. The basinward thinning is most pronounced along the Idaho-Montana border, where the isopached unit thins basinward from about 800 m to about 400 m in a distance of only 20 km. This prounced thinning suggests topographic relief along the foreslope of perhaps 200 to 400 m, assuming that no compensatory isostatic adjustment occurred for the thicker shelf margin sediment load during Meramecian time and that correlation surfaces are in fact isochronous. Further, if the carbonate ridge crest was essentially at sea level as the shallow-water carbonates suggest, water depth in the starved basin to the west was about 200-400 m.

4. A profound change in facies coincides with the basinward thinning (Fig. 6). Schematic regional cross sections (Fig. 8A, B, C) show that shelf carbonates of the upper regressive phase (Mission Canyon and equivalents) thin abruptly and disappear, apparently grading into much thinner, dark silty carbonates of the lower part of the Woodman, the Little Flat, and their equivalents. These silty sediments represent craton-derived terrigenous clastics swept across the shelf during periods of exposure and gentle cratonic uplift. (See section "The Carbonate Depositional Model Through Time.") They are genetically related to the sandy carbonates of the Humbug Formation of eastern and central Utah, which accumulated in an east-west trough that seems to have served as a broad conduit for transport of terrigenous clastics from the shelf into the starved basin primarily after the close of Mission Canyon deposition, when the carbonate shelf was subaerially exposed. The transgressive Lodgepole (and equivalent) carbonates seem relatively unaffected by any thickness or lithologic changes, in contrast to the shelf-to-basin changes of the Mission Canyon and Little Flat. The previous characterization of the Mission Canyon as regressive is compatible with the earlier discussion of the accretionary process of shelf-margin construction. (See section "The Carbonate Depositional Model Through Time.") Although he was the first to recognize this combination of facies change and basinward thinning, Sando (1967b, 1975, 1976) ascribed the relationship not to reciprocal sedimentation along a shelf-margin belt, but rather to Meramecian uplift and erosion, which removed all the Mission Canyon strata in the Idaho Province and which was followed by subsidence that accommodated Little Flat deposits (Fig. 3C). However, Sando's original interpretation is modified in a forthcoming paper (Sando et al., 1976) to coincide essentially with the concept presented herein.

5. The consistency of these thickness and facies changes along a regional curvilinear trend is similar to that of other well-documented shelf margins, such as the Lower Cretaceous Edwards reef trend of the Gulf Coast region and the Guadalupian shelf margin in the Permian Basin of West Texas and New Mexico.

In summary, the lower depositional complex represents a complete miogeosynclinal sequence with a wide cratonic carbonate shelf thickening gradually to the west. This carbonate shelf was terminated on the west by an accretional barrier-type shelf margin having seaward topographic relief of several hundred metres. Behind this stratigraphic reef accumulated shelf-interior evaporites and low-energy, very shallow water dolomites and limestones. Seaward of this shelf margin, deposition occurred slowly in a starved basin, where dark silty limestones and shales formed in waters as much as 400 m deep. Madison-equivalent basinal sediments thin westward toward the thick Lower Mississippian flysch deposits derived from the Antler orogenic belt (Poole, 1974). The lower depositional complex can therefore be characterized as a transgressive-regressive carbonate couplet (Fig. 5); shelf-interior evaporites, constructional skeletal shelf-margin carbonates, and dark starved-basin sediments were generated during the upper regressive phase.

If this postulated shelf margin is so prominent, why was it not recognized long ago, inasmuch as the Madison and its equivalents have been studied so widely and for so many years? There are probably two reasons.

1. Most of the mountain ranges in which the shelf margin occurs are controlled by Laramide thrust faults, which trend generally north-south. Thus these mountain ranges tend to be narrow features also elongated north-south. Unfortunately, the shelf margin is also a linear north-south feature. Accordingly, the presence of extended east-west outcrops is severely limited, and the possibility of viewing the outcropping shelf margin in an extended depositional dip direction, which would trend east-west, is extremely rare. On the other hand, the presence of spectacular, clear outcrops of the Guadalupian shelf margin, oriented parallel to depositional dip and showing steep accretionary reef foreset beds, was of great help in identifying and understanding the Permian Reef of West Texas.

2. Carbonate stratigraphy has undergone a revolution in the past 20 years. Many previous workers were simply unfamiliar with carbonate-style stratigraphy. Traditional stratigraphic concepts and methods, particularly those involving submarine depositional topography, simply did not lend themselves readily to working out the true nature of carbonate-shelf and starved-basin relationships.

UPPER DEPOSITIONAL COMPLEX

The upper depositional complex contains strata of middle Meramecian through middle late Chesterian age. It specifically exludes the thick lens of Manning Canyon Shale of latest Chesterian and Morrowan age that filled the foundering incipient Oquirrh basin of central Utah; also it excludes thinner contemporaneous black shales of the so-called "Chainman Shale" that overlie Great Blue shelf carbonates in western and northern Utah.

The upper depositional complex reflects sedimentation in a relatively narrow, rapidly subsiding miogeosyncline confined between an emergent craton on the east and the active Antler orogen on the west.

Thin deposits of red beds and sandstones form an apron along the emergent craton to the east (Fig. 9); these are the basal deposits of the Amsden Formation. Two east-west depressions indent this upper shelf. The northernmost is the Big Snowy trough of central Montana, filled by estuarine sands, silty limestones, and black carbonaceous shales of the Big Snowy Group. To the south, in northeastern Utah and northwestern Colorado, is the counterpart of the Big Snowy trough, here called the Doughnut trough (Uinta basin of Sando and others, 1975). This east-west depression, precursor to the Pennsylvanian Oquirrh basin, began to form near the end of Deseret deposition. During deposition of the upper carbonate complex, this depression was first the conduit for transport of cratonic sands of the Humbug Formation into the starved basin west of the old lower shelf margin (see item 4 on "Evidence for Lower Shelf Margin"). It later received thick deposits of euxinic black shales and limestones of the Doughnut and Great Blue formations. Regional mapping shows clearly, however, that the Doughnut trough was a cratonic depression behind the upper shelf margin, separated from the Late Mississippian Cordilleran seaway by a wide belt of thick shallow-marine shelf limestones of the Great Blue and Ochre Mountain Limestones.

Scattered islands or ridges of Madison-age carbonate rocks projected above the depositional surface of the upper shelf interior and were never covered. Prominent among these was a north-south ridge along the Wyoming-Idaho border, representing the exposed crest of the lower shelf margin (Fig. 9). This feature was called the Bannock Highland by Richardson (1941). The highest observed average porosity of outcropping Madison-age carbonate rocks in the entire study area is present here, possibly owing to Late Mississippian subaerial exposure and leaching.

As the upper depositional complex thickens westward across the shelf, the eastern red beds grade into sandy dolomites that in turn grade into thick-bedded, skeletal shelf limestones, which reach a maximum thickness of at least 1,500 m in west-central Utah and more than 1,000 m in northeastern Idaho (Fig. 10). This thick carbonate succession has been called the White Knob Limestone and Scott Peak, South Creek, and Surrett Canyon formations in east-central Idaho, Monroe Canyon Limestone or upper part of the so-called Brazer Limestone in southeastern Idaho, Ochre Mountain Limestone in westernmost Utah, Great Blue Limestone throughout most of central Utah, and, where it is much thinner in southern Nevada, the Battleship Wash Formation of Langenheim and Langenheim (1965).

The conspicuous transgressive-regressive character of the lower carbonate depositional complex is somewhat less apparent in the upper carbonate depositional complex. Instead, a basal terrigenous clastic succession, the upper parts of the silty Woodman, Peers Spring, Little Flat, and Middle Canyon formations previously described, is ubiquitous beneath the upper carbonate succession. I believe most of this terrigenous clastic sequence represents a cratonic sand sheet swept across the exposed and eroding lower shelf and dropped into the basin after the close of Madison deposition. Much of the middle or upper silty facies, therefore, is correlative with the unconformity at the top of the Madison and has no recognized shelf equivalents, except for the Humbug Formation, the Darwin Member of the Amsden Formation, and sandstone fillings of post-Madison karst topography (see section "The Cabonate Depositional Model through Time"). The thick upper depositional succession appears to consist of several thick carbonate cycles, rather than the simple transgressive-regressive couplet of its lower counterpart. Huh (1967), however, emphasized the overall accretionary offlap nature of the entire upper carbonate succession (White Knob Limestone) in northeastern Idaho.

Farther west, the carbonate rocks of the upper depositional complex grade into thin, dark, silty limestones and calcareous siltstones, just as the lower shelf carbonate rocks grade into comparable dark, basinal, silty carbonate deposits. These Upper Mississippian basinal deposits are included in the upper part of the Woodman, Peers Spring, and Chainman formations in Utah and Nevada and in part of the Deep Creek Formation of eastern Idaho. These sediments in turn grade westward into flysch deposits of the Antler foreland basin, which were derived the Antler orogenic highland.

In western Utah and eastern Nevada, the Upper Mississippian basinal silty carbonates occur in a starved-basin setting, where they are generally 300-500 m thick; equivalent shelf carbonate rocks to the east are three or four times thicker (Fig. 8, cross section B-B'). Northward in Idaho, however, the depositional trough narrows markedly, the Antler-derived flysch deposits are adjacent to shelf limestone deposits, and no starved basin is present (Fig. 8, cross section A-A').

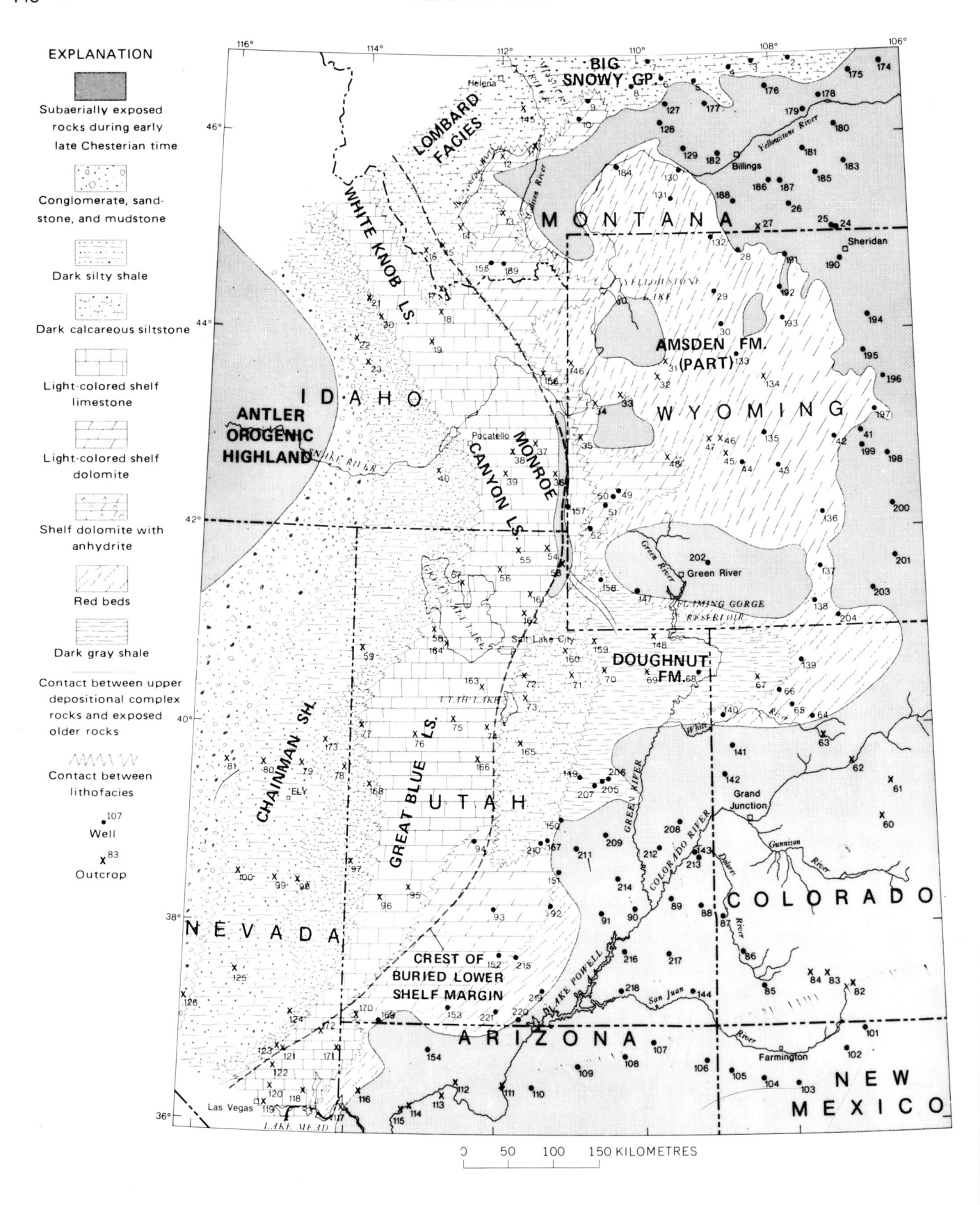

Fig. 9 — Lithofacies map of the upper depositional complex in early late Chesterian time showing provincial formation names.

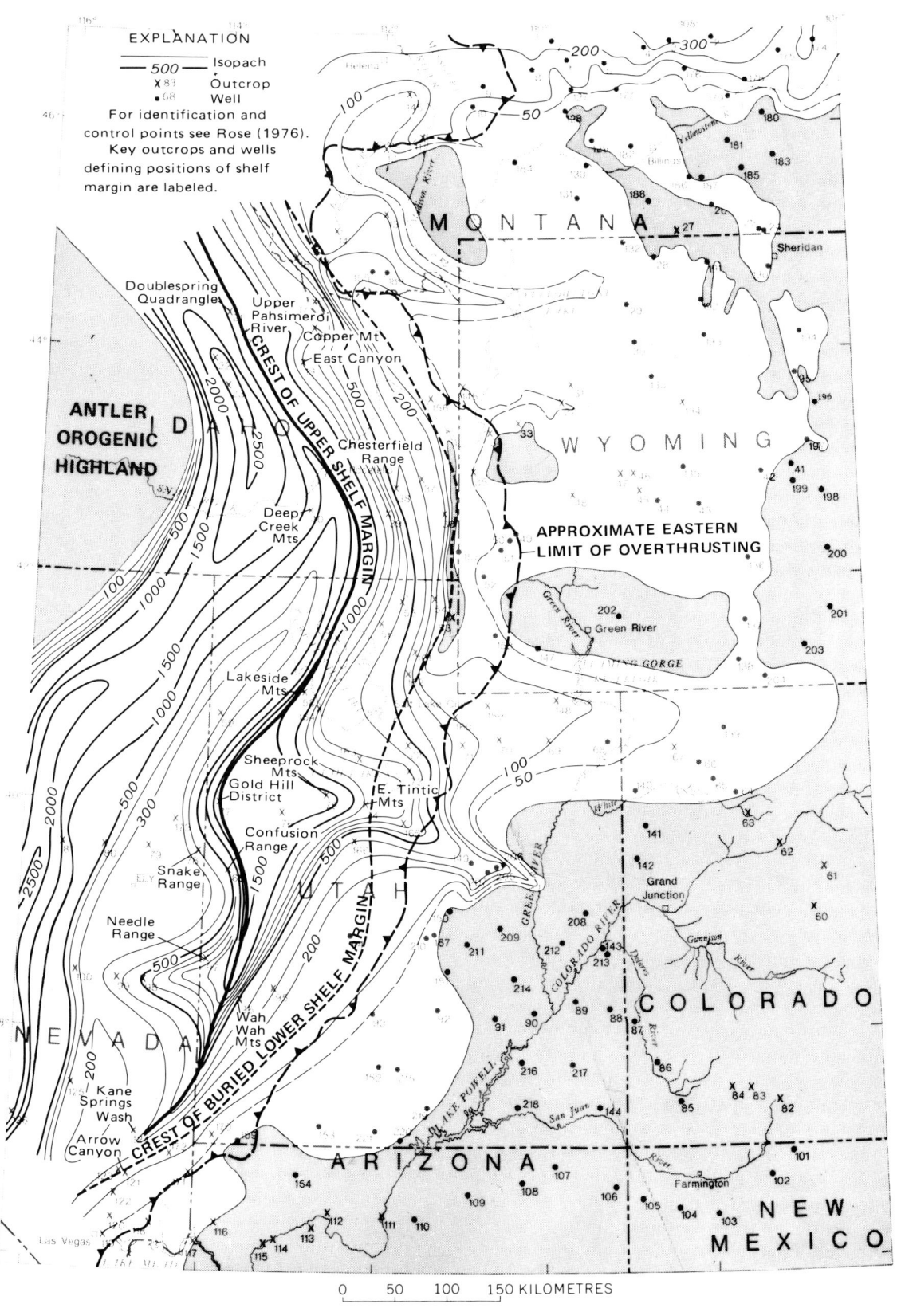

Fig. 10 — Isopach map of the upper depositional complex. Flysch trough east of Antler orogenic highland modified from Poole (1974). Eastern edge of Antler orogenic highland in Idaho restored to compensate for eastern overthrusting. Patterned areas indicate nondeposition or erosion of upper depositional complex. Hachures indicate areas of thinning. Isopach interval 50 (dashed), 100, and 500 m.

South of the thick carbonate accumulation in west-central Utah, the Great Blue Limestone apparently thins gradually, converges upon, and finally pinches out against the old lower shelf margin in southern Nevada (Fig. 8, cross section C-C'). Here the thin Battleship Wash Formation appears to be equivalent to all the thicker Great Blue carbonate rocks to the north.

EVIDENCE FOR UPPER SHELF MARGIN

Available evidence suggests that the upper shelf margin may have been more complex than its lower counterpart. Nevertheless, the same five main elements of evidence exist:

1. Isopach mapping (Fig. 10) suggests the existence of a constructional barrier that could have restricted circulation of marine waters in the shelf interior.

2. The presence of dolomites and red beds indicates restricted circulation in the shelf interior.

3. Isopach mapping shows abrupt basinward thinning in Utah immediately seaward of the barrier, suggesting topographic relief of 500 to 1,000 m. Proximity of Antler flysch deposits to the shelf margin in central Idaho apparently precluded formation of a deep starved-basin there, however, and so distal orogenic and carbonate deposits are intertongued along the shelf margin.

4. Massive skeletal limestone beds change facies to dark silty carbonates coincident with the aforementioned basinward thinning (Fig. 9).

5. All these features occur in a consistent regional pattern analogous to other ancient and modern shelf margins, and also similar to the immediately underlying lower shelf margin.

In summary, the upper depositional complex represents carbonate shelf sedimentation in a miogeosynclinal setting between an emergent craton on the east, which shed terrigenous clastics into the shelf interior, and an active orogen on the west, with its accompanying flysch trough. This north-south carbonate shelf was crossed by two elongate east-west troughs filled with cratonic sands and euxinic muds. Substantial topographic relief of perhaps as much as 1,000 m developed where the carbonate shelf margin declined into the starved basin. In Idaho, however, where the flysch trough converges upon the carbonate shelf, little topographic relief probably developed at the edge of the upper carbonate shelf.

PERTINENT FUTURE RESEARCH

Three lines of future research on Mississippian shelf margins of the western United States are suggested:

1. Detailed field investigations in appropriately located mountain ranges are needed for both the lower and upper shelf-carbonate successions, to determine if shelf-margin features can be recognized in outcrop and to compare such features with other ancient counterparts, as well as with features in different parts of the same Mississippian shelf. Such investigations may document the nature of the lower and upper shelf margins and add new dimensions to existing shelf-margin concepts. Preliminary investigations in some parts of the report area have already been initiated by F. G. Poole, C. A. Sandberg, and myself. Continuing detailed biostratigraphic studies by W. J. Sando, J. T. Dutro, Jr., Mackenzie Gordon, Jr., and B. L. Mamet in Wyoming, Montana, Idaho, and Utah will also provide important data.

2. Using the shelf-margin concept, eastern Great Basin stratigraphy should be re-evaluated, specifically in regard to possible petroleum source rocks and phosphate-rich units, as well as preferentially mineralized rock units or facies. Such a study is underway by Gutschick (1976) and Sandberg (1975).

3. The distribution of porosity and permeability in the lower carbonate depositional complex should be examined in light of the shelf-margin model to see if useful generalizations can be made regarding primary carbonate facies, early diagenetic history, and development and present occurrence of reservoir rocks and seals.

REFERENCES CITED

Adams, J. E., Frenzel, H. N., Rhodes, M. L., and Johnson, D. P., 1951, Starved Pennsylvanian Midland basin: Am. Assoc. Petroleum Geologists Bull., v. 35, no. 12, p. 2600-2607.

Andrichuk, J. M., 1955, Mississippian Madison group stratigraphy and sedimentation in Wyoming and southern Montana: Am. Assoc. Petroleum Geologists Bull., v. 39, no. 11, p. 2170-2210.

Coogan, A. H., 1969, Recent and ancient carbonate cyclic sequence, in Cyclic sedimentation in the Permian Basin — Symposium, Midland, Texas, 1967: West Texas Geol. Soc. Pub. 69-56, p. 5-16.

Craig, L. C., 1972, Mississippian System, in Mallory, W. W., ed., Geologic atlas of the Rocky Mountain region: Rocky Mtn. Assoc. Geologists, p. 100-110.

Dunham, R. J., 1970, Stratigraphic reefs versus ecologic reefs: Am. Assoc. Petroleum Geologists Bull., v. 54, no. 10, p. 1931-1932.

Eardley, A. J., 1962, Structural geology of North America [2d ed.]: New York, Harper and Row, 743 p.

Enos, Paul, 1974, Reefs, platforms and basins of Middle Cretaceous in northeast Mexico: Am. Assoc. Petroleum Geologists Bull., v. 58, no. 5, p. 800-809.

Fisher, W. L., and Rodda, P. U., 1969, Edwards Formation (Lower Cretaceous), Texas — Dolomitization in a carbonate platform system: Am. Assoc. Petroleum Geologists Bull., v. 53, no. 1, p. 55-72.

Ginsburg, R. N., 1956, Environmental relationships of grain size and constituent particles in some south Florida carbonate sediments: Am. Assoc. Petroleum Geologists Bull., v. 40, no. 10, p. 2384-2427.

Gutschick, R. C., 1976, Preliminary reconnaisance study of Lower and lower Upper Mississippian strata across northwestern Utah: U.S. Geol. Survey Open-File Rept. 76-200, 40 p.

Hintze, L. F., 1973, Geologic history of Utah: Brigham Young Univ. Geology Studies, v. 20, pt. 3, 181 p.

Huh, O. K., 1967, The Mississippian System across the Wasatch line, east-central Idaho, extreme southwestern Montana, in Montana Geol. Soc. Guidebook 18th Ann. Field Conf., 1967, Centennial basin of southwest Montana: p. 31-62.

Illing, L. V., Wells, A. J., and Taylor, J. C. M., 1965, Penecontemporary dolomite in the Persian Gulf, in Pray, L. C., and Murray, R. C., eds., Dolomitization and limestone diagenesis — a symposium: Soc. Econ. Paleontologists and Mineralogists Spec. Pub. 13, p. 89-111.

Imbrie, John, and Purdy, E. G., 1962, Classification of modern Bahamian carbonate sediments, in Classification of carbonate rocks — A symposium: Am. Assoc. Petroleum Geologists Mem. 1, p. 253-272.

Irwin, M. L., 1965, General theory of epeiric clear water sedimentation: Am. Assoc. Petroleum Geologists Bull., v. 49, no. 4, p. 445-459.

Jackson, W. E., 1964, Depositional topography and cyclic deposition in west-central Texas: Am. Assoc. Petroleum Geologists Bull., v. 48, no. 3, p. 317-328.

Kay, Marshall, 1951, North American geosynclines: Geol. Soc. America Mem. 48, 143 p.

King, P. B., 1948, Geology of the southern Guadalupe Mountains, Texas: U.S. Geol. Survey Prof. Paper 215, 183 p. [1949].

_____, 1959, The evolution of North America: Princeton, N. J., Princeton Univ. Press, 190 p.

Langenheim, R. L., Jr., 1963, Nomenclature of the Late Mississippian White Pine Shale and associated rocks in Nevada: Illinois Acad. Sci. Trans. 1962, v. 55, no. 2, p. 133-145.

_____, Carss, B. W., Kennerly, J. B., McCutcheon, V. A., and Waines, R. H., 1962, Paleozoic section in Arrow Canyon Range, Clark County, Nevada: Am. Assoc. Petroleum Geologists Bull., v. 46, no. 5, p. 592-609.

Langenheim, V. A. M., and Langenheim, R. L., Jr., 1965, The Bird Spring Group, Chesterian through Wolfcampian at Arrow Canyon, Arrow Canyon Range, Clark County, Nevada: Illinois Acad. Sci. Trans., v. 58, no. 4, p. 225-240.

Laporte, L. F., 1967, Carbonate deposition near mean sea-level and resultant facies mosaic — Manlius Formation (Lower Devonian) of New York State: Am. Assoc. Petroleum Geologists Bull., v. 51, no. 1, p. 73-101.

Meissner, F. F., 1972, Cyclic sedimentation in Middle Permian strata of the Permian Basin, west Texas and New Mexico, in Cyclic sedimentation in the Permian Basin — Symposium, Midland, Texas, 1967 [2d ed.]: West Texas Geol. Soc. Pub. 72-60, p. 203-232.

Neumann, A. C., and Land, L. S., 1975, Lime mud deposition and calcareous algae in the Bight of Abaco, Bahamas — A budget: Jour. Sed. Petrology, v. 45, no. 4, p. 763-786.

Poole, F. G., 1974, Flysch deposits of Antler foreland basin, western United States, in Dickinson, W. R., ed., Tectonics and sedimentation: Soc. Econ. Paleontologists and Mineralogists Spec. Pub. 22, p. 58-82.

Rich, J. L., 1951, Three critical environments of deposition, and criteria for recognition of rocks deposited in each of them: Geol. Soc. America Bull., v. 62, no. 1, p. 1-20.

Rich, Mark, 1963, Petrographic analysis of Bird Spring Group (Carboniferous-Permian) near Lee Canyon, Clark County, Nevada: Am. Assoc. Petroleum Geologists Bull., v. 47, no. 9, p. 1657-1661.

Richardson, G. B., 1941, Geology and mineral resources of the Randolph quadrangle, Utah-Wyoming: U.S. Geol. Survey Bull. 923, 54 p.

Rose, P. R., 1976, Key wells and outcrops for regional analysis of Mississippian rocks, Western United States: U.S. Geol. Survey Open-File Rept. 76-242.

Sandberg, C. A., 1975, McGowan Creek Formation, new name for Lower Mississippian flysch sequence in east-central Idaho: U.S. Geol. Survey Bull., 1405-E, 11 p.

Sando, W. J., 1967a, Madison Limestone (Mississippian), Wind River, Washakie, and Owl Creek Mountains, Wyoming: Am. Assoc. Petroleum Geologists Bull., v. 51, no. 4, p. 529-557.

_____, 1967b, Mississippian depositional provinces in the northern Cordilleran region, in Geological Survey research 1967: U.S. Geol. Survey Prof. Paper 575-D, p. D29-D38.

_____, 1972, Madison Group (Mississippian) and Amsden Formation (Mississippian and Pennsylvanian) in the Beartooth Mountains, northern Wyoming and southern Montana, in Montana Geol. Soc. Guidebook, 21st Ann. Field Conf., Crazy Mountains Basin, 1972: p. 57-63, 1 pl., 9 figs.

_____, 1974, Mississippian history of the northern Rocky Mountains (abs.): Geol. Soc. America Abs. with Programs, v. 6, no. 7, p. 938-939.

_____, 1975, Diastem factor in Mississippian rocks of the northern Rocky Mountains: Geology, v. 3, no. 11, p. 657-770.

_____, 1976, Mississippian history of the northern Rocky Mountains region: U.S. Geol. Survey Jour. Research, v. 4, no. 3, p. 317-338.

_____, and Dutro, J. T., Jr., 1960, Stratigraphy and coral zonation of the Madison group and Brazer dolomite in northeastern Utah, western Wyoming, and southwestern Montana, in Wyoming Geol. Assoc. Guidebook, 15th Ann. Field Conf., Overthrust belt of southwestern Wyoming and adjacent areas 1960: p. 117-126.

_____, Dutro, J. T., Jr., Sandberg, C. A., and Mamet, B. L., 1976, Revision of Mississippian stratigraphy, eastern Idaho and northeastern Utah: U.S. Geol. Survey Jour. Research, v. 4, no. 4, in press.

_____, Gordon, Mackenzie, Jr., and Dutro, J. T., Jr., 1975, Stratigraphy and geologic history of the Amsden Formation (Mississippian and Pennsylvanian) of Wyoming: U.S. Geol. Survey Prof. Paper 848-A, 78 p., 11 pl., 23 figs.

Shinn, E. A., Ginsburg, R. N., and Lloyd, R. M., 1965, Recent supratidal dolomite from Andros Island, Bahamas, in Pray, L. C., and Murray, R. C., eds., Dolomitization and limestone diagenesis — A symposium: Soc. Econ. Paleontologists and Mineralogists Spec. Pub. 13, p. 112-123.

Stockman, K. W., Ginsburg, R. N., and Shinn, E. A., 1967, The production of lime mud by algae in south Florida: Jour. Sed. Petrology, v. 37, no. 2, p. 633-648.

Van Siclen, D. C., 1958, Depositional topography — Examples and theory: Am. Assoc. Petroleum Geologists Bull., v. 42, no. 8, p. 1897-1913.

Wilson, J. L., 1967, Cyclic and reciprocal sedimentation in Virgilian strata of southern New Mexico: Geol. Soc. America Bull., v. 78, no. 7, p. 805-818.

_____, 1974, Characteristics of carbonate-platform margins: Am. Assoc. Petroleum Geologists Bull., v. 58, no. 5, p. 810-824.

photo by Peter J. Varney

North Side of Little Cottonwood Canyon near Alta and Snowbird. The dark colored unit on the left is the Upper Precambrian Mineral Fork Formation. The light colored unit on the right and in the canyon is the Mississippian Madison Limestone.

RELATIONSHIPS OF PENNSYLVANIAN-PERMIAN STRATIGRAPHY TO THE LATE MESOZOIC THRUST BELT IN THE EASTERN GREAT BASIN, UTAH AND NEVADA

by

John E. Welsh[1]

ABSTRACT

Exploration for oil and gas in the late Paleozoic rocks along the Mesozoic thrust belt in the eastern Great Basin has tested the Callville-Pakoon-Queantoweap-Kaibab platform with discoveries at the Anderson Junction, Upper Valley, and Last Chance fields. Wildcat tests in Clark County, Nevada and southwestern Utah are few and widely scattered. The possibility for future discoveries is excellent in the platform sequence. Allochthonous rocks of the Bird Spring and Oquirrh basins and time-equivalent rocks in the western Utah and eastern Nevada synclinoria have not been seriously tested. These basinal sequences have organic-rich source beds interbedded with possible reservoir rocks and deserve wildcat tests.

Stratigraphic relations of the Pennsylvanian and Permian rocks illustrate west to east beveling of the Permian Kaibab Group by the Triassic, wedge-edge pinch-out of Pennsylvanian carbonates on the Emery-Kaibab uplift and Ely-Ferguson Mountain shelf beneath Wolfcampian carbonates, and eustatic sea level control on evaporite-carbonate distribution. All of these situations have the effect of increasing porosity in the carbonates beneath unconformities.

Subsurface interpretation of wildcat tests along the "hingeline" in southeastern Utah is reviewed and compared with surface outcrop sections. Differences between the Laramide thrust belt and the wedge-edge of the Paleozoic depo-prism, the Wasatch-Las Vegas "hingeline", are discussed for the area between Salt Lake City and Las Vegas.

INTRODUCTION

This paper will illustrate a few specific stratigraphic studies of the writer's concerning the Pennsylvanian and Permian sections and wildcat tests along the "hingeline" between Salt Lake City and Las Vegas. Reference to other areas in Utah and Nevada will be made in the discussion. The reader is referred to Figure 1, for a generalized location map of the late Paleozoic depositional areas and the late Mesozoic tectonic elements. Correlation Table 1 lists the relative time-stratigraphic positions of rock units as used by the writer. This paper will make no specific reference to the many excellent regional published papers on the upper Paleozoic since they are readily available (Bissell, 1962). A few of the more obscure references are cited at the end of the paper. The statements made are based either upon field observations or laboratory examination of well samples by the writer. There has been no collaboration with other investigators on the interpretation of the data presented.

[1] Economic geologist, 4780 Bonair St., Holladay, Utah 84117.

The writer wishes to acknowledge A. W. Grier, formerly of AMAX Petroleum, for supporting the surface studies, Jock Campbell, Utah Geol. Survey for providing logs and samples, and Mountain Fuel Supply Co. for loan of the Shurtz Creek samples. Janet R. Welsh assisted in the laboratory and typed the manuscript.

TECTONIC SETTING

Two independent criteria may be used to define the Wasatch-Las Vegas "hingeline" along the edge of the eastern Great Basin. One method is to use the stratigraphic thinning of the Paleozoic depo-prism and the other is to use the leading edge of the overthrust belt (Fig. 1). The thrust belt, in a regional sense, can be demonstrated to cut obliquely to the depo-prism edge.

Eastward thinning of the Cambrian through Devonian depo-prism is principally by wedge-edge unconformities. Generally, near the "hingeline", the Upper Devonian carbonates or clastics overlie unconformably the Upper Cambrian carbonates. Recycled clastics were eroded from the Late Devonian Stansbury high in northwestern Utah. Upper Paleozoic sequences thin markedly in a southeasterly direction across the Las Vegas line in southern Nevada and across the Wasatch line in the Nephi to Salt Lake City area of central Utah. The Bird Spring basin received three times and the Oquirrh basin eight times the thickness of sediments than did the platform to the east in Pennsylvanian and Permian times. However, in the St. George to Nephi area stratigraphic thinning is minimal and the thrust belt is oblique to and east of the late Paleo-

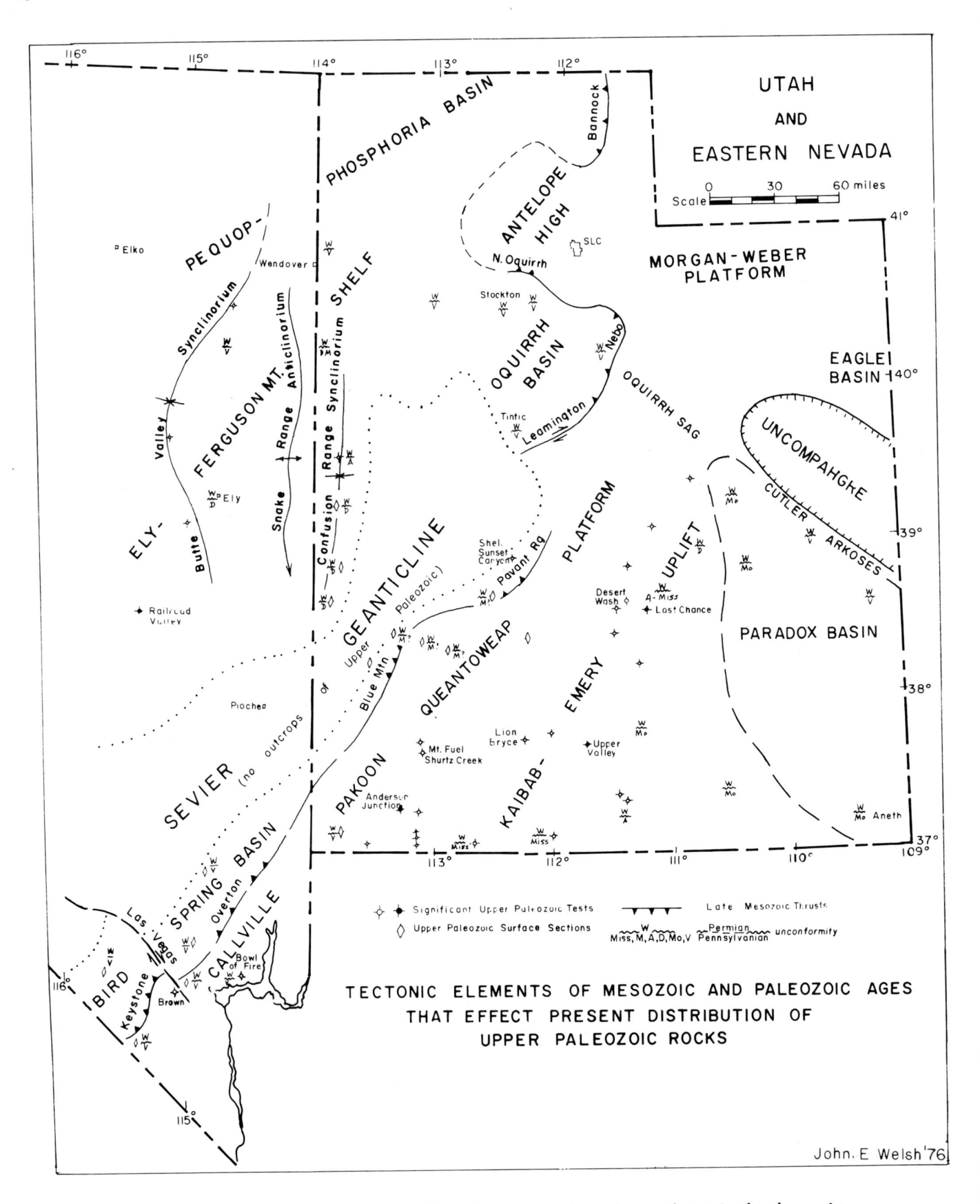

Fig. 1 — Late Paleozoic depositional areas and late Mesozoic tectonic elements.

zoic depo-prism edge. The Confusion Range has 7,000 ft of Pennsylvanian-Permian section compared to 5,000 ft in the Beaver Dam Mountains. The Pennsylvanian section does increase approximately two times from the east side of the Sevier geanticline to the west side but the Permian section is essentially unchanged. The thrust belt actually cuts across the late Paleozoic platform in the southwestern Utah area.

Laramide thrusting shows a continuous northeast trend from the Keystone thrust west of Las Vegas, Nevada to the Charleston-Nebo thrust east of Provo, Utah. The thrusts moved the Cambrian to Triassic sequences eastward either as intact sheets or as large, asymmetrical parallel folds. The thrust belt is subparallel with the Paleozoic "hingeline" but may break obliquely across basinal and platform boundaries. When the thrust plates break obliquely across the basinal margins, tear faults separate the two sedimentary facies. The Leamington Canyon tear separates the thick Oquirrh section from the thin Callville-Pakoon-Queantoweap section. The Las Vegas tear juxtaposes the Bird Spring basinal sections to the northeast against the platform sections south of the Keystone thrust. The Pennsylvanian and Permian sections in the Bird Spring basin in southern Nevada are inbricated in the ranges northeast of Las Vegas because the plate is ramping southeastward approximately perpendicular to the Paleozoic Las Vegas hingeline. The drag on either side of the northwest-trending Las Vegas valley is related to the Las Vegas shear and friction at the base of the thrust plates as they ramp the late Paleozoic platform.

A similar situation may be seen in the Oquirrh basin between Leamington and Tooele. Large north-trending asymmetrical folds in the Tintic and Oquirrh Mountains abruptly bend westward in the Stockton area. The Ophir anticline illustrates this right-angle bend in its axis near Stockton. The right-angle change in fold-axis direction is caused by drag at the thrust-plate bottom against the pre-existing Antelope high basement. The north Oquirrh thrust is a south-directed ramping overthrust from the Antelope high which moved in an opposite direction to the other overturned asymmetrical folds in the Oquirrh Mountains.

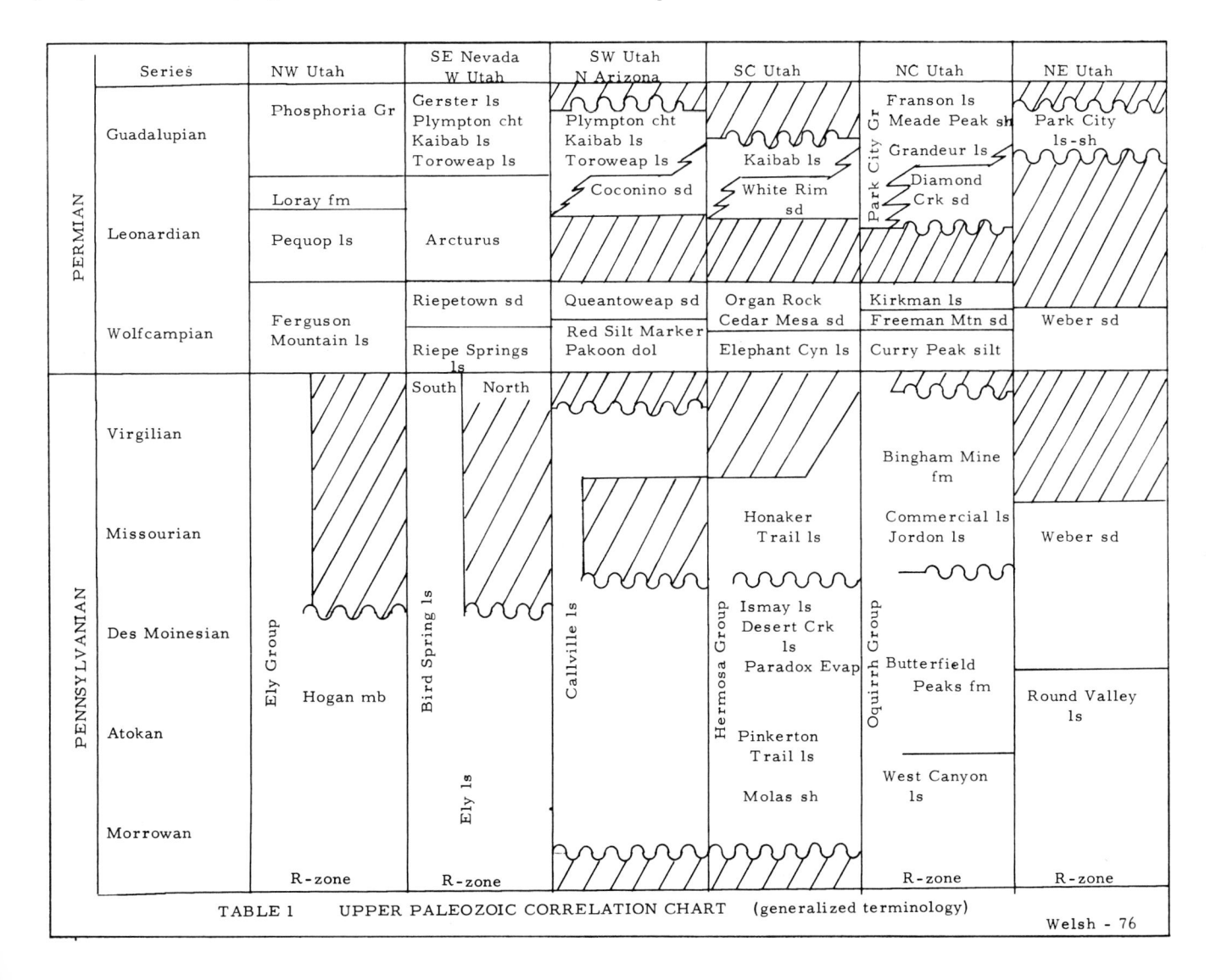

TABLE 1   UPPER PALEOZOIC CORRELATION CHART   (generalized terminology)

Welsh - 76

The folds and imbricate plates of the Bird Spring and Oquirrh basins have not been tested by drilling for oil and gas. Several tests have been drilled into the platform sections.

PERMIAN-TRIASSIC UNCONFORMITY

Evidence that the Triassic red beds and limestones rest unconformably upon the Permian carbonates is present in surface sections and in the subsurface. Generally this boundary is marked by a few feet of chert-pebble conglomerate. The limestones of the Triassic Timpoweap and Sinbad members are non-cherty, fine-crystalline, algal, and pelletal. The Permian limestone and dolomite invariably contains chert. If these criteria are rigorously applied it is usually possible to select the position of the unconformity. There is at present no consistency in the commercial stratigraphic logs but there is less confusion in the published outcrop sections. In the Mountain Fuel Supply Company Desert Wash No. 1A wildcat, the chert-pebble conglomerate can be readily identified at 4420-4430 ft. The Mountain Fuel Supply Company Shurtz Creek No. 1 wildcat near Cedar City is an example of Timpoweap Limestone directly overlying Plympton chert at 950 ft.

Regionally the Triassic Thaynes Limestone overlies unconformably the Permian Gerster Limestone northwest of the Sevier geanticline, the Moenkopi Formation overlies unconformably the Plympton dolomites and chert in the St. George to Cedar City area and the Kaibab Limestone in the Lion Bryce No. 1 and Shell Sunset Canyon No. 1 wildcat tests. Whether the "Kaibab" in the Last Chance and Upper Valley fields is actually Kaibab or Toroweap has not been confirmed by this writer. Regardless, there is a regional erosional truncation of the Upper Permian rocks from west to east across the "hingeline" in southern Utah, (Campbell, 1969).

UPPER PERMIAN

The thickest preserved Upper Permian sequence is in the Phosphoria depo-prism of northwestern Utah. Here, the cherty shale, Rex Chert, Franson Limestone, Meade Peak Shale, and Grandeur Limestone members can be recognized. The Confusion Range area of western Utah has Gerster Limestone, Plympton Formation, Kaibab, and Toroweap limestones (Hose and Repenning, 1964). The southwestern Utah area has only the Plympton, Kaibab, and Toroweap lithostratigraphic units. The type Coconino Sandstone of Arizona is a dune field equivalent to the Toroweap Formation (Rawson and Turner, 1975).

There are several excellent surface sections in the Star Range, southern Mineral Mountains, Pavant Range, and the Beaver Dam Mountains. The Shivwits section in the Beaver Dam Mountains has well exposed outcrops which correlate with the wells in Washington County, Utah. The subsurface section (Fig. 2) of the Mountain Fuel Shurtz Creek No. 1 shows the lithostratigraphic units. In the Shurtz Creek well the Kaibab Group may be subdivided into 460 ft of Plympton Formation, 320 ft of Kaibab Limestone, 320 ft of Toroweap evaporite, and 310 ft of Toroweap Limestone. A lower Toroweap evaporite member is absent although it is present in many of the outcrop sections. The usage of the Plympton Formation in southwestern Utah is consistent with lithostratigraphic characteristics in the Confusion Range of western Utah.

The lithologies of the Kaibab Group indicate a sequence of vertically and laterally changing environments from dune, beach, and sabkha to restricted and open marine. The underlying beach and coastal dune complex of the Queantoweap Sandstone was transgressed by the Toroweap marine invasion after a Middle Permian hiatus. Eustatic sea level changes are believed to have caused the vertical transitions illustrated in Figure 2. The formations are transitional and lithologic changes represent relatively rapid transgressions and regressions. The feasibility of correlating lithologies for a hundred miles in a northwest to southeast direction as well as northeast to southwest, both parallel and perpendicular to the "hingeline", indicates widespread tectonic stability. The beveling of the Upper Permian carbonates to the east and the variations in the Toroweap evaporite members are the main lithologic differences in the sections.

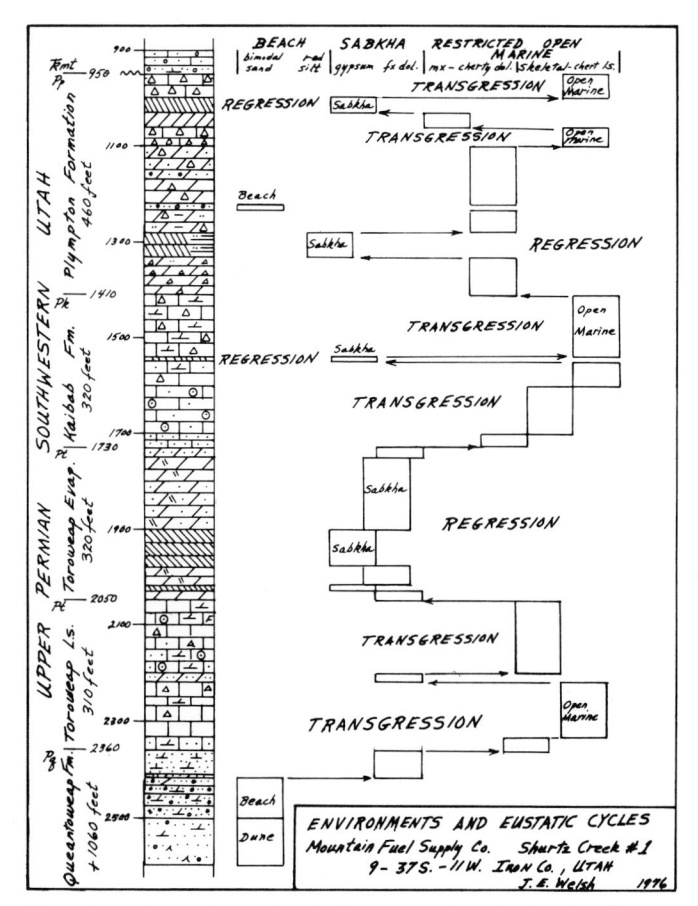

Fig. 2 — Mountain Fuel Supply Co. Schurtz Creek No. 1 illustrating environments and eustatic cycles in the Upper Permian.

The 1400 ft outcrop section of the Kaibab Group in the Beaver Dam Mountains and in the Shurtz Creek No. 1 test near Cedar City and the 1200 ft measured in the Star and Confusion Ranges indicates lateral uniformity of the lithostratigraphic units in southwestern Utah. The thin eastern edge of the Kaibab Group in the western Paradox basin is an erosional wedge edge. The Park City Group in the Oquirrh basin and Wasatch Mountains shows the same eastward thining from 1,200 to 1,500 ft in the Cedar and Tintic Mountains to 600 ft at Park City.

MIDDLE PERMIAN

The Middle Permian rocks are not present in southwestern or central Utah. On the west side of the Sevier geanticline the Arcturus Formation, and in northwestern Utah and northeastern Nevada the Pequop Formation, represent the Middle Permian sequence. Possibly the Diamond Creek Sandstone of the Oquirrh basin is correlative. It is also probably correct that there is an hiatus between the Toroweap Formation and the Queantoweap Sandstone in southwestern Utah.

LOWER PERMIAN

The Lower Permian platform (Fig. 1) has Pakoon dolomite and gypsum overlain by Queantoweap Sandstone. Thick gypsum beds in the Pakoon Formation in southern Nevada are not well developed in Utah. The Queantoweap reaches its maximum thickness of 1,900 ft in the South Mormon Mountains, southern Nevada and 1,750 ft in the Shivwits section in the Beaver Dam Mountains, Utah. Only 500 ft are present on Bradshaw Mountain in the Mineral Mountains and in the Pavant Range. The Mountain Fuel Shurtz Creek No. 1 well intersected 1,000 ft of Queantoweap before crossing a fault, but the Shell Sunset Canyon No. 1 well had only 300 ft of Queantoweap. The source of the Queantoweap Sandstone is principally from the southeast. The eolian Cedar Mesa Sandstone is the equivalent lithostratigraphic unit in the Paradox basin although the Queantoweap Sandstone is a coastal beach deposit.

The Bird Spring basin and Ely-Ferguson Mountain shelf have essentially the same lithostratigraphic units for the Lower Permian. The Pakoon Dolomite is equivalent to the Riepe Springs Limestone while the Riepetown Sandstone is partially the marine equivalent of the Queantoweap Sandstone. In northeastern Nevada the Ferguson Mountain Formation and Strathern Formation are partial equivalents of these units.

In the Oquirrh basin, in the Oquirrh Mountains type area, over 5,000 ft of the Kirkman Limestone, Freeman Peak (Clinker) Sandstone, and Curry Peak Siltstone are equivalents of the less than 1,000 ft of Queantoweap Sandstone and Pakoon Dolomite south of Leamington Canyon (Welsh and James, 1961). There are no transitional surface sections between the Oquirrh basin section and the platform section south of Leamington Canyon.

PENNSYLVANIAN-PERMIAN UNCONFORMITY

Usually the Lower Permian rocks rest disconformably upon beveled Middle and Upper Pennsylvanian rocks throughout the eastern Great Basin. Missourian to Missippian age limestones and dolomites underlie the Wolfcampian Elephant Canyon Formation in the Kaibab-Emery uplift area. Virgilian rocks underlie the Wolfcampian in the Oquirrh and Bird Spring basin and over much of the Callville-Pakoon platform. There is beveling of the underlying Virgilian to Atokan rocks from both the north and south sides of the Ely-Ferguson Mountain shelf. Commonly the unconformity is marked by a few inches or feet of chert-pebble conglomerate which is even present in the Oquirrh basin section in the Oquirrh Mountains (Welsh and James, 1961).

Continuous marine deposition across the Pennsylvanian-Permian time boundary possibly occurred in the Bird Spring basin at the Apex Siding section. The youngest Virgilian and oldest Wolfcampian fusulinid zones are present here and at the North Arrow Canyon section. The Strathern Formation in northeastern Nevada also crosses the time boundary but the chert-pebble beach congolmerates are not necessarily indicative of continuous marine sedimentation. The Weber Sandstone also crosses the time boundary.

UPPER PENNSYLVANIAN

Upper Pennsylvanian rocks are represented by complete sections of Missourian and Virgilian time-stratigraphic sequences in both the Bird Spring and Oquirrh basins. The maximum thickness in the Bird Spring basin is approximately 800 ft of limestone in the Las Vegas Range while there are 6,500 ft of limestone and sandstone in the Bingham Mine Formation of the Oquirrh Group. On the Callville platform there are only 200-300 ft of Virgilian limestone overlying disconformably the Des Moinesian limestone. At Frenchman Mountain there are well exposed Virgilian clastic clinoform limestones deposited on the northwest edge of the deposlope into the Bird Spring basin. In the western Paradox basin Missourian limestones rest disconformably upon Des Moinesian limestones and below Wolfcampian Elephant Canyon Limestone. In the depocenter of the Paradox basin there are also limestones of Virgilian age.

Upper Pennsylvanian rocks were eroded from the Ely shelf in eastern Nevada and western Utah. Beveled wedge edges of these rocks are truncated either by the overlying Wolfcampian Riepe Springs Limestone or a dolomite equivalent to the Pakoon which occurs locally beneath the Riepe Springs Limestone (Steele, 1959).

POST-MIDDLE PENNSYLVANIAN UNCONFORMITY

A pre-Missourian unconformity is indicated in all marine Pennsylvanian sections by detailed faunal correlation. The physical break is seldom discernible, but a fuanal break is present except in the most complete sections in the Bird Spring and Oquirrh basins. A generic change in fusulinids across this gap is independent evidence for a continent-wide hiatus. Megafossil changes also show this chronostratigraphic break. This disconformity is a useful time-stratigraphic datum in the Paradox basin where it can be correlated with accuracy. This datum, which is immediately above the Ismay and Desert Creek limestone, is a valuable tool in lithostratigraphic studies for selecting potentially favorable reservoir areas because it allows correlation of individual limestone members.

MIDDLE PENNSYLVANIAN

Atokan and Des Moinesian lithostratigraphic units are the most widespread of all Pennsylvanian rocks. The middle Des Moinesian limestones represent the maximum invasion of the open marine environment and also contain the most prolific oil producing strata. If the Middle Pennsylvanian rocks are missing or thin it was caused by erosion on the later unconformities, not by non-deposition. In the Oquirrh Mountains these rocks were named the Butterfield Peaks Formation and consist of 7,550 ft of limestone and sandstone. The correlation between two sections of lower and middle Pennsylvanian rocks in the Oquirrh Mountains is shown in Figure 3. *Chaetetes* biostromes are common in both the Atokan and Des Moinesian limestones. The megafossil and fusulinid zones support the lithologic correlation of individual limestone beds as members in ths deeper water facies of the Pennsylvanian.

LOWER PENNSYLVANIAN

A thickness of 2,130 ft for the West Canyon Formation in the southern Oquirrh Group, Figure 3, equals the total thickness of the Pennsylvanian sequence anywhere else in the eastern Great Basin. Rocks of Morrowan age are widespread indicating that this was a time of major submergence of the Bird Spring, Oquirrh, and Paradox basins and the Callville platform and Ely shelf.

MISSISSIPPIAN-PENNSYLVANIAN BOUNDARY

Everywhere on the Callville platform and within the Paradox basin, the lower Pennsylvanian limestones rest disconformably upon the Redwall or Deseret limestones. Commonly there is a regolith or karst surface at this contact. In the Bird Spring and Oquirrh basins and the Ely shelf there apparently was continuous sedimentation from Chesterian to Morrowan time. The *Rhipidomella nevadensis* brachiopod zone commonly marks this transition as is shown in the Oquirrh Mountains (Fig. 3). In eastern Nevada these beds are commonly cut out by thrusting but they range from 10 to 200 ft thick in the Bird Spring and Oquirrh basins and in western Utah. These Mississippian clastic limestones generally map most conveniently with the overlying Pennsylvanian rocks but grade downward into the Mississippian Chainman or Manning Canyon shales.

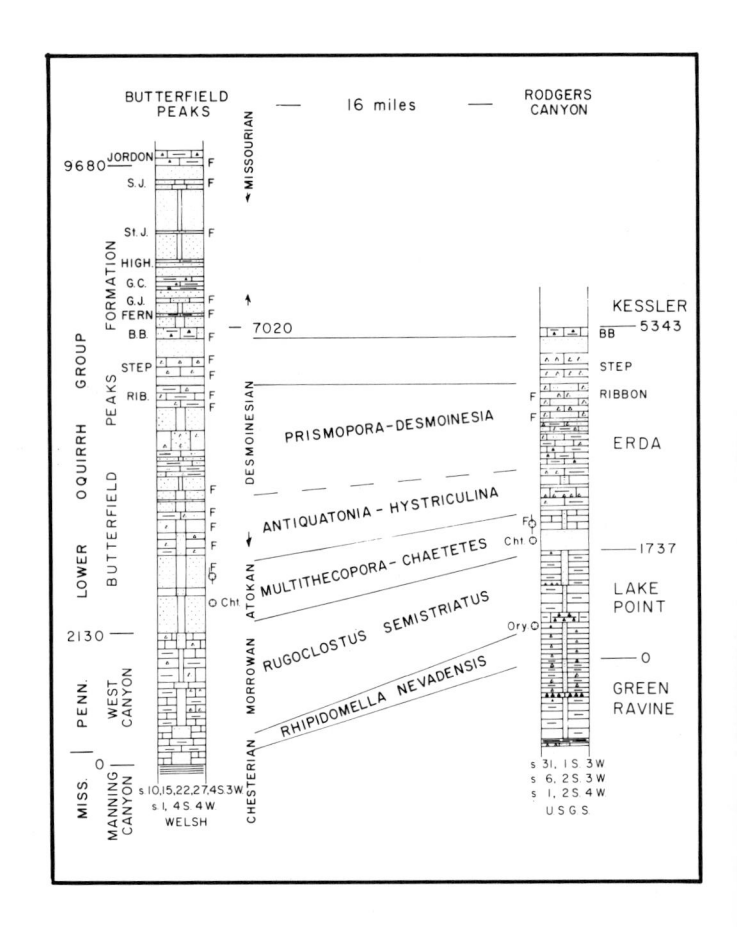

Fig. 3 — Lower Oquirrh Group correlation, megafauna, and fusulinid zonation, Oquirrh Range, Utah.

SELECTED SURFACE SECTIONS AND WELLS

In order to make stratigraphic sense out of the Pennsylvanian and Permian rocks along the "hingeline" between Salt Lake and Las Vegas, the reader may wish to command the initative to look at a few surface sections. A traverse over the Butterfield Peaks section (Fig. 3) and the South Mountain section west of Stockton, Utah will suffice for the Oquirrh basin. The South Pavant Range, in T. 24 S., R. 6 W. Millard County, Utah, has well exposed outcrops across an overturned Devonian through Permian section. The Shivwits section in the Beaver Dam Mountains, Utah and the Frenchman Mountain section outside of Las Vegas, Nevada adequately illustrate the platform sequences. A hike across the Spring Mountain sections or through the North Arrow Canyon section in southern Nevada will detail the Bird Spring basin. With these sections measured and examined the subsurface cuttings from wildcat wells can be intelligently logged.

The tops and comments on the following four wildcat holes should be useful in interpreting the transition zone across the "hingeline" between Salt Lake City and St. George, Utah. The Shell Bowl of Fire No. 1 well and the Joe Brown No. 1 well in Clark County, Nevada are also examples of platform sections which compare with the many tests in the St. George, Utah area (Heylmun, 1961).

Shell Oil Company
Sunset Canyon (W) No. 1
C SW SE Sec. 21, T. 22 S., R. 4 W.
Millard County, Utah

TRIASSIC	Moenkopi with limestone members	125	-3410
	— UNCONFORMITY —		
PERMIAN	Kaibab Limestone	3410	-3530
(Leonardian)	Toroweap evaporite member	3530	-3760
	Toroweap limestone member	3760	-4000
(Wolfcampian)	Queantoweap Sandstone	4000	-4310
	Pakoon Dolomite	4310	-4510?
	— UNCONFORMITY —		
PENNSYLVANIAN	Callville Formation	4510?	-4800
	— UNCONFORMITY —		
MISSISSIPPIAN	Deseret Limestone	4800	-5590
	Gardison Limestone	5590	-6000
	Fitchville Formation	6000	-6200
DEVONIAN	Pinyon Peak Formation	6200	-6310
	Crystal Pass Limestone	6310	-6450
	Simonson Dolomite	6450	
		8962	T. D.

Discussion. — The writer has not logged the samples of this well, but by using a measured section in the Pavant Range and studying the lithology of the drillers and commercial logs a preliminary interpretation is made on the formational tops. These tops differ significantly from those of other investigators and it is hoped that the samples can be examined in the future.

Mountain Fuel Supply Company
Desert Wash Unit No. 1A
NW NW SE Sec. 14, T. 25 S., R. 5 E.
Sevier County, Utah

TRIASSIC	Moenkopi Formation	3150-4430
Trms	Shnabkaib Member	3150-4080
Trmsb	Sinbad Limestone	4080-4230
Trml	Lower Red Siltstone Member	4230-4420
	Chert-pebble conglomerate	4420-4430
	— UNCONFORMITY —	
PERMIAN		
Pk	Kaibab Dolomite	4430-4570
Pq	Queantoweap Sandstone	4570-4590 T. D.

Discussion. — The Triassic-Permian boundary is marked by a chert-pebble conglomerate with a coarse sand matrix. Which part of the Kaibab Group is represented by the cherty, white, fine-crystalline dolomite is problematical. There are no fossil casts to suggest that the dolomite is replacing limestone and it is interpreted to be primary dolomite of the sabkha environment. Further study is needed to correlate the Kaibab of the Desert Wash-Last Chance area with the eastern Great Basin sections.

Lion (Monsanto) Oil Co.
Bryce No. 1, Bryce Canyon (W)
SE NW NW Sec. 10, T. 36 S., R. 4 W.
Garfield County, Utah

TRIASSIC		8130-10190
Trc	Chinle Formation	8130-8730
Trs	Shinarump Formation	8730-8860
Trm	Moenkopi Formation	8860-10190
Trmsb	Shnabkaib Member	8860-9980
Trmt	Timpoweap Member	9980-10190
	— UNCONFORMITY —	
PERMIAN		
Pk	Kaibab carbonate member	10190-10443
	Kaibab sandstone member	10443-10480
Pt	Toroweap dolomite member	10480-10760
	Toroweap dolomitic sandstone member	10760-10987
Pq	Queantoweap Sandstone	10987-11,221 T. D.

Discussion. — The interval 10,190 to 10,220 ft may contain a basal conglomerate with chert and cherty limestone regolith. The first cherty limestone of the Kaibab appears at 10,190 ft, while on the radioactive log the top of the massive carbonate appears to be lower. Comparison of the Bryce No. 1 with the Shurtz Creek No. 1, in Sec. 9, T. 37 S., R. 11 W., indicates that the Plympton Formation is missing at Bryce Canyon and the Triassic Timpoweap Member rests upon the Kaibab carbonate member. Also the evaporite and limestone members of the Toroweap Formation grade into a sandy dolomite to dolomitic sandstone facies and bimodal beach sandstone is a common component. Algal buildup in the Kaibab calcareous dolomites from 10,250 to 10,350 ft is an indication of favorable reservoir rock. Oil shows in the Toroweap dolomites at 10,500 to 11,000 ft indicate that the Bryce area has untested hydrocarbon potential. Which member of the Kaibab Group is present at the Upper Valley and Johns Valley areas to the east has not been determined.

**Mountain Fuel Supply Company
Shurtz Creek No. 1
NE NE Sec. 9, T. 37 S., R. 11 W.
Iron County, Utah**

QUATERNARY		
Qal	Quaternary Alluvium	0-110
	— UNCONFORMITY —	
TRIASSIC	Moenkopi Formation	110-950
Trms	Shnabkaib Member	110-530
Trmv	Virgin (?) Limestone Member	170-190
Trmt	Timpoweap Limestone Member	530-950
	main limestone	730-950
	— UNCONFORMITY —	
PERMIAN		
Pp	Plympton Formation	950-1400
Pk	Kaibab Limestone	1400-1700
Pte	Toroweap evaporite member	1700-2050
Ptl	Toroweap limestone member	2050-2360
Pq	Queantoweap Sandstone	2360-3420
	biotite monzonite @	3290-3320
	FAULT @ 3420	
TRIASSIC	Moenkopi Formation	3420-4970
Trms	Shnabkaib Member	3420-4710
Trmv	Virgin (?) Limestone Member	4360-4420
Trmt	Timpoweap	4710-4970
	FAULT @ 4970	
PERMIAN		4970
Pk	Kaibab Limestone	4970-5330
Pte	Toroweap evaporite member	5330-5580
Ptl	Toroweap limestone member	5580-5840
Pq	Queantoweap Sandstone	5840-5995 T. D.

Discussion. — The Triassic-Permian boundary in the upper block is an unconformity, while in the lower structural block it is a fault, with the Plympton Formation having been omitted. The Permian units correlate closely with measured sections in the Beaver Dam Mountains and the Mineral Range. A common error is to include part of the Timpoweap limestone in the Kaibab Group but the unconformity is clearly at 950 ft in the upper structural block. Environmental analysis of the Kaibab group is illustrated on Figure 2.

EXPLORATION POSSIBILITIES

The Pennsylvanian and Permian sequences have a wide variety of potential reservoir rocks in the different environmental and structural settings. The erosional wedge edges of the Kaibab and Toroweap carbonates under the High Plateaus of Utah, the Callville and Hermosa limestones on the flanks of the Kaibab-Emery uplift, and the Ely Limestone and Pakoon Dolomite against the Ely shelf are areas of truncated potential reservoir rocks. Porosity and permeability changes may be expected both below the unconformities and in response to eustatic sea level changes. The eustatic sea level changes were important in causing the cyclical lithologies of the carbonates of shelf and platform environments both in the Pennsylvanian and Permian. Coarse-clastic clinoform limestones of the Bird Spring and Oquirrh basins may have well developed intraclast porosity. Variations in the crystallinity of dolomite in the Pakoon Formation and Kaibab Group will effect the porosity. Most of the Pennsylvanian and Permian sandstones are fine-grained with either calcite or dolomite cement and locally silica cement. Thick sandstones in the Oquirrh basin and on the Ely-Ferguson Mountain shelf are untested in the subsurface. The Arcturus and Riepetown sandstones in Nevada, the Freeman Mountain and Bingham Mine sandstones of the Oquirrh basin and the Queantoweap Sandstone of the southwest Utah platform are potential reservoirs. The Pequop sandstones in northeastern Nevada are also interesting because they are interbedded with source beds.

Limestones of the maximum marine invasion have proven to be prolific oil producers in the Paradox basin. The Des Moinesian limestones in all areas have potential where porosity is developed. The Virgilian limestones of the Callville platform in Utah and the Wolfcampian Riepe Springs Limestone in Nevada are also examples of major marine invasions. The Toroweap and Kaibab limestones are other examples.

Source beds are present in the algal carbonates of the Callville, Toroweap, and Kaibab limestones. Many of the basinal calcilutites and calcisiltites in the Oquirrh, Bird Spring, Pequop, and Phosphoria depo-prisms are organic rich, as are the underlying Mississippian black shales and siltstones. The Riepetown Sandstone equivalents at Ferguson Mountain in northeast Nevada are organic-saturated. The Oquirrh basin should not be overlooked as a source area and fracture porosity in the fine sediments could markedly increase reservoir potential.

Where are the best places for exploration? This writer's selections are first, the Butte Valley synclinorium of Nevada; second, the Kaibab wedge edge beneath the Utah High Plateaus; third, the overthrust belt of the Bird Spring basin; fourth, the Oquirrh basin; and fifth, the erosional truncations adjacent to the Kaibab-Emery high and the Ely platform.

CONCLUSION

The Pennsylvanian and Permian sequences of the eastern Great Basin have excellent potential for oil production. Wildcatting to this date has barely provided adequate stratigraphic control, the best information is still in the surface stratigraphic sections. Future wildcat drilling should select areas with the best source bed and reservoir sequences.

REFERENCES CITED

Bissell, H. J., 1962, Pennsylvanian and Permian rocks of Cordilleran area, in Pennsylvanian System in the United States: Am. Assoc. Petroleum Geologists, pp. 188-263.

Campbell, J. A., 1969, Upper Valley oil field, Garfield County, Utah: Four Corners Geol. Soc. Guidebook, 5th Ann. Field Conf., p. 195-200.

Heylmun, E. B., 1961, Results of deep drilling in southwestern Utah: Am. Assoc. Petroleum Geologists Bull., v. 45, no. 2, pp. 252-255.

Hose, R. K., and Repenning, C. A., 1964, Geologic map and sections of the Cowboy Pass SW Quadrangle, Confusion Range, Millard County, Utah: U.S. Geol. Survey Map I-390.

Rawson, R. R. and Turner, C. E., 1975, The Toroweap: a new look: Dept. of Geology, Northern Arizona University.

Steele, Grant, 1959, Stratigraphic interpretation of the Pennsylvanian-Permian Systems of the eastern Great Basin: unpub. Ph.D. dissertation, Univ. of Washington.

Welsh, J. E., 1959, Biostratigraphy of the Pennsylvanian and Permian Systems in southern Nevada: unpub. Ph.D. dissertation, Univ. of Utah.

—————, and James, A. H., 1961, Pennsylvanian and Permian Stratigraphy of the Central Oquirrh Mountains, Utah: Utah Geol. Soc. Guidebook No. 16, p. 1-16.

STRATIGRAPHY AND PETROLEUM POTENTIAL OF THE PERMIAN KAIBAB BETA MEMBER, EAST CENTRAL UTAH

by

L. Clark Kiser[1]

ABSTRACT

The Beta Member of the Permian Kaibab Formation in east-central Utah has the potential for large reserves of stratigraphically trapped hydrocarbons. The Beta Member is locally very porous and permeable, frequently carries oil shows, and is productive within the study area. The member contains calcarenitic and bioclastic dolomites, indicative of relatively shallow-water, high energy shelf environments of deposition.

Trapping mechanisms present in the Beta Member in east-central Utah are (a) truncation traps, (b) stratigraphic traps formed by updip loss of porosity and permeability, and (c) anticlinal traps, particularly where favorable hydrodynamic conditions exist.

The potential for large scale truncation traps is present in the Beta Member, due to erosion and downcutting by the Permian-Triassic unconformity. Transgression of the Triassic seaway over the erosion surface in-filled valleys and channels in the Beta Member with tight, impermeable, basal Triassic sediments, forming truncation traps prior to Laramide tectonism and hydrocarbon migration and accumulation in the San Rafael Swell area.

INTRODUCTION

Permian stratigraphy of the Colorado Plateau and the San Rafael Swell area of east-central Utah has been the subject of regional study dealing with the complex stratigraphic relationships, major facies changes, and regional correlation problems present within the Permian System. The Kaibab Formation is of particular interest, due to the uncertainty as to its exact age and the nature of the overlying Permian-Triassic unconformity.

Hydrocarbon shows at outcrop localities and in exploratory wells in the shallow-water shelf carbonates of the Kaibab occur throughout the Colorado Plateau. Regional geologic analysis of a large, relatively unexplored region of east-central Utah was conducted in an attempt to identify areas with potential for large hydrocarbon reserves and favorable conditions for development of stratigraphic traps. Lithologic and stratigraphic data from published material, including both surface and subsurface data, were compiled within the 7000 sq mi (18,200 sq km) surrounding the San Rafael Swell. One of the long-standing stratigraphic problems regarding the Kaibab in east-central Utah is the nature and magnitude of relief on the Permian-Triassic unconformity

which occurs at the top of the formation. Subsurface data has been incorporated into a reconstruction of the pre-Triassic topography of the San Rafael area. The resulting interpretation indicates the presence of large-scale truncation traps in the Kaibab Beta Member as a result of erosional relief on the Permian-Triassic unconformity surface.

Geographic Setting

The study area (Fig. 1) lies immediately east of the Paleozoic-early Mesozoic Cordilleran geosyncline. The eastern margin of the geosyncline is generally referred to as the "hingeline", and marks the point of abrupt westward thickening of many Paleozoic and early Mesozoic sediments into this depositional basin.

Centrally located within the study area is the San Rafael Swell, a prominent structural and topographic feature trending northeast-southwest for 90 mi (145 km), with a maximum width of approximately 50 mi (91 km). The gently dipping west flank of the feature terminates against the Wasatch Plateau and the Last Chance anticlinal trend, while the steeply dipping east flank is bounded by the Nequoia arch and the Uinta basin. The axis of the structure plunges northward into the Uinta basin; the south plunge terminates near the north end of the Henry Mountain basin. The oldest rocks exposed by post-Laramide erosion on the swell are Pennsylvanian (?) (Hallgarth, 1962), with Permian and Triassic rocks forming the majority of out-

---

[1] Energetics, Inc., Englewood, Colorado.

The writer wishes to thank Mr. C. L. Thetford and Mr. Jim Meunier for drafting the illustrations and Ms. Jo Hansen for typing the manuscript. Appreciation is also extended to the many helpful colleagues who offered their critique and suggestions. The writer also acknowledges a small beneficial interest within certain areas covered by this paper.

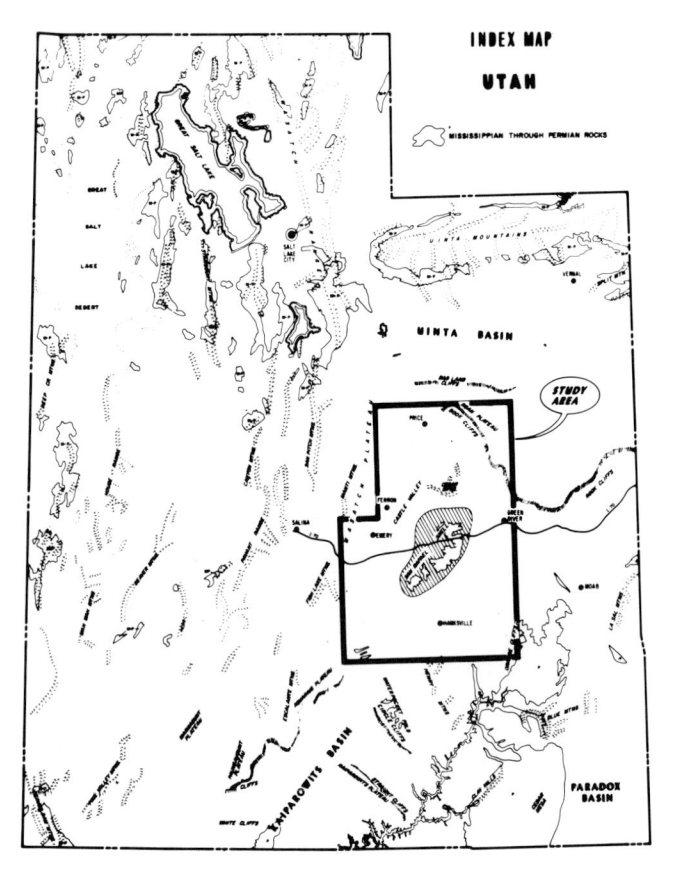

Fig. 1 — Index map of Utah showing study area.

crops in the core of the anticline. The flanks of the swell are in Cretaceous, Jurassic, and Late Triassic rocks. The broad, gentle nature of this large fold feature is indicated by the structure map (Fig. 2). The west flank and north plunge of the San Rafael Swell exhibit homoclinal dips of less than 1 degree to 5 degrees. The climate of the region is high desert, with 7 in (17.8 cm) annual precipitation. Accessibility of the area has been greatly improved with completion of Interstate Highway 70 across the top of the swell, and numerous secondary roads provide access to the west flank and north plunge of the uplift.

Intrusive igneous dikes and sills present in the southwestern portion of the study area appear to be confined to the Jurassic Entrada and Carmel Formations, and have had no significant effect upon exploratory tests.

KAIBAB FORMATION

The division of the Kaibab Formation into the lowermost Gamma, middle Beta, and uppermost Alpha Members was suggested by McKee (1938) from his work in the Grand Canyon area. As interpreted by McKee, these members represent the initial transgressive phase of the Kaibab seaway (Gamma Member), deposition during maximum transgression of the seaway (Beta Member), and a regressive phase (Alpha Member).

Regional subsurface correlation by Irwin (1971) from northern Arizona northward to the vicinity of the San Rafael Swell demonstrates the continuity of the Gamma and Beta Members from the general area of McKee's surface work into the study area, based on correlation of lithologic changes and gamma-ray log characteristics.

The age of the Kaibab in the San Rafael Swell area has not been conclusively established on the basis of faunal evidence, due in part to widely separated outcrops, uncertainty as to the amount of Kaibab rocks removed by pre-Moenkopi erosion in many areas, and dissimilarity of certain fossil assemblages between central Utah and the northern Arizona-Grand Canyon area. Baars (1962) noted that the *Bellerophon* fossil assemblage (gastropods, cephalopods, tribolites, brachiopods and scaphopods) present in the Alpha Member in the Grand Canyon area is also present in uppermost Kaibab rocks on the San Rafael Swell. This relationship suggests that the Beta Member of east-central Utah, which represents the youngest Kaibab rocks present under the Permian-Triassic unconformity, may be age equivalent to the Alpha Member in the Grand Canyon area. Orgill (1971) states that the Kaibab of southeastern Utah (including the area of this paper) is younger than the type Kaibab (Leonardian) in northern Arizona, based on faunal evidence. Fauna collected by Orgill included several genera which displayed a close affinity to those from the Phosphoria Formation (Guadalupian) of Wyoming and its correlative Franson, Gerster, Meade Peak, and Rex formations of the upper Park City Group of northern Utah and Nevada.

These data tentatively establish the age of the Beta Member of the Kaibab as Lower Guadalupian in east-central Utah. The Beta Member may be slightly time-transgressive from south to north, and therefore younger in the San Rafael area than at the Kaibab type section of northern Arizona.

Gamma Member

The Gamma Member of the Kaibab Formation exhibits a west to east facies change from thin-bedded, anhydritic, and fossiliferous micritic dolomite to the fine- to coarse-grained marginal marine and eolian sandstones of the upper White Rim Sandstone, as shown on cross-section A-A' (Fig. 6). East of this facies line, the Beta Member overlies White Rim Sandstone. The marine carbonate to nonmarine sandstone facies change records the earliest eastward transgression of the Kaibab seaway, as noted by McKee (Fig. 3).

The contact between the Gamma Member and the overlying Beta Member is everywhere a sharp radioactive "kick" which can be readily identified on gamma-ray logs. This marker may represent a period of stillstand or minor regression in the overall transgressive pattern of deposition of the Beta and Gamma members, although no evidence of unconformity can be observed. Irwin (1971) described the relationship of the two

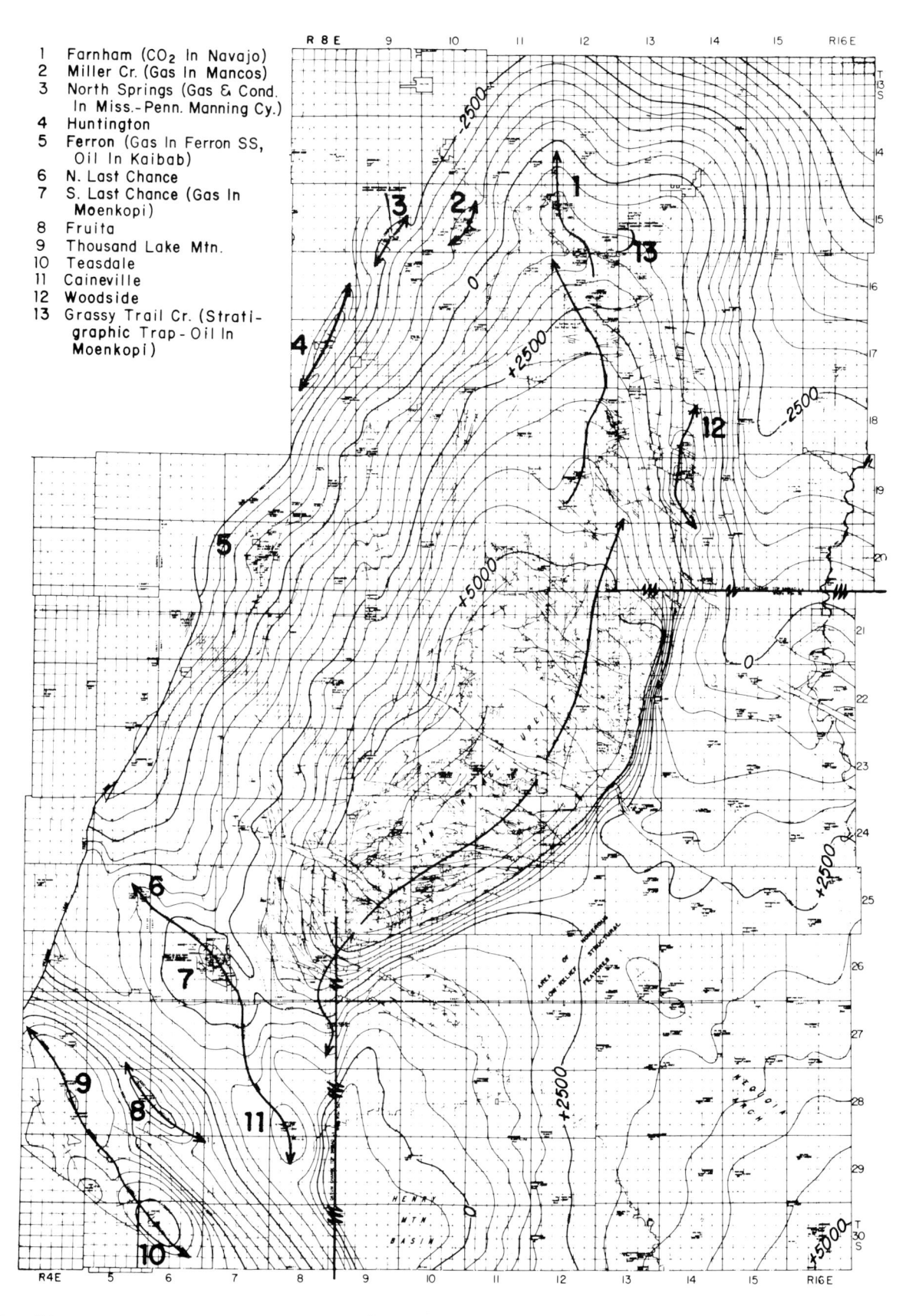

Fig. 2 — Structure contour map on top of the Sinbad Limestone (Triassic) and White Rim Sandstone (Permian). Contour interval 500 ft.

1. Farnham (CO₂ In Navajo)
2. Miller Cr. (Gas In Mancos)
3. North Springs (Gas & Cond. In Miss.-Penn. Manning Cy.)
4. Huntington
5. Ferron (Gas In Ferron SS, Oil In Kaibab)
6. N. Last Chance
7. S. Last Chance (Gas In Moenkopi)
8. Fruita
9. Thousand Lake Mtn.
10. Teasdale
11. Caineville
12. Woodside
13. Grassy Trail Cr. (Stratigraphic Trap-Oil In Moenkopi)

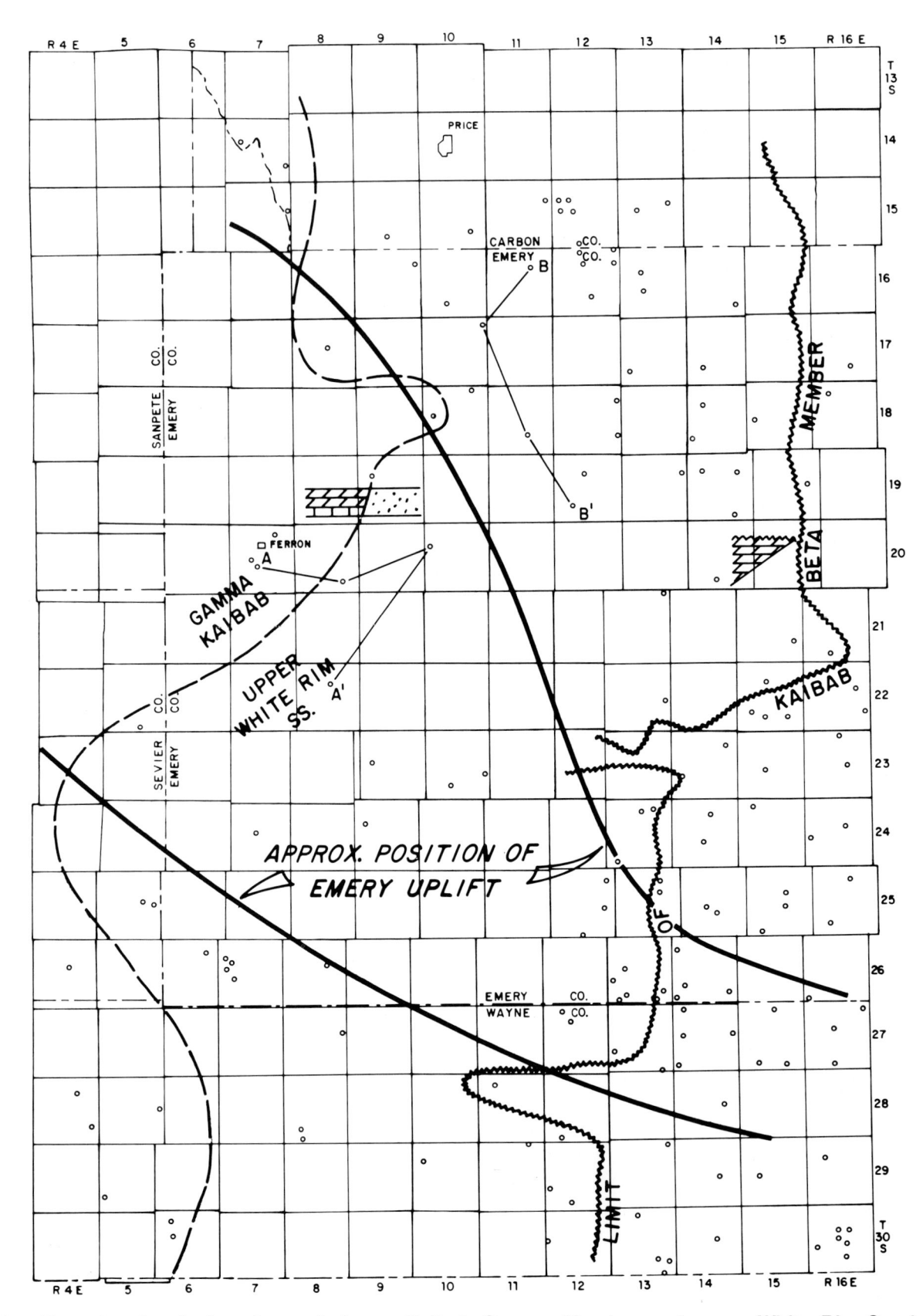

Fig. 3 — Map showing facies change between Kaibab Gamma Member and upper White Rim Sandstone, approximate limit of the Beta Member, and approximate position of Pennsylvanian Emery uplift. Circles indicate Kaibab-White Rim subsurface control.

members at this contact as merely that of carbonate strata (Beta Member) overriding a transgressive beach deposit.

Beta Member

The subsurface contact between the top of the Beta Member and the overlying Triassic Moenkopi can usually be determined at the sharp lithologic break between the dolomite of the Beta Member and the shaley, silty lower Moenkopi. In local instances, however, the Kaibab erosion surface has been weathered, reworked, and redeposited as basal marine beds of the transgressive Moenkopi seaway. This conglomeratic, cherty carbonate unit has Kaibab-type log characteristics, and can make the exact positioning of the Permian-Triassic unconformity questionable on the basis of mechanical log control only.

Surface and subsurface descriptions of Beta Member lithology in the San Rafael area report gray to buff calcarenitic, bioclastic, and micritic dolomite, with minor amounts of interbedded gray chert, fossiliferous limestone, anhydrite, sandstone and siltstone stringers, and occasional glauconite and pyrite. The calcarenitic and bioclastic dolomites are locally very porous. Indistinct marine megafossils, oolites, and large vugs are frequently described. The chert present in the Beta Member appears to be especially abundant near the top of the unit and is present as a lag gravel at the Permian-Triassic unconformity, as a result of its erosion from the upper Beta Member and subsequent redeposition.

The Beta Member was cored in the Energetics, Inc. #23-7 Reserve Federal well in NE SW Sec. 7, T26S-R7E, Emery County, Utah, near the crest of the South Last Chance anticline. Two cores were cut in the upper 73 ft (22.2 m) of the member, which provide the only core data of the upper Beta Member available on the western side of the swell. The top of the Kaibab in this well occurs at 3035 ft. The well was air drilled, and the core point was picked when gray chert pebbles were encountered at the unconformity. The total core recovered spans the interval from 3035 to 3108 ft. The thickness of the member at this location is 150 ft (45.7 m).

Porosity of the Beta Member over the cored interval averages 17.7 percent while air permeability averages 22 md. Due to the extremely fine-grained nature of the calcarenite, visual porosity is difficult to observe in core specimens, although the high porosity recorded on the core analysis is confirmed by a gamma-ray neutron log run across the interval. Fine-grained calcarenitic dolomite is the dominant carbonate rock type in the cored interval, with occasional large vugs lined with calcite crystals, indistinct megafossils, interbedded gray anhydrite and chert stringers, and occasional pyrite.

Sedimentary structure in the upper portion of the Beta Member consists primarily of contorted bedding at dolomite-anhydrite interfaces.

The writer has observed bioclastic zones in the lower portion of the Beta Member, in outcrops on the San Rafael Swell, which consist primarily of small pelecypods and brachiopods. Fossilmoldic porosity can be seen in bioclastic dolomite which is cemented with earthy, micritic dolomite. The calcarenitic and bioclastic dolomites suggest deposition in a relatively shallow-water shelf environment with zones of high-energy wave action. In the subsurface, these dolomitic rock types grade laterally into zones of dense, impermeable, micritic dolomite, suggesting variations of sea level and sea floor topography during the overall eastward transgression of the Kaibab seaway.

Porosity within the Beta Member appears to be continuous along north-south stratigraphic strike, but is laterally discontinuous in an east-west direction. This condition creates the potential for large trends of stratigraphically-trapped hydrocarbons along the west flank of the San Rafael Swell, due to updip intraformational loss of porosity and permeability.

The basal 40-50 ft (12 to 15 m) of the Beta Member along the west flank and north plunge of the swell is almost always a tight, impermeable sequence of dense, fossiliferous, micritic dolomite and limestone stringers, interbedded with anhydrite, chert, and shaley siltstones. Occasional fine to coarse floating quartz grains and sandstone stringers are also found in this interval. The importance of this basal tight sequence is that it forms a "floor" for potential hydrocarbon accumulations.

Beta Member Isopach. — The isopachous map of the Beta Member (Fig. 4) shows regional thinning from 200 ft (61 m) in the western portion of the study area to a zero edge in the eastern portion. This eastward thinning is accomplished by depositional onlap and truncation by pre-Moenkopi erosion.

The isopach also suggests the influence of the ancient Emery uplift, a northwest-trending Late Pennsylvanian feature which occupied much of the present day San Rafael area. Pennsylvanian rocks are absent from the crestal portion, and basal Permian sediments rest unconformably upon eroded Mississippian carbonates along the top of this feature (Fig. 3).

Moulton (1975) described the influence of the Emery uplift in the hingeline area of central Utah. His Permian isopachous map shows a pronounced thin over the feature, indicating its continuing influence during the deposition of oldest Permian rocks. By the time of Beta Member deposition, however, the uplift apparently had very little relief, as its influence on the thickness of the member is much less pronounced than on older Permian rocks. Triassic isopachous maps show no significant thinning over the Emery uplift.

Maximum erosion of the Beta Member did, however, occur in the vicinity of the ancient uplift, as evidenced by the complete removal of the member over a large area of the north-central San Rafael Swell. In many localities, basal Moenkopi sediments rest directly on upper White Rim Sandstone (Gamma Kaibab).

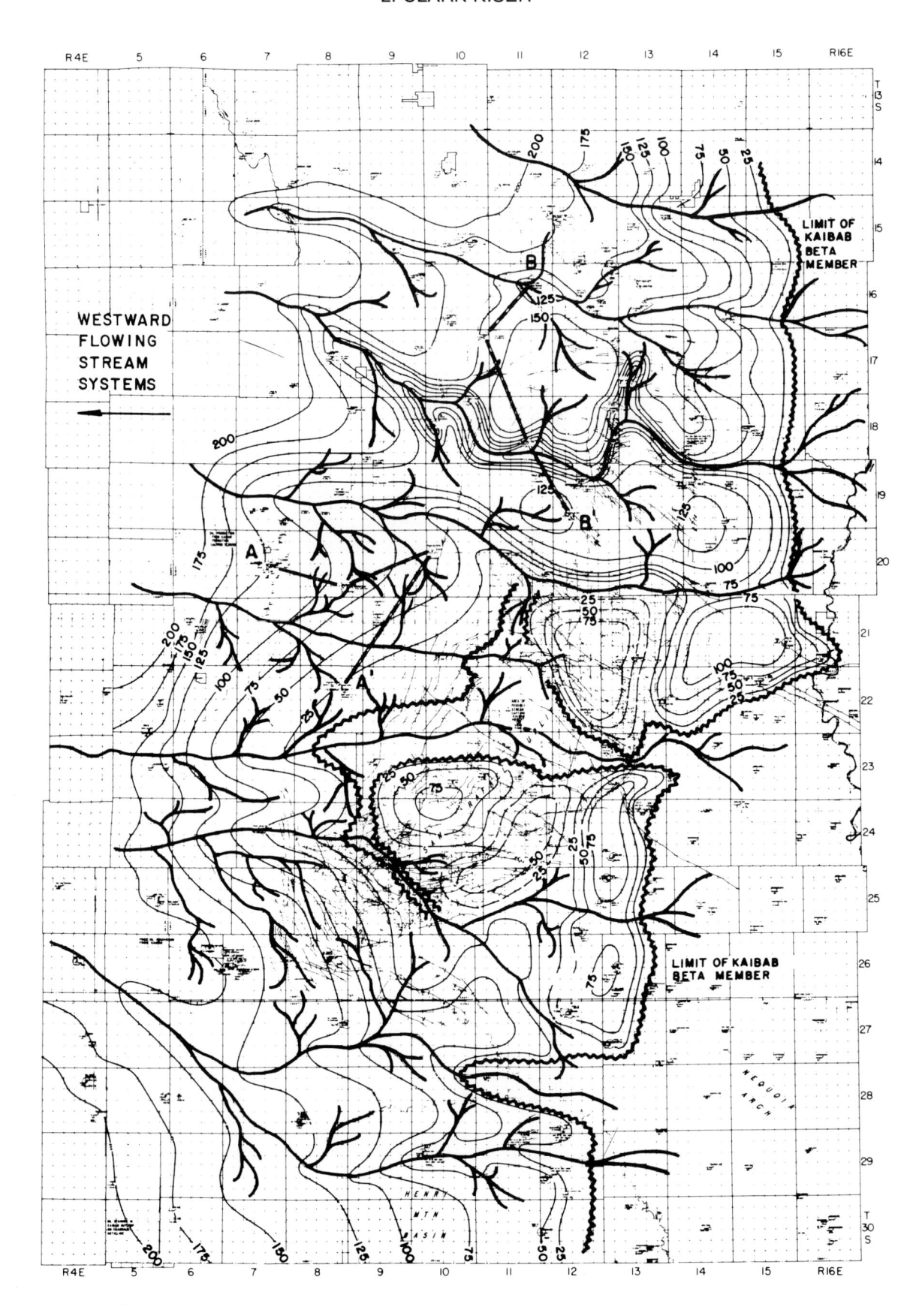

Fig. 4 — Isopach map of Kaibab Beta Member, with interpretation of drainage patterns on pre-Moenkopi erosion surface. Contour interval 25 ft.

PERMIAN-TRIASSIC UNCONFORMITY

The extensive erosional surface at the top of the Beta Member represents a hiatus between rocks of Middle Permian age and overlying beds of middle Lower Triassic. Baars (1962) observed that the physical evidence of the unconformity in the Colorado Plateau area suggests a period of epiorogenic uplift. In the San Rafael area, this uplift began the removal of the Alpha Member (assuming that this unit was originally deposited) and erosion of the Beta Member by an extensive stream system. The exact time of uplift cannot be demonstrated, although Baars (1962) noted that it may have been in the Upper Permian. The erosional period is herein referred to as pre-Moenkopi.

The magnitude and extent of erosion and relief on the unconformity surface at the outcrop has been described in several published studies of the San Rafael area. Gilluly (1928) described the absence of the Kaibab in areas of the San Rafael Swell, and noted the basal Moenkopi conglomerate and the irregular surface of the Kaibab as evidence of unconformity. Other writers who have discussed the unconformity in the San Rafael area include: Baker, 1946; Stewart and Smith, 1954; McKee, 1938, 1954; Herman and Sharps, 1956; Hawley, et al., 1968; Orgill, 1971; Irwin, 1971.

Observations as to the magnitude of relief on the unconformity surface range from only 12 ft (3.66 m) (McKee, 1938) to 102 ft (31 m) in the Green River Desert (Baker, 1946).

The most convincing evidence of erosional relief on the Kaibab Beta Member is found in the subsurface. Cross-section B-B' (Fig. 6) demonstrates removal of at least 100 ft (30.5 m) of Kaibab section by pre-Moenkopi erosion. Although the Beta Member has been completely removed in an area on top of the San Rafael Swell, subsurface evidence indicates that in most cases erosion did not progress beyond the tight basal portion, which may have acted as an erosion-resistant "floor" preventing further downcutting into the Gamma Member.

The Beta Member isopach on the west flank of the San Rafael Swell suggests development of a cuesta-like topography, while the north plunge of the swell developed into steep canyons and cliffs, similar to the present-day topography seen in the vicinity of the San Rafael River in Triassic and Jurassic outcrops. The crestal portion of the swell and the east flank consists of erosional outliers and re-entrants of carbonate tongues of the Beta Member near its zero edge.

In addition to the post-depositional modification of thickness in the Beta Member, it is probable that pre-Beta paleotopography, local structural uplift during deposition, and minor intraformational unconformities have played a part in isopachous variations seen in this unit. At present, the paucity of well control makes it difficult to differentiate between depositional and structurally-controlled variations and those due purely to erosion.

TRIASSIC MOENKOPI FORMATION

Triassic sedimentation in the San Rafael area began with a southward marine transgression out of the Triassic basin, which was located initially in northern Utah, southern Idaho, and southeastern Nevada. The basal Triassic unit deposited by this transgressive phase has been informally identified as the "lower slope-forming member" of the Moenkopi by Blakey (1973) and again by the same writer in a later publication (Blakey, 1974) as the Black Dragon Member. This unit consists of yellowish-gray to red, micaceous shales, mudstones, and siltstones; thin carbonates; and sandstones. The basal portion contains conglomerates of weathered and eroded Kaibab sediments, primarily cherty and sandy lag gravels. The Black Dragon Member thins from 375 ft (114 m) in the northern portion of the study area to a zero edge in the Circle Cliffs area south of the San Rafael Swell.

A local isopachous "thick" to the west of the buried Uncompahgre uplift indicates that this feature was shedding sediment into the Triassic seaway during deposition of the Black Dragon Member.

Marine Triassic sediments filled in topographic lows, as evidenced by local "thicks" over major stream systems mapped on the Beta Member in the northern San Rafael area (Fig. 5). This relationship is very similar to the infilling of Pennsylvanian Minnelusa erosional channels by Opeche Shale in the Powder River basin of Wyoming.

Blakey (1974) maps a pronounced Black Dragon Member "thick" in the north-central portion of the swell on the basis of surface data. This "thick" generally coincides with the area of widespread removal of the Beta Member, and is further evidence of erosional relief prior to deposition of Triassic sediments.

OCCURRENCE OF OIL IN THE BETA MEMBER

The Pan American #3 Ferron Unit (Sec. 21-T20S-R7E, Emery County, Utah) was completed in 1964 for an IPF of 43 bbls of oil, 1107 mcf gas, and 34 bbls of water per day from the Beta Member. The oil is 40 deg. API gravity, asphalt based, low sulfur crude. The gas is composed of 64 percent inert constituents, primarily nitrogen and carbon dioxide, and has a Btu rating of 450. This well has produced approximately 34,000 bbls of oil to date.

The only other significant production from the Kaibab Formation in Utah is south of the study area in Garfield County, where Upper Valley field (T36, 37S-R1, 2E) produces from 25 wells in the Timpoweap (Triassic) and Kaibab formations on a tightly folded anticline. The oil accumulation on this feature

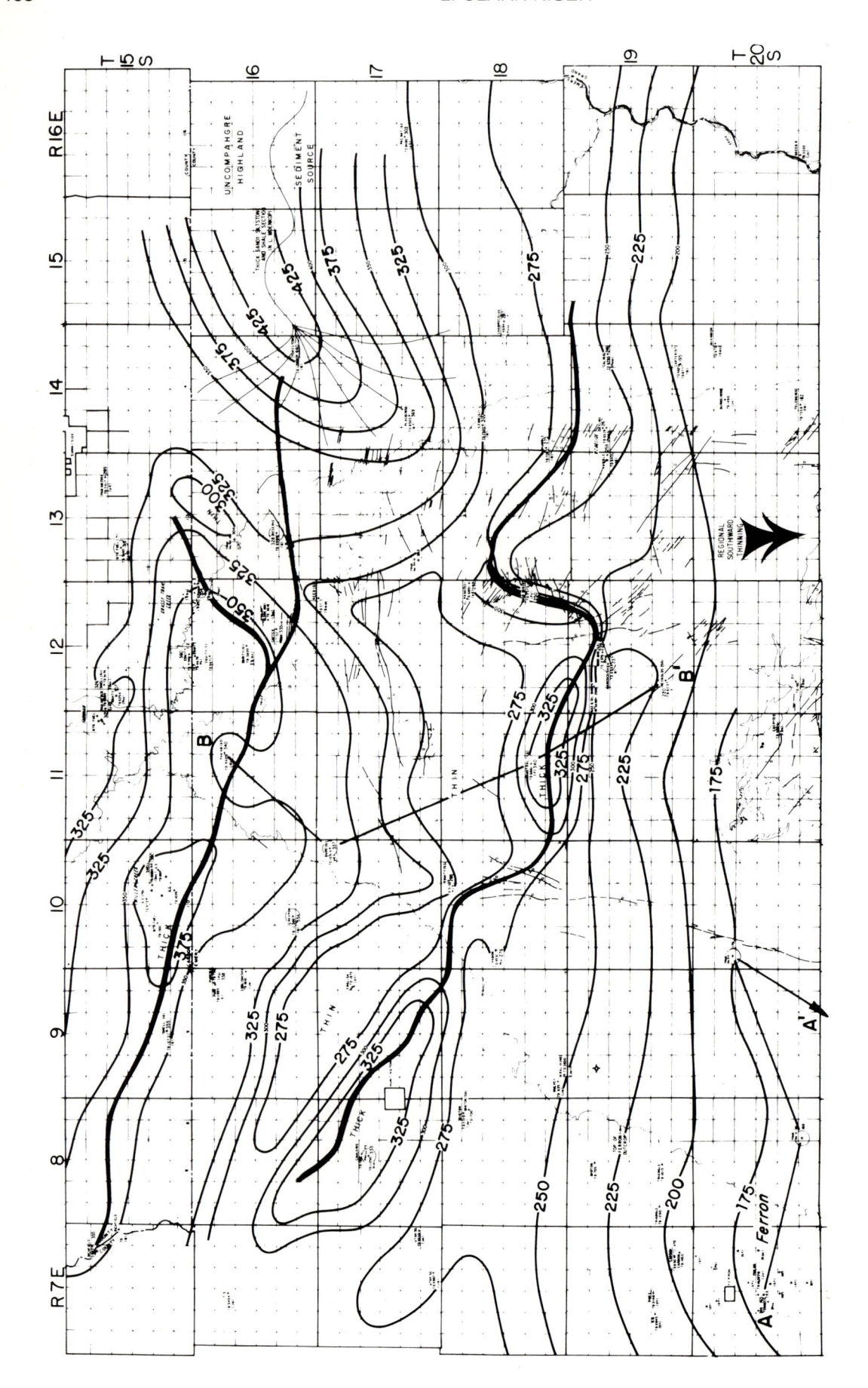

Fig. 5—Isopach map of Moenkopi Black Dragon Member ("lower slope-forming member") showing local isopach "thicks" in vicinity of Kaibab Beta Member channels. Note depositional "thick" in northeast portion of map area, indicating sediment derived from nearby Uncompahgre uplift. Circles around wells indicate oil shows in Black Dragon Member. Contour interval 25 ft.

occurs along the south plunge and west flank, due to the presence of a strong southwest-dipping potentiometric surface gradient. The field has produced over 14 million bbls of oil, and may ultimately produce 25 to 30 million bbls.

Almost every well in the study area has yielded oil shows in the Kaibab, ranging from spotty oil staining and fluorescence reported on sample logs and in well reports to free oil in air-drilled and cable-tool holes on the west flank of the swell. The core described previously from the South Last Chance anticline had residual oil saturation in the upper Beta Member of up to 36.2 percent. Oil staining in the Kaibab at the outcrop on the San Rafael Swell has been reported by Ritzma (1968) and described by Orgill (1971).

The presence of oil shows in the Kaibab is analogous to the frequent occurrence of oil in the Permian Phosphoria Formation in the Wyoming shelf area and the San Andres Formation of eastern New Mexico and West Texas. In the Big Horn basin, the source of the oil in all Paleozoic formations which produce on anticlinal closure is believed to be the Phosphoria (Stone, 1967).

The tremendous capability of Phosphoria rocks to generate and accumulate oil in the Wyoming shelf area is due primarily to two factors: (1) presence of a western phosphatic-shale source facies, and (2) since its deposition, the regional relationship of downdip source rocks to updip reservoir beds and sealing facies in the Phosphoria was not significantly disturbed or reversed prior to, or in many cases during, Laramide tectonism. Laramide uplift merely served to refocus the oil from the existing regional stratigraphic trap into local anticlinal traps.

A similar relationship occurs in the Kaibab Beta Member on the eastern edge of the "hingeline" along which most Paleozoic and lower Mesozoic formations thicken westward into the Cordilleran geosyncline. Due to the sparcity of subsurface data in the area of the geosyncline, Permian stratigraphy is not well understood. Source beds in the Beta Member and upper White Rim Sandstone zones should logically have been deposited west of the San Rafael Swell, with oil migrating updip into the shelf carbonates of the Kaibab Beta Member. Although no eastern redbed and shale equivalent of the Kaibab is present in east-central Utah, as is present in the Phosphoria of Wyoming, the eastern updip wedgeout of the Beta Member and its overlying Triassic caprock served as an effective barrier to further migration of hydrocarbons. Laramide uplift redistributed oil accumulations in this regional stratigraphic trap into local truncation and anticlinal traps on the west flank and north plunge of the San Rafael Swell. Although many such traps have been breached along the crestal portion of the swell by post-Laramide erosion, many may remain undiscovered throughout the study area.

Migration and Accumulation of Oil

Very little data are available in the study area to indicate time of migration and accumulation of oil within the Beta Member. Gussow (1955) suggested that probably 2000 ft (610 m) of overburden is necessary to initiate expulsion of hydrocarbons from source beds. Regional isopachous studies indicate that this thickness of sediment had accumulated in the hingeline area of the Cordilleran geosyncline by Middle to Late Triassic time.

Stone (1967) observed that the depth of burial of Permian Phosphoria source rocks in the miogeosyncline west of the Wyoming shelf probably was sufficient to begin expulsion and migration before Late Triassic time and the primary flush migration of Phosphoria oil on a regional scale was complete by Early Jurassic time. Without evidence to the contrary, similar times of migration and accumulation may be assumed for the Kaibab oil in east-central Utah.

Truncation Traps

The nature and extent of the unconformity surface at the top of the Beta Member has been described in the preceding discussion along with the essential parameters for formation of truncation traps:

(a) an oil-bearing stratigraphic unit containing reservoir-quality rocks,

(b) an impermeable caprock (basal beds of the Triassic Moenkopi) which effectively prevents the updip migration of hydrocarbons from truncated reservoirs, and

(c) a "floor" of basal impermeable strata, which prevents escape of hydrocarbons from under the trapping facies due to regional tilt or flushing, and allows development of an oil column substantially in excess of the reservoir thickness.

Cross section B-B' (Fig. 6) demonstrates the presence of these conditions along the north plunge of the swell, the area with the best potential for development of the truncation-type trap within the region. Truncation has also played an important part in isolating a large portion of the Beta Member on the west flank of the San Rafael Swell from the outcrop.

Intraformational Porosity and Permeability Barriers

An example of this trapping mechanism can be seen in the vicinity of Ferron field. Although ± 150 ft (46 m) of closure can be mapped on this feature on the Upper Cretaceous Ferron Sandstone, stratigraphic variation is the controlling factor for oil accumulation in the Beta Member. A well drilled by True Oil Co. in Sec. 35, T20S-R8E, 8 mi (12.9 km) updip from Kaibab production in the Pan American #3 Ferron Unit (cross-section A-A', Fig. 6) encountered a tight Beta Member section, which yielded a slight amount of free oil from a fracture zone.

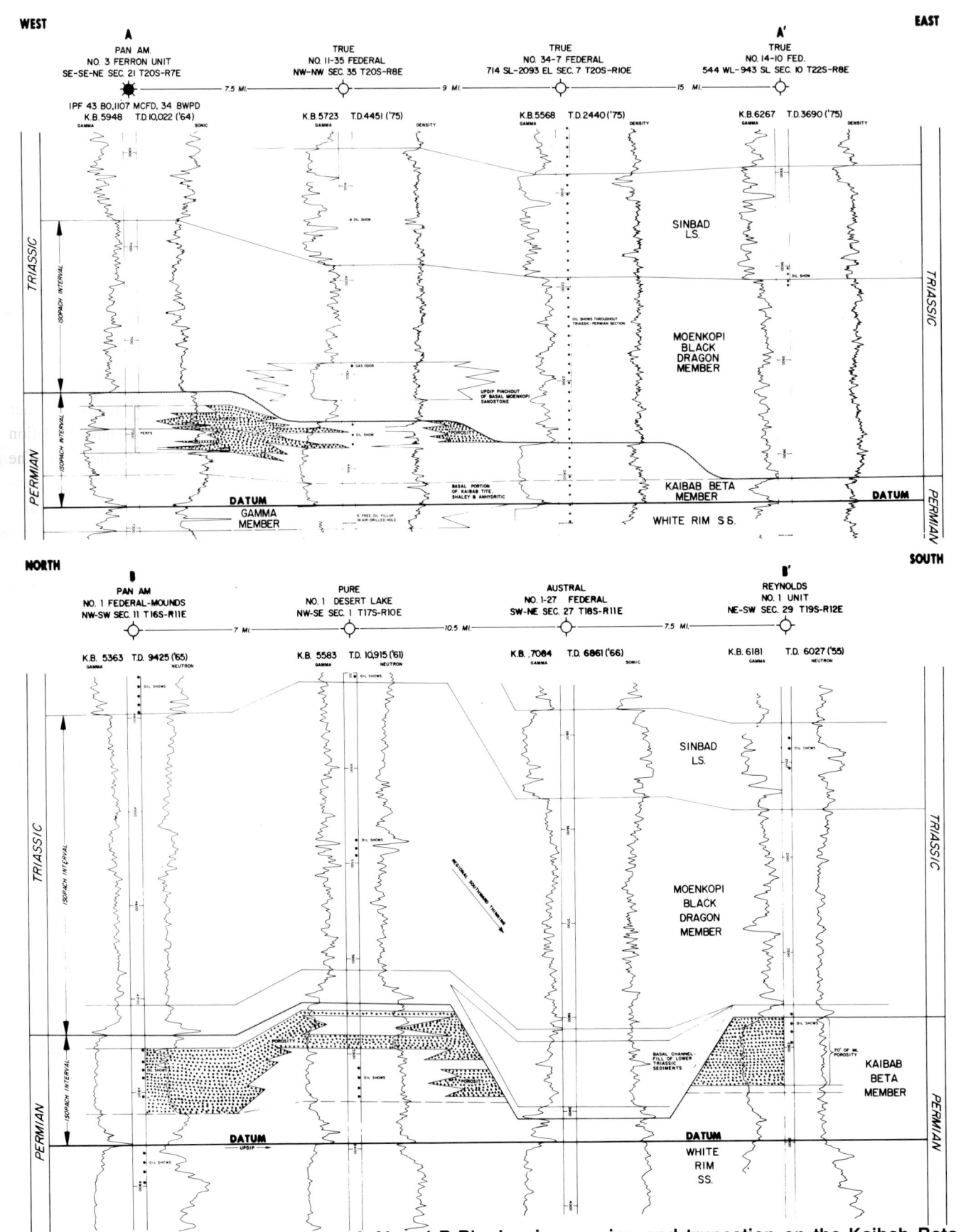

Fig. 6 — Stratigraphic cross-sections A-A' and B-B', showing erosion and truncation on the Kaibab Beta Member at the pre-Moenkopi unconformity. Stipple pattern indicates porosity within Beta Member.

The lack of subsurface control along the west flank of the swell (Fig. 3) makes it difficult to project porosity trends. Additional drilling is needed to provide the necessary control to locate intraformational stratigraphic traps in the Beta Member.

Anticlinal Traps

Several large anticlinal structures, other than the breached San Rafael Swell, occur within the area. The South Last Chance anticline (T26S-R6, 7E) has 800 ft (244 m) of closure, a gas cap in the lenticular sands of the Moenkopi "ledge-forming member", and oil saturation in the Kaibab, yet the structure does not produce oil. Shows are present in Triassic through Mississippian sediments, possibly indicating the presence of oil accumulation in these rocks at some previous time.

Analysis of pressure data from wells on South Last Chance anticline indicates the presence of a southwest-dipping potentiometric surface across the structure, which has effected the removal of most, if not all, of the pre-existing oil accumulation. Apparently the excellent porosity and permeability present in the Beta Member on South Last Chance anticline is connected to porosity in the southern San Rafael Swell, and only further exploration will determine if an oil accumulation is present on the feature as a result of hydrodynamic trapping conditions. The Triassic gas cap on the anticline has remained undisturbed, due to the great difference in relative density of gas and water, and the lenticular nature of the gas-bearing sandstones.

The Caineville anticline (T28, 29S-R8E) and North Last Chance anticline (T25S-R5E) both have similar problems with regard to the potentiometric surface. The Fruita and Thousand Lake Mountain structures in the southwestern portion of the study area encounter the Kaibab at structural elevations of +5000 ft (+1524 m) or higher, with extremely low formation pressures. The Kaibab crops out on the Teasdale feature, at a surface elevation of about +7500 ft (+286 m).

Woodside anticline, in T18-20S, R13-14E, has at least 800 ft (244 m) of closure, and yet contains only noncombustible gases and water in the Kaibab and White Rim. The explanation for this occurrence is probably related to the age of the structure. The steep west flank assymetry of Woodside is in direct contrast to the steep east flank configuration of the San Rafael Swell, which suggests different tectonic origins for these structures. The absence of oil on Woodside may be an indication that the San Rafael fold was the earlier structural event, and caused hydrocarbons in the west-dipping regional stratigraphic trap to be refocused toward the structure, and into existing truncation traps prior to formation of the Woodside feature.

Careful analysis of the "plumbing system" in this area of large anticlinal closures may result in the discovery of fields with hydrodynamic trapping conditions similar to Upper Valley field in southern Utah.

CONCLUSIONS

1. The Kaibab Beta and Gamma members of east-central Utah are lithologically continuous with the Beta and Gamma members of the Kaibab of southern Utah-northern Arizona, as indicated by subsurface correlation. The Beta Member in the San Rafael area may be slightly younger than at the northern Arizona type section. The age of the member is placed as Lower Guadalupian in east-central Utah.

2. The marine carbonates of the Gamma Member, in the westernmost part of the study area, interfinger eastward with marginal marine and eolian sandstones of the upper White Rim. This facies change represents the first transgressive phase of the Kaibab seaway into the central Utah area.

3. The Beta Member on the west flank and north plunge of the San Rafael Swell contains bioclastic and calcarenitic dolomite, which is locally very porous, and was deposited under shallow-water, high-energy shelf conditions. Lateral gradation into tight, micritic dolomite occurs throughout the area. The basal portion of the member is composed of impermeable, thin-bedded carbonates which form a "floor" for potential hydrocarbon accumulations in the upper Beta Member.

4. Isopachous mapping of the Beta Member indicates slight influence of the ancient Emery uplift on Beta Member thickness in the vicinity of the San Rafael Swell.

5. The extensive erosion surface present at the top of the Beta Member has created the potential for large-scale truncation traps in the Kaibab, where porosity zones have been cut and subsequently infilled with impermeable basal Triassic sediments. On the west flank of the swell, the tight basal section of the Beta Member may have acted as an erosion-resistant "floor" preventing downcutting into the Gamma Member. Several major westward-flowing erosional stream systems can be mapped throughout the study area.

6. The Triassic seaway transgressed over the erosion surface from the north, and infilled valleys and channels in the Beta Member with compensating thicknesses of basal Moenkopi marine shales and siltstones. In areas where well control is of sufficient density, these "thicks" can be mapped in the subsurface overlying thin Beta Member sections, and are further evidence of pre-Moenkopi erosion.

7. Oil production from the Kaibab occurs in the study area at Ferron field, and at Upper Valley field in southern Utah. The Beta Member bears many similarities to the Permian Phosphoria Formation of Wyoming, including the regional relationship of downdip source beds to updip reservoir and sealing facies, and the widespread occurrence of oil saturation in surface and subsurface sections. In both the Wyoming shelf and east-central Utah, regional west dip persisted until the time of Laramide tectonism. Oil in the regional stratigraphic traps was refocused into local structural traps and, in the case

of east-central Utah, truncation traps similar to those present in the Minnelusa Formation of the Powder River basin in Wyoming.

8. Three types of hydrocarbon traps in the Beta Member exist in the east-central Utah area:

(a) truncation traps

(b) stratigraphic traps formed by intraformational porosity and permeability barriers

(c) anticlinal traps, particularly on those anticlines whose flanks may have untested potential for hydrodynamically controlled hydrocarbon accumulations.

SELECTED BIBLIOGRAPHY

Baars, D. L., 1962, Permian System of Colorado Plateau: Am. Assoc. Petroleum Geologists Bull., v. 46, no. 2, p. 149-218.
Baker, A. A., 1946, Geology of the Green River Desert-Cataract Canyon region, Emery, Wayne, and Garfield Counties, Utah: U.S. Geol. Survey Bull. 951, 122 p.
Blakey, R. C., 1973, Stratigraphy and Origin of the Moenkopi Formation (Triassic) of Southeastern Utah: Mountain Geologist, v. 10, no. 1, p. 1-17.
_____, 1974, Stratigraphic and Depositional Analysis of the Moenkopi Formation, southeastern Utah: Utah Geol. and Mineral Survey Bull. 104, 81 p.
Davidson, E. S., 1967, Geology of the Circle Cliffs area, Garfield and Kane Counties, Utah: U.S. Geol. Survey Bull. 1229, 140 p.
Gilluly, J., 1929, Geology and oil and gas prospects of part of the San Rafael Swell, Utah: U.S. Geol. Survey Bull. 806-C, p. 69-130.
_____, and Reeside, J. B., Jr., 1928, Sedimentary rocks of the San Rafael Swell and some adjacent areas in eastern Utah: U.S. Geol. Survey Prof. Paper 150-D, p. 61-110.
Girdley, W. A., 1974, Kaibab Limestone and associated strata, Circle Cliffs, Utah: Utah Geology, v. 1, no. 1, p. 5-20.
Gussow, W. C., 1954, Differential entrapment of oil and gas: a fundamental principle: Am. Assoc. Petroleum Geologists Bull., v. 38, no. 5, p. 816-853.
Hallgarth, W. F., 1962, Upper Paleozoic rocks exposed in Straight Wash Canyon, San Rafael Swell, Utah: Am. Assoc. Petroleum Geologists Bull., v. 46, no. 8, p. 1494-1501.
Hawley, C. C., Robeck, R. C., and Dyer, H. B., 1968, Geology, altered rocks and ore deposits of the San Rafael Swell, Emery County, Utah: U.S. Geol. Survey Prof. Paper 1239, 113 p.
Herman, G., and Sharps, S. L., 1956, Pennsylvanian and Permian stratigraphy of the Paradox salt embayment: Intermtn. Assoc. Petroleum Geologists 7th Ann. Field Conf. Guidebook, p. 77-84.
Irwin, C. D., 1971, Stratigraphic analysis of Upper Permian and Lower Triassic strata in Southern Utah: Am. Assoc. Petroleum Geologists Bull., v. 55, no. 11, p. 1976-2007.
McKee, E. D., 1938, The environment and history of the Toroweap and Kaibab formations of northern Arizona and southern Utah: Carnegie Inst. Washington Pub. 492, 268 p.
_____, 1954, Permian stratigraphy between Price and Escalante, Utah: Intermtn. Assoc. Petroleum Geologists 5th Ann. Field Conf. Guidebook, p. 21-24.
_____, and Oriel, S. S., *et al.*, 1967, Paleotectonic Investigations of the Permian System in the United States: U.S. Geol. Survey Prof. Paper 515, 271 p.
Moulton, F. C., 1975, Lower Mesozoic and Upper Paleozoic petroleum potential of the hingeline area, central Utah *in* Symposium on Deep Drilling Frontiers in the Central Rocky Mountains: Rocky Mtn. Assoc. Geologists, p. 87-97.
Orgill, J. R., 1971, The Permian-Triassic unconformity and its relationship to the Moenkopi, Kaibab, and White Rim formations in and near the San Rafael Swell, Utah: Brigham Young Univ. Geol. Studies, v. 18, part 3, p. 131-179.
Ritzma, H. R., 1968, Preliminary location map, oil-impregnated rock deposits of Utah: Utah Geol. & Min. Survey Map #25.
Stewart, J. H., and Smith, F. J., Jr., 1954, Triassic rocks in the San Rafael Swell, Capitol Reef and adjoining parts of southeastern Utah: Intermtn. Assoc. Petroleum Geologists 5th Ann. Field Conf. Guidebook, p. 25-33.
_____, Poole, F. G., and Wilson, R. F., 1972, Stratigraphy and origin of the Triassic Moenkopi Formation and related strata in the Colorado Plateau region: U.S. Geol. Survey Prof. Paper 691, 195 p.
Stokes, W. L., and Cohenour, R. E., 1956, Geologic Atlas of Utah: Utah Geol. & Mineral Survey Bull. 52, 92 p.
Stone, D. S., 1967, Theory of Paleozoic oil and gas accumulation in Big Horn Basin, Wyoming: Am. Assoc. Petroleum Geologists Bull., v. 51, no. 10, p. 2056-2114.

A CITY, BY ANY OTHER NAME . . .

It was the practice of writers in the employ of the Union Pacific Railroad during its early years (they would be called Public Relations Counselors today) to attach grandiose and flattering, yet empty, descriptive phrases to frontier towns along their route. Denver, for example, is referred to in early railroad guidebooks, as "The Queen City of the Plains." Cheyenne was "The Magic City of the Plains." Laramie was "The Gem City of the Mountains."

Utah managed to escape this nonsense. The writers had no nickname at all for Ogden. The best they could do for Salt Lake City was to refer to it as "The City of the Saints," a designation it had acquired elsewhere years earlier.

— W. Lyle Dockery

PERMIAN PHOSPHORIA CARBONATE BANKS, IDAHO-WYOMING THRUST BELT

by
Marvin D. Brittenham[1]

ABSTRACT

Carbonate lentils within the Rex Chert Member were examined to define stratigraphic relations and genesis of the chert and carbonates. This facies represents shelf-edge equivalents of Franson Member shelf carbonates on the east in Wyoming. Lateral and vertical stratigraphic relations of the lentils and enclosing chert are gradational. Fossils and lithologies of the lentils indicate that they originated as carbonate banks with interbank areas of spicular siliceous muds. Main biotic constituents of the banks include crinoids, brachiopods, and bryozoa. Rugose corals occur in moderate abundance at several localities. Siliceous monaxial sponge spicules comprise the major biotic element in the cherts. Lithologically, the banks consist mainly of crinoid and brachiopod biograinstones with moderate to high degrees of textural maturity; back-bank lithologies consist primarily of bryozoan biopackstone-wackestone of relatively low textural maturity. Interbank cherts are primarily bituminous spicular chert biowackestones of very low textural maturity. From facies relations of the chert and carbonate, chert deposition is interpreted to have occurred in water depths of less than 100 ft. Carbonate bank development was initiated when a transgression of the Phosphoria sea introduced open circulation resulting in the deposition of Franson carbonates. During the pioneer phase of carbonate bank development, ramose bryozoa colonized areas of slight topographic relief along the shelf edge. Laterally, siliceous sponges were disarticulated to form spicular chert flanking the banks. As the transgression continued, circulation increased and more larvae migrated basinward establishing a mature community consisting largely of brachiopods, crinoids and bryozoa. During the climactic stage of carbonate deposition; foraminifers, algae, molluscs, and brachiopods occupied a carbonate shoal flanked by a winnowed sponge spicule bar.

INTRODUCTION

Carbonate lenses in the Rex Chert Member of the Phosphoria Formation were first noted by geologists of the U.S. Geological Survey in the early 1900's when the importance of the phosphates of the Phosphoria was realized and mapping of the phosphate reserves was initiated by Mansfield and Richards (Richards and Mansfield, 1910; Mansfield, 1927). At that time several localities of occurrence of the carbonate lenses were noted, but no detailed description was made, although some of the fossils collected were described by Girty (Mansfield, 1927).

After World War II the importance of the Phosphoria, not only as a source of phosphates but also as a suspected source of radioactive minerals, influenced the U.S. Geological Survey to initiate further study of the stratigraphy, paleontology, petrology, and geochemistry of the Phosphoria Formation, as well as more detailed mapping of the phosphate reserves (McKelvey, et al., 1956). Most of the detailed mapping was concentrated in the southeast Idaho area, which is the area of maximum phosphate occurrence. During this mapping project the carbonate lenses of the Rex Chert Member were again noted. Fossils were collected from the lenses and described by E. L. Yochelson (1968), whose data have formed a basis for most of the paleontological aspects of this study and also provided a listing of most of the localities of lens occurrence. W. C. Gere initially suggested a study of the lenses and located many of them. Most of the field work for the project was completed in the summer of 1969 and a brief portion of the spring of 1970. The study included the stratigraphic analysis of eight localities with thirty-five stratigraphic sections.

The area of study (Fig. 1) was extreme southeast Idaho near the Idaho-Wyoming and Idaho-Utah borders. Broad north-south trending folds and thrusts of the Idaho-Wyoming thrust belt are expressed throughout the area as high rolling ridges and broad valleys. Most of the carbonate lenses studied

---

[1] Impel Corporation, Denver, Colorado

I wish to especially acknowledge the assistance of Professor James A. Peterson from the University of Montana whose suggestions and criticisms greatly contributed to the completion of this project. The National Science Foundation provided funds for the field work and laboratory expenses under Professor Peterson's grant, NSF GA-986, which deals with a more regional study of the carbonates of the Phosphoria Formation.

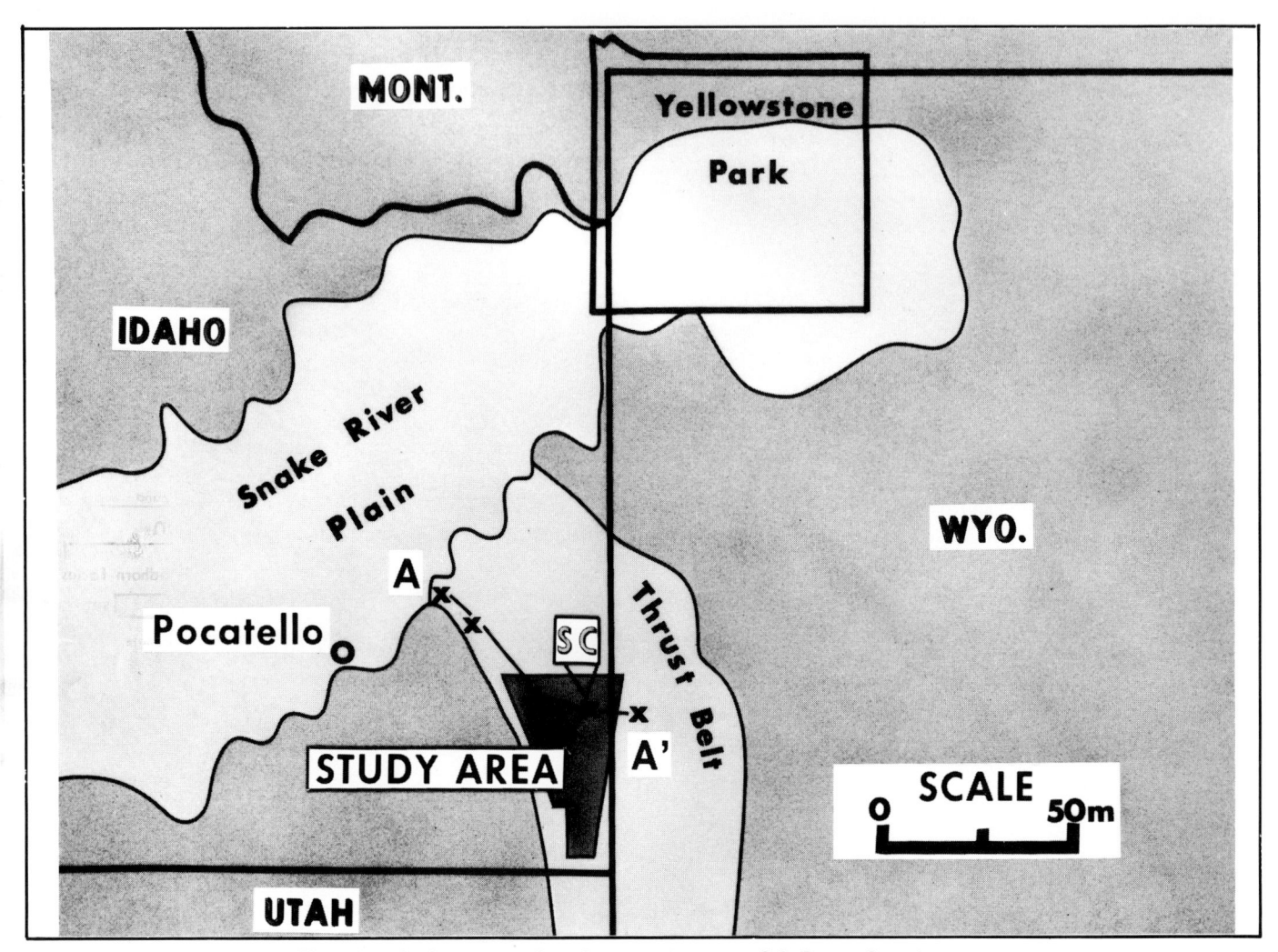

Fig. 1 — Index map showing study area, SC-Sage Creek area.

(Table 1) were within the Bannock thrust sheet east of Soda Springs, Idaho. The primary locality studied, because of the quality of the exposure, was the Sage Creek area close to the Idaho-Wyoming border at the edge of the Bannock sheet.

Table 1. Localities of Carbonate Lenses, Southeast Idaho

1. Sage Creek, Idaho; Secs. 3 and 10, T9S, R45E, Caribou County.
2. South Sage Creek, Idaho; Secs. 1, 6, 13 and 24, T9S, R45E and Sec. 7, T9S, R46E, Caribou County.
3. Deer Creek, Idaho (See Yochelson, 1968).
4. Timber Creek, Idaho; Secs. 21 and 22, T8S, R45E, Caribou County.
5. Stewart Canyon, Idaho; Secs. 30 and 31, T8S, R45E, Caribou County.
6. Dry Valley, Idaho; NW¼ Sec. 31, T4S, R44E, Caribou County.
7. Wood Canyon, Idaho; Sec. 6, T9S, R43E and Secs. 30 and 31, T8S, R43E, Caribou County.
8. Hot Springs, Idaho; Sec. 13, T15S, R44E, Bear Lake County.

REGIONAL STRATIGRAPHY

The basic regional stratigraphic relationships of the Phosphoria Formation are well known as a result of the definitive work of the U.S. Geological Survey (McKelvey, *et al.*, 1967; Sheldon, 1963; Cressman and Swanson, 1964; and McKee, *et al.*, 1967). The complex intertonguing nature within the unit, although it is fairly well defined, adds difficulty to any discussion of regional stratigraphy because of the resultant problems of stratigraphic nomenclature.

For the purpose of clarity in this brief discussion, the nomenclature diagrammed in Figure 2 will be used. The use of the term, Phosphoria Formation, in this manner is widespread in the petroleum and mining industries. This figure illustrates the stratigraphic relationships of rock units in the study area to those in western Wyoming. A similar scheme was used by Cole

(1969) in central Wyoming where the Phosphoria shelf carbonates (Park City facies) intertongue with equivalent continental sediments.

The complex of intertonguing units will be referred to collectively as the Phosphoria Formation, with each separate unit referred to as member, tongue, or lentil, wherever applicable (Fig. 2). The eastern Idaho units, dominated by shales, cherts, and phosphorites, will be informally referred to as the Phosphoria facies: Meade Peak, lower chert, Rex Chert, Retort, Tosi Chert, and cherty shale members. The western and central Wyoming units, dominated by carbonates, will be informally referred to as the Park City facies: Grandeur, Franson, and Ervay Members. The units in southern Montana and eastern Wyoming, dominated by redbeds, sandstone, and evaporites, will be informally referred to as the Shedhorn-Goose Egg facies: Shedhorn, Opeche, Glendo, and Freezeout Members.

The regional stratigraphic section (Fig. 3) illustrates in more detail the relationships shown in Figure 2. The carbonate lentils studied are equivalent to the Franson Member carbonates of western Wyoming as evidenced by: their stratigraphic position, their lithology, and their fauna. They are an integral part of the second major transgressive Permian cycle, the Franson cycle.

The area of study is very near the western limit of the Franson Member carbonates as they intertongue with the cherts of the Rex Chert Member on the west (Fig. 3). Westward from the area of study near the Fort Hall Indian Reservation, the Rex Chert Member grades laterally and vertically into the siliceous shale of the cherty shale member. The cherty shale member, in turn, grades laterally to the southwest in the Sublette Range of south-central Idaho to a thick section dominated by chert (Mansfield, 1927; Warner, 1956; McKelvey, 1959).

In the area of the Lost River Range in central Idaho, Bostwick (1955) reports sandstones of Lower Permian age, however, all of the younger Permian and Mesozoic rocks are eroded. In north-central Idaho volcanic rocks of Permo-Triassic age (Seven Devils Volcanics) are exposed (Bissel, 1959; McKee, et al., 1967).

Figure 4 is a schematic representation of the paleogeography at the time of deposition of the Franson cycle sediments as interpreted from the data discussed above. The following genetic relationships within the Franson cycle may be interpreted from the regional stratigraphic evidence.

Clastic Shelf Sediments

The Shedhorn Sandstone in southwestern Montana and northwestern Wyoming represents the littoral sediments of the north margin of the Franson sea. These sands were derived from older sandstones in the uplifted area to the north in Montana and were transported southward into the northwest portion of the diagrammed area.

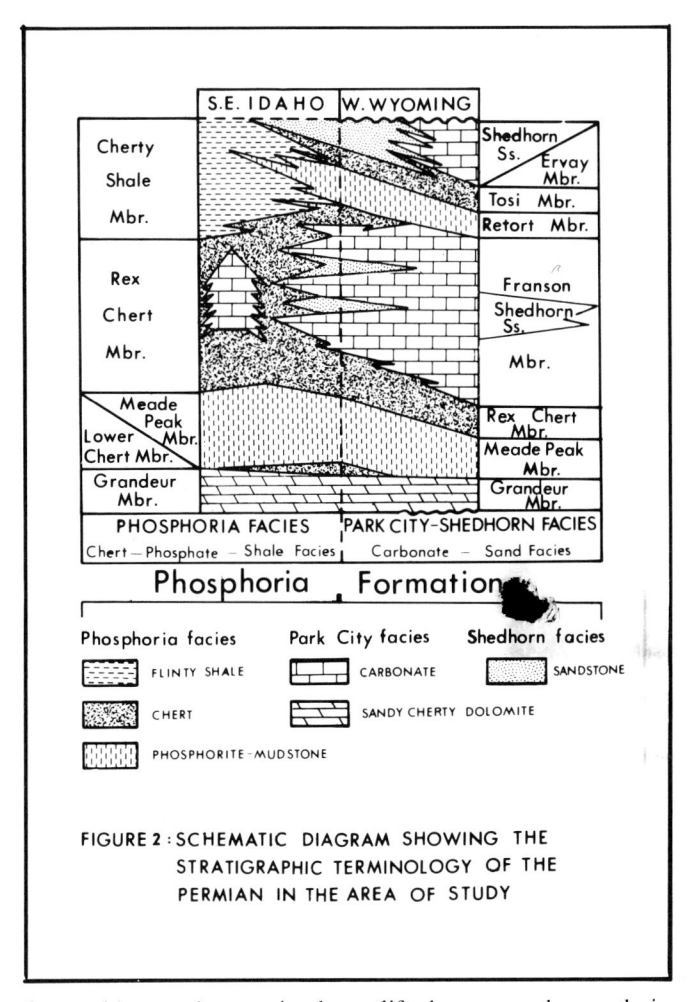

The shoreline to the east in Wyoming (Cole, 1969) consisted of a more gradual change from subtidal and tidal-flat carbonates to restricted marine and continental redbeds and evaporites of the Goose Egg facies.

Farther to the southeast of the study area the shoreward equivalents of the Franson are eroded, but may have originally consisted of the littoral and aeolian sandstones of the White Rim Sandstone and the continental redbed-feldspathic sandstone sequence of the Cutler Formation on the flanks of the Ancestral Rockies and Uncompahgre uplift (McKee, et al., 1967).

Inner Shelf Carbonates

Most of the Franson Member carbonates were deposited offshore in shoal conditions resulting in the deposition of biostromes and bioherms. Both local and regional paleotopography probably affected the position of bioherms relative to the more normal biostromal nature of the carbonates. Locally, sponge spicular cherts were deposited in the topographic lows adjacent to the carbonate bioherms.

FIGURE 2: SCHEMATIC DIAGRAM SHOWING THE STRATIGRAPHIC TERMINOLOGY OF THE PERMIAN IN THE AREA OF STUDY

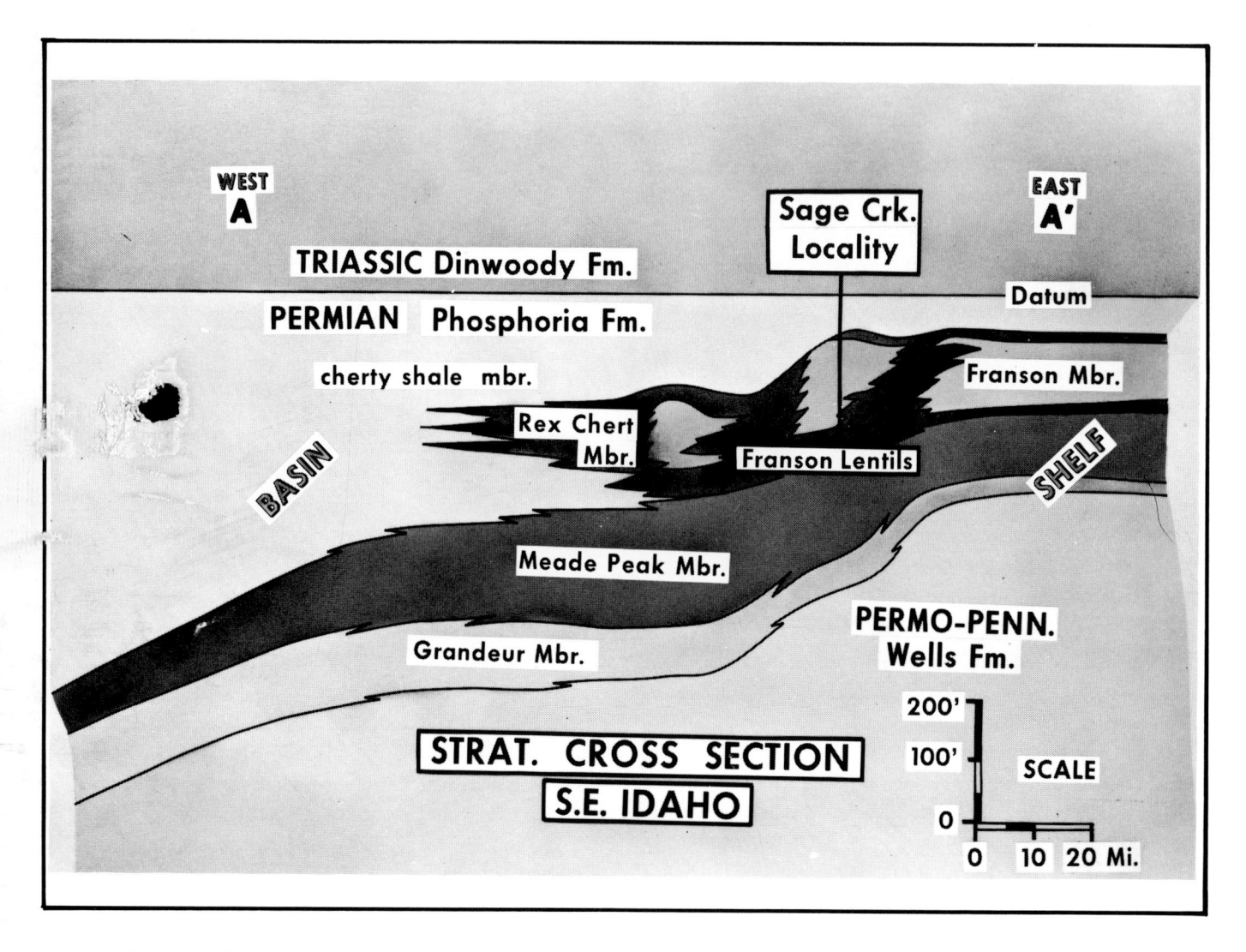

Fig. 3 — Regional stratigraphic cross section showing relationship of Phosphoria members.

Outer Shelf Chert and Carbonates

Carbonate deposition was largely restricted to the shallow-shelf areas of western and west-central Wyoming and southwestern Montana, while deposition of spicular cherts of the Rex Chert Member dominated the lower portions of the shelf in the study area. However, lentils of Franson-equivalent carbonate bioherms occur westward from the main body of the Franson up to and on the "hinge" or shelf edge, representing local shoals on paleotopographic highs.

Shelf Slope and Basinal Cherts and Shales

The nature of the slope and basinal sediments of the Phosphoria Formation is very poorly understood. From what is known to date (Mansfield, 1927; Warner, 1956; McKelvey, 1959; McKee, et al., 1967) the change from slope to basin is represented only by a facies change from chert on the east to siliceous shales basinward. This facies change may not have represented a sharp topographic boundary, but rather the outer limits of the average depth at which siliceous sponges were present to form the sediment supply for the cherts. The thick section of chert present in the Sublette Range presumably near the Phosphoria basinal axis is non-spicular; therefore, a different genesis is implied for these cherts as opposed to the shelf and eastern shelf-slope cherts.

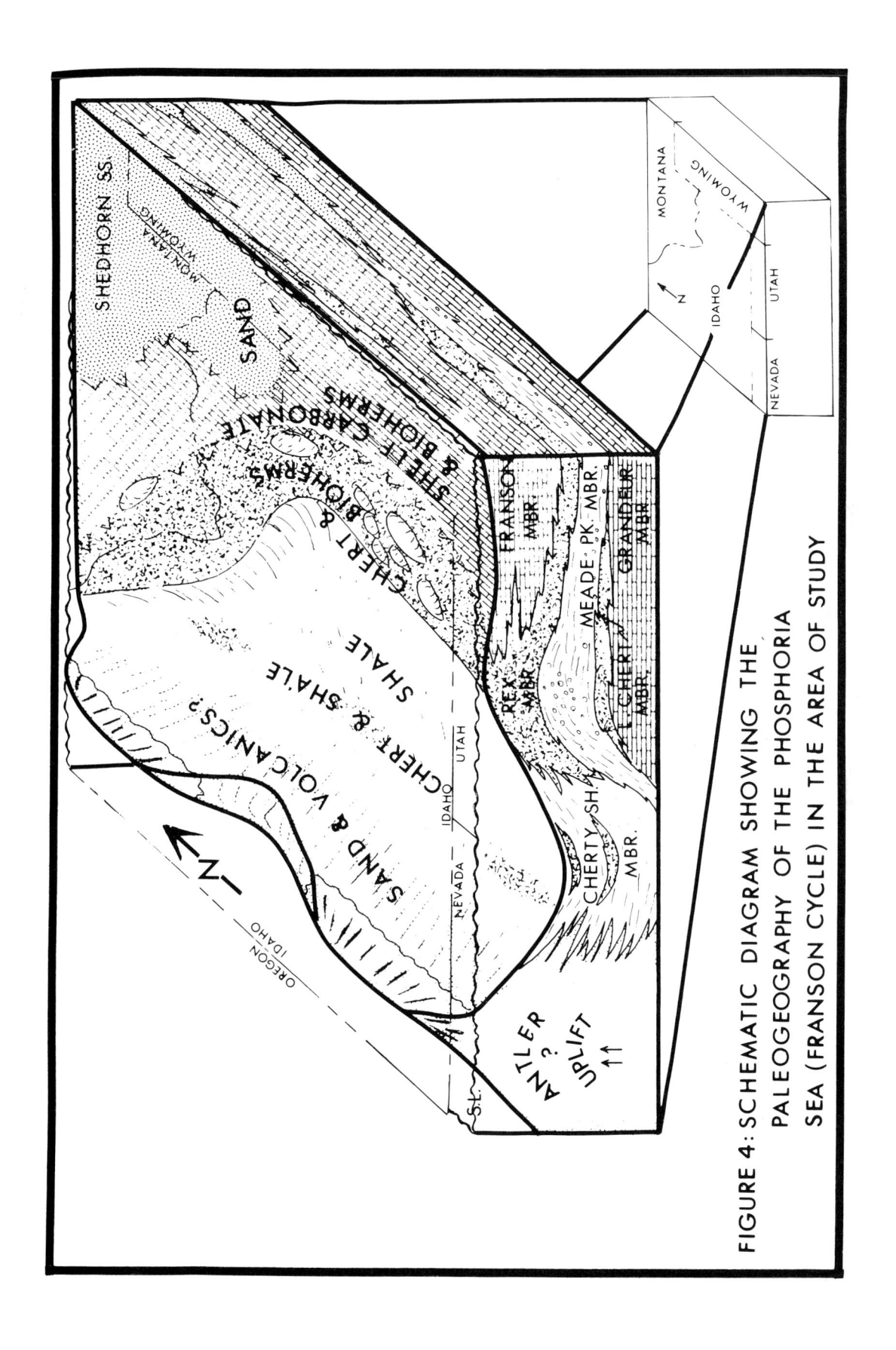

FIGURE 4: SCHEMATIC DIAGRAM SHOWING THE PALEOGEOGRAPHY OF THE PHOSPHORIA SEA (FRANSON CYCLE) IN THE AREA OF STUDY

Orogenic Belt Clastics and Volcanics

The presence of an orogenic belt through central to western Idaho is assumed from scattered localities in which Permian rocks outcrop. Present distributions of the clastics and volcanics of the Permian System in western Idaho, relative to marine limestones and shales in northwestern Washington (McKee, et al., 1967) indicate that this was a narrow orogenic belt.

LOCAL STRATIGRAPHY

Details of the Permian stratigraphy of southeast Idaho are somewhat incompletely described in the geologic literature. Most of what is known is a result of the mapping of the U.S. Geological Survey by Richards and Mansfield in the early 1900's (Mansfield, 1927) and in more detailed quadrangle mapping in the 1950's (Cressman and Gulbrandsen, 1955; Gulbrandsen, et al., 1956; McKelvey, et al., 1959; Cressman, 1964; and Cheney and Montgomery, 1967). Stratigraphic sections of various parts of the Phosphoria Formation were published as circulars in the early 1950's, however, they generally included detailed descriptions of the Meade Peak Member only. Figure 5 is a generalized stratigraphic section for Permian rocks in the area of study. Complete Permian sections are exposed in many localities in the area, where phosphate exploration trenches add greatly to the quality of exposure.

**FIGURE 5
Generalized Stratigraphic Section
Southeast Idaho**

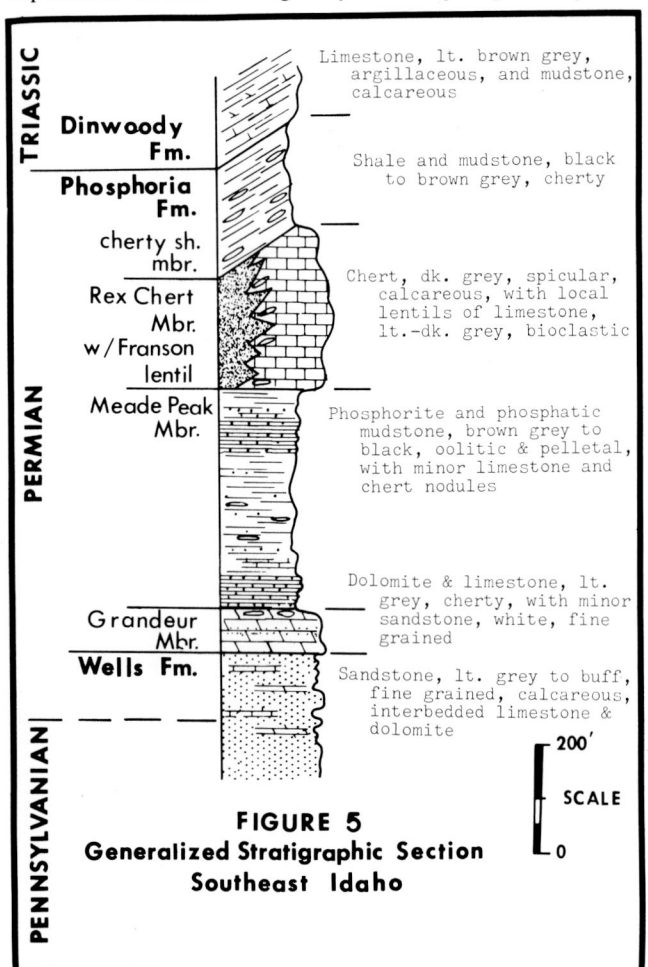

PALEONTOLOGY

Biota of the carbonate lenses studied (Table 2) are quite similar to those of the Franson shelf carbonates to the east in Wyoming as described by Yochelson (1968). The major differences of these lenses from those of the Franson shelf carbonates are the presence of rugose corals and the relative abundance of foraminifera in the lenses studied.

Table 2. Biota of the Rex Chert Member and enclosed bioherms, southeast Idaho.

(After Yochelson, 1968, and from personal observations)

Algae
  ?Schizophyte (Blue-green)
  Dasyclad (Green)
    Epimastapora
    Vermiporella?

Foraminifera
  Textularid indet.
  Globivalvulina
  Encrusting foram

Porifera
  *Monaxial siliceous spicules

Coelenterates
  Rugose corals
    Rugose coral indet.
    Bradyphyllum sp.

Bryozoa
  Ramose bryozoa
    (Cryptostomes)
  Fenestrate bryozoa
    (Trepostomes)

Brachiopods
  Inarticulate
    Orbiculoidea sp. indet.
  Rhynchonellids
    Leiorhynchus sp. indet.
    Rhynchopora sp. indet.
  Strophomenid
    Derbyia sp. indet.
  Productids
    Echinauris sp. indet.
    "*Liostella*" sp. indet.
    Kochiproductus c.f.
      K. longus
    Bathymyonia c.f.
      nevadensis
    Antiquatonia c.f.
      A. sulcatus
    Anidanthus eucharis
    Cancrinella sp.
    Muirwoodia multistriatus
    Sphenosteges sp. indet.
    dictyoclostid indet.
  Spiriferids
    Composita sp. indet.
    Spiriferina sp. indet.

Gastropods
  Babylonites sp.
  indet.

Scaphopods

Pelecypods
  Nuculopsis sp.
  indet.
  Polidevica obesa
  Aviculopectin sp.
  indet.
  Girtypectin sp.
  indet.
  Streblochondria
    c.f. *montpelierensis*

Echinoderms
  Echinoid spines and
    fragments
  Crinoid stem plates
    and fragments

Arthropods
  Ostracodes

*Fish scales and bone
  fragments

Incerte sedis
  *Organic walled
    fossils
  Spindle-shaped
    borings
  Burrows

* Biotic elements occurring primarily in a chert matrix.

Since few of the organisms in the southeast Idaho lenses exhibit the ecologic potential to erect a rigid wave- or current-resistant structure and since most of the organic productivity involved in creating the bioherms was related to brachiopods and crinoids, the genetic term bank is used to describe these bodies in preference to the term reef (Nelson, Brown, and Brindman, 1962). It is fairly unlikely that the few rugose corals present within the bioherms could have contributed to building a rigid wave-resistant structure.

The faunal associations interpreted in the exposure of the carbonate bank at South Sage Creek are summarized in Figure 6 for comparison with the faunal elements of a normal Franson cycle in western Wyoming as described by Yochelson (1963, 1968) shown in Figure 7. Only the lateral sequence is shown in Figure 6 since a cyclic sequence comparable to Yochelson's model does not occur within the bank. Yochelson's model involves a biostromal sequence where faunas and lithologies migrated in response to transgressions and regressions of the Franson sea.

The southeast Idaho banks on the other hand, were initiated on and restricted to topographic highs surrounded by a toxic, siliceous mud. The paleoenvironments, then, in terms of both the fauna and lithology within these banks were relatively narrowly restricted. In response to the transgression of the Franson sea the bank grew vertically with increasing productivity and gradually spread laterally to the north. At the point of maximum transgression the lateral sequence was as shown in Figure 6. This sequence is similar to the model (Fig. 7) except that the area of decreasing energy regime and increasing depth occurs behind the bank zone of high energy. The deeper, low energy regime environment in front of the bank was occupied by siliceous sponges living on a substratum of siliceous mud which was toxic to the fauna living within the bank.

Biotic relationships were determined from detailed thin section petrographic study as well as hand specimen identification. Identification of bioclasts in thin sections was accomplished through comparison with hand samples and general data from the geologic literature (Majewske, 1969).

FACIES DISTRIBUTION — SOUTH SAGE CREEK BANK

The Sage Creek and South Sage Creek localities were studied in detail with respect to paleontology and petrology for the following reasons:
  a. Quality of the exposures.
  b. Lateral stratigraphic relationships observable in the field.
  c. Apparent lateral facies changes.
  d. The possibility that the two localities represent exposures of a single bank or bank complex.

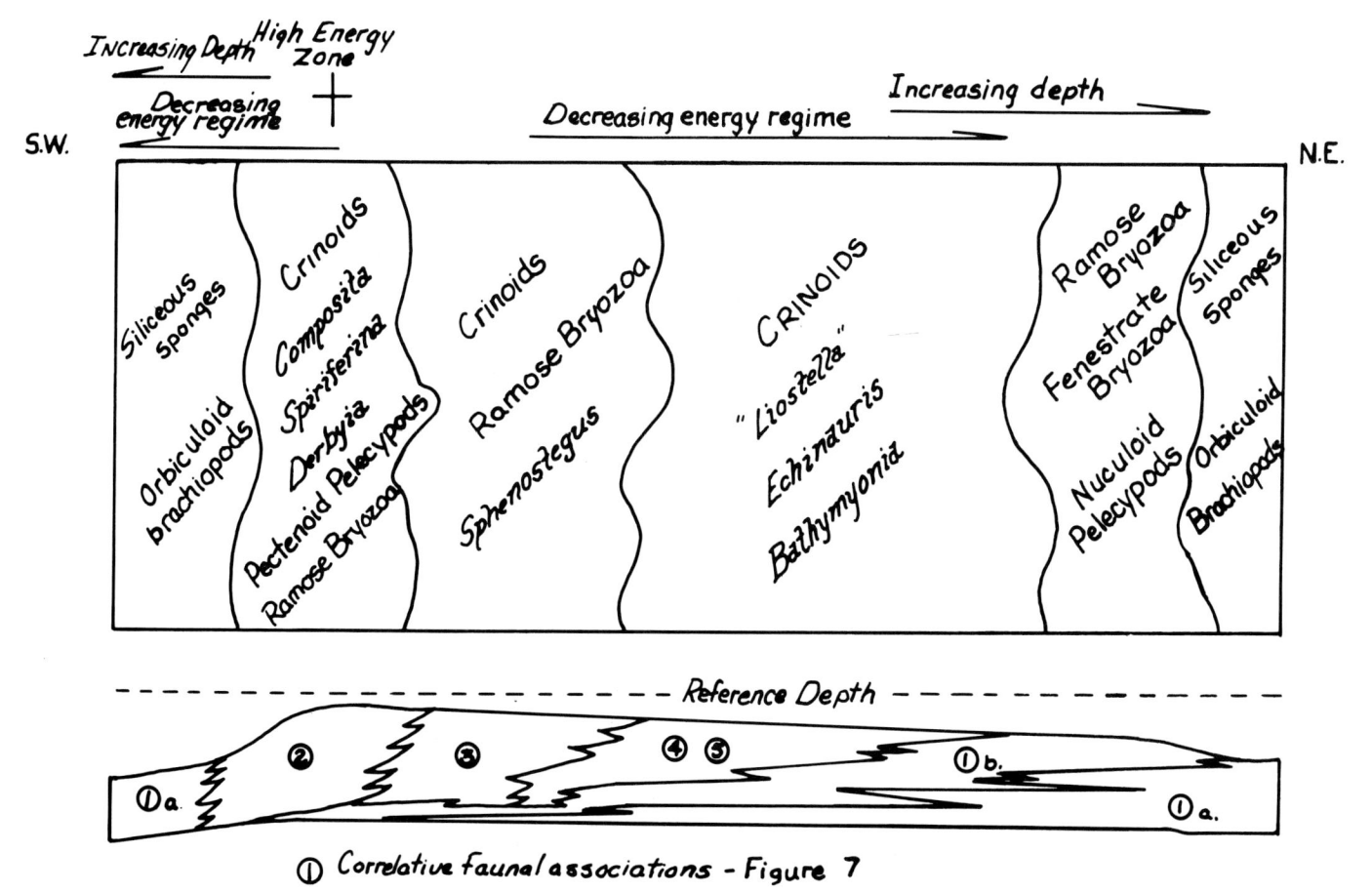

Fig. 6 — Interpreted faunal associations — South Sage Creek, Idaho.

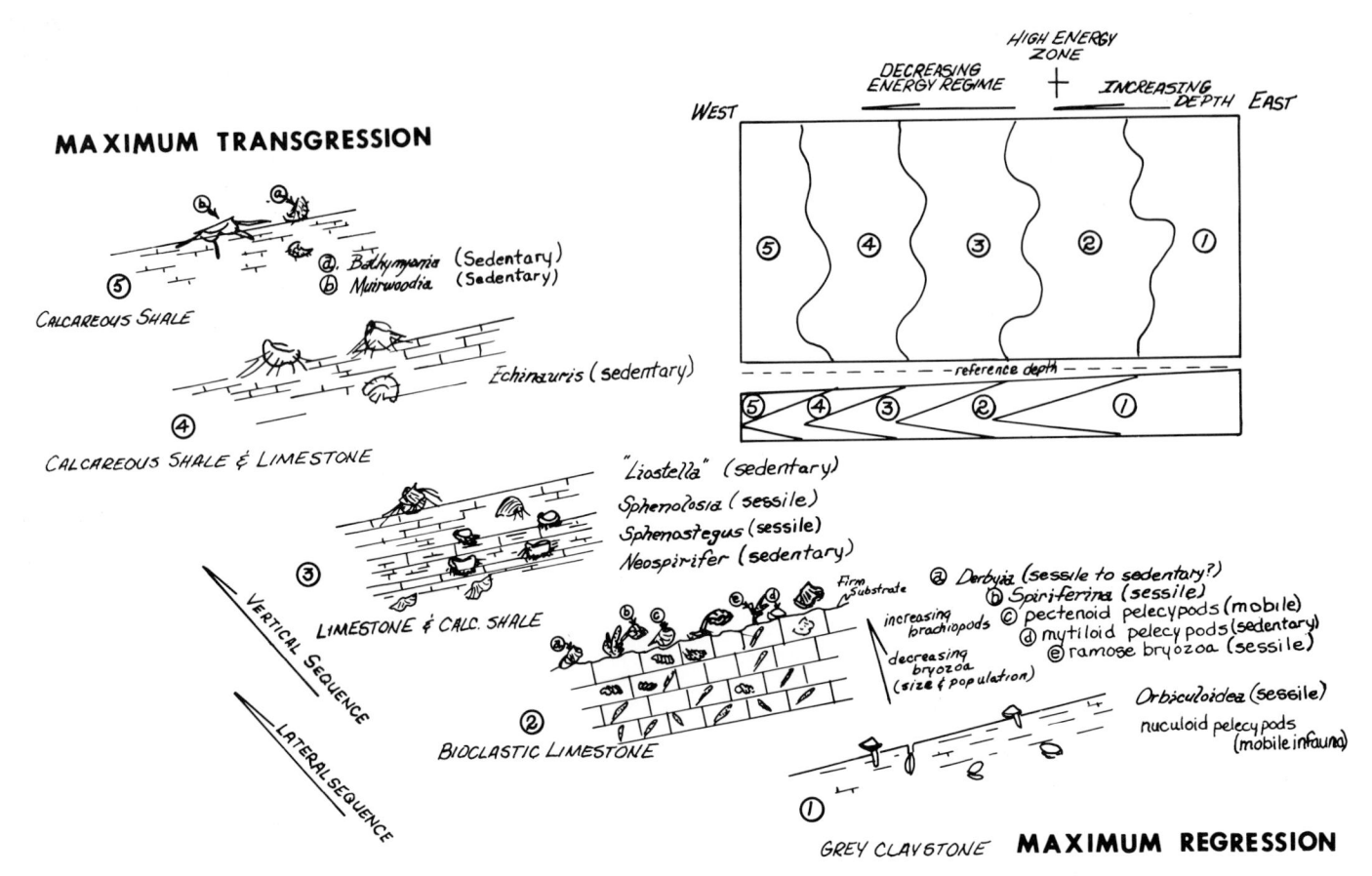

Fig. 7 — Faunal elements — Franson cycle, West Wyoming. (Data from Yochelson, 1963, 1968).

The Sage Creek and South Sage Creek localities are on the west and east limbs respectively of the Webster syncline (Fig. 8). There is no evidence that the two carbonate exposures are separate or unrelated. The Sage Creek carbonate occurrence is terminated on the west flank of the Snowdrift anticline by an outcrop of chert. The facies change is eroded from the crest of the anticline (Fig. 8). At the South Sage Creek locality the eastern limit of the carbonate occurrence is poorly defined. The east flank of the asymmetric Boulder Creek anticline is overturned and faulted to the east of the South Sage Creek locality (Mansfield, 1927, Plate VII). In the one poor outcrop of the Rex Chert Member observed immediately to the east (i.e. beneath) the reverse fault, no carbonate was exposed. For this reason, the bank is interpreted to occur only to the eastern boundary of Section 18, T9S, R46E.

The carbonate isolith (Fig. 8) shows the interpreted extent of the South Sage Creek bioherm, projecting outcrop thicknesses into the subsurface beneath the Webster syncline. Thicknesses of total carbonate are shown for each measured section. The maximum carbonate occurrence at Sage Creek is breached and partially covered with alluvium. That portion of the isolith therefore, is restored, as are the portions west of the Sage Creek locality and east of the South Sage Creek locality where the Rex Chert Member has been removed by erosion.

The stratigraphic cross section (Fig. 9) illustrates the generalized lithofacies distributions for the South Sage Creek locality and was generalized from the detailed lithotypes described petrologically for South Sage Creek. The lithotypes (e.g. 2A, 1B, etc.) included in each lithofacies are listed for comparison with Tables 3 and 4. An explanation of the terminology for various rock types is given in Table 5. The same lithotypes were recognized during the petrologic study of the Sage Creek locality.

Interpretation of time-stratigraphic relationships is difficult at both of the localities, however, the datums used in construction of Figure 9 are felt to be closely isochronous. The use of the multiple datum system at South Sage Creek is necessitated by the lack of exposure of the base of the Rex Chert Member. These datums are arbitrary and only used for interpretations of time-stratigraphic relationships and probably approach isochroniety only very locally.

The interpreted time-stratigraphic relationships at various stages of development of the bank are shown on Figure 9 and are illustrated in the facies distribution maps (Fig. 10 to 14). The maps include data synthesized from petrologic analysis, field relationships, interpreted faunal associations, and experience gained from observations at other localities.

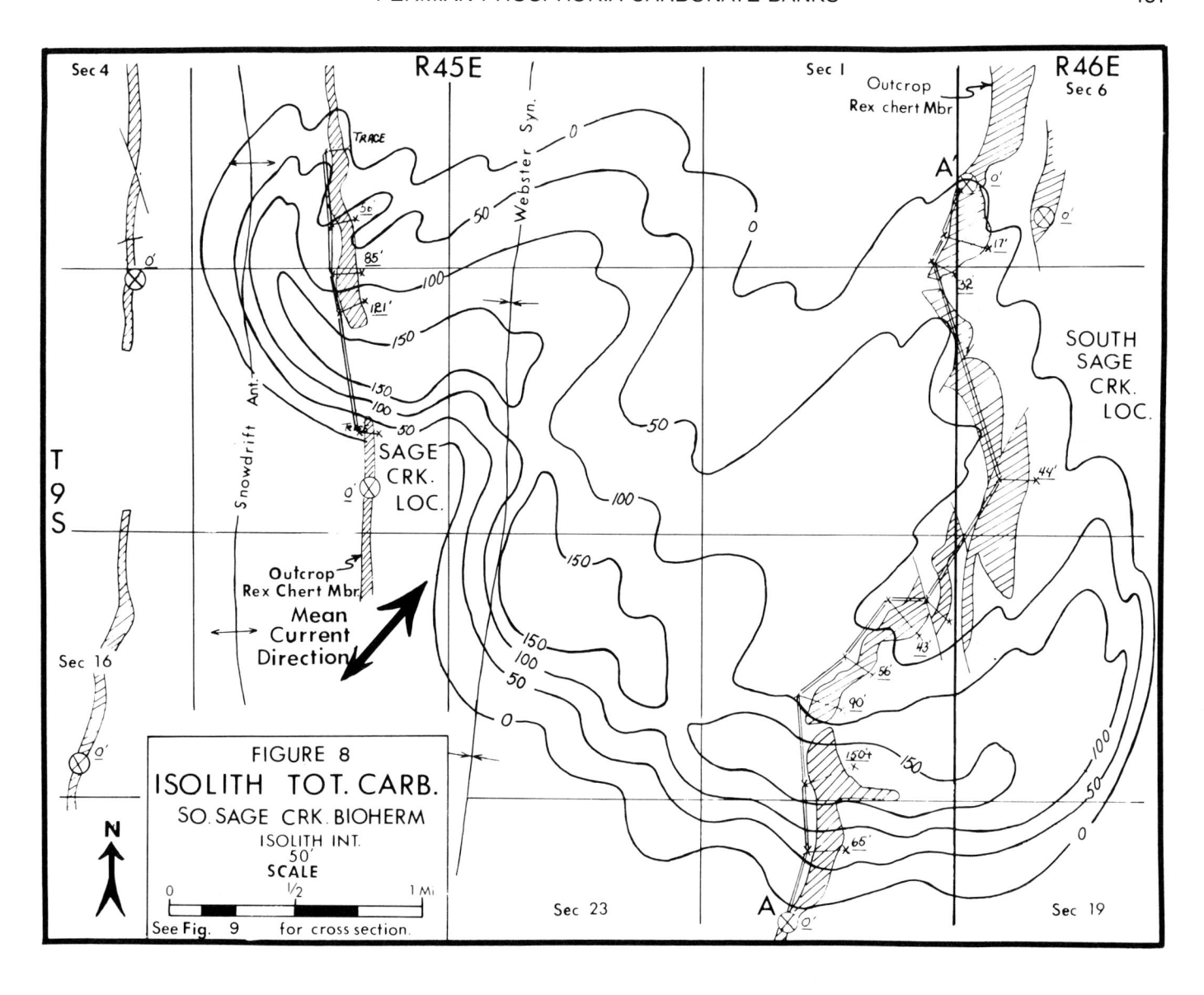

Figure 8. ISOLITH TOT. CARB. SO. SAGE CRK. BIOHERM. ISOLITH INT. 50'. See Fig. 9 for cross section.

Table 3. Relative Textural Maturity — Carbonate Lithotypes, South Sage Creek Bank

LITHOTYPES	ENERGY REGIME		
	HIGH	MODERATE	LOW
2A	←——————→		
2C	←——————→		
2B	←——————→		
2E	←——————→		
2D		←——————→	
2F		←——————→	
2G		←——————→	
2I		←——————→	
2H			←——————→

(RELATIVE TEXTURAL MATURITY INCREASING ↑)

Table 4. Relative Textural Maturity — Chert Lithotypes, South Sage Creek Bank

LITHOTYPES	ENERGY REGIME		
	HIGH	MODERATE	LOW
1B			←——→
1A			←——→
1C			←——→
1F			←——→
1D			←——→
3			←——→
1E (Replacement)	←————————————→		

(RELATIVE TEXTURAL MATURITY INCREASING ↑)

Fig. 9 — Cross section A-A', South Sage Creek bioherm.

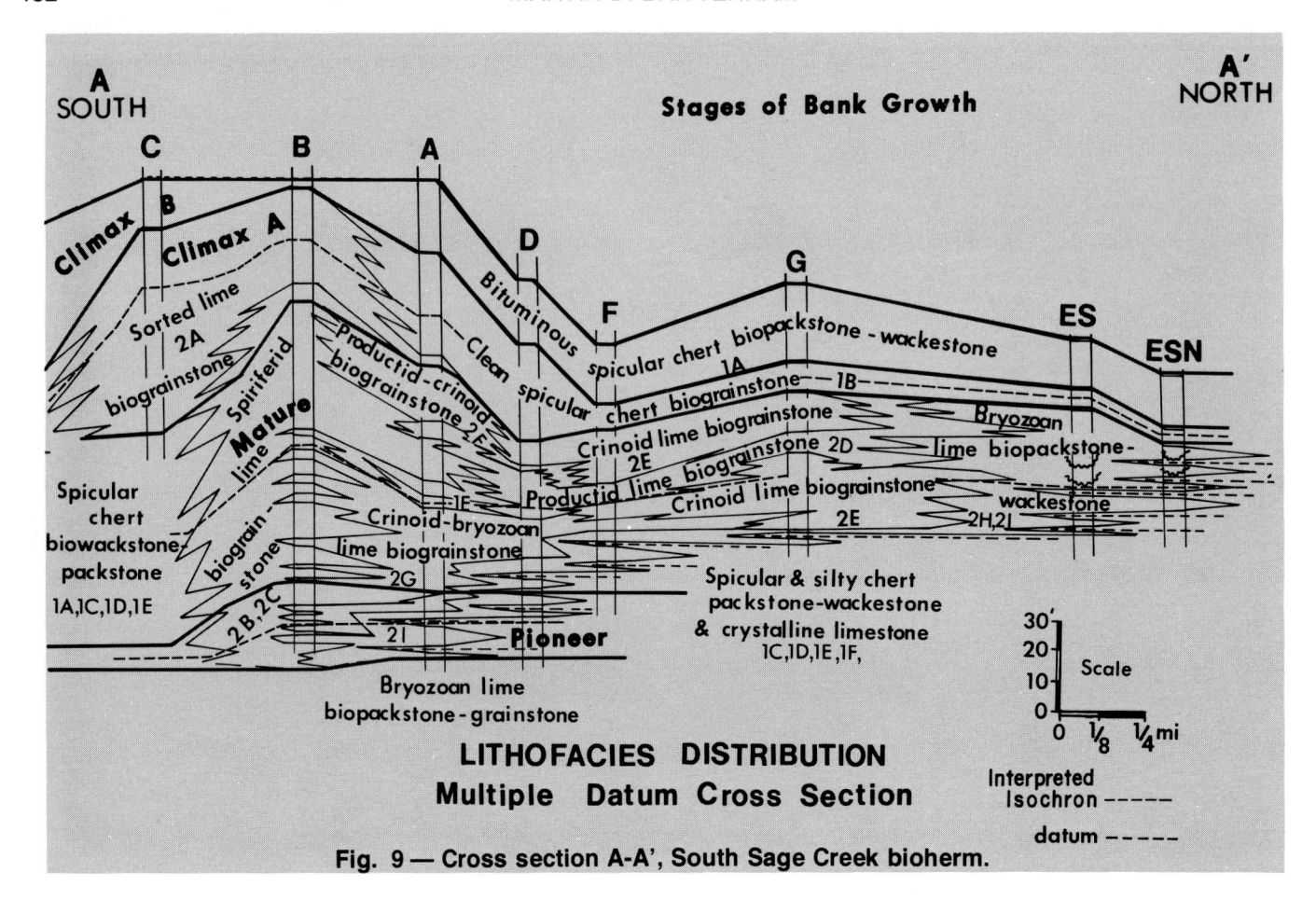

Table 5. Lithologic Classification (After Dunham, 1962)

I. TYPES WITH GREATER THAN 10% CLASTS

A. Texture (See chart below)

MATRIX		
MUD (>4 μ)		**SPAR**
Mudstone — (Ms)* < 10% grains		Crystalline types (Cry)* < 10% grains
Wackestone — (Ws)* mud support > 10% grains		Grainstone (Gs)* grain support > 10% grains
Packstone — (Ps)* grain support		

(TEXTURES)

B. **Prefixes**
 1. Mineralogy
    a. Lime — Limestone (L)*
    b. Dolo — Dolomite (D)*
    c. Chert — Chert (C)*
 2. Clast type-major constituent as prefix; others as modifiers
    a. Bio — > 10% bioclasts (Bc)*
    b. Intra — > 10% intraclasts (Ic)*
    c. Pel — > 10% pellets (P)*
    d. Oo — > 10% oolites (Oo)*
    e. Litho — > 10% lithoclasts (Lc)*
 3. Fossil types
 4. Clast grain-size
    a. Coarse — > 50% 2mm
    b. Medium — > 50% 1/16 — 2mm
    c. Fine — > 50% 4μ — 1/16 mm
 5. Clast sorting (qualitative)
    a. Unsorted
    b. Moderately sorted
    c. Sorted
 6. Rounding or angularity
 7. Terrigenous clasts
 8. Other modifiers — porosity, diagenesis, organic bitumen content, minor constituents, lamination, structures, etc.

II. **CRYSTALLINE TYPES**
 A. Crystalline limestone (CryL)*
 B. Crystalline dolomite (CryD)*
 C. Crystalline chert (CryC)*

* Abbreviations

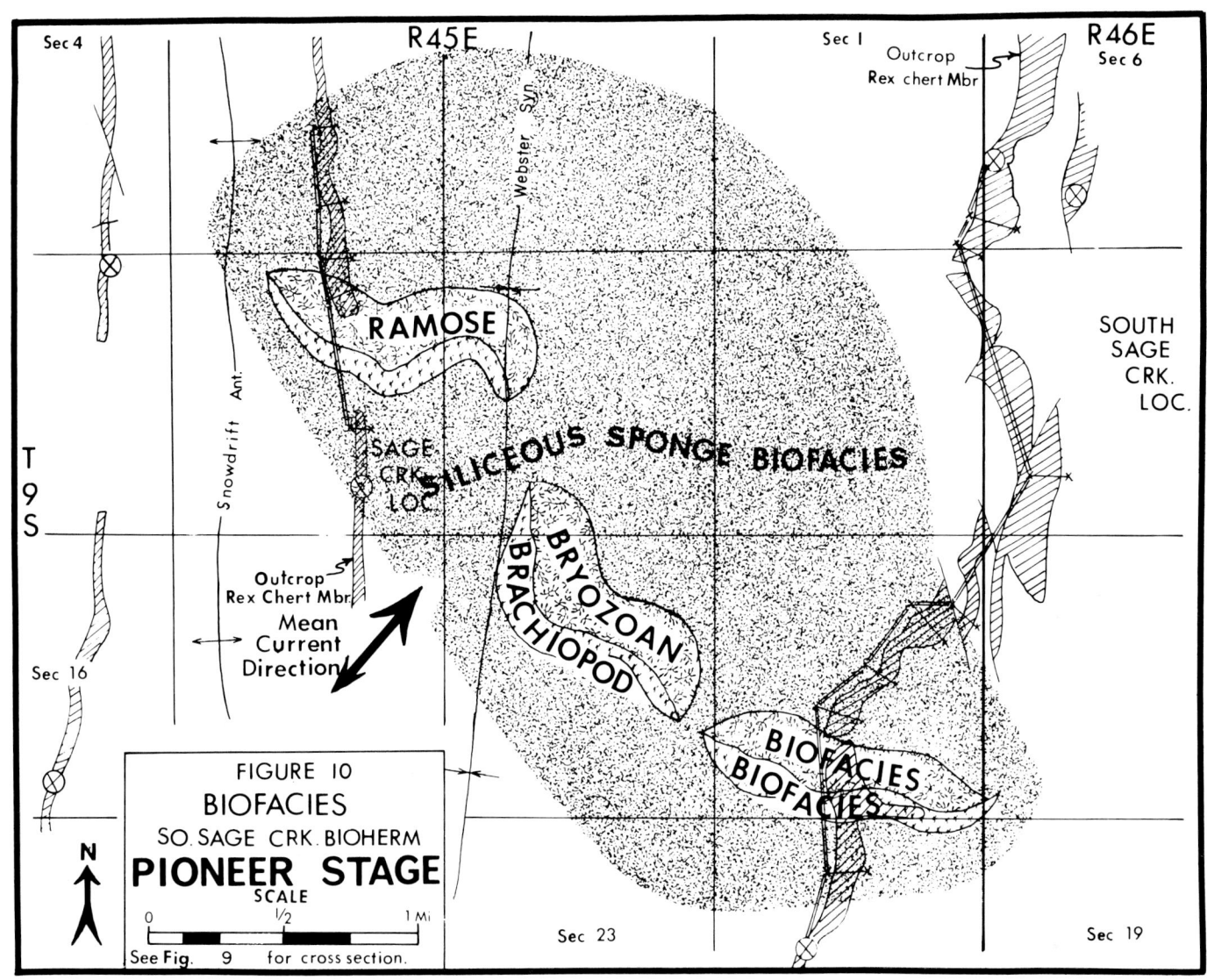

FIGURE 10
BIOFACIES
SO. SAGE CRK. BIOHERM
PIONEER STAGE
SCALE
0   1/2   1 Mi
See Fig. 9 for cross section.

The vertical stacking of lithofacies, as defined in the stratigraphic cross section (Fig. 9) is considered to be indicative of paleotopographic control of carbonate productivity. In the case of the South Sage Creek bank the highest degree of productivity occurred in the area of Section B, South Sage Creek and immediately south of Section B, Sage Creek, and was initiated in that area and restricted to it by pre-existing topography.

The spiriferid brachiopods and crinoids which inhabited the main bank area received the highest degree of current action and circulation of nutrients. In the face of sea level changes the productivity of the fauna rose and fell in compliance with the restrictions of the environment, however, because of the paleotopographic control, they did not migrate from the area they initially occupied.

Not all lithofacies are stacked as is the spiriferid-lime biograinstone facies (Fig. 9). The bryozoan-lime biopackstone-grainstone facies (2I), for instance, transgresses from southeast to northeast from the area of initial bank development to the northeast margin of the bank where the environment was stabilized to form the stacked bryozoan lime biopackstone-wackestone facies. The general distribution of facies, defined by cross section A-A' is interpreted to represent a genetic sequence as illustrated in the following facies maps. (Figs. 10-14)

Pioneer Stage

As an initial stage of carbonate sedimentation in the area, ramose bryozoa occupied small topographic highs (Fig. 10). The bryozoa were originally restricted to small microcommunities surrounded by spicular siliceous muds, but rapidly spread to form larger mounds. With increased circulation, in response to the transgressing sea, brachiopods inhabited the areas of greatest circulation.

Field evidence from all localities observed in the study area supports a similar mode of bank initiation. Also, the northward transgression of the South Sage Creek bank (Fig. 9) may be directly attributed to pioneering communities of bryozoa which initiated carbonate deposition in protected areas on the lee side of the bank.

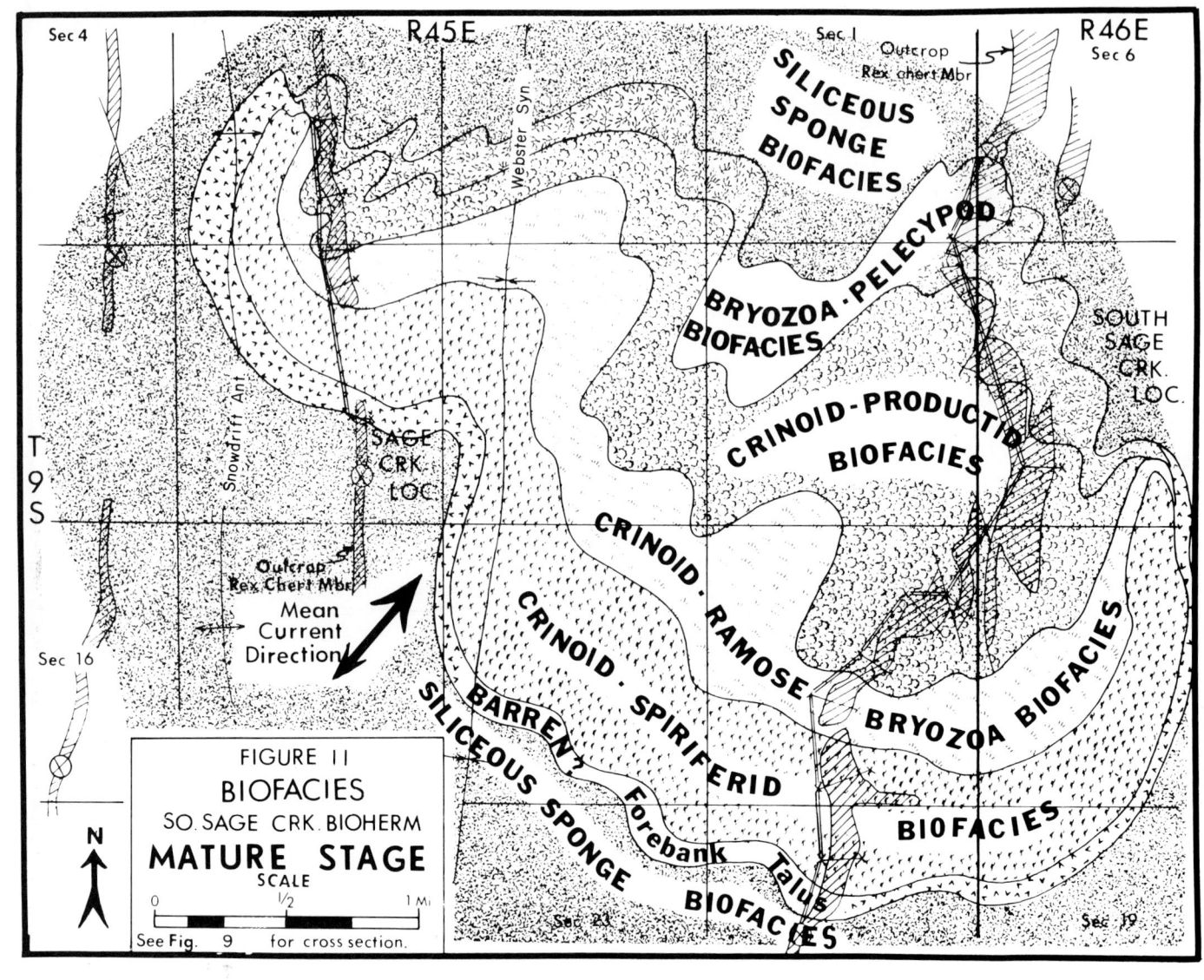

FIGURE 11
BIOFACIES
SO. SAGE CRK. BIOHERM
MATURE STAGE

Mature Stage

With continued transgression of the sea, maximum carbonate productivity was achieved and diverse environments within the bank were occupied by a variety of faunal elements. The kinds of sediment deposited within an environment (Fig. 12) were affected by the organisms living within or immediately adjacent to that environment (Fig. 11) and were modified by the processes active within the environment (Fig. 13).

BIOFACIES

The interpreted biofacies distribution for the South Sage Creek bank (Fig. 11) was generalized from detailed petrologic data and other paleontological evidence previously described (Fig. 6). The faunal elements included within each biofacies are summarized as follows (listed in order of abundance):

A. Siliceous Sponge Biofacies:
   1. Siliceous Sponges
   2. Orbiculoid Brachiopods
   3. Scaphopods

B. Crinoid-Spiriferid Biofacies
   1. Spiriferid Brachiopods
      Spiriferina, Composita
   2. Crinoids
   3. *Derbyia*
   4. Ramose Bryozoa
   5. Pectinoid Pelecypods

C. Crinoid-Ramose Bryozoa Biofacies
   1. Crinoids
   2. Ramose Bryozoa
   3. Productid Brachiopods
      mostly *Sphenostegus*(?)
   4. Spiriferid Brachiopods
   5. Fenestrate Bryozoa
   6. Pectinoid Pelecypods
   7. Foraminifera
   8. Echinoids
   9. Rugose Corals

D. Crinoid-Productid Biofacies
   1. Crinoids
   2. Productid Brachiopods
      "Liostella"
      Echinauris
      Bathymyonia
   3. Ramose Bryozoa
   4. Pectinoid Pelecypods
   5. Echinoids
   6. Rugose Corals (other localities)
E. Bryozoa-Pelecypod Biofacies
   1. Ramose Bryozoa
   2. Fenestrate Bryozoa
   3. Nuculoid Pelecypods
   4. Encrusting Bryozoa
   5. Foraminifera
   6. Echinoids
   7. Algae

The carbonate debris shed off the steeper southern flank of the bioherm encroached upon the toxic bottom muds of the siliceous sponge biofacies. Existence within such an environment would have been very difficult for organisms with carbonate shells; therefore, the environment is considered to have been relatively barren of living carbonate-secreting organisms.

Lithofacies

The lithofacies distribution for the South Sage Creek bank was generalized from detailed petrologic data. When considered in context with the biofacies described, the distribution of lithofacies is a good measure of depositional processes and environments.

A. Coarse Lime Biograinstone Lithofacies:

This lithofacies contains the texturally most mature lithotypes (2B, 2C) of the bank complex. It also represents the maximum accumulation of carbonate sediment within the bank. It was certainly deposited under optimum conditions for carbonate sedimentation.

B. Medium to Coarse Lime Biograinstone-Packstone Lithofacies:

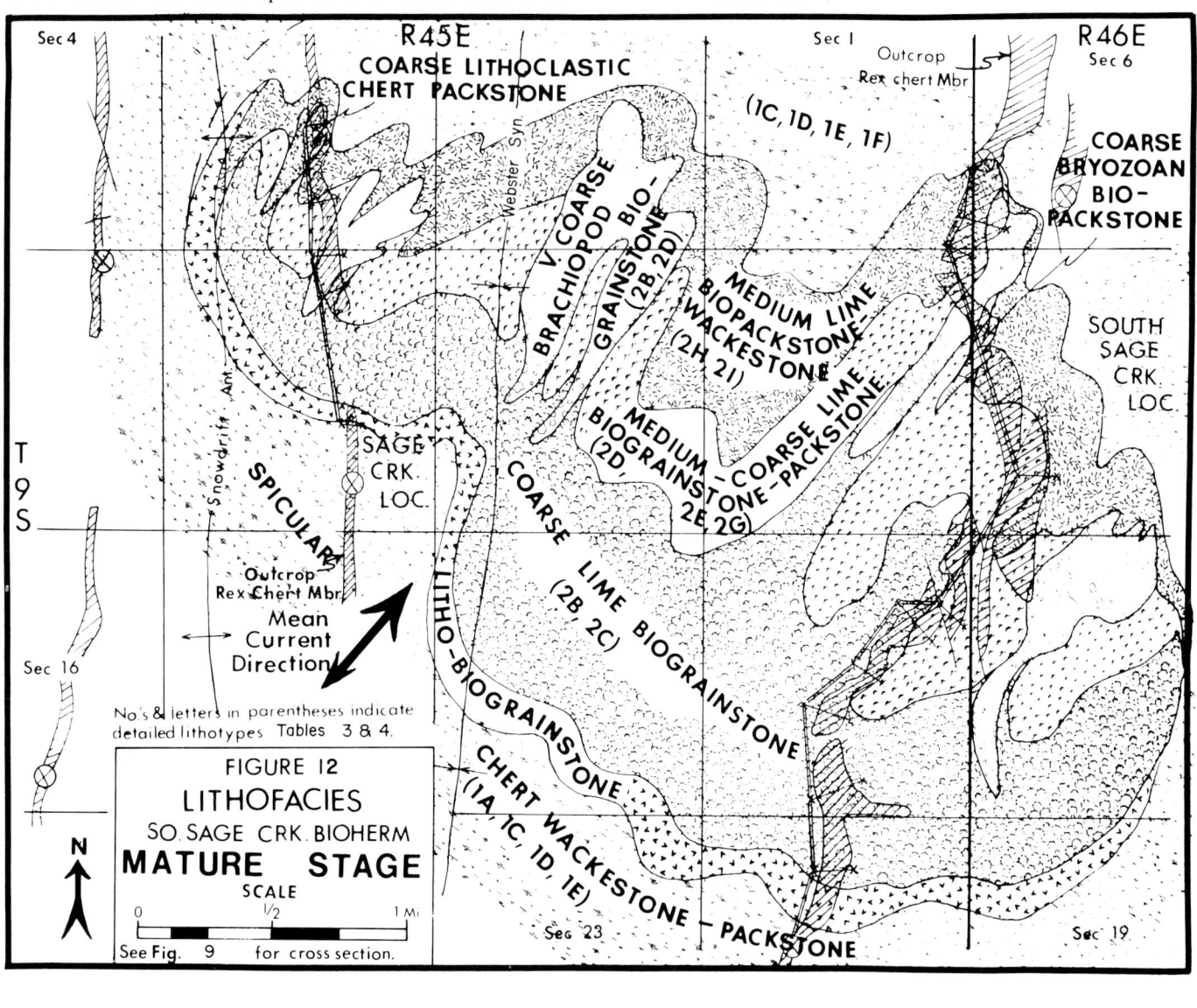

FIGURE 12
LITHOFACIES
SO. SAGE CRK. BIOHERM
MATURE STAGE

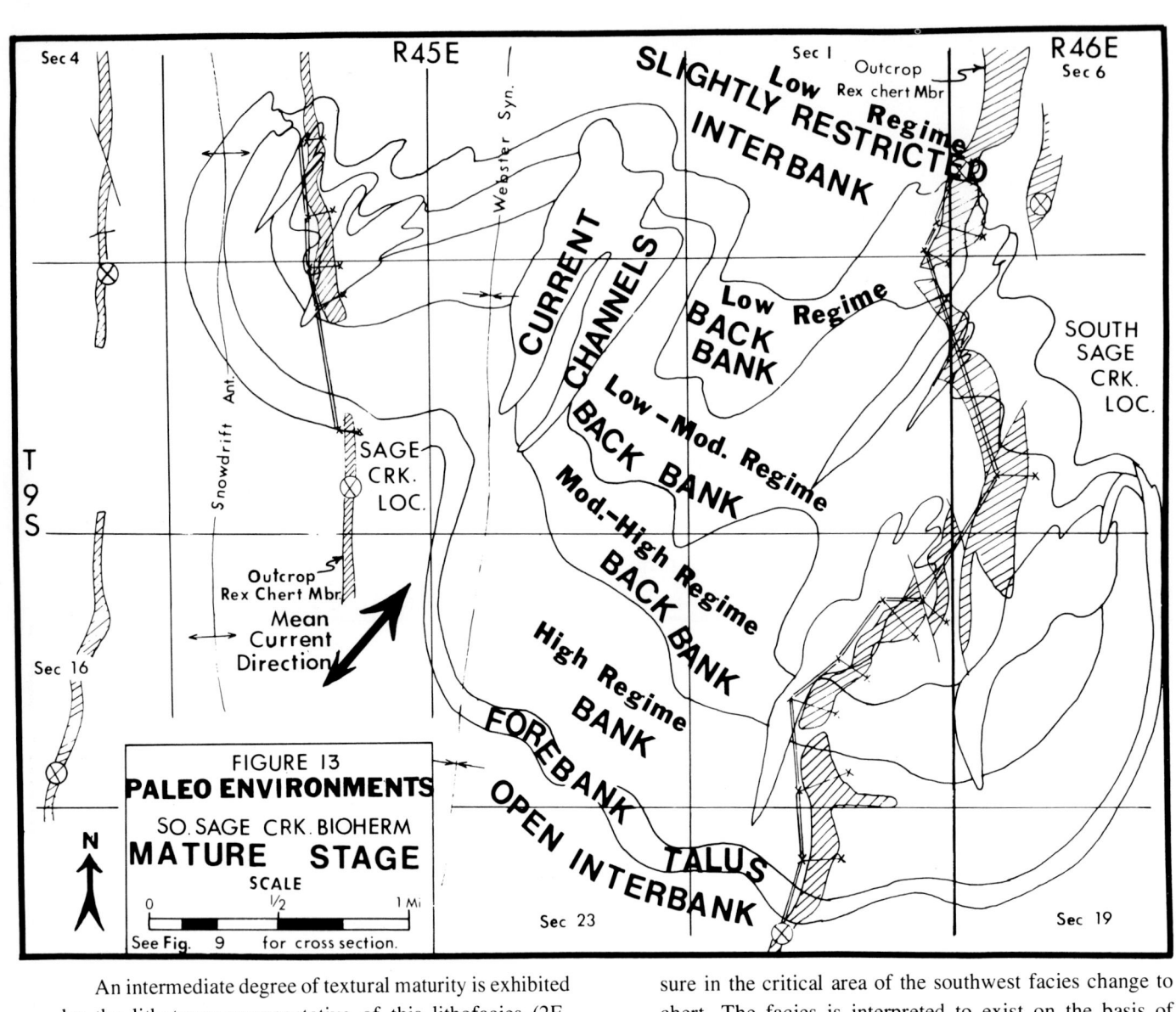

FIGURE 13
PALEO ENVIRONMENTS
SO. SAGE CRK. BIOHERM
MATURE STAGE

An intermediate degree of textural maturity is exhibited by the lithotypes representative of this lithofacies (2E, 2D, 2G; Table 3. The presence of minor amounts of mud matrix is indicative of the lower degree of textural maturity. However, locally some lithotypes, especially the crinoid grainstones (2E), exhibit a high degree of textural maturity, which along with field evidence of cross bedding and mounding suggests the presence of small local carbonate shoals leeward of the main bank.

C. Medium Lime Biopackstone-Wackestone:

Of the carbonate lithofacies, this facies includes the texturally least mature lithotypes (2H, 2I; Table 3). Mud matrix occurs in moderate abundance in this lithofacies. The presence of articulated ramose bryozoa as well as the rest of the fauna within this facies is compatible with the low degree of textural maturity.

D. Litho-Biograinstone Lithofacies:

This facies was not actually observed at the Sage Creek or South Sage Creek localities because of the lack of exposure in the critical area of the southwest facies change to chert. The facies is interpreted to exist on the basis of evidence at other localities (especially Timber Creek, Table 1) where that facies change is better exposed. At other localities this lithofacies is most often highly silicified.

E. Very Coarse Brachiopod Biograinstone, Coarse Lithoclastic Chert Packstone and Coarse Bryozoan Biopackstone Lithofacies:

These lithofacies represent anomalous occurrences of more mature lithotypes in areas occupied by much less mature lithotypes. All lines of evidence: scoured bases, discontinuous nature, aligned bioclasts, coarse grain size, etc., suggest transportation to the areas of deposition in current channels.

F. Chert Wackestone-Packstone Lithofacies:

Two subfacies may be recognized dependent upon the lithotypes found within this species. The presence of the most texturally mature spicular chert lithotype (1A) within the facies (Table 4) as opposed to the presence of the silty chert lithotype (2F) is one of the criteria for recognition of the subfacies. The former occurs southwest of the bank and the latter immediately to the northeast.

Paleoenvironments — Processes

The depositional processes and paleoenvironments for the South Sage Creek bank (Fig. 13) were interpreted from the lithofacies and biofacies distribution. Depositional processes are described only in terms of energy regimes interpreted from the textural maturity, faunal content, sedimentary structures, and stratigraphic relationships of the rocks:

A. **Bank — High Regime:** This depositional environment represents the highest degree of carbonate productivity as evidenced by the volume of sediment produced within it. All aspects of the environment are indicative of a high energy regime. The fauna were especially well adapted and thrived in this area of high circulation and resultant abundant nutrients. The main bank environment was most probably the highest topographic expression of the South Sage Creek bank (Figs. 8 and 9).

B. **Back Bank** — Moderate to High Regime, Low to Moderate Regime, and Low Regime: The back bank environments are interpreted to represent decreasing energy regimes from southwest to northeast as a result of the protection afforded by the main bank from wave or current action. Locally, energy regimes were higher where small crinoidal shoals (lithotype 2E, Fig. 9) developed or where current channels transversed the bank.

C. **Current Channels:** Varying energy regimes were active in the current channel environment as indicated by the diversity of rock types within them, ranging from low (lithoclastic chert packstone), or moderate (bryozoan biopackstone) to high (brachiopod biograinstone).

D. **Forebank Talus:** The energy regime of this environment was also probably quite variable. Much of the deposition within this environment could have occurred as a simple downslope gravity process, but a certain amount of winnowing must also have been involved. Interpretation of the textural maturity of the sediments is difficult (even at other localities where exposures are better) because of the degree of silicification.

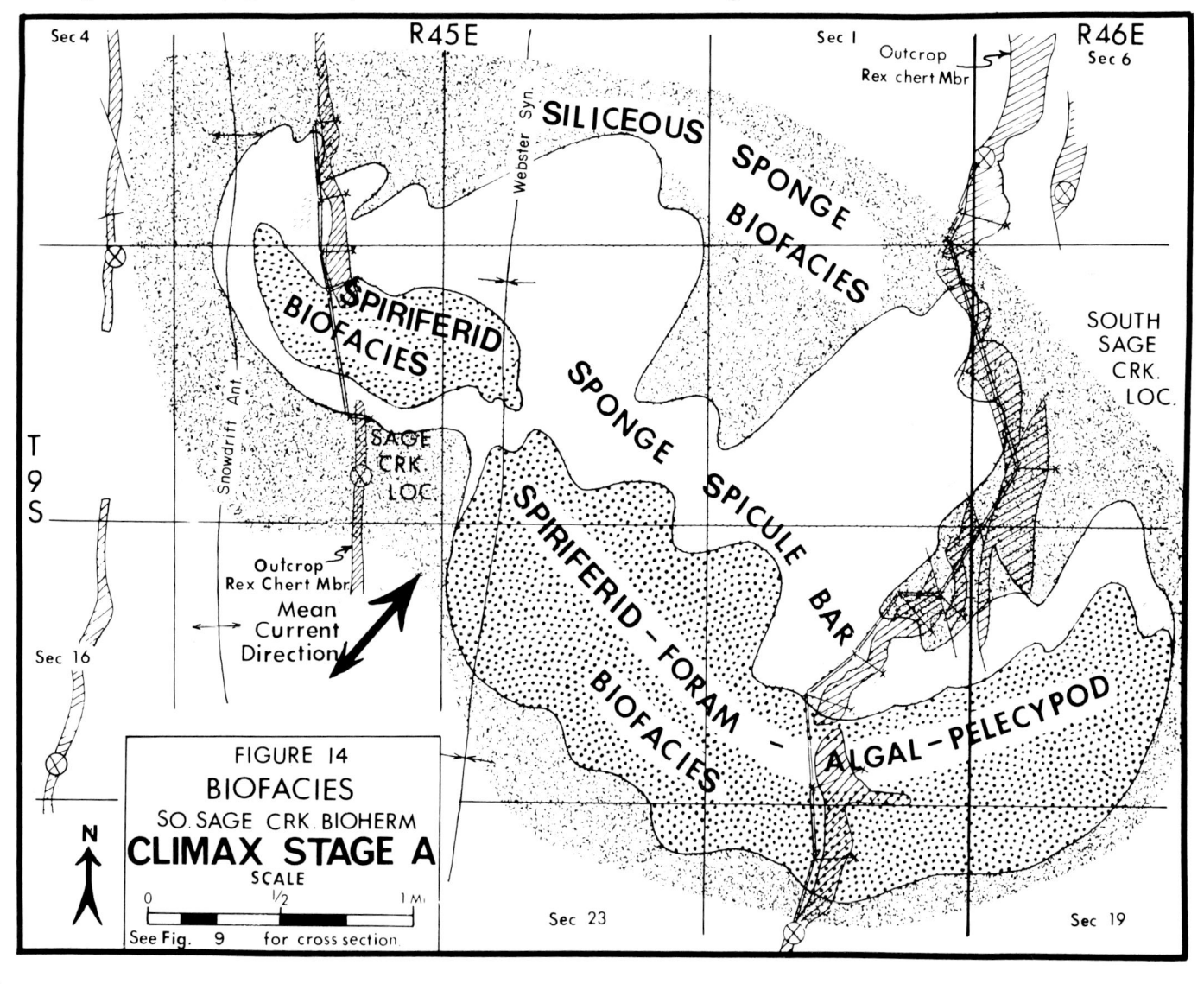

FIGURE 14
BIOFACIES
SO. SAGE CRK. BIOHERM
CLIMAX STAGE A

E. **Open Interbank:** In the areas between carbonate banks where siliceous spicular muds were deposited, only a very low energy regime existed. The siliceous sponges thrived with only slight current action. The open interbank environment exhibits a slightly higher regime than the restricted interbank environment as evidenced by the presence of lithotype 1A (spicular chert packstone-wackestone).

F. **Slightly Restricted Interbank:** The lowest degree of textural maturity and, consequently, the interpreted lowest energy regime is typical of this environment. The difference in energy regime in relation to the more normal interbank environment was afforded by the protection from current activity by the carbonate bank.

Energy Modes

The primary energy mode for the mature stage of carbonate development is interpreted to have been current action, most probably upwelling currents, as evidenced by:

A. The regional position of the banks relative to the interpreted basin geometry and the shelf edge.

B. Evidence for deposition primarily, if not entirely, in subtidal conditions at moderate depths.

C. Evidence for moderate winnowing over long periods of time.

D. Presence of current channels.

E. Moderate development of cross-bedding.

The interpreted mean current direction is shown on all the map figures (Figs. 8 and 10 to 14) and was interpreted primarily from the geometry of the bank. Cross-bedding directions were not determinable in the field; however, alignment of articulated ramose bryozoan fronds on bedding planes in a roughly north-south orientation supports the interpreted current direction.

Climax Stage A

In the last stage of carbonate deposition with regression of the Phosphoria Sea, a shoal developed on the unprotected flank of the South Sage Creek bank (Fig. 14). The regressive or progradational nature of the unit is illustrated in the stratigraphic cross section for that locality (Fig. 9). The textural maturity of the carbonate lithotype (2A; Table 3) representative of this stage of deposition, the massive bedding, evidence of cross-bedding and the fauna which inhabited the environment of deposition all indicated deposition in shallow subtidal waters (i.e. possibly less than 60 ft).

The spiriferid-foram-algal-pelecypod biofacies of the climactic stage of deposition includes the following faunal elements.

A. Spiriferid Brachiopods
B. Ramose Bryozoa
C. Foraminifera
D. Algae
   Dasyclad — *Epimastapora*
   Blue green?
E. Pelecypods
F. Echinoids
G. Gastropods
H. Encrusting Bryozoa
I. Ostracodes

Other bioclasts found within lithotype 2A are not considered to be indigenous to the depositional environment, but to have been reworked from pre-existing sediments.

On the leeward side of the carbonate shoal, sponge spicule bars were formed from the winnowing of siliceous sponge spicules. Chert lithotype 1B (Table 4) deposited in this environment exhibits a high degree of textural maturity. Other than siliceous sponge spicules, bioclasts are rarely found within lithotype 1B and it is doubtful that any faunal element occupied the environment in abundance.

Laterally, in more protected waters, siliceous sponges thrived and were disarticulated to form spicular cherts (lithotype 1A). The energy regime of the depositional environment of lithotype 1A was very low relative to carbonate lithotype 2A and chert lithotype 1B, but slightly higher than the lithotypes deposited in the interbank environments during the mature stage of bank development.

Carbonate deposition was culminated as the basin became restricted and adequate circulation for carbonate productivity was not achieved. At this time spicular cherts of lithotype 1A (Climax B Stage, Fig. 9) and finally, due to a slight transgression of the Phosphoria Sea and the resultant influx of fine clastics, the basinal siliceous shales of the cherty shale member were deposited in the study area.

CHERT ORIGIN

In the area of study the origin of the siliceous material from which the Rex cherts were lithified is not difficult to establish. The extreme volume of siliceous spicules indicates that biotic processes, mainly that of siliceous sponges, were primarily responsible for the initial precipitation of the silica from sea water. Only normal amounts of silica concentrations would be needed for precipitation, but in an area where upwelling currents are thought to have existed, such as southeast Idaho, conditions would certainly have been optimum for the high volumes of silica deposited in the Rex Chert Member.

Actual lithification of that sediment under conditions of normal marine pH pore water is somewhat more difficult to explain. Silica solubility increases with increasing pH (Hauskopf, 1959, 1967) and at normal marine pH and temperature silica would be soluble. Lithification of the Rex cherts in the area of study must have involved a change in pore fluid geochemistry and/or temperature.

The presence of differing geochemical environments below the sediment-water interface is known to exist in modern sediments (Sharma, 1965). Also, the petrologic evidence of silicified or corroded carbonate bioherms and the lack of burrowing infauna is suggestive of a more acidic environment which was toxic to organisms with calcareous tests within the sediment. The solubility of silica decreases with decreasing pH which would be conducive to chert lithification with depth below the sediment-water interface.

The interpreted method of lithification of the Rex Cherts within the study area is summarized in Fig. 15, after Sharma. The most noticeable difference between the chert lithification model proposed (Fig. 15) and that of Sharma (1965) is the deletion of Sharma's Zone D (i.e. carbonate replacement of silica). This is justifiable since there is no evidence within the system studied for replacement of chert by carbonate. Actually, replacement of carbonate by silica occurs much more commonly.

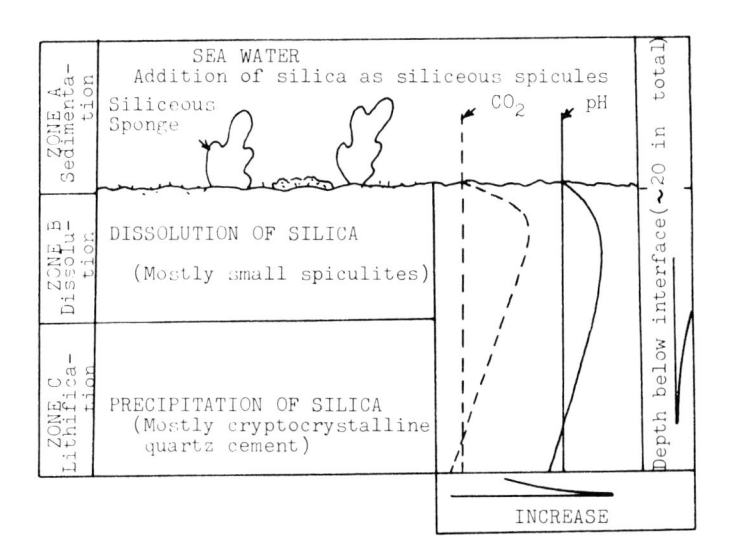

Fig. 15 — Chert Lithification Model (after Sharma, 1965)

Most of the petrologic evidence supports deposition of the chert lithotypes described (Fig. 9) with the exception of 1B in a very low energy regime environment. Textural maturity of the chert lithotypes is summarized in Table 4 relative to the carbonate lithotypes (Table 3), however, most of the chert lithotypes can be demonstrated to be direct facies equivalents of the carbonate bank lithotypes. Lack of evidence for any abrupt paleotopographic variation from bank to interbank deposits would therefore preclude chert deposition occurring in very deep water within the area of study.

The genesis of the Rex Chert Member has been interpreted variously by Keller (1941), Warner (1956), Bissell (1959), and McKelvey et al. (1959). Most of the differences of opinion may be resolved as a matter of the area in which the Rex Chert Member was studied. The cherts of the Rex Chert Member are most certainly polygenetic.

Keller, for instance, studied the Rex Chert Member near the Idaho-Wyoming border where the influence of clastic sediments of the Shedhorn facies (Fig. 8) lead to the supposition of a clastic origin for most of the silica present. Warner was partially misled by his own interpretation of silica solubility (opposite that which is presently accepted) to conclude that the Rex cherts were primarily inorganically precipitated. Bissell has studied the miogeosynclinal facies where there is a possible influence of volcanism. McKelvey et al. studied the shelf facies where, as in the area of this study, the origin of the cherts was primarily due to organic activity (i.e. mostly siliceous sponges).

On the basis of examination of samples collected by J. A. Peterson in 1970 from the Rex Chert Member in the Sublette Range of southcentral Idaho, west of the study area, the basinal facies of the Rex Chert appears to have formed primarily from different mechanisms than those active within the study area. There is a notable paucity of siliceous spicules in those samples as previously noted by Warner (1956). The abundance of carbonate minerals and siliceous crystalline dolomitic limestone is most compatible with a replacement origin for this area. Similar rock types were noted in the study area (lithotypes 1D and 1C).

Lithotype 1F within the study area is similar to rock types described by Keller (1941) in that it contains moderately abundant quartz silt. The greater distance from the clastic influx of the Shedhorn Sandstone (Fig. 4) precludes an abundance of silt in the chert within the area of study.

In conclusion, most of the hypotheses of the origin of the Rex Cherts are not only acceptable, but when considered in context with the regional stratigraphic setting, allow additional understanding of the paleoenvironments of the Rex Chert Member.

EXPLORATION POTENTIAL

The most important parameters within the Phosphoria bioherms in southeast Idaho relative to petroleum exploration potential were silicification, carbonate cementation, and subsequent porosity destruction (Fig. 16). Cementation by sparry calcite is the most common form of diagenesis as can readily be seen by the predominance of grainstones. The complete degree of calcite cementation resulted in almost total lack of porosity in a rock which must have had a moderate to high degree of original porosity. Also notably lacking is the dolomitization and resultant porosity enhancement typical of productive Phosphoria carbonates farther to the east in Wyoming. In addition, the high risks of exploring for a stratigraphic trap in an extremely complex structural area would appear to preclude hopes for exploration success at this time.

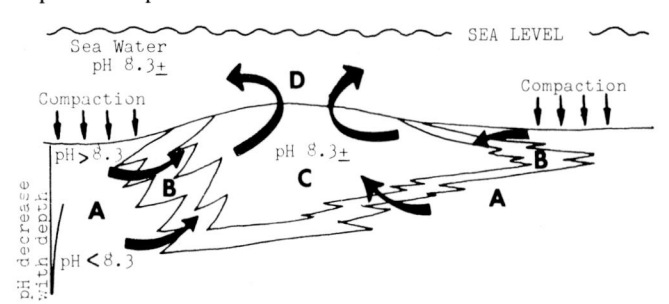

Fig. 16 — Silicification and Carbonate Cementation

Explanation:

A. Siliceous spicular muds — a pH greater than normal sea water (D) is necessary for mobilization of the silica in the sediments (Krauskopf, 1959; 1967). A relatively more acidic pH is needed for lithification of the chert. Pore waters probably vary with depth from relatively more basic, just below the sediment-water interface, to relatively more acidic with depth (Sharma, 1965).

B. Silicified carbonate — (arrows indicate migration of pore fluids). The high pH fluids saturated with respect to silica, intermix with the more normal pH pore fluids of the carbonate and the more acidic fluids of the buried siliceous muds resulting in the precipitation of siliceous cement and silicification of carbonate material.

C. Sparry calcite cemented limestone — Fluids migrating from the silicified carbonate zone are overly-saturated with respect to calcium carbonate. After mixing of these fluids with the normal pore water of the porous carbonate sediment the sparry calcite cement is precipitated.

D. Sea water — Excess pore waters were probably expelled where the originally porous carbonates were not capped by the less permeable siliceous muds.

The romance, however, of exploring for "reefs" covering six or more sq mi with up to 150 ft thickness combined with intense reservoir fracturing, abundant dead oil stain, fetid petroliferous odor in almost all of the carbonate rocks, surface trend indicators, and encasement in the richest source rock in the Rocky Mountain area should eventually lead adventuresome explorationists to this play.

SELECTED REFERENCES

Bissel, H. S., 1959, Silica in sediments of the Upper Paleozoic of the Cordilleran area: in Ireland, H. A., ed., Silica in sediments, a symposium: Soc. Econ. Paleontologists and Mineralogists Spec. Pub. 7, p. 150-185.

Bostwick, D. A., 1955, Stratigraphy of the Wood River Formation, south-central Idaho: Jour. Paleontology, v. 29, no. 6, p. 941-951.

Branson, C. C., 1930, Paleontology and stratigraphy of the Phosphoria Formation: Missouri Univ. Studies, v. 5, no. 2, p. 1-99.

Brittenham, M. D., 1973, Permian Phosphoria bioherms and related facies, southeastern Idaho: Unpub. Master's thesis, University of Montana.

_____, 1974, Permian Phosphoria bioherms and related facies, southeastern Idaho: abs. Am. Assoc. Petroleum Geologists and Soc. Econ. Paleontologists and Mineralogists Ann. Meetings, San Antonio, Texas, v. 1, p. 11.

Cheney, T. M., and Montgomery, K. M., 1967, Geology of the Stewart Flat Quadrangle, Caribou Conty, Idaho: U.S. Geol. Survey Bull. 1217, 63 p.

Chilingar, G. V., 1958, Sponge spicule deposits as indicators of physical-chemical environments of deposition: Compass of Sigma Gamma Epsilon, v. 35, p. 215-219.

Churkin, Michael, Jr., 1962, Facies across Paleozoic miogeosynclinal margin of central Idaho: Am. Assoc. Petroleum Geologists Bull., v. 46, p. 569-591.

Cole, G. P., 1969, Permian carbonate facies, southern Big Horn Basin-Owl Creek Mountains area, north-central Wyoming: Unpub. Master's thesis, Univ. of Montana.

Cramer, H. R., 1971, Permian rocks from the Sublette Range, southern Idaho: Am. Assoc. Petroleum Geologists Bull., v. 55, no. 10, p. 1787-1801.

Cressman, E. R., 1955, Physical stratigraphy of the Phosphoria Formation in part of southwestern Montana: U.S. Geol. Survey Bull. 1027-A, p. 1-31.

_____, 1964, Geology of the Georgetown Canyon-Snowdrift Mountain area, southeastern Idaho: U.S. Geol. Survey Bull. 1153, 105 p.

_____, and Gulbrandsen, R. A., 1955, Geology of the Dry Valley Quadrangle, Idaho: U.S. Geol. Survey Bull. 1015-I, p. 257-270.

_____, and Swanson, R. W., 1964, Stratigraphy and petrology of the Permian rocks of southwestern Montana: U.S. Geol. Survey Prof. Paper 313-C, p. 275-569.

Davidson, D. F., Smart, R. A., Pierce, H. W. and Weiser, J. D., 1953, Stratigraphic sections of the Phosphoria Formation in Idaho, pt. 2: U.S. Geol. Survey Circ. 305, 28 p.

Dunham, R. J., 1962, Classification of carbonate rocks according to depositional texture: in Ham, W. E., ed., Classification of carbonate rocks — a symposium: Am. Assoc. Petroleum Geologists Mem. 1, p. 108-121.

Finks, R. M., Yochelson, E. L., and Sheldon, R. P., 1961, Stratigraphic implications of a Permian sponge occurrence in the Park City Formation of western Wyoming: Jour. Paleo., v. 35, p. 554-556.

Girty, G. H., 1927, Descriptions of Carboniferous and Triassic fossils, in Mansfield, G. R., Geography, geology, and mineral resources of part of southeastern Idaho: U.S. Geol. Survey Prof. Paper 152, p. 78-81.

Gulbrandsen, R. A., McLaughlin, K. P., Honkala, F. S., and Clabaugh, S. E., 1956, Geology of the Johnson Creek Quadrangle, Caribou County, Idaho: U.S. Geol. Survey Bull. 1042-A, p. 1-23.

Heckel, P. H., 1972, Recognition of ancient shallow marine environments, in Rigby, J. K., and Hamblin, W. K., eds., Recognition of ancient sedimentary environments: Soc. Econ. Paleontologists and Mineralogists Spec. Pub. 16, p. 226-286.

Keller, W. D., 1941, Petrology and origin of the Rex Chert: Geol. Soc. America Bull., v. 52, p. 1279-1298.

Krauskopf, K. B., 1959, The geochemistry of silica in sedimentary environments: in Ireland, H. A., ed., Silica in sediments — a symposium: Soc. Econ. Paleontologists and Mineralogists Spec. Pub. 7, p. 4-18.

_____, 1967, Introduction to Geochemistry: New York, McGraw-Hill, 721 p.

Lowell, W. R., 1952, Phosphate rocks in the Deer Creek-Wells Canyon area, Idaho: U.S. Geol. Survey Bull. 982-A, p. 1-52.

Majewske, O. P., 1969, Recognition of invertebrate fossil fragments in rocks and thin sections: Leiden, E. J. Brill, 101 p., 106 pl.

Mansfield, G. R., 1920, Geography, geology, and mineral resources of the Fort Hall Indian Reservation, Idaho: U.S. Geol. Survey Bull. 713, 152 p.

_____, 1927, Geography, geology and mineral resources of part of southeastern Idaho: U.S. Geol. Survey Prof. Paper 152, 453 p.

_____, 1929, Geography, geology, and mineral resources of the Portneuf Quadrangle, Idaho: U.S. Geol. Survey Bull. 803, 110 p.

_____, 1952, Geography, geology, and mineral resources of the Ammon and Paradise Valley Quadrangles: U.S. Geol. Survey Prof. Paper 238, 92 p.

McKee, E. D., Oriel, S. S., et al., 1967, Paleotectonic investigations of the Permian system in the United States: U.S. Geol. Survey Prof. Paper 515, 271 p.

McKelvey, V. E., Davidson, D. F., O'Malley, F. W., and Smith, L. E., 1953a, Stratigraphic sections of the Phosphoria Formation in Idaho, 1947-48, pt. 1: U.S. Geol. Survey Circ. 208, p. 1-49.

_____, Armstrong, F. C., Gulbrandsen, R. A., and Campbell, R. M., 1953b, Stratigraphic sections of the Phosphoria Formation in Idaho, 1947-48, pt. 2: U.S. Geol. Survey Circ. 301, p. 1-58.

_____, Swanson, R. W., and Sheldon, R. P., 1953, Phosphoria Formation in southeastern Idaho and western Wyoming: Intermtn. Assoc. Pet. Geol. Guidebook, Fourth Ann. Field Conf., p. 41-47.

_____, Williams, J. S., Sheldon, R. P., Cressman, E. R., Cheney, R. M., and Swanson, R. W., 1956, Summary description of Phosphoria, Park City and Shedhorn Formation in western phosphate field: Am. Assoc. Petroleum Geologists Bull., v. 40, p. 2826-2863.

_____, 1959, Relation of upwelling marine waters to phosphorite and oil: Geol. Soc. America Bull., v. 70, p. 1783.

_____, Everhart, D. L., and Garrells, R. M., 1959, The Phosphoria, Park City and Shedhorn Formations in the western phosphate field: U.S. Geol. Survey Prof. Paper 313-A, p. 1-47.

_____, Williams, J. S., Sheldon, R. P., Cressman, E. R., Cheney, T. M., and Swanson, R. W., 1967, The Phosphoria, Park City and Shedhorn Formations in western phosphate field, Intermtn. Assoc. of Geologists 15th Ann. Field Conf. Guidebook, p. 15-33.

Nelson, H. F., Brown, C. Wm., and Brineman, J. H., 1962, Skeletal limestone classification: in Ham, W. E., ed., Classification of carbonate rocks — a symposium: Am. Assoc. Petroleum Geologists Mem. 1, p. 224-252.

Richards, R. W., and Mansfield, G. R., 1910, Preliminary report of a portion of the Idaho phosphate reserve: U.S. Geol. Survey Bull. 470-H, p. 371-481.

_____, and Mansfield, G. R., 1914, Geology of the phosphate deposits northeast of Georgetown, Idaho: U.S. Geol. Survey Bull. 577, 76 p.

Rioux, R. L., Hite, R. J., Dyni, J. R., and Gere, W. C., 1966, Preliminary geologic map of the Upper Valley Quadrangle, Caribou County, Idaho: U.S. Geol. Survey Open File Report.

Roberts, R. J., et al., 1965, Pennsylvanian and Permian basins in northwestern Nevada and south-central Idaho: Am. Assoc. Petroleum Geologists Bull., v. 49, p. 1926-1956.

Ross, C. P., 1962, Upper Paleozoic rocks in central Idaho: Am. Assoc. Petroleum Geologists Bull., v. 46, p. 384-387.

Scholten, R., 1957, Paleozoic geosynclinal margin north of the Snake River Plain, Idaho-Montana: Geol. Soc. America Bull., v. 68, p. 151-170.

Sharma, G. D., 1965, Formation of silica cement and its replacement by carbonates: Jour. Sed. Petrology, v. 35, no. 3, p. 733-745.

Swanson, R. W., Cressman, E. R., Jones, R. S., and Replogle, B. K., 1953, Stratigraphic sections of the Phosphoria Formation, 1953 (Idaho-Montana-Utah): U.S. Geol. Survey Circ. 385, p. 1-30.

_____, Carswell, L. D., Sheldon, R. P., and Cheney, T. M., 1956, Stratigraphic sections of the Phosphoria Formation, 1953 (Idaho-Montana-Utah): U.S. Geol. Survey Circ. 375, 30 p.

Shannon, J. P., 1961, Upper Paleozoic stratigraphy of east-central Idaho: Geol. Soc. America Bull., v. 72, p. 1829-1836.

Sheldon, R. P., Warner, M. A., Thompson, M. E., and Peirce, H. W., 1953, Stratigraphic sections of the Phosphoria Formation in Idaho, 1949, pt. 1: U.S. Geol. Survey Circ. 304, p. 1-30.

_____, 1957, Physical stratigraphy of the Phosphoria Formation in northwestern Wyoming: U.S. Geol. Survey Bull. 1042-E, p. 105-185.

_____, 1963, Physical stratigraphy and mineral resources of Permian rocks in western Wyoming: U.S. Geol. Survey Prof. Paper 313-B, p. B47-B271.

Smart, R. A., Waring, R. G., Cheney, T. M., and Sheldon, R. P., 1954, Stratigraphic sections of the Phosphoria Formation in Idaho, 1950-51: U.S. Geol. Survey Circ. 327, p. 1-22.

Thomasson, M. R., 1959, Late Paleozoic stratigraphy and paleotectonics, central and eastern Idaho: Unpub. PhD Thesis, Univ. of Wisconsin, 274 p.

Warner, M. A., 1956, The origin of the Rex Chert: Unpub. PhD Thesis, Univ. of Wisconsin.

Wilkerson, B. H., 1967, Paleoecology of the Permian Ervay Member of the Park City Formation in north-central Wyoming: Unpub. Thesis, Univ. of Wyoming.

Williams, J. S., 1959, Fauna, age, and correlation of rocks of Park City age: in McKelvey, V. E., et al., The Phosphoria, Park City and Shedhorn Formations in the western phosphate field: U.S. Geol. Survey Prof. Paper 313-A, p. 36-40.

Yochelson, E. L., 1963, Paleoecology of the Permian Phosphoria Formation and related rocks: Geological Survey Research 1963: U.S. Geol. Survey Prof. Paper 475-B, p. B123-B124.

_____, 1968, Biostratigraphy of the Phosphoria, Park City, and Shedhorn Formations: U.S. Geol. Survey Prof. Paper 313-D, p. 571-660.

THE TALKING WIRE COMES TO THE ROCKIES

Samuel F. B. Morse graduated from Yale in 1810. Although he had studied under Benjamin Silliman and Jeremiah Day, for twenty years he thought of himself as an artist and pursued art as a profession. It was not until the 1830s that he became involved in the perfection of an instrument that would transmit messages over long distances. He completed a working model in 1836 and applied for a patent in 1838. The commercial telegraph was made possible by the invention of the electromagnet by the great physicist and first Secretary of the Smithsonian Institution, Joseph Henry, who is memorialized today by physicists in the henry, the unit of inductance, and by geologists in the Henry Mountains, which were named in his honor.

Morse attempted to sell his invention to the government for $100,000, but was turned down by the Postmaster General, who was uncertain that the telegraph would ever generate $100,000 in revenue. A telegraph line was built between Washington, D.C., and Baltimore. The first message, "What hath God wrought!", was sent on May 24, 1844. The message was the inspiration of Miss Annie Ellsworth, daughter of the Commissioner of Patents, who had found it in the Old Testament in the Book of Numbers.

Within two years Washington, D.C., was connected by telegraph with New York City. New Orleans was connected to New York City in 1848. By 1851, there were over 50 telegraph companies operating in the United States.

The telegraph came to the Rocky Mountains in 1861, almost eight years ahead of the railroad. It was built by two companies whose lines met in a two-story building on Main Street in Salt Lake City. The eastern company was named, somewhat inappropriately, the Pacific Telegraph Company. Their line extended from Omaha up the Platte River to Fort Kearney and Fort Laramie, and up the Sweetwater to South Pass, and from there to Salt Lake City.

The western company was the Overland Telegraph Company, described as "a puppet of the California State Telegraph Company." Their section of line extended eastward from Carson City to Ruby Valley, Egan Canyon, and Deep Creek, and skirted the south side of the Great Salt Lake.

A wagon train consisting of 25 wagons loaded with equipment, 228 oxen, 18 mules and horses, and 50 men, left Sacramento on May 27, 1861. The expedition was so large as to be unwieldy. It had to be broken into smaller units in order to cross the mountains, and did not reach Carson City until late in June. The first post hole on the Pacific Telegraph section of line was dug on the Fourth of July.

The route of the telegraph was essentially that of the Pony Express. Work was pushed both to the east and west from Salt Lake City, and from both ends of the line. As the unfinished gaps shortened, telegraph operators were stationed at the ends of the lines and moved with the advancing line every day. Messages were relayed across the unfinished gaps by Pony Express riders. The Pacific Telegraph Company completed their line on October 19. The last stretch of line to be finished was the Overland Telegraph Company's section between Ruby Valley and Schell Creek, Nevada. It was completed on October 24.

The Pony Express went out of business two days later.

— W. Lyle Dockery

PERMIAN AND LOWER TRIASSIC RESERVOIR ROCKS OF CENTRAL UTAH

by

C. Dennis Irwin[1]

ABSTRACT

Permian and Lower Triassic reservoir rocks throughout central Utah consist of beach and shallow water sandstones and shallow marine carbonates. These reservoirs are the Permian-age Cedar Mesa Sandstone, Toroweap Formation, White Rim Sandstone, Kaibab Formation, and the Lower Triassic-age Sinbad-Timpoweap carbonate member of the Moenkopi Formation.

Depositional patterns of these stratigraphic units are controlled by the relative positions of the Oquirrh and Bird Spring basins and the Emery uplift which separates the two.

Hydrocarbon shows are found in all of the reservoirs but generally increase in abundance upward in the stratigraphic sequence and the upper two, the Kaibab and Sinbad, both produce commercial oil.

INTRODUCTION

Permian-age source rock and reservoir beds deposited along the eastern shelf edge of the Cordilleran geosyncline have yielded some of the most prolific oil fields of the United States in the Big Horn and Wind River basins of Wyoming. All logic seems to dictate that since the Cordilleran geosyncline and its attendant eastern shelf continued south from central Wyoming through central Utah and into the Grand Canyon area that Permian hydrocarbon accumulations should also be found in these areas. To date the results have been disappointing but commercial production from the Kaibab Formation at the Upper Valley field, noncommercial oil from the Kaibab at the Ferron field and the giant dead oil accumulation in the White Rim Sandstone at Elaterite Basin show the potential of the area.

Presented here are a set of cross sections and maps showing the correlation and distribution of Upper Permian and Lower Triassic reservoir rocks in central Utah. These maps and cross sections were part of my dissertation (Irwin, 1969) and part were published in a more abbreviated form (Irwin, 1971). The maps have been revised to include the latest well data in the "Hingeline play" and the study has been extended slightly northward into the Uintah Basin.

This study is generally terminated on the west where the Sevier orogenic belt is encountered. It is highly probable that the western-most well on cross section D-D' (Fig. 15, in pocket), the Shell Sunset Canyon No. 1, Sec. 21, T22S, R4W is located on an overthrust sheet. The horizontal displacement in this case is unknown and the contour pattern on the maps assumes no movement. This well is located south of the Emery uplift, and, if the Emery does continue west of the Pavant Range, considerable eastward fault transport would probably not result in significant telescoping of differing sedimentary facies in the rocks under discussion as the depositional strike of the rocks would be east-west. The Beta Kaibab Member in this well is limestone in contrast to 100% dolomite in the Beta Kaibab Member in the Anschutz well 19 mi northnortheast. In the Sunset Canyon well, an initial depositional position further west would be beneficial in contouring the dolomite percentage map. It is doubtful, however, that the rocks were transported more than several tens of miles.

The loss of the reservoir rock by facies change, truncation, and non-deposition terminates this study on the east.

Figure 1 is a paleotectonic map which shows the major tectonic features that controlled a large part of the Paleozoic sedimentation in Utah. The Emery uplift is shown as that area over which no Pennsylvanian sediments are present. Hallgarth (1962) questioned whether the lowest carbonate rocks encountered in Straight Wash Canyon on the San Rafael Swell were Pennsylvanian or Permian in age and finally settled on questionable Pennsylvanian. Roger L. Billings (personal communication) in a regional Mississippian study for Humble Oil Company measured this section and states that the lowermost rocks are Mississippian and that no Pennsylvanian is present. This information on the east plus additional drilling to the

---

[1] Consulting Geologist, Denver, Colorado.

Acknowledgement

The cross sections presented here are part of my dissertation prepared at the University of New Mexico in 1969. I would like to thank American Stratigraphic Company again for permission to use their sample descriptions in these cross sections.

west in Sevier County shows the Emery uplift to be a substantial paleotectonic feature extending laterally some 100 mi from the eastern edge of the San Rafael Swell to the Pavant Range on the west. Moreover, the strong possibility exists that this feature may have extended westward to the Nevada state line.

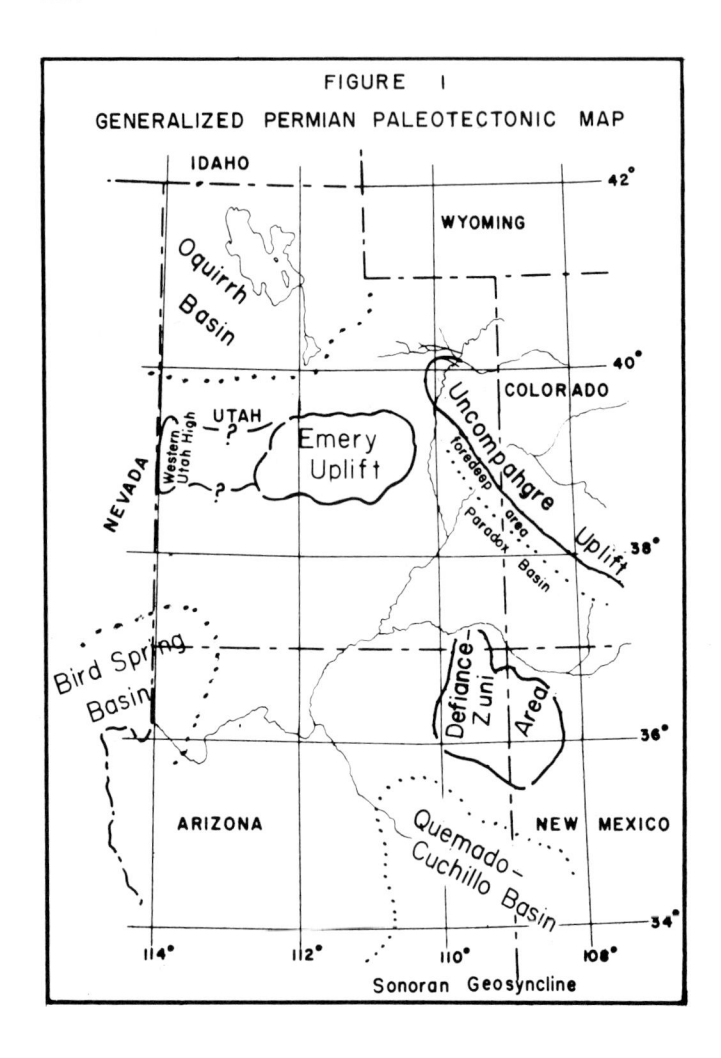

FIGURE 1
GENERALIZED PERMIAN PALEOTECTONIC MAP

RESERVOIR ROCKS

Cedar Mesa Sandstone

The lowermost reservoir presented here is the Cedar Mesa Sandstone Member of the Cutler Formation. Underlying the Cedar Mesa south of the latitude of the Abajo Mountains (37°45′) is the Halgaito Shale Member of the Cutler Formation. North of the Abajo Mountains underlying and eventually exhibiting a lateral facies relationship to the Cedar Mesa is the Elephant Canyon Formation of Baars (1962).

The Cedar Mesa Sandstone is a medium- to fine-grained quartz sandstone with scattered grains of chert, feldspar, and mica that was deposited in a shallow-water marine to beach environment with some back-beach dune development.

The Cedar Mesa changes facies rapidly eastward from the type section on the Monument uplift into an interbedded sequence of red gypsiferous shale and siltstone. This zone of facies change is geographically narrow; occurring within a horizontal distance of less than two mi. An isopach map of the Cedar Mesa west of the zone of rapid facies change can be considered a sandstone isolith within the mapped area (Fig. 3). The most noticeable features of this map are the east-west strike of the maximum sandstone thickness and the northward thinning to zero. The thinning is caused both by facies change into the Elephant Canyon Formation and thinning upon the ancient Emery uplift. Over this feature the Elephant Canyon is quite thin and in places absent.

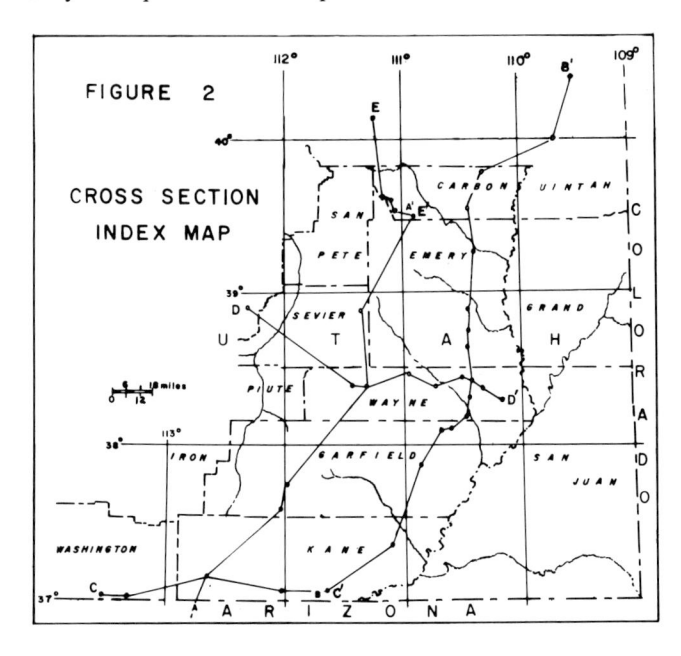

FIGURE 2
CROSS SECTION INDEX MAP

The paleotectonic map (Fig. 1) indicates that the accessway between the Oquirrh basin to the north and the Paradox basin to the south was restricted to a rather narrow area between the Emery and Uncompahgre uplifts. The north-south cross sections show that sediment communication between these two basins did exist during Cedar Mesa time and that the Cedar Mesa Sandstone does extend laterally as an integral part of the Weber Formation. This is not unexpected and has been postulated by previous geologists because Wolfcampian age fusulinids are present in both the Cedar Mesa and Weber sandstones. Cross section B-B' (Fig. 13, in pocket) shows the correlation between the Emery uplift, the Uncompahgre uplift and the Uintah segment of the Oquirrh basin. At Ashley Creek (Sec. 36, T2S, R20E) about 10 mi north of the Atlantic Oil Maeser Federal, Sec. 12, T4S, R20E, the northern-most well on the cross section, Bissell (1964) found Wolfcampian fusulinids 160 ft below the top of the Weber. It is tempting to correlate a larger part of the upper Weber with the Cedar Mesa but this is not compatible with fusulinid data that

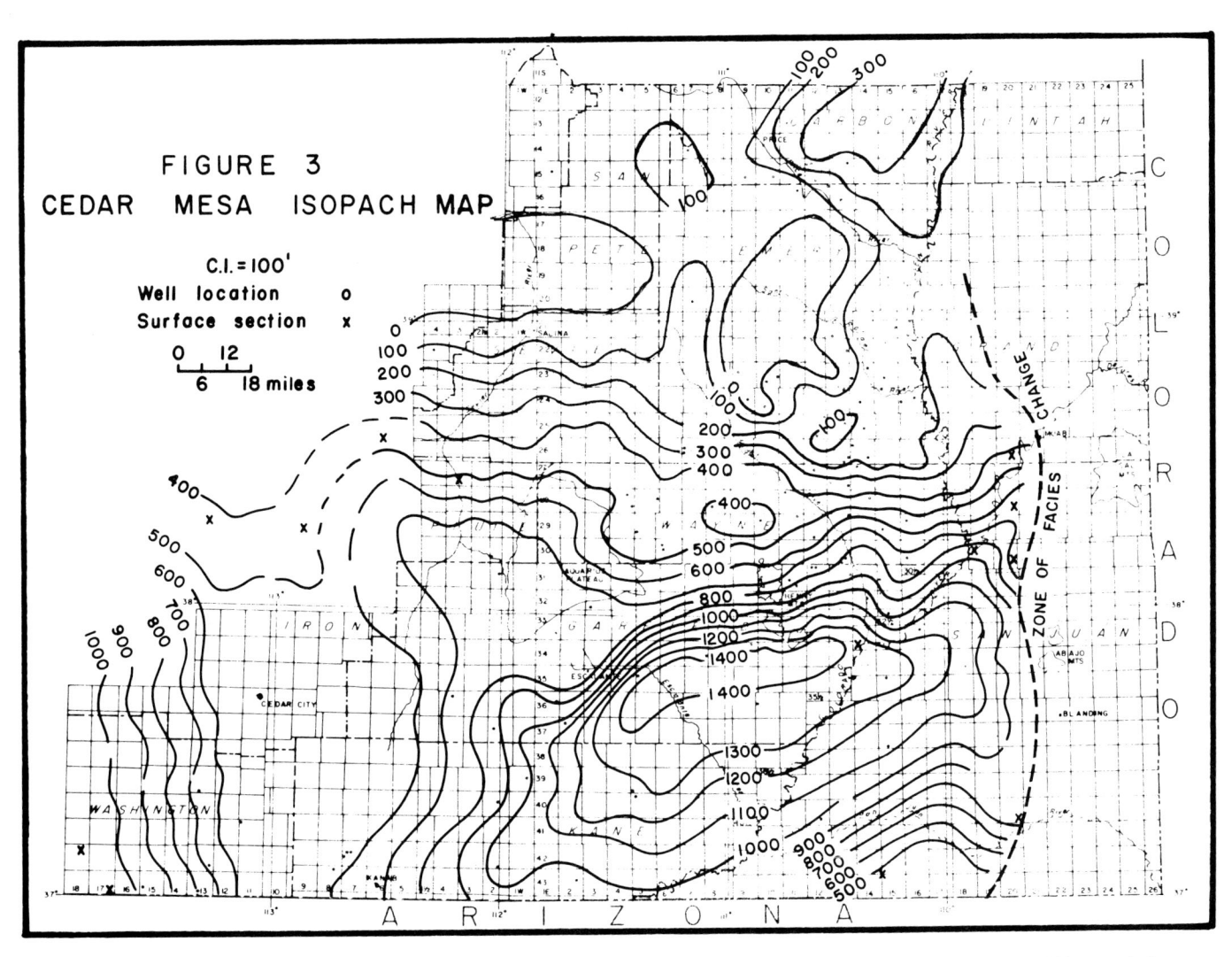

**FIGURE 3
CEDAR MESA ISOPACH MAP**
C.I. = 100'
Well location o
Surface section x

shows most of the Weber to be Pennsylvanian in age. The disparity between Wolfcampian thicknesses north and south of the Emery and Uncompahgre uplifts indicates strongly differing sedimentation rates for the two areas. Apparently this portion of the Oquirrh basin was almost completely filled and perhaps its clastic sediment source eroded to near base level before comparable sedimentation was well under way in central Utah.

Only one oil show that I know of has been credited to the Cedar Mesa. This show occurs in the subsurface where about 30 ft of oil saturated core was recovered from this formation in the discovery well (Tenneco Oil Co., Upper Valley No. 2) at the Upper Valley field.

The intertonguing relationship between the Cedar Mesa and the marine Elephant Canyon Formation suggests that it should not be entirely barren of hydrocarbons, however, the lack of well-defined source beds adjacent to this reservoir makes its future as a significant oil producer rather suspect.

Overlying the Cedar Mesa is the Organ Rock Shale. The cross sections demonstrate the northward and westward thinning and eventual loss of this unit. It is restricted on the north by the Emery and Uncompahgre uplifts, and thus not present in the Oquirrh basin.

Toroweap Formation and White Rim Sandstone

The Toroweap Formation is composed of a transgressive-regressive sequence of intertonguing sandstone, carbonate, and anhydrite that grades from beach sandstones to marine carbonates from east to west. The beach sandstones are known throughout central Utah as the White Rim Formation. Over a large part of the mapped area the upper part of the White Rim is not a lateral equivalent of the Toroweap but instead is the beach facies of the Gamma Member of the Kaibab which overlies the Toroweap. This relationship is shown on the cross sections. Thus the term White Rim includes the stacked beach facies of two major Upper Permian depositional cycles.

The sandstones of the White Rim-Toroweap are generally fine- to medium-grained, white in color, and locally are slightly to highly anhydritic. The limestones of the Toroweap are light-colored, thin-bedded, discontinuous, and locally carry a shallow-water marine invertebrate fauna. Over a large part of southwestern Utah the limestones have been completely dolomitized. However, indistinct shell outlines, shell-shaped voids, and pellet outlines are evidence of the original shallow-marine nature of the carbonates. Anhydrite can be found scattered throughout the beach-marine transition area but it is most common in the upper parts of the formation and microscopic examination reveals that anhydrite metasomatism is common.

Figure 4 is an isopach of the Toroweap-White Rim formations exclusive of that portion of the White Rim assigned to the Gamma Kaibab Member. Noted on this map by a patterned line is the zero carbonate line. Rocks to the east of this line are entirely composed of the White Rim Sandstone.

The north-south cross section B-B' (Fig. 13, in pocket) shows the northward thinning of the White Rim and its correlation into the uppermost part of the Weber. It must be pointed out that a time problem exists with this correlation because the White Rim is Leonardian in age and no fossils younger than Wolfcampian age have yet been found in the Weber, thus it may be that: (1) no fossils have been found but the Weber is Leonardian at its top, (2) the basal White Rim is Wolfcampian (most likely improbable), or (3) the White Rim equivalent is absent by erosion or nondeposition in the areas of Weber Sandstone deposition and the correlation is wrong.

In terms of petroleum exploration one important fact is established; that the Weber, Cedar Mesa, and White Rim are in lithologic continuity and hence undoubtedly in fluid communication with each other in the southwestern Uintah Basin. This assumes significance if one accepts long-distance migration of hydrocarbons.

The White Rim thins to the east and eventually intertongues with the red beds of the Cutler. It is along this eastern zero line that the giant heavy-oil accumulation at Elaterite Basin occurs. Baars and Seager (1970) state their belief that an ancient sand bar with as much as 200 ft of relief contributed significantly to the oil entrapment. Large stratigraphic anomalies such as bars of this size can now be identified with sophisticated seismic techniques and this avenue of exploration deserves further investigation.

Figure 5 is a carbonate isolith map of the Toroweap

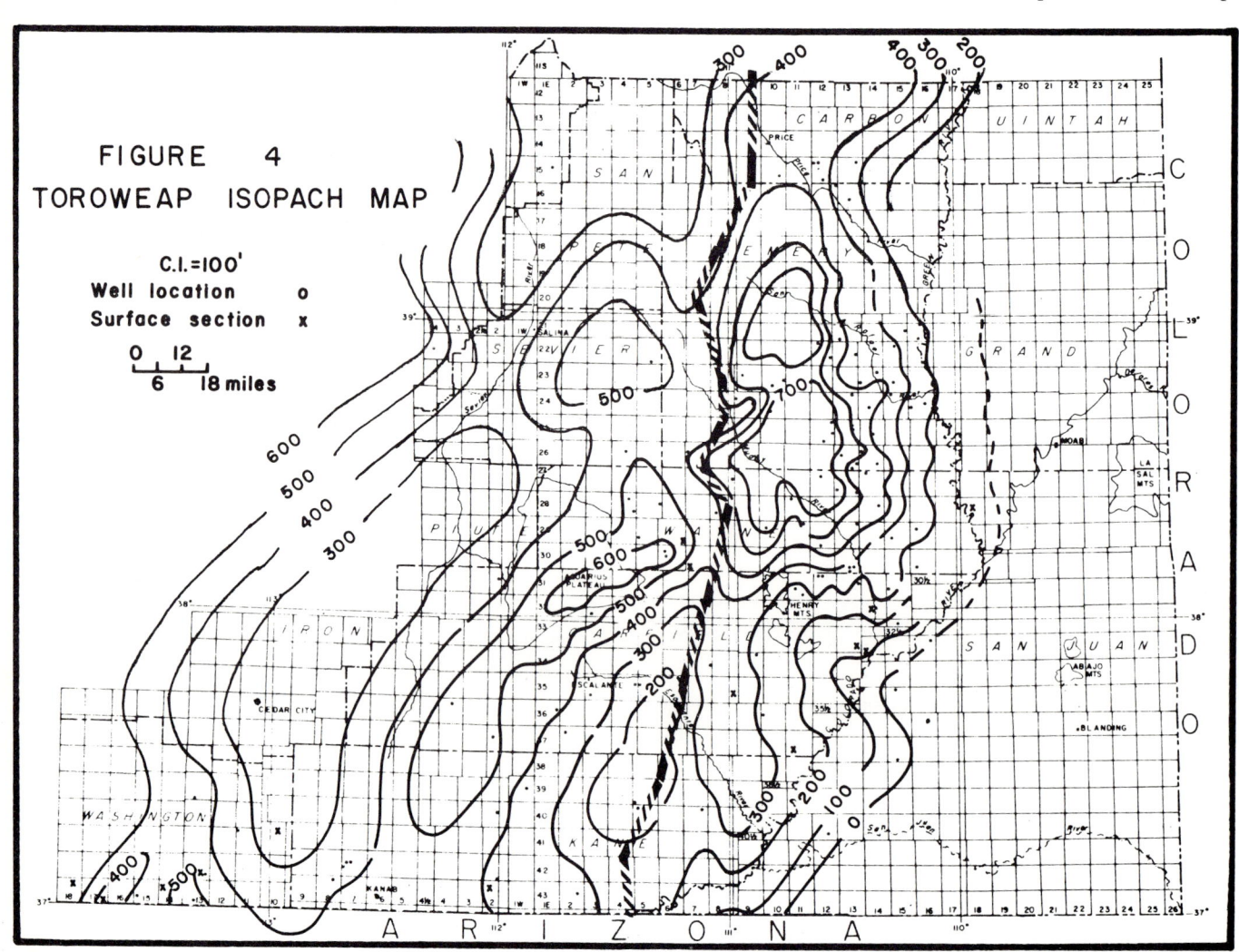

FIGURE 4
TOROWEAP ISOPACH MAP

C.I.=100'
Well location   o
Surface section   x

0   12
6   18 miles

**FIGURE 5
TOROWEAP CARBONATE ISOLITH**
C.I = 50'
Well location ○
Surface section ×

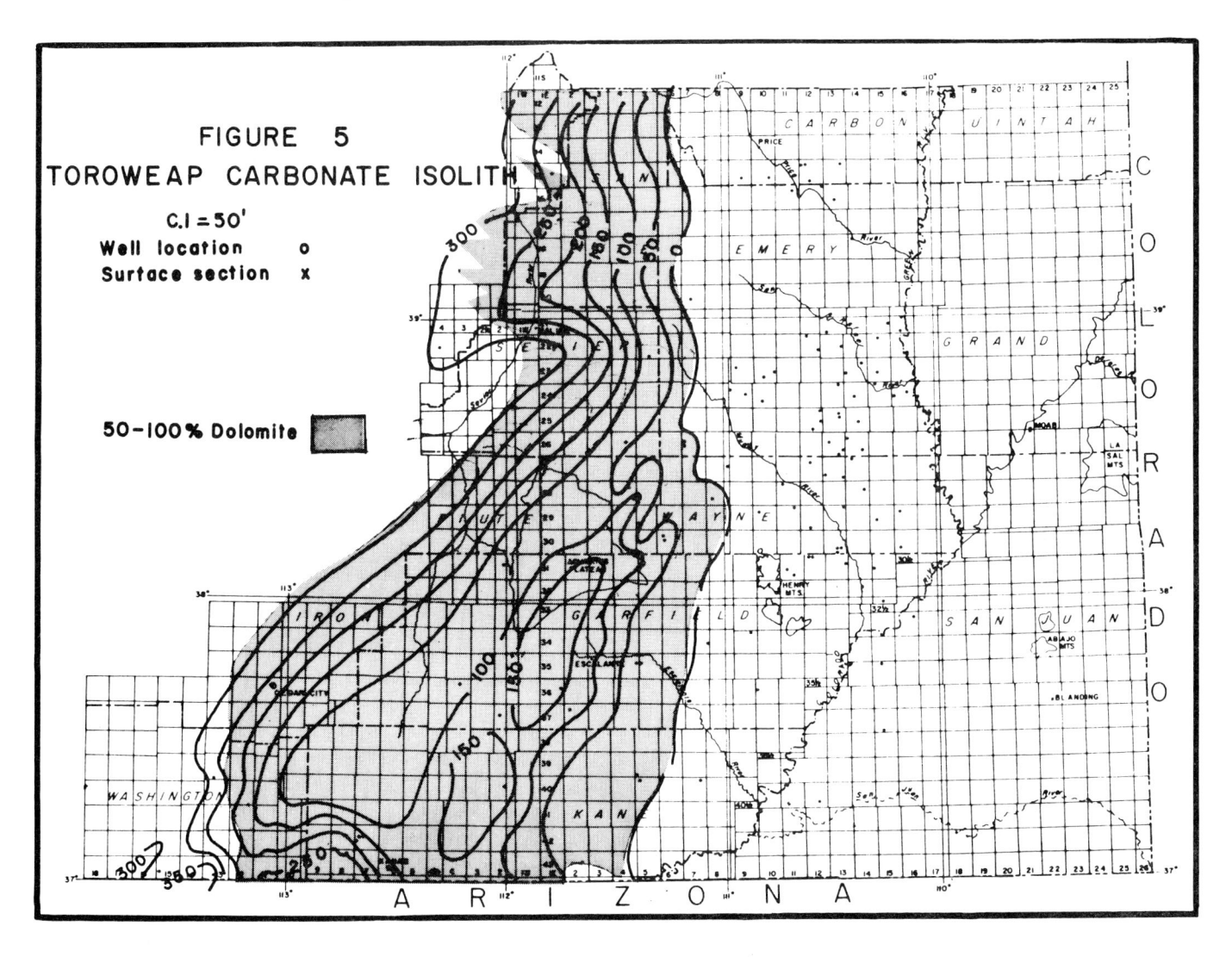

50-100% Dolomite ▨

Formation. The rapid increase of carbonate in the western part of the map may possibly reflect the position of an ancient break in depositional slope. Shaded on the Toroweap carbonate isolith map is the area of greater than 50% dolomite, however, throughout most of this area the carbonates are in reality almost 100% dolomite.

Cross section E-E' (Fig. 14, in pocket) shows the correlation of the Toroweap northward into the Diamond Creek—Thistle dome area. The correlation of this unit northward from Clear Creek is not exact. Correlation by stratal position would tie the Toroweap to the Grandeur Member of the Park City Formation, however a considerable thickness of Leonardian age rocks exists to the west of this area and many geologists have indicated a preference to correlate the White Rim-Toroweap with the Diamond Creek Sandstone which underlies the Grandeur. The widespread disconformity between the Toroweap and Kaibab may represent the time of Grandeur deposition, or all of the Grandeur and Diamond Creek strata could be represented in the Toroweap-White Rim sequence to the east. Examination of this cross section (E-E') shows several mechanical log breaks that appear to correlate internally within the Toroweap-White Rim and these could be the thin shelf-edge tongues of the much thicker basinal sediment sequence of both the Grandeur and Diamond Creek.

No production has been established from the Toroweap but excellent oil shows and some free oil recoveries have been obtained from dolomites of this formation in the Kaiparowits basin. The intertonguing of sandstone and carbonate in this formation may be somewhat detrimental in that oil may be free to migrate into the White Rim Sandstone, thus precluding stratigraphic traps near the beach facies. Possibly exploration in this formation should be targeted to areas of maximum carbonate facies change, this may occur to the west or southwest.

Kaibab Formation

Overlying and separated from the Toroweap by a widespread disconformity surface is the Kaibab Formation. In southwestern Utah the Kaibab Formation, like the Toroweap, was divided by McKee (1938) into three members which represent transgressive, maximum transgression, and regressive phases of a marine cycle. These are named Gamma, Beta and Alpha in ascending order. The isopach of the total Kaibab is shown on Figure 6.

The Gamma Kaibab, locally called the sandy Kaibab by many geologists, consists of sandstone, sandy limestone, limestone, and dolomite and clearly is a transgressive marine deposit. It is mostly sandstone along its eastern limits where it makes up the upper portion of the White Rim Sandstone and grades westward to arenaceous marine carbonates. Northwestward this unit thins and, as shown by the north-south cross sections B-B' and E-E' (Fig. 13 & 14, in pocket) overlies the Meade Peak phosphatic shale. In the Thistle area (T8S, R5E) the Gamma Kaibab also becomes somewhat phosphatic. Williams (1969) traces the outcrop of the Mead Peak the length of the Uintah Mountains; thus the Kaibab correlation can now be carried to the west with a sense of security.

Figure 7 is an isopach map of the Gamma Kaibab. The patterned line on the map represents the eastern limit of carbonate occurrence. The Emery uplift continues to demonstrate its effect upon sedimentation patterns as shown by the absence of carbonate over the crest of the uplift in contrast to the presence of carbonate in the basins to the north and south.

Immediately overlying and conformable on the Gamma Member is the Beta Member. The Beta Kaibab, which is almost all carbonate, represents the time of maximum Upper Permian marine transgression onto the shelf.

The Beta Kaibab throughout most of the study area is mainly composed of cherty dolomite but locally may contain some impure cherty limestones. In western Washington County and southwestern Kane County, Utah, limestone does become the dominant lithology however.

Figure 8 is an isopach map of the Beta Kaibab Member. Continuous carbonate deposition occurred across the Emery uplift and into the Bird Spring and Oquirrh basins.

The rapid thinning of this unit on the east reflects the effects of post-Kaibab erosion. This map may also be considered a carbonate isolith map because the clastic content of the Beta Kaibab is so slight and scattered that it is insignificant. Viewed as a carbonate isolith map it becomes apparent

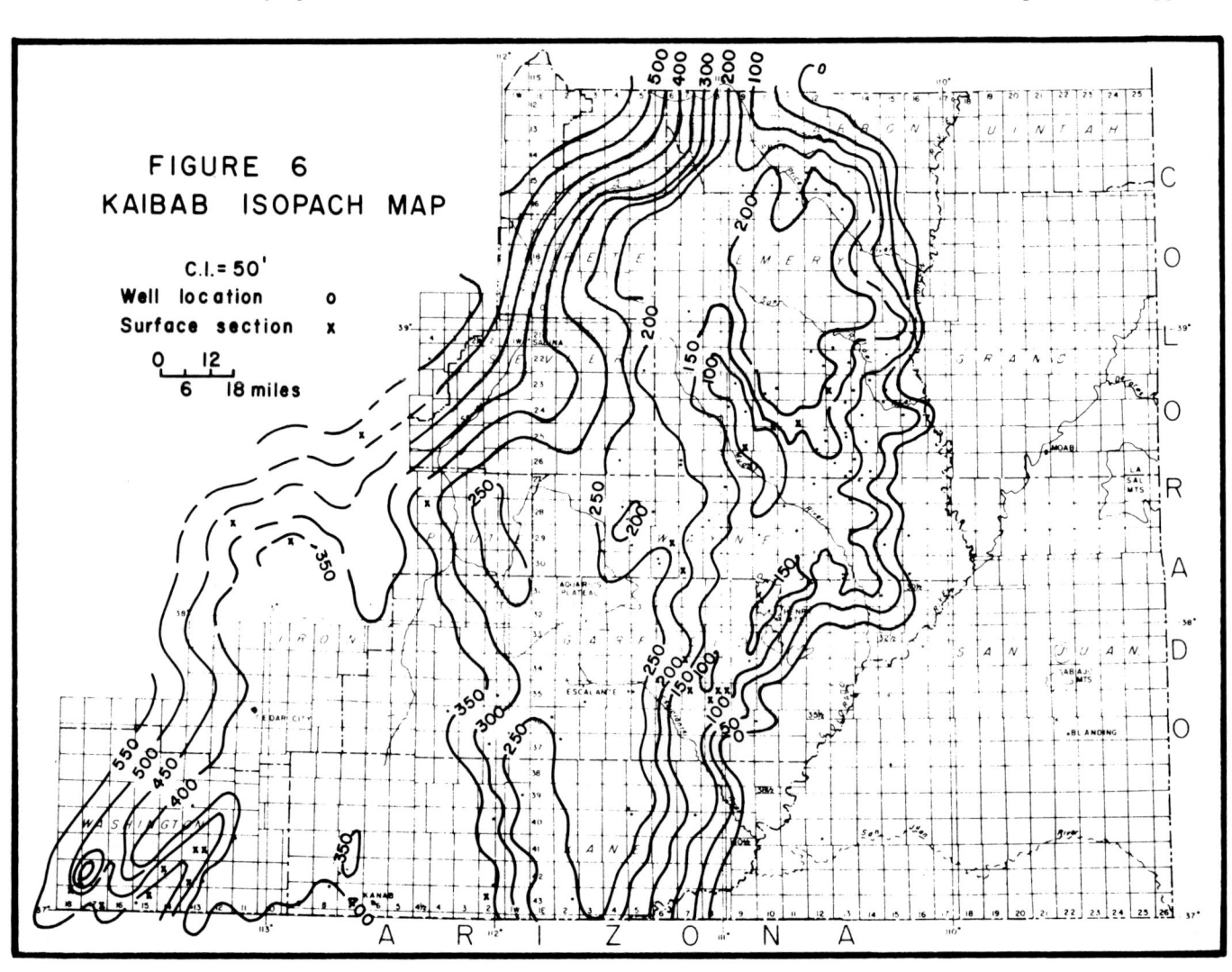

FIGURE 6
KAIBAB ISOPACH MAP
C.I.= 50'
Well location   o
Surface section x
0   12
  6   18 miles

FIGURE 7
GAMMA KAIBAB ISOPACH MAP

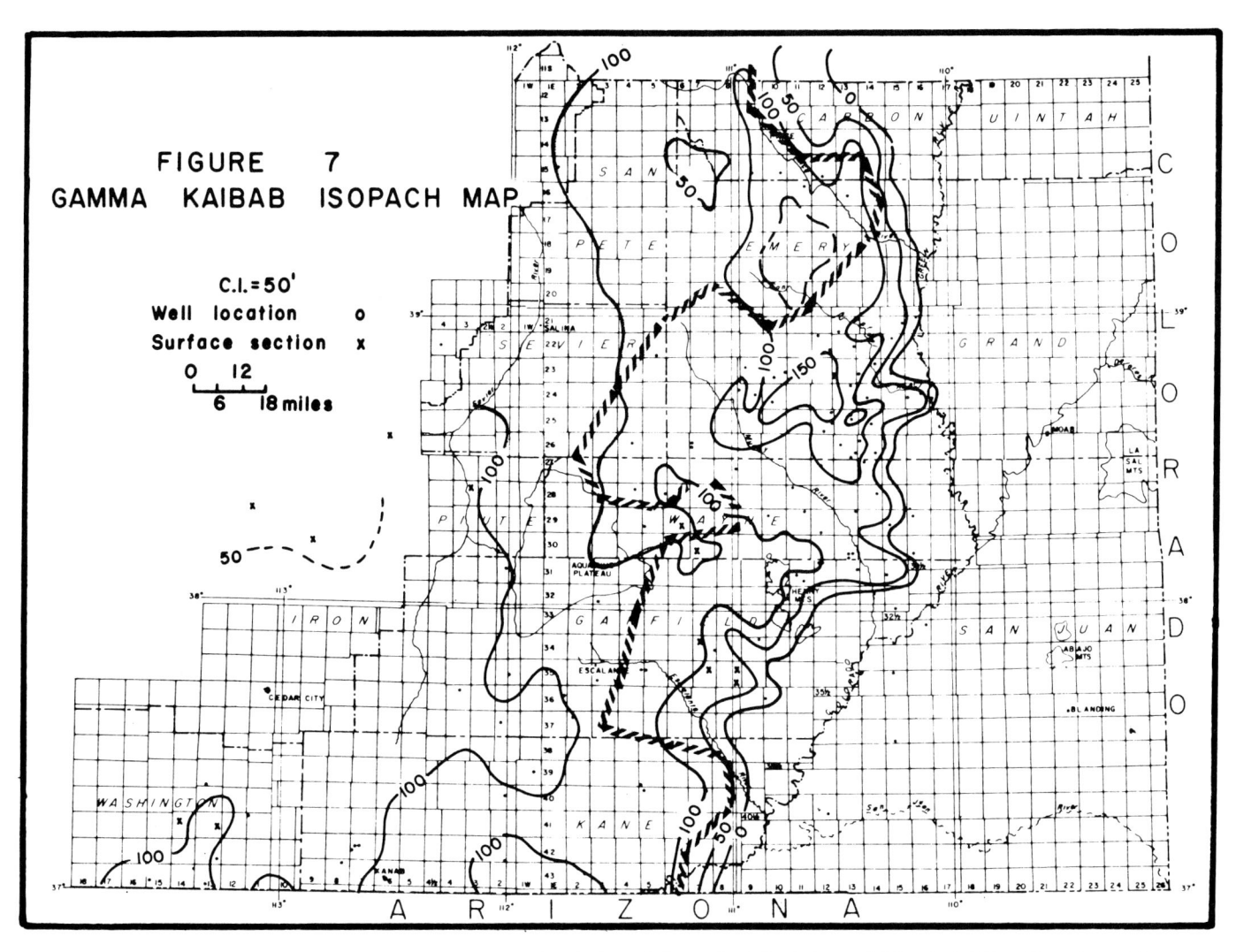

C.I.=50'
Well location o
Surface section x
0 12
6 18 miles

there is no eastern beach sandstone facies present as in the underlying formations. The absence of a beach facies may be considered a combination of little original beach development as well as its subsequent removal by erosion.

Shaded on this map is the area where the Beta Kaibab consists of over 75% dolomite. Excellent leached and intercrystalline porosity is usually present in the dolomites, thus there is a large area over which effective porosity can be expected in the Kaibab.

The uppermost member of the Kaibab is termed the Alpha Member. Also locally called the Harrisburg Member; this lithologic unit is the regressive phase of the Kaibab and is composed mostly of anhydrite and dolomite with lesser amounts of shale and sandstone. Figure 9 is an isopach map of this unit.

Cross section E-E' (Fig. 14, in pocket) shows the correlation of the Kaibab northward from North Springs into the Thistle Dome—Diamond Creek area and shows the position of the Grand Canyon Kaibab in the Park City sequence. The D.D. Feldman Oil & Gas Diamond No. 1 at the north end of this cross section is in the upper plate of the Wasatch Mountain Allochthon of Bissell (1959). Therefore this stratal sequence is not in place but has been transported eastward by the Charleston thrust. Estimates of the distance vary from about 16 to 40 mi. One of the interesting features of this well is the presence of an apparent Alpha Kaibab section. This may imply a lateral continuity of Alpha Kaibab around the west end of the Emery uplift or marine withdrawal and evaporation in separate basins to the north and south of the uplift thus creating two lithologically similar but independent "Alpha" Kaibab units. The continued deposition of carbonate above the Kaibab in the Diamond No. 1 well shows the continued presence of the Oquirrh basin. The absence of additional Permian rocks above the Kaibab across central Utah probably dates the beginning of the Kaibab—Lower Triassic hiatus.

Over 12 million bbl of oil have been produced from the Beta Kaibab at the Upper Valley field, noncommercial oil is found at Ferron field, and shows of oil have been encountered in this unit in almost every wildcat well drilled in central Utah. The porosity in the Beta Member is well developed over much of the mapped area and can be considered sheet

**FIGURE 8
BETA KAIBAB ISOPACH MAP**

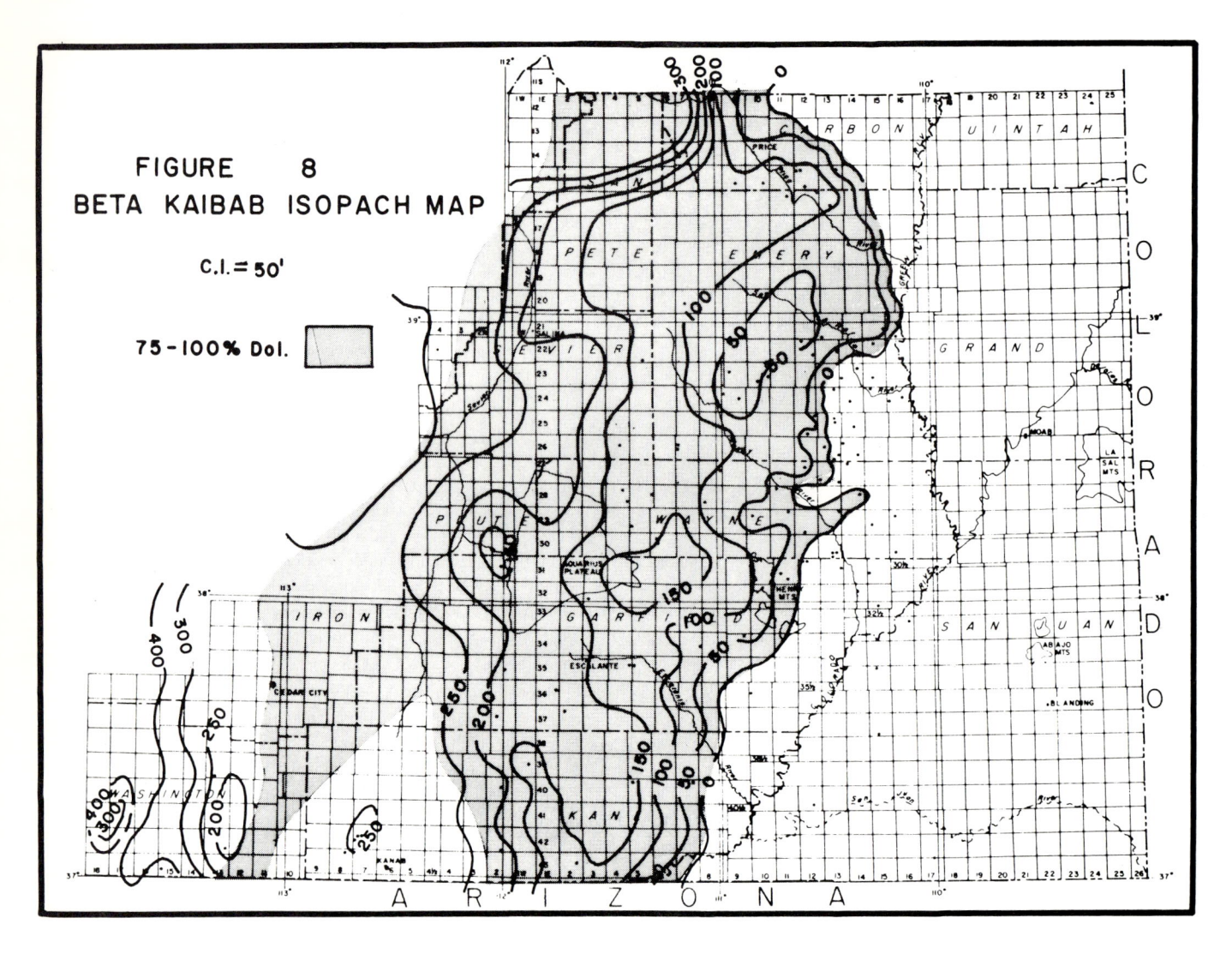

C.I. = 50'

75-100% Dol.

type, thus structural closure is undoubtedly necessary before hydrocarbon accumulation can take place. Most of the structures drilled to date are Laramide in age and are probably post-migration features. A detailed study of the Upper Valley field in this context is necessary.

A wide spread unconformity surface between the Kaibab and overlying Moenkopi Formation exists throughout central and southern Utah. Sedimentation was apparently continuous across the system boundaries in the Oquirrh Basin but elsewhere the Kaibab was subjected to active erosion.

Sinbad Member of the Moenkopi Formation

Triassic sediments onlap the Kaibab from north to south. In the North Springs-Gordon Creek area (T14-15S, R7-9E) the Sinbad Limestone Member of the Moenkopi is separated from the Kaibab by about 350' of Woodside Shale. To the south in southern Iron and Garfield counties the Sinbad is in depositional contact with the Kaibab, the contact being evident by the debris of the intervening erosional surface.

The Sinbad is composed of varying mixtures of limestone, dolomite, and shale. The carbonates generally are oolitic and commonly are porous. The porosity consists of microcrystalline, vuggy, and leached oolitic types and often contains considerable dead oil residue.

Figure 10 is a clean carbonate (less than 25% argillaceous material) isolith map of the Sinbad Formation. Shaded on this map is the area where the carbonates are more than 50% dolomite. From the 100 ft. thickness line eastward the Sinbad is almost entirely dolomite. Cross section D-D' (Fig. 15, in pocket) shows the basinward thickening of the Sinbad with the cleaner carbonate occurring mostly at the top and bottom of the member. Basinward the Sinbad contains considerably more limestone.

The Sinbad produces oil at the Upper Valley and Virgin fields and is saturated with dead oil in outcrops on the San Rafael Swell and at Capitol Reef. Porosity and permeability developments in the Sinbad are less uniform than in the Kaibab, also the Sinbad is encased in shales on its eastern wedge-edge, hence the opportunity exists for stratigraphic traps to

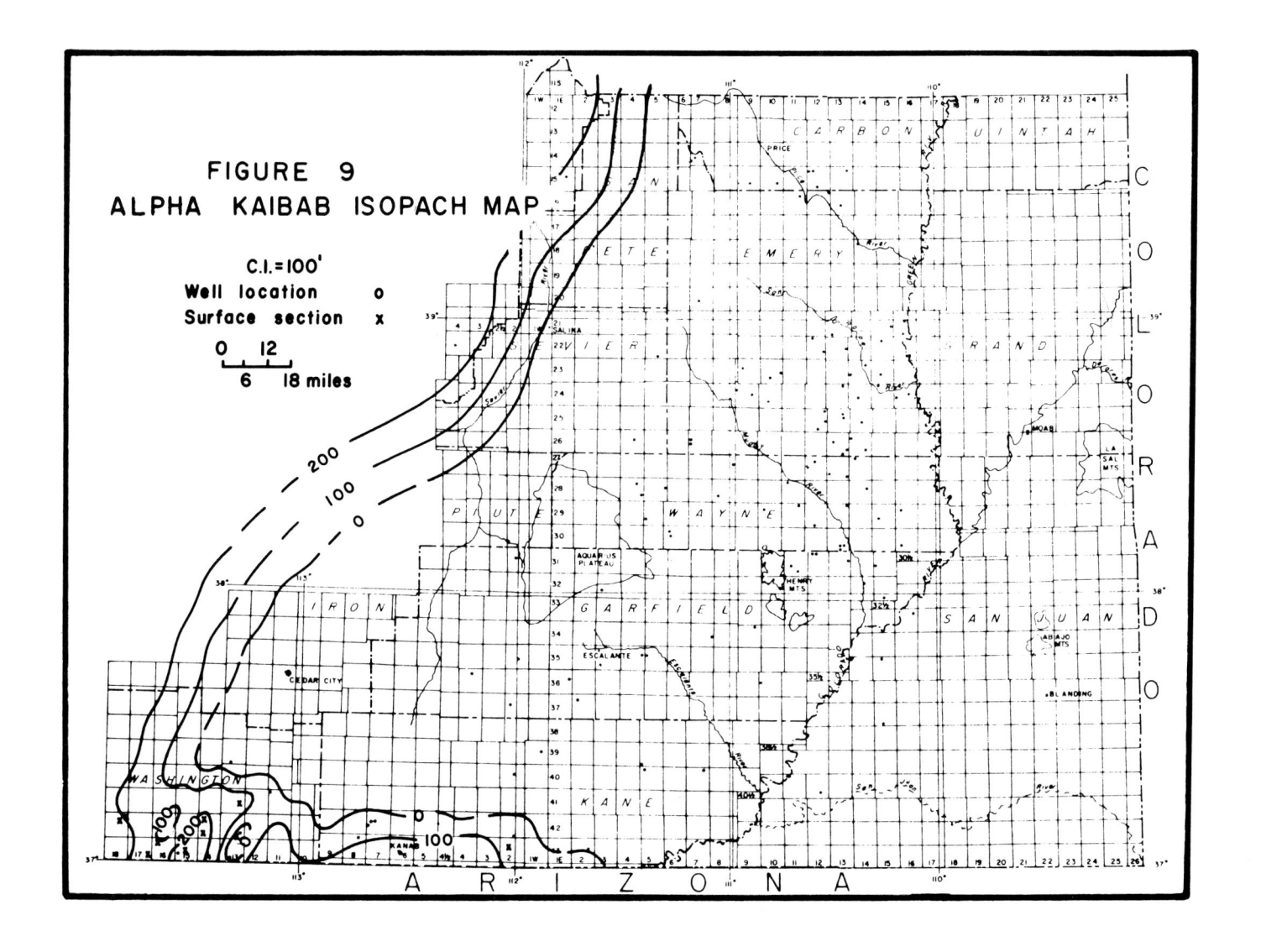

FIGURE 9
ALPHA KAIBAB ISOPACH MAP

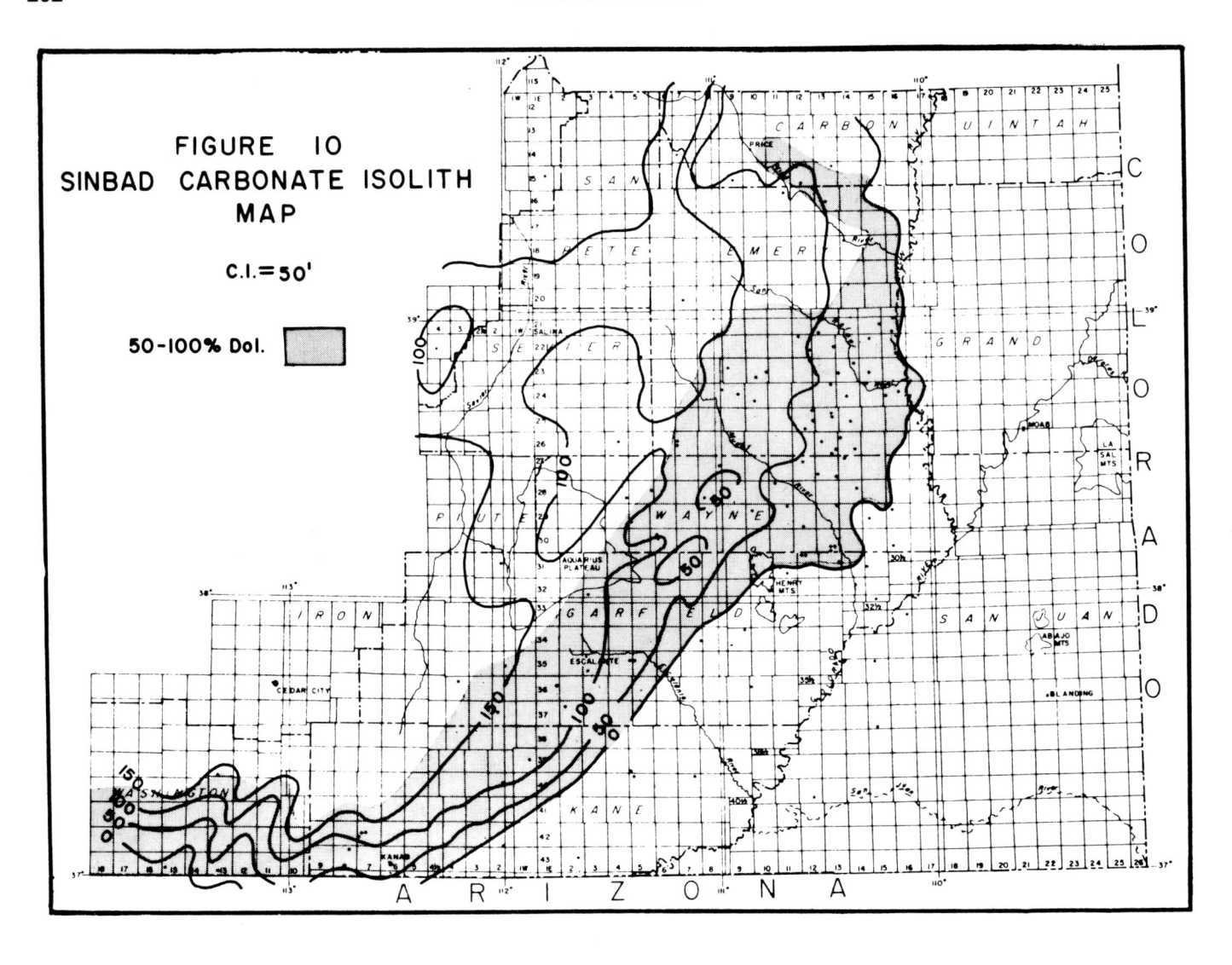

FIGURE 10
SINBAD CARBONATE ISOLITH MAP
C.I.=50'
50-100% Dol.

occur within this unit.

Figure 11 (in pocket) is a regional structure map of central Utah contoured on top of the Kaibab Formation. When this map is compared to the isopach maps it becomes apparent that most of the structure is younger than Permian and probably is Laramide in age. Integrating topographic, hydrodynamic, and reservoir pressure relationships into this structural picture explains why paleofeatures such as the Emery uplift and Thousand Lake Mountain are nonproductive.

It is difficult to believe that Upper Valley is or will be the only significant field found from this section of rocks in central Utah. Hydrodynamic tilts as at Upper Valley, extreme canyon dissection as in the San Rafael Swell, Circle Cliffs, Monument uplift, and the Grand Canyon, and extensive volcanic cover require the exploration geologist and geophysicist to put more effort into exploration techniques than merely searching for a simple undrilled high.

REFERENCES

Baars, D.L., 1962, Permian System of Colorado Plateau: Am. Assoc. Petroleum Geologists Bull., v. 46, no. 2, p. 149-218.

_____, and Seagar, W.R., 1970, Stratigraphic control of petroleum in White Rim Sandstone (Permian) in and near Canyonlands National Park, Utah: Am. Assoc. Petroleum Geologists Bull., v. 54, no. 5, p. 709-18.

Bissell, H. J., 1959, North Strawberry Valley sedimentation and tectonics: Intermtn. Assoc. Petroleum Geologists Guidebook, 10th Ann. Field Conference, p. 159-65.

_____, 1964, Lithology and petrography of the Weber Formation in Utah and Colorado: Intermtn. Assoc. Petroleum Geologists Guidebook, 13th Ann. Field Conference, p. 67-91.

Hallgarth, W. E., 1962, Upper Paleozoic rocks exposed in Straight Wash Canyon, San Rafael Swell, Utah: Amer. Assoc. Petroleum Geologists Bull., v. 46, no. 8, p. 1494-1501.

Irwin, C.D., 1969, Stratigraphic analysis of the Upper Permian and Lower Triassic in southern Utah: Unpublished PH.D. dissertation, Univ. of New Mexico, 157 p.

_____, 1971, Straigraphic analysis of Upper Permian and Lower Triassic strata in southern Utah: Am. Assoc. Petroleum Geologists Bull., v. 55, no. 11, p. 1976-2007.

McKee, 1938, The environment and history of the Toroweap and Kaibab formations of northern Arizona and southern Utah: Carnegie Inst. Washington Pub. 492, 268 p.

Williams, J. S., 1969, The Permian System in the Uintah Mountain area: Intermtn. Assoc. Petroleum Geologists Guidebook, 16th Ann. Field Conference, p. 153-68.

LOWER TRIASSIC FACIES IN THE VICINITY OF THE CORDILLERAN HINGELINE: WESTERN WYOMING, SOUTHEASTERN IDAHO AND UTAH

by

W. Jerry Koch[1]

ABSTRACT

Lower Triassic strata (Dinwoody, Woodside, Thaynes, Moenkopi) in the vicinity of the Cordilleran Hingeline were deposited in a great variety of continental to marine environments. They form a wedge-shaped body that is over 5,000 ft thick in southeastern Idaho and thins eastward to less than 1,000 ft in Wyoming and eastern Utah. A dramatic thinning of section as well as a facies change from dominantly gray siltstones, shales, and limestones to red-beds occurs along the hingeline. The distribution of environments of deposition, and consequently facies development and early diagenesis, was strongly controlled by three major tectonic elements: 1) the miogeosyncline, 2) the hingeline, and 3) the platform. Lower Triassic strata in the vicinity of the Cordilleran Hingeline contain the basic elements necessary for hydrocarbon entrapment: 1) mature source rocks, 2) reservoirs, 3) seals, and 4) structural and stratigraphic trap configurations. With increased exploration activity in the vicinity of the Cordilleran Hingeline, it is likely that Lower Triassic strata will become a productive exploration target.

INTRODUCTION

The purpose of this paper is to provide an overview of the development of Lower Triassic facies and their distribution in the vicinity of the Cordilleran Hingeline. The paper includes discussions of correlation, tectonic setting, palinspastic reconstruction, facies and paleogeography and the hydrocarbon potential of the Lower Triassic. In addition to focusing on details of the Lower Triassic, the criteria used to interpret facies can be applied to stratigraphic sequences of other ages.

CORRELATION

Because of the tremendous facies changes and the essentially unfossiliferous nature of the red-beds, it is difficult to correlate all of the stratigaphic units of the Lower Triassic over the area studied. Fortunately, however, the miogeosynclinal deposits contain a distinct sequence of ammonite faunal zones, and some of the faunal zones are present in marine limestone and shale tongues interbedded with the platform red-beds to the east. In many cases, the unfossiliferous stratigraphic units can be correlated because of their relationship to overlying and underlying dated units.

The most complete sequence of ammonite faunas, as documented by Kummel (1954), occurs in the vicinity of Bear Lake in southeastern Idaho. In ascending order, the zones are *Otoceras*, *Genodiscus*, *Meekoceras*, *Anasibirites*, *Tirolites*, *Columbites*, and *Prohungarites*. The *Otoceras* and *Genodiscus* zones which are confined to the Dinwoody Formation in the area are rarely present, and for this study have been combined and called the pre-*Meekoceras* Triassic zone.

Table 1 is a correlation chart for Lower Triassic formations in this area. The stratigraphic position of the ammonite faunal zones identified are indicated for key areas. This provides a rigid framework for the correlation of stratigraphic units that do not contain distinctive faunal elements. Some conjecture remains about the correlation of formations not bracketed by faunal zones. The scheme presented here is one of many that could be proposed, based on techniques of correlation by physical stratigraphy.

TECTONIC SETTING

In order to interpret the development of Lower Triassic facies, it is important to understand the regional tectonic setting during Lower Triassic time (synsedimentary) and during the Sevier-Laramide orogenic events (post-sedimentary). The rates of subsidence and influx of clastic sediment, for instance, strongly influenced the environments of deposition. Later tectonic movements such as transportation and superposition of laterally equivalent facies, particularly in southeastern Idaho and western Wyoming, must also be considered in reconstructing and interpreting the origin of the facies.

---

[1] Filon Exploration Corporation, 1700 Broadway, Denver, Colorado 80202.

Financial support for this study was provided by the Department of Geological Sciences, Harvard University.

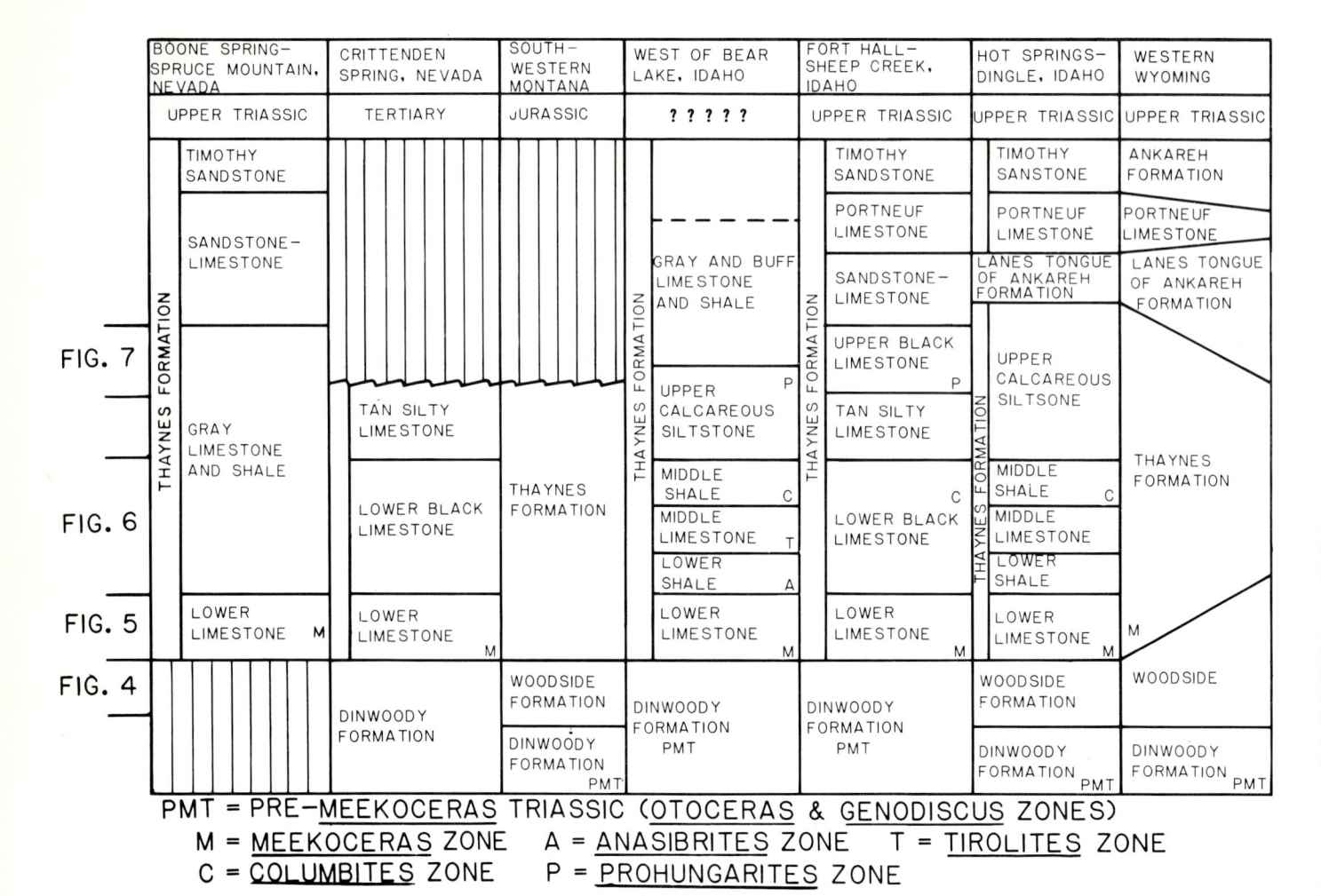

BOONE SPRING-SPRUCE MOUNTAIN, NEVADA	CRITTENDEN SPRING, NEVADA	SOUTH-WESTERN MONTANA	WEST OF BEAR LAKE, IDAHO	FORT HALL-SHEEP CREEK, IDAHO	HOT SPRINGS-DINGLE, IDAHO	WESTERN WYOMING
UPPER TRIASSIC	TERTIARY	JURASSIC	?????	UPPER TRIASSIC	UPPER TRIASSIC	UPPER TRIASSIC
TIMOTHY SANDSTONE (Thaynes Formation)			GRAY AND BUFF LIMESTONE AND SHALE (Thaynes Formation)	TIMOTHY SANDSTONE (Thaynes Formation)	TIMOTHY SANSTONE (Thaynes Formation)	ANKAREH FORMATION
SANDSTONE-LIMESTONE				PORTNEUF LIMESTONE	PORTNEUF LIMESTONE	PORTNEUF LIMESTONE
GRAY LIMESTONE AND SHALE	TAN SILTY LIMESTONE	THAYNES FORMATION	UPPER CALCAREOUS SILTSTONE — P	SANDSTONE-LIMESTONE	LANES TONGUE OF ANKAREH FORMATION	LANES TONGUE OF ANKAREH FORMATION
			MIDDLE SHALE C	UPPER BLACK LIMESTONE — P	UPPER CALCAREOUS SILTSONE	THAYNES FORMATION
	LOWER BLACK LIMESTONE		MIDDLE LIMESTONE T	TAN SILTY LIMESTONE	MIDDLE SHALE C	
			LOWER SHALE A	LOWER BLACK LIMESTONE C	MIDDLE LIMESTONE	
					LOWER SHALE	
LOWER LIMESTONE M	LOWER LIMESTONE M		LOWER LIMESTONE M	LOWER LIMESTONE M	LOWER LIMESTONE M	M
		WOODSIDE FORMATION			WOODSIDE FORMATION	WOODSIDE
	DINWOODY FORMATION	DINWOODY FORMATION PMT	DINWOODY FORMATION PMT	DINWOODY FORMATION PMT	DINWOODY FORMATION PMT	DINWOODY FORMATION PMT

PMT = PRE-MEEKOCERAS TRIASSIC (OTOCERAS & GENODISCUS ZONES)
M = MEEKOCERAS ZONE   A = ANASIBRITES ZONE   T = TIROLITES ZONE
C = COLUMBITES ZONE   P = PROHUNGARITES ZONE

Table 1. Correlation chart for Lower Triassic Formations in the vicinity of the Cordilleran Hingeline

Synsedimentary Tectonic Activity

The Cordilleran Miogeosyncline was a generally negative area with respect to the hingeline and platform during the Lower Triassic. The entire section is thicker (over 5,000 ft) and consists of predominantly marine facies, some apparently deposited in relatively deep water. The negative aspect of the area (most negative in southeastern Idaho) facilitated the accumulation of the thick section, as well as establishing rather continuous marine conditions.

During the Lower Triassic, the hingeline between the relatively negative miogeosyncline and relatively stable platform was a zone of transition rather than a distinct line. The zone is recognized by a rapid thickening in section and is a locus of dramatic facies changes — in particular the change from red beds to non-red units. The apparent location of the hingeline shifted in an east-west direction during deposition of various Lower Triassic units. The hingeline can be shown to have occupied a zone over 50 mi wide during the Paleozoic, Triassic, and Jurassic.

The platform was relatively stable during the Lower Triassic. Slow subsidence precluded the accumulation of thick deposits (> 2,500 ft) and, though seas flooded the area, sedimentation was dominantly under continental or very shallow marine conditions. Even in this relatively stable region, some areas were generally positive (eg. Uncompahgre Highland) and others were negative, as evidenced by increased thicknesses of Lower Triassic strata and the location of sources of coarse clastic sediment. On the platform, early diagenesis of the sediments is dominated by dolomitization and moldic porosity development in carbonate units. Diagenesis is greatly different from that in the miogeosyncline. The differences are related to the shallow water evaporative conditions and subareal exposure on the platform contrasting with the marine to deep marine conditions in the miogeosyncline.

Post-Sedimentary Tectonic Activity

Much of western Wyoming, southeastern Idaho, and western Utah were involved in major eastward overthrust movements during the Jurassic-Cretaceous Sevier-Laramide

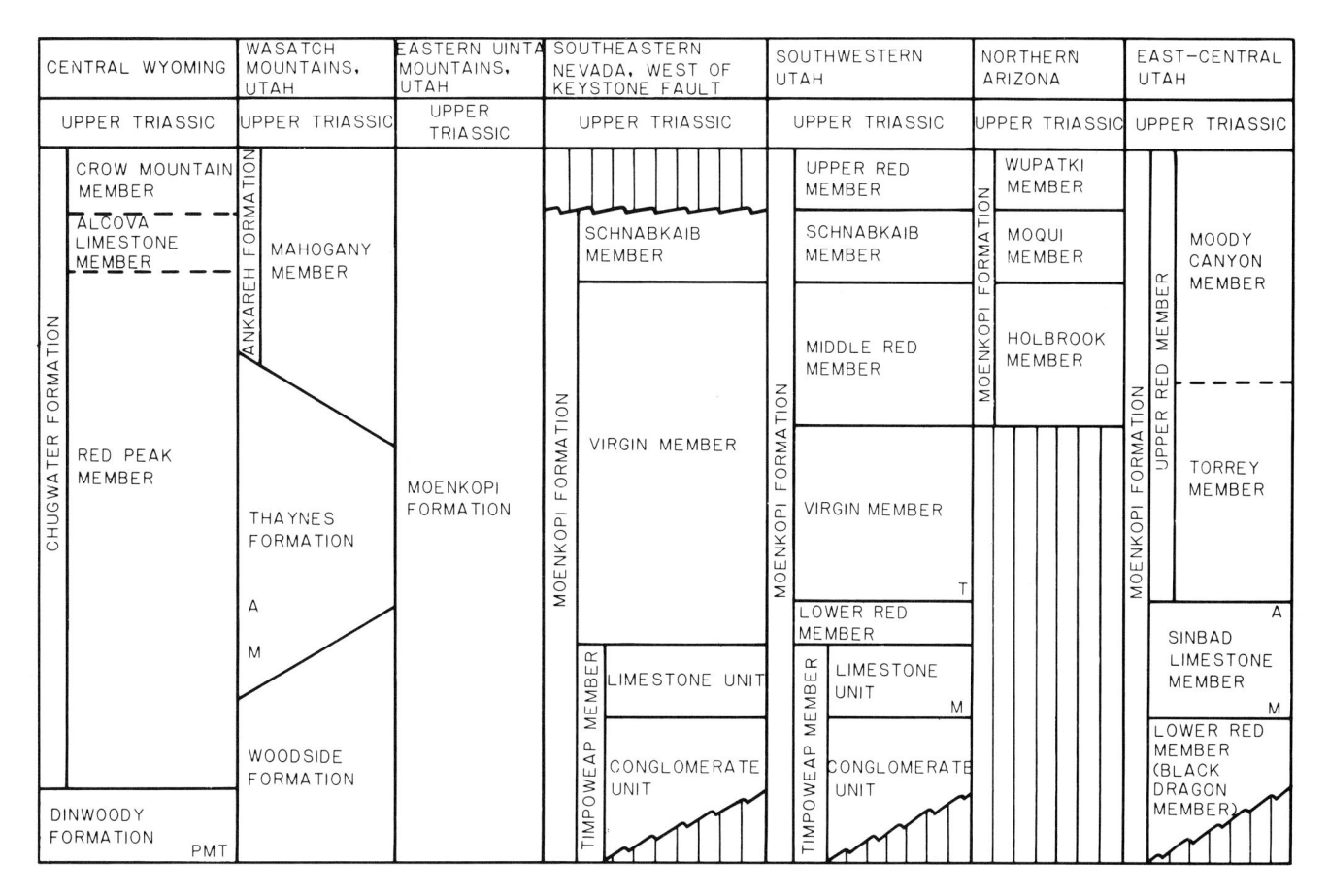

CENTRAL WYOMING	WASATCH MOUNTAINS, UTAH	EASTERN UINTA MOUNTAINS, UTAH	SOUTHEASTERN NEVADA, WEST OF KEYSTONE FAULT	SOUTHWESTERN UTAH	NORTHERN ARIZONA	EAST-CENTRAL UTAH
UPPER TRIASSIC	UPPER TRIASSIC	UPPER TRIASSIC	UPPER TRIASSIC	UPPER TRIASSIC	UPPER TRIASSIC	UPPER TRIASSIC
CHUGWATER FORMATION: CROW MOUNTAIN MEMBER / ALCOVA LIMESTONE MEMBER / RED PEAK MEMBER; DINWOODY FORMATION (PMT)	ANKAREH FORMATION: MAHOGANY MEMBER; THAYNES FORMATION (A, M); WOODSIDE FORMATION	MOENKOPI FORMATION	MOENKOPI FORMATION: SCHNABKAIB MEMBER; VIRGIN MEMBER; TIMPOWEAP MEMBER (LIMESTONE UNIT / CONGLOMERATE UNIT)	MOENKOPI FORMATION: UPPER RED MEMBER; SCHNABKAIB MEMBER; MIDDLE RED MEMBER; VIRGIN MEMBER; LOWER RED MEMBER; TIMPOWEAP MEMBER (LIMESTONE UNIT / CONGLOMERATE UNIT) (T, M)	MOENKOPI FORMATION: WUPATKI MEMBER; MOQUI MEMBER; HOLBROOK MEMBER	MOENKOPI FORMATION: UPPER RED MEMBER (MOODY CANYON MEMBER / TORREY MEMBER); SINBAD LIMESTONE MEMBER; LOWER RED MEMBER (BLACK DRAGON MEMBER) (A, M)

deformations. The original spatial relations of the Lower Triassic facies were thus greatly altered. In order to view the distribution of facies in their proper perspective, it is necessary to make a palinspastic reconstruction by mentally restoring the thrust sheets to their pre-thrusting position. Palinspastic reconstruction may be accomplished using structural or stratigraphic approaches or a combination of the two. All techniques involve numerous assumptions and should be viewed as "best guess" approximations and used accordingly.

The structural approach, as aptly illustrated by Royse, et al., (1975), consists of an analysis of the lateral movements along all of the thrust faults in the system using "rules" of thrust mechanics and geometry. As a rule of thumb, net shortening is generally about 50 percent.

In the stratigraphic approach to palinspatic reconstruction data on facies and thicknesses in the various thrust sheets are compiled and fit into a "reasonable" set of facies maps, isopachs, etc., to produce a palinspastic map based on stratigraphic principles. The displacements on the faults are then determined from the offsets necessary to shorten the palinspastic reconstruction to fit the present day location of the stratigraphic sections. If the ultimate goal of the palinspastic map is use in facies reconstruction, the stratigraphic approach using only the formation of interest obviously involves a circular argument.

The Lower Triassic isopach map (Fig. 2) and the paleogeography maps (Fig. 4-7) are palinspastic. The reconstruction is based largely on structural techniques, including the work of Royse, et al., (1975). Some of the movements have also been inferred using stratigraphic reconstructions of units older and younger than Lower Triassic. The restoration is a "best guess" at this time, and a more accurate reconstruction awaits additional detailed structural and stratigraphic work over the whole area. If anything, the estimate of offset or telescoping of facies and isopachs used here is probably conservative (i.e., a minimum displacement is assumed).

FACIES AND PALEOGEOGRAPHY

The Lower Triassic facies in the vicinity of the Cordilleran Hingeline were deposited in a spectrum of environments from continental to deep marine. A detailed discussion of the origin of all stratigraphic units in the area is obviously beyond the scope of a paper of this length. Paleogeograhic maps for four increments of time have been constructed and palinspastically restored. The stratigraphic intervals selected for the maps correlate with 1) the Upper Dinwoody, 2) the Lower Limestone, 3) the Lower Black Limestone, and 4) the Upper Black Limestone units of the Fort Hall-Sheep Creek sections (Table 1).

A tremendous volume of literature exists on sedimentary

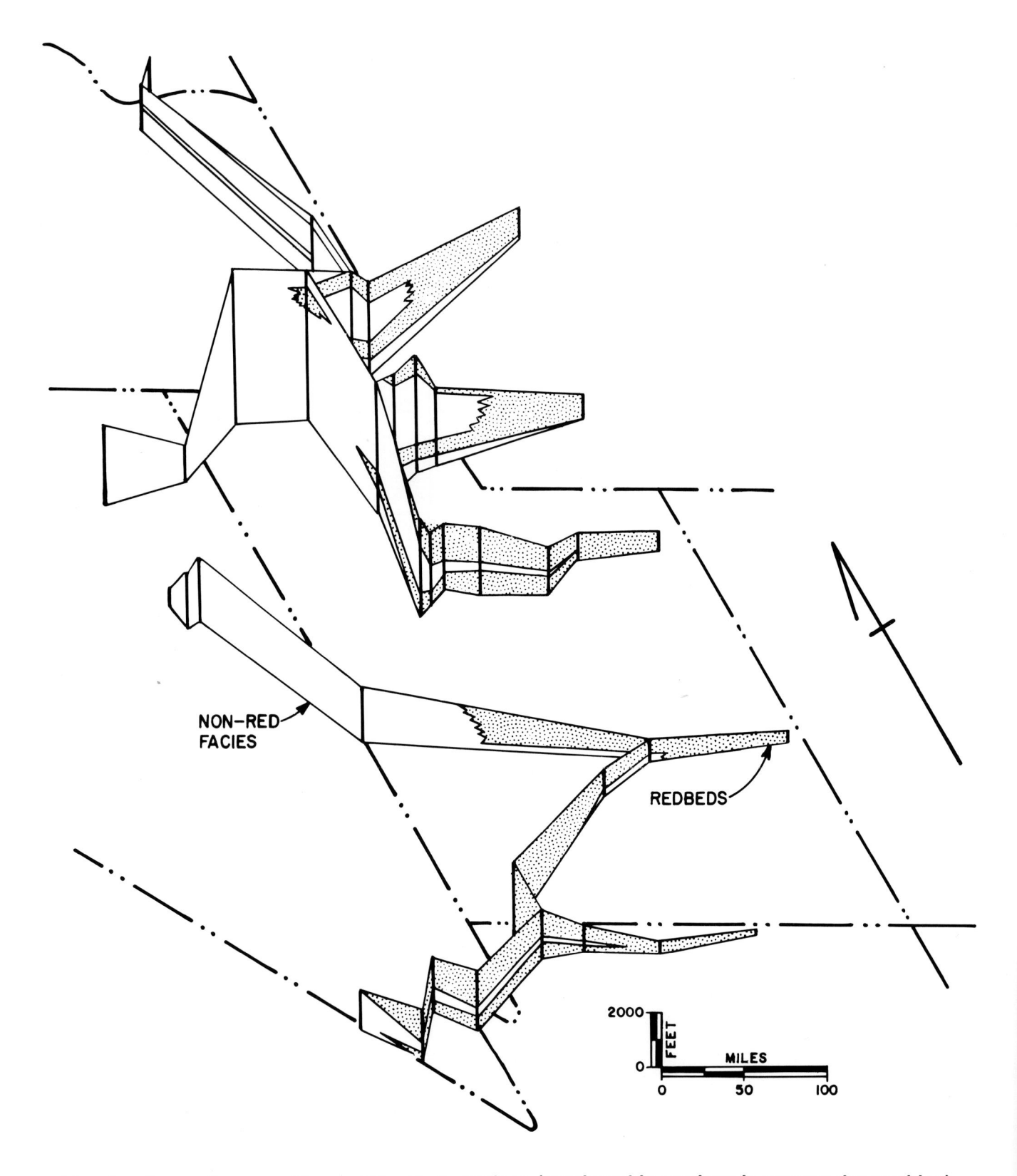

Fig. 1 — Fence diagram of Lower Triassic formations (stratigraphic sections in present-day position).

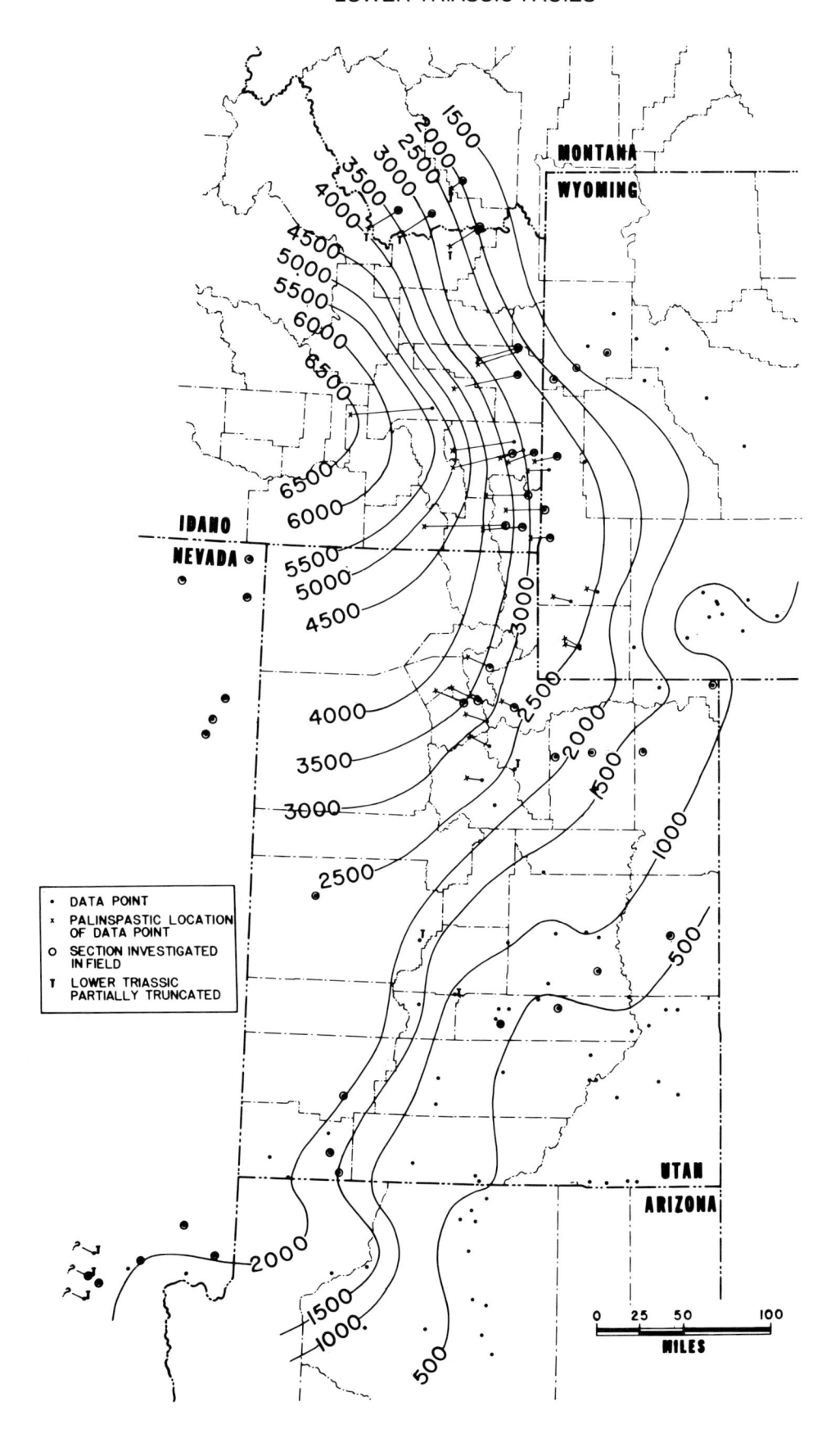

Fig. 2 — Palinspastic isopach map of Lower Triassic.

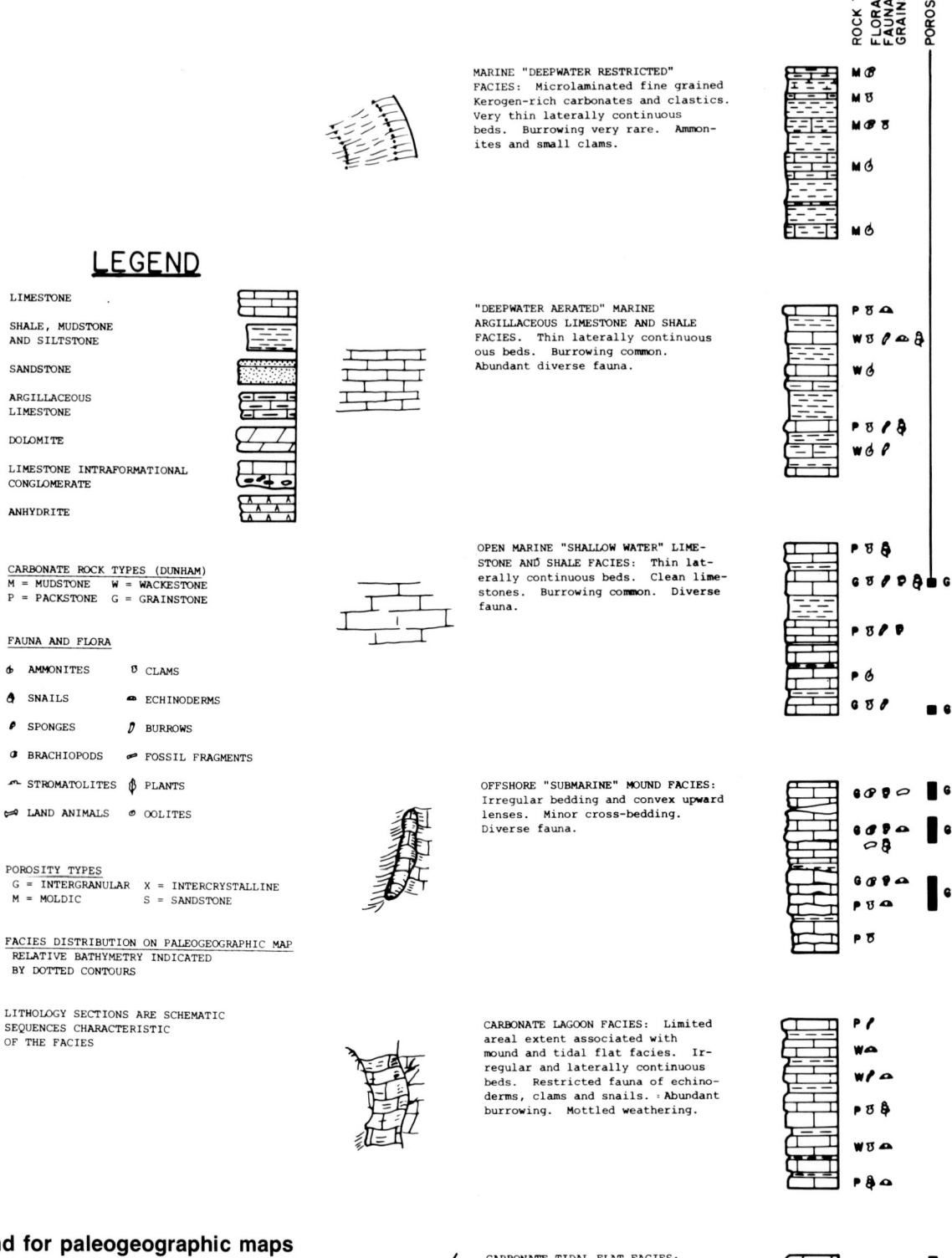

MARINE "DEEPWATER RESTRICTED" FACIES: Microlaminated fine grained Kerogen-rich carbonates and clastics. Very thin laterally continuous beds. Burrowing very rare. Ammonites and small clams.

LEGEND

LIMESTONE

SHALE, MUDSTONE AND SILTSTONE

SANDSTONE

ARGILLACEOUS LIMESTONE

DOLOMITE

LIMESTONE INTRAFORMATIONAL CONGLOMERATE

ANHYDRITE

CARBONATE ROCK TYPES (DUNHAM)
M = MUDSTONE   W = WACKESTONE
P = PACKSTONE G = GRAINSTONE

FAUNA AND FLORA

- AMMONITES
- CLAMS
- SNAILS
- ECHINODERMS
- SPONGES
- BURROWS
- BRACHIOPODS
- FOSSIL FRAGMENTS
- STROMATOLITES
- PLANTS
- LAND ANIMALS
- OOLITES

POROSITY TYPES
G = INTERGRANULAR   X = INTERCRYSTALLINE
M = MOLDIC   S = SANDSTONE

FACIES DISTRIBUTION ON PALEOGEOGRAPHIC MAP
RELATIVE BATHYMETRY INDICATED BY DOTTED CONTOURS

LITHOLOGY SECTIONS ARE SCHEMATIC SEQUENCES CHARACTERISTIC OF THE FACIES

"DEEPWATER AERATED" MARINE ARGILLACEOUS LIMESTONE AND SHALE FACIES. Thin laterally continuous beds. Burrowing common. Abundant diverse fauna.

OPEN MARINE "SHALLOW WATER" LIMESTONE AND SHALE FACIES: Thin laterally continuous beds. Clean limestones. Burrowing common. Diverse fauna.

OFFSHORE "SUBMARINE" MOUND FACIES: Irregular bedding and convex upward lenses. Minor cross-bedding. Diverse fauna.

CARBONATE LAGOON FACIES: Limited areal extent associated with mound and tidal flat facies. Irregular and laterally continuous beds. Restricted fauna of echinoderms, clams and snails. Abundant burrowing. Mottled weathering.

CARBONATE TIDAL FLAT FACIES: Dolomities, intraformational conglomerates and thin bedded limestones. Irregular bedding. Cross-bedding in channels. Stromatolites and relatively restricted fauna (normal marine fauna near base of sequence). Leached oolites, pellets and possibly some anhydrite. Sucrosic dolomite.

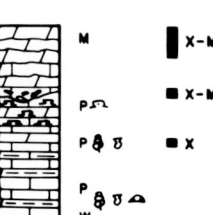

Fig. 3 — Legend for paleogeographic maps and summary of characteristics used for facies interpretation.

LOWER TRIASSIC FACIES

OOLITE BAR FACIES: Lenticular irregular layers of oolitic limestone and sucrosic dolomite. Festoon cross-bedding abundant. Oolites and coated rounded grains. Normal marine fauna near base of sequence.

CLASTIC BEACH AND BAR FACIES: Irregular to laterally continuous beds. Coarsening upward sequence. Cross-bedding common. Restricted marine fauna and burrowing most common near base of sequence. Many fossils abraded.

DELTA FACIES: Irregular to laterally continuous bedding. Coarsening upward sequence. Lobate sand distribution. Association of deltaic facies. Channels with cross-bedding. May become beach deposits along depositional strike. Terrestrial flora and fauna near top and restricted marine fauna near base of sequence.

DELTA CHANNEL, LEEVE AND OVERBANK FACIES: Irregular to laterally continuous bedding. Fining upward sequence in channels cutting delta plain sequence. Terrestrial flora and fauna.

"UNCLASSIFIED REDBED FACIES". Laterally continuous to irregular bedding. Dominantly terrestrial flora and fauna but in some areas a restricted marine fauna may be present. Mud cracks, animal tracks present.

CONTINENTAL FACIES: Irregular channel bedding. Cross-beds and signs of sub areal exposure (mud cracks, animal tracks). Terrestrial flora and fauna.

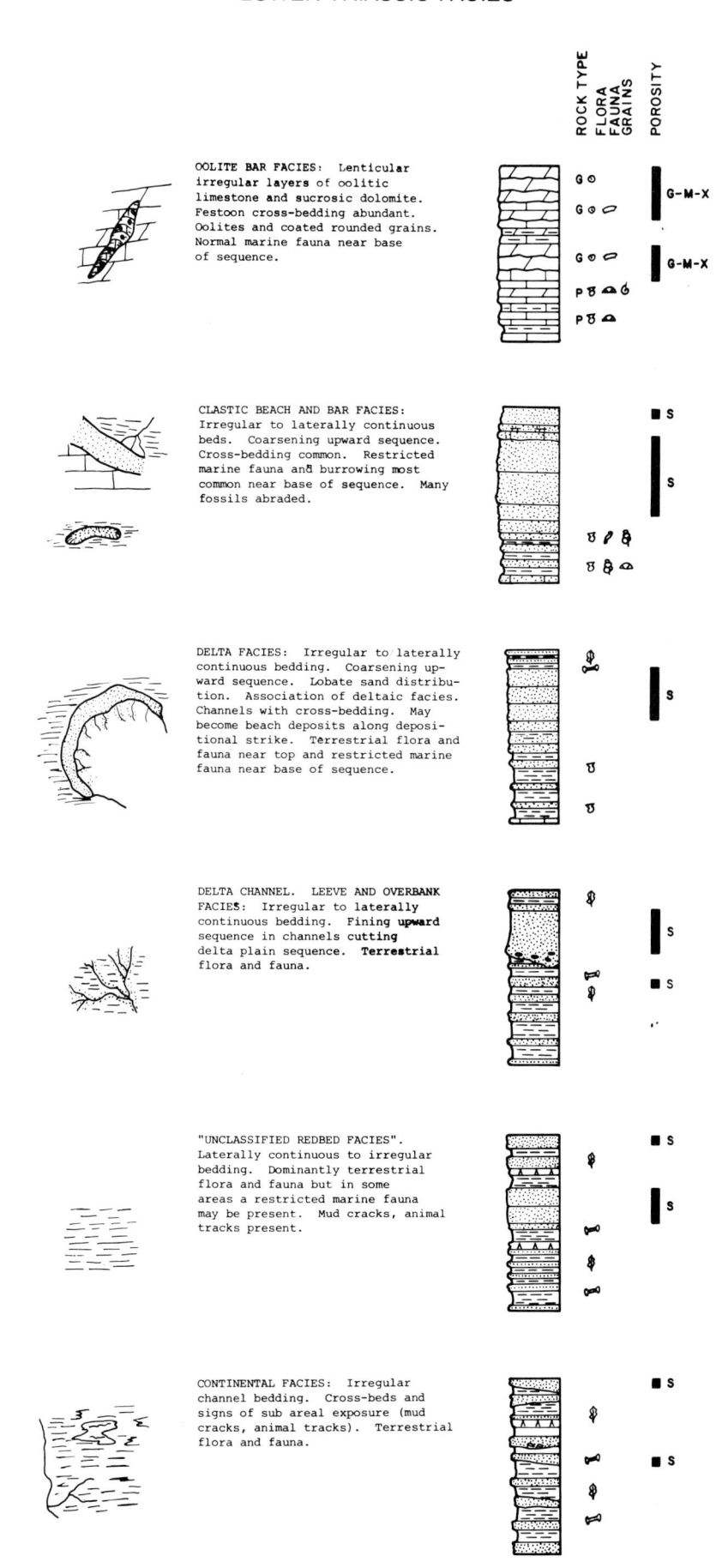

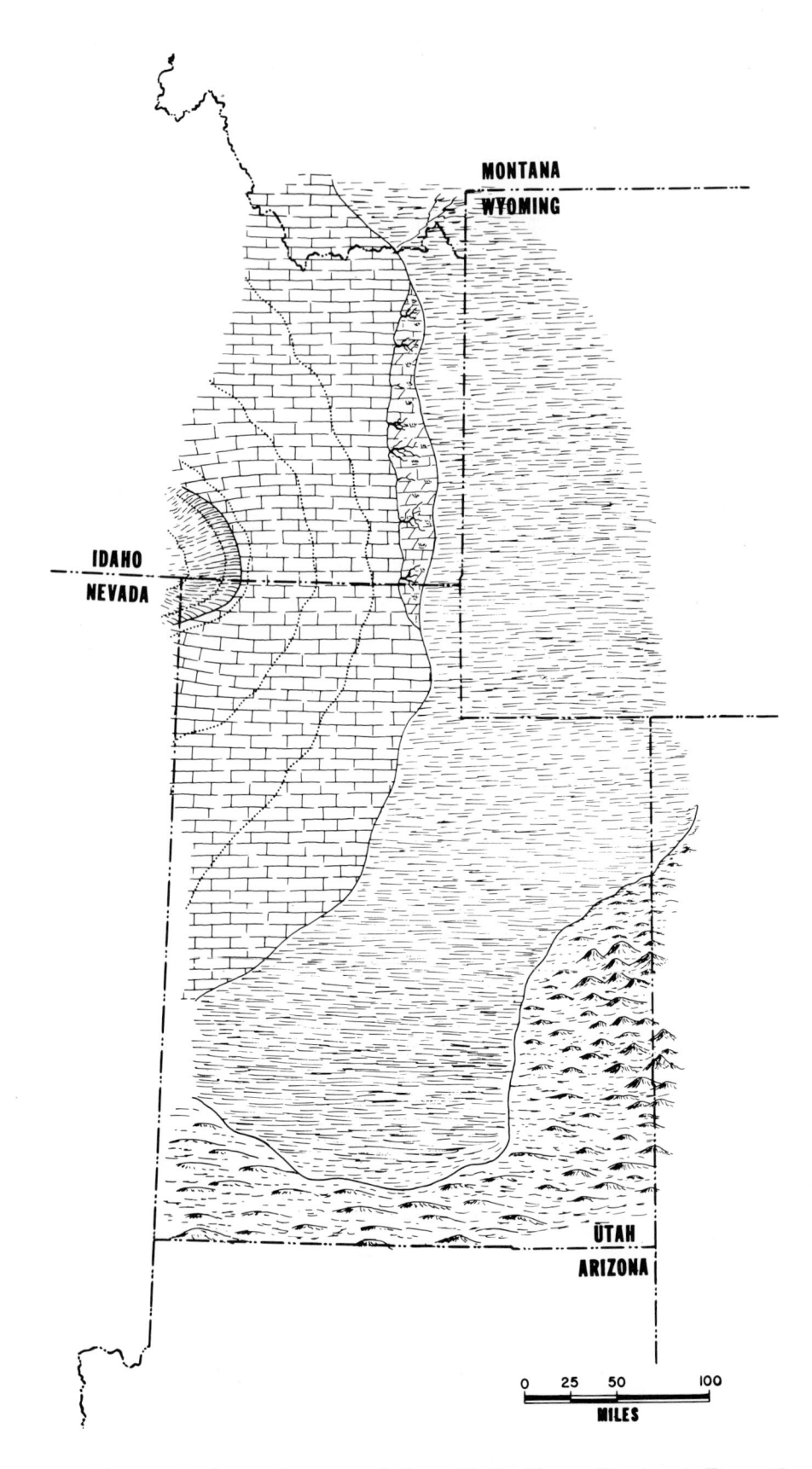

Fig. 4 — Paleogeographic map for formations correlative with the Upper Dinwoody Formation in southeastern Idaho (see correlation chart).

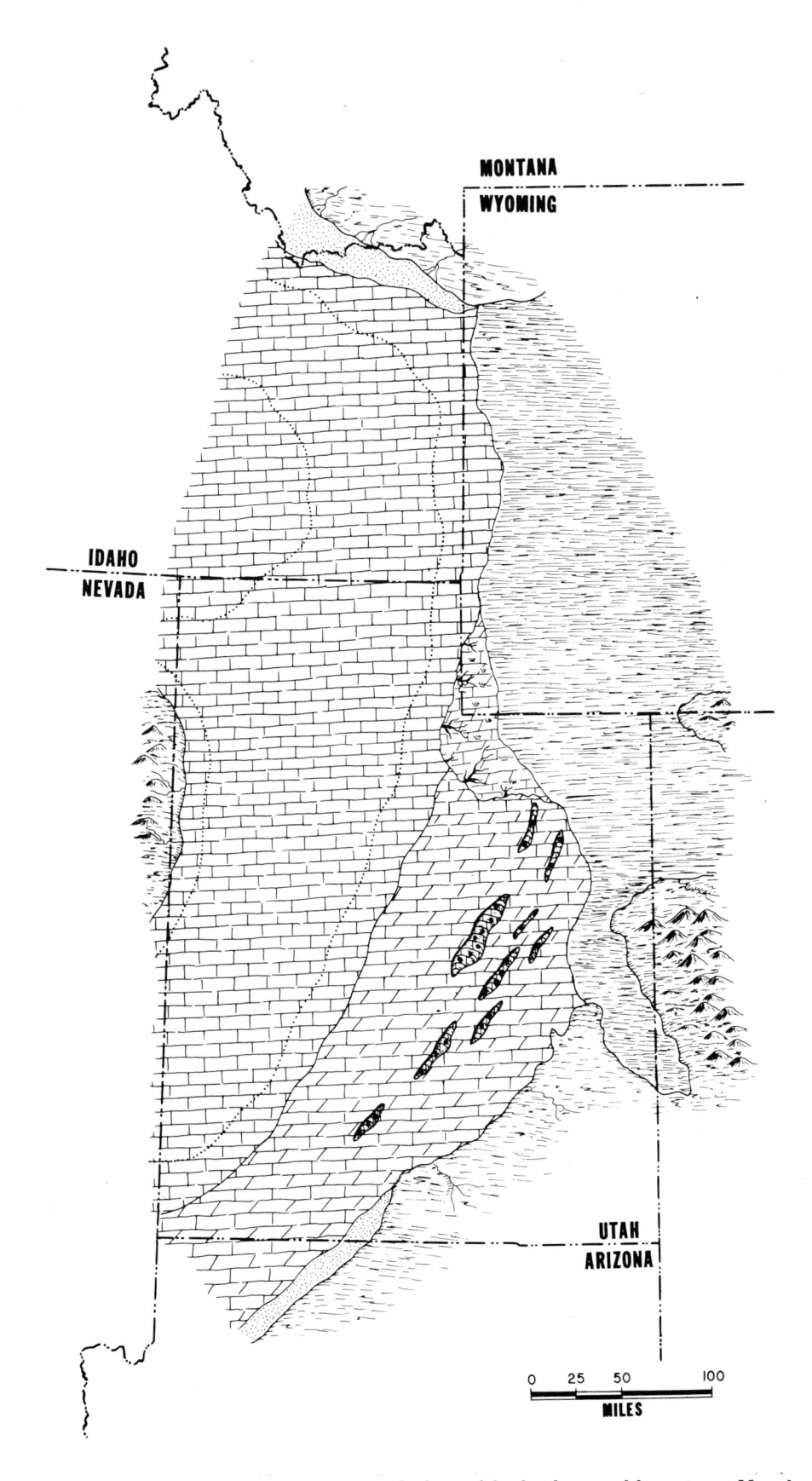

Fig. 5 — Paleogeographic map for formations correlative with the Lower Limestone Member of the Thaynes Formation in southeastern Idaho (see correlation chart).

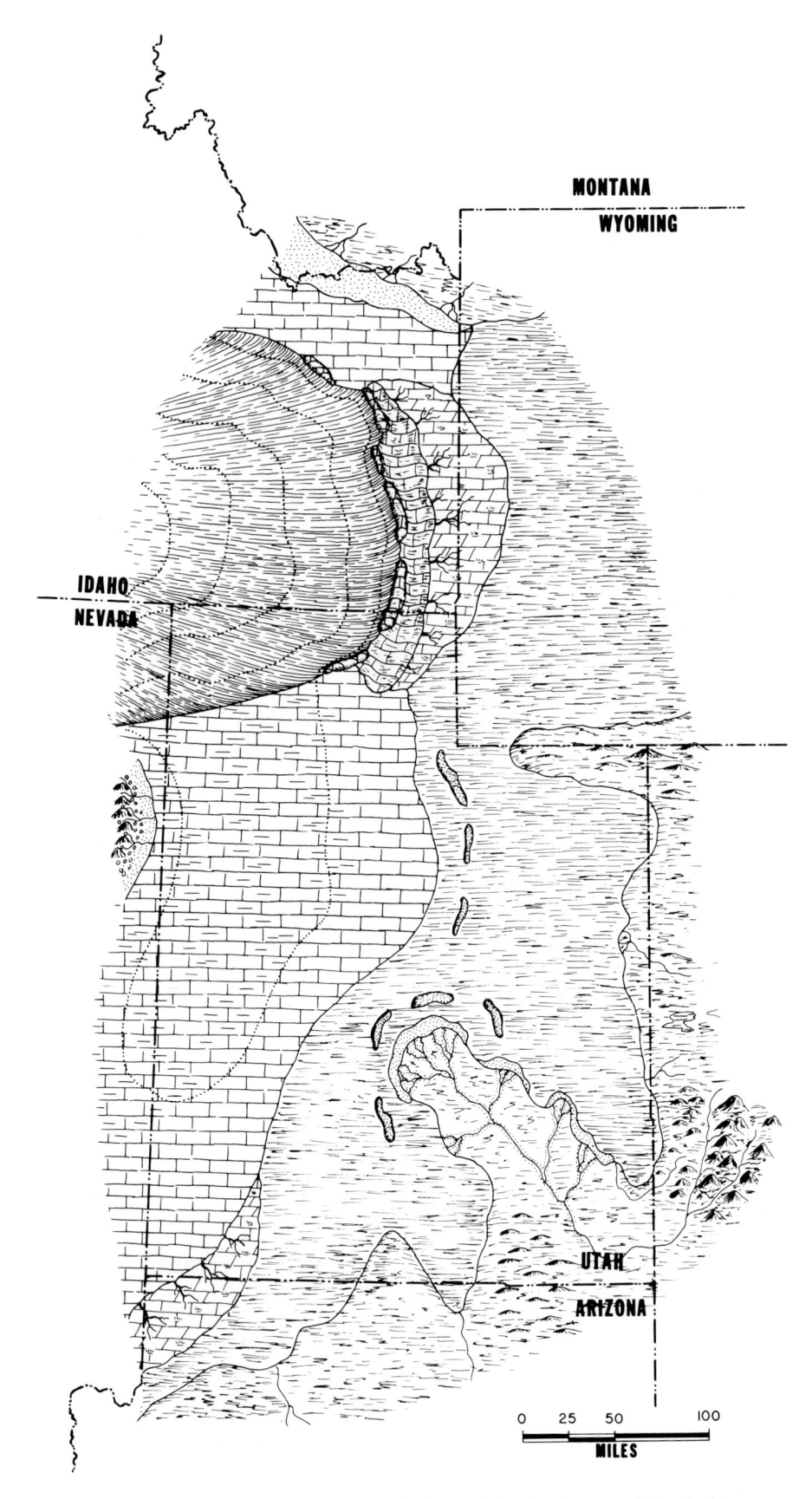

Fig. 6 — Paleogeographic map for formations correlative with the Lower Black Limestone Member of the Thaynes Formation in the Fort Hall-Sheep Creek sections, southeastern Idaho (see correlation chart).

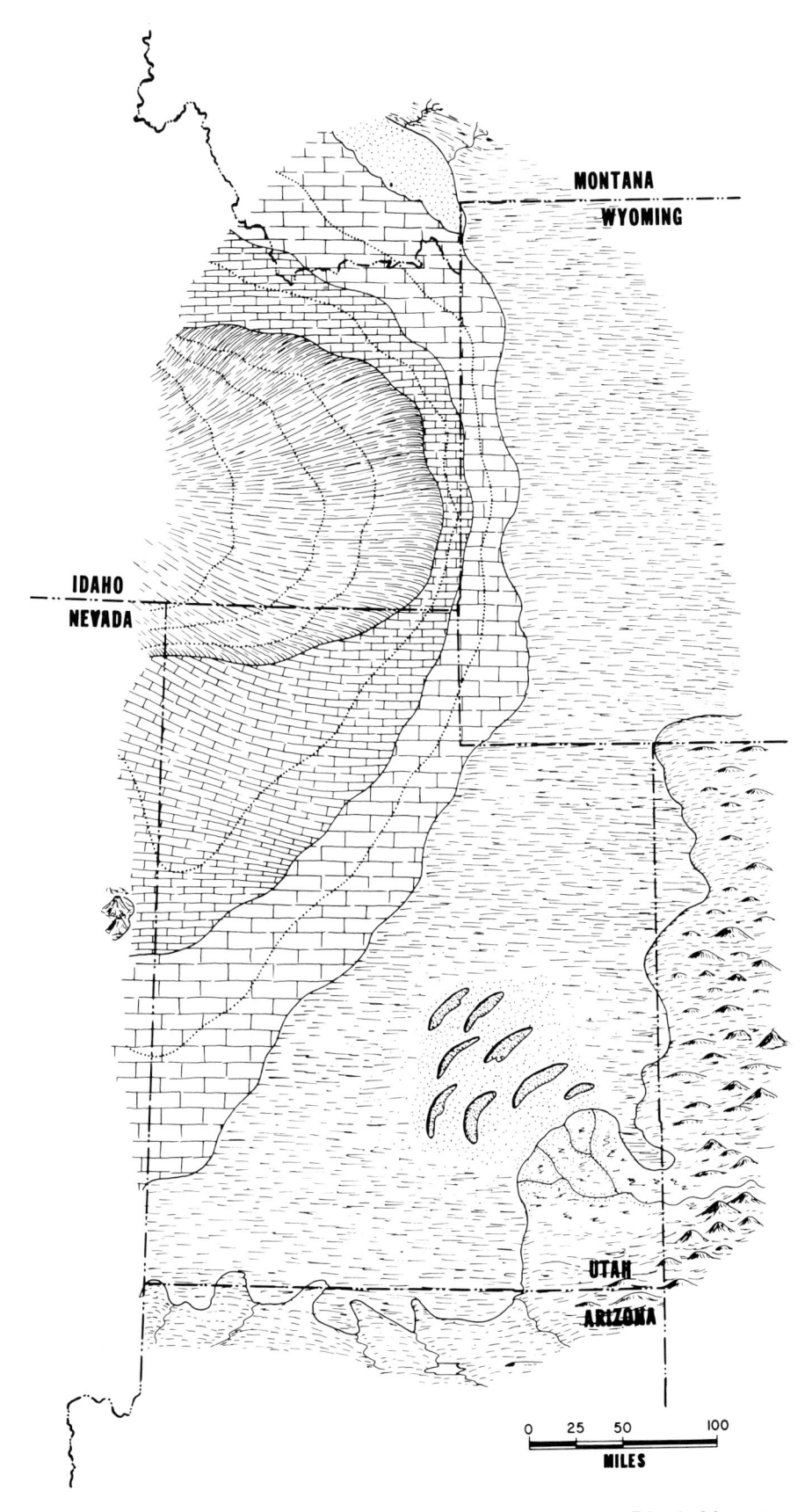

Fig. 7 — Paleogeographic map for formations correlative with the Upper Black Limestone Member of the Thaynes Formation in the Fort Hall-Sheep Creek sections, southeastern Idaho (see correlation chart).

facies and criteria for their interpretation. Some of the facies are very distinct and environments of deposition are easy to interpret, such as carbonate tidal flats, oolite bars, carbonate buildups, delta lobes, or laminated deepwater limestones. The origin of others, such as some of the red-bed facies, are much more problematic and leave a great deal of room for interpretation. The sedimentologic characteristics used to interpret the facies are summarized in Figure 3. Some of the most characteristic lithologies are illustrated in Figures 8-12. The generalized lithologic sections illustrate the association of rock types, and in some cases, vertical sequences representative of specific environments of deposition. These actually fit models based on investigations of modern environments and other ancient deposits. Thus, the facies distributions shown on the paleogeographic maps (Fig. 4-7) were interpreted by classifying isolated outcrops within correlative stratigraphic units (using the facies criteria in Fig. 3) and fitting them into a paleogeographically reasonable scheme.

For the sake of brevity, the discussion of origin of facies will focus on the red-beds. Interpretations of the other facies are covered by the paleogeographic maps and Figure 3. Some of the papers that have been relied upon most for the facies interpretations are Blakey (1974), Koch (1969), Kummel (1957, McKee, et al., (1959), Picard, et al., (1969), and Picard (1975).

The origin of red beds has long elicited discussion and controversy. Some units of the Moenkopi or Chugwater are readily interpreted as deltaic or continental in origin (McKee, 1954; Picard, 1967; and Blakley, 1974). Many of the evenly bedded, laterally continuous red siltstones and mudstones show evidence of intermittent subaerial exposure. The extremely rare faunal evidence includes vertebrate and invertebrate remains of continental and marine origin (McKee, 1954). Picard (1967) and Picard, et al., (1969), interpreted much of the Red Peak Member of the Chugwater as being deposited under shallow marine and tidal flat conditions. The association of rock types, petrography, sedimentary structures, and the fact that relatively thin mudstone and siltstone units can be correlated over great distances strongly support the shallow marine-tidal flat interpretation. The enigmatic faunal content, however, is more suggestive of a lacustrine or continental environment. The deposits have many of the characteristics of lacustrine accumulations as documented by Picard and High (1972). The facies also cover about the same area as the younger Triassic lacustrine Popo Agie (High and Picard, 1965; Picard and High, 1972), and are similar in extent to lacustrine deposits of the Jurassic Todilto Formation of Utah, Colorado, and New Mexico (Tanner, 1974). For the purpose of this paper, the red beds that are not obviously of continental, deltaic, lacustrine, or shallow marine origin have been grouped into one category — "unclassified red beds". More refined paleogeographic maps

thus await even more detailed studies of the red-beds, so that they may be placed in their proper regional setting in relation to the other more readily interpretable facies.

HYDROCARBON POTENTIAL

The Lower Triassic strata in the vicinity of the Cordilleran Hingeline have been associated with only limited hydrocarbon production (Timpoweap in the Virgin and Upper Valley fields, southwestern Utah, and the Sinbad in the shut-in South Last Chance gas well; Moulton, 1975). There have been numerous hydrocarbon shows in Lower Triassic strata throughout the Colorado Plateau and along the hingeline (Bissel, 1970; Irwin, 1971; and Picard, 1975). It also appears that the San Rafael Swell is an exhumed fossil oil field because of the ubiquity of asphalt impregnation in the Sinbad Limestone. The Permian White Rim and Kaibab also are impregnated locally and possibly regionally on the San Rafael Swell.

The Lower Triassic includes the basic elements for hydrocarbon entrapment (mature source-rocks, reservoirs, seals, and structural and stratigraphic trap configurations).

Source rock facies are present in southeastern Idaho in the Lower Black Limestone and Upper Black Limestone members of the Thaynes Formation. Kerogenous material is visible in thin sections of microlaminated fettid limestones and shales. Organic material contained in at least some of the section appears to be thermally mature. Live oil oozes from small vugs in concretions in the *Columbites* beds of the Lower Black Limestone in Webster Canyon (Fig. 8B). The Phosphoria Formation is also a proven source rock unit (yields of 5 to 20 gallons per ton) and could logically charge some Lower Triassic reservoirs.

Some of the more obvious potential reservoirs are carbonate mounds, carbonate tidal flats, oolite bars (primary intergranular porosity and leached oolites and pellets, Fig. 11), carbonate or clastic beach deposits, and sucrosic dolomites below the red bed units. The distribution of porosity types is summarized in the generalized stratigraphic sections in Figure 3. Most facies, even with relatively low matrix porosities, may also develop fracture porosity. The quality of such reservoirs has been vividly demonstrated by recent production from fractured Twin Creek Limestone (Jurassic) in the Pineview area.

Numerous trapping configurations can be visualized for Lower Triassic reservoirs. Simple stratigraphic traps should be present, particularly in porosity pinchouts on the flanks of structures, in areas along the hingeline not subsequently deformed by intense tectonic movements. The ideal location for this type of stratigraphic trap is where regional dip has been continuously westward. With burial and subsequent thermal maturation of the Triassic and Permian source rocks, hydrocarbons migrated updip to the east. Porosity pinchouts should

Fig. 8A — Microlaminated kerogen-rich dark gray argillaceous limestone of the Marine "Deepwater Restricted" Facies. Lower Black Limestone Member of the Thaynes Formation; Paris Canyon (West of Bear Lake), Idaho.

Fig. 8B — Fossiliferous calcite concretion in laminated kerogen-rich calcareous shale of the Marine "Deepwater Restricted" Facies. Live oil bleeds from vugs in some concretions. Lower Black Limestone Member of the Thaynes Formation; Webster Canyon, Idaho.

Fig. 9A — Light gray grainstones and layers of brachiopods *Pugnoides triassica* in the offshore "Submarine" Mound Facies. Middle Gray Limestone Member of the Thaynes Formation; Raymond Canyon, Wyoming.

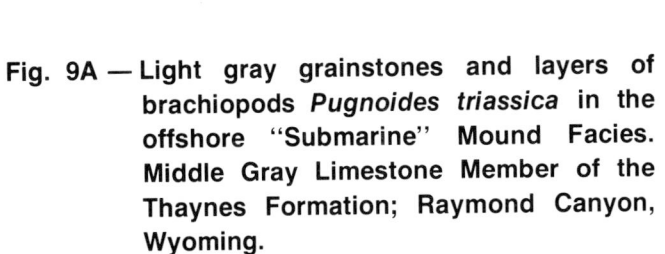

Fig. 9B — Brown and gray burrowed mottled lime mudstone of the Carbonate Lagoon Facies. Thaynes Formation; Big Cottonwood Canyon (East of Salt Lake City), Utah.

Fig. 10A — Algal stromatolite in carbonate tidal flat sequence. Upper Dinwoody Formation; Raymond Canyon, Wyoming.

Fig. 10B — Intraformational limestone conglomerate in channel cut into thin bedded algal limestones and bioclastic packstones and wackestones of the Carbonate Tidal Flat sequence. Thaynes Formation; Weber River, Utah.

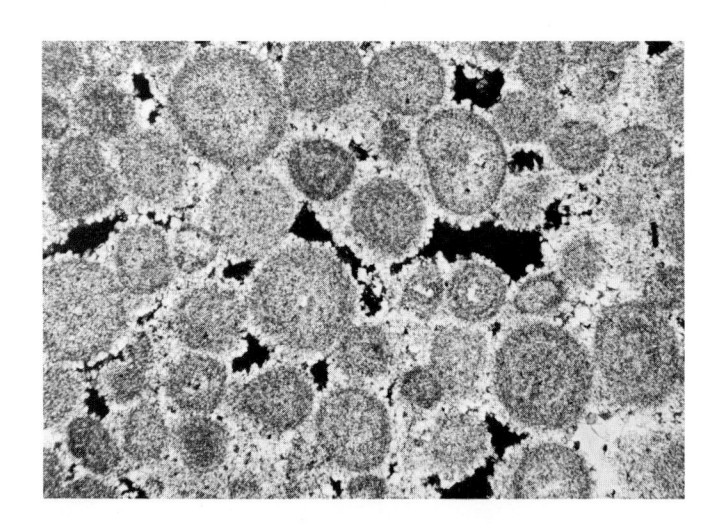

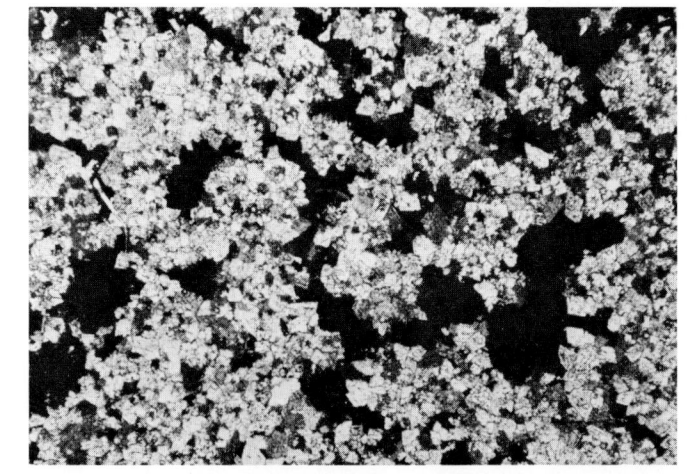

Fig. 11A — Oolite and pellet grainstone with some primary porosity preserved in Oolite Bar Facies. Thaynes Formation; Lake Fork, Utah.

Fig. 11B — Dolomitized oolite and pellet grainstone with intercrystalline and leached-pellet moldic porosity from the Oolite Bar Facies. Sinbad Limestone Member of the Moenkopi Formation; San Rafael Swell, Utah.

have trapped significant volumes of hydrocarbons along the hingeline. It is likely that much of the hydrocarbon impregnation of Lower Triassic exposures on the Colorado Plateau (San Rafael Swell, etc.) reached there through relatively continuous porosity zones linked to the "hydrocarbon cooking pot" to the west.

A great variety of structural configurations exist in which porous Lower Triassic facies should form structural traps in the Wyoming-Idaho overthrust belt, as well as other places in the vicinity of the hingeline.

With the increased exploration activity in the vicinity of the Cordilleran Hingeline, it is likely that the Lower Triassic strata will become a productive exploration target.

REFERENCES CITED

Bissell, H. J., 1970, Realms of Permian tectonism and sedimentation in western Utah and eastern Nevada: Am. Assoc. Petroleum Geologist Bull., v. 54, p. 285-312.

_____, 1969, Permian and Lower Triassic transition from shelf to basin, (Grand Canyon, Arizona to Spring Mountains, Nevada): Four Corners Geol. Soc. Guidebook, p. 135-169.

Blakley, R. C., 1974, Stratigraphic and depositional analysis of the Moenkopi Formation, southeastern Utah: Utah Geol. and Mineral Survey Bull. 104, 81 p.

Boutwell, J. M., 1907, Stratigraphy and structure of the Park City Mining District, Utah: Jour. Geology, v. 15, p. 434-458.

Clark, D. L., 1957, Marine Triassic stratigraphy in eastern Great Basin: Am. Assoc. Petroleum Geologists Bull., v. 41, p. 2192-2222.

Collinson, J. W., 1968, Permian and Triassic biostratigraphy of the Medicine Range, northeastern Nevada: Earth Science Bulletin, Wyoming Geol. Assoc., v. 1, no. 4, p. 25-44.

_____, Kendal, C. G. St. C. and Mercantel, J. B., 1976, Permian-Triassic boundary in eastern Nevada and west-central Nevada: Geol. Soc. America Bull., v. 87, p. 821-824.

Davidson, E. S., 1967, Geology of the Circle Cliffs area, Garfield and Kane Counties, Utah: U.S. Geol. Survey Bull. 1229, 140 p.

Dunham, R. J., 1962, Classification of carbonate rocks according to depositional texture: in Classification of carbonate rocks (Ham, W. E., editor), Am. Assoc. Petroleum Geologists Mem. 1, p. 108-121.

Gilluly, J., 1929, Geology and oil and gas prospects of part of the San Rafael Swell, Utah: U.S. Geol. Survey Bull. 806, p. 69-130.

_____, and Reeside, J. B., Jr., 1928, Sedimentary rocks of the San Rafael Swell and some adjacent areas in eastern Utah: U.S. Geol. Survey Prof. Paper 150, p. 61-110.

Gregory, H. E., 1917, Geology of the Navajo Country: U.S. Geol. Survey Prof. Paper 93, p. 23.

Hose, R. K., and Repenning, C. A., 1959, Stratigraphy of Pennsylvanian, Permian and Lower Triassic rocks of the Confusion Range, west-central Utah: Am. Assoc. Petroleum Geologists Bull., v. 43, p. 2167-2196.

Irwin, C. D., 1971, Stratigraphic analysis of Upper Permian and Lower Triassic strata in southern Utah: Am. Assoc. Petroleum Geologists Bull., v. 55, p. 1976-2007.

Koch, W. J., 1969, Lower Triassic lithofacies of the Cordilleran Miogeosyncline in the western United States: Ph.D. dissertation, Harvard University.

Kummel, B., 1954, Triassic stratigraphy of southeastern Idaho and adjacent areas: U.S. Geol. Survey Prof. Paper 254 H, p. 165-194.

_____, 1957, Paleoecology of Lower Triassic formations of southeastern Idaho and adjacent areas: Geol. Soc. America Mem. 67, p. 437-468.

Larson, A. R., 1966, Stratigraphy and paleontology of the Moenkopi Formation of southern Nevada: Univ. of California at Los Angeles, Ph.D. dissertation.

Mansfield, G. R., 1927, Geography, geology and mineral resources of part of southeastern Idaho: U.S. Geol. Survey Prof. Paper 152, 409 p.

McKee, E. D., 1954, Stratigraphy and history of the Moenkopi Formation of Triassic age: Geol. Soc. America Mem. 61, 133 p.

McKee, E. D., et al., 1959, Paleotectonic maps of the Triassic System: U.S. Geol. Survey Misc. Geol. Inv. Map I-300, 32 p.

Moritz, C. A., 1951, Triassic and Jurassic stratigraphy of southwestern Montana: Am. Assoc. Petroleum Geologists Bull., v. 35, p. 1781-1814.

Moulton, F. C., 1975, Lower Mesozoic and Upper Paleozoic petroleum potential of the Hingeline area, central Utah: in Deep drilling frontiers of the Rocky Mountains: Rocky Mtn. Assoc. Geologists Symposium, p. 87-97.

Newell, N. D. and Kummel, B., 1942, Lower Eo-Triassic stratigraphy, western Wyoming and southeast Idaho: Geol. Soc. America, v. 53, p. 937-996.

Picard, M. D., 1966, Petrography of the Red Peak Member, Chugwater Formation (Triassic), west-central Wyoming: Jour. Sedimentary Petrology, v. 36, p. 904-926.

_____, 1967, Stratigraphy and depositional environments of the Red Peak Member of the Chugwater Formation, west-central Wyoming, Wyoming Univ. Contrib. Geology, v. 6, p. 39-67.

_____, 1975, Shelf edge carbonate-redbed transitions. Red Peak and Thaynes formations (Triassic), western Wyoming and adjacent Idaho: in Deep drilling frontiers of the Rocky Mountains: Rocky Mtn. Assoc. Geologists Symposium, p. 99-107.

_____, and High, R. H., Jr., 1968, Shallow marine currents on the Early Triassic Wyoming Shelf: Jour. Sedimentary Petrology, v. 38, no. 2, p. 411-423.

_____, Aadland, A., and High, L. R., 1969, Correlation and stratigraphy of the Red Peak and Thaynes formations, western Wyoming and adjacent Idaho: Am. Assoc. Petroleum Geologists Bull., v. 53, p. 2274-2289.

_____, and High, R. H., Jr., 1972, Criteria for recognizing lacustrine rocks: in Recognition of ancient sedimentary environments: Soc. Econ. Paleontologists and Mineralogists Spec. Pub. 16, p. 108-145.

Poborski, S. J., 1954, Virgin Formation (Triassic) of the St. George, Utah area: Geol. Soc. America Bull., v. 65, p. 971-1006.

Reeside, J. B., Jr., and Bassler, H., 1922, Stratigraphic sections in southwestern Utah and northwestern Arizona: U.S. Geol. Survey Prof. Paper 129 D, p. 52-77.

Royse, F., Jr., Warner, M. A., and Reese, D. L., 1975, Thrust belt structural geometry and related stratigraphic problems, Wyoming-Idaho-northern Utah: in Deep drilling frontiers of the Rocky Mountains: Rocky Mtn. Assoc. Geologists Symposium, p. 41-54.

Scott, W. F., 1954, Regional physical stratigraphy of the Triassic in part of the eastern Cordillera: Univ. of Washington, Ph.D. dissertation.

Smith, H. P., 1969, The Thaynes Formation of the Moenkopi Group, north-central Utah: University of Utah, Ph.D. dissertation.

Stephenson, G. R., 1961, Stratigraphy of the Thaynes Formation in southeast Idaho: Washington State Univ., M.S. thesis, 113 p.

Tanner, W. F., 1974, History of Mesozoic lakes of northern New Mexico: New Mexico Geol. Soc. Field Conf. Guidebook, p. 219-223.

Thomas, H. D., and Kruger, M. L., 1946, Late Paleozoic and early Mesozoic stratigraphy of the Uinta Mountains, Utah: Am. Assoc. Petroleum Geologists Bull., v. 30, p. 1255-1293.

HILLIARD

The town of Hilliard was an interesting sight for the traveler on the Union Pacific in the 1880s. It had more evident industrial activity than any town along the railroad in Wyoming Territory, although it was never as large as nearby Evanston.

First, there was the elevated timber flume of the Hilliard Flume & Lumber Company which crossed over the railroad, and could be seen disappearing in the distance far to the south toward the Uinta Mountains. The flume was 24 miles in length, and had a drop of 2,000 feet between its head and Hilliard. At the head of the flume there was a company sawmill with a capacity of 40,000 board feet per day, powered by a steam engine of 40 horsepower. (It is scarcely necessary to point out that horsepower was calculated differently at Hilliard than later at Detroit.) All the lumber used in the construction of the flume was sawed by this mill, and the flume alone required more than two million board feet. The flume carried crossties, logs, lumber, and cordwood down to the railroad.

The second attraction at Hilliard was the charcoal kilns, twenty-nine of them, in which were burned 2,000 cords of wood per month to yield 100,000 bushels of charcoal. The charcoal was used commercially in smelting operations where coke is now employed, in the manufacture of gunpowder, and in shop and foundry applications. Give a competent blacksmith a good bellows, an anvil, a heavy hammer, some simple tools, and a few sacks of charcoal and he could work wonders.

Hilliard was founded in 1873. It was located on a section of the railroad where the track was winding and the grades were steep. The Union Pacific decided to relocate the line farther to the north. They laid track up Pioneer Hollow and tunneled through Aspen Ridge. The new section of track cost $12,000,000, and eliminated 10 miles of steep winding grade. It was completed in 1901. Piedmont, Old Aspen, and Hilliard were abandoned.

A. C. Veatch, however, reported in U.S. Geological Survey Professional Paper 56 in 1907 that several tenacious operators were still manufacturing charcoal at Hilliard, and hauling it in wagons to the railroad at Altamont.

— W. Lyle Dockery

LOWER MESOZOIC AND UPPER PALEOZOIC PETROLEUM POTENTIAL OF THE HINGELINE AREA, CENTRAL UTAH

by

Floyd C. Moulton[1]

ABSTRACT

A Late Pennsylvanian uplift in central Utah experienced extensive erosion which removed Pennsylvanian and most Mississippian rocks. The segment of this ancient arch where Pennsylvanian sediments are missing has been named the Emery uplift. The western part of the uplift was affected by a north-south trending Middle Jurassic depression herein named the Sanpete-Sevier rift. The eastern growth fault defining this depression is herein named the ancient Ephraim fault. More than 8,000 feet of Jurassic Carmel-Arapien marine shale and evaporites filled this rift. The Sanpete-Sevier rift has been partly defined by extensive seismograph surveys.

Recent wildcat drilling has encountered 2,038 feet of salt in the Jurassic Carmel-Arapien sequence and limits of these salt deposits can now be postulated.

The west part of the Emery uplift, still covered by the thick Jurassic evaporite sediments, has excellent oil and gas potential in the lower Mesozoic and upper Paleozoic rocks.

INTRODUCTION

The Hingeline Area of central Utah is located along the western margin of the Colorado Plateau Province. It has long been a puzzle to exploration geologists. Until Late Pennsylvanian time this area was the hingeline along which most lower Paleozoic formations thicken westward into the Cordilleran geosyncline. Late Pennsylvanian uplift and erosion exposed the Mississippian rocks, which later were buried by onlapping Permian sediments.

The suface structure is complex due in part to compression forming thrust faults and later relaxation causing some tensional faulting typical of the Basin and Range Province to the west. Additional complexity caused by diapirism involving Jurassic salt and shale, now evident on the surface, has made this area unique and different.

Some resolution of the puzzle is now possible with the accumulation of data from exploratory wells and from extensive seismograph surveys carried out in recent years. Systemic and formation isopach maps, together with panel cross sections, help to define the best places to anticipate commercial oil and gas fields in central Utah.

Figure 1 shows the area of this study. It is situated along the western part of the Colorado Plateau Province and adjacent to discontinuous, eastwardly-directed thrust segments which mark the eastern limit of the Basin and Range Province. These thrusts are mapped as the southern extension of the Thrust Belt Province of northern Utah, western Wyoming and southeastern Idaho. The Pavant Range and Mount Nebo thrusts are within the study area. The axis of the Sevier orogenic belt lies west of the area but is referred to later in the discussion of Jurassic sediments.

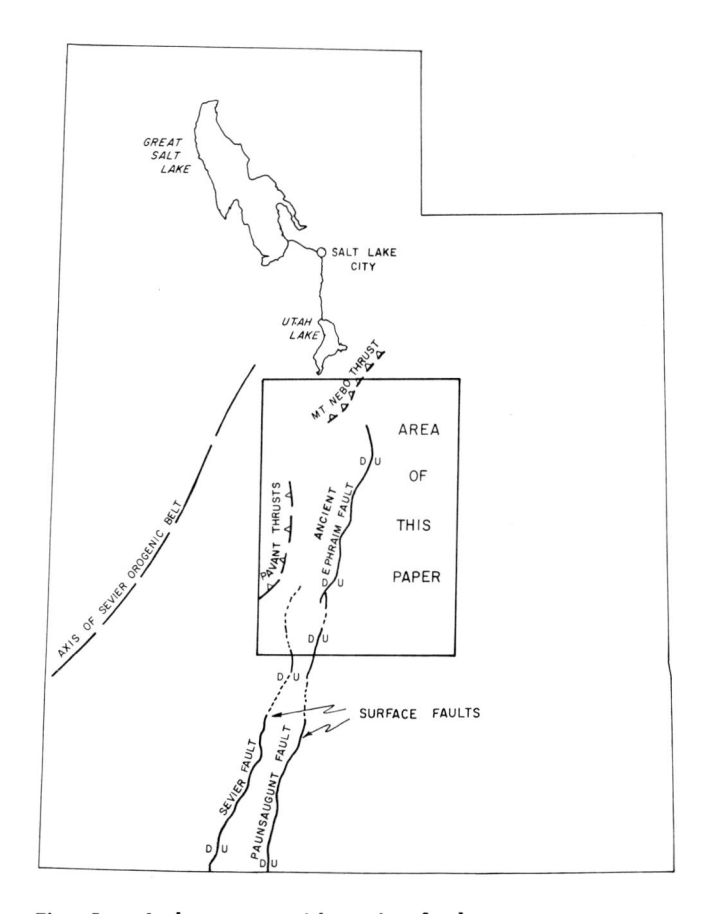

Fig. 1 — Index map with major faults.

[1]Geologist, Phillips Petroleum Company, Denver, Colorado.

Reprinted from: Symposium on deep drilling frontiers in the central Rocky Mountains; Rocky Mtn. Assoc. Geologists, 1975.

The ancient Ephraim fault (Fig. 1) is located on the eastern side of a deep depression which formed in Early Jurassic time. This is a buried fault with no direct surface expression. The Sevier and Paunsaugunt faults, which are surface faults south of the study area, may have some genetic relationship to the ancient Ephraim fault.

SEDIMENTARY AND TECTONIC HISTORY

The sedimentary and tectonic history of the Mississippian through Jurassic systems are summarized below. Figure 2 shows the stratigraphic nomenclature and the position of the isopach maps and panel cross sections in the order they are discussed.

Fig. 2 — Generalized correlation chart. Circles indicate positions of Figs. 3-18.

Mississippian

Figure 3 is an isopach map of the preserved Mississippian sediments (not including the Manning Canyon Formation). The area of thin Mississippian rocks represents the highest part of the northwest trending uplift. It was exposed during a Late Pennsylvanian uplift and erosion cycle (Fig. 2).

The Mississippian carbonates are generally porous because of solution during this Late Pennsylvanian erosion. These carbonates constitute a major potential oil and gas reservoir. Oil staining and some gas shows have been observed in many wells.

The Manning Canyon Formation includes shale, carbonate and thin sandstone. It is Late Mississippian to Early Pennsylvanian in age. The northeast part of Figure 4 shows what previously has been named the "Manning Canyon Embayment". This should not be called an embayment since the isopachs represent only the preserved thickness of the Manning Canyon. The Manning Canyon was deposited over a much larger area, probably with an east-west depositional trend, and was subsequently reduced to its present extent by Late Pennsylvanian erosion.

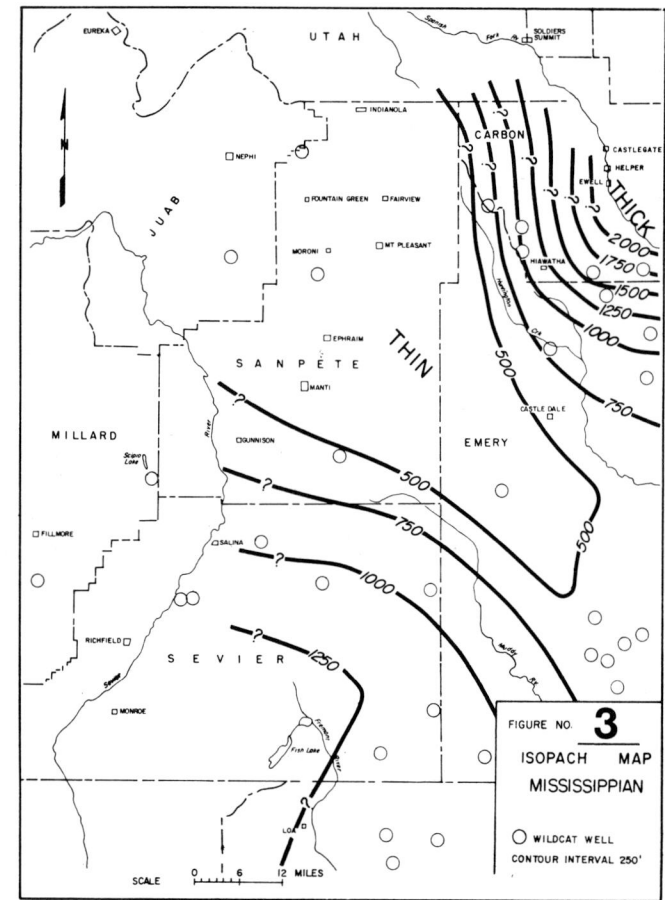

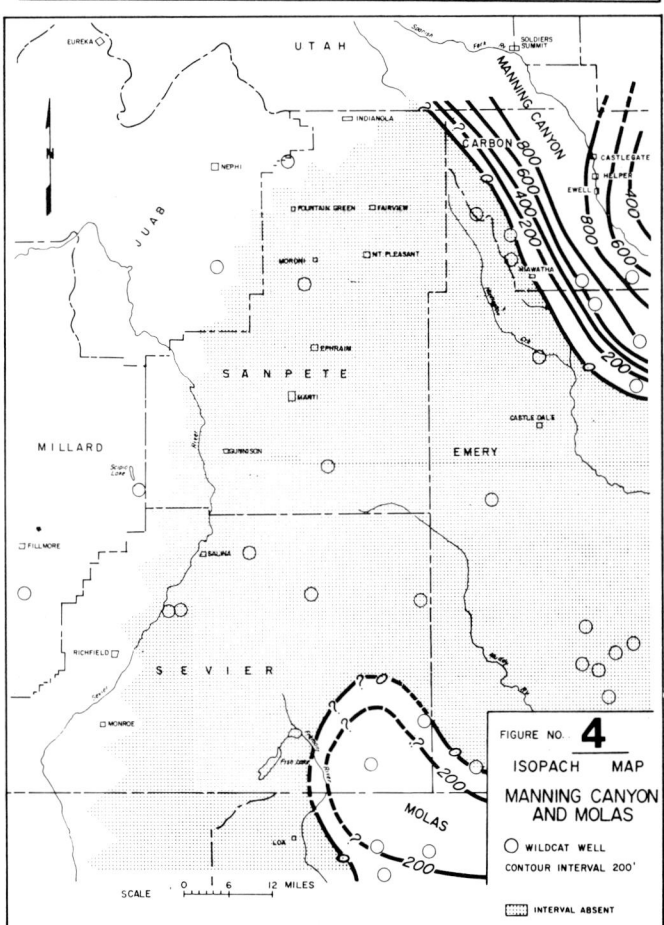

The southern part of the map shows the preserved Molas shale sequence. Originally the Molas may have covered a much larger area, but the formation was removed from part of the area by Late Pennsylvanian erosion. The shaded area on Figure 4 represents the exposed Mississippian carbonates.

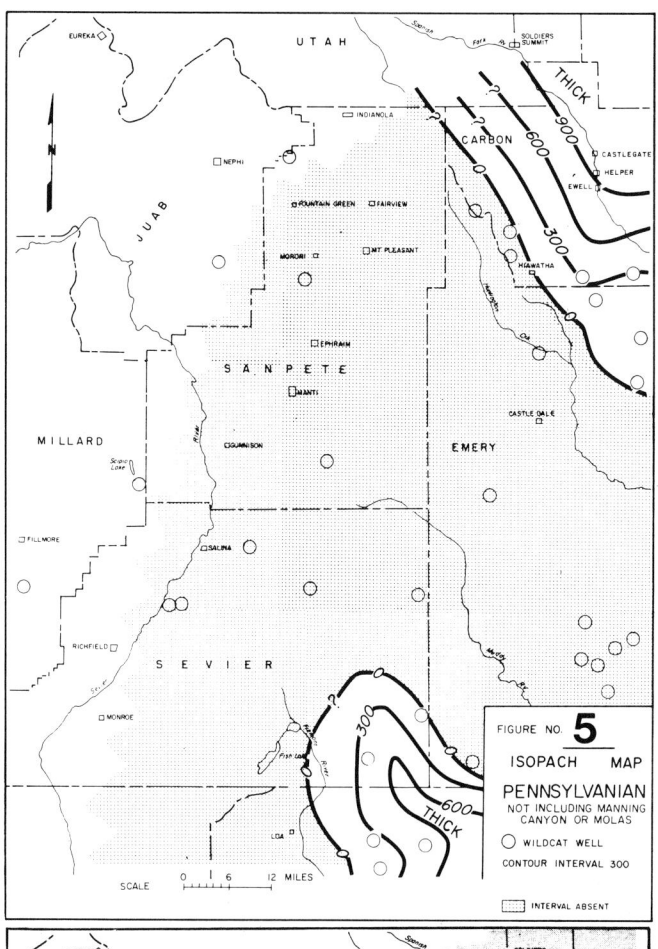

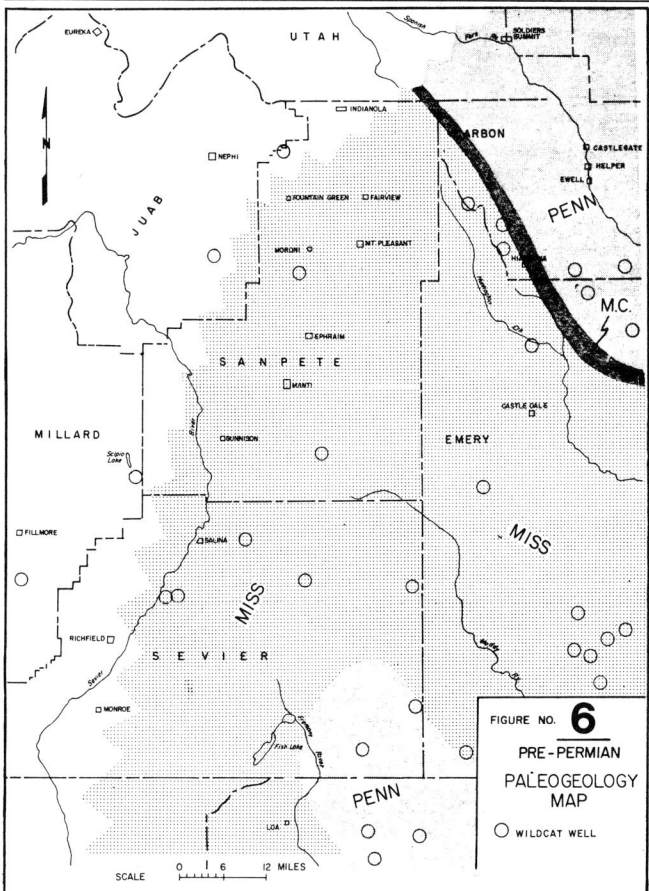

Oil staining and gas shows have been recorded in the Manning Canyon sandstone within the study area.

Pennsylvanian

The thickness of the Pennsylvanian carbonate and shale sequence is shown in Figure 5. This map shows the Pennsylvanian deposits that were preserved after the Late Pennsylvanian uplift and erosion of the Emery uplift. All the shaded area represents the exposed Mississippian carbonates.

Pennsylvanian rocks have not had significant oil and gas shows within the study area.

Permian

The Permian rests unconformably on older rocks (Fig. 6), except possibly in the extreme northeast part of the area where the Permian is believed to be conformable on Pennsylvanian. The Permian consists of carbonates, evaporites, sandstone and shale.

The Permian isopach thin (Fig. 7) defines the highest part of the ancient Emery uplift. The northwest trend is again obvious and may represent uplift during Permian sedimentation.

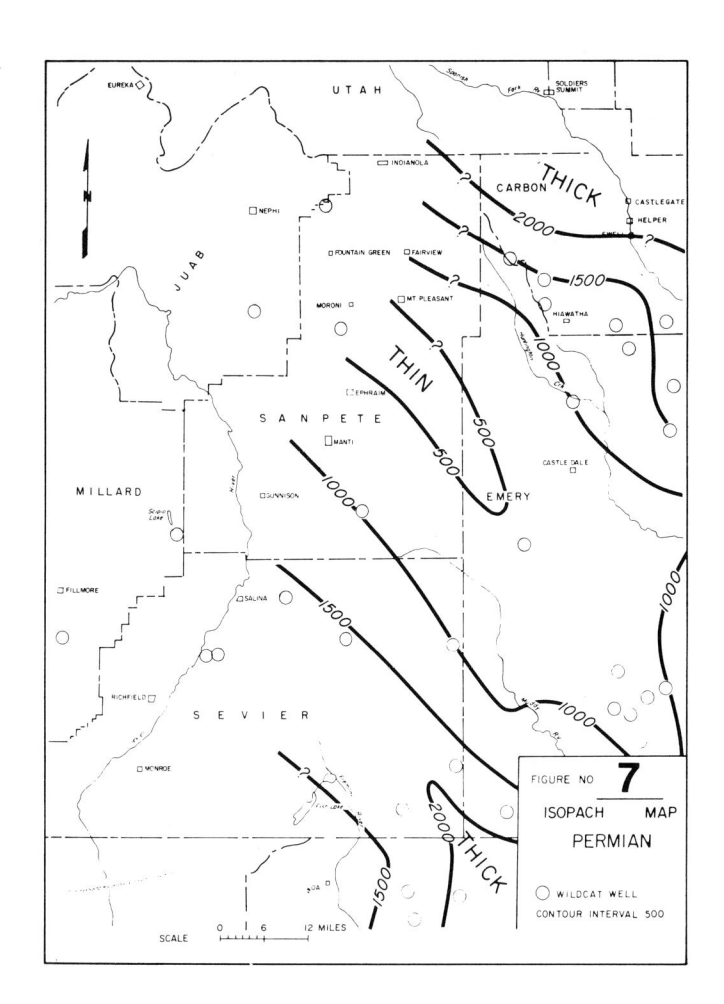

FIGURE NO. 5 — ISOPACH MAP PENNSYLVANIAN NOT INCLUDING MANNING CANYON OR MOLAS. WILDCAT WELL. CONTOUR INTERVAL 300. INTERVAL ABSENT.

FIGURE NO. 6 — PRE-PERMIAN PALEOGEOLOGY MAP. WILDCAT WELL.

FIGURE NO. 7 — ISOPACH MAP PERMIAN. WILDCAT WELL. CONTOUR INTERVAL 500.

Permian carbonates are porous and contain good live oil shows in many wells on and near the high part of the Emery uplift. The Permian reservoirs constitute a major objective throughout most of the area. In the southeast corner of the area, however, they are exposed at the surface. Upper Valley oil field, located in Garfield County south of the map area, has produced more than twelve million barrels of oil from the Kaibab limestone and dolomites.

Triassic

Early Triassic marine shale and thin carbonates were deposited over the Emery uplift with no significant reflection of the old high. The late Triassic rocks are dominantly continental and thicken to the west and southwest. As shown in Figure 8, the regional thickening to the west followed the early Paleozoic pattern.

The Navajo Sandstone, a major objective reservoir, is present across the buried Emery uplift. It thickens from 400 feet on the north to more than 1,100 feet (estimated) on the south (Fig. 9).

The early Triassic Sinbad carbonates have yielded live oil shows and combustible gas. South Last Chance (T.26S., R.7E.) is a shut-in gas field capable of producing from the Sinbad. The Navajo Sandstone is stained in some places and contains live oil shows in a few wells.

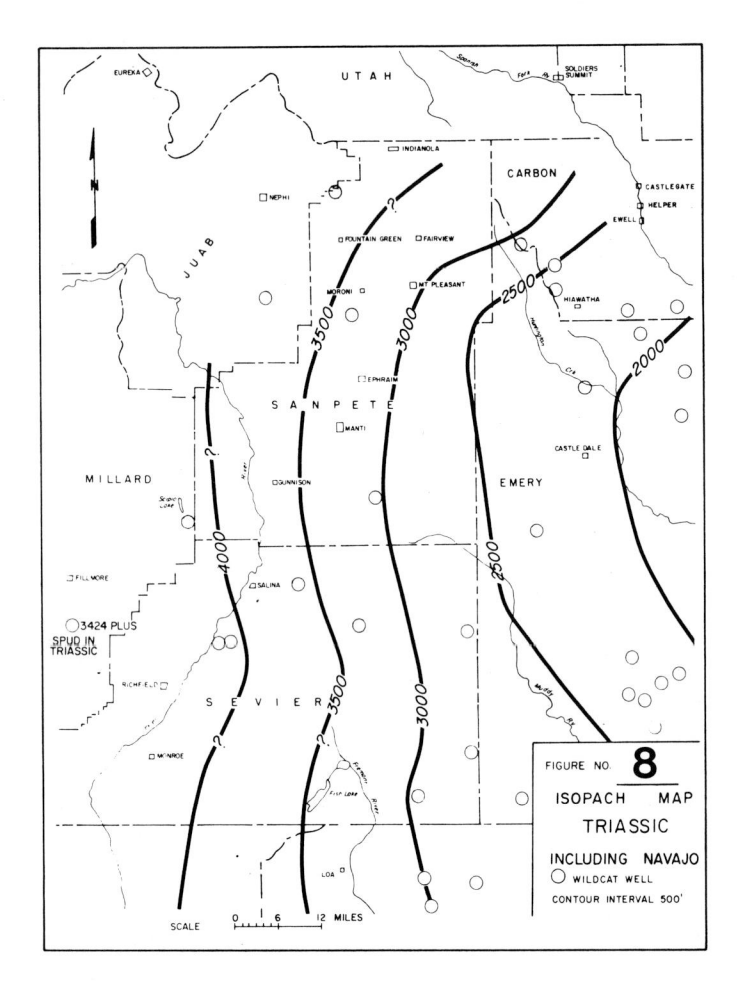

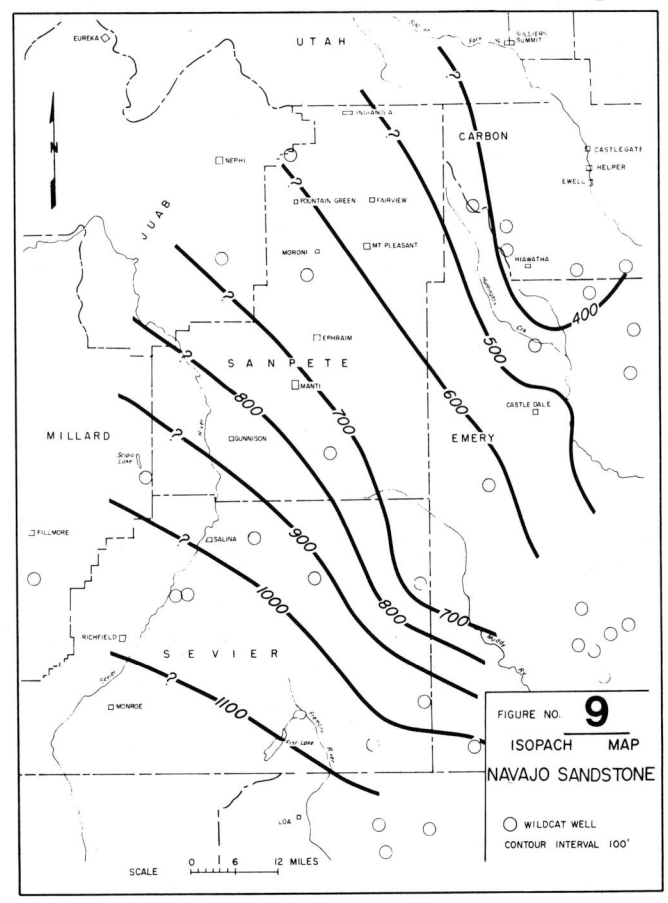

FIGURE NO. 9
ISOPACH MAP
NAVAJO SANDSTONE
○ WILDCAT WELL
CONTOUR INTERVAL 100'

Jurassic

The general configuration of the Emery high at the end of Navajo Sandstone deposition is shown on Figure 10. This panel cross section depicts the broad Emery uplift. At the center of the Emery high, a thin Permian sequence rests unconformably on thin Mississippian rocks. As will be shown later the western side of this uplift was severely faulted and modified during the Jurassic.

Figure 11 shows the preserved thickness of all Jurassic sediments where they can be mapped on the east side of the study area. In the Sanpete-Sevier rift (shaded), the Jurassic thickness values are posted on the map for reference.

The lowest Jurassic sediments of the Sanpete-Sevier rift area are thin carbonates which seem to correlate with thin carbonates on the shelf to the east. These carbonates may represent a shallow marine invasion from the northwest during an early stable period before the deep rift was formed. Subsequently, in about Middle Jurassic time, the Sevier uplift in western Utah was elevated as a broad arch (see Fig. 1). The position of the present Sanpete-Sevier rift was downwarped and filled with evaporites, carbonates, shale and thin sandstones called the Carmel-Arapien.

The Sanpete-Sevier rift area is bounded by growth faults with displacements up to 4,000 feet or more along its east side. The ancient Ephraim fault system, along the

FIGURE NO. 8
ISOPACH MAP
TRIASSIC
INCLUDING NAVAJO
○ WILDCAT WELL
CONTOUR INTERVAL 500'

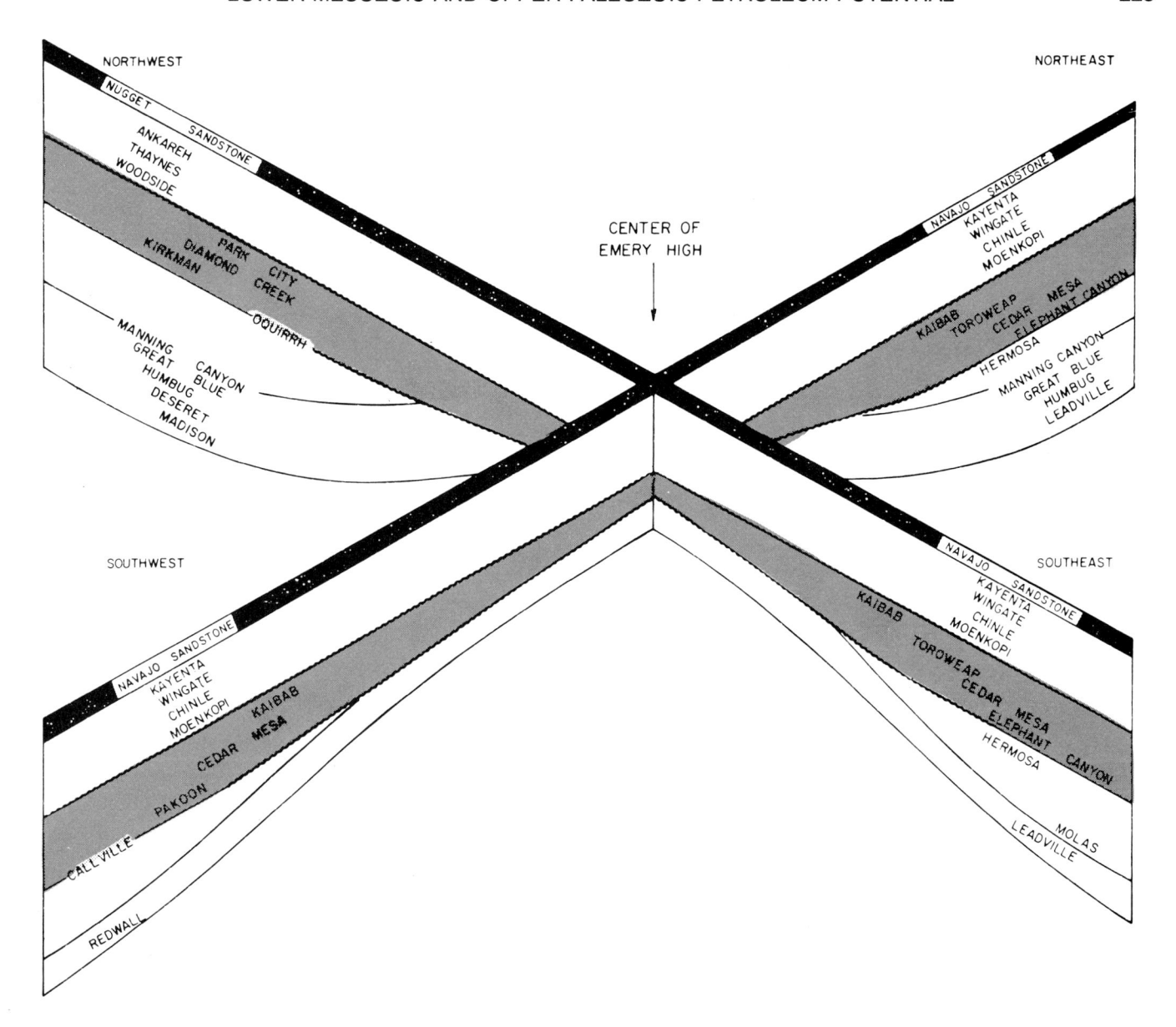

Fig. 10 — Generalized panel cross section showing the "Emery High" at end of Navajo deposition.

east side of the Sanpete-Sevier rift, may be genetically related to faults of similar trend both north and south of the map area. To the south the Paunsaugunt and Sevier surface faults (Fig. 1) may have a similar middle Jurassic history of movement. These two surface faults have moved in recent history, but the ancient Ephraim fault has been relatively inactive since Middle Jurassic time.

Within the Sanpete-Sevier rift deep structures both related and unrelated to the ancient faulting are now buried under Jurassic salt, other evaporites and shale which locally exceed 11,000 feet in thickness. Excessive thicknesses could be a result of flowage of salt, other evaporites and shale into "salt cells" or broad diapirs.

The Phillips Petroleum Company No. 1 Price "N" wildcat well (Sec. 29, T.15S., R.3E.), in Sanpete County south of the town of Moroni, was drilled to a total depth of 12,332 feet. Salt 2,038 feet thick was penetrated from 9,072 to 11,110 feet; 1,222 feet of Jurassic shale was drilled below the salt. A total of 5,832 feet of Jurassic rocks was drilled.

The Jurassic Carmel-Arapien isopach map (Fig. 12) is based on direct correlation of typical Carmel shale and thin carbonate beds in the eastern part of the map area with most of the Arapien as defined in central Utah by previous writers. These original investigations had only the surface outcrops to study and assigned much of the Arapien shales and evaporites to a younger age.

Within the thick Sanpete-Sevier rift belt, most of the San Rafael Group (Carmel, Entrada, Curtis and Summerville Formations) can be distinguished in recently drilled wells. This new information proves that most of the Arapien strata in the rift belt correlates with the Carmel. It is significant to future exploration to note here that in the

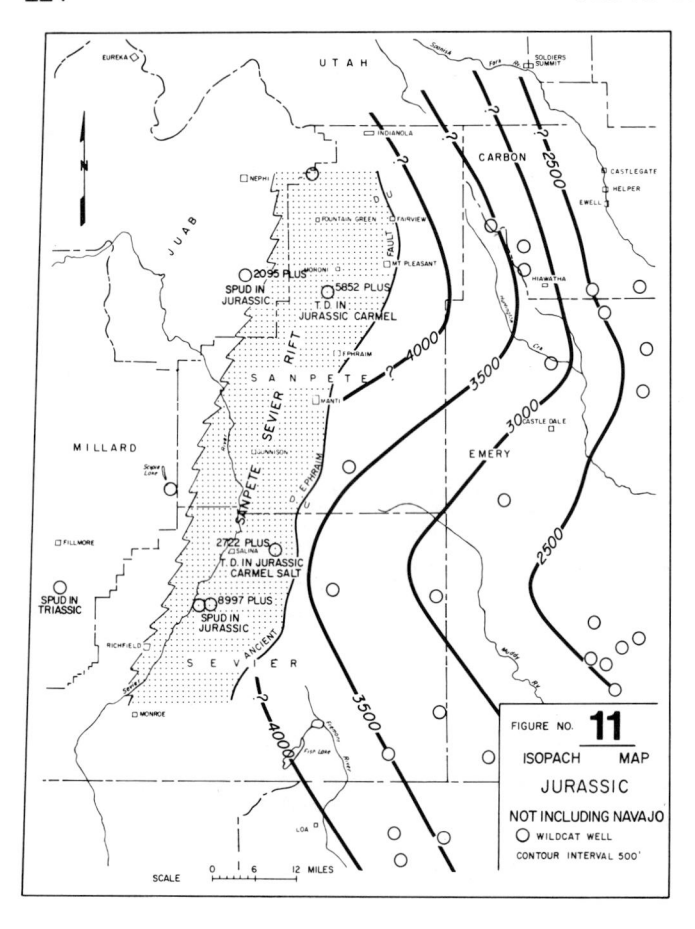

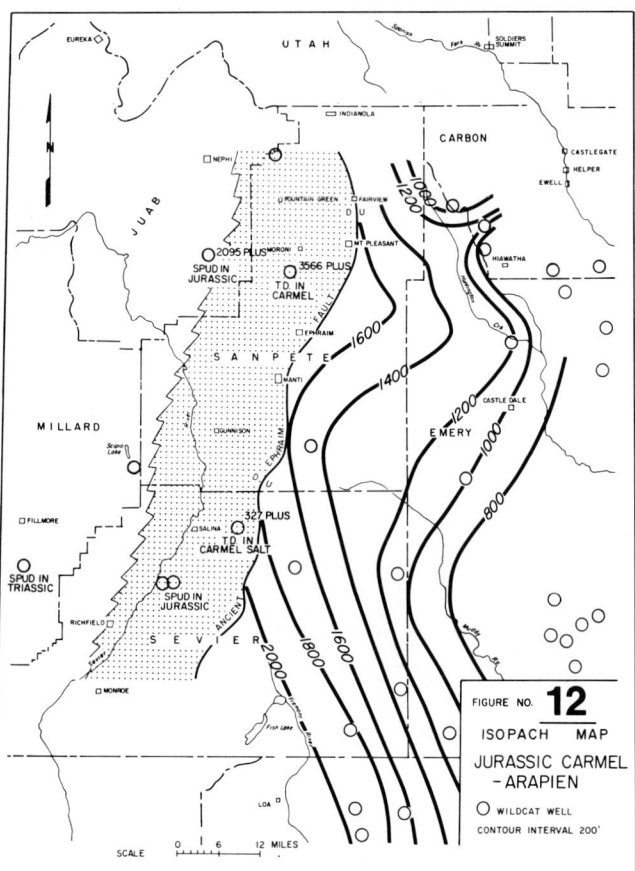

Phillips No. 1 Price "N" a salt bed, 16 feet thick, was encountered at 6,897 feet in the Summerville. Some surface outcrops containing salt are Summerville and not Carmel as mapped by some surface geologists.

Recent exploration work and deep drilling indicate that the Carmel-Arapien is more than 8,000 feet thick in local areas. These thick shale and salt sequences may be due in part to flowage into broad thick diapiric masses and smaller salt cells. Thrust faults have moved Paleozoic rocks eastward over the Mesozoic rocks more than ten miles along the west sector of the study area. Compressive forces affected the low density salt, evaporites and shale sequences of the Jurassic, locally forming excessive thicknesses in front of these thrusts.

The possible limits of the Carmel salts are shown in Figure 13. The eastern limits are generally defined by data from wells, but the western limits are more speculative. All wells that encountered Carmel salt are noted. The Phillips No. 1 Price "N" is labeled with a salt interval of 2,038 feet.

The rift belt with probable maximum salt tihckness is the shaded area on the map. Within the shaded area is a Jurassic surface outcrop belt which has been named the Sanpete-Sevier Valley anticline. This anticline may define the area of maximum thickness of the Jurassic and also the Carmel salt.

The Sanpete-Sevier Valley anticline is confined within the rift belt and is probably a result of vertical diapir movement of Jurassic sediments only. The Triassic Navajo Formation probably served as the floor. The Navajo Formation was folded and faulted to some extent below the Jurassic sediments but was not included in the vertical diapir mass.

Overturned beds on the surface have been interpreted as being formed from large scale overthrusts in some areas along the Sanpete-Sevier Valley anticline. However, the same evidence also can be explained as erosional remnants of recumbent or mushroom-shaped folds formed by diapiric movement.

The vertical diapir movement can also be explained by differential sediment loading along early compressional folds formed within the Jurassic sequence. Vertical diapir growth during Tertiary time is documented in some areas.

Algal reefs are associated with evaporite deposits in many basins. A logical place to expect these reefs in the Carmel-Arapien would be within and along the edges of the Sanpete-Sevier rift belt. Future drilling will help to define the depressed areas where the salt and evaporites were concentrated and also the optimum shelf localities for algal reef growth near normal marine water. Some dark to black shale sequences in the Carmel are believed to be excellent source beds of hydrocarbon because they have very high contents of kerogen.

The panel cross section (Fig. 14) shows the thick Carmel-Arapien trough formed during the major fault movements. This depression, or rift, developed as the Sevier

FIGURE NO. 13
ISOPACH MAP
CARMEL SALTS

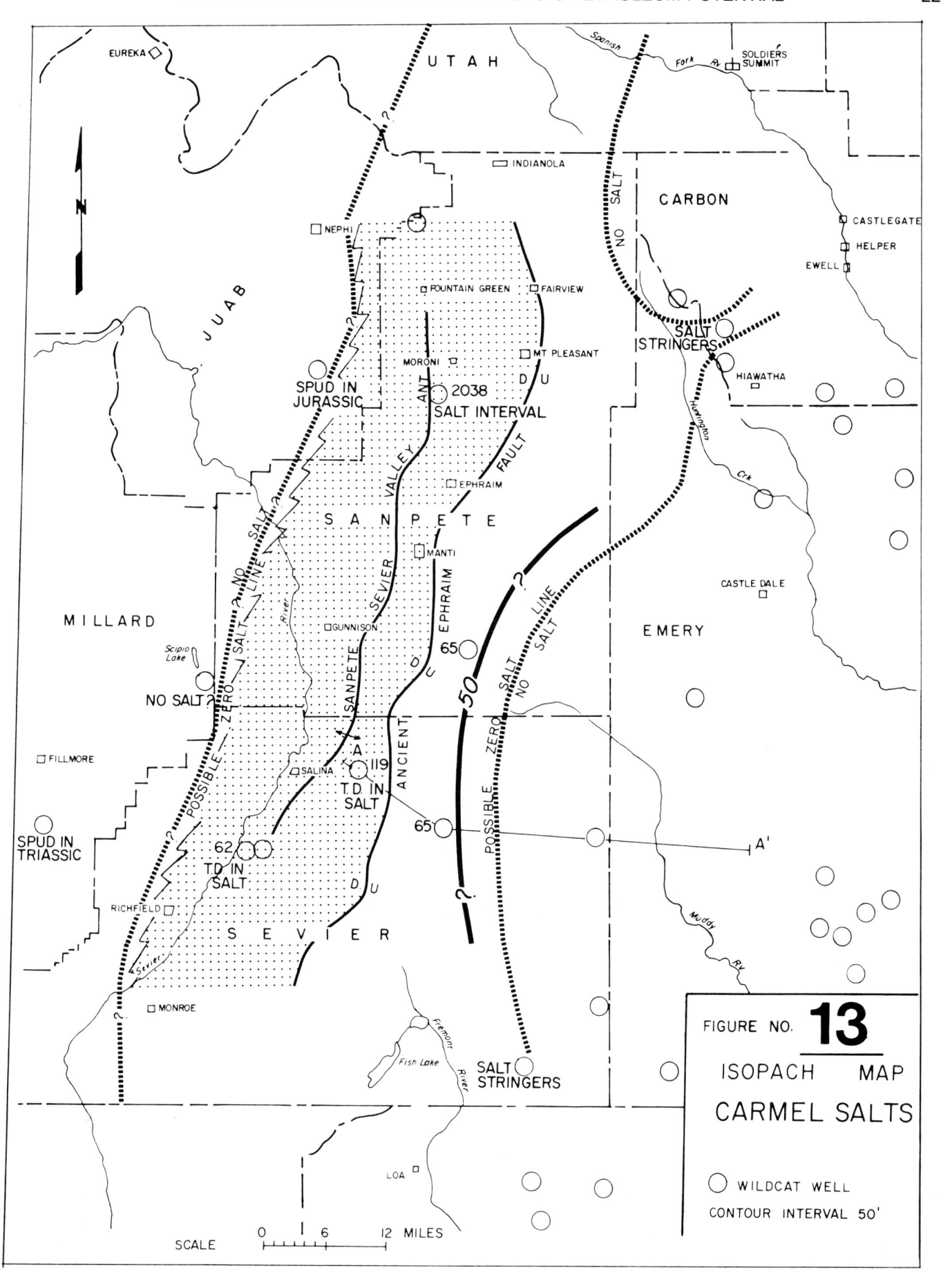

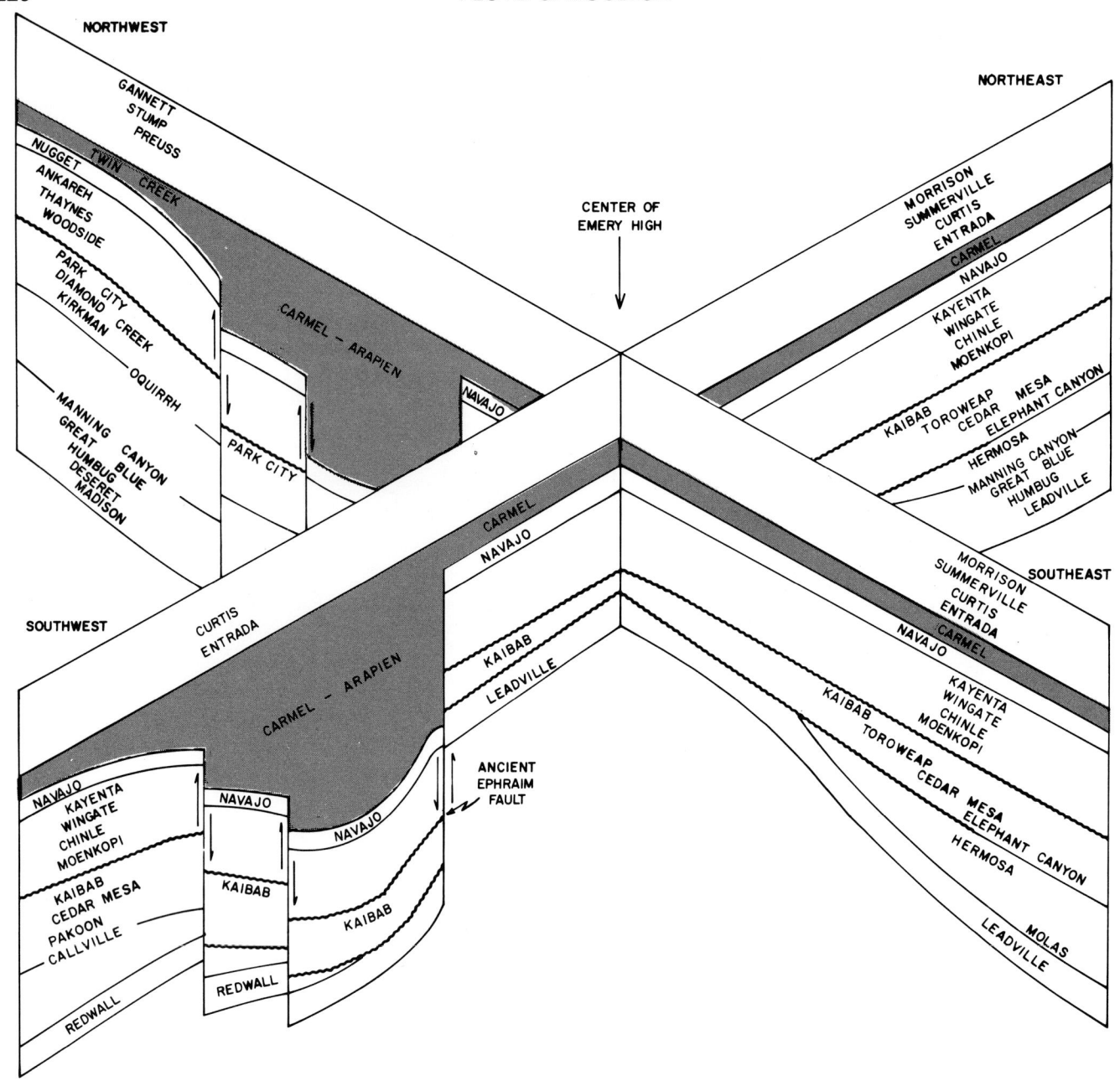

Fig. 14 — Generalized panel cross section showing the "Emery High" after Jurassic deposition.

orogenic belt was elevated in western Utah and southeastern Nevada. The tectonic instability which caused this elongate rift occurred mostly during the Jurassic Carmel-Arapien sedimentary cycle. The presence of salt and anhydrite within a dominantly marine shale sequence indicates that downwarping of the depressed area exceeded the supply of available clastics.

As shown in this panel cross section, the Entrada was not significantly affected by the previous depression.

Some faults, including the ancient Ephraim fault, *have not* had any significant movement since Middle Jurassic time. Structural traps, fault traps and facies traps could be preserved below and within the Carmel-Arapien sediments.

Figures 15, 16, 17 and 18 show the relationship of the Jurassic evaporite basin to the older formations on the Emery uplift. The northwest portion of the uplift is covered by thick evaporites, which have provided an impervious seal over pre-Jurassic sedimentary rocks.

The Carmel carbonates are oil stained and have had gas shows in some wells in the Sanpete-Sevier rift belt. Thick oolitic limestone sequences in the Carmel have scattered oil staining and could be important objectives in some areas.

STRUCTURE

The present structural relationship of the eastern part of the study area is shown by a schematic cross section (Fig.

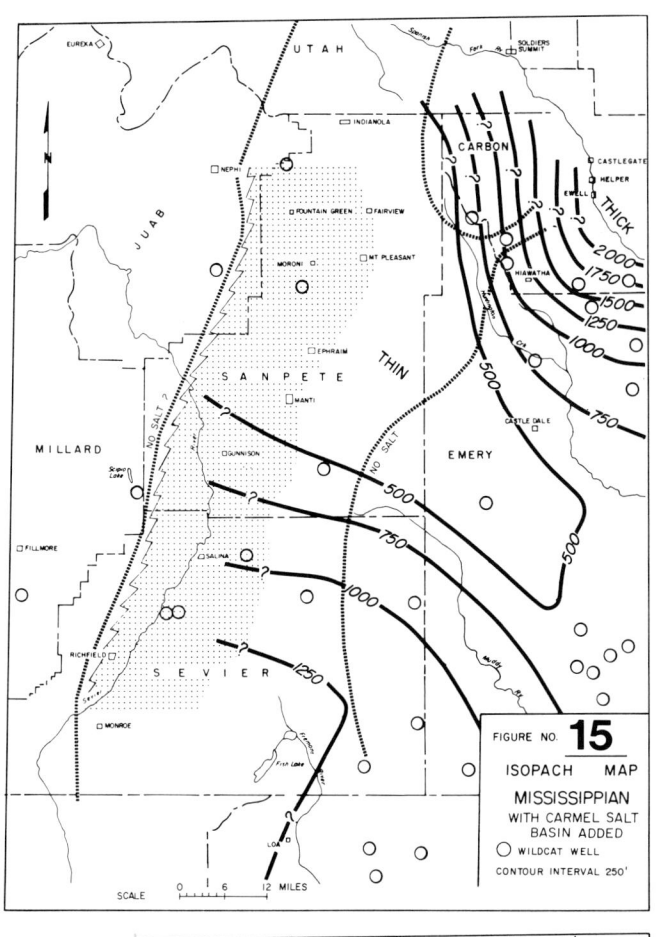

LOWER MESOZOIC AND UPPER PALEOZOIC PETROLEUM POTENTIAL

FIGURE NO. 15 — ISOPACH MAP, MISSISSIPPIAN WITH CARMEL SALT BASIN ADDED. Wildcat well. Contour interval 250'.

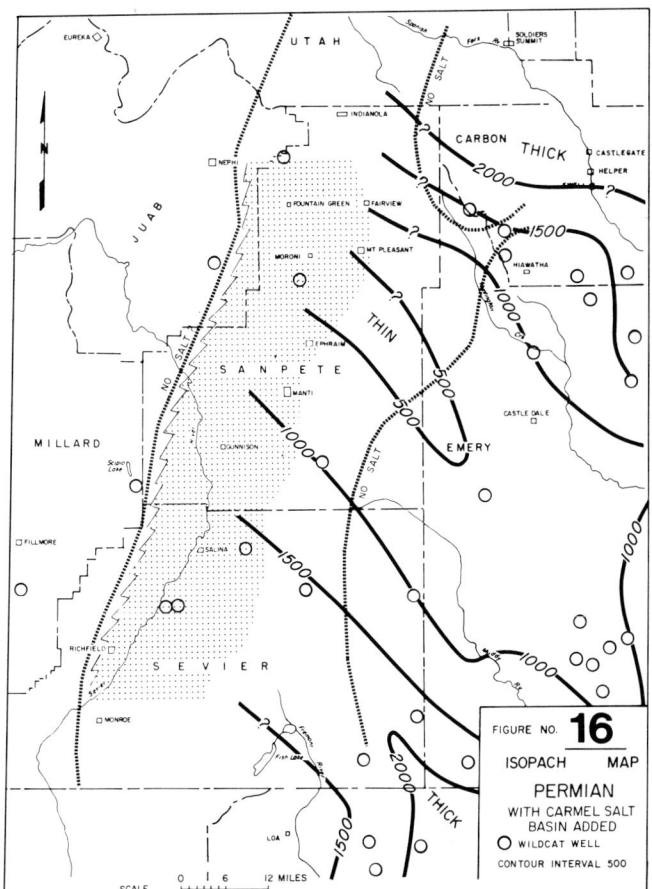

FIGURE NO. 16 — ISOPACH MAP, PERMIAN WITH CARMEL SALT BASIN ADDED. Wildcat well. Contour interval 500'.

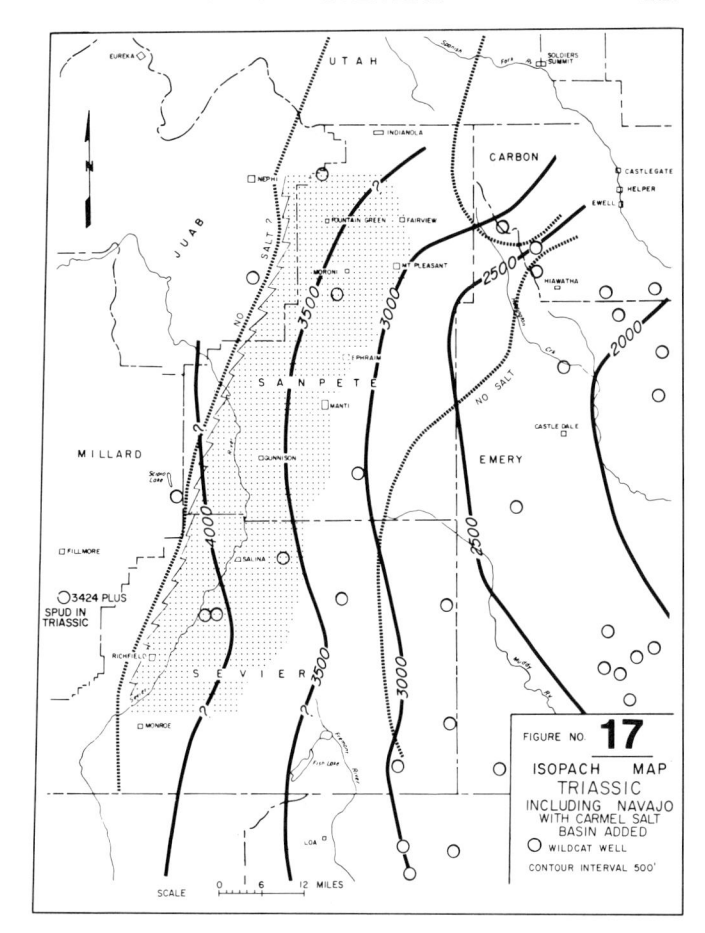

FIGURE NO. 17 — ISOPACH MAP, TRIASSIC INCLUDING NAVAJO WITH CARMEL SALT BASIN ADDED. Wildcat well. Contour interval 500'.

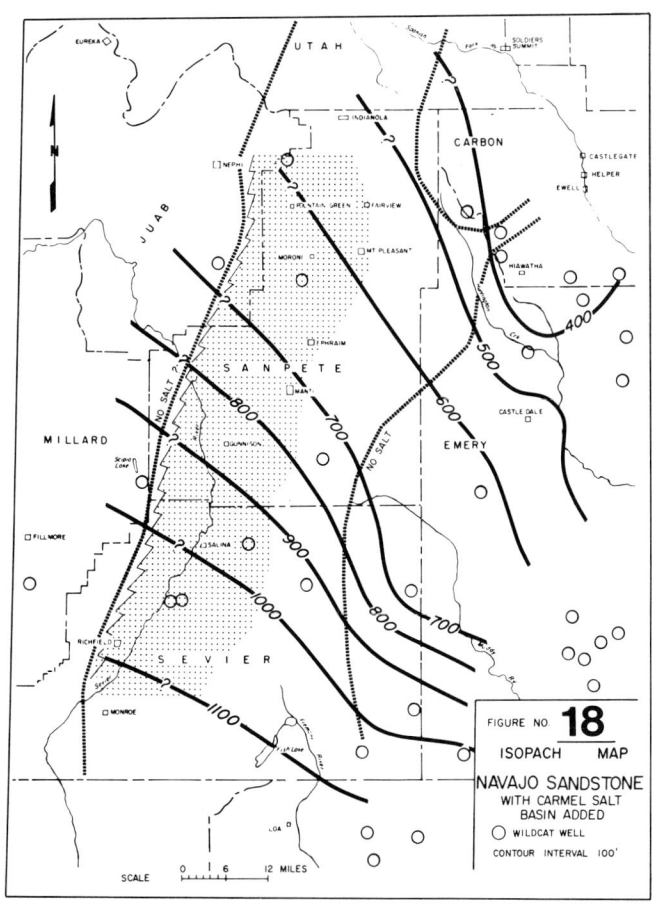

FIGURE NO. 18 — ISOPACH MAP, NAVAJO SANDSTONE WITH CARMEL SALT BASIN ADDED. Wildcat well. Contour interval 100'.

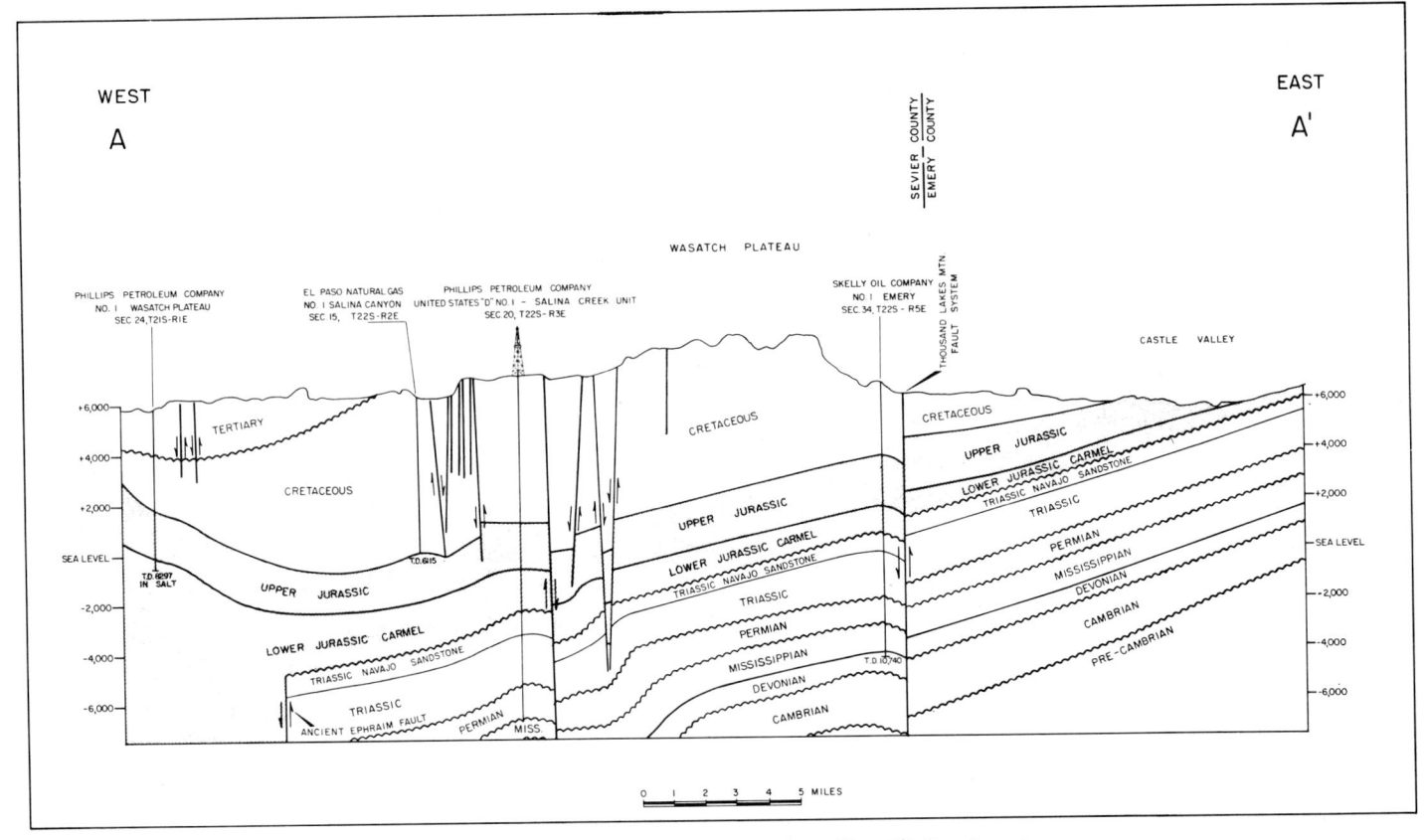

Fig. 19 — Schematic cross section (see Fig. 13 for location.)

19). The location of this cross section is shown on Figure 13. The location of the Phillips Petroleum Company No. 1 U.S. "D" is shown on a structural fault trap. On the west end of this cross section the ancient Ephraim fault can be seen with the Jurassic Carmel sediments down-faulted against the lower Mesozoic and upper Paleozoic rocks.

OIL AND GAS POSSIBILITIES

The central Utah area is covered by thick Jurassic strata and has excellent oil and gas possibilities in lower Mesozoic and Paleozoic formations. Hydrocarbon source beds and reservoir rocks are present in the Jurassic, Triassic, Permian, Pennsylvanian and Mississippian Systems. The older Paleozoic formations also could have oil and gas possibilities because they are in fault contact with all formations up to and including the Jurassic.

Good oil and gas shows and some production are found in all formations discussed in this paper. Shows of live oil and gas have been found in the Carmel carbonates, the Navajo Sandstone, and the Sinbad Timpoweap. Producible gas occurs in the Sinbad at South Last Chance field (shut-in). The Sinbad had an excellent oil and gas show in the Phillips Petroleum Company No. 1 U.S. "E" (Sec. 27 T. 19S., R.3E.), Sanpete County. As previously discussed the Permian Kaibab carbonates have produced more than twelve million barrels of oil at the Upper Valley oil field south of the study area. Other Permian formations which have had oil stains in exploratory wells include the Cedar Mesa, White Rim, Toroweap and Elephant Canyon. The Lower Pennsylvanian and Upper Mississippian Manning Canyon is oil stained in some wells, and subcommercial gas production was found in one well. Some exploratory wells have found live oil shows in the Madison limestones.

Bissell (1970) has presented a detailed discussion of oil and gas shows and production south of the study area.

CONCLUSIONS

Recent seismic surveys and exploratory drilling have provided better definition of the Paleozoic Emery uplift. Some ancient structures with repeated growth are associated with this old high. The upper Paleozoic and lower Mesozoic sequences include many good reservoirs. Source beds of hydrocarbons also have been identified; portions of the Jurassic are especially important. Thick evaporite and shale deposits of Early Jurassic age provide impervious seals which should have prevented the escape of hydrocarbons from deeper structures and stratigraphic traps and from algal mounds or reefoid carbonates within the Jurassic sequence.

The geologic history of the area covered by thick Jurassic evaporites indicates that a large segment of central Utah should be highly prospective for oil and gas. Numerous shows, including some production, appear to confirm this conclusion. The lower Paleozoic formations, which have not been considered in this report, also have potential in this deep drilling frontier. It is the writer's opinion that very large oil and gas reserves will be found in the central Utah Hingeline Area.

SELECTED BIBLIOGRAPHY

Armstrong, Richard Lee, 1968, Sevier Orogenic Belt in Nevada and Utah: Geol. Soc. America Bull., v. 79, p. 429-59.
_____ 1968, The Cordilleran Miogeosyncline in Nevada and Utah: Utah Geological and Mineralogical Survey, Bull. 78.
Bissell, H. J., North Strawberry Valley sedimentation and tectonics: Intermountain Assoc. Petrol. Geologists, 10th Ann. Field Conf. Guidebook, p. 159-165.
_____, 1963, Pennsylvanian and Permian Systems of southwestern Utah: Intermountain Assoc. Petrol. Geologists 12th Ann. Field Conf. Guidebook, p. 42-58.
_____ 1970, Realms of Permian tectonism and sedimentation in western Utah and eastern Nevada: A.A.P.G. Bull., v. 54, no. 2, p. 285-312.
Black, B. A., 1965, Nebo Overthrust, southwestern Wasatch Mountains, Utah: Brigham Young Univ. Geology Studies, v. 12, p. 58-89.
Burchfiel, B. C. and Hickcox, C. W., 1972, Structural development of central Utah: Utah Geological Association Pub. 2., p. 55-66.
Gilliland, W. N., 1963, Sanpete-Sevier Anticline of central Utah: Geol. Soc. America Bull., v. 74.
Gunderson, Wayne Campbell, 1961, An isopach and lithofacies study of the Price River, North Horn, and Flagstaff Formations of central Utah: Thesis for Master of Science, Department of Geol., Univ. of Nebraska, Lincoln, Nebraska.

Hardy, Clyde T., 1952, Eastern Sevier Valley, Sevier and Sanpete Counties, Utah, with reference to formations of Jurassic age: Utah Geol. Mineral Survey Bull. 43.
_____ 1953, Evidence for variable vertical movement in central Utah: Geol. Soc. America Bull., v. 64, p. 245-247.
_____ and Zeller, Howard D., 1953, Geology of the west-central part of the Gunnison Plateau, Utah: Geol. Soc. America Bull., v. 64, p. 1261-1278.
Harris, H. D., 1959, A late Mesozoic positive area in western Utah: A.A.P.G. Bull., v. 43, no. 11, p. 2636-2652.
Hintze, Lehi F., 1973, Geologic history of Utah: Brigham Young Univ. Geology Studies, v. 20, part 3.
Hite, Robert J. and Cater, Fred W., 1972, Pennsylvanian rocks and salt anticlines, Paradox Basin, Utah and Colorado, in Geologic Atlas of the Rocky Mountain Region: Rocky Mtn. Assoc. Geologists, p. 133-138.
Mallory, William W., 1972, Regional synthesis of the Pennsylvanian System, in Geologic Atlas of the Rocky Mountain Region: Rocky Mtn. Assoc. Geologists, p. 111, 127.
Peterson, J. A., 1972, Jurassic System, in Geologic Atlas of the Rocky Mountain Region: Rocky Mountain Assoc. Geologists, p. 177-189.
Sales, John K., 1968, Crustal mechanics of Cordilleran foreland deformation: a regional and scale-model approach: A.A.P.G. Bull., v. 52, no. 10, p. 2016-2044.
Spieker, Edmund M., 1946, Late Mesozoic and Early Cenozoic history of central Utah: U.S. Geol. Survey Prof. Paper 205-D.
_____ 1949a, Sedimentary facies and associated diastrophism in the Upper Cretaceous of central and eastern Utah: Geol. Soc. America Mem. 39, p. 55-82.
_____ 1954, Structural history: Intermountain Assoc. Petrol. Geol., 5th Ann. Field Conf.. p. 9-14.
Stokes, W. L., 1963b, Triassic and Jurassic formations of southwestern Utah: Intermountain Assoc. Petrol. Geol., 12th Ann. Field Con., Guidebook, p. 60-64.

photo by Peter J. Varney

View of Twin Peaks on southside of Little Cottonwood Canyon above Snowbird. Peaks are capped with Upper Precambrian Big Cottonwood Formation which is underlain in this location by Tertiary quartz monzonite like that used in the construction of the "Mormon" Temple in Salt Lake City.

PETROLOGY OF ENTRADA SANDSTONE (JURASSIC), NORTHEASTERN UTAH

by

E. P. Otto[1] and M. Dane Picard[2]

ABSTRACT

In northeastern Utah, the Entrada Sandstone contains two lithologic facies: a lower pale yellowish-orange, fine- to medium-grained sandstone (sandstone facies), and an upper moderate reddish-orange, very fine-grained silty sandstone (silty sandstone facies). The silty sandstone facies is present only in the western part of the area where it generally interfingers with the sandstone facies.

Based mainly on lithology and sedimentary structures, informal lower and upper units are differentiated in the sandstone facies. Locally within the sandstone facies, it is possible to recognize subaerial and subaqueous deposits that are distinguished on the basis of sedimentary structures.

The lower unit of the sandstone facies accumulated in an arid or semi-arid, inland dune environment in the eastern part of the area, and as coastal dunes and shallow-water marine deposits in the western part of the area. The upper unit probably represents a reworking of the Entrada during a minor transgression of the Preuss sea. The silty sandstone facies was deposited in a shallow-water marine environment.

Based on 44 modal analyses, the sandstone of both facies is dominantly subarkose. Orthoclase is the most abundant feldspar; metamorphic rock fragments and chert are the most abundant rock fragments. Hematite, magnetite, and mica (mostly biotite) are the most common accessory minerals. Post-depositional alteration of biotite and magnetite contributed the red pigment, which stains the silty sandstone facies, and the yellowish color that stains the sandstone facies. Clay minerals present include montmorillonite, kaolinite, and illite.

Eolian beds in the sandstone facies exhibit an average porosity of 15.9 percent and offer the best potential as reservoir rocks. Where eolian beds are close to favorable source rocks, low porosities in the overlying silty sandstone facies (1.9 percent) and the subaqueous deposits (9 percent) may provide stratigraphic traps for the accumulation of oil and gas.

INTRODUCTION

In northeastern Utah the Entrada Sandstone of Jurassic age crops out on the north and south flanks of the eastern end of the Uinta Mountains (Fig. 1). In this area, the Entrada is a well-exposed sequence of clastic sedimentary rocks that reaches a maximum thickness of 234 ft (71.3 m). Although the Entrada Sandstone does not attain as great a thickness as other eolian sandstone formations in the Colorado Plateau region, it has the widest areal extent (Poole, 1963, p. 395; Tanner, 1965, p. 564). The Entrada is also one of the youngest eolian sandstones in this region, representing the termination of a long period of wind deposition that began in the Permian and ended in the Jurassic.

The Entrada Sandstone of northeastern Utah was deposited east of the Cordilleran Hingeline. Therefore the formation does not show the rapid thickness and facies changes that characterize many of the Paleozoic and Mesozoic formations deposited across the hingeline. The westward equivalent of the Entrada Sandstone, the Preuss Formation, does exhibit this rapid thickening.

---

[1] Sunmark Exploration Company, 12850 Hillcrest Road, Dallas, Texas. This investigation was done while at the University of Utah.

[2] Department of Geology and Geophysics, University of Utah, Salt Lake City, Utah.

We thank T. A. Larson and R. O. Johnson for critically reading the manuscript and offering suggestions for its improvement. S. E. Otto drafted the illustrations. Richard S. Tousley, superintendent, Dinosaur National Monument, permitted the collection of samples from within the monument.

Acknowledgement is made to the donors of the Geological Research Fund, administered by the Department of Geology and Geophysics, University of Utah, for partial support of this research. NASA Grant NGR 45-003-095 also helped support this study.

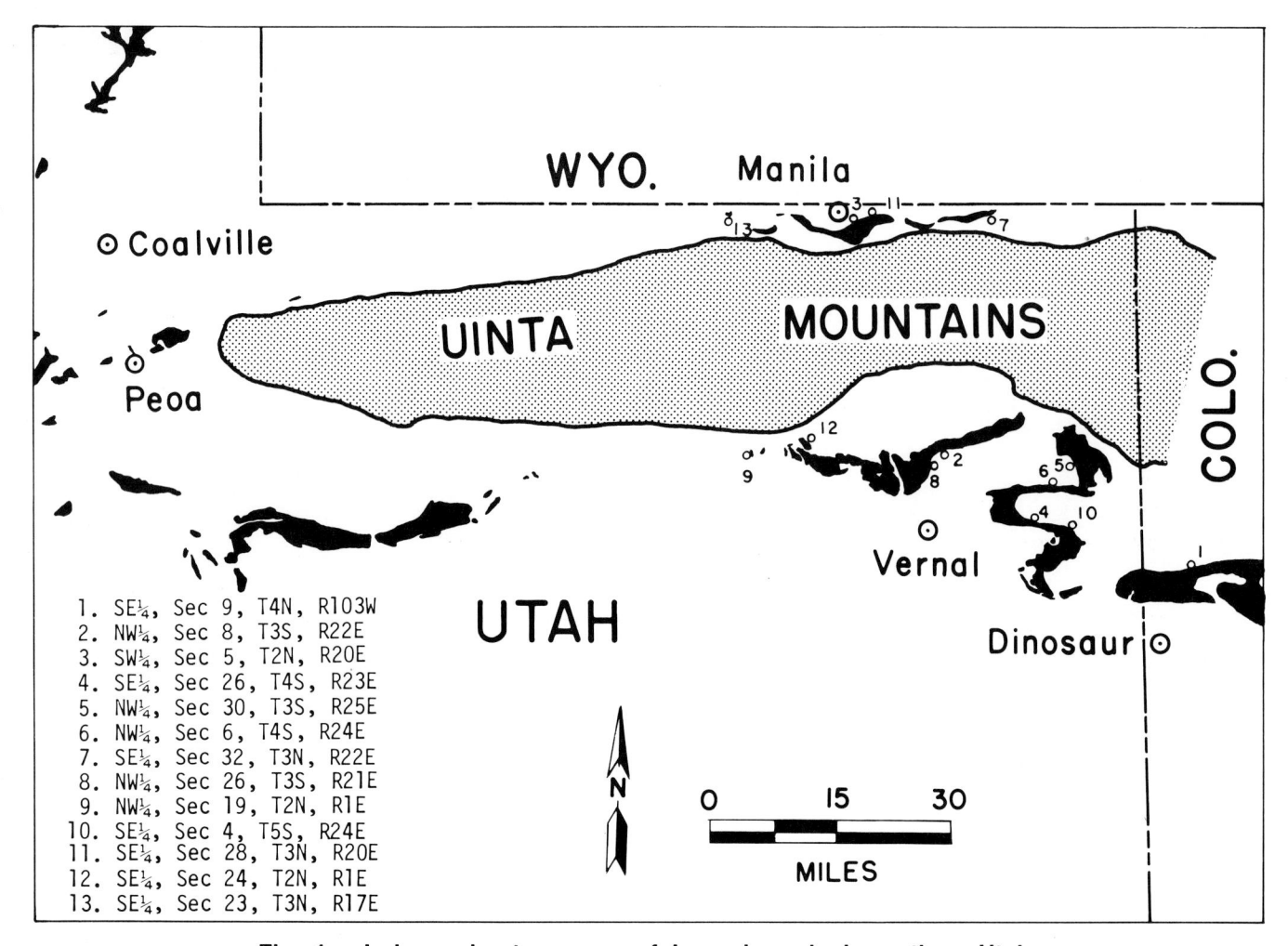

Fig. 1 — Index and outcrop map of Jurassic rocks in northern Utah.

PREVIOUS STUDIES

Gilluly and Reeside (1928, p. 76) named the Entrada Sandstone from exposures at Entrada Point in the northern part of the San Rafael Swell, Utah. They included these "earthy" sandstones and subordinate shales in the San Rafael Group and assigned them a Late Jurassic age.

The Entrada Sandstone was first recognized in northeastern Utah by Baker et al. (1936, p. 14, 15, 27, 41) who differentiated it from the Nugget Sandstone with which it earlier had been grouped. Through later work by Thomas and Krueger (1946, p. 1276-1278), Untermann and Untermann (1954, p. 50-51), Kinney (1955, p. 81-85), Stokes et al. (1955), Hansen (1965, p. 79-81), and Otto and Picard (1975), knowledge of the Entrada Sandstone in northeastern Utah has been expanded.

Published information on the petrography of the Entrada Sandstone is sparse. Dane (1935, p. 93-95) described 5 thin sections from the Entrada in Grand County, Utah and San Miguel County, Colorado. Shawe (1968, p. 9-11) examined 7 thin sections from the Slick Rock Member of the Entrada Sandstone in San Miguel and Dolores Counties, Colorado. The only other petrographic analysis published is the brief petrographic descriptions of the Entrada Sandstone in northeastern Utah by Otto and Picard (1975, p. 133-134).

GEOLOGIC SETTING AND GENERAL STRATIGRAPHY

In northeastern Utah, the Entrada Sandstone contains two lithologic facies (Otto and Picard, 1975, p. 131): a lower pale yellowish-orange, fine- to medium-grained sandstone (sandstone facies), and an upper moderate reddish-orange, very fine-grained silty sandstone (silty sandstone facies, Fig. 2). The silty sandstone facies is present only in the western part of the study area where it generally interfingers with the sandstone facies. Based mainly on lithology and sedimentary structures, informal lower and upper units are differentiated in the sandstone facies. Locally, within the sandstone facies, it is possible to recognize subaerial and subaqueous deposits that are distinguished on the basis of sedimentary structures (Otto and Picard, 1975). Figure 3 is a generalized stratigraphic column of the Entrada Sandstone showing these divisions.

Fig. 2 — Outcrop character of the two lithologic facies of the Entrada, with the Curtis (Jcu) above and the Carmel (Jca) below. Thickness of upper slope-forming silty sandstone facies (2) is 75 ft (23 m); sandstone facies (1) is 56 ft (17 m) thick. Silty sandstone facies occurring below the sandstone facies is 39 ft (12 m) thick. Contact between Entrada and Carmel is not visible from this angle. Location 3.

The lower unit of the sandstone facies accumulated in an arid or semi-arid, inland-dune environment in the eastern part of the study area, and as coastal dunes and shallow-water marine deposits in the western part of the area. Within the inland-dune fields, in flat, low-lying areas, deposition in fresh water playas occurred. The upper unit probably represents a reworking of the Entrada during a minor transgression of the Preuss sea. The silty sandstone facies was deposited in a shallow-water marine environment. The sandstone and silty sandstone facies, in the western part of the area, were situated near and possibly on a shelf edge that bordered the Preuss seaway to the west. A more detailed discussion of the stratigraphy, sedimentary structures, and depositional environments of the Entrada Sandstone in northeastern Utah is given in Otto and Picard (1975).

Jurassic formations in northern Utah display rapid facies and thickness changes associated with the Cordilleran Hingeline. Figure 4 shows correlations of the Entrada Sandstone and other Jurassic formations in northeastern Utah with their equivalents in north-central Utah. At location 8, just north of Vernal (Fig. 1), the Entrada is only 100 ft (30.5 m) thick, pale yellowish-orange in color, and dominantly very fine- to medium-grained sandstone. One hundred miles (161 km) to the west, just north of Peoa, the Entrada thickens to 1,196 ft (364.5 m; Thomas and Krueger, 1946, p. 1282) and changes to red silty sandstone and siltstone called the Preuss Formation. The silty sandstone facies of the Entrada in northeastern Utah

LEGEND

LITHOLOGY
- SILTSTONE
- VERY FINE GRAINED SILTY SANDSTONE
- VERY FINE TO FINE-GRAINED SANDSTONE
- MEDIUM-GRAINED SANDSTONE

MODIFIERS
- GLAUCONITIC
- OOLITIC
- LIMY
- PEBBLY

GODDARD COLORS
- PALE YELLOWISH ORANGE
- MODERATE REDDISH ORANGE
- MODERATE REDDISH BROWN
- WHITE, BLUISH WHITE, VERY LIGHT GRAY
- LIGHT GRAY, MEDIUM LIGHT GRAY, MEDIUM DARK GRAY

STRATIFICATION
- SMTP SMALL-TO MEDIUM-SCALE TABULAR-PLANAR
- LTP   LARGE-SCALE TABULAR-PLANAR
- SMWP SMALL-TO MEDIUM-SCALE WEDGE-PLANAR
- LT   LARGE-SCALE TROUGH
- H   HORIZONTAL
- S   STRUCTURELESS
- R   RIPPLE MARKS
- M   MULTIPLE PARALLEL-TRUNCATION BEDDING PLANES

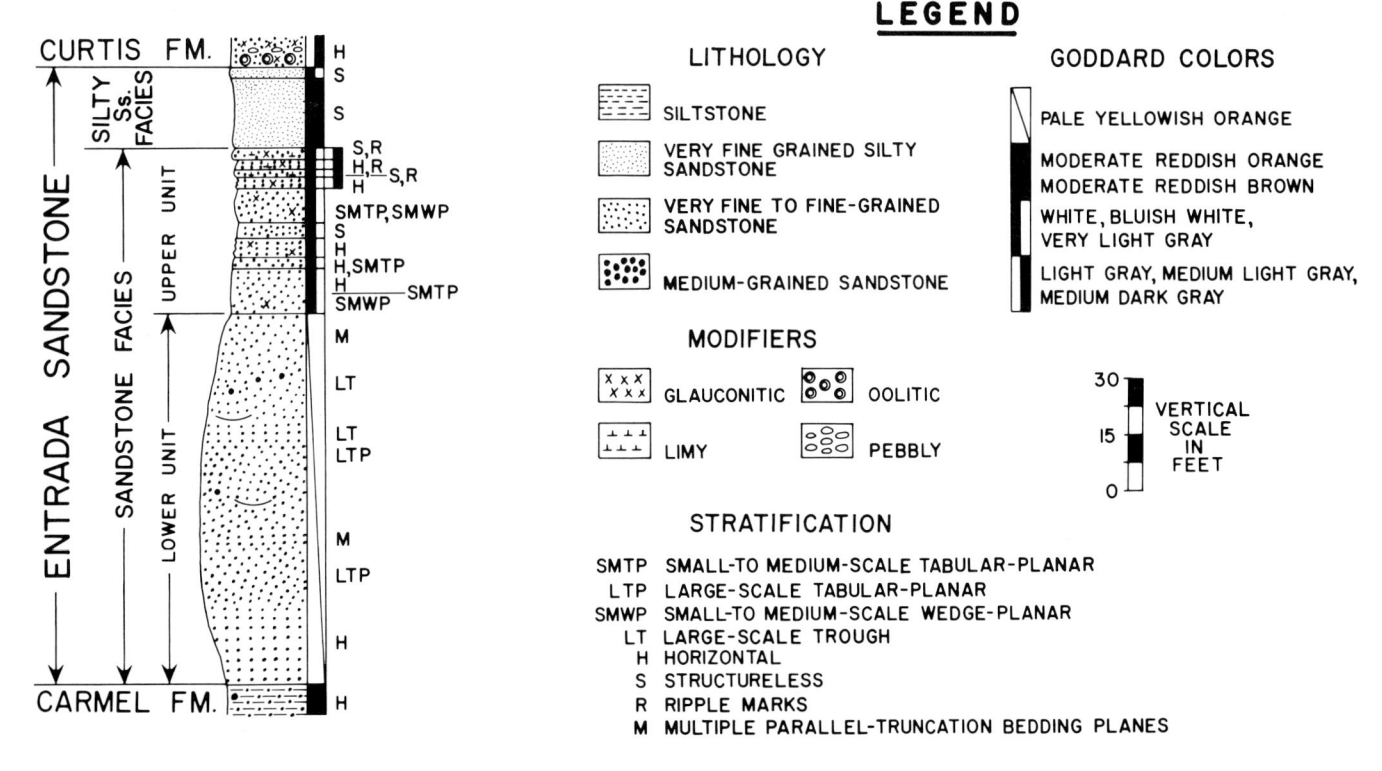

Fig. 3 — Generalized stratigraphic column of the Entrada Sandstone in northeastern Utah.

		NORTH-CENTRAL UTAH	NORTHEASTERN UTAH
JURASSIC	UPPER	MORRISON FM	MORRISON FM
		STUMP FM	CURTIS FM
	MIDDLE	PREUSS FM	ENTRADA FM
		TWIN CREEK FM	CARMEL FM
	LOWER	NUGGET FM	NUGGET FM

Fig. 4 — Correlation chart showing Jurassic nomenclature for northeastern Utah (shelf edge) and north-central Utah (within Cordilleran Hingeline).

actually is an eastward extension of the Preuss Formation (Heaton, 1939, p. 1172-1175; Thomas and Krueger, 1946, p. 1277-1278; Otto and Picard, 1975, p. 136-137). The Nugget, Carmel, and Curtis formations of northeastern Utah also display pronounced stratigraphic changes from east to west as they cross the hingeline.

METHODS OF STUDY

Outcrops of the Entrada Sandstone were studied and sampled at 13 principal locations (Fig. 1). To minimize bias, geographic location, stratigraphic position within the section, and stratification type of individual samples were ignored during all analyses.

The nomenclature used to describe cross-stratification and stratification follows that of McKee and Weir (1953). Grain size is according to Wentworth's (1922) scale. Sorting and roundness were visually estimated with binocular and petrographic microscopes and classified according to Compton (1962, p. 214) and Powers (1953, p. 118), respectively. Rock color is according to Goddard (1948).

Forty-six representative samples were chosen and thin sections were cut perpendicular to bedding. Twenty-seven thin sections were stained for potassium feldspar with sodium cobaltinitrite as suggested by Bailey and Stevens (1960). Six thin sections were stained for calcite with Alizarine Red S according to Friedman's (1959) method. At least 200 points in each thin section were counted to determine composition (Tables 1, 2, 3).

Of the 46 modal analyses; 38 are from the sandstone facies of the Entrada, 6 from the silty sandstone facies, and 2 from the Preuss Formation (sample locations — Parleys and Emigration Canyons, east of Salt Lake City). The 38 thin sections from the sandstone facies consist of 14 samples from large scale, subaerial trough and tabular-planar cross-stratification, 4 samples from small- to medium-scale tabular- and wedge-planar cross-stratification, and 20 samples from horizontal stratification. The last two stratification types are considered subaqueous (playa and marine) in origin, and the large-scale trough and tabular-planar cross-stratification are considered subaerial in origin in the diagrams that follow. In the text, the term subaqueous is qualified as marine or playa deposits.

X-ray analyses of several samples were helpful in determining gross mineralogy and especially the clay mineralogy.

PETROGRAPHY

Texture. — Grain size in the sandstone facies of the Entrada ranges from very fine- to coarse-grained. Most sandstone, however, is fine to medium grained and slightly bimodal (Fig. 5). Sorting within the sandstone facies ranges from very well to poorly sorted. Most samples are moderately sorted, reflecting the bimodality.

Typical sandstones of the silty sandstone facies are very fine-grained, silty, and poorly sorted. The quartz grains present are subangular to angular.

Roundness of the quartz grains in both facies varies with grain size. Very fine to fine quartz grains are subangular to subrounded; medium to coarse grains are generally rounded or well-rounded.

Almost all quartz grains have a frosted surface. Some grain surfaces are corroded. True polished grains are rare.

Composition. — All rocks were classified according to Folk's (1968) classification (Fig. 6). The quartz pole includes all types of monocrystalline and polycrystalline quartz, except chert. All single feldspars and feldspar-rich rock fragments (granite and gneiss) are combined to form the feldspar pole. The rock fragment pole is composed of all other rock fragments including chert. Accessory minerals such as hematite, magnetite, and mica, along with all types of cement and matrix, are ignored in the classification of the sandstones.

All 44 modal analyses from both facies of the Entrada Sandstone exhibit more than 75 percent quartz; two samples contain greater than 95 percent quartz. Only three samples contain more rock fragments than feldspar. Thirty-nine sandstones are therefore classified as subarkose, two as quartzarenite, and three as sublitharenite. Two of the sublitharenites

Table 1. Modal analyses of subaerial deposits in the sandstone facies of the Entrada. (two-hundred points counted per thin section; T — trace.) Note: modal analyses in all tables are rounded off to the nearest whole percent.

Loc. No.	Quartz	K-feld.	Plagio-clase	Micro-cline	Rock Frags.	Acces. Min.	Pore Space	Q/F Ratio	Authi-genic Carb.	Silica	Matrix (<1/16 mm)
4	66	5	1	T	7	0	15	11.0	2	0	4
4	64	4	2	1	5	2	16	9.1	0	2	4
6	66	3	3	0	4	0	19	11.0	0	2	3
Colo. Nat. Mon.	73	4	4	0	5	1	6	9.1	2	0	5
4	66	6	0	T	2	0	21	11.0	0	1	4
4	60	6	0	0	1	0	23	10.0	T	T	9
4	68	6	0	0	1	T	15	11.3	3	T	6
4	68	8	1	1	1	0	17	6.8	0	3	2
4	69	10	T	0	3	T	13	6.9	1	T	4
4	68	8	0	0	3	T	16	8.5	T	2	3
4	66	9	0	0	T	0	19	7.3	1	2	3
4	68	7	T	0	3	T	14	9.7	T	2	6
10	77	4	1	1	T	T	15	12.8	T	T	1
10	64	4	0	0	T	T	14	16.0	16	1	T
Mean:	67.4	6.0	0.9	0.2	2.5	0.2	15.9	10.0	1.8	1.1	3.9
Standard Deviation:	4.1	2.1	1.3	—	2.2	—	4.0	2.5	4.2	1.1	2.2

Fig. 5 — Photomicrograph of a well-cemented subaqueous (marine) sandstone in the sandstone facies of the Entrada. Rock contains about 17 percent cement. Note bimodal grain distribution. View is under crossed nicols. Sample location 7.

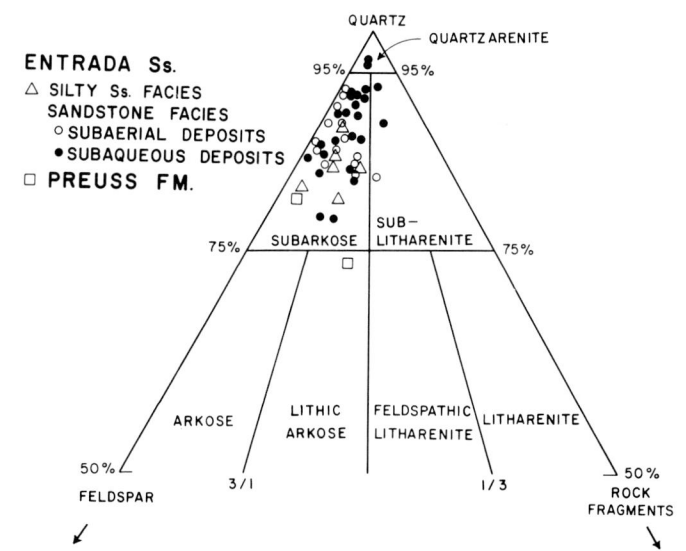

Fig. 6 — Ternary plot of quartz, feldspar, and rock fragments in the sandstone and silty sandstone facies of the Entrada Sandstone and in the Preuss Formation (classification from Folk, 1968).

Table 2. Modal analyses of subaqueous deposits in the sandstone facies of the Entrada. (two-hundred points counted per thin section; m - marine, p - playa.)

Loc. No.	Quartz	K-feld.	Plagio-clase	Micro-cline	Rock Frags.	Acces. Min.	Pore Space	Q/F Ratio	Authi-genic Carb.	Silica	Matrix (<1/16 mm)
3m	62	2	1	1	2	2	11	15.5	13	1	5
3m	66	4	0	0	1	T	21	16.5	7	0	1
2m	63	2	1	0	1	1	3	21.0	26	T	3
3m	57	1	1	0	2	2	2	28.5	32	T	3
3m	63	4	3	T	2	1	1	9.0	17	T	8
4p	52	6	3	1	4	0	24	5.2	2	0	8
5p	75	1	1	0	3	T	10	37.5	1	5	4
5p	58	3	3	0	4	T	18	9.7	5	3	6
5p	77	2	2	0	2	0	13	19.2	0	2	2
6p	65	1	2	T	5	T	4	21.7	16	T	6
6p	60	2	5	T	5	T	1	8.6	21	0	5
6p	61	4	4	T	2	T	8	7.6	16	1	3
6p	61	2	3	T	4	0	7	12.2	21	0	2
7m	66	2	4	0	1	1	4	11.0	18	0	4
7m	67	4	2	T	3	2	1	11.2	17	0	4
7m	70	T	1	0	1	4	5	70.0	17	0	2
1p	64	1	0	T	1	0	2	64.0	30	T	1
1p	68	4	2	0	2	0	12	11.3	8	T	4
1p	72	2	1	T	2	T	4	24.0	14	2	2
1p	36	5	2	0	2	2	3	5.1	41	3	6
4p	68	8	0	1	1	0	18	8.5	0	2	2
4p	67	4	0	0	2	1	21	16.8	T	2	3
10p	56	9	0	0	T	1	10	6.2	14	2	8
10p	60	9	0	0	2	T	18	6.7	6	1	4
Mean:	63.1	3.4	1.7	0.1	2.2	0.7	9.2	18.6	14.2	1.0	4.0
Standard Deviation:	8.2	2.5	1.4	—	1.3	1.0	7.4	16.9	10.9	1.4	1.7

Table 3. Modal analyses of 6 samples from the silty sandstone facies of the Entrada and 2 samples from the Preuss Formation. (two-hundred points counted per thin section.)

Loc. No.	Quartz	K-feld.	Plagio-clase	Micro-cline	Rock Frags.	Acces. Min.	Pore Space	Q/F Ratio	Authi-genic Carb.	Silica	Matrix (<1/32 mm)
11	47	4	3	0	2	1	2	6.7	26	0	15
11	45	7	2	0	1	1	T	5.0	36	0	8
11	40	3	1	T	3	2	T	10.0	30	0	20
3	53	3	1	T	1	3	0	13.2	14	2	23
3	49	4	2	T	2	3	4	8.2	19	T	16
3	54	4	4	T	4	T	4	6.8	15	T	14
Mean:	48	4.2	2.2	—	2.2	1.7	1.7	8.3	23.3	0.3	16
Standard Deviation:	5.3	1.5	1.2	—	1.2	1.2	2.0	2.9	8.8	—	5.2
Preuss Fm.	44	3	4	2	1	2	4	4.9	29	4	7
	41	4	4	T	6	2	2	5.1	32	T	9
Mean:	42.5	3.5	4	—	3.5	2	3	5.0	30.5	—	8

can be further classified as phyllarenites and the remaining one as a chertarenite.

There is no significant regional variation in the three main compositional parameters (quartz, feldspar, and rock fragments). Also, there is no significant variation in the relative abundance of these constituents between the sandstone and silty sandstone facies.

Accessory minerals in the sandstone of both facies compose less than 5 percent of any sample; the mean is 0.7 percent. Hematite, magnetite, and mica (mostly biotite) are the most common accessory minerals. Authigenic hematite composes about 60 percent of all hematite; rounded, detrital hematite grains make up the remaining fraction. Magnetite grains are well-rounded and detrital in origin. In the sandstone facies, most mica flakes appear abraded and detrital in origin. Mica flakes are generally the same size or smaller than associated quartz and feldspar grains.

Biotite in the silty sandstone facies of the Entrada generally is present in long (up to 0.3 mm), bent plates. These plates are pleochroic and are very faintly green or pale gold. Extinction is wavy because of the bent nature of the plates. Hematite and magnetite in the silty sandstone facies is not different from that in the sandstone facies except for its smaller grain size.

The silty sandstone facies contains more biotite and hematite and less magnetite than the sandstone facies. In the sandstone facies, the subaqueous (marine) deposits contain a greater percent of these three minerals than the subaqueous (playa) and the subaerial deposits.

Other accessory minerals present in rare amounts include: goethite, zircon, tourmaline, topaz, glauconite, and ilmenite.

Quartz. — All quartz grains were classified according to extinction angle, crystallinity, and inclusions (Folk, 1968, p. 73-74). Quartz types from the sandstone facies average 74 percent nonundulatory (monocrystalline), 18 percent undulatory (monocrystalline), and 8 percent polycrystalline. The average quartz types from the silty sandstone facies are 65 percent nonundulatory (monocrystalline), 24 percent undulatory (monocrystalline), and 11 percent polycrystalline.

There is considerable variability in quartz types within individual sandstone samples from the sandstone and silty sandstone facies. However, the relationship of quartz type to any other parameter, except grain size, is not readily apparent.

In contrasting the two facies, the amount of nonundulatory quartz decreases, while the amount of undulatory and polycrystalline quartz increases with decreasing grain size. This relationship is opposite to observations of Conolly (1965), Blatt (1967), and Andersen and Picard (1971), and may indicate two completely different source areas for the sandstone and silty sandstone facies.

Krynine (1946), Folk (1968), and Basu *et al.* (1975) stressed the genetic significance of quartz types. Blatt and Christie (1963) and Blatt (1967) believe that quartz types are of limited use in determining source areas. However, most geologists agree that the amount of undulose and polycrystalline quartz decreases, while nonundulose quartz increases with decreasing grain size for samples from a single formation. The opposite relationship observed for the two facies of the Entrada Sandstone suggests two different source areas for these facies.

Quartz with vacuoles, needle-like inclusions (rutile), and microlites (mostly carbonate and hematite) compose less than one percent of all quartz present. Overgrowths of authigenic silica on quartz grains are rare. Most quartz grains are not altered.

Feldspar. — Feldspar is the second most abundant framework constituent in the sandstone in both facies and ranges between 1.5 and 10.0 percent of the total sample. The mean is 6.0 percent. The relative order of feldspar abundance in most samples from the Entrada Sandstone is: orthoclase > plagioclase > microcline. Shawe (1968, p. 9) found the same relative order of feldspar abundance in his modal analyses of 7 samples from the Slick Rock Member of the Entrada Sandstone in San Miguel and Dolores Counties, Colorado. Forty-three samples from both facies of the Entrada Sandstone in northeastern Utah contain one percent or more orthoclase; 27 samples contain one percent or more plagioclase; and only three samples contain one percent or more microcline. In 10 samples, plagioclase is equal to or greater than orthoclase. Granite and gneiss fragments are present in trace amounts in 24 samples, but total one percent or more in only 6 samples.

Orthoclase was identified by staining and general appearance (Folk, 1968, p. 83). It is almost always untwinned, although Carlsbad twins are present in rare amounts.

Plagioclase species were distinguished based on the maximum extinction angle of albite twins cut normal to (010). Average extinction angles of left and right twins generally range between 13 and 15 degrees, indicating that most plagioclase feldspar is probably andesine (Kerr, 1959, p. 258). The plagioclase is generally twinned by the albite law; pericline twins are rare. In several samples, however, untwinned plagioclase is present and can be more abundant than twinned plagioclase.

Polysynthetic twinning is present in almost all microcline. Because the potassium in microcline will "pick up" the sodium cobaltinitrite stain, care was exercised not to confuse untwinned microcline with orthoclase.

Granite and gneiss fragments consist primarily of orthoclase and microcline, and compose less than 2 percent of any one thin section.

Alteration of the feldspar is variable. Most grains are clear and apparently unaltered, although some grains are cloudy in appearance because of incipient alteration. Two main types of alteration were observed; sericitization and kaolinization. Sericite is present around grain edges and yields a straw yellow color when observed parallel to cleavage under crossed nicols. Grains, which have altered to kaolinite (or clays in the kaolin group), yield a "velvet fog" or white "silky" appearance. Authigenic feldspar overgrowths are present in trace amounts.

No regional variations exist between feldspar content or type. However, one thin section analyzed from Colorado National Monument, southwest of Grand Junction, Colorado, showed abnormally large (0.5 mm) plagioclase and microcline grains. The percent of feldspar present in this thin section is similar to samples from northeastern Utah.

Variations in feldspar content or type do not exist between the sandstone and silty sandstone facies. However there is a subtle difference in the percent of feldspar present in the subaqueous (playa) and subaerial deposits of the sandstone facies. Tables 1 and 2 show that the percent of feldspar is generally higher in the subaerial deposits. Thus the feldspars may have been destroyed in the subaqueous (playa) environment by fresh (nonsaline) water, and preserved in an arid or semi-arid, subaerial (eolian) environment.

Rock fragments. — Rock fragments in sandstone in both facies compose between 0.5 and 7.0 percent of the total sample. The mean is 2.3 percent. Three main types of rock fragments were distinguished; metamorphic, chert, and sedimentary rock fragments (carbonate, mudstone, and sandstone). Metamorphic rock fragments and chert are the most abundant.

Metamorphic rock fragments present are composed of microcrystalline quartz and may contain inclusions of mica, clay, or detrital quartz grains. They are rarely foliated (grains elongated), and generally are gray to black in color.

Chert is composed of microcrystalline quartz, which forms a pinpoint-birefringent aggregate, dominantly consisting of equidimensional grains. Grain diameters may reach 20 μ, but generally range between 1 and 5 μ, and are coarser than the microcrystalline quartz in metamorphic rock fragments. The chert is generally brown to reddish-brown, and many grains contain inclusions of hematite and carbonate. Chalcedonic quartz is rare.

Chert and metamorphic rock fragments are generally better rounded than associated quartz grains, and commonly are the same or smaller in grain size. The percent of chert and metamorphic rock fragments increases with decreasing grain size.

The relative order of abundance of sedimentary rock fragments is: carbonate > mudstone > sandstone. Carbonate rock fragments generally are composed of rounded micrite grains. Mudstone fragments generally are well-rounded, brown to reddish-brown, and may contain detrital quartz grains. Sandstone fragments generally are subangular to rounded and consist of quartz grains cemented by carbonate.

The percent and type of rock fragments do not vary geographically or within individual sample locations. However, very large (1.0 mm) subangular chert grains are present in a thin section from the Colorado National Monument, southwest of Grand Junction, Colorado (Fig. 7).

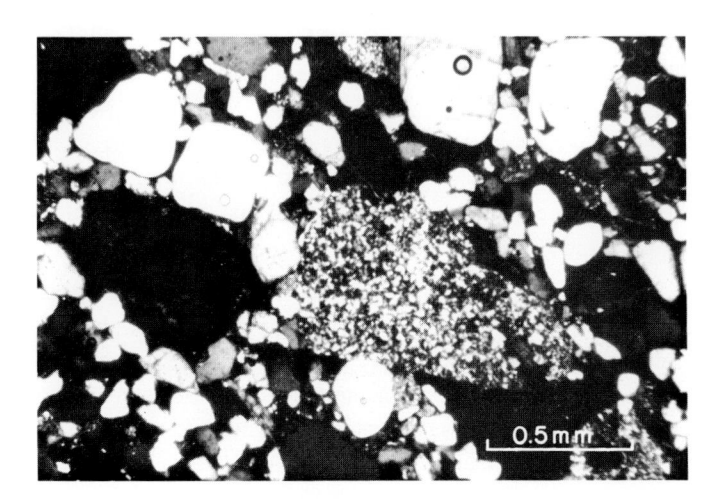

Fig. 7 — Photomicrograph of large subangular chert fragment. Note large detrital hematite (dark grain) to left of chert. Rock contains about 4 percent chert. View is under crossed nicols. Sample location is Colorado National Monument.

Matrix and cement. — Figure 8 is a ternary diagram of grains, matrix, and cement for all 46 modal analyses. The most important matrix constituents are quartz, feldspar, and clay minerals (mostly montmorillonite and kaolinite).

Fine-grained sandstone from the sandstone facies contains an average of 4.0 percent matrix, ranging between 0.5 and 9.0 percent. The very fine-grained silty sandstone from the silty sandstone facies contains an average of 16.0 percent matrix, ranging between 8.5 and 23.0 percent.

Carbonate cement is the most abundant type; silica and chert cement are rare. Calcite is the dominant carbonate cement; dolomite and siderite are present in trace amounts. Almost all calcite cement is spar. However micrite is present in trace amounts in the form of ooliths and carbonate rock fragments.

In the sandstone facies, the subaqueous (playa and marine) deposits contain more cement than the subaerial deposits. Cement in the subaqueous (playa and marine) deposits ranges between 1.5 and 44.0 percent; the mean is 15.0 percent. Cement in the subaerial deposits ranges between 1.0 and 17.5 percent; the mean is 3.0 percent.

Cement in the silty sandstone facies ranges between 15.0 and 36.0 percent; the mean is 24.0 percent.

Silica cement and quartz overgrowths are rare. Most carbonate cement has formed from direct precipitation, not by replacement. Because of the rarity of quartz overgrowths and the lack of carbonate replacement of quartz grains, it is doubtful if these rocks underwent significant silica cementation. Carbonate cementation and the lack of silica cement and quartz overgrowths may have been partly controlled by a low pH (below pH 9) of the depositional pore water as suggested by Walker (1960). Cementation of the Entrada Sandstone probably is not related to depth of burial and higher temperatures as suggested by Siever (1959, p. 76-77).

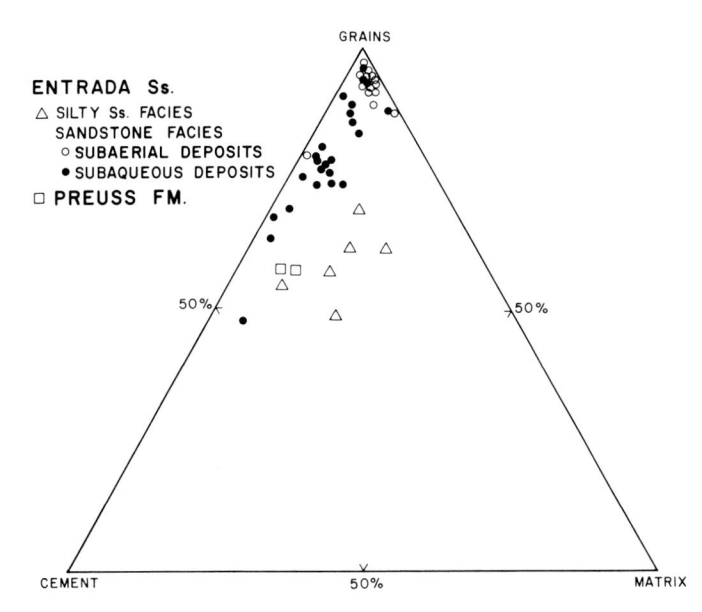

Fig. 8 — Ternary plot of grains, matrix, and cement in the sandstone and silty sandstone facies of the Entrada and in the Preuss Formation.

Pigmentation. — The sandstone facies of the Entrada is pale yellowish-orange in color. This color is derived largely from varying amounts of stain (hematite and goethite) present in the cement and as coatings on grains. Several thin sections contain local patches of hematite cement. Hematite is also present as rounded detrital grains (Fig. 7), and as dust that also darkens the color of the rock. Many "detrital" hematite grains may actually have been detrital magnetite that has altered to hematite. Ilmenite locally also has altered to hematite.

Many biotite and some opaque grains (hematite, magnetite, and ilmenite) contain red pigment radiating outward from the grain margins. The intensity of the pigmentation generally decreases outward from the altered grains.

The silty sandstone facies of the Entrada is moderate reddish-orange in color. Its darker color is primarily related to the presence of more hematite than is present in the sandstone facies (Fig. 9). Three processes are responsible for the

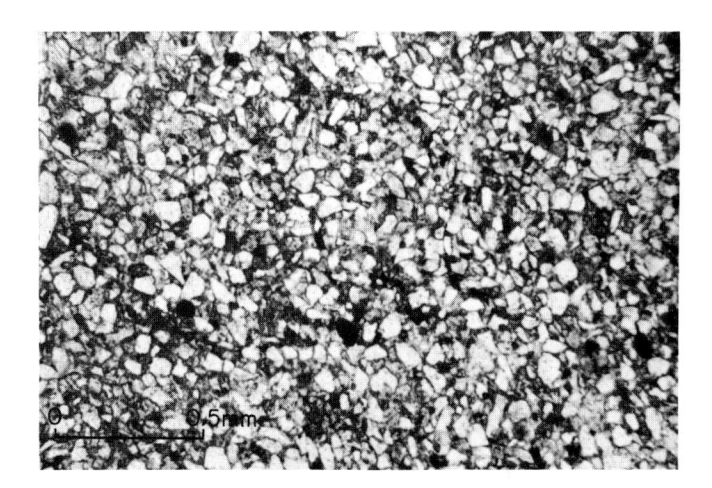

Fig. 9 — Photomicrograph of a very fine-grained silty sandstone from the silty sandstone facies of the Entrada. Note abundance of detrital hematite (dark grains). Rock contains about 3 percent detrital hematite. View is in plane light. Sample location 3.

greater concentration of hematite in the silty sandstone facies. Picard (1965, p. 468-469) noted that the percent of total iron increased with decreasing grain size in his study of the Chugwater Formation in Wyoming. The darker color of the silty sandstone facies is partly related to its finer grain size. The presence of less magnetite in the silty sandstone facies may indicate that most of the magnetite, which was once present, has altered to hematite. The silty sandstone facies also contains more biotite than the sandstone facies (Fig. 10). Most biotite is pale gold in color, which according to Folk (1968, p. 87), indicates a loss of iron (oxidization). Some biotite, however, has a very faint green tint, possibly suggesting that some reduction occurred after a more extensive period of oxidization. The alteration of biotite not only provides iron oxide for pigment, but also may yield authigenic clay for matrix and calcite for cement (Walker, 1967, p. 359).

The absence of hornblende in the Entrada may indicate that this unstable iron-bearing mineral has been entirely altered to hematite. According to Walker (1967, p. 365) the paucity of hornblende where it should be a common accessory mineral is evidence of an authigenic origin of the hematite pigment.

The presence of red pigment in cement and surrounding

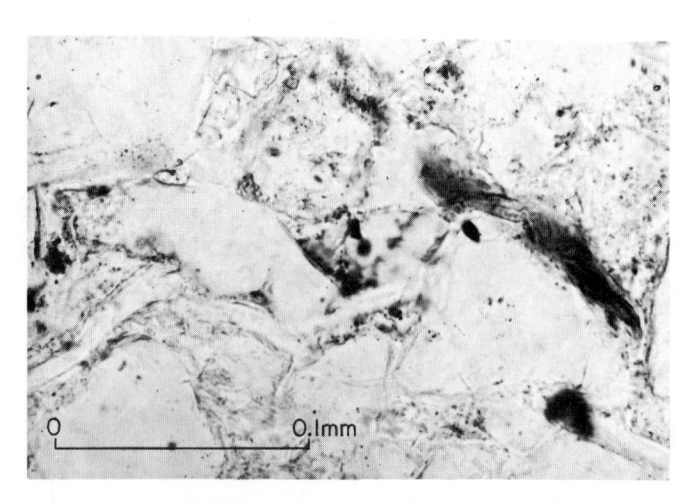

Fig. 10 — Photomicrograph of bent biotite plates from the silty sandstone facies of the Entrada. Note pigment in cement and moving outward from biotite on right. View is in plane light. Sample location 3.

quartz grains, but its absence at grain contacts (Fig. 11), further indicates that the pigment was available during diagenesis. The red pigment in both facies apparently is primarily post-depositional in origin. However, a few grain contacts in the sandstone facies are stained, possibly indicating that erosion of pre-existing red beds contributed a small part of the stain to this facies.

Pore space. — Porosity as determined by modal analyses

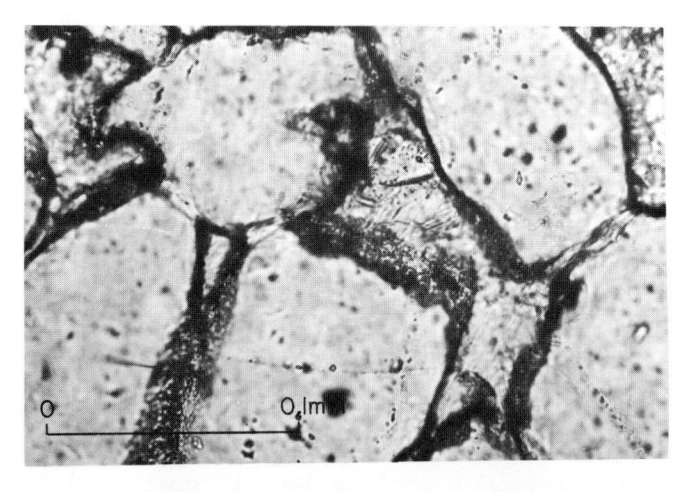

Fig. 11 — Photomicrograph of post-depositional hematite stain in the sandstone facies of the Entrada. Note presence of stain in cement and surrounding quartz grains, but its absence at grain contacts. View is under plane light. Sample location 1.

must be considered a maximum value because of possible grain plucking during thin section preparation.

Porosity in the subaqueous (playa and marine) deposits of the sandstone facies ranges between 1.0 and 24.0 percent; the mean is 9.0 percent. Values for the subaerial deposits range between 5.5 and 22.5 percent; the mean is 15.9 percent. The differences in porosity in the sandstone facies reflect the different degree of cementation between the subaqueous and subaerial deposits. The mean initial porosity (addition of the mean percent of cement and pore space) for the subaqueous (playa and marine) deposits is 24 percent and for the subaerial deposits is 19 percent. Thus, the initial porosity of the sediments in these environments was about equal before cementation.

Porosity in the silty sandstone facies of the Entrada ranges from 0 to 4.3 percent; the mean is 1.9 percent. The low porosity observed in the silty sandstone facies is mainly related to the well cemented nature of the rock.

X-ray analysis. — X-ray analysis of several samples was helpful in determining the gross mineralogy before modal analysis. Five samples from the sandstone facies and two samples from the silty sandstone facies of the Entrada were analyzed for clay minerals. The predominant clay minerals present in both facies are, in order of decreasing abundance: montmorillonite, kaolinite, and illite. Montmorillonite was identified by its 12 Å spacing, which collapsed to 10 Å when heated to 550°C for 1 hour, and expanded to 17 Å when glyconated. Kaolinite was identified by its 7 Å and 3.5 Å spacings that were destroyed when heated to 550°C for 1 hour. Illite or K-mica was identified by its 10 Å and 5 Å spacings which were unaffected by glyconation or heat treatment. The 3.3 Å illite spacing was obscured by alpha quartz.

Except for a greater amount of clay in the silty sandstone facies, no variation in clay mineralogy exists between the sandstone and silty sandstone facies. Samples of equal mass from the subaqueous and subaerial deposits of the sandstone facies were prepared identically to determine if differences in clay mineralogy exist. No difference in clay mineralogy exists between rocks from the two environments. However, the subaqueous (playa) deposits contain a larger amount of clay than the subaerial deposits. The destruction of feldspars by fresh (nonsaline) water may have, in part, produced the greater amount of clay observed in rocks from the subaqueous (playa) environment. Because there is less feldspar present in the subaqueous (playa) deposits, a possible direct relationship may exist between the destruction of feldspar and generation of clay minerals in the subaqueous (playa) environment. Another possible explanation is that the clays were winnowed out of the eolian environment and deposited in the subaqueous (playa) environment.

Comparison with the Preuss Formation. — Samples of the Preuss Formation from Parleys and Emigration Canyons, east of Salt Lake City, are poorly sorted, very fine- to medium-grained with an abundance of silt sized grains present. Chert is also abundant. According to Folk's (1968) sandstone classification, the two samples are classified as subarkose and lithic arkose, reflecting the abundance of chert and feldspar present (Fig. 6; Table 3). Plagioclase feldspar is more abundant than orthoclase which is more abundant than microcline. Many of the plagioclase feldspars and some of the orthoclase and microcline are rounded.

Important petrographic differences between the Preuss and Entrada are: the sandstone of the Preuss Formation is generally coarser grained than the silty sandstone facies of the Entrada; also the plagioclase feldspars are generally rounded and more abundant than orthoclase, in contrast to both facies of the Entrada; and the Preuss contains more chert than the silty sandstone facies of the Entrada.

PROVENANCE

Sandstone facies. — Paleocurrent azimuths, based on 275 measurements, indicate a dominant wind transport direction toward the west and southwest (Otto and Picard, 1975, p. 134-135). This suggests that source areas for the sandstone facies were on the east and northeast.

At several locations in Colorado, the Entrada rests unconformably on the ancestral Front Range and Uncompahgre uplifts, indicating that these areas were positive features during Entrada time. At these locations the Entrada occasionally contains a basal conglomerate with angular fragments of underlying Precambrian rock (Holmes, 1956, p. 34, 37). These areas were probably one source of sediment for the sandstone facies of the Entrada.

The abundance of feldspar in the Entrada suggests source areas that were relatively close or contained abundant feldspar. Upper Paleozoic and Triassic sandstones in Alberta and Montana may have been too distant and of improper composition to have supplied much of the feldspar observed in the Entrada. However, erosion of these Paleozoic and Triassic sandstones could have supplied large amounts of quartz sand.

Based on modal analyses, most of the quartz in the sandstone facies is nonundulatory to slightly undulatory (common quartz) and much may be recycled from sedimentary sources. Evidence for this is the well-rounded nature of some grains that are associated with angular grains. The remainder of the common quartz may indicate erosion of plutonic, granitic, or granite-gneiss source areas (Folk, 1968, p. 71, 73; Basu *et al.*, 1975, p. 879). The rarity of polycrystalline quartz suggests only minor contributions from metamorphic sources (Folk, 1968, p. 73-74; Basu *et al.*, 1975, p. 879). The rare presence of metamorphic rock fragments further suggest only minor contributions from metamorphic sources. The presence of chert in almost all Entrada samples substantiates the presence of pre-existing sedimentary rocks in the source areas.

The Precambrian complex of the Colorado Front Range consists of 70 to 75% granitic rocks (Boos and Boos, 1957, p. 2635). According to Stokes (1961, p. 154), the Ancestral Rockies consist mainly of crystalline and granitic rocks rich in quartz, feldspar, clay-rich mica, silt, and pre-Pennsylvanian sedimentary rocks. The ancestral Front Range and Uncompahgre uplifts contain rocks of the proper composition to have supplied sediment to the Entrada.

In summary, paleocurrent azimuths and the detrital grains in the sandstone facies of the Entrada in northeastern Utah suggests multiple and varied source areas. Plutonic, granitic, and granite-gneiss rocks of the ancestral Front Range and Uncompahgre uplifts probably were the main sources. Weathering of pre-existing sedimentary rocks and metamorphic rocks in these uplifts probably contributed only minor amounts of detritus to the Entrada. Upper Paleozoic and Triassic sandstones in Alberta and Montana may have supplied a major portion of the quartz sand. Varied sources are inferred from the composition of the sandstone facies and the vast areal extent of the Entrada. Reconstruction of the source areas for the sandstone facies certainly deserves further study.

Silty sandstone facies. — Heaton (1939, p. 1172-1175) first suggested that the entire Entrada Sandstone in the Uinta Mountains grades westward into the Preuss Formation of southern Idaho and the central Wasatch Mountains of Utah. Thomas and Krueger (1946, p. 1277-1278) demonstrated that the silty sandstone facies of the Entrada is an eastward extension of the Preuss Formation. According to Neely (1937, p. 752-753), the Preuss is a great deltaic deposit derived from a land mass in western Utah and eastern Nevada (Mesocordilleran highland). Kinney (1955, p. 85) suggested that the silty sandstone facies of the Entrada Sandstone probably also originated from this source.

According to Stokes (1961, p. 154), the Mesocordilleran highland was composed primarily of sedimentary rocks that contributed relatively more chert and finer grained detritus than the Ancestral Rockies. Volcanic activity on the west also supplied debris at various times during the existence of the Mesocordilleran highland (Stokes, 1961, p. 154). The silty sandstone facies of the Entrada is very fine grained and contains abundant chert and biotite (possibly volcanic), suggesting a possible western source. The Preuss Formation is generally coarser grained than the silty sandstone facies, further supporting a western source.

Another possible source area for the Preuss and the silty

sandstone facies of the Entrada is the Big Belt and Little Belt Mountains in west-central Montana (Imlay, 1952, p. 1745). Imlay based his conclusions partly on the coarser grain size of the Preuss at its northernmost areas of outcrop in Wyoming and Idaho and its siltier, finer grain size southward. According to Imlay (1952), the Preuss was deposited in a series of highly saline lagoons connected on the north and west with normal marine waters.

Contrasting lithologies and the presence of greater amounts of polycrystalline quartz in the silty sandstone facies than is present in the sandstone facies suggests different source areas for the two facies of the Entrada Sandstone. The presence of the silty sandstone facies only in the western part of the area and its lateral continuity with the Preuss Formation suggests western or possibly northern sources.

Several source areas for the silty sandstone facies probably existed, however, a western source (Mesocordilleran highland) is likely for some of the fine grained detritus, chert, and biotite present in the silty sandstone facies. A more detailed study of this facies, the Preuss Formation, and their mutual relationship might be helpful in further delineating the source areas.

OIL AND GAS

General. — Oil and gas production from the Entrada Sandstone in Utah has been meager. Nevertheless, there has been sufficient established production and showings of oil and gas to stimulate considerable interest in the formation.

Production and good shows of gas have been established from the Entrada at Bar-X, West Bar-X, San Arroyo, Westwater, and Harley dome in northeastern Utah (Fig. 12; Table 4). Minor amounts of oil have been produced from the formation at Ashley Valley. In 1974, Champlin's Entrada discovery, 14 Deep Brady Unit (sec. 4, T.17N., R.100W.), in Wyoming, created further interest in this formation. More recently, Amoco's No. 1 Champlin well (sec. 5, T.2N., R.8E.) in Summit County, Utah encountered gas shows from the Preuss Formation. Hopefully, this well, located adjacent to the Cordilleran Hingeline, will stimulate interest in the Preuss in this region.

San Arroyo field. — Production and shows at San Arroyo have been found in several sandstone reservoirs: the Cretaceous Cedar Mountain, Dakota, and Mancos formations, and the Jurassic Entrada and Morrison formations. According to Monsalve (1972), productive sandstones are elongate along the San Arroyo anticline, and most entrapment is related both to stratigraphic changes and structure. However, Williams (1961) and Monsalve (1972) believe that the gas in the Entrada is a structural accumulation.

The San Arroyo anticline is an asymmetrical doubly-plunging fold with the steepest limb on the south. As mapped at the Dakota silt marker, two areas of structural closure are present: a vertical closure of 350 ft (106.7 m) on the east and a lesser vertical closure of about 150 ft (45.7 m) on the west. Entrada gas is present in the larger eastern closure.

The Entrada at San Arroyo is a cross-stratified, white to brownish red, fine-grained sandstone approximately 225 ft (68.6 m) thick (Williams, 1961). Porosity in the Entrada at San Arroyo is 16 percent and the permeability is 30 MD (Table 4).

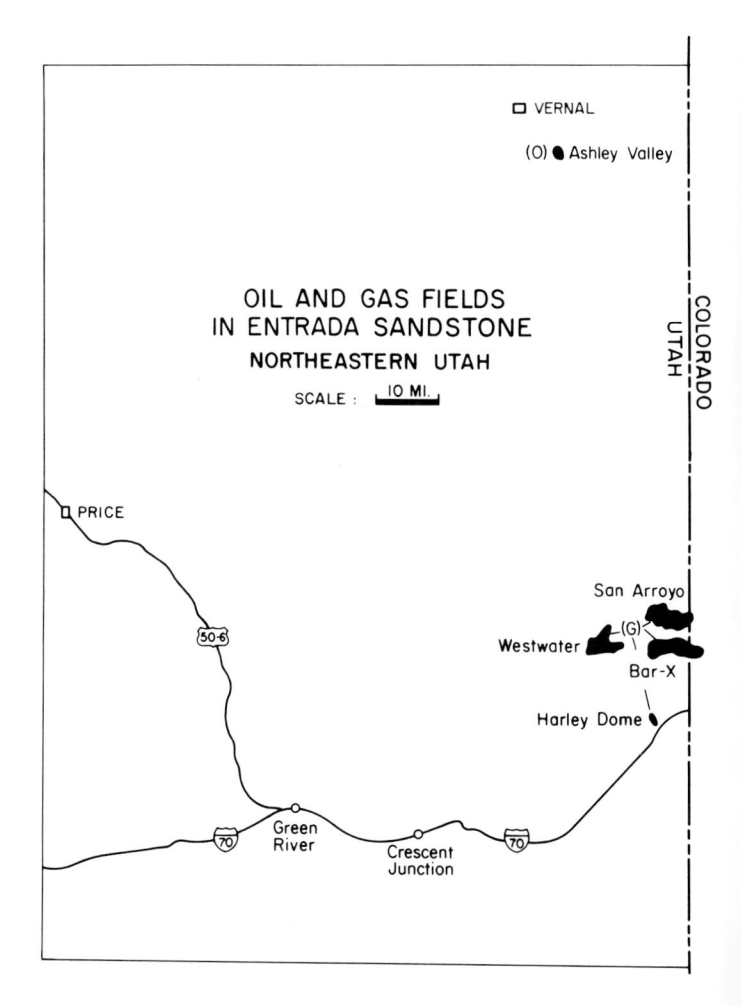

Fig. 12 — Index map of oil and gas fields that produce from the Entrada in Utah.

The Btu value of gas at San Arroyo is low (Table 5), and the gas must be scrubbed to remove noncombustible gases and raise the Btu value for pipeline transmission.

Bar-X field. — The Bar-X anticline is about 3 mi (4.8 km) to the south, and occupies a position en echelon with the San Arroyo anticline. Bar-X anticline contains two separate structural highs and many normal faults that are transverse to the main axis (Owen and Whitney, 1956, p. 195).

Gas in the Entrada at Bar-X is trapped at the top of the

Table 4. Summary of geology of oil and gas fields in Entrada Sandstone (Information from Schuh, 1961; Williams, 1961; and Oil and Gas Fields of Utah, 1961)

Fields	Location	Year of Discovery	Average Depth (ft)	Net Pay (ft)	Porosity (%)	Permeability (MD)	Initial Pressure (psi)	Btu (ft³)
Ashley Valley	5S-22E				17.8	93		
Bar-X	17S-25E	1948	3616	31	22.5		1274	535
San Arroyo	16S-25,26E	1955	5250	80	16	30	1600	631
Westwater	17S-23,24E	1955	5700	118	24	762	1860	800
Harley Dome	19S-25E	1926	900	85			155	

Table 5. Composition of gas in Entrada Sandstone, northeastern Utah (Information from Utah Geological Survey and U.S. Bureau of Mines Information Circulars)

Field Name	Methane	Ethane	Higher Fractions	Nitrogen	Carbon Dioxide	Helium	Btu (ft³)
Bar-X	49.6	2.1	0.8		9.6	0.5	589
	45.0	0.9	2.0		24.3	0.8	530
	44.7	0.9	0.4	28.6	25.2		486
Mean	46.4	1.3	1.1		19.7		535
San Arroyo	47.5	3.5	5.1		26.1	0.6	703
	47.9	2.2	2.5		19.6	0.9	604
	45.1	2.4	2.4		24.5	0.9	572
	46.8	2.1	2.9		23.4	1.0	610
	46.8	2.0	2.9		22.7	0.9	601
	47.2	2.2	3.1		19.7	1.0	620
	47.2	2.3	3.0		19.7	0.5	619
	51.4	2.3	4.3	17.1	23.9	0.5	721
Mean	47.5	2.4	3.3		22.4	0.8	631
Westwater	75.1	1.9	0.6	18.0	1.0	0.6	819
	68.2	2.8	1.2		2.0	0.6	780
Mean	71.6	2.3	0.9		1.5	0.6	800
Mean	51.0	2.1	2.4		18.6	0.7	635

formation where the net pay is about 31 ft (9.4 m) and the porosity is 22.5 percent. The mean Btu value of the gas is very low (535), and the gas contains large amounts of nitrogen (28.6 percent) and carbon dioxide (mean is 19.7 percent). The methane content averages 46.4 percent.

Westwater field. — The Westwater field is a local structural closure on a long westward-plunging nose commonly called the Westwater anticline (Schuh, 1961), located about 2 miles (3.2 km) southwest of the Bar-X anticline. The reservoirs in the Westwater field are sandstone in the Entrada, Morrison, Dakota, and Castlegate formations. The Entrada is the best reservoir, producing from sandstone that exhibits prominent cross-bedding (Schuh, 1961). Both the porosity (24 percent) and the permeability (762 MD) are excellent. The mean Btu value of the gas is 800. The methane content averages 71.6 percent and the ethane content averages 2.3 percent.

Harley dome field. — Harley dome is a northwestward-plunging anticline in Grand County, Utah. The axes of Seiber dome on the south and Cottonwood dome on the southwest parallel the axis of Harley dome and also are northwestward-plunging anticlines. Normal faults are present in the area and generally strike to the northwest.

Production from the Cottonwood-Harley dome area has been from the Dakota Formation and the Brushy Basin Member of the Morrison Formation. The Entrada Sandstone and the Salt Wash Member of the Morrison contain helium in the Harley dome structure.

In 1926, the No. 2 Government (sec. 4, T.19S., R.25E.)

at Harley dome encountered a gas flow of 5000 MCFPD (1,420,000 cubic m) from the Entrada at a depth of 860 to 945 ft (262.1 to 288.0 m). The helium content ranged from 2 to 7 percent; the total inert gas present was 92.6 percent (Keebler, 1956, p. 190). In 1932, the area surrounding the helium-producing wells was incorporated into a helium reserve for the United States Government.

Ashley Valley field. — Oil is present in the Entrada at the Ashley Valley field. However, production in the field has been almost entirely from the Pennsylvanian Weber Sandstone and the Permian Phosphoria Formation.

The structure at Ashley Valley is a northwest-trending anticline that is situated on the west-plunging Blue Mountain uplift. At the top of the Weber Sandstone, anticlinal and fault closure is about 300 ft (91.4 m). The anticline is cut by many normal faults; the largest displacement observed is about 150 ft (46 m). Only one period of faulting has been recognized in the field and involves the youngest rocks exposed (Mancos Shale; Peterson, 1961).

Log information indicates that porosity in the Entrada at Ashley Valley is good. On the basis of 20 measurements, the porosity ranges between 7.5 and 22.1 percent; the mean is 17.8 percent. Permeability ranges from 0 to 600 MD; the mean is 93 MD.

Petroleum potential of Entrada. — There is not a significant oil or gas field producing from the Entrada Sandstone in Utah. San Arroyo is the best gas field and was productive during 1974. Production figures for the Entrada are difficult to obtain because production from the Entrada is not large enough to merit a separate listing in production statistics and therefore is combined with the production from other formations. Probably very little oil or gas was produced during 1974 from the Entrada in Utah.

Drilling to the formation has not been extensive in many areas because of excessive depths. This is unfortunate because the Entrada is likely to have reservoir quality throughout most of its subsurface extent (Table 6).

Table 6. Porosity in different facies and in oil and gas fields of the Entrada Sandstone

Facies or fields in Entrada	Range	Mean
Subaerial (eolian) beds of sandstone facies	5.5-22.5	15.9
Subaqueous beds of sandstone facies	1.0-24.0	9.0
Beds of silty sandstone facies	0.0- 4.3	1.9
Ashley Valley field	7.5-22.1	17.8
San Arroyo field		16.0
Bar-X field		22.5
Westwater field		24.0

Porosity and permeability are important factors in developing Entrada exploration targets. Where the silty sandstone facies or the subaqueous deposits of the sandstone facies are present, their low porosity may provide stratigraphic or combination structural-stratigraphic traps for the more porous eolian deposits below. Prediction of the distribution of these different facies would be helpful to exploration.

Although most Entrada production has been from structural traps, variations in porosity, the presence of shoreline conditions in part of the area during Entrada time, and an unconformity at the top of the Entrada suggests that subtle stratigraphic traps could exist. As more and deeper subsurface information becomes available through increased drilling and seismic surveys, exploration for stratigraphic and structural-stratigraphic traps will become more precise.

Compositions of gas in the Entrada Sandstone and the Salt Wash Member of the Morrison Formation at Bar-X field are different. Gas in the Entrada and the Dakota Sandstone at San Arroyo also differ considerably. Dakota gas at San Arroyo contains 87.4 percent methane, 1.0 percent carbon dioxide, and has a Btu value of 1045. In contrast, Entrada gas contains an average of 47.5 percent methane, 22.4 percent carbon dioxide, and has an average Btu value of 631 (Table 5). The Entrada gas also contains large amounts of nitrogen. Different sources for the gases may be involved, thus enhancing the possibilities for future successful exploration in the Entrada.

Shallow-water marine deposits of the silty sandstone facies offer the possibility that hydrocarbons may have originated within the formation. It is unlikely, however, that the playa deposits in the lower unit of the sandstone facies were capable of generating much petroleum. The scarcity of favorable source rocks seems to be the main element preventing prolific production from the Entrada. The location of possible source rocks should be investigated further.

The Entrada may be more economically attractive than has previously been supposed. Some oil and gas production and shows are known, although no large fields have been discovered. With increasingly higher prices for petroleum and improved drilling, completion and recovery techniques, deeper exploration is becoming more economical. The Entrada's wide extent and capacity as a reservoir provide attractive exploration targets for innovative geologists.

REFERENCES CITED

Andersen, D. W., and Picard, M. D., 1971, Quartz extinction in siltstone: Geol. Soc. America Bull., v. 82, no. 1, p. 181-186.

Bailey, E. H., and Stevens, R. E., 1960, Selective staining of K-feldspar and plagioclase on rock slabs and thin sections: The Amer. Min., v. 45, p. 1020-1025.

Baker, A. A., Dane, C. H., and Reeside, J. B., Jr., 1936, Correlation of the Jurassic formations of parts of Utah, Arizona, New Mexico and Colorado: U.S. Geol. Survey Prof. Paper 183, 66 p.

Basu, Abhijit, Young, Steven W., Suttner, Lee J., James, W. Calvin, and Mack, Greg H., 1975, Re-evaluation of the use of undulatory extinction and polycrystallinity in detrital quartz for provenance interpretation: Jour. Sed. Petrology, v. 45, no. 4, p. 873-882.

Blatt, Harvey, and Christie, J. M., 1963, Undulatory extinction in quartz of igneous and metamorphic rocks and its significance in provenance studies of sedimentary rocks: Jour. Sed. Petrology, v. 33, no. 3, p. 559-579.

_____, 1967, Original characteristics of clastic quartz grains: Jour. Sed. Petrology, v. 37, no. 2, p. 401-424.

Boos, C. Maynard, and Boos, Margaret Fuller, 1957, Tectonics of eastern flank and foothills of Front Range, Colorado: Am. Assoc. Petroleum Geologists Bull., v. 41, no. 12, p. 2603-2676.

Compton, R. R., 1962, Manual of field geology: New York, John Wiley and Sons, Inc., 378 p.

Conolly, J. R., 1965, The occurrence of polycrystallinity and undulatory extinction in quartz in sandstones: Jour. Sed. Petrology, v. 35, no. 1, p. 116-135.

Dane, C. H., 1935, Geology of the Salt Valley Anticline and adjacent areas Grand County, Utah: U.S. Geol. Survey Bull. 863, 184 p.

Folk, R. L., 1968, Petrology of sedimentary rocks: Austin, Hemphill's, 170 p.

Friedman, G. M., 1959, Identification of carbonate minerals by staining methods: Jour. Sed. Petrology, v. 29, no. 1, p. 87-97.

Gilluly, James, and Reeside, J. B., Jr., 1928, Sedimentary rocks of the San Rafael Swell and some adjacent areas in eastern Utah: U.S. Geol. Survey Prof. Paper 150-D, p. 61-110.

Goddard, E. N. (chm.), 1948, Rock-color chart: Washington, National Research Council, 6 p.

Hansen, W. R., 1965, Geology of the Flaming Gorge Area Utah-Colorado-Wyoming: U.S. Geol. Survey Prof. Paper 490, 196 p.

Heaton, R. L., 1939, Contribution to Jurassic stratigraphy of Rocky Mountain Region: Am. Assoc. Petroleum Geologists Bull., v. 23, no. 8, p. 1153-1177.

Holmes, C. N., 1956, Tectonic history of the Ancestral Uncompahgre Range in Colorado, in Geology and economic deposits of east central Utah: Intermtn. Assoc. Petroleum Geol. Guidebook, 7th Ann. Field Conf., p. 29-37.

Imlay, R. W., 1952, Marine origin of Preuss Sandstone of Idaho, Wyoming, and Utah: Am. Assoc. Petroleum Geologist Bull., v. 36, no. 9, p. 1735-1753.

Keebler, W. E., 1956, Cottonwood-Harley Dome area, Grand County, Utah, in Geology and economic deposits of east central Utah: Intermtn. Assoc. Petroleum Geol. Guidebook, 7th Ann. Field Conf., p. 190-194.

Kerr, P. F., 1959, Optical mineralogy, 3rd ed.: New York, McGraw-Hill Book Co., 442 p.

Kinney, D. M., 1955, Geology of Uinta River-Brush Creek area, Duchesne and Uintah Counties, Utah: U.S. Geol. Survey Bull. 1007, 185 p.

Krynine, P. D., 1946, Microscopic morphology of quartz types: Pan-Am. Cong. Mining and Geol. Engineers, Annals of 2nd Comm., p. 36-49.

McKee, E. D., and Weir, G. W., 1953, Terminology for stratification and cross-stratification in sedimentary rocks: Geol. Soc. America Bull., v. 64, no. 4, p. 381-390.

Monsalve, O. A., 1972, Geology of the San Arroyo gas field, Grand County, Utah: Unpub. M. S. thesis, Univ. of Utah, Salt Lake City, Utah, 58 p.

Neely, Joseph, 1937, Stratigraphy of the Sundance Formation and related Jurassic rocks in Wyoming and their petroleum aspects: Am. Assoc. Petroleum Geologists Bull., v. 21, no. 6, p. 715-770.

Otto, E. P., and Picard, M. Dane, 1975, Stratigraphy and oil and gas potential of Entrada Sandstone (Jurassic), northeastern Utah, in Deep drilling frontiers in the central Rocky Mountains: Rocky Mtn. Assoc. of Geologists, 1975 Symposium, p. 129-139.

Owen, A. E., and Whitney, G. W., 1956, San Arroyo Bar-X area, Grand County, Utah and Mesa County, Colorado, in Geology and economic deposits of east central Utah: Intermtn. Assoc. Petroleum Geol. Guidebook, 7th Ann. Field Conf., p. 195-198.

Peterson, V. E., 1961, Ashley Valley oil field, Uintah County, Utah, in A symposium of the oil and gas fields of Utah: Intermtn. Assoc. Petroleum Geol.

Picard, M. D., 1965, Iron oxides and fine-grained rocks of Red Peak and Crow Mountain sandstone members, Chugwater (Triassic) Formation, Wyoming: Jour. Sed. Petrology, v. 35, no. 2, p. 464-479.

Poole, F. G., 1963, Palaewinds in the western United States, in A. E. M. Nairn, ed., Problems in palaeclimatology: London, Interscience Publ., p. 393-405.

Powers, M. C., 1953, A new roundness scale for sedimentary particles: Jour. Sed. Petrology, v. 23, no. 2, p. 117-119.

Schuh, E. J., 1961, Westwater Field, Grand County, Utah, in A symposium of the oil and gas fields of Utah: Intermtn. Assoc. Petroleum Geol.

Shawe, D. R., 1968, Petrology of sedimentary rocks in the Slick Rock district, San Miguel and Dolores Counties, Colorado: U.S. Geol. Survey Prof. Paper 576-B, 34 p.

Siever, Raymond, 1959, Petrology and geochemistry of silica cementation in some Pennsylvanian sandstones, in H. A. Ireland, ed., Silica in sediments: Soc. of Econ. Paleontologists and Mineralogists, Special Publ. 7, p. 55-79.

Stokes, W. L., Peterson, J. A., and Picard, M. D., 1955, Correlation of Mesozoic formations of Utah: Am. Assoc. Petroleum Geologists Bull., v. 39, no. 10, p. 2003-2019.

_____, 1961, Fluvial and eolian sandstone bodies in Colorado Plateau, in J. A. Peterson and J. C. Osmond, eds., Geometry of sandstone bodies: Tulsa, Am. Assoc. Petroleum Geologists, p. 151-178.

Tanner, W. F., 1965, Upper Jurassic paleogeography of the Four Corners region: Jour. Sed. Petrology, v. 35, no. 3, p. 564-574.

Thomas, H. D., and Krueger, M. L., 1946, Late Paleozoic and Early Mesozoic stratigraphy of Uinta Mountains, Utah: Am. Assoc. Petroleum Geologists Bull., v. 30, no. 8, p. 1255-1293.

Untermann, G. E., and Untermann, B. R., 1954, Geology of Dinosaur National Monument and vicinity, Utah-Colorado: Utah Geol. and Mineralog. Survey Bull. 42, 221 p.

Walker, T. R., 1960, Carbonate replacement of detrital silicate minerals as a source of authigenic silica in sedimentary rock: Geol. Soc. America Bull., v. 71, no. 2, p. 145-151.

_____, 1967, Formation of red beds in modern and ancient deserts: Geol. Soc. America Bull., v. 78, no. 3, p. 353-368.

Wentworth, C. K., 1922, A scale of grade and class terms for clastic sediments: Jour. Geology, v. 30, no. 5, p. 377-392.

Williams, F. E., 1961, San Arroyo gas field, Grand County, Utah, in A symposium of the oil and gas fields of Utah: Intermtn. Assoc. Petroleum Geol.

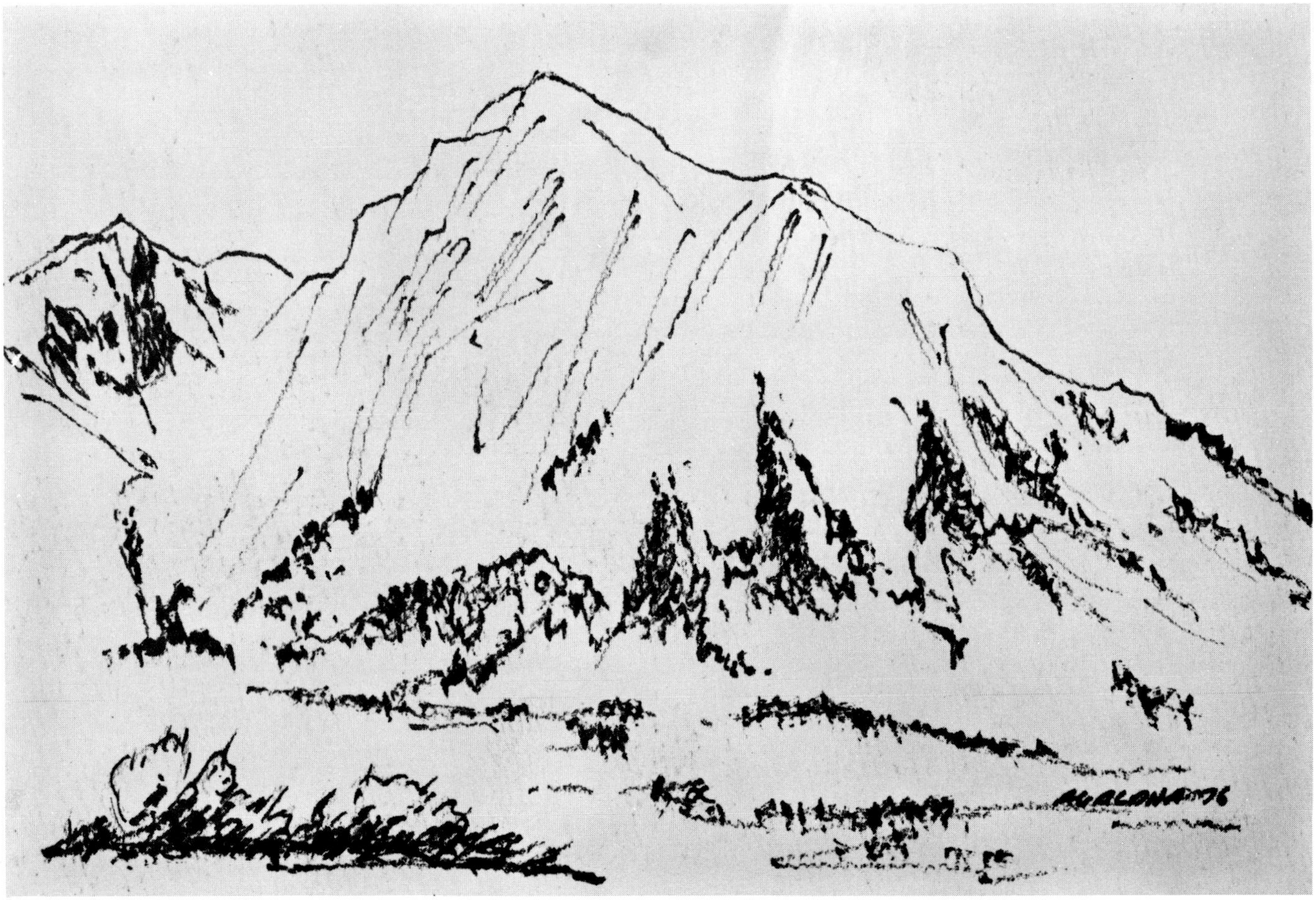

… ROCKY MOUNTAIN ASSOCIATION OF GEOLOGISTS — 1976 SYMPOSIUM

REGIONAL STRATIGRAPHY AND DEPOSITIONAL ENVIRONMENTS OF THE GLEN CANYON GROUP AND CARMEL FORMATION (SAN RAFAEL GROUP)

by

William E. Freeman[1]

ABSTRACT

Recent paleogeographic studies suggest there was no Mesocordilleran geanticline during Early Jurassic time and deposition typical of the eastern miogeosyncline extended uninterrupted across Nevada and Utah.

Re-examination of Navajo Sandstone textures, sedimentary structures, paleontologic evidence, and stratigraphic position indicates processes more consistent with a tidal-dominated marine shelf than an interior-desert sand-sea.

Vertical sequence and facies patterns of the Glen Canyon Group and Carmel Formation suggest a transgression of the Navajo and Carmel marine deposits eastward from the Cordilleran geosyncline onto continental deposits of the Kayenta, Moenave, Wingate, and Chinle formations.

INTRODUCTION

The study concentrates on the depositional origins of the Triassic-Jurassic Glen Canyon Group and Jurassic Carmel Formation (San Rafael Group) of the central and southwestern Colorado Plateau in southern Utah and northern Arizona (Fig. 1). Figure 2 shows a generalized stratigraphic column of the sedimentary rocks within the Glen Canyon Group and Carmel Formation in the southwestern Colorado Plateau. This paper is derived from a thesis submitted by the author in 1973 to the Graduate School, University of Tulsa, Tulsa, Oklahoma and a subsequent paper by Freeman and Visher (1975).

PALEOGEOGRAPHIC SETTING

The Triassic-Jurassic Glen Canyon Group and Jurassic Carmel Formation of the San Rafael Group were deposited on the eastern flank of the Cordilleran geosyncline in an area which is now the Colorado Plateau. Some Early Jurassic paleogeographic interpretations for the western United States (McKee, et al., 1956; Eardley, 1962; Dunbar and Waage, 1969) assumed two sedimentation troughs, one marine and the other continental, separated by a high Mesocordilleran geanticline (Stanley, et al., 1971). The postulated persistence of the Mesocordilleran geanticline through Early Jurassic time in central Nevada and Idaho is based on the apparent absence of Triassic and Jurassic rocks in that region. However, Lower Jurassic lithofacies patterns in Idaho, Utah, western Nevada, and Oregon fail to indicate the presence of an Early Jurassic geanticline in central Nevada or Idaho. The presence of marine Triassic rocks in north-central Nevada and Jurassic mature quartz sandstones in

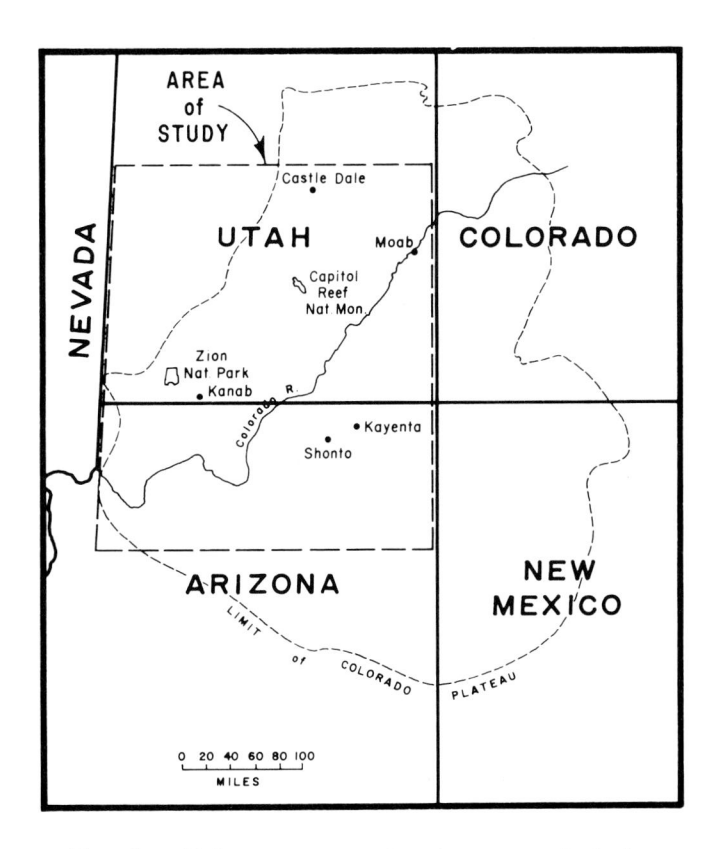

Fig. 1 — Reference map showing area of study.

western Nevada indicates that no continuous Mesocordilleran geanticline separated the miogeosynclinal and eugeosynclinal belts during early Mesozoic time (Stanley, et al., 1971). Early Jurassic sedimentary conditions typical of the eastern Cordilleran miogeosyncline extended uninterrupted across Nevada. The

[1] Apache Corporation, Oil and Gas Division, Denver, Colorado.

scarcity of Triassic and Jurassic strata on the postulated geanticline in central Nevada is due to post-Lower Jurassic erosion rather than non-deposition (Stanley, 1971).

STRATIGRAPHY

Chinle Formation (Mid (?) — Late Triassic)

The Chinle Formation consists of non-marine conglomerate, sandstone, siltstone, claystone, and limestone. The thickness of the Chinle is about 300 m near the southeastern corner off Utah, increasing westward to 400 m at Monument Valley, Utah, and thinning northward to 100 m in east-central Utah.

The lower portion of the Chinle is an alluvial plain deposit extending over the southern Colorado Plateau represented by widespread sandstones and conglomerates, i.e., braided stream deposits (Shinarump and Moss Back members) and argillaceous siltstones and sandstones, i.e., meandering stream and lake deposits (Monitor Butte and Petrified Forest members) (Stewart, 1961). Further evidence of the fluvial origin of these sediments includes carbonized and silicified wood fragments (Stewart, 1956) and non-marine vertebrates, i.e., amphibian, fish, and reptile (phytosaur) fossils (Stewart and Smith, 1954). Stream directions were north to northwest (Poole, 1961), the source being the Mogollon highland in southern Arizona (Stewart, 1961).

The upper part of the Chinle was deposited as a widespread alluvial plain containing many lakes which extended from northwest New Mexico to northwest Colorado. Deltas spread out into a large lake which covered most of northeastern Arizona, northwestern New Mexico, southeastern Utah, and southwestern Colorado (Stewart, 1956; 1961). The lithology of the lower member (Owl Rock Member) includes structureless mudstone and horizontally bedded siltstone interstratified with thin limestone beds (Stewart, 1961); the lithology of the upper member (Church Rock Member) consists of siltstone with interbedded cross-stratified sandstone (Stewart, 1961). The upper Chinle is characterized by small non-marine gastropods (Stewart and Smith, 1954) and freshwater pelecypods (*Unio*) (Baker, *et al.*, 1936; Stewart and Smith, 1954). Paleocurrent indicators of stream directions for the upper Chinle suggest that the streams probably flowed northwest (Stewart, 1956; 1961). Lithofacies — isopach maps for the Upper Triassic confirm the presence of a major basin centered south of the "four corners" (MacLachlan, 1972). Uncompahgre and Front Range igneous and metamorphic terrain were the main source areas (Stewart, 1961).

GLEN CANYON GROUP

Wingate Formation (Late Triassic)

The Wingate Sandstone consists of two mappable units. The lower member (Rock Point Member) is thickest (248 m) in the south-central part of the Navajo Indian Reservation, thins abruptly toward the western border of the reservation and gradually thins northward into Utah and eastward into New Mexico. The Rock Point was deposited in a shallow basin that plunged to the south (Harshbarger, *et al.*, 1957). The upper member (Lukachukai Member) maintains a thickness of about 100 m throughout the southern Colorado Plateau; it has a maximum thickness of 180 m in extreme northwest New Mexico and thins to a feather edge in western Colorado and central New Mexico.

Thin parallel-bedded siltstone and sandstone units characterize the Rock Point Member. Fluviatile sandstone lenses at the base of the Rock Point suggest an initial fluvial stage followed by silty, horizontally bedded, quiet-water deposition (Harshbarger, *et al.*, 1957). Sedimentary structures (mud cracks, ripple marks), color, composition, fossil evidence (worm trails), and regional trends of sorting and grain size suggest a semi-restricted lagoon (Harshbarger, *et al.*, 1957).

The Lukachukai Member is comprised of fine-grain sandstone with large-scale cross-stratification and according to Harsharger, *et al.*, represents a coastal dune facies which spread southwestward during the regression of the Rock Point lagoon. Intertonguing of the Lukachukai Member within the Rock Point

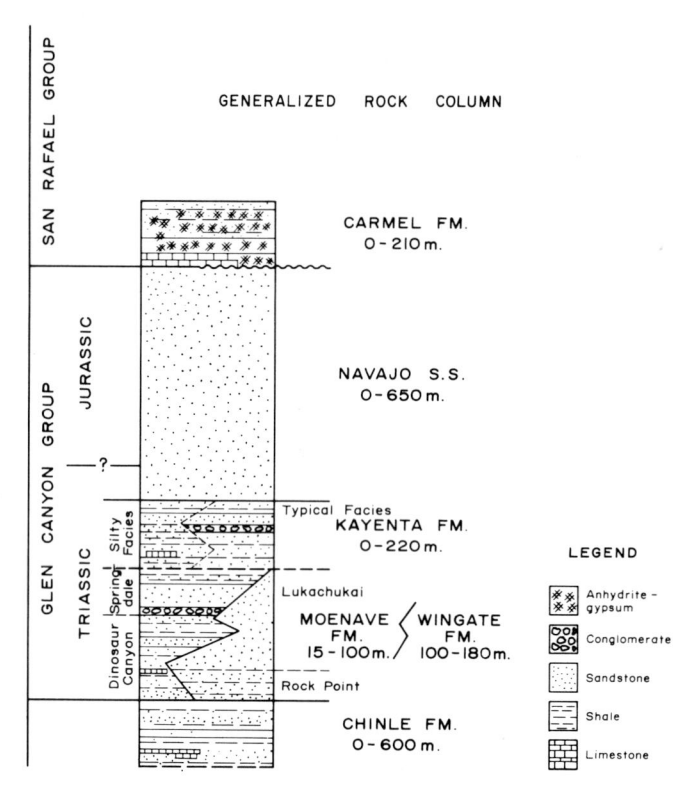

Fig. 2 — **Generalized stratigraphic column of Glen Canyon Group and Carmel Formation (San Rafael Group) of the southwestern Colorado Plateau.**

Member indicates shoreline oscillation. The cross-bedded sandstone tongues coalesce with the main body of the Lukachukai Member to the northeast and pinch-out toward the Rock Point Member to the southwest (Harshbarger, *et al.*, 1957). The Lukachukai Member interfingers with the underlying Chinle and with the overlying basal Kayenta Formation (Jordan, 1965). Paleocurrent directions for the Wingate (Poole, 1962) indicate a northwesterly wind direction.

Kayenta and Moenave Formations (Late Triassic (?))

The "typical facies" of the Kayenta Formation contains trough cross-bedded coarse- to very fine-grain sandstone and horizontally stratified siltstone, mudstone, and shale. Channeling, lenticular sand bodies, intraformational conglomerates, clay flakes, freshwater mollusks, and gastropods indicate a fluvial or flood plain environment. The Kayenta's "typical facies" undergoes a change to a "silty facies" in the southwestern portion of the Colorado Plateau (Harshbarger, *et al.*, 1957; Wright and Dickey, 1958; Stokes and Holmes, 1954; Baker, *et al.*, 1936).

The underlying Moenave Formation, which is similar to the Kayenta in lithology and depositional environment (Poole, 1961), consists of the Dinosaur Canyon Sandstone Member (lower Moenave) and the Springdale Sandstone Member (upper Moenave); the extent of the Moenave is confined to the southwestern Colorado Plateau. The contact between the Dinosaur Canyon Sandstone and Springdale Sandstone Members is gradational. Initial deposition of the Kayenta appears to have been continuous with deposition of the Moenave. Fish remains are found throughout the Moenave and primitive crocodile bones (*Protosuchus*) have been found in the Dinosaur Canyon Sandstone Member (Harshbarger, *et al.*, 1957). Quiet water deposition is suggested by minor amounts of fine-grain limestone in the Dinosaur Canyon Sandstone and silty facies of the Kayenta. The Dinosaur Canyon Sandstone is differentiated from the overlying Springdale Sandstone Member by an increased amount of silty units and clay matrix in the sandstones. The amount of fine-grain material increases from east to west indicating an easterly source. The contact between the Kayenta and the underlying Moenave is conformable, becoming more gradational to the southwest (Harshbarger, *et al.*, 1957). The thickness of the Moenave ranges from about 15 m in south-central Utah increasing westward to over 100 m in northwestern Arizona.

The Kayenta formation occupies southeastern and south-central Utah and northeastern Arizona and is usually about 65 m thick, although it thickens abruptly along its western edge to 220 m at Zion Canyon, Utah (Averitt, *et al.*, 1955). The eastern pinchout is probably depositional but some pre-San Rafael Group erosion is suggested by increased thinning wherever the Kayenta lies directly beneath the San Rafael Group in southeastern Colorado (Craig and Dickey, 1956). The northern boundary of the Kayenta is the Uinta Basin. It is unresolved whether the Kayenta wedges-out beneath the Uinta Basin (Stokes and Holmes, 1954) or is lost in a facies change within the Glen Canyon Group of the Uinta Mountains (MacLachlan, 1957). In southwestern Utah, the Kayenta and Navajo intertongue on a large scale (Averitt, *et al.*, 1955; Wright and Dickey, 1963; Stokes, 1963). Elsewhere on the Colorado Plateau the Kayenta — Navajo contact is usually gradational. The contact is conformable in that exact boundaries are difficult to select due to intertonguing and the lack of a widespread erosional surface. The facies change of the Kayenta from predominantly sandstone in southeast Utah to predominantly siltstone in southwest Utah and northwest Arizona (Averitt, *et al.*, 1955) suggests transport from a source area located to the northeast in western Colorado (Uncompahgre highland) (Craig and Dickey, 1956). Poole's (1961) paleocurrent map indicates that Kayenta and upper Moenave streams flowed west and southwest.

The abrupt thickening of the silty facies and Dinosaur Canyon Sandstone along their depositional margin in northwestern Arizona and southwestern Utah suggests that the Kayenta and Moenave streams formed a large deltaic system as they flowed southwest into the basin.

Navajo Formation (Late Triassic (?) — Early Jurassic)

The Navajo Sandstone (Late Triassic (?) — Early Jurassic) is a homogenous feldspathic quartz arenite presently cropping out in an area approximately delineated by the boundaries of the Colorado Plateau in southern Utah, southwestern Colorado, and northeastern Arizona. The Navajo Formation comprises the uppermost portion of the Glen Canyon Group, overlies the Kayenta Formation (Late Triassic (?)), and is overlain by the Carmel Formation (Middle Jurassic) of the San Rafael Group.

The Navajo has been interpreted as desert-eolian in origin principally based upon "eolian-type" cross-bedding; well rounded, well sorted, and frosted grains; and lack of fossil evidence. (Gregory, 1917; Reeside, 1929; Baker, 1946; Kiersch, 1950; Harshbarger, *et al.*, 1957; Stokes, 1961; Poole, 1962; and others).

However, in light of reinvestigation of the Navajo, these criteria are insufficient to conclude an eolian interpretation. Other authors (Keller, 1945; Grater, 1948; Digert, 1955; Jordan, 1965; Marzolf, 1969; Stanley, *et al.*, 1971; Freeman and Visher, 1975) have hypothesized a marginal marine rather than desert-eolian environment.

The "eolian-type" cross-bedding is the primary basis for the desert-eolian interpretation for the Navajo Sandstone; however, no single type of cross-bedding typifies present-day eolian dunes (Pryor, 1971). Trough (festoon)-type cross-bedding is relatively uncommon in modern eolian dunes (McKee, 1966); whereas, trough cross-bedding predominates in the Navajo

(Jordan, 1965). Cross-stratification similar in size and morphology to that found in the Navajo is presently found in marine environments; trough and planar cross-stratification are produced by tidal channels and marine bed forms, i.e., megaripples, sand waves, and tidal current ridges (Hoyt, 1962, 1967; Off, 1963; Houbolt, 1968).

Depositional dips of high-angle marine cross-bedding range from 20° up to the maximum repose angle of subaqueous sand (30°) as discussed by Hoyt (1967); such dip angles are compatible with measurements of Navajo cross-bed foresets made by Kiersch (1950). Depositional dips of Holocene eolian dune foresets are generally steeper, usually ranging from 30° to 34° with some dips up to 42° (McKee, 1957, 1966; Bigarella, et al., 1969). Glennie (1972) explained the large discrepancy between Navajo and modern dune cross-bed dips by assuming that the Navajo originally possessed eolian dips and has undergone 25-27 percent compaction of sediment. Such a high degree of compaction seems unlikely since the amount of compaction of sand grains achieved by rearrangement under load is small and accounts for a decrease in porosity of only a few percent (Pettijohn, et al., 1972).

The high degree of sorting found within the Navajo is inconclusive in determining environment of deposition as both eolian and tidal current processes produce comparably high sorting coefficients.

Although the modifiers "well rounded" and "frosted" have been used to describe the eolian nature of Navajo sand grains, roundness lacks environmental significance in ancient-modern analogue models and the evidence is that the frosting on Navajo grains is a result of diagenetic solution and precipitation of quartz rather than mechanical action (Freeman and Visher, 1975).

There are no marine fossils found in the Navajo; however, this does not preclude a marine origin: (1) a high energy, tidal dominated sandy shelf subjected to strong bidirectional currents is not conducive to the support of marine benthonic forms; (2) biogenic structures, such as tracks and trails, when preserved, are difficult to distinguish in a homogenous sand medium; (3) water conditions could have been abnormal (highly saline or alkaline) and unsuitable for the support of marine organisms; and (4) the high porosity and permeability of Navajo deposits are unfavorable for the preservation of hard parts.

The body fossils of two bipedal reptiles and one primitive crocodile from northeastern Arizona have been cited as further proof of the Navajo's terrestrial origin, but the rare occurrence of these vertebrates does not substantiate such an interpretation. The bodies of these small dinosaurs possibly were carried out to sea by Chinle, Kayenta-Moenave streams and rivers flowing from the eastern upland terrain where vertebrates were common.

Freeman and Visher (1975) found that comparisons of log-probability curve shapes of samples from Navajo and modern tidal-current environments show they were deposited by similar processes. The Navajo possesses sedimentary structures, such as current lineation, bioturbation, and large-scale convolute bedding, characteristic of subaqueous conditions of deposition. Pelletal glauconite occurs in the Nugget Formation of the Wind River Basin (a Navajo correlative) and is indicative of a marine environment.

The intertonguing relationship of the Navajo with the overlying marine Carmel Limestone (Wright and Dickey, 1963) casts serious doubt upon an eolian-desert origin for the Navajo. A broad shallow shelf is suggested by increases in Navajo thickness and sand-shale ratios toward a probable shelf margin to the northwest. Similarly, the lithofacies of the Lower Jurassic (Fig. 3) shows the initial stage of transgression from west and northwest by the Navajo Sandstone and its correlatives (Nugget and Aztec formations), followed by the decidedly marine lithofacies of the Middle and Upper Jurassic (Carmel and Curtis-Summerville formations) shown in Figure 4.

The Navajo represents the shelf sands of an overall transgression across a shallow tidal-dominated shelf floor. These marine sands encroached eastward upon continental deposits (deltaic, fluvial, desert, and lacustrine) and were a prelude to the sea in which the Carmel Formation was deposited.

SAN RAFAEL GROUP

Carmel Formation (Mid Jurassic)

The Carmel Formation attains a maximum thickness in central Utah of 210 m and thins to a feather edge near the western Colorado border. The lower Carmel Formation includes interbedded oolitic and fossiliferous limestone and calcareous shale ("limestone facies"). The fossils are mostly marine pelecypods and ammonites of Middle Jurassic and Late Jurassic age (Bajocian-Callovian Stages) (Wright and Dickey, 1958). The lower Carmel grades upward into thick bedded gypsum and anhydrite ("gypsum facies") and eastward into gypsiferous sandstone and siltstone ("red bed facies") (Baker, et al., 1936; Wright and Dickey, 1958). To the west, the limestone facies is correlated with the Twin Creek Limestone (Baker, et al., 1936; Richards, 1958; Wright and Dickey, 1958).

The limestone facies and gypsum facies are interpreted as normal marine and restricted lagoon, respectively; the red bed facies is characterized by gypsum stringers, mud cracks, ripple marks, and bioturbation (worm trails) and is interpreted as representing tidal flat and lagoonal environments under high-evaporation conditions (Wright and Dickey, 1958; Peterson, 1957).

In southeastern Utah and northeastern Arizona, the Carmel rests unconformably on the Navajo as indicated by the sharp

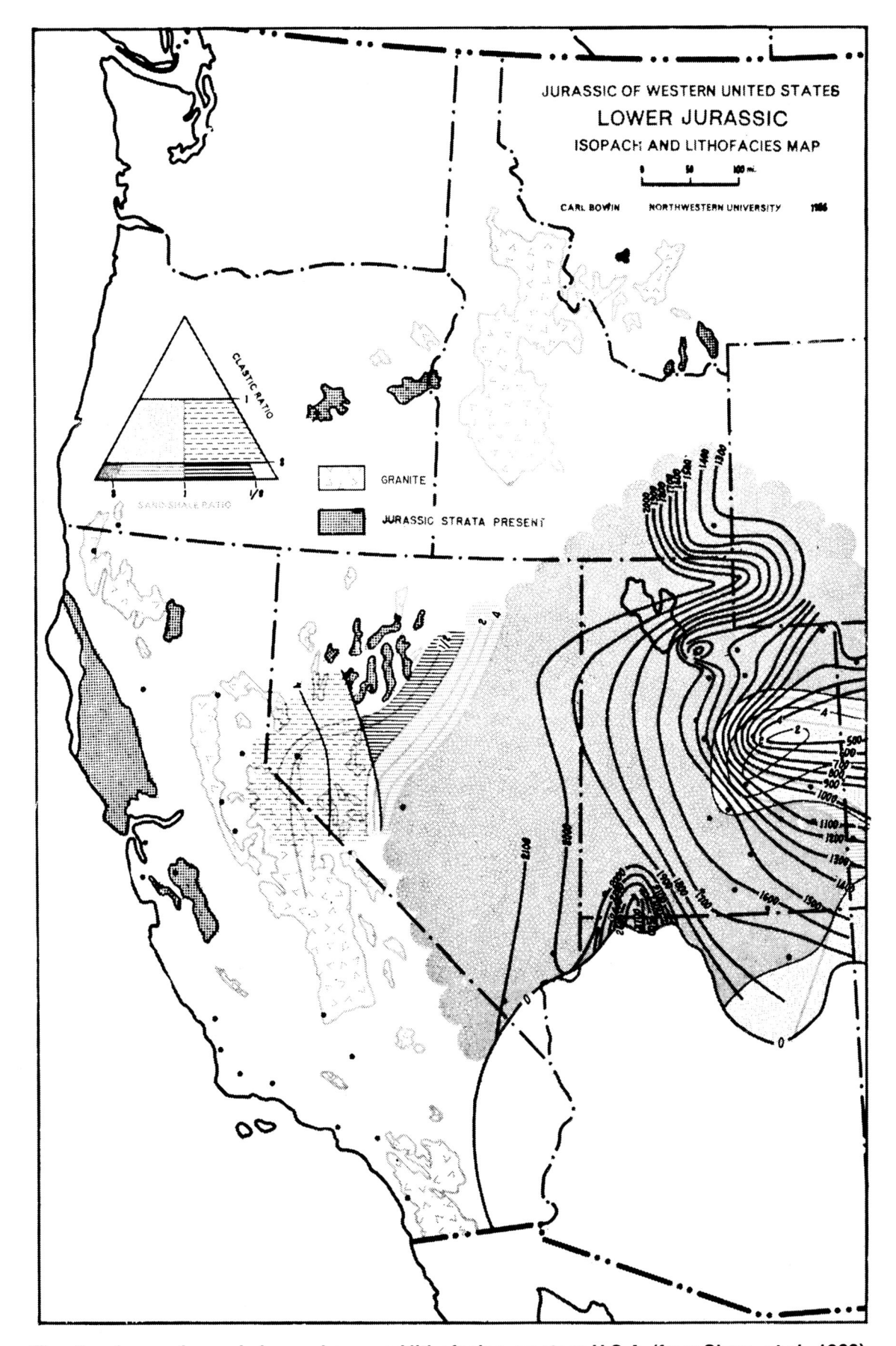

Fig. 3 — Lower Jurassic isopachous and lithofacies, western U.S.A. (from Sloss, et al., 1960).

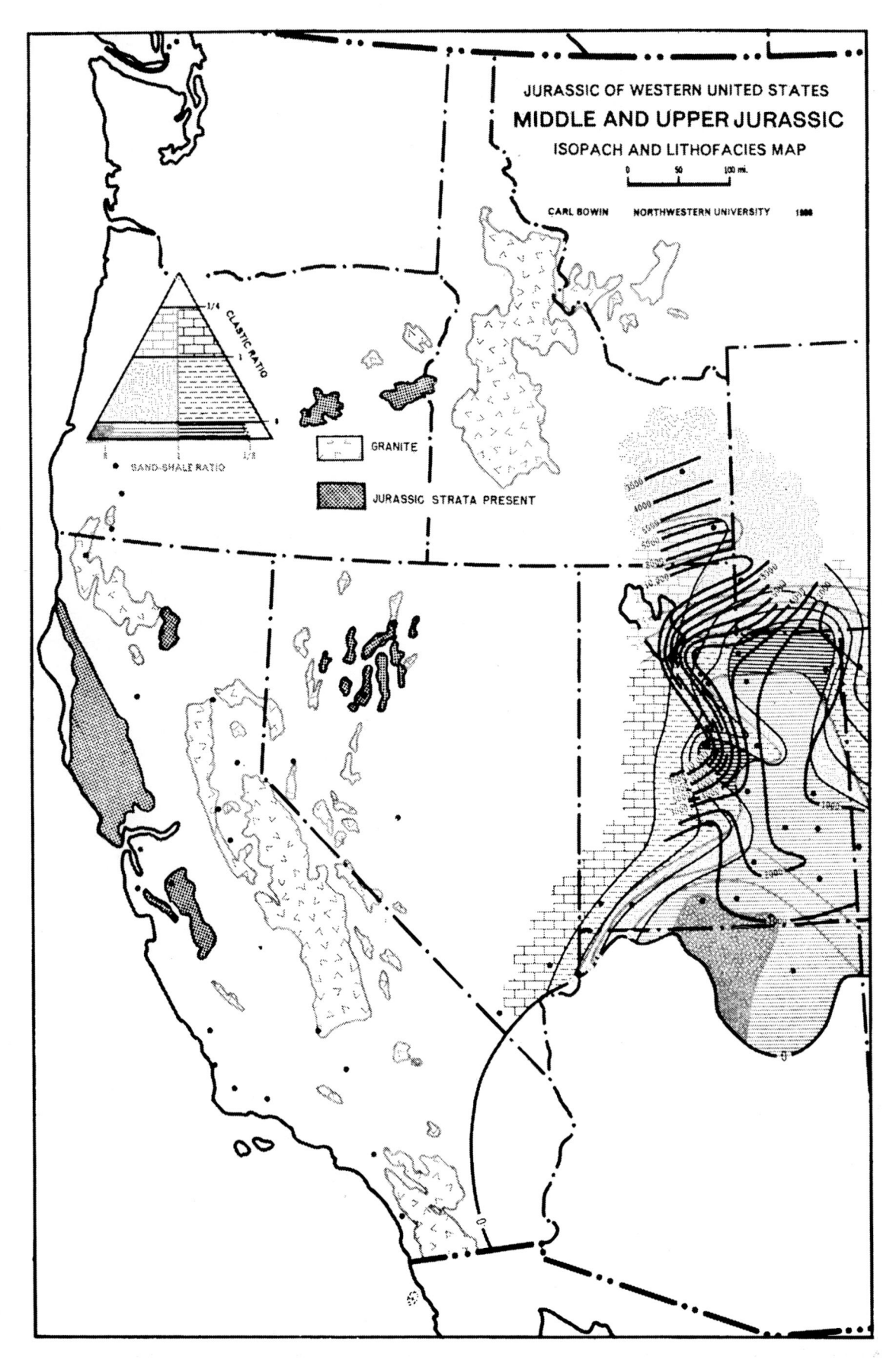

Fig. 4 — Middle and Upper Jurassic isopachous and lithofacies, western U.S.A.

truncation of the Navajo cross-bedding, abrupt lithologic change, and accumulations of clastic chert at the Carmel-Navajo contact (Dane, 1935; McKnight, 1940; Baker, 1946, Harshbarger, *et al.*, 1957; Wright and Dickey, 1958; Pipiringos and O'Sullivan, 1975). The unconformity diminishes westward where the Carmel and Navajo intertongue in southwestern Utah (Wright and Dickey, 1963).

SUMMARY

A summary of the lithologies and environments of deposition of the Glen Canyon Group (Chinle, Wingate, Moenave-Kayenta, and Navajo Formations) and the Carmel Formation (San Rafael Group) is found in Table 1.

DEPOSITIONAL HISTORY OF THE GLEN CANYON GROUP AND CARMEL FORMATION

Problems in correlation and environmental reconstruction in the Colorado Plateau are (1) the paucity of fossil evidence resulting in vague time-rock relationships, (2) the lack of consistent, regional unconformable surfaces, and (3) the absence of widespread isochronous marker-beds. Figure 5 is a southwest-northeast cross-section through southern Utah and central Colorado. This cross-section shows the rock units and their major facies changes (including environmental interpretations) within the Glen Canyon Group and the overlying Carmel Formation as previously discussed.

Time lines were drawn parallel to the Carmel-Navajo unconformity in eastern Utah (Dane, 1935; McKnight, 1940; Baker, 1946; Kinney, 1955; Harshbarger, *et al.*, 1957; Wright and Dickey, 1958; Pipiringos and O'Sullivan, 1975). A sequence of paleoenvironmental interpretations based upon the facies occurrence within particular time intervals is shown in Figures 6 through 9.

Stage I

Figure 6 is a paleoenvironmental map as interpreted from the facies pattern during the assumed time interval T_1-T_2.

Note the eastward transition from shelf sandstone (Navajo Formation) to continental facies, i.e., lagoonal-deltaic (Rock Point Member of Wingate Formation), fluvio-deltaic (Moenave and Kayenta formations), and fluvio-lacustrine (upper part of Chinle Formation). The source of the shelf sands was streams which flowed generally westward from the ancestral Rocky Mountain highland and continental source areas to the east. These streams with paleochannels in the upper part of the Chinle and in the Kayenta and Moenave developed deltaic deposits out

Table 1 Stratigraphic—Environmental Summary

Rock Unit	Environment	Predominant Lithology
Carmel Formation		
Red Bed Facies	tidal flat	gypsiferous sltstn., ss.
Gypsum Facies	restricted lagoon	bedded gyp., anhyd.
Limestone Facies	shallow marine	fossiliferous ls.
Navajo Formation	transgressive marine	cross-bedded ss.
Kayenta Formation		
"Typical" Facies	fluvial	ss., sltstn., conglom.
Silty Facies	fluvial-deltaic(?)	sltstn., ss., cherty ls.
Moenave Formation		
Springdale Sandstone Member	fluvial	ss., sltstn., conglom.
Dinosaur Canyon Sandstone Member	fluvial-deltaic(?)	sltstn., ss., cherty ls.
Wingate Formation		
Lukachukai Member	eolian	cross-bedded ss.
Rock Point Member	lagoonal-deltaic	sltstn., ss.
Chinle Formation		
Upper Part	fluvio-lacustrine	sltstn., ss., freshwater ls.
Lower Part	fluvial	ss., sltstn., conglom.

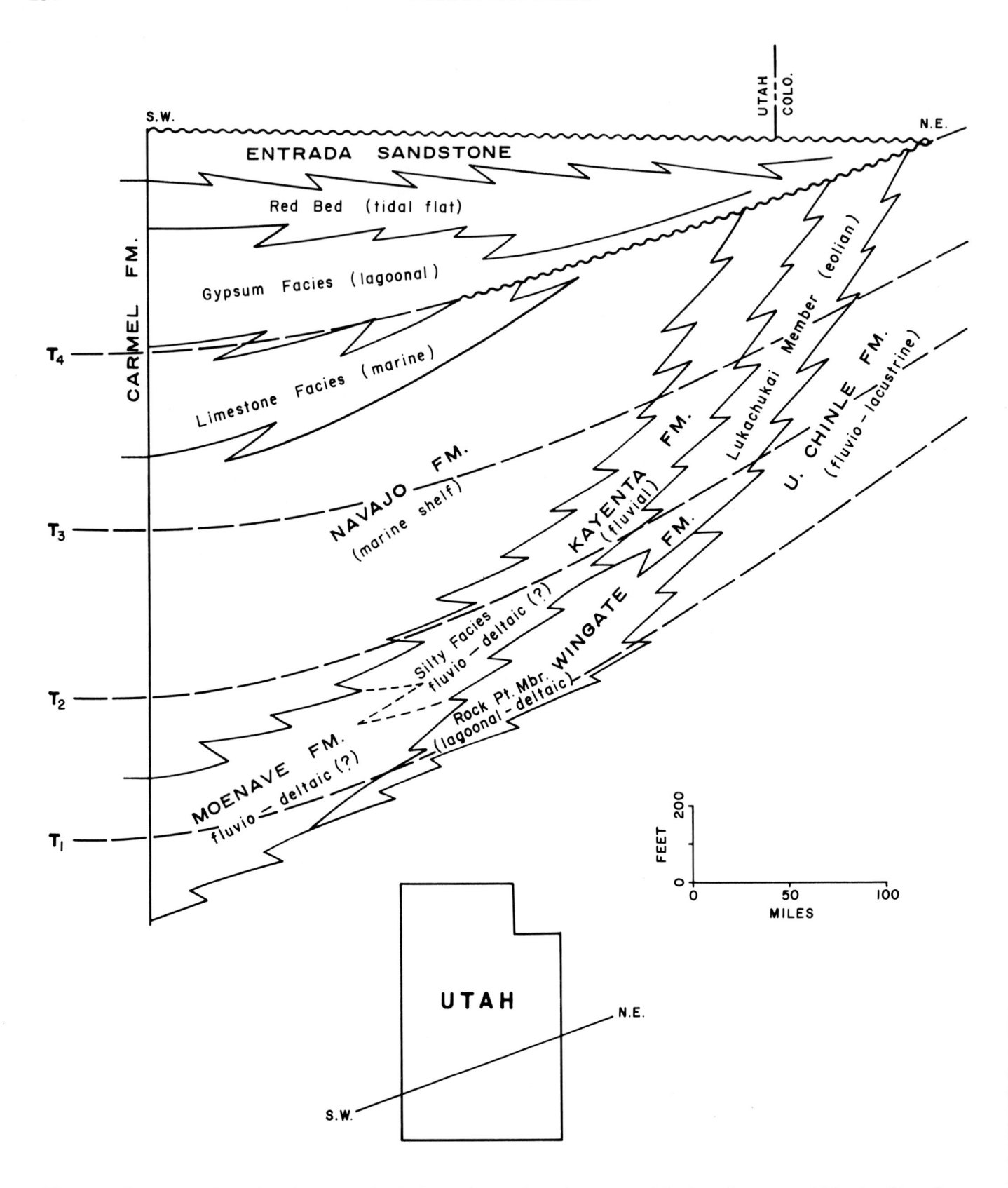

Fig. 5 — Cross-section showing stratal relationships and environmental facies changes within the Glen Canyon Group and Carmel Formation. Assumed time lines are drawn parallel to the Carmel-Navajo unconformity (drawn from Peterson, 1972).

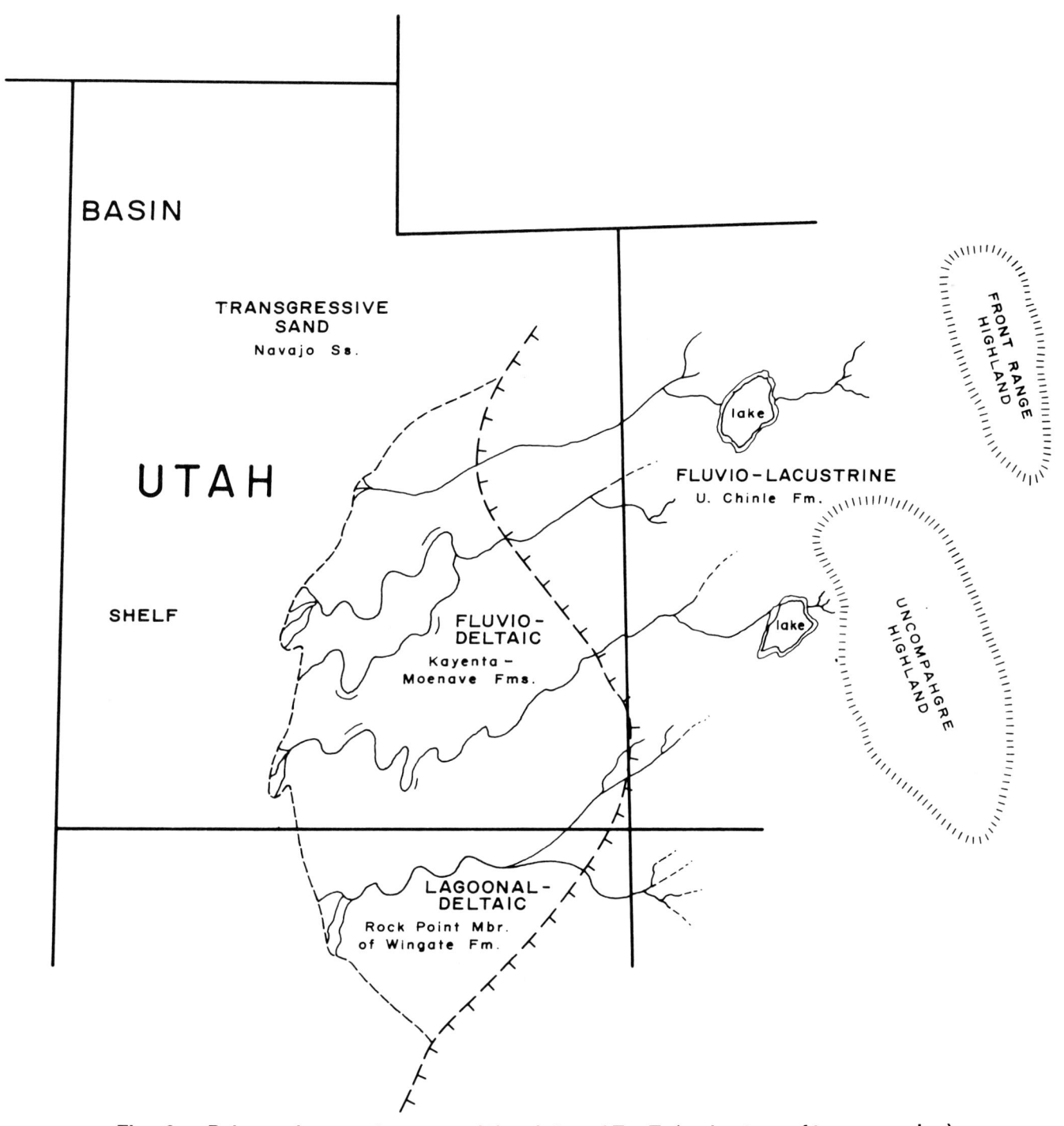

Fig. 6 — Paleoenvironment: assumed time interval T_1-T_2 (early stage of transgression).

into the basin (Dinosaur Canyon Sandstone Member of Moenave Formation, Rock Point Member of Wingate Formation, and silty facies of Kayenta Formation). The fluvial sands were reworked and redeposited as tidal ridge and sand wave fields on the shallow, tidal-dominated shelf floor (Navajo Formation). A narrow belt of coastal dunes (Navajo Formation) may have marked the forefront of the transgressive advance, but most of these dune structures had little preservation potential and probably were destroyed by the transgression.

Stage II

Figure 7 is a paleoenvironmental map as interpreted from the facies deposited during the assumed time interval T_2-T_3.

The transgressive sands overlapped the fluvio-deltaic facies of the Kayenta-Moenave and lagoonal-deltaic facies of the lower part of the Wingate (Rock Point Member) and lapped onto the Kayenta alluvial plain. Along the eastern margin of the

Kayenta alluvial plain, interior eolian dunes (Wingate Formation) developed due to prevalent northwesterly winds and arid conditions (Poole, 1962; Harshbarger, et al., 1957), the source of the sand being the Kayenta alluvial plain deposits. Farther east, the eolian dune deposits of the Kayenta-Wingate basin grade into upland fluvio-lacustrine deposits of the upper part of the Chinle.

Stage III

Figure 8 is a paleoenvironmental map based upon interpretation of the facies deposited during the assumed time interval $T_3 - T_4$.

The figure shows the continuation of the transgression and the development of "reefal" limestone (limestone facies of Carmel Formation) to the northwest. Within this stage, the supply of sand from streams diminished as the transgression covered the continental interior.

Stage IV

Figure 9 shows the paleoenvironmental interpretation and

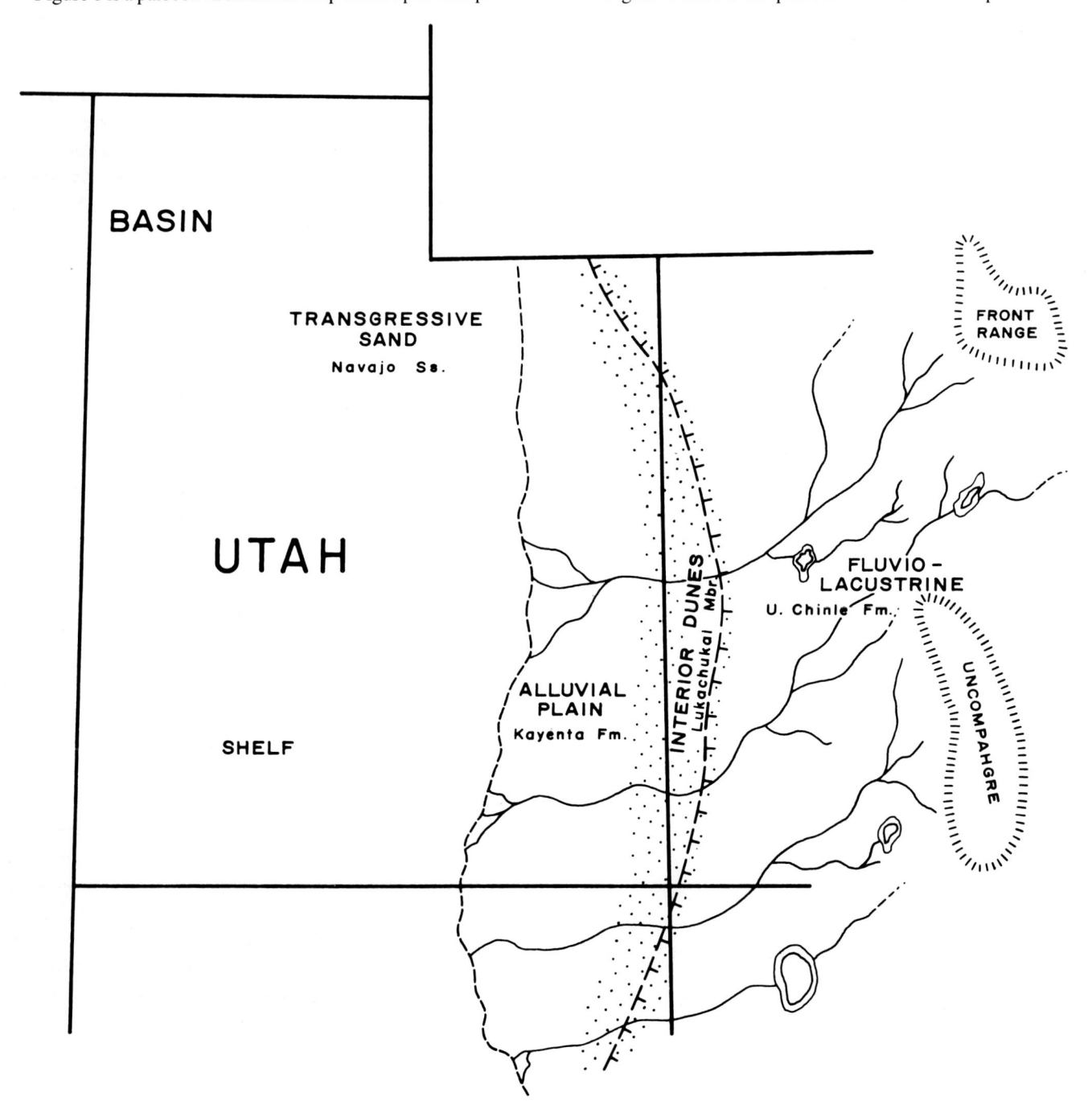

Fig. 7 — Paleoenvironment: assumed time interval T_2-T_3 (continued transgression).

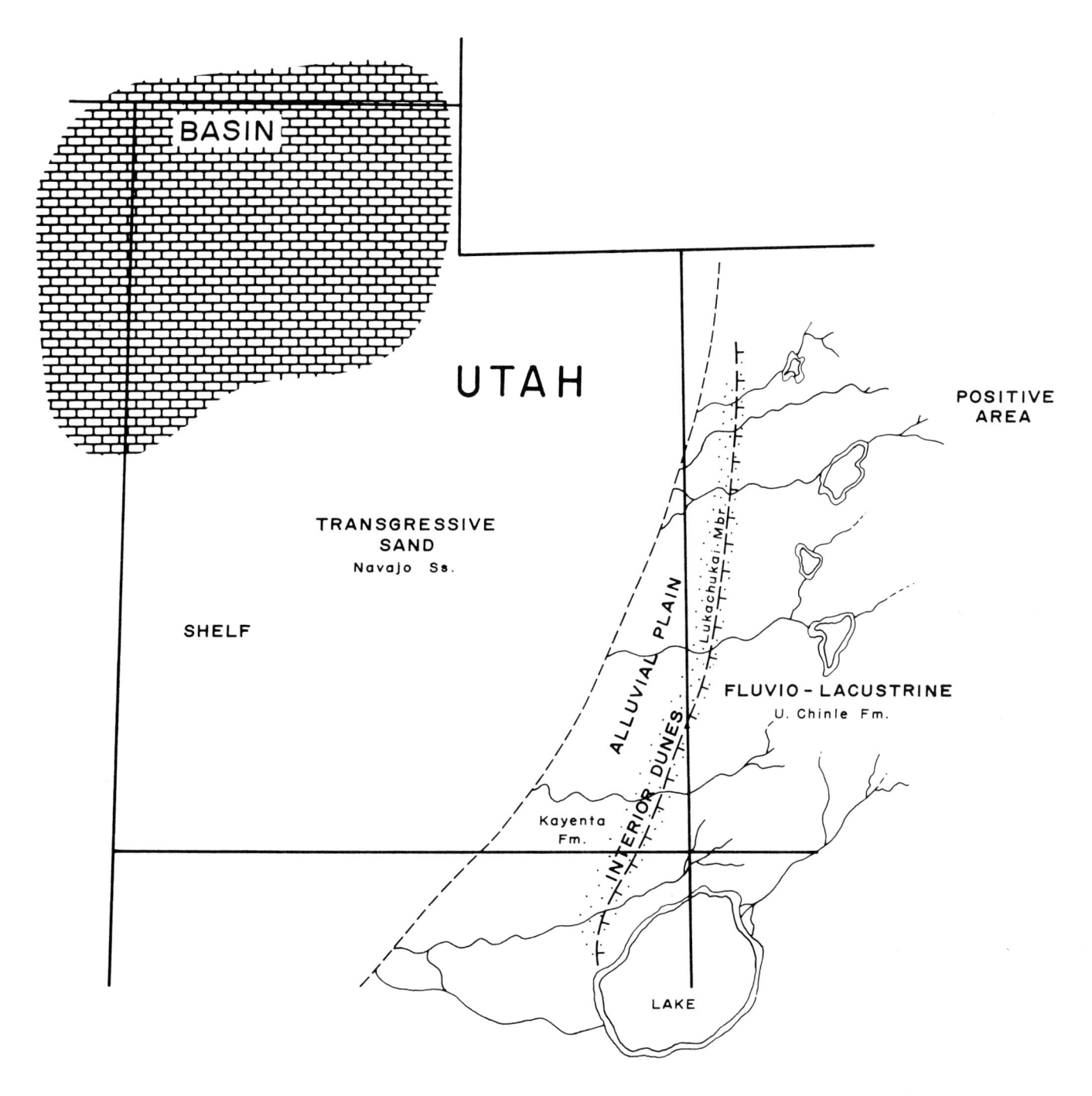

Fig. 8 — Paleoenvironment: assumed time interval T_3-T_4 (continued transgression).

facies patterns occurring after assumed time-line T_4 (i.e., during the time of deposition of the Carmel Formation).

The transgression attained its ultimate phase. The Carmel Formation onlapped the eastern shore. Combined factors of cessation of sand supply via streams and erosion to wave base by the sea in which the Carmel was deposited produced a marine unconformity between the Carmel Formation and the Glen Canyon Group. Facies patterns indicate tidal flat deposits (red bed facies) near shore, and restricted lagoon (gypsum facies) and normal marine deposition (limestone facies) basinward; the basin filled with sediments and the waters shallowed.

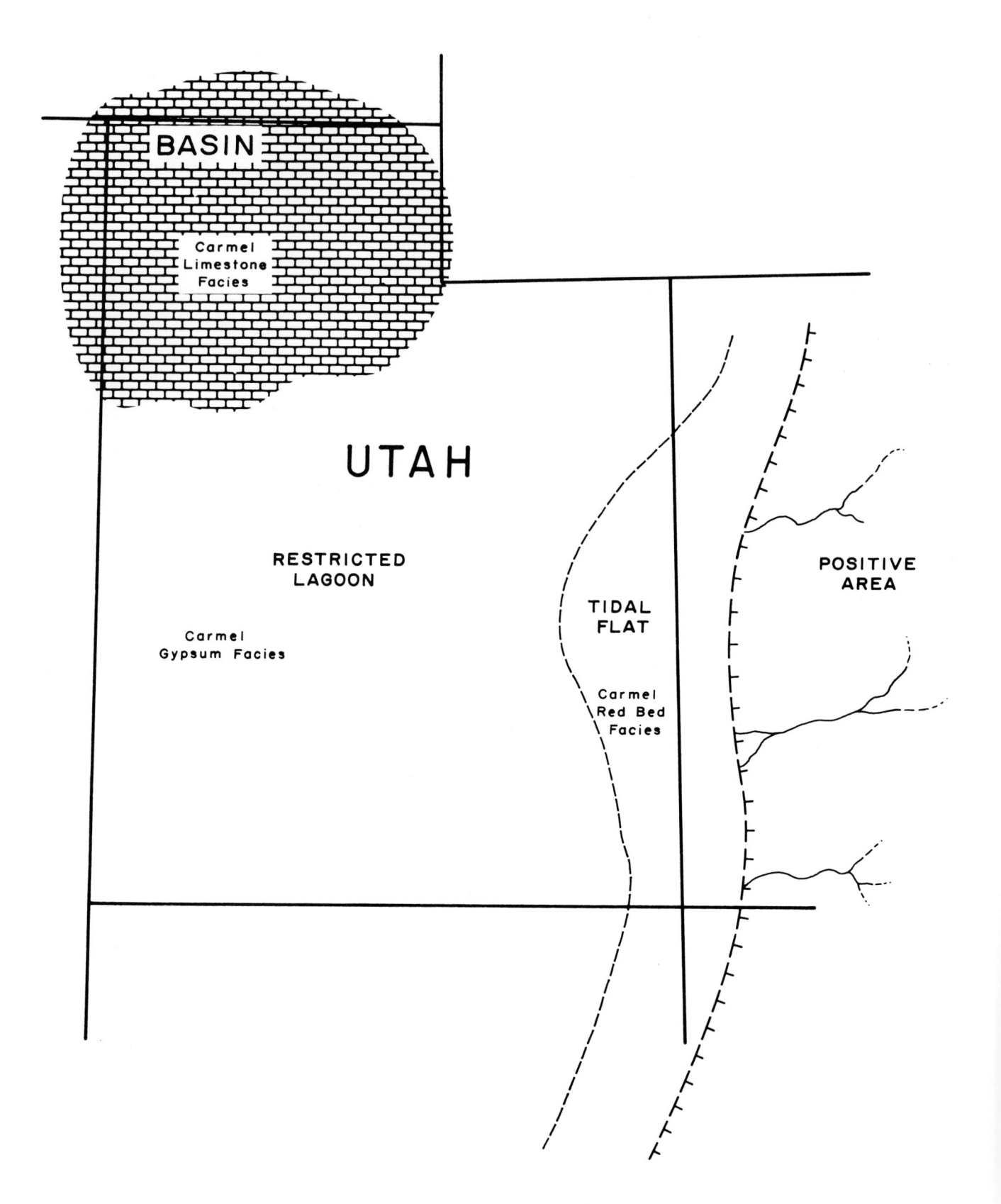

Fig. 9 — Paleoenvironment: assumed time interval T_4 plus (time of deposition of Carmel Formation) (advanced stage of transgression).

REFERENCES CITED

Averitt, P., Detterman, J. S., Harshbarger, J. W., Repenning, C. A. and Wilson, R. F., 1955, Revisions in correlation and nomenclature of Triassic and Jurassic formations in southwestern Utah and northern Arizona: Am. Assoc. Petroleum Geologists Bull., v. 39, p. 2515-2525.

Baker, A. A., 1946, Geology of the Green River Desert-Cataract Canyon region, Emergy, Wayne, and Garfield Counties, Utah: U.S. Geol. Survey Bull. 951, 122 p.

_____, Dane, C. H., and Reeside, J. B., Jr., 1936, Correlation of the Jurassic formations of parts of Utah, Arizona, New Mexico, and Colorado: U.S. Geol. Survey Prof. Paper 183, 66 p.

Bigarella, J. J., Becker, R. D., and Duarte, G. M., 1969, Coastal dune structures from Parana, Brazil: Marine Geology, v. 7, p. 5-55.

Craig, L. C., and Dickey, D. D., 1956, Jurassic strata of southeastern Utah and southwestern Colorado: in Intermtn. Assoc. Petroleum Geologists Guidebook, 7th Ann. Field Conf., p. 93-104.

Dane, C. H., 1935, Geology of the Salt Valley anticline and adjacent areas, Grand County, Utah: U.S. Geol. Survey Bull. 863, 184 p.

Digert, F. E., 1955, The depositional environment of the Navajo Sandstone of the southwestern Colorado Plateau: unpub. M.A. thesis, Univ. of Wisconsin, 55 p.

Dunbar, C. O., and Waage, K. M., 1969, Historical geology: New York, John Wiley and Sons, 556 p.

Eardley, A. J., 1962, Structural geology of North America, 2nd ed.: New York, Harper and Bros., 743 p.

Freeman, W. E., and Visher, G. S., 1975, Stratigraphic analysis of the Navajo Sandstone: Jour. Sedimentary Petrology, v. 45, no. 3, p. 651-668.

Glennie, K. W., 1970, Desert sedimentary environments: New York, Elsevier Publishing Company, 222 p.

Grater, R. K., 1948, Some features of the Navajo Formation in Zion National Park, Utah: Am. Jour. Sci., no. 246, p. 311-318.

Gregory, H. E., 1917, Geology of the Navajo country, a reconnaissance of parts of Arizona, New Mexico, and Utah: U.S. Geol. Survey Prof. Paper 93, 161 p.

Harshbarger, J. W., Repenning, C. A., and Irwin, J. H., 1957, Stratigraphy of the uppermost Triassic and the Jurassic rocks of the Navajo country: U.S. Geol. Survey Prof. Paper 291, 74 p.

Houbolt, J. J. H. C., 1968, Recent sediments in the southern bight of the North Sea: Geologie en Mijnbouw, v. 47, no. 4, p. 245-273.

Hoyt, J. H., 1962, High-angle beach stratification, Sapelo Island, Georgia: Jour. Sed. Petrology, v. 32, p. 309-311.

_____, 1967, Occurrence of high angle stratification in littoral and shallow neritic environments, central Georgia coast, U.S.A.: Sedimentology, v. 8, p. 229-238.

Jordan, W. M., 1965, Regional environmental study of the early Mesozoic Nugget and Navajo Sandstones: unpub. Ph.D. dissertation, Univ. of Wisconsin, 206 p.

Keller, W. D., 1945, Size distribution of sand in some dunes, beaches, and sandstones: Am. Assoc. Petroleum Geologists Bull., v. 29, p. 215-223.

Kiersch, G. A., 1950, Small scale structures and other features of the Navajo Sandstone, northern part of San Rafael Swell, Utah: Am. Assoc. Petroleum Geologists Bull., v. 34, p. 923-942.

Kinney, D. M., 1955, Geology of the Uinta River-Brush Creek area, Duchesne and Uinta Counties, Utah: U.S. Geol. Survey Bull. 1007, 185 p.

MacLachlan, M. E., 1957, Triassic stratigraphy in parts of Utah and Colorado: in Intermtn. Assoc. Petroleum Geologists Guidebook, 8th Ann. Field Conf., p. 82-91.

_____, 1972, Triassic system: in Geologic atlas of the Rocky Mountain region, Rocky Mtn. Assoc. Geologists, p. 172.

Marzolf, J. E., 1969, Regional stratigraphic variations in primary features of the Navajo Sandstone, Utah (abs.): Geol. Soc. America Abstracts with Programs for 1969, pt. 3, p. 40.

McKee, E. D., et al., 1956, Paleotectonic maps of the Jurassic System: U.S. Geol. Survey Misc. Geol. Inv. Map I-275, 6 p.

_____, 1957, Primary structures in some Recent sediments: Am. Assoc. Petroleum Geologists Bull., v. 41, p. 1704-1747.

_____, 1966, Dune structures: Sedimentology, v. 7, p. 1-69.

McKnight, E. T., 1940, Geology of area between Green and Colorado Rivers, Grand and San Juan Counties, Utah: U.S. Geol. Survey Bull. 908, 185 p.

Off, T., 1963, Rhythmic linear sand bodies caused by tidal currents: Am. Assoc. Petroleum Geologists Bull., v. 47, p. 324-341.

Peterson, J. A., 1957, Marine Jurassic of northern Rocky Mountains and Williston Basin: Am. Assoc. Petroleum Geologists Bull., v. 41, p. 399-440.

_____, 1972, Jurassic System: in Geologic atlas of the Rocky Mountain region, Rocky Mtn. Assoc. Geologists, Denver, Colorado, p. 188.

Pettijohn, F. J., Potter, P. E., and Siever, R., 1972, Sand and sandstone: New York, Springer-Verlag, 618 p.

Pipiringos, G. N., and O'Sullivan, R. B., 1975, Chert pebble unconformity at top of the Navajo Sandstone in southeastern Utah: in Four Corners Geol. Soc. Guidebook, 8th Ann. Field Conf., p. 149-156.

Poole, F. G., 1961, Stream directions in Triassic rocks of the Colorado Plateau: U.S. Geol. Survey Prof. Paper 424C, p. 139-141.

_____, 1962, Wind directions in late Paleozoic to middle Mesozoic time on the Colorado Plateau: U.S. Geol. Survey Prof. Paper 450D, p. 147-151.

Pryor, W. A., 1971, Petrology of the Permian yellow sand of northeastern England and their North Sea basin equivalents: Sedimentary Geology, v. 6, p. 221-254.

Reeside, J. B., Jr., 1929, Triassic-Jurassic "red beds" of the Rocky Mountain region; A discussion: Jour. Geology, v. 37, p. 47-63.

Richards, H. B., 1958, Cyclic deposition in the Jurassic Carmel Formation of eastern Utah: Jour. Sed. Petrology, v. 28, no. 1, p. 40-45.

Sloss, L. L., Dapples, E. C., and Krumbein, W. C., 1960, Jurassic of western United States: in Lithofacies Maps and Atlas of the United States and southern Canada: New York, John Wiley and Sons, Inc., p. 57.

Stanley, K. O., 1971, Tectonic and sedimentologic history of Lower Jurassic Sunrise and Dunlap Formations, west-central Nevada: Am. Assoc. Petroleum Geologists Bull., v. 55, no. 3, p. 454-477.

_____, Jordan, W. M., and Dott, R. H., 1971, New hypothesis of Early Jurassic paleogeography and sediment dispersal for western United States: Am. Assoc. Petroleum Geologists Bull., v. 55, no. 1, p. 10-19.

Stewart, J. H., 1956, Triassic strata of southwestern Utah and southwestern Colorado: in Intermtn. Assoc. Petroleum Geologists Guidebook, 7th Ann. Field Conf., p. 85-92.

_____, 1961, Stratigraphy and origin of the Chinle Formation on the Colorado Plateau: unpub. Ph.D. dissertation, Stanford University, 247 p.

_____, and Smith, J. F., 1954, Triassic rocks in the San Rafael Swell, Capitol Reef and adjoining parts of southwest Utah: in Intermtn. Assoc. Petroleum Geologists Guidebook, 5th Ann. Field Conf., p. 25-33.

Stokes, W. L., 1961, Fluvial and eolian sandstone bodies in the Colorado Plateau: in Geometry of Sandstone Bodies, Am. Assoc. Petroleum Geologists, p. 151-178.

_____, 1963, Triassic and Jurassic formations of southwestern Utah: in Intermtn. Assoc. Petroleum Geologists Guidebook, 12th Ann. Field Conf., p. 60-64.

_____, and Holmes, C. N., 1954, Jurassic rocks of south-central Utah: in Intermtn. Assoc. Petroleum Geologists Guidebook, 5th Ann. Field Conf., p. 34-41.

Wright, J. C., and Dickey, D. D., 1958, Pre-Morrison Jurassic strata of southwestern Utah: in Intermtn. Assoc. Petroleum Geologists Guidebook, 9th Ann. Field Conf., p. 172-181.

_____, and Dickey, D. D., 1963, Relations of the Navajo and Carmel Formations in southwest Utah and adjoining Arizona: U.S. Geol. Survey Prof. Paper 450E, p. 63-67.

JURASSIC SALTS OF THE HINGELINE AREA SOUTHERN ROCKY MOUNTAINS

by

Alan R. Hansen[1]

ABSTRACT

Jurassic-age salt is known to exist in ancient graben valleys from southeast Idaho to northwest Arizona. The salt occurs as diapiric intrusions and bedded salt, and has been dated as Middle and Upper Jurassic.

Salt domes in central and southeast Arizona are apparently associated with graben valleys marginal to the Colorado Plateau, but their ages have not been positively identified.

The composite trend of salt occurrences along the western and southwest margins of the Colorado Plateau are structurally and stratigraphically similar, suggesting a "mother salt" of Jurassic age from southeast Arizona to southeast Idaho. This trend is now at a position coincident with the Paleozoic "hingeline" and the present eastward limit of the Rocky Mountain overthrust belt.

INTRODUCTION

Recent oil and gas discoveries in Jurassic age rocks along the "hingeline" areas of Utah and Wyoming has focused increased attention on similar areas in Idaho and Arizona.

GEOLOGIC SETTING

Regionally, the hingeline extends north-south, representing a zone of transition between the geosynclinal facies to the west and continental shelf deposits on the east (Fig. 1). Except for occasional eastward transgressions of the sea onto the continental shelf and a Pennsylvanian uplift in central Utah, the position of this transition zone remained rather closely confined during Cambrian through Middle Jurassic time. The hingeline of central Utah is part of this overall Rocky Mountain orogenic complex. Reference should be made to the diagrammatic cross section (Fig. 2) as this text attempts to reconstruct the Mesozoic depositional history of the hingeline area.

Triassic sedimentation in western Utah represented a continuation of the widespread marine carbonate deposition which lasted from Permian until Early Jurassic when marine waters were drained from central and eastern Utah for the last time, receding to the north (Armstrong, 1968, p. 432). To the east the Ancestral Rockies of western Colorado were once again topographic highs and supplied major volumes of clastic sediments into the Utah hingeline (Stokes, 1972). A time-rock chart for the Jurassic is presented in Figure 3.

[1] Consulting Geologist, Denver, Colorado

The writer is indebted to Floyd C. Moulton for many suggestions that improved the manuscript and illustrations.

Early in Jurassic time the miogeosyncline in western Utah was uplifted regionally, becoming a subdued intermittant source area (Stokes, 1972). Simultaneous with these movements and later in Jurassic time, the eastern miogeosyncline probably was subjected to tensional forces and north-south graben valleys began to develop in central Utah and beyond. In this subsiding, faulted trough more than 8,000 ft of Jurassic sediments accumulated, including 2,000 ft of salt (Moulton, 1975; See Fig. 4 and 5).

During Early Cretaceous time the eastern margin of the miogeosyncline in western Utah was beset by thrusting (Sevier orogeny) and vast amounts of clastic debris were carried eastward from the elevated thrust sheets into the subsiding trough (Armstrong, 1968, p. 432). This thrusting, which continued intermittantly into the Paleocene, was not just a local feature but occurred from northwest Arizona to western Wyoming, Idaho and beyond (Fig. 4). As we study the thrusting and rifting of northern Utah and southeast Idaho, one should be aware that 40 or more mi of eastward salt displacement probably occurred on the thrust sheets. Subsequent volcanism, plutonic intrusions, normal faulting, and some thrusting followed by erosion has served to define the present day topographic and geologic expressions.

JURASSIC TROUGH IN UTAH AND IDAHO

Our attempt to reconstruct the events of Jurassic time have been aided by recently available deeper well control and remote sensing devices such as gravity, magnetics, and seismic. Using these tools and published material, we can become increasingly convinced of an extensive north-south Jurassic-age trough and

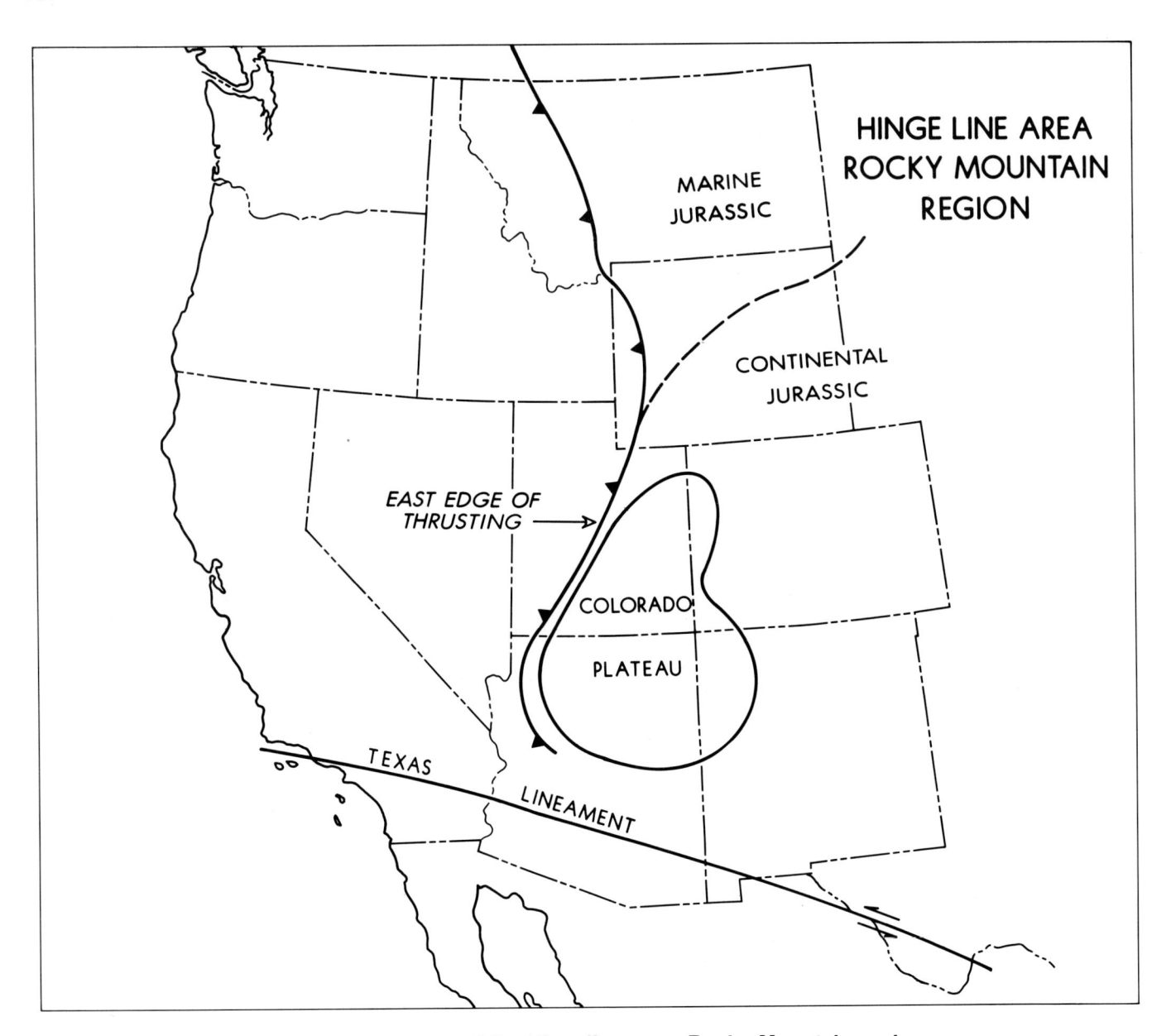

Fig. 1 — Index map of the Hingeline area, Rocky Mountain region.

associated Jurassic-dated faulting in the southern Rocky Mountain area. Evidence for this ancient faulting is presented by Moulton (1975), who discusses the Jurassic rift valley and associated evaporites of central Utah.

Aided by recent well control, surface evidence, and the "filtering down" of seismic data, the writer is reasonably sure of at least two salt diapirs in the northern Utah and southwest Wyoming area. Figure 4 shows the various Jurassic salt thicknesses, and their probable limits.

JURASSIC (?) TROUGH AND SALT IN ARIZONA

Of particular interest is the occurrence of salt domes in Arizona which have been reported on by Koester (1971). Each of these domes appear to be located in graben-like valleys along the southwest flank of the Colorado Plateau (Fig. 4). The limits indicated for the potential salt accumulation in Arizona is not well supported and can only be inferred. This trend is, however, marginal to the Colorado Plateau and occupies a position of Paleozoic overthrusting (Crosby, 1968, p. 2001), which has now been identified as the "Texas lineament" in the language of plate tectonics (Sales, 1968, p. 2017 and p. 2035). In addition, isopach studies of the Triassic, having greater supporting data, indicate Triassic thickening along the Jurassic salt trend of this paper.

The salt masses in Arizona show evidence of continuous upward movement to the present time. Their shape and total mass, where known, suggest diapiric action similar to salt

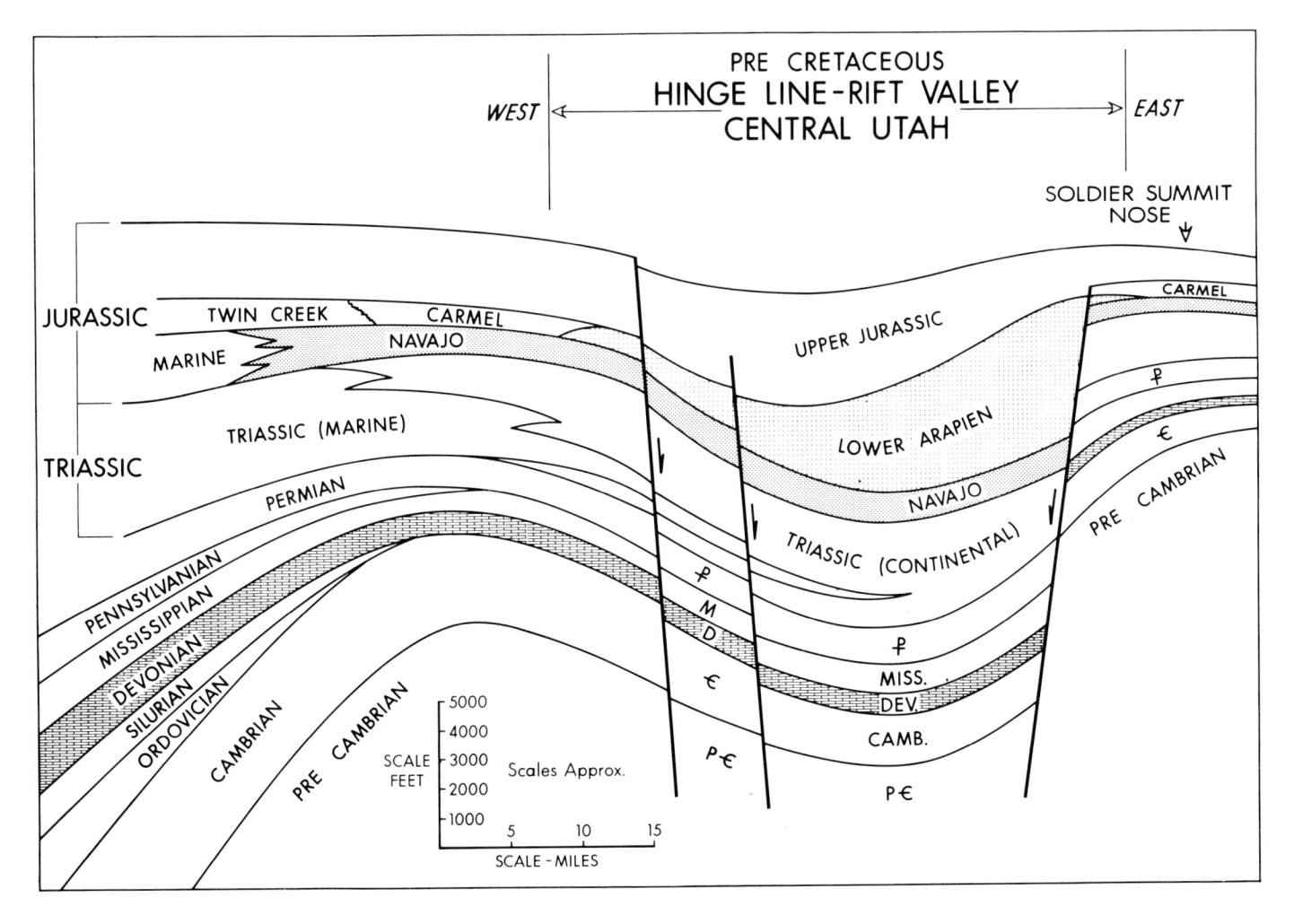

Fig. 2 — Diagrammatic cross-section of the Hingeline rift valley of central Utah showing the area of major salt deposition (cross-hatched pattern).

domes in the Gulf Coast (Koester, 1971).

A Bouguer gravity study in south-central Arizona (Peterson, 1968) shows many gravity lows along a northwest trend. Two of these lows, the Luke and Picacho salt domes, have been drilled, and massive salt is present. Other salt domes may therefore exist at depth in association with the remaining gravity anomalies.

Regarding the age of the salt and associated beds in Arizona, Koester (1971) states that no dating is available at the Detrital Valley, Luke, or Pima locations, whereas the Overton Beach and Picacho areas are Pliocene or older. The age of the salts at Red Lake have been established by radiometric dating and palynology as Triassic-Jurassic age.

As a final speculative thought about the salt occurrences in Arizona, Koester (1971) discusses a potential "mother salt" and suggests it may be of Mesozoic age. Such a conclusion would support the data gathered for this paper.

CONCLUSIONS

Since Jurassic time, orogenic activity including rifting, thrusting, and normal faulting have both complicated and refined the "hingeline" geology for oil and gas exploration. Many of these present-day structural features have been influenced by Jurassic salt that has dissolved or relocated causing collapse and major linear faulting. Such responses are due to zones of weakness developed during Jurassic time. Some surface faults that may be related to Jurassic rift zones are: the Wasatch fault near Salt Lake City and the Paunsaugunt, Sevier, Hurricane, and Grand Wash faults of southwest Utah.

Finally this paper attempts to document the extent of the Jurassic salts and define the vast area where the salt and other evaporites can serve as seals to trap oil and gas.

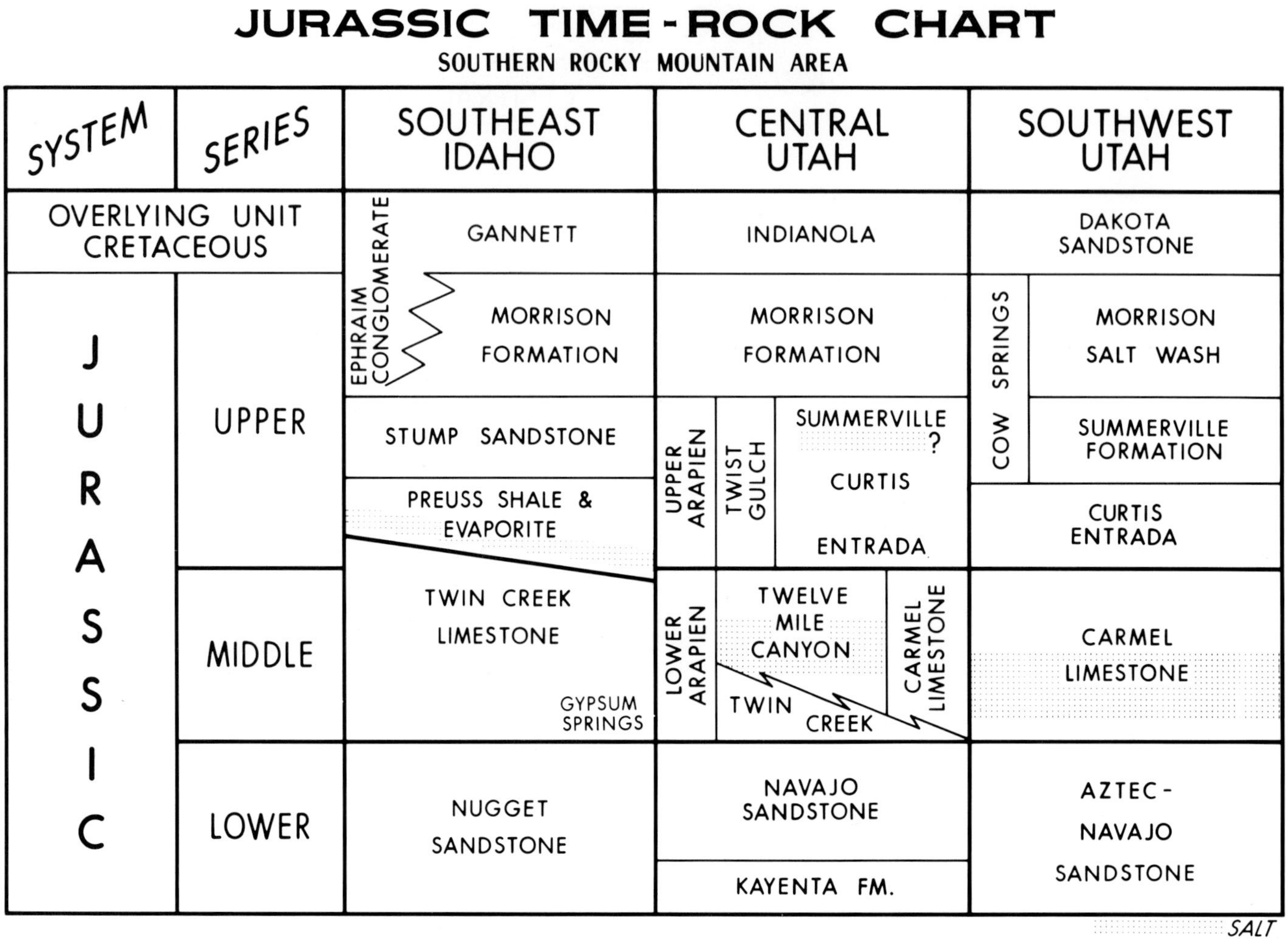

JURASSIC TIME-ROCK CHART
SOUTHERN ROCKY MOUNTAIN AREA

SYSTEM	SERIES	SOUTHEAST IDAHO	CENTRAL UTAH	SOUTHWEST UTAH
OVERLYING UNIT CRETACEOUS		GANNETT	INDIANOLA	DAKOTA SANDSTONE
JURASSIC	UPPER	EPHRAIM CONGLOMERATE / MORRISON FORMATION	MORRISON FORMATION	COW SPRINGS / MORRISON SALT WASH
JURASSIC	UPPER	STUMP SANDSTONE	UPPER ARAPIEN / TWIST GULCH / SUMMERVILLE ? / CURTIS / ENTRADA	SUMMERVILLE FORMATION / CURTIS ENTRADA
JURASSIC	UPPER	PREUSS SHALE & EVAPORITE		
JURASSIC	MIDDLE	TWIN CREEK LIMESTONE / GYPSUM SPRINGS	LOWER ARAPIEN / TWELVE MILE CANYON / TWIN CREEK / CARMEL LIMESTONE	CARMEL LIMESTONE
JURASSIC	LOWER	NUGGET SANDSTONE	NAVAJO SANDSTONE / KAYENTA FM.	AZTEC-NAVAJO SANDSTONE

Fig. 3 — Time-rock correlation chart of Jurassic units, southern Rocky Mountain area.

EXPLORATION METHODS AND CONCEPTS — CORDILLERAN HINGELINE

Those of us associated with the oil and gas industry should and must direct most of our professional effort toward finding many large fields. To this end, the "hingeline" fairway of the Rocky Mountain area is prime for our exploration efforts.

As we analyse the oil and gas potential of the hingeline, the total geologic column must be considered, and one should be aware of the following facts and comments.

1. All sediments below the salt basins are more prospective because of the salt and other related "top seals."
2. Pinchouts of the Ordovician, Silurian and Pennsylvanian-age rocks occur on the hingeline (Fig. 2).
3. Good source rocks of the Mesozoic era include Triassic marine shales and the Jurassic Twin Creek, Carmel, and Arapien sediments (Moulton, 1976, personal communication). The most obvious Mesozoic reservoir bed is the Jurassic Nugget-Navajo Sandstone which attains thicknesses of over 1000 ft.
4. Many Jurassic-age salt diapirs are believed to occur in a Jurassic trough from Arizona to Idaho. These salt plugs can generate the tilting and sealing necessary for oil and gas accumulations.
5. Jurassic salt is considered the most likely "grease" for bedding plane thrusts on the hingeline. As thrusting developed some of the salt was "scooped" along and probably concentrated at the toe of eastward directed thrusts. Some thrust plates remain buried and we need geophysics to locate them.
6. Geophysical tools are important in any exploration program. Of particular value is the gravity survey to help define thick salt formations and buried thrust plates. Our confidence in seismic work can be greatly improved by the integration of proper salt velocities in the interpretations.

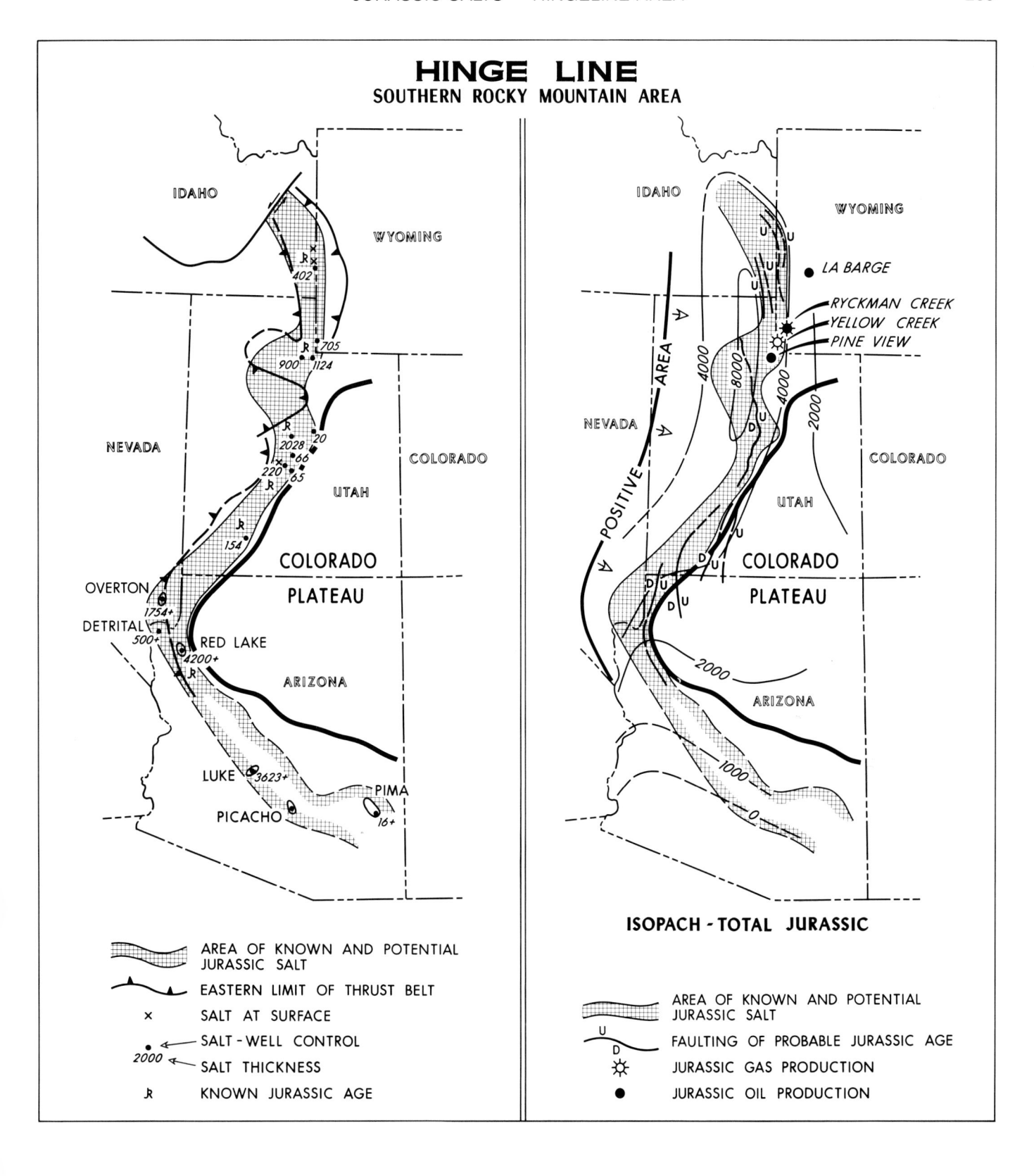

Fig. 4 — Distribution of known and potential Jurassic age salt, southern Rocky Mountain Hingeline area.

Fig. 5 — Isopach map, total Jurassic, southern Rocky Mountain Hingeline area.

Pineview Field — The Example

Pineview field in northern Utah (T2N, R7E) appears to be a major oil discovery of the 100 million bbl recoverable class from the Nugget Sandstone. The cap rock is probably the Preuss salt, with the Jurassic Twin Creek carbonate rocks as a source. Salt thicknesses at Pineview vary from 20 to over 1000 ft indicating probable gouging, faulting, and solution activity along a thrust plane.

REFERENCES CITED

Armstrong, F. C. and Oriel, S. S., 1965: Tectonic development of Idaho-Wyoming thrust belt: Am. Assoc. Petroleum Geologists Bull., v. 49, no. 11, pp. 1847-1866.

Armstrong, R. L., 1968, Sevier orogenic belt in Nevada and Utah: Geol. Soc. America Bull., v. 79, pp. 429-458.

Crosby, G. W., 1968, Vertical movements and isostasy in western Wyoming overthrust belt: Am. Assoc. Petroleum Geologists Bull., v. 52, no. 10, p. 2000-2015.

Hintze, L. F., 1963, Geologic history of Utah: Brigham Young Univ. Geol. Studies, v. 20, part 3.

Koester, E. A., 1971, Salt domes in Arizona: Oil and Gas Conservation Comm., State of Arizona, Geol. Dept., no. 1, pp. 1-20.

Moulton, F. C., 1975, Lower Mesozoic and upper Paleozoic petroleum potential of the Hingeline area: Rocky Mtn. Assoc. Geologists, Symposium on deep drilling frontiers in the central Rocky Mountains, pp. 87-98.

Peterson, D. L., 1968, Bouguer gravity map of parts of Maricopa, Pima, Pinal, and Yuma counties, Arizona: U.S. Geol. Survey Geophys. Inv. Map GP-615.

Sales, J. K., 1968, Crustal mechanics of Cordilleran foreland deformation: A regional scale-model approach: Am. Assoc. Petroleum Geologists Bull., v. 52, no. 10, p. 2016-2044.

Stokes, W. L., 1972, Stratigraphic problems of the Triassic and Jurassic sedimentary rocks of central Utah: Plateau — Basin and Range transition zone, central Utah, 1972: Utah Geol. Assoc. Pub. 2., p. 21-28.

STRATIGRAPHY, SEDIMENTOLOGY, AND PETROLEUM POTENTIAL OF DAKOTA FORMATION, NORTHEASTERN UTAH

by

R. L. Vaughn[1] and M. Dane Picard[2]

ABSTRACT

The Lower Cretaceous Dakota Formation in northeastern Utah is a fluvial deposit containing channel and overbank facies. The channel facies consist of multistoried sandstone bodies deposited as point-bars in sinuous streams and rivers. Each point-bar displays an upward fining sequence of grain sizes and sedimentary structures that indicate an upward decrease in flow density.

Through time the fluvial system changed from relatively large streams to smaller, more numerous streams that underwent repeated channel avulsions. Deposits of both stream types display a distinctive outcrop morphology. This stream system flowed across a low-lying coastal plain not far from the advancing shoreline of the Mowry sea.

The lower contact of the Dakota with the Cedar Mountain Formation is a scoured erosional surface. The upper contact of the floodplain facies with the overlying Mowry Formation is also an unconformity marking the common boundary between a continental and a transgressive marine deposit. Above the channel facies the contact is conformable. Deposition was probably continuous within the channel facies during late Dakota and early Mowry time.

Paleocurrent measurements of individual channel sandstones are strongly unimodal. However, successive channel deposits at a particular location show a wide scatter of transport directions through time. Based on 349 measurements, the net regional dispersal direction was N 3° E.

Sedimentary rocks of the Dakota Formation predominantly consist of grains recycled from pre-existing sandstone and limestone beds. Sandstone and siltstone of the Dakota are quartzarenite or quartz-rich varieties of sublitharenite and subarkose. The source area probably was the Mesocordilleran Geanticline of west-central Utah and adjacent parts of Nevada.

INTRODUCTION

The Dakota Formation of the Western Interior has long been of interest to geologists because of its important record of Cretaceous history, and because it is a frequent objective in the exploration for oil and gas. This paper extends the collective knowledge of the Dakota in both of these areas.

---

[1] Getty Oil Company, Houston, Texas.

[2] Department of Geology and Geophysics, University of Utah, Salt Lake City, Utah.

We gladly acknowledge the help of Everett R. Sharp and Jonathan H. Goodwin for critically reviewing the manuscript, which improved its form and content. Charles L. Rutherford permitted samples to be taken from within the Dinosaur National Monument. Figures were drafted by D. L. Olson. Several versions of the manuscript were typed by Marion Vaughn and V. R. Picard.

Acknowledgement is made to the donors of the Geological Research Fund, administered by the Department of Geological and Geophysical Sciences, University of Utah, for partial support of this research.

The Dakota Formation is well exposed on the north and south flanks of the Uinta Mountains where 11 Dakota sections were measured in detail (Fig. 1). In each section detailed descriptions were made of the beds and sedimentary structures. Special attention was given to the identification and characteristics of the cyclicity of fluvial channels. A total of 349 paleocurrent azimuths were taken, and evaluated at different interpretative levels. Thin sections of 50 rock samples were studied in detail, 35 of which were analyzed by modal analysis. One hundred-fifty rock samples were examined under a binocular microscope to determine modal grain size, frosting, sorting, color, and roundness. Sixteen rock samples were selected for X-ray analysis.

Information from these observations is used here to briefly discuss the following topics: 1) stratigraphy, 2) types of outcrops, 3) formational contacts, 4) internal sedimentary structures, 5) petrography, 6) paleocurrents, and 7) provenance.

In addition, oil and gas possibilities of the Dakota Formation are discussed briefly.

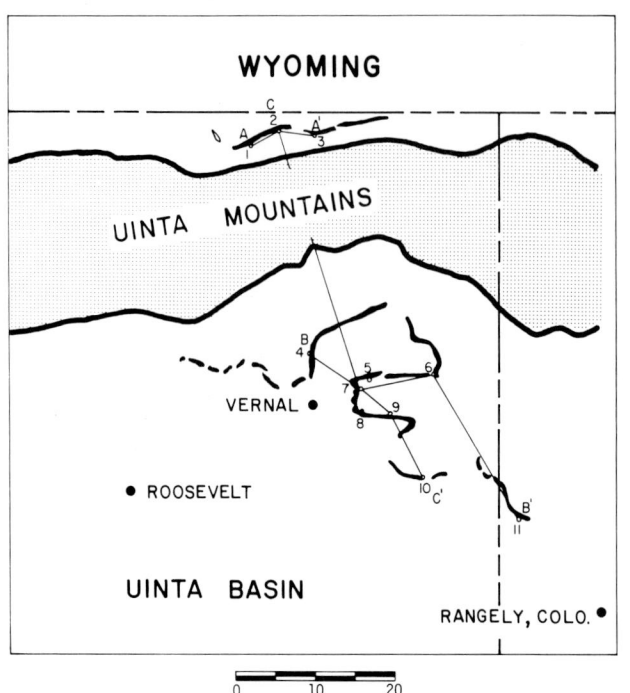

Fig. 1 — Index map showing location of cross-sections in study area.

1 - Manila, Utah area, (M)
2 - Finch Draw (F)
3 - Chokecherry Draw (C)
4 - Steinaker Draw (S)
5 - Split Mountain area, (SM-A)
6 - Split Mountain area, (SM-D)
7 - Split Mountain area, (SM-B)
8 - Split Mountain area, (SM-C)
9 - Quarry area, (QA)
10 - Green Mountain area, (GM)
11 - Dinosaur, Colorado area, (DC)

PREVIOUS WORK

Meek and Hayden (1861) were the first to study the Dakota Formation from exposures near the town of Dakota, Nebraska. Since then the Dakota has been the subject of many papers and theses whose study areas ranged from the western Great Plains to the western slope of the Rocky Mountains. Depending on the geographic location, the Dakota has been reported to represent several marine and nonmarine depositional environments that are usually closely associated with shoreline conditions. Some of the more recent contributions dealing with Dakota stratigraphy are those of Kinney (1955), Haun (1959, 1963), Young (1960, 1973, 1975), Haun and Barlow (1962), MacKenzie and Poole (1962), Weimer (1962, 1970), Lane (1963), Hale and Van De Graff (1964), Suttner (1969), Haverfield (1970), and Furer (1970).

STRATIGRAPHY
General

The Dakota Formation, as here described, is bounded below and above by the Cedar Mountain and Mowry formations respectively. The Cedar Mountain Formation is continental in origin, and is recognized by its slope-forming habit and variegated colors of red, purple, and gray. Claystone, siltstone, and sandstone lenses are dominant in the Cedar Mountain. The Mowry is a dark gray to black, siliceous marine shale, easily identified in the field by its color, fissility, diagnostic fish scales, and slope-forming topography.

Dakota Stratigraphy

The Dakota Formation in northeastern Utah (Fig. 1) is a fluvial deposit consisting of channel and overbank facies. The channel facies consist of locally thick deposits of sandstone, pebbly sandstone, and conglomerate. The overbank facies are mainly siltstone, mudstone, and thin beds of sandstone. Exposures of the channel facies are common, and form prominent sandstone cliffs. Thickness of the channel deposits is variable, reaching a maximum of about 100 feet (30m) on the south flank of the Uintas and thickening to about 240 feet (72m) on the north flank. Two east-west cross-sections (Figs. 2 and 3) and a north-south cross-section (Fig. 4) summarize the stratigraphy and the various sedimentary structures of the channel facies.

Figures 2 and 3 are oriented generally perpendicular to the paleoslope and are situated across ancient Dakota fluvial valleys. The substantial thinning and thickening of the channel deposits is a reflection of the paleotopography as well as the erosive nature of the Dakota streams. Within the thicker channel deposits, it is evident that the Stage I and Stage II deposits (Figs. 2 - 4) represent deposition in a dynamic fluvial system. Beds of Stage I can be correlated surprisingly well although they were deposited in relatively large streams and are typical point-bar deposits. This was a degrading stream system, as shown by its erosional base. The energy level was relatively high compared to later streams. The banks were cohesive enough to minimize lateral migration and allow a substantial thickness of overbank material to accumulate and be preserved. The ease of correlation suggests that the fluvial system was operating under stable geomorphic conditions, and sedimentary processes were controlling deposition uniformly across northeastern Utah.

Subsequent to Stage I, the fluvial system evolved into smaller streams in which the thinner channel deposits of Stage II were deposited. The lower boundary surfaces of Stage II channels are smooth (not undulatory), indicating that the streams approached "grade" and that downward erosion was negligible. In contrast to Stage I deposits, a general lack of floodplain material suggests that the banks of the streams were not resistant to erosion, which in turn allowed rapid lateral migration across the floodplain and frequent reworking of sediments.

The coarse-grained unit contains pebbly sandstone and conglomerate and minor amounts of sandstone. Its internal texture is generally massive. In some places, however, it is crudely graded and contains unusually large ripple marks. The north-south cross-section (Fig. 4) indicates that the coarse-grained unit thickens and becomes finer grained northward.

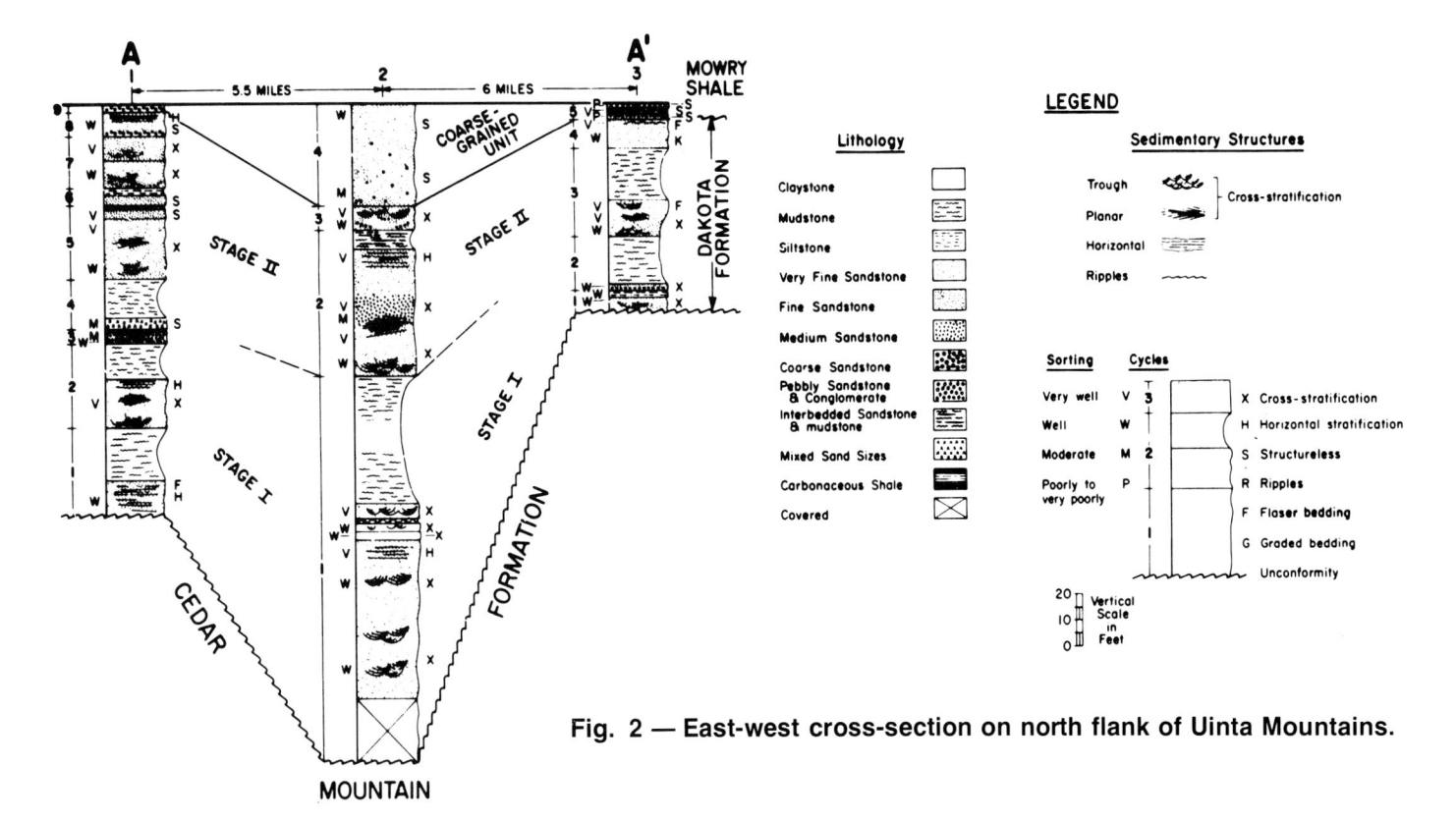

Fig. 2 — East-west cross-section on north flank of Uinta Mountains.

The north-south cross-section (Fig. 4) is oriented approximately parallel to the paleoslope and shows an over-all thickening of the Dakota north of section 9, which is the seaward direction.

Types of outcrops. — Outcrop morphologies of the Dakota channel facies are of two types and are useful in analyzing the nature of this ancient fluvial system. The change in outcrop morphology indicates that the fluvial system changed substantially through time. The two types of outcrops vary in gross thickness, thickness of individual channels, sand/shale ratios, lateral continuity of individual channels, and width of outcrops. Figures 5 and 6 are photographs of the two types of outcrops.

The first type (Fig. 5) is a thick cliff-forming outcrop that contains point-bar deposits formed in the large meandering streams (Stages I and II). Outcrops of this type range in thickness from approximately 100 feet (30m) on the south side of

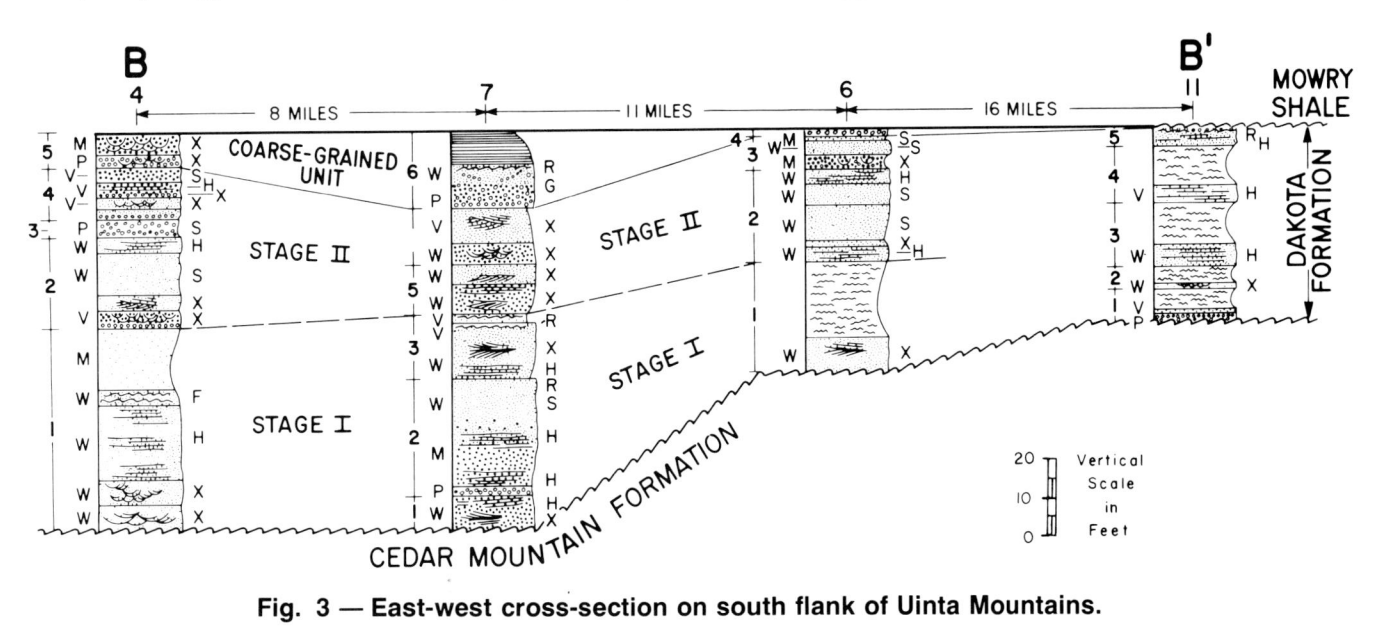

Fig. 3 — East-west cross-section on south flank of Uinta Mountains.

the Uintas to approximately 240 feet (72m) on the north side. They have high sand/shale ratios, ranging from 2.3 to 13.0. These are cyclic or multistoried sands, with the basal erosional surface of a sandstone commonly resting directly on the underlying channel deposit with no intervening floodplain shale. Individual channels of the larger outcrops may have lateral dimensions of over 1000 feet (300m).

The second type of channel outcrop (Fig. 6) contains deposits of the smaller, more numerous, meandering streams. These deposits are thinner than those of type one, ranging in outcrop thickness from 50 to 80 feet (15 to 24m). The sand/shale ratio varies only slightly, and is about 1. Individual channels are less extensive laterally than those of type one, commonly pinching out within the outcrop. These channel sandstones are often enclosed by fine-grained overbank beds.

Type two outcrops contain a cluster of isolated channel deposits at various stratigraphic levels, and have lateral dimensions of approximately 200 to 400 feet (60 to 120m). This outcrop pattern probably resulted from deposition in aggrading streams where repeated avulsions formed new channels in nearby low areas. A few deposits that probably represent crevasse-splay deposition were noted.

Contacts. — The basal contact of the Dakota with the Cedar Mountain Formation is unconformable. Dakota streams incised channels into the underlying Cedar Mountain Formation, as indicated by undulating, erosional, channel bottoms. Local relief on this erosional surface is as much as 50 feet (15m) according to Hansen (1965). The magnitude of the hiatus or the thickness of beds removed was not determined. The basal Dakota contact is a regional disconformity.

The upper contact of the Dakota with the Mowry Formation is probably unconformable over the floodplain facies and conformable over the channels. The upper contact marks a common boundary between a continental deposit and an overlying marine deposit. However, complications arise when the nature of the marine transgression is considered in detail.

As the Mowry sea transgressed over the Dakota, it failed to deposit a beach or nearshore sand along its leading edge. This is probably one reason why the Dakota has been called a transitional deposit between the Cedar Mountain and Mowry formations.

However, during this investigation, it became increasingly apparent that the coarse-grained unit, which is present above the channel facies, is genetically related to the transgression of the Mowry sea and represents, at least in part, the "missing link" in the geologic record. The coarse-grained unit was deposited in channel bottoms near the mouths of streams and rivers as they entered into the advancing Mowry sea. These topographically low entry points along the shoreline were subjected to very strong currents, probably alternating tidal currents.

Summary. — The stratigraphic information indicates that the Dakota fluvial system consisted of meandering streams that probably flowed sluggishly across a low-relief coastal plain not far from the advancing shoreline of the Mowry sea. The system was a dynamic one, and the streams changed in size and character in order to reach equilibrium as the sea transgressed.

FEATURES OF DAKOTA FLUVIAL CHANNELS

The basic element of a fluvial facies is the channel deposit. Knowledge of the physical parameters of a fluvial facies is useful in petroleum exploration, especially in terms of predicting sandstone geometry. Some of the salient features of channels analyzed in this study are rock types, thicknesses, internal structures, vertical profiles, and flow regimes.

Each channel deposit of the Dakota Formation consists of a coarse, poorly-sorted basal lag deposit that is overlain by finer grained sandstone. The lag consists of pebbly sandstone and/or conglomerate. Large clasts of bank material are rare. The sandstone portion, which constitutes as much as 95 percent of the channel deposits, is dominated by fine-grained sand.

Channel thicknesses, which are indicative of stream size, were compared. In the larger, meandering, stream channels of Stages I and II, 73 percent are 20 feet (6m) thick or less. The remaining 27 percent ranges from just over 20 feet (6m) to more than 40 feet (12m). In the smaller stream system, the majority of channels are less than 10 feet (3m) thick. Only about 30 percent are over 10 feet (3m), and none exceed 20 feet (6m).

Sedimentary structures are abundant in the Dakota and are useful for interpreting the environments of deposition. Based on the terminology of McKee and Weir (1953), the most common types of cross-stratification are trough and planar. Horizontal stratification is also common. Many bedding plane features were noted, including current lineation marks, sole marks, and rib-and-furrow structures. Ripple marks, ripple stratification, and flaser bedding are present in the uppermost parts of some channels.

One of the most conclusive indicators of the fluvial origin of the Dakota sandstone is the vertical sedimentary structure profile. From base to top, the sequence of sedimentary structures is: 1) cross-stratified channel lag deposit (conglomeratic sandstone), 2) trough cross-stratification, 3) planar cross-stratification, 4) horizontal stratification, and 5) ripple stratification. These structures indicate an upward decrease in flow intensity and are a sequence of primary sedimentary structures

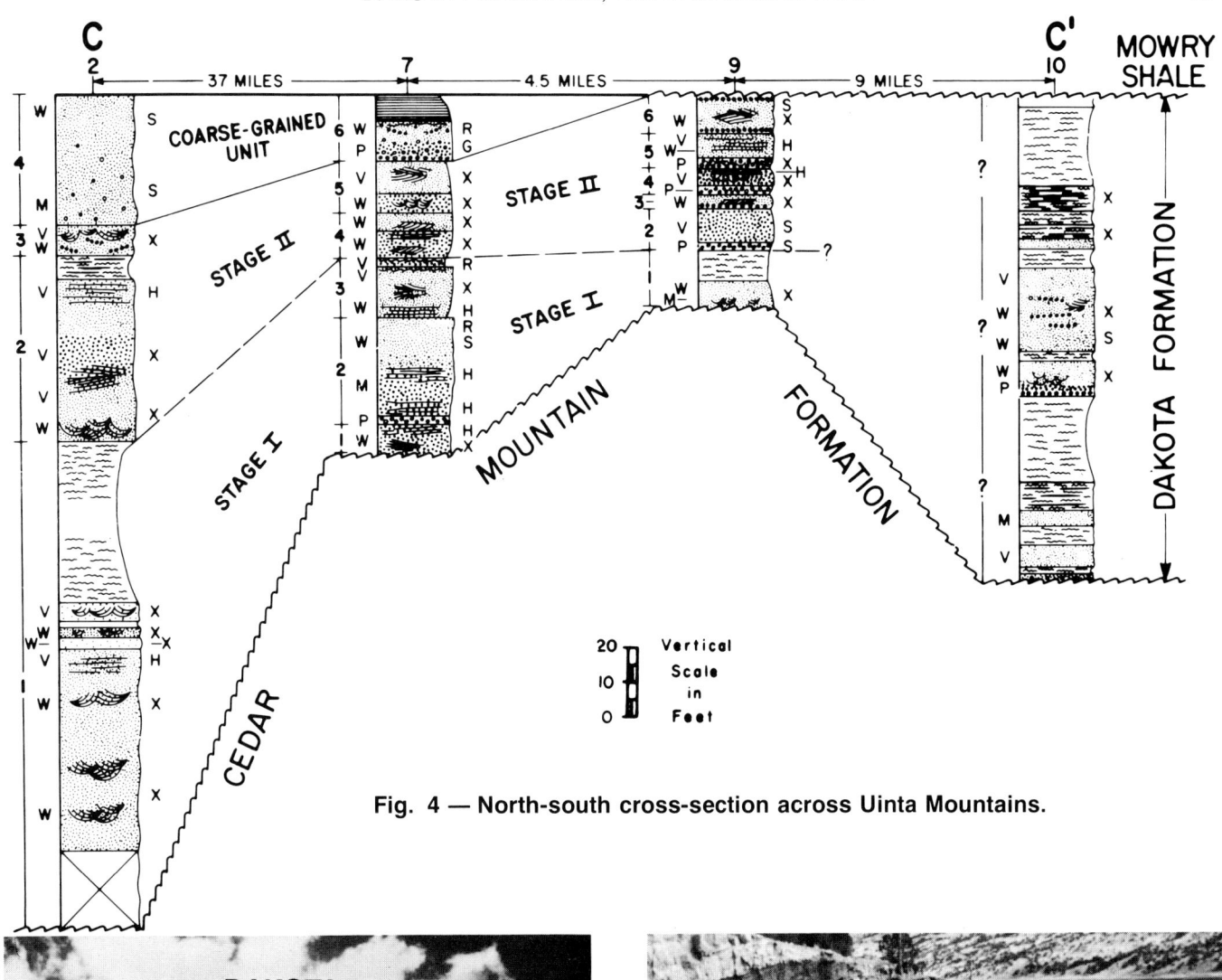

Fig. 4 — North-south cross-section across Uinta Mountains.

Fig. 5 — Outcrop of large meandering stream deposit (section SM-B) showing cyclic deposition of fluvial channels. Outcrop thickness about 112 feet.

Fig. 6 — Outcrop of alluvial plain stream deposit (section M) showing low sand/shale ratio, pinching out of channels and crevasse-splay deposits.

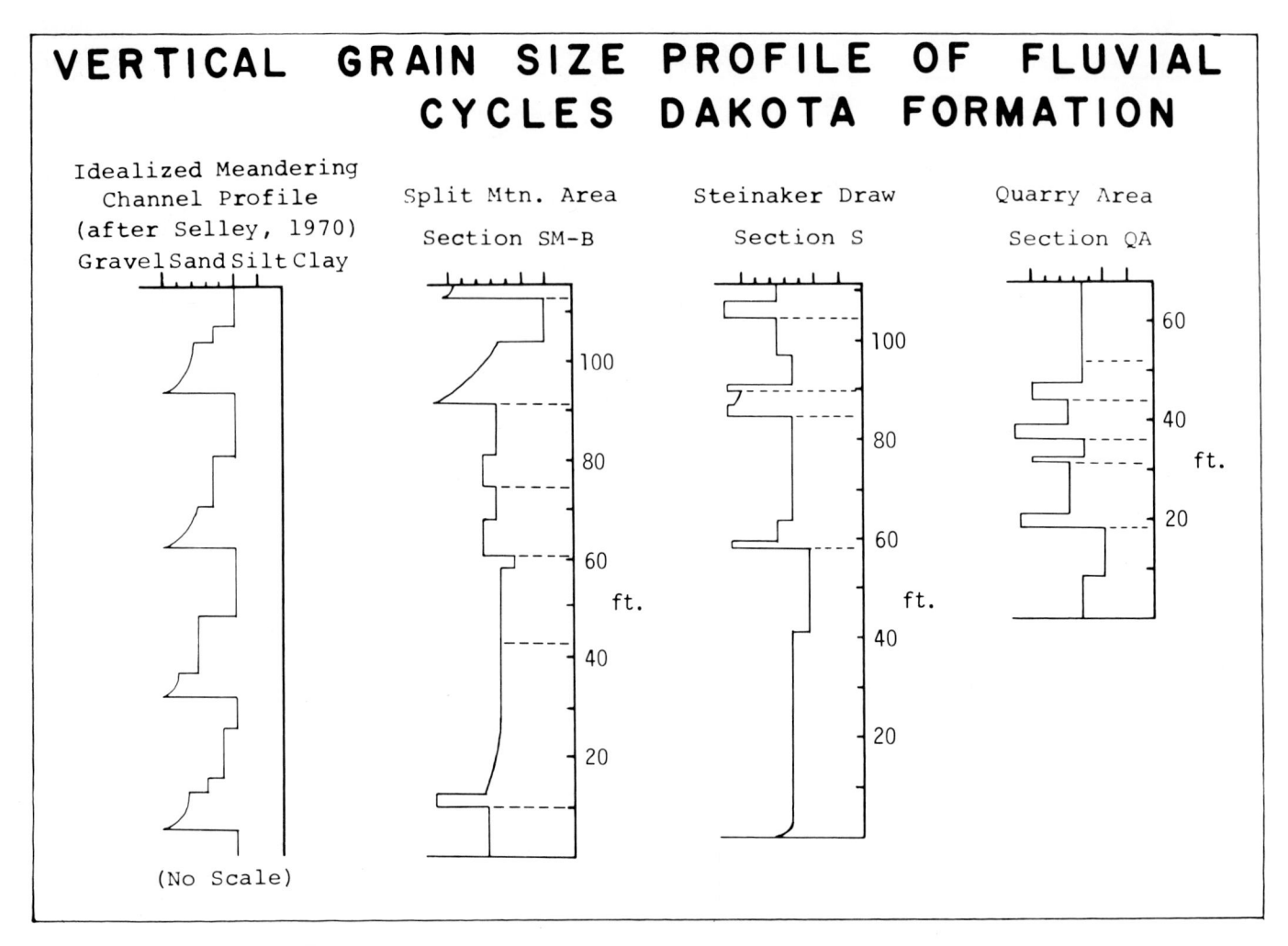

Fig. 7 — Shows vertical repetition of channels by grain size.

commonly found in point-bar deposits (Allen, 1970; Bernard and Major, 1963; and Harms, et al., 1963).

A distinctive feature of the channel facies is the vertical repetition of channels, each channel resulting from a new episode of fluvial deposition. This pattern imparts a cyclicity to the outcrops, which is also shown in a vertical grain-size profile chart (Fig. 7). The upward fining sequence of grain sizes within each cycle is indicative of point-bar deposits in which lateral accretion was the mode of deposition (Allen, 1965 and 1970).

PETROGRAPHY

Pebbly Sandstone and Conglomerate

The pebbly sandstones and conglomerates that constitute the channel lag deposits (Folk, 1968) are litharenites (Fig. 8). Rock fragments consist of chert plus metamorphic and sedimentary rock fragments. The matrix is composed primarily of authigenic clays. X-ray analyses indicate that kaolinite is the dominant clay mineral. Montmorillonite is also present, but in lesser amounts.

Feldspar grains are rare, generally forming less than one percent of the rocks. These beds have a characteristic reddish-brown color because of hematite stain emanating from sedimentary rock fragments. Porosity values range from less than 1 to 13 percent and average 4.7 percent. The principal bonding agent is the clay matrix. Silica is also present as a bonding agent, but is insignificant compared with the clay. Carbonate cement is rare. However, this may be partly a weathering phenomenon and could change substantially in the subsurface.

Sandstone and Siltstone

Based on modal analyses, the sandstone and siltstone is quartzarenite and sublitharenite. The quartz grains are mainly monocrystalline and have straight extinction. Silica overgrowths are present on many quartz grains and are of two types, post-depositional and reworked. The first type is recognized by straight crystal faces that occasionally form an interlocking network of grains. Reworked silica overgrowths are recognized by nuclei of rounded grains whose overgrowths are very irregular in outline and have been "chipped" away in places down medium-grained, with the fine-grained sandstone most common. Eighty-one percent of the sandstone and siltstone ranges from well- to very well sorted. Well-sorted samples are the largest group, accounting for 40 percent of all samples. Plots of grain roundness values show a strong grouping of grains in the subangular and subrounded classes.

Diagenesis

Thin section examination suggests that the clay matrix did not form from the alteration of original grains. Silica overgrowths are commonly coated by the matrix showing that silica was precipitated first. The presence of matrix within the pore spaces, its absence at grain contacts, and the lack of feldspar grains that could alter into clays indicate that the matrix formed from formation waters. Complete alteration of the feldspar grains is not likely because the existing grains show almost no deterioration.

PALEOCURRENT PATTERN

A total of 349 paleocurrent measurements were made from small- and medium-scale trough and planar cross-stratification. The number of measurements per locality ranged from 10 to 64 and averaged 35. Measurements were corrected for tectonic tilt where necessary. The information was evaluated quantitatively according to the methods described in Potter and Pettijohn (1963, p. 264), and was grouped at different interpretative levels to determine: 1) paleocurrent patterns for individual cycles, 2) variation of dispersal direction through time by comparing results of all the cycles at each location, 3) the net dispersal direction at each location, and 4) the net regional dispersal direction.

Results. — Paleocurrent azimuths for individual cycles are strongly unimodal. The larger meandering stream deposits have a slightly greater concentration of azimuths than the smaller channel deposits. When paleocurrent information from all the cycles at a particular outcrop is compared, it is seen that the streams transported sediment in widely scattered directions through time (Fig. 9). This variation in the dispersal direction is indicative of the meandering nature of the streams. The net paleocurrent direction at each location is shown in figure 10, and the net regional paleocurrent direction is shown in figure 11.

PROVENANCE

As seen from the paleocurrent analysis, the general direction of sediment dispersal was to the north. Source areas were therefore on the south. The most likely source areas were in west-central and southern Utah (MacKenzie and Ryan, 1962; Young, 1970). This region was occupied by the Mesocordilleran Geanticline, a highland area from which the dispersal system transported sediment north and east into Colorado and Wyoming.

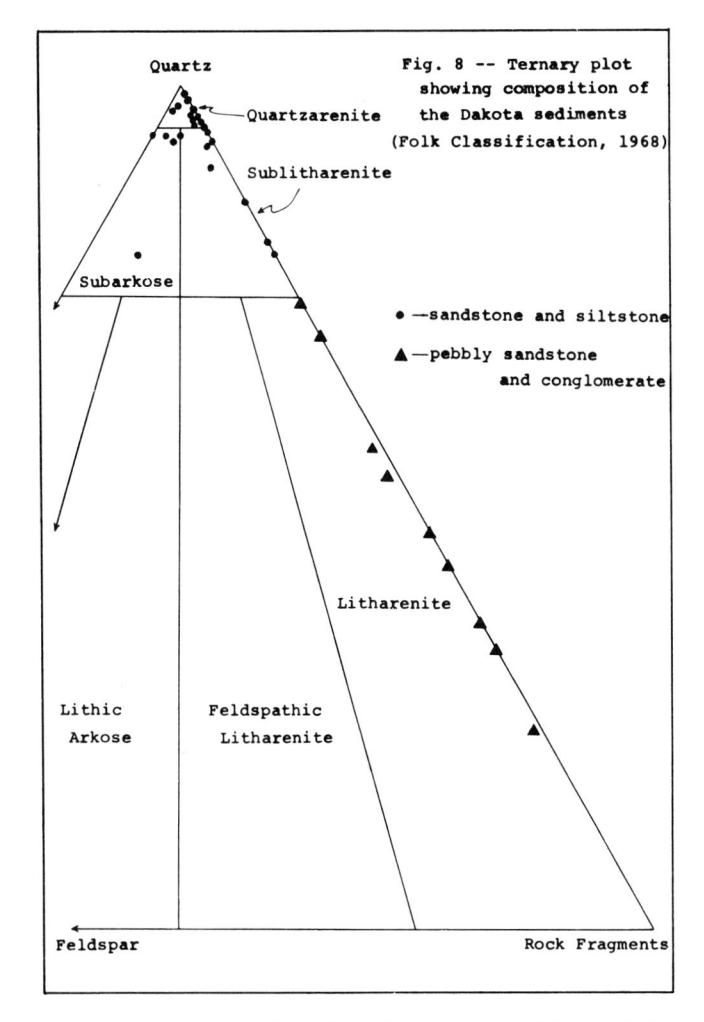

Fig. 8 — Ternary plot showing composition of the Dakota sediments (Folk Classification, 1968).

to the surface of the original grain. In some grains, the "chipped" surfaces penetrate through the overgrowth and into the original grain.

Feldspar grains are rare, ranging from 0 to a maximum of 9 percent. The relative abundance of feldspar varieties is orthoclase>Na-plagioclase>microcline. The rock fragments consist of chert and metamorphic rock fragments that average 2 percent and 1.5 percent respectively. Sedimentary rock fragments are virtually absent.

The matrix of the sandstone and siltstone is almost exclusively authigenic kaolinite with lesser amounts of montmorillonite and illite. The clay matrix is the primary bonding agent. Silica cement is also present, but only in small amounts. Porosity values for the sandstone and siltstone range from less than 1 to 22 percent, averaging 10.6 percent.

Textural parameters indicate that 83 percent of the sandstone and siltstone is unimodal in grain size. Furthermore, 96 percent of the unimodal samples range from very fine to

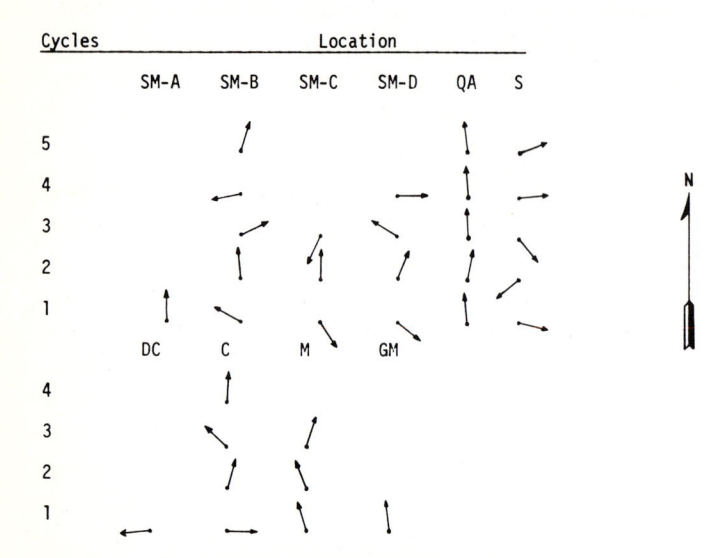

Fig. 9 — Shows the cyclic variation of transport direction at each section.

Petrographic analysis indicates that pre-existing sandstone and limestone were the source rocks for most of the Dakota sediment. Metamorphic rocks contributed minor amounts of material. Recycling of pre-existing sedimentary rocks is suggested by the predominance of quartzarenite or quartz-rich varieties of sublitharenite or subarkose in the Dakota (Fig. 8).

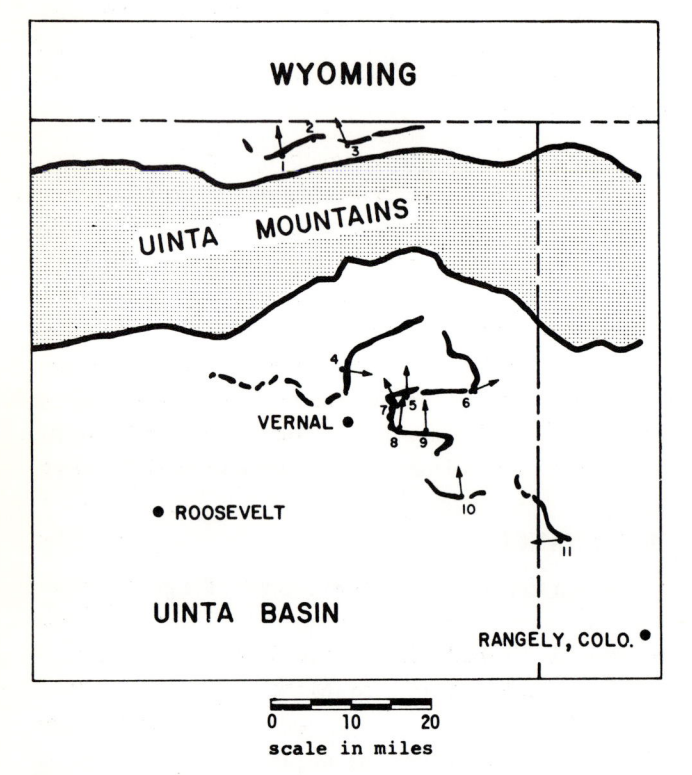

Fig. 10 — Shows net direction of sediment movement at each locality.

NET REGIONAL PALEOCURRENT DISPERSION PATTERN

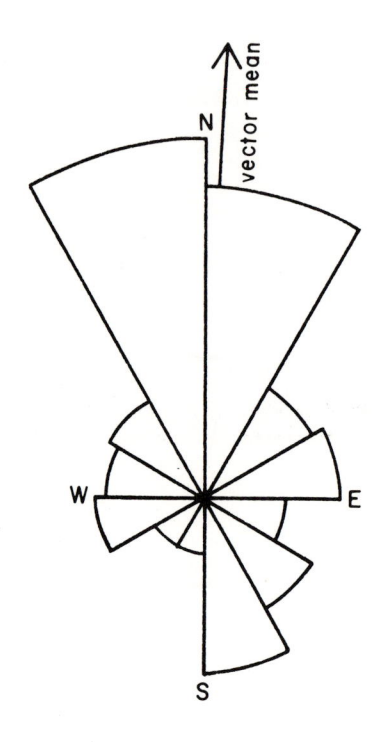

N = 349

$\bar{x}$ = N 2.7° E

Fig. 11 — Shows the net regional direction of sediment movement in northeast Utah.

Easily decomposed particles are rare in rocks of the Dakota Formation. Some scattered quartz grains have silica overgrowths but others do not. Chert is ubiquitous in the Dakota, and is a good diagnostic indicator of carbonate source terranes. Pebbly sandstone and conglomerate are common, and rock fragments in them are also indicative of sedimentary sources.

Sedimentary sources are further indicated by textural features. The Dakota is characterized by fairly well-rounded quartz grains, which suggests that they have been recycled at least once and possibly several times. Although textural inversions are not abundant, they were observed, and consist of fairly well-rounded grains that are not well-sorted, and well-sorted bimodal grains. Both of these inversion types probably represent multiple sedimentary source rocks (Folk, 1968, p. 106).

PETROLEUM POTENTIAL

Oil and gas production from the Dakota Formation in northeastern Utah is limited. However, there has been sufficient production at Bridger Lake, Clay Basin, and other fields in Emery and Grand counties of eastern Utah (Fig. 12) to sustain interest in the formation.

The following sections briefly discuss the geology of some important Dakota producing fields. Some ideas concerning the origin of the hydrocarbons are outlined. Information summarized on the composition of oil and gas in the Morrison, Cedar Mountain, and Dakota formations (Tables 2 - 6) is from the Symposium on Oil and Gas Fields of Utah (1961), Bureau of Mines Information Circulars, and Stowe (1972).

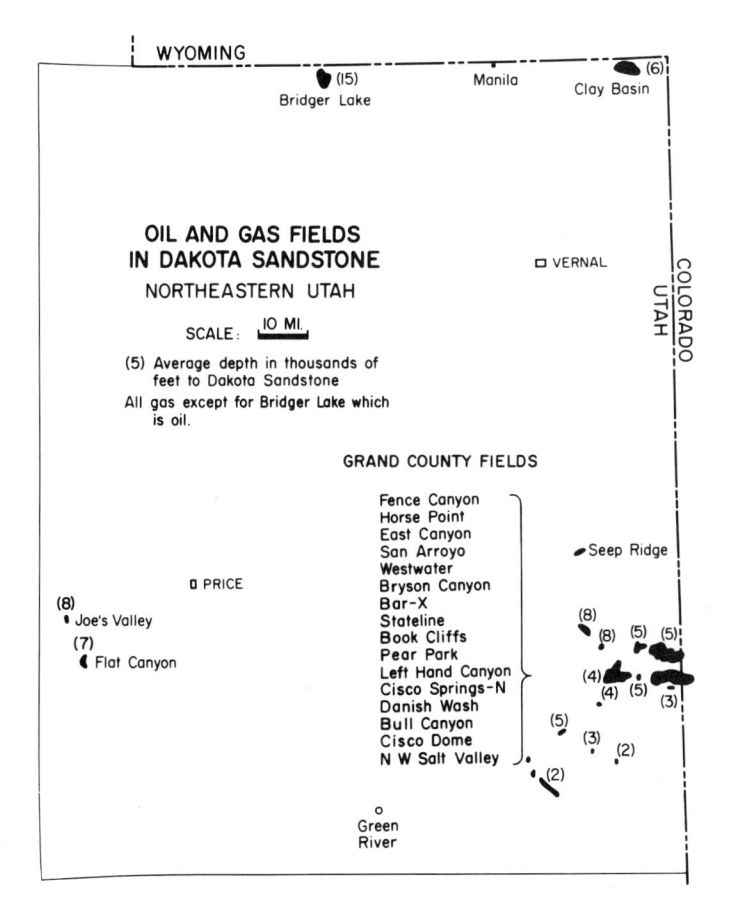

Fig. 12 — Index map showing location of Dakota oil and gas fields.

Oil and Gas Fields

Bridger Lake field. — Significant production of oil from the Dakota Formation has come from the Bridger Lake field immediately north of the Uinta Mountain uplift. The geology of this field is discussed in papers by Garvin (1969) and Peterson (1973).

Production at Bridger Lake is from lower sandstone beds of the Dakota on a south-plunging, faulted, anticlinal nose near the south end of the Church Buttes-Moxa arch. South of the field, the anticlinal trend is cut by the Uinta Mountain fault that separates the Uinta Mountain uplift from the Green River Basin. Lithologic studies indicate that the field is mainly a stratigraphic trap. Two dry holes on the north side of the field are structurally higher than several producers, but do not contain adequate reservoir beds.

According to Peterson (1973), the wells at Bridger Lake are the deepest producers in Utah, and the deepest Cretaceous oil producers in the Rocky Mountains. Estimated original oil in place is 63 million barrels (9 million metric tons), of which 40 million barrels (5.7 million metric tons) are expected to be recovered. The estimated original gas in solution is 54,000 million cubic feet (14,600 million cubic meters).

Clay Basin field. — Gas was first discovered in the Dakota Formation at Clay Basin in 1935. Structure contour maps drawn at the top of the Dakota show an asymmetric anticlinal axis trending west with about 450 feet (135m) of structural closure.

The Clay Basin anticline closely parallels the Uinta Mountain fault, which is immediately south of the field. A tightly folded to slightly overturned syncline separates the Clay Basin anticline from the fault zone. Structural closure on the south flank of the field is believed (Hummel, 1969) to be formed by the lowest closing contour rather than by faulting.

The Dakota sandstones at Clay Basin are considered by Hummel (1969) to be mainly fluvial deposits that were partially reworked by an advancing marine transgression. The average sandstone thickness per well is 40 feet (12m), ranging from 15 feet (4.5m) to 60 feet (18m). Reservoir properties vary considerably, but generally deteriorate northward across the field.

Flat Canyon and Joe's Valley fields. — The Flat Canyon gas field is on the Flat Canyon anticline midway across the north-south trending Wasatch Plateau. The anticline is a closed structure (75m of closure at the Ferron Sandstone) trending northeast. It is bounded and partly truncated on the west by the steeply-dipping Joe's Valley graben (Seeley, 1961). The structure apparently is terminated on the east by a north-south fault complex that is downthrown to the east.

Gas in both the Ferron and Dakota formations is very similar. Seeley (1961) suggested that a common source for both accumulations is possible. His preference is for Lower Cretaceous source beds, probably marine beds below the Ferron Sandstone.

Joe's Valley gas field, northwest of Flat Canyon and southwest of the Clear Creek gas field, also contains gas in both

Table 1. Summary of geology of oil and gas fields in Dakota Formation, northeastern Utah (Information from Oil and Gas Fields of Utah, 1961; Garvin, 1969; and Hummel, 1969)

Field Name	Location	Year of Discovery	Average Depth (ft.)	Net Pay (ft.)	Porosity (%)	Permeability (MD)	Initial Pressure (psi)	Btu (ft.³)	Trap
Bridger Lake	3N-14E	1966	15,500	32	13	80	7,230		faulted anticlinal nose
Clay Basin	3N-24E	1935	5,760	40	16	24	2,100	1080	faulted anticline
Joe's Valley	15S-6E	1956	7,700						faulted anticline
Flat Canyon	16S-6E	1953	7,020	44	4		1,514	1151	faulted anticline
Bar-X (Colorado)	8S-104W	1953	2,844	29	13.5		925	1045	stratigraphic
Fence Canyon	15S-22E	1960	8,126	9			1,600	1067	stratigraphic
Horse Point	16S-23E	1962	7,958	9					stratigraphic
East Canyon	16S-24,25E	1962	5,500						stratigraphic
San Arroyo	16S-25,26E		4,500	35	15	50	1,100	1085	stratigraphic-structural
Westwater	17S-23,24E	1957	4,400	24	19		1,140	1085	stratigraphic
Bryson Canyon	17S-24E	1960	4,600					983	stratigraphic
Stateline	17S-25,26E	1963							stratigraphic
Book Cliffs	18S-22E	1961	5,350					1075	stratigraphic
Pear Park	18S-23E	1961							
Cisco Springs-N	19S-23E	1954	2,820	20					stratigraphic
Danish Wash	19S-24E	1955	1,965	15	17.5		770	1013	stratigraphic
Cisco Dome	20S-21,22E	1925	2,000	30	16.5		880	1083	stratigraphic-structural
NW Salt Valley	21S-19E	1956	3,166						

Table 2. Characteristics of oil in Dakota and Morrison formations, northeastern Utah

	Bridger Lake (Dakota)	Bar-X (Morrison)	Seiber Nose (20S-24E) (Morrison)	Cisco Townsite (21S-23E) (Morrison)	Agate (20S-24E) (Morrison)
Gravity (°API)	40	40.9	34.8	33.8	40.4
Gravity (specific)	0.826	0.821	0.851	0.856	0.823
Pour point	50°F	55°F	30°F	< 5°F	< 5°F
Color	brn-grn	brn-grn	brn-blk	brn-blk	grn-blk
Gas-oil ratio	859 to 1				
Base	paraffin				
Viscosity	100°F, 40 sec.	100°F, 40 sec.	100°F, 44 sec.	100°F, 48 sec.	100°F, 38 sec.
Sulfur (%)	0.05	0.12	0.96	1.07	0.61
Nitrogen (%)	0.047	0.013	0.055	0.070	0.034

the Dakota and Ferron formations. The trap at Joe's Valley is a faulted anticline in an area of complex normal faulting. Johnson (1961) suggested that the gas in both reservoirs originated in the Mancos Shale.

The wells at Flat Canyon and Joe's Valley are now plugged and abandoned.

Grand County fields. — All 16 Dakota discoveries (Fig. 12) in Grand County, Utah are closely related to lithologic and stratigraphic variables. Although well defined structural features are present, few accumulations are without stratigraphic components.

Most of the Dakota discoveries in Grand County have been abandoned or are shut-in. Calculation of reserves, based on present production information, is speculative because many of the wells are completed in several reservoirs (Entrada, Morrison, Cedar Mountain, and Dakota).

Characteristics of Oil

At Bridger Lake, oil in the Dakota is high gravity, low in sulfur and nitrogen, and characterized by a paraffin base (Table 2). Oil with similar characteristics was found at the Bar-X field in the Morrison Formation of northeastern Utah. Three other fields in the Morrison (Table 2) contain oil that has lower gravities and pore points, higher nitrogen and sulfur contents, and no paraffin base.

Characteristics of Gas

Gas of the Dakota Formation is of good quality. The amounts of noncombustible gases are generally low (Table 3). Gas in the Cedar Mountain and Morrison formations (Table 4) has characteristics similar to gas in the Dakota (Table 5). Gas found in the Morrison contains more nitrogen and helium than gas from the Dakota and Cedar Mountain formations, but the differences are small (Table 6). Morrison gas has a lower

Table 3. Composition of gas in Dakota Formation, northeastern Utah

	Methane	Ethane	Higher Fractions	Nitrogen	Carbon Dioxide	Helium	Btu (ft.³)
Bridger Lake	71.3	11.4	12.6	2.0	0.8		1357
Clay Basin	99.3						1080
	92.4	3.9	3.2		0.3	trace	1074
Joe's Valley	86.0	4.6	5.6		0.1	trace	1129
Flat Canyon	86.5	6.4		0.6	0.2		1151
Bar-X (Colorado)	90.3	4.1	1.8	2.8	0.4		1045
	91.6	4.1	1.9		0.3	0.10	1060
Fence Canyon	95.1	2.4	0.6	0.7	1.0	trace	1024
	92.7	4.1	1.8	0.4	1.1	0.10	1067
San Arroyo	89.9	5.1	1.9		0.8	0.10	1059
	91.1	4.4	1.8	0.8	0.9	0.10	1064
	90.2	5.4	2.5		0.9	0.10	1089
	78.6	5.1	2.3	10.3	1.2	0.10	969
Westwater	94.5	2.2					1086
	91.1	5.2	2.3	0.7	0.5	0.08	1089
	94.0	3.1	1.0	0.5	1.2	0.06	1041
Bryson Canyon	87.1	3.7	1.1	5.7	1.0	0.10	983
Book Cliffs	88.9	5.7	2.5		0.2	0.20	1081
	87.7	5.5	2.6	3.4	0.7	0.20	1070
	89.4	5.5	2.3	2.4	0.2	0.20	1077
Harley Dome	84.2	0.7	0.3	11.4	0.5	0.20	877
Cisco Dome	92.4	3.6	2.3		0.1	0.10	1074
Mean	89.2	4.4	2.7	3.2	0.6	0.12	1070

Table 4. Composition of gas in Cedar Mountain Formation, northeastern Utah

Field Name	Methane	Ethane	Higher Fractions	Nitrogen	Carbon Dioxide	Helium	Btu (ft.³)
Evacuation Creek (12S-25E)	90.2	5.3	2.9	0.6	1.0	trace	1100
East Canyon	89.0	5.9	3.2	0.4	1.4	0.1	1107
Bar-X (Buckhorn Cgl.)	88.8	4.7	2.1		0.5	0.3	1049
Segundo Canyon (16,17S-21E)	96.6	1.4	0.4	0.6	0.8	trace	1017
	89.5	6.1	2.3	0.6	1.3	trace	1083
Book Cliffs	86.9	6.7	4.3		1.0	0.1	1132
Gravel Pile (20S-24E)	92.4	1.8	1.0	3.6	0.7	0.3	1001
Cisco Townsite (Buckhorn Cgl.)	82.1	2.5	1.4	13.1	trace	0.7	922
Mean	89.4	4.3	2.2	2.4	0.8	0.2	1051

Table 5. Composition of gas in Morrison Formation, northeastern Utah

Field Name	Methane	Ethane	Higher Fractions	Nitrogen	Carbon Dioxide	Helium	Btu (ft.³)
10-10S-24E	73.4	1.1	0.9	23.2		1.3	795
Fence Canyon	92.8	3.9	1.1		1.1	0.1	1045
San Arroyo	92.7	3.6	1.7	0.9	1.0	0.1	1058
Westwater	77.9	4.3	4.7		0.5	0.2	1027
	93.4	3.3	1.9		0.8	0.1	1073
Bryson Canyon	82.4	3.6	1.7	9.3	0.5	0.7	958
Cisco Dome	89.5	2.1	0.6		3.0	0.1	967
	89.9	4.0	2.5		0.1	0.2	1063
Agate (20S-24E)	88.7	1.0	1.3	8.4	trace	0.5	958
	76.7	2.9	2.8	16.7	0.1	0.7	913
Harley Dome (19S-25E)	83.2	1.4	0.5	14.9			
Mean	85.5	2.8	1.8		0.6	0.8	986

Table 6. Comparison of average compositions of gas in Morrison, Cedar Mountain, and Dakota formations (Number of analyses in average)

	Dakota Ss. (22)	Cedar Mountain Fm. (8)	Morrison Fm. (11)
Methane	89.2	89.4	85.5
Ethane	4.4	4.3	2.8
Higher Fractions	2.7	2.2	1.8
Nitrogen	3.2	2.4	
Carbon Dioxide	0.6	0.8	0.6
Helium	0.12	0.2	0.8
Btu (ft.^{3})	1070	1051	986

Btu value because of slightly larger contents of noncombustible gases.

Picard (1962) found a general increase in the percentage of nitrogen and helium with depth in the Pennsylvanian and Mississippian rocks of the Four Corners region. He attributed their origin to degassing of Precambrian basement rocks. A similar origin for inert gases in the Jurassic and Cretaceous seems likely.

Origin of Oil and Gas

Young (1975) stressed that Lower Cretaceous beds of northeastern Utah and northwestern Colorado have not been prolific producers of oil and gas, yielding only 14.6 million barrels (2.1 million metric tons) of oil and 164.8 billion cubic feet (44.5 billion cubic meters) of gas by the end of 1974. However, Lower Cretaceous beds in Wyoming and eastern Colorado are highly productive. Young (1975) suggested several reasons for this difference: 1) the predominantly nonmarine Lower Cretaceous beds in this area, 2) thinner Mowry Shale, which is probably a source bed, and 3) drilling confined to basin margins near existing pipelines.

In addition to Young's suggestions, another possibility for the origin of oil and gas in the Dakota, Morrison, and Cedar Mountain formations in northeastern Utah is that hydrocarbons originated in predominantly nonmarine environments and migrated short distances from local source beds into lenticular sandstone reservoirs. Organic material was not as abundant in these depositional settings as in the marine conditions of Wyoming and eastern Colorado. This organic deficiency may be responsible for the generally small accumulations of hydrocarbons in this area. Although the Mancos and Mowry shales contain possible source beds, migration of hydrocarbons from these beds into nonmarine reservoirs is believed to be minimal.

Petroleum Potential of the Dakota

Through January 1976, the Bridger Lake Field has produced 8,675,608 barrels of oil (1,239,372 metric tons) and 50,369,948 MCF of gas (about 13.6 billion cubic meters) from the Dakota. Such quantities of production are very encouraging even at deep drilling depths (Table 1).

Recently, the American Quasar Petroleum No. 35-1, UPRR, in section 31, T. 2 N., R. 6 E. tested 2 million cubic feet (0.54 million cubic meters) of gas per day from the Dakota at depths of almost 17,000 feet (5,000m).

In contrast to deep production, the Clay Basin field has produced 129,910,099 MCF of gas (about 35 billion cubic meters) and 299,125 barrels of oil (42,732 metric tons) through January 1976 (including Frontier production) from moderate depths. This field is nearing its economic limit and may be used as a gas storage reservoir.

Except for Grand County, the Dakota has not been fully tested because of drilling depths and complex structural conditions. Much of the drilling has been confined to locations close to present pipelines.

In view of the production records of the Bridger Lake and Clay Basin fields, recent discoveries, and the generally sparse testing of the Dakota, we feel that the full potential of the formation has not been realized. Further stratigraphic and sedimentologic studies coupled with aggressive exploration programs could very possibly locate significant reservoirs in the future.

REFERENCES CITED

Allen, J. R. L., 1965, Fining-upwards cycles in alluvial successions: Liverpool Manchester Geol. Jour., v. 4, p. 229-246.

_____, 1970, Studies in fluviatile sedimentation: a comparison of fining-upwards cyclothems, with special reference to coarse-member composition and interpretation: Jour. Sed. Petrology, v. 40, p. 298-324.

Bernard, H. H., and Major, C. F., 1963, Recent meander belt deposits of the Brazos River: an alluvial "sand" model (abs.): Am. Assoc. Petroleum Geologists Bull., v. 47, p. 350.

Folk, R. L., 1968, Petrology of Sedimentary Rocks: Hemphill's, Austin, 170 p.

Furer, L. C., 1970, Petrology and stratigraphy of nonmarine upper Jurrasic — lower Cretaceous rocks of western Wyoming and southeastern Idaho: Am. Assoc. Petroleum Geologists, v. 54, p. 2282-2302.

Garvin, R. F., 1969, Bridger Lake field, Summit County, Utah: in Geologic Guidebook of the Uinta Mountains, Intermtn. Assoc. Geologists, 16th Annual Field Conference, p. 109-115.

Hale, L. A., and Van De Graff, F. R., 1964, Cretaceous stratigraphy and facies patterns — northeastern Utah and adjacent areas: in Guidebook to the Geology and Mineral Resources of the Uinta Basin, Intermtn. Assoc. Geologists, 13th Annual Field Conference, p. 115-138.

Hansen, W. R., 1965, Geology of the Flaming Gorge Area, Utah — Colorado - Wyoming: U.S. Geol. Survey Prof. Paper 490, 196 p.

Harms, J. C., MacKenzie, D. B., and McCubbin, D. G., 1963, Stratification in modern sands of the Red River, Louisiana: Jour. Geology, v. 71, p. 566-580.

Haun, J. D., 1959, Lower Cretaceous stratigraphy of Colorado: in Symposium on Cretaceous Rocks of Colorado and Adjacent Areas, Rocky Mountain Assoc. Geologists, 11th Annual Field Conference, p. 1-8.

_____, 1963, Stratigraphy of Dakota group and relationship to petroleum occurrence, northern Denver Basin: in Geology of the Northern Denver Basin and Adjacent Uplifts, Rocky Mountain Assoc. Geologists, 14th Annual Field Conference, p. 119-134.

_____, and Barlow, J. A., Jr., 1962, Lower Cretaceous stratigraphy of Wyoming: in Symposium on Early Cretaceous Rocks of Wyoming and Adjacent Areas, Wyom. Geol. Assoc., 17th Annual Field Conference, p. 15-22.

Haverfield, J. J., 1970, Lower Cretaceous sediments of North Park Basin: The Mountain Geologist, v. 7, p. 139-149.

Hummel, J. M., 1969, Anatomy of a gas field — Clay Basin, Daggett County, Utah: in Geologic Guidebook of the Uinta Mountains, Intermtn. Assoc. Geologists, 16th Annual Field Conference, p. 117-126.

Johnson, E., 1961, Joe's Valley gas field: in Oil and Gas Fields of Utah, Intermtn. Assoc. Petrol. Geologists.

Kinney, D. M., 1955, Geology of the Uinta River-Brush Creek Area, Duchesne and Uinta Counties, Utah: U.S. Geol. Survey Bull. 1007, 180 p.

Lane, D. W., 1963, Sedimentary environments in Cretaceous Dakota sandstone in northwestern Colorado: Am. Assoc. Petroleum Geologists, v. 47, p. 229-256.

MacKenzie, D. B., and Poole, D. M., 1962, Provenance of Dakota group sandstone of the western interior: in Symposium on Early Cretaceous Rocks of Wyoming and Adjacent Areas, Wyom. Geol. Assoc., 17th Annual Field Conference, p. 62-71.

MacKenzie, F. T., and Ryan, J. D., 1962, Cloverly-Lakota and Fall River paleocurrents in the Wyoming Rockies: in Symposium on Early Cretaceous Rocks of Wyoming and Adjacent Areas, Wyom. Geol. Assoc., 17th Annual Field Conference, p. 44-60.

McKee, E. D., and Weir, G. W., 1953, Terminology for stratification and cross-stratification in sedimentary rocks: Geol. Soc. America, v. 64, p. 381-390.

Meek, F. B., and Hayden, F. V., 1861, Descriptions of new Lower Silurian (primordial), Jurassic, Cretaceous, and Tertiary fossils, collected in Nebraska, ..; with some remarks on the rocks from which they were obtained: Acad. Nat. Sci. Phila. Proc., p. 415-447.

Peterson, P. R., 1973, Bridger Lake Field (Unitized): Utah Geol. and Mineralogical Survey, Oil and Gas Field Studies No. 5.

Picard, M. D., 1962, Occurrence and origin of Mississippian gas in Four Corners region: Am. Assoc. Petroleum Geologists, v. 46, p. 1681-1700.

Potter, P. E., and Pettijohn, F. J., 1963, Paleocurrent and Basin Analysis: Academic Press Inc., New York, 296 p.

Seeley, de Benneville K., Jr., 1961, Flat Canyon field: in Oil and Gas Fields of Utah, Intermtn. Assoc. Petrol. Geologists.

Selley, R. C., 1970, Ancient Sedimentary Environments: Cornell Univ. Press, Ithaca, New York, 237 p.

Stowe, C., 1972, Oil and gas production in Utah to 1970: Utah Geol. and Min. Survey, Bull. 94, 179 p.

Suttner, L. J., 1969, Stratigraphic and petrographic analysis of upper Jurassic — lower Cretaceous Morrison and Kootenai Formations, southwest Montana: Am. Assoc. Petroleum Geologists, v. 53, p. 1391-1410.

Weimer, R. J., 1962, Late Jurassic and early Cretaceous correlations, south central Wyoming and northwestern Colorado: in Symposium on Early Cretaceous Rocks of Wyoming and Adjacent Areas, Wyom. Geol. Assoc., 17th Annual Field Conference, p. 124-130.

_____, 1970, Dakota Group (Cretaceous) stratigraphy, southern Front Range, South and Middle Parks, Colorado: The Mountain Geologist, v. 7, p. 157-184.

Young, R. G., 1960, Dakota Group of Colorado Plateau: Am. Assoc. Petroleum Geologists Bull., v. 44, p. 156-194.

_____, 1970, Lower Cretaceous of Wyoming and southern Rockies: in 22nd Annual Field Conference Guidebook, Wyoming Geol. Assoc., p. 147-159.

_____, 1973, Depositional environments of basal Cretaceous rocks of the Colorado Plateau: in Memoir on Cretaceous and Tertiary Rocks of the Southern Colorado Plateau, Four Corners Geol. Soc., p. 10-27.

_____, 1975, Lower Cretaceous rocks of northwestern Colorado and northeastern Utah: in Symposium on Deep Drilling Frontiers in the Central Rocky Mountains, Rocky Mountain Assoc. Geologists, p. 141-147.

Ogden to Salt Lake — Wasatch Range

TERTIARY TECTONICS AND SEDIMENTARY ROCKS ALONG THE TRANSITION: BASIN AND RANGE PROVINCE TO PLATEAU AND THRUST BELT PROVINCE, UTAH

by

Robert E. McDonald[1]

Abstract

There are numerous areas of extensive fluvio-lacustrine Tertiary sediments throughout the western United States. These units sometimes reach several thousand ft in thickness, but with a few notable exceptions, such as the Green River Formation, they are poorly exposed and frequently occupy undrilled or sparsely drilled basinal areas. As a result very little is known concerning their economic potential.

The Great Basin is largely delineated by Tertiary tectonism that readily falls into a system of wrench-fault classification. The peripheral areas appear to be controlled by 1st-order right- and left-lateral systems, while 2nd- and 3rd-order systems dominate the interior.

In southwestern Utah the eastern edge of the Sevier thrust belt closely coincides with the Las Vegas-Wasatch Line, but in west-central Utah the "Wasatch Front" bends northward and northwestward, while the thrust belt continues in a more north-northeasterly direction. It is believed that collapse of the Great Basin along the Las Vegas-Wasatch Line commenced by Eocene time and perhaps as early as Paleocene time. This may also be true along the Wasatch Front in northern Utah.

Geophysical evidence indicates as much as 10,000 ft of Tertiary-age sediment is present in the Great Salt Lake and Sevier Desert basins. Seismic data in the Sevier Desert basin are believed to reveal a prominent detachment plane along which major thrusting occurred. Above the detachment, apparent Tertiary-age structure is extensional and typical of Gulf Coast tectonics. It is postulated that as the area made the transition from compression to tension, overthrust Paleozoics retreated episodically westward along the same glide plane over which they were originally thrust eastward.

INTRODUCTION

Tertiary sedimentary strata covered by this paper include southwestern to central Utah along the transition zone between the Basin and Range and Colorado Plateau provinces. From central Utah into northern Utah it will deal with Tertiary sediments in the area of collapse of the thrust belt, roughly paralleling the Wasatch Front. An attempt will be made to summarize published literature to date, and also introduce a few new ideas utilizing this summarization and published and unpublished recent geophysical data.

SIGNIFICANT TERTIARY FLUVIO-LACUSTRINE STRATA: WESTERN UNITED STATES

A brief discussion of major areas of lacustrine to intermittent-lacustrine Tertiary sediments in the western U.S. is helpful in gaining an understanding of hingeline Tertiary deposition. All important Tertiary units are not necessarily lacustrine, but since Tertiary sedimentary history in this area is dominated by shifting areas of vast, shallow, fresh-water lakes, this approach quickly gives some insight into the tectonism along the hingeline and adjacent areas.

Paleocene and Eocene Lacustrine Deposits. (Fig. 1.)

Area 1: Central and west-central Washington.—These strata include both Paleocene and Eocene continental sediments. These lower Tertiary strata were first described from several localities and are represented by a varied nomenclature, including Chuckanut, Cowlitz, Guye, Manatash, Roslyn, Skookumchuck, Swauk, and Tootle formations. There are also essentially volcanic Eocene formations not included in the foregoing. The present occurence of Paleocene and Eocene sediments is primarily on the eastern slopes of the Cascades, and does not represent their original areal extent. The irregular distribution and relative thicknesses at various localities indicate that the present Cascade Mountains in central Washington originally represented a

---

[1] Wolf Energy Company, Denver, Colorado

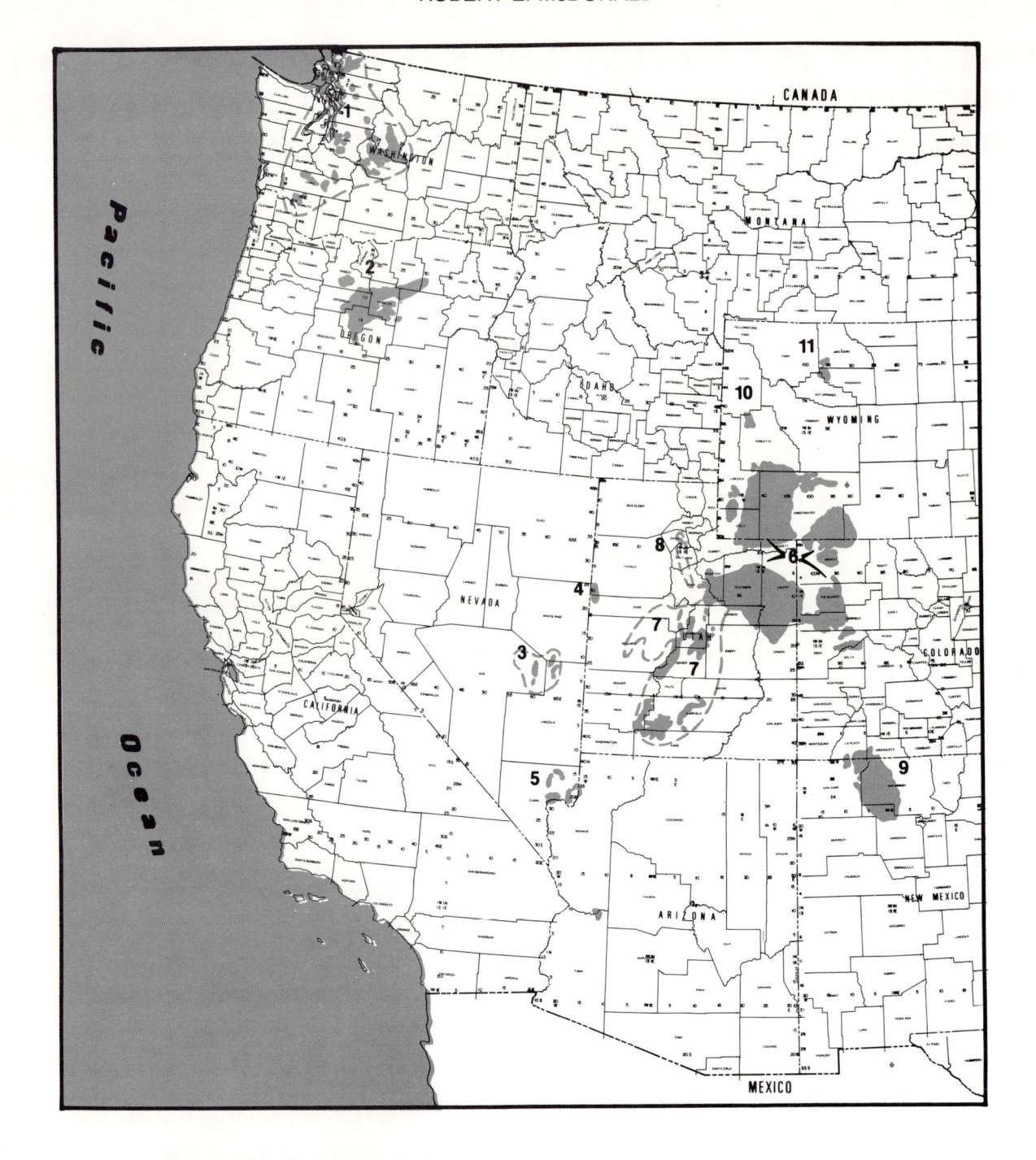

AREAS OF SIGNIFICANT LACUSTRINE DEPOSITION
during
PALEOCENE AND EOCENE

1. **PALEOCENE & EOCENE**
   Chuckanut, Cowlitz, Guye, Manastash, Roslyn, Skookumchuck, Swauck & Tootle Fms.
2. **EOCENE**
   Clarno Fm
3. **EOCENE**
   Sheep Pass Fm
4. **EOCENE?**
   White Sage Fm
5. **PALEOCENE - EOCENE?**
   Horse Springs Fm
6. **PALEOCENE - EOCENE**
   Fort Union, Wasatch & Green River Fms
7. **PALEOCENE - EOCENE**
   North Horn, Flagstaff & Green River Fms
8. **EOCENE?**
   Unnamed
9. **EOCENE**
   San Jose Fm
10. **PALEOCENE**
    Hoback Fm
11. **EOCENE**
    Tatman Fm

FIG. 1

topography of low and irregular relief, not too much above sea level. There were probably hills and valleys, with westerly-flowing rivers. Differential upwarping and downwarping caused some damming of the westward-flowing rivers and temporary lacustrine conditions. These Paleocene and Eocene strata were very extensive and ranged upwards of several thousand ft in thickness; however, subsequent uplift of the Cascade Mountains has resulted in erosion of the major portion of the deposits.

Area 2: John Day area, north-central Oregon.—As much as 6000 ft of Eocene strata are present in the Clarno Formation which outcrops extensively in north-central Oregon and underlies Miocene-Pliocene sediments in the John Day basin. Although the Clarno is in part lacustrine, measured sections suggest that a great abundance of volcanic materials make it unattractive as a significant reservoir or source rock.

Area 3: East-central Nevada.—A lacustrine sequence of unique Eocene limestones, reaching a thickness of 3500 ft is present in some of the mountain ranges and beneath parts of valleys in east-central Nevada. The Sheep Pass Formation produces oil in the Eagle Springs field, Railroad Valley, Nevada, and has had significant oil shows in other wells. Petroliferous limestones have been reported in surface outcrops. There is no question, of course, that the Sheep Pass is a potentially productive oil horizon, and it continues to receive close scrutiny. Currently known occurrences of Sheep Pass would indicate a lake of at least 1200 sq mi in area, occupying portions of White Pine, Lincoln and Nye Counties, Nevada. The lack of volcanic rocks in this Eocene lacustrine sequence is unusual in the Great Basin, and the unit usually contains nonmarine mollusks, ostracodes, and occasional fossil vertebrates.

Area 4: Gold Hill area, Nevada-Utah.—The White Sage Formation is a predominantly lacustrine unit believed to be Eocene in age that is exposed extensively along the Nevada-Utah state line. It is undoubtedly an erosional remnant of Tertiary lacustrine and fluvial strata that covered a much larger area, and it has been suggested that the lacustrine facies of the White Sage Formation may have been deposited in the same extensive Eocene lake in which the Sheep Pass Formation was deposited. On outcrop its present maximum thickness is 600 ft, and there is no geologic reason to project its presence with any great continuity into adjacent valleys.

Area 5: Muddy Mountains area, Nevada.—The Horse Springs Formation has been reported in the literature as being possibly Paleocene or Eocene in age. Recent potassium-argon dating from biotites near the base indicate an age of 23 m.y., giving this formation a probable lower Miocene age. It will be included later in discussions of the Miocene-Pliocene lacustrine deposits.

Area 6: Green River, Washakie, Uinta, and Piceance basins, Wyoming, Utah, and Colorado.—Oil and gas production from fluvial and lacustrine Paleocene and Eocene-age strata occurs in all of these basins and includes major oil and gas accumulations. These well-known early Tertiary units are extensively covered in the literature.

Area 7: Wasatch Plateau-Sevier Desert area, Utah.—This vast area underwent persistent and recurring lacustrine conditions from Cretaceous through Eocene time, and during the Eocene, Lake Uinta covered much of the Wasatch Plateau and lapped high onto the flanks of the mountains of the Sevier orogenic belt. The dashed line around area 7 (Fig. 1) indicates a general projected extent of important Paleocene-Eocene lacustrine units across areas where they either have been removed by erosion or are believed to be covered by more recent sediments. It is not meant to imply that the entire area within the outline was a single lake at any given time—although much of it was during middle Eocene—but extensive fluvial and lacustrine deposits did accumulate over the entire area during varying portions of Paleocene and Eocene time. Over most of area 7, namely the large eastern lobe that is now the Wasatch Plateau and other high plateaus of southern and central Utah, the Paleocene and Eocene sections crop out and are largely removed by erosion. For this reason their potential for oil and gas accumulation has generally been destroyed. The western lobe that may exist beneath the Sevier Desert basin is believed to be entirely covered by Oligocene through Quaternary-age strata. The extension of this lobe beneath the Sevier Desert basin is not a certainty, but believed logical on the basis of the geologic history, nearby stratigraphy, and interpretation of the Gulf Oil Company Gronning No. 1 well.

Area 8: Jordan-Utah Valley, Utah.—This area of lacustrine deposition is postulated but believed possible on the basis of apparent Eocene outcrops in the Salt Lake salient and the geologic history of the area. A lacustrine limestone unit north of Salt Lake City has yielded a fossil snail collection that could be dated anywhere from late Paleocene to late Eocene. Based on relationships with underlying and overlying units and the fact that the limestone is tuffaceous, it is believed by the writer that the unit is probably middle to late Eocene. The Sevier orogenic thrusting responsible for emplacement of the Willard thrust to the north and the Charleston-Nebo thrust system to the south appears to have been completed by the end of Cretaceous. In about middle Eocene, or earlier, north-south trending folds, along with associated normal faulting, began forming the north-south trending mountain ranges and valleys as are now seen. Thus, extensive lake formation may have commenced immediately west of the Wasatch Front by late Eocene time.

Area 9: San Juan basin, New Mexico.—The San Juan basin is included in this discussion only because the Eocene San Jose Formation has been referred to in some of the literature as being partly lacustrine. There were undoubtedly minor ponding and laking; however, this formation as well as the underlying Nacimento is essentially of fluvial origin throughout the basin.

Area 10: Hoback Formation, Northern Green River Basin and Jackson Hole areas, western Wyoming.—Several thousand ft of Paleocene Hoback Formation accumulated in a rapidly subsiding trough between the Gros Ventre-Wind River Mountains on the east, and the "disturbed belt" to the west. The section is essentially fluvial, but with significant periods of lacustrine deposition, and is certainly prospective for hydrocarbons. Much of the Paleocene area of hydrocarbon potential has been complicated by thrusting and post-Paleocene tectonic activity that complicates the geology and make much of the terrain very rugged.

Area 11: Big Horn basin, Wyoming.—The Tatman Formation is an Eocene lacustrine unit located in Wyoming's Big Horn basin. Its stratigraphic and areal extent is too limited to consider in a search for major oil and gas reserves.

Oligocene, Miocene and Pliocene Lacustrine Deposits. (Fig. 2)

Area 1: Washington and northern Idaho.—There are several scattered intermontane basins that contain Oligocene through Pliocene lacustrine and fluvial deposits, including the Hammer Bluff, Ellensburg, Latah, and Wilkes formations. In all instances, however, the areal and stratigraphic extent of any potential source and reservoir beds are too limited to consider the areas as sources of major hydrocarbon accumulations.

Area 2: North-central Oregon.—There are very extensive outcroppings and adequate thicknesses of Oligocene through Pliocene strata present in this area, particularly in the John Day basin. The thickest and most widespread is the John Day Formation of Oligocene and Miocene age. Some very limited exploratory work has been undertaken in this area. Although much of the John Day Formation is lacustrine and the abundance of leaves and plant remains indicate possible presence of adequate organic hydrocarbon source material, potential reservoir beds appear to be lacking. In surface sections examined, the bulk of the material was pyroclastics, ignimbrites, flows of reworked tuffs, and tuffaceous breccias.

Area 3: Upper Kalamath River basin, Oregon.—The Pliocene Yonna Formation possesses an interesting depositional environment, with a combination of extensive lacustrine, marsh, and fluvial conditions represented. Published stratigraphy and geologic history indicates that maximum thicknesses reach about 2000 ft and there appears to be no probable reason to project greater thicknesses of older buried strata deposited under similar depositional settings.

Area 4: Malheur and Harney basins, southeastern Oregon.—The Pliocene strata in this area are a combination of fluvial and lacustrine environments, and were originally named the Danforth Formation. Subsequently, the name Idaho Formation was extended into this area from the Snake River downwarp further east. The Malheur-Harney area has some reported shows of gas in shallow water wells and the Idaho Formation does have some petroleum potential, particularly for natural gas. Depth of this Tertiary basin is unknown, but is probably considerably less than that of the Snake River downwarp. It is not known whether the basin contains any Miocene age strata. Petroleum potential is not believed to be major. So far as can be determined, there were no periods of high clastic input by major drainages to provide the high energies desirable for deltaic buildups in the lacustrine environments, nor deposit well-sorted fluvial bar and point bar deposits through the marshes and playas.

Area 5: Northeastern California-northwestern Nevada.—The Alturas Formation and Cedarville series offer an interesting, partially lacustrine stratigraphy, but their thickness is commonly less than 1000 ft and not adequate for petroleum generation.

Area 6: Western Snake River downwarp, southwestern Idaho and southeastern Oregon.—There have been numerous shows of gas and several shows of oil in the Pliocene Idaho Formation and Miocene Payette Formation, but very little exploration work has been carried out. There have been two periods of extensive lacustrine depositional conditions, one in Miocene time and another in Pliocene time. There is fossil evidence of periods of lush vegetation, and high energy clastic conditions have been adequate from the Snake River and Idaho batholith drainages to provide reservoir beds. Regional gravity data indicate the western portion of the Snake River downwarp is a great graben basin and there have been estimates of as much as 18,000 to 20,000 ft of Tertiary sediments within the graben. The downwrap may have begun sagging before Miocene—perhaps as early as Eocene—and deep drilling may encounter Tertiary sediments older than those observed on the surface or encountered in the shallow drilling. The big deterent to exploration has been the large volumes of volcanic flow material, which has rendered seismic data nearly useless to date.

Area 7: Intermontane Valleys, western Montana.—There are fluvial and lacustrine deposits of Oligocene and Miocene age in the intermontane valleys of western Montana. In most cases they are either of inadequate thickness or too limited in areal extent to be very prospective. In a couple of instances, they do reach thicknesses of 3000 to 5000 ft, but not over large areas. The strata may have some petroleum possibilities, but do not appear to have potential for major fields.

Area 8: Southeastern Idaho.—This area of outcrops represents several intermontane valleys which contain sometimes quite thick sections of the Salt Lake Group—as much as 8000 ft. The dating of the Salt Lake Group is quite inexact, but it is

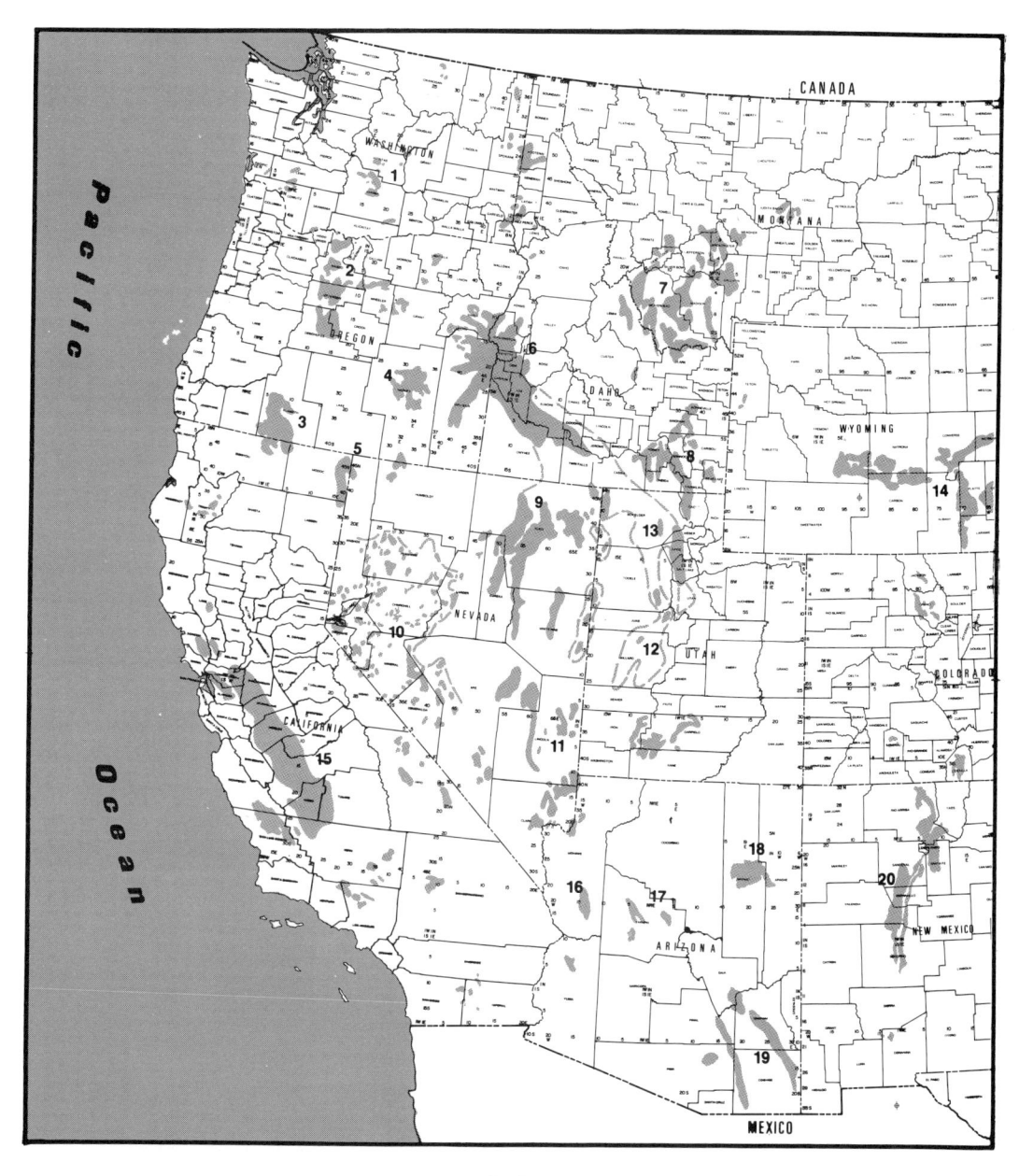

FIG. 2
AREAS of SIGNIFICANT LACUSTRINE DEPOSITION during OLIGOCENE, MIOCENE, and PLIOCENE

1. OLIGOCENE - PLIOCENE
   HAMMER BLUFF, ELLENSBURG, LATAH, LATOUR, & WILKES FMS.

2. OLIGOCENE - PLIOCENE
   DALLES, JOHN DAY, & MASCAL FMS.

3. PLIOCENE
   YONNA FM.

4. PLIOCENE
   DANFORTH or IDAHO FMS.

5. PLIOCENE
   ALTURAS FM. & CEDARVILLE SERIES

6. MIOCENE - PLIOCENE
   IDAHO & PAYETTE FMS.

7. OLIGOCENE and MIOCENE
   TERTIARY LAKE BEDS

8. OLIGOCENE? - PLIOCENE
   SALT LAKE GROUP

9. MIOCENE - PLIOCENE
   HUMBOLDT FM.

10. MIOCENE - PLIOCENE
    TRUCKEE, COAL VALLEY, ESMERELDA, DESERT PEAK FM.
    WASSUK MTS. GROUP

11. MIOCENE - PLIOCENE
    MUDDY CREEK & HUALPAI FM.

12. SEVIER DESERT, UTAH
    OLIGOCENE - PLIOCENE
    SALT LAKE GROUP

13. GREAT SALT LAKE
    MIOCENE - PLIOCENE
    SALT LAKE GROUP

14. SE WYOMING - NW NEBRASKA
    WHITE RIVER, CHADRON, & BRULE FM.

15. CENTRAL VALLEY, CALIFORNIA
    PLIOCENE - TULARE & SAN JOAQUIN FMS.

16. WESTERN ARIZONA
    MIDDLE TO LATE TERTIARY, CHAPIN WASH FM.

17. WEST CENTRAL ARIZONA
    PLIOCENE
    CHINO & CHERRY VALLEY BEDS
    VERDE FM.

18. NORTHEASTERN ARIZONA
    PLIOCENE
    BIDAHOCHI FM.

19. SOUTHEASTERN ARIZONA
    PLIOCENE
    PLANTANA FM., SAN PEDRA, SAN CARLOS, GILA, SAN SIMON VALLEY BEDS

20. RIO GRANDE VALLEY, NEW MEXICO
    PLIOCENE
    SANTE FE FM.

thought to be primarily Pliocene in age. It may very well include Miocene and even Oligocene strata, however. There have been significant shows of gas in the Salt Lake Group in this general area. The depositional settings favorable for hydrocarbons in these valley areas are localized and limited, and are not believed to offer potential for major accumulations.

Area 9: Northeastern Nevada.—In northeastern Nevada, the Humboldt Formation crops out extensively, but not as continuously as shown on Figure 2. The stratigraphy of the Humboldt Formation is in part quite favorable for the occurrence of oil and gas, it reaches a maximum known thickness of 8000 ft. The Humboldt is regarded as essentially Miocene, although it includes strata of Late Eocene to Pliocene ages. The Humboldt Formation is essentially lacustrine, but of course its lithologies and depositional environments are controlled by proximity to mountainous source areas and periods of volcanism. In the northeasternmost exposures, just before the Humboldt Formation disappears beneath some of the younger Snake River basalts at the junction of Utah-Idaho-Nevada, it consists of carbonaceous and brittle bituminous shale and petroliferous limestone assigned to the upper Miocene and lower Pliocene. Toward the south end of the outcrop area in Huntington Valley, the Humboldt consists of an upper vitric tuff unit, underlain by aphanitic ostracod- and snail-bearing limestone, together with some mudstone and a few layers of platy chert. The basal beds are reddish-brown Paleozoic pebble conglomerate and sandy limestone, overlain by dark, platy petroliferous limestone and subordinate amounts of crumbly dark carbonaceous mudstone. The well-known oil shales of the Elko, Nevada area are included in the Humboldt Formation, and recoveries from these oil shales has ranged as high as 85 gal/ton. Tectonicly, northeastern Nevada was subjected to extensive uplift and breakup subsequent to deposition of the Humboldt Formation, which complicates exploratory efforts. The structure of the Humboldt Formation is often complex and its presence in the subsurface unpredictable.

Area 10: Carson Desert basin, Nevada.—The Carson Desert-Buena Vista-Dixie Valley areas of west-central Nevada were major late Tertiary basins, filled with a fluvial-lacustrine sedimentary and volcanic rock sequence. Fluvial and lacustrine sediments crop out extensively in the surrounding mountains and hills, with thicknesses of sedimentary units reaching 8200 ft and the combined Tertiary section, including both sediments and volcanics, exceeding 12,000 ft. The age of Tertiary units reaches back into the Oligocene, with Miocene and Pliocene strata providing the bulk of the exposed section. Tectonic history and regional relationships suggested that the Walker Lane strike-slip rift-faulting continues along the west side of Carson Desert, bending back northward and northeastward through the Pyramid Lake area, thus forming an orocline in a major right-lateral wrench-fault zone complimentary to the San Andreas fault. There is considerable evidence for this, including alignment of earthquake epicenters. Rifting commenced by at least mid-Miocene, with the downthrown side of the rift system being most usually along the east, toward the center of the orocline now occupied by the Carson Sink. This indicates a strong probability of a section of fluvio-lacustrine Miocene and Pliocene sedimentary rocks several thousand ft thick through the central Carson Desert, Buena Vista, and Dixie Valley areas. The full extent of the Miocene-Pliocene lake cannot be appreciated from the plentiful but scattered outcrops. A dashed line indicates the inferred extent of laking, and the western Nevada lake may have joined the Humboldt lake at times.

Area 11: Southern Nevada.—The northern portion of this group of lacustrine outcrops is the Currant Creek tuff, which is a series of fresh water limestones and tuffs. Fragments of snails and ostracodes are present, but deposits are neither extensive enough nor have adequate clastic source for hydrocarbon potential. The lacustrine strata in the south part of this section includes the Miocene (?) Muddy Creek Formation and the Pliocene Hualpai. Both formations are a combination of fluvial and lacustrine deposits, with some interesting chemical precipitates. These sediments were deposited in a chain of intermontane lakes, however, and did not reach adequate areal extent or thickness to be included as an objective for large oil and gas accumulations.

Area 12: Sevier Desert, Utah.—This area is positioned in a major desert basin, immediately adjacent to a rift system that is part of the Las Vegas-Wasatch Line, an extensive 1st-order left-lateral wrench-fault zone. It is also situated in an orocline similar to the Carson Sink, only there is a change from a northeast-trending to a northwest-trending rift system. The shear system is complete with a zone of earthquake epicenters, hot springs, sag ponds, and most indicators of rifts. Thick fluvio-lacustrine sections are apparently present in this persistently negative area, with at least Oligocene through Recent and possibly Eocene through Recent strata represented. There have been several shows of oil and gas at shallow depths, and the writer believes the Tertiary possesses hydrocarbon potential.

Area 13: Great Salt Lake, Utah.—The Great Salt Lake is also a good exploratory area, with many similarities to the Sevier Desert basin. It is currently under lease by Amoco Production Co., which has done extensive seismic exploration and has two wildcat tests in the planning stage.

Area 14: Southwestern Wyoming, northwestern Nebraska.—The White River, Chadron and Brule formations lie in areas of limited lacustrine deposition — mostly local ponding — in an open fluvial system. Thickness, source, and reservoir rocks are not conducive to hydrocarbon accumulation. Ages of these formations are Oligocene and Miocene.

Area 15: Central Valley, California.—The area is covered extensively in the existing literature and is productive of oil and gas. It is mentioned here only to complete the western picture.

Area 16: Western Arizona.—The area contains the Chapin Wash Formation and unnamed middle to late Tertiary valley deposits. There are some extensive areas of relatively attractive lacustrine and fluvial sediments, but maximum reported thickness is 1500 ft and most of the basinal areas have lesser thicknesses.

Area 17: West-central Arizona.—Chino and Cherry Valley beds of the Verde Formation are present in this area. These deposits of Pliocene and Pleistocene lacustrine strata are due to volcanic flow material damming-up valleys. The basins contain thick lacustrine limestones, sulphates, halites, and gypsum, but the largest basin covers only about 300 sq mi and total Tertiary and Quaternary strata are not known to exceed 200 ft in thickness.

Area 18: Northeastern Arizona.—The Bidahochi Formation is Pliocene age and covers a relatively large area; however, it reaches a maximum thickness of only about 250 ft.

Area 19: Southeastern Arizona.—This area includes the Plantano Formation, San Pedro Valley beds, Gila and San Carlos Valley beds, and San Simon Valley beds. The deposits are thin, with a maximum thickness of 1250 ft. The stratigraphy is mostly red beds with abundant evaporties, and considered representative of playa depositional environments.

Area 20: Rio Grande Valley, New Mexico.—The Rio Grande Valley is underlain by the Pliocene Santa Fe Formation. In the literature its maximum reported thickness is 4100 ft, but it may be much thicker in places. The Santa Fe Formation was deposited in a subsiding structural depression, consisting of a number of basins separated end-to-end by structural constrictions. Playa deposits are the main depositional environment and just southwest of Albuquerque about 400 ft of Santa Fe was described as tan, buff, and brown sand, and brown and red gypsiferous silt and clay. There may be places along the Rio Grande trough that the Santa Fe Formation attains thicknesses that would be attractive for hydrocarbon exploration; however, the total depositional setting is not believed attractive for major accumulations. The Rio Grande trough might be compared to a miniature Snake River downwarp that never achieved similar depth or areal extent. So far as is known, it was never dammed to form immense lakes and marshes. Lacustrine conditions appear to have been limited to small, temporary, closed basins along the depression, and the trough as a whole has remained open.

Pleistocene Lakes, (Fig. 3)

Pleistocene sediments are not discussed in this paper, but Fig. 3 is included as a matter of interest to complete the evolution of major lake development in the western United States. For more information the reader is referred to Feth (1964).

CENOZOIC TECTONIC FRAMEWORK

Fig. 4 is a physiographic map of the Great Basin and environs. The basin outline and its internal structure appears to be largely delineated by wrench-fault zones, and wrench-fault tectonics can be used to explain nearly all the major structural lineaments. The western Snake River downwarp, Carson Desert, and Sevier Desert are indicated, and it is perhaps significant that two of these fall adjacent to major, known 1st-order wrench-faults, and a third, the western Snake River downwarp, is in all probability adjacent to or within a 1st-order wrench-fault. While there have been periods of extensive lacustrine deposition and important Tertiary sedimentary deposits in areas dominated by 2nd-order wrench-faults, in every instance their existence as a negative area has not been persistent and they have been subjected to later breakup.

In Southwestern Nevada, the Walker Lane and Las Vegas shear have long been recognized as 1st-order right-lateral zones of wrench-faulting, and several geologists have suggested their connection. As can be seen on Fig. 4, there appears to be no question that the physiography strongly suggests this. Trending slightly southeastward from the Walker Lane are two extensive fault zones — the Death Valley and Owens Valley faults — that do not easily fit the wrench-fault pattern, although their northern extremities appear to curve northwesterly into the 1st-order left-lateral pattern. Across the Sierras, the San Andreas fault is indicated, which is undoubtedly the most famous 1st-order right-lateral wrench-fault in American geology and folklore.

Just northwest of the Carson Desert, the margins of the Great Basin make a bend and take on a northeasterly trend. This bend has been recognized and named the Mendocino-Idaho orocline. Southeast of this northwesterly margin, the basins and ranges have definite patterns typical of Great Basin physiography, but to the northwest — an area dominated by volcanic plateaus — the mountain ranges are not as pronounced and take on no particular structural lineaments. The physiographic map suggests that this northwestern margin of the Basin and Range Province is bounded by a series of short 1st-order right- and left-lateral wrench-faults.

Commencing just north of the Carson Desert basin and extending in a northeasterly direction across northern Nevada, the primary orientation of the mountain ranges suggest an extensive zone of 2nd-order left-lateral wrench-faults. This is not to the exclusion of 2nd-order right-laterally controlled ranges, but the left-lateral trends dominate and point toward the eastern portion of the Snake River downwarp in southeastern Idaho.

Through central Nevada and into western Utah, an extensive area of north-south to north northeast-south southeast trending ranges appear to nearly all fit the pattern of 2nd-order right-lateral wrench-faults.

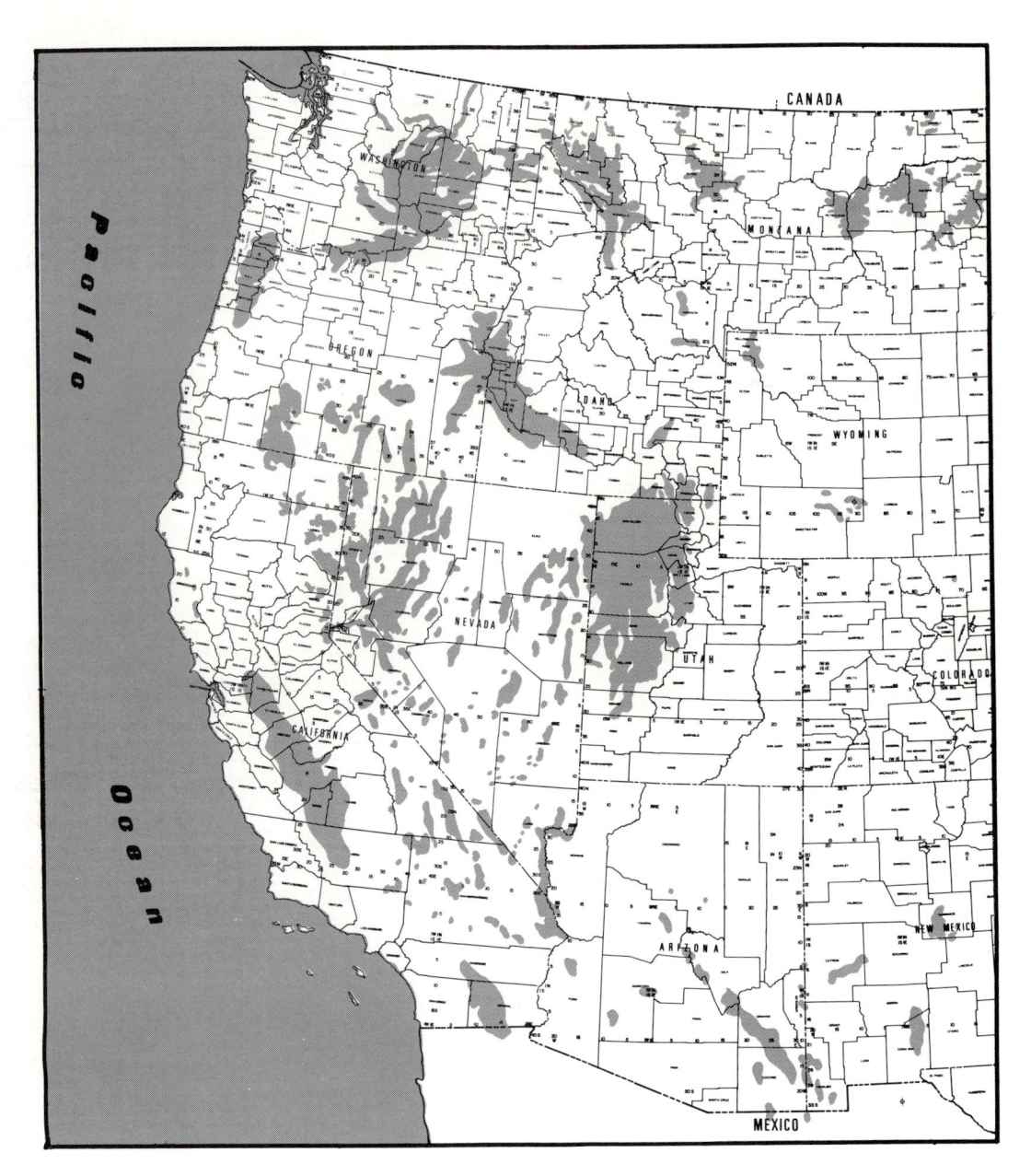

FIG. 3

MAXIMUM EXTENT of KNOWN and INFERRED PLEISTOCENE LAKES, WESTERN UNITED STATES
From: Feth, John H., USGS Bull. 1080, 1964.

In southern Nevada and southwestern Utah, the southeastern margins of the Great Basin Province are delineated, in general, by the Las Vegas-Wasatch Line, along which left-lateral movement has been reported, and it is shown as a 1st-order left-lateral wrench-fault on Fig. 4. The Las Vegas-Wasatch Line roughly coincides with the southeastern margins of the Sevier orogenic belt.

There are a series of major faults in southwestern Utah trending north-south to northeast-southwest, such as the Hurricane, Sevier, and Paunsagunt faults. Their angular relationship with the Las Vegas-Wasatch Line is a mirror image to that of the Owens and Death Valley faults to the Walker Lane. It is not known whether the Sevier and Paunsagunt faults reflect lateral displacement; however, it has been reported from gravity

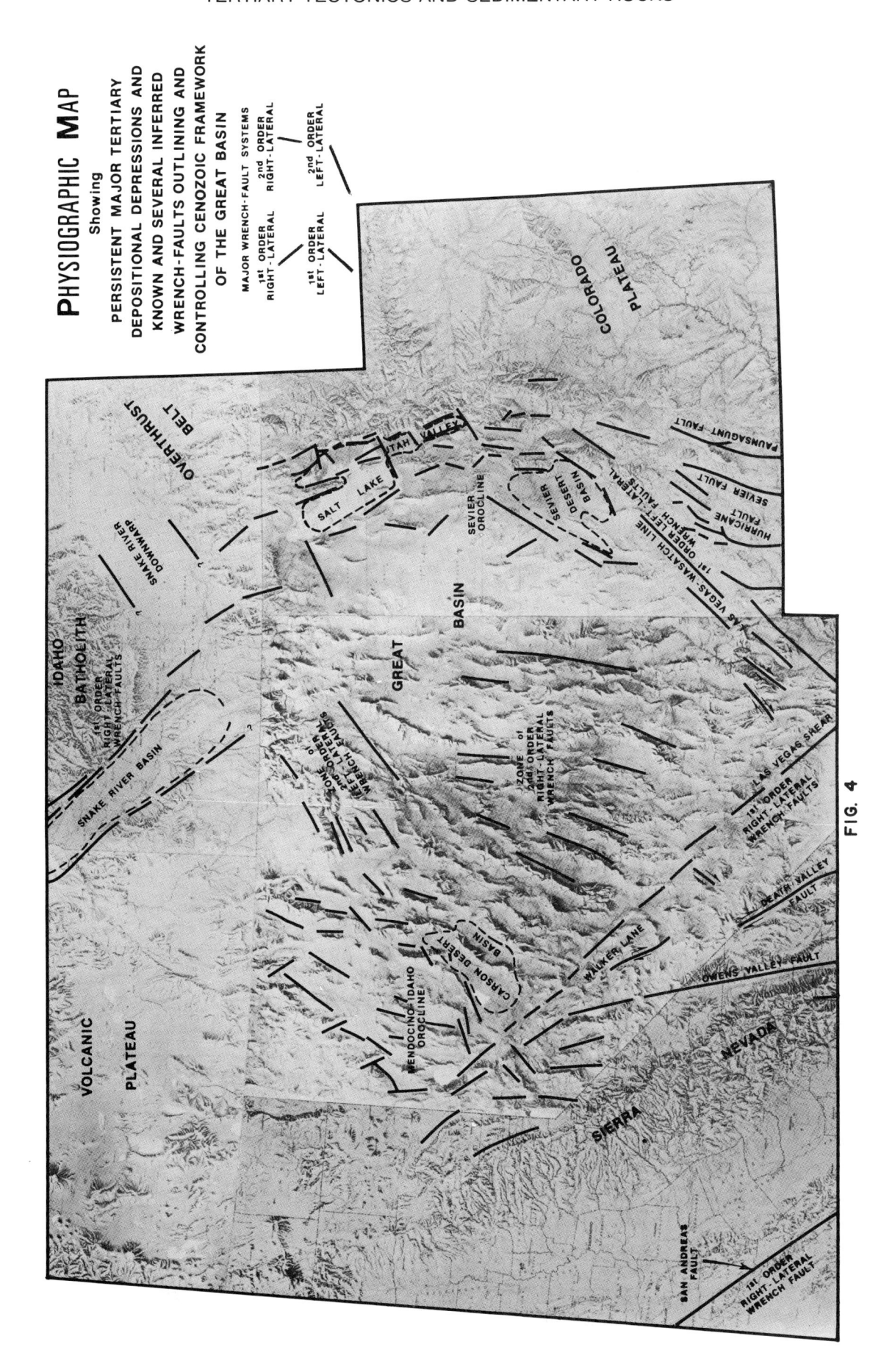

FIG. 4. PHYSIOGRAPHIC MAP showing PERSISTENT MAJOR TERTIARY DEPOSITIONAL DEPRESSIONS AND KNOWN AND SEVERAL INFERRED WRENCH-FAULTS OUTLINING AND CONTROLLING CENOZOIC FRAMEWORK OF THE GREAT BASIN

evidence that lateral offsetting of structural lineaments east and west of the Hurricane fault does occur.

In west-central Utah, there is a northward and northwestward bending of the Wasatch Front just northeast of the Sevier Desert basin forming another orocline which the writer has called the Sevier orocline (Fig. 4). North of this bend the gross trend of the Wasatch Front is more northerly than would be anticipated from a 1st-order right-lateral wrench-fault. When the faulting along the Wasatch Front is examined in detail, however, it is not a single fault but a series of en echelon faults that individually reflect a more northwesterly trend. The Front also is offset often by northeast-southwest trending faults that fit the pattern of 1st- and 2nd-order left-lateral wrench-faults. Immediately northeast of the Sevier orocline, the major rifting and collapse portion of the Wasatch Front cuts the Nebo-Charleston thrust sheet and no longer parallels the overthrust belt.

Around the common corners of Utah and Nevada where they join Idaho, is a large area of intersection between dominantly northwest- and northeast-trending structural lineaments. Significantly, the Snake River downwarp makes a dramatic bend at this point and the southeasterly-flowing Snake River abruptly makes a 90° change of direction, flowing off to the northwest.

Gravity work on the eastern side of the Snake River downwarp suggests it is possibly just that — a downwarp. Faulting exists, but it could not be called a major graben basin. Conversely, gravity work suggests that the western portion of the downwarp is a great graben basin, delineated on either side by a series of faults that trend northwest-southeast. To the writer's knowledge, lateral movement has not been demonstrated on faults that delineate the western side of the downwarp, but it appears quite probable that it is adjacent to or within a major zone of 1st-order right-lateral wrench-faults.

In all cases, the 1st-order wrench-fault systems are accompanied by a swarm of earthquake epicenters, hot springs, high temperature gradients, sag ponds, and centers of volcanic activity. In addition to hydrocarbon potential in certain selected areas, there is a proliferation of geothermal prospects along these systems.

TERTIARY SEDIMENTARY ROCKS: SOUTHWESTERN UTAH

Southwestern Utah is defined for purposes of this paper as that portion bounded roughly by the Colorado Plateau province on the east, the Sevier Desert on the north and the San Francisco, Beaver and Cricket Mountains on the west.

The only Tertiary sedimentary unit in this area with significant area of outcrop is the Claron Formation, which is a fluvial and lacustrine unit predominately Paleocene and Eocene in age. No faunal dating of the Claron has been published and as Spieker (1949) emphasized long ago, correlation of Tertiary units purely by lithology in the plateau country of Utah is frequently very misleading. Threet (1952) found it very difficult to distinguish between Claron and Cretaceous conglomerates on the Markagunt Plateau, and apparent Claron-type conglomerates were traced by Wiley (1963) beneath a Laramide thrust sheet in the northern Beaver Dam Mountains. The Claron unconformably overlies formations from the Cretaceous Kaiparowits down to the Jurassic Navajo Sandstone (Cook, 1960), and is most commonly overlain by Oligocene ignimbrites of the Needles Range Formation. This upper contact can be either unconformable, or conformable and intertonguing; thus, some upper Claron may be lower Oligocene in age.

Cook (1960) gives a good generalized description of the Claron: "In most places the Claron is a sequence of fresh-water limestones with subordinate calcareous sandstone, limestone pebble conglomerate, and shale above a basal sandstone and quartzite-cobble conglomerate. The beds of the Claron show a wide range of composition, texture, and color. The general color of the unweathered limestone is pale red, yellow, grey, and white, but weathering produces a strong pink tone, especially in the lower part of the formation."

The Claron is a homolog of the Pink Cliffs Wasatch of the High Plateaus, and the upper limestones are probably correlative both in time and environment with the Flagstaff Limestone. A great deal of the Claron outcrop area is around the Pine Valley Mountains north of St. George, Utah, where it maintains a remarkably uniform 460 ft in thickness (Cook, 1957). Cook also observed that the Claron thickens to over 1000 ft and grades rapidly into conglomerates as a thrust sheet is approached northwest of Gunlock, Utah. This thickening is maintained in a northeasterly direction along the thrust front, leading Cook to postulate that this linear zone may represent a shallow syncline extending from the Iron Springs district to the Bull Valley district. East of the Pine Valley Mountains and just northwest of Pintura, Utah, the Claron is over 2000 ft thick and perhaps as much as 3400 ft in thickness (Lovejoy 1973). East of the Hurricane fault on the Markagunt Plateau, the Claron is about 1300 ft thick (Cook, 1957). Lovejoy interprets the abnormally thick Claron at Leap Creek (northwest of Pintura) as part of a reverse-drag zone along the west side of the Hurricane fault, and this represents one of the evidences cited in his hypothesis that 85 percent of the displacement along the Hurricane fault occurred prior to the end of Paleocene time.

Exposures of Paleocene and Eocene rocks are not adequate in southwestern Utah to reconstruct the Claron-Flagstaff lake geometry, but there is little question that the immense Flagstaff playa-lake complex of central Utah extended into southwestern Utah also. This essentially lacustrine unit indicates synclinal warping and movement of fault blocks concurrent with sedimentation.

A well drilled by Amoco Production Company in the middle

of Escalante Valley, Section 1, T. 34 S., R. 15 W., to a total depth of 6762 ft should have yielded a valuable Tertiary section. Unfortunately, severe lost circulation plagued the test almost continuously and much of it was drilled dry with no cuttings sample returns.

Interpretations by Amoco from limited cuttings samples and mechanical logs indicate the top of the Tertiary ignimbrites at 810 ft and the base at 3250 ft. The Claron Formation top is at 3250 ft and its base is at 3890 ft for a total thickness of 640 ft. The Claron may be in fault contact with the underlying Paleozoic units. Cores taken at 5013 to 5027 ft and 6742 to 6762 ft recovered dolomite identified paleontologically as Devonian age.

In the mountains immediately west and northwest of the Amoco location, the exposed ignimbrite sequences commonly range from 1000 to 1500 ft in thickness and rest unconformably on Paleozoic rocks with as much as 500 ft of local relief on the erosion surface (Mackin, 1963). To the east and southeast of Escalante Valley the ignimbrite sequence is commonly thinner and rests unconformably on Mesozoic units or on the Claron Formation, both unconformably and with intertongueing relationship. Mackin (1963) points out that persistently thicker sections of Needles Range Formation (ignimbrites) occur west of Escalante Valley than in areas to the east and southeast, possible indicating foundering along the Wasatch-Las Vegas Line during deposition of the ignimbrite in early Oligocene time.

TERTIARY SEDIMENTARY ROCKS: WEST-CENTRAL UTAH

West-central Utah is here defined as that portion of the transition zone bounded by the high plateaus on the east, the Sevier Desert on the west, the Nebo-Charleston thrust on the north, and the Tushar Mountains on the south. Much has already been published on the Cenozoic stratigraphy of west-central Utah east of the Sevier orogenic front, as manifested in the overthrust Pavant and Canyon Ranges, and will not be repeated here. For someone not familiar with the stratigraphy, there are several good references in the bibliography and a starting point should be Spieker (1949). The most recent good summary is by Doelling (1972). This paper is confined to some general observations regarding Cenozoic stratigraphy and tectonics both east and west of the Sevier orogenic front, and engages in some speculation regarding what may be underlying the immense Sevier Desert basin.

There have been some shallow shows of oil and gas in the Sevier Desert and northernmost Escalante Valley area although this area remains virtually unexplored. Some geologists believe the Sevier Desert basin and environs is filled with ignimbrites and volcanic debris, and its only economic potential is geothermal resources. The writer believes, however, that significant sagging may have occurred in the Sevier Desert basin as early as Eocene, and portions of the basin may contain as much as 10,000 ft of Tertiary strata, including the possibility of Eocene lacustrine strata, and extensive Miocene-Pliocene lacustrine sediments.

SEVIER DESERT, UTAH

The earliest verified shows of oil and gas in the Sevier Desert occurred in water wells drilled in the early 1900's. One test was drilled by the Union Pacific Railroad at Neels, Utah (west side of the basin just east of the Cricket Range) in 1906 to a total depth of 1998 ft. This well encountered several shallow water horizons, but all the water was hot and too poor in quality for drinking purposes. There was steam in some zones, and it is reported that "steam escaped freely from the well while drilling." Lee (1908) reports, "A slight amount of oil was found at several horizons in this well, and gas under great pressure was encountered at a depth of 1802 ft." The drillers log states: "Gas under pressure sufficient to raise 6200 lbs of tools 400 ft in the air." This is noted opposite the 1802 ft depth. In reviewing the drillers log, the most prevalent lithology is "blue waxy clay," with "grey clays" being the second most common lithology. There is frequent mention of interbedded sandstones, and some red and yellow clays. Two igneous flows of 12 ft and 14 ft were penetrated, and the well probably bottomed in an igneous flow.

Meinzer (1911) states that the State of Utah drilled a well 5 mi northwest of Fillmore, Utah (Sec 36., T. 20 S., R. 5 W., Fig. 17), searching for artesian water. No definite record is available on this test, but Meinzer states that "a small amount of oil was struck" and the total depth was about 900 ft...the upper 100 ft are said to consist of clay and sand, below which the drill penetrated shale and red sandstone." The section penetrated was probably Quaternary. Subsequently a Westwood Oil Co. did additional drilling around this well. A drillers log records penetrating "petroliferous shales and oil-bearing sands." It would almost appear they were drilling in an oil seep.

South of the Sevier Desert in Sec. 28, T. 29 S., R. 7 W., another early oil show was recorded in Quarternary lake beds. Lee (1908) reports on a 200 ft water well, "The well is not now in use, but is of interest especially in showing the character of the Beaver Lake beds and for the reason that an oil stratum was penetrated at a depth of 140 ft. The evidences of oil were not sufficient, however, to indicate the presence of any large quantity." It was not stated when the well was drilled; however, this paper was published in 1908.

In 1910, Tom Kearns, prominent in mining in Salt Lake City, drilled a 1920 ft hole in either Sec. 12 or 13, T. 20 S., R. 5 W. The well was reportedly drilled because of oil and/or gas shows in a nearby water well, and apparently remained in Quaternary to total depth.

In 1920, a Kaminski drilled an 800 ft test in the northern end of Escalante Valley in Sec. 31, T. 25 N., R. 9 W., which reported shows of gas. A Beaver Oil Company followed up on these shows, drilling two wells in Sec. 19 and 20, T. 25 N., R. 9 W., to a total depths of 1545 ft and 3500 ft respectively in 1930. The operator reported shows of natural gas in these wells.

Following the Kaminski and Beaver Oil Company wells, no drilling was done until 1950 when El Capitan Drilling Company tried the north end of Escalante Valley once again. Their well offset the deeper Beaver Oil test, being located in SESW, Sec. 17, T. 25 S., R. 9 W., and was drilled to a total depth of 3682 ft, reporting shows of both oil and gas. El Capitan also drilled a test to 3379 ft in Sec. 29, T. 25 S., R. 10 W., and reported shows of gas. Drillers records indicate that none of this cluster of tests penetrated beyond Quaternary-late Tertiary strata at total depth. An electric log was run to TD on the well in Sec. 17 and reveals a series of well-developed sands below 3000 ft.

The first test drilled in the Sevier Desert by an oil company was the Gulf Oil Company, Gronning No. 1, Sec. 24, T. 16 S., R. 8 W. This was a joint-venture stratigraphic test financed by Gulf Research, Standard of California, Shell, and Union Oil. No seismic was conducted and the well was drilled solely as a stratigraphic test. This well is the only control point in the entire Sevier Desert area, and since reports of the age and nature of the section penetrated vary significantly, the samples were obtained from the Utah Geological and Mineral Survey and examined. The latest drilling activity in the area occurred in 1960, when Shell Oil Company drilled an 8962 ft Paleozoic test in Sec. 21, T. 22 S., R. 4 W. This well really has no bearing on the Tertiary stratigraphy or petroleum potential of the Sevier Desert.

Tertiary Stratigraphy of the Sevier Desert

The Sevier Desert basin is an area of roughly 3000 sq mi in which only one significant test has been drilled, and the only outcrops of the prospective sedimentary section lie across mountain ranges to the east and northeast. Fig. 6 illustrates several line sections and gives notations regarding outcrops. These will, for the most part, be referred to in the section on "Tectonics." This summary of stratigraphy will be primarily devoted to a discussion of the Gulf Oil, Gronning No. 1 test, and a few nearby outcrops, and the regional stratigraphic relationships will be discussed later.

The Gronning No. 1 wildcat was drilled in Sec. 24, T. 16 S., R. 8 W., to a total depth of 8064 ft without problem. The well was spudded February 1, 1957, and plugged March 20, 1957, and remains a little-studied stratigraphic outpost in the geologic "no man's land" of the Sevier Desert. Gulf simply reported that the well bottomed in Tertiary red beds. American Stratigraphic Company ran the samples in 1963, and concluded that Gulf was in Eocene strata at total depth. Subsequently, Ed Heylmun, a geologist formerly with the Utah Geological and Mineral Survey, reported in two publications that he believed that Triassic red beds were penetrated by the test. The conclusion by the writer is that the Gulf well bottomed in Tertiary strata, possibly Eocene, primarily on the basis of relative amounts of intercalated volcanic material and sand grain constituents. It is believed the maximum geologic range of time that could be represented at total depth is uppermost Cretaceous to Miocene.

Fig. 5 is a generalized sample description and electric log of the Gulf well. The first 2120 ft have been tentatively assigned to the Pliocene-Pleistocene Sevier River Formation. In samples, this section is a series of tan to grey claystones, interbedded with sandstones and a few conglomerates, fresh water limestones, and lignites. The log has numerous highly resistive kicks accompanied by low SP curves that suggest lignites or limestones, but only a few examples of these lighologies were seen in the samples. The first 2000 ft undoubtedly drilled extremely rapidly, and possibly with little supervision over sample catching. The gross lithologies are probably valid, but it is very possible that much important detail is missing. It is probable, however, that this interval represents a fluvio-lacustrine sequence.

A change in character on the electric log at 2120 ft is also reflected in the samples, initially in the color and induration of the claystones. The tans become darker or light brown in color, greys become more common, and green colors appear. A little further into the second unit there are also some red claystones. This sedimentary unit continues to a depth of 3500 ft on the basis of similar lithologies and apparent depositional environments, with a total thickness of 1380 ft. The sandstones are primarily very fine- to medium-grained, but occasionally grade to coarse-grained. They are sometimes clayey, some appear porous, and in most instances they are friable. There is surprisingly little tuffaceous material, and perhaps much of it was altered to clays. A small amount of ostracodal limestone was penetrated.

There are igneous flows at 2520 ft (80 ft thick), 2630 ft (80 ft thick), 2820 ft (100 ft thick), 3050 ft (20 ft thick), and 3300 feet (50 ft thick). The first three flows are definitely basalts, and are difficult to assign an age. There are a few early Tertiary basalts in southwestern Utah, but the writer is unaware of any in proximity to the Sevier Desert where the basalts are commonly late Tertiary or Quaternary. The lower two flows are distinctly different, with red to brown colors, green crystals that appear to be augite, and vesicles apparent in the cuttings. The writer is not qualified to classify extrusive igneous rocks, but in appearance these flows are very similar to basaltic andesites. Basaltic andesites are common in the Oligocene-Miocene portion of the Tushar volcanic pile to the southeast.

Reducing environments dominate over oxidizing depositional environments in the sedimentary rocks, and it is believed the sediments represent shallow-water lacustrine to playa and low-energy fluvial-type strata. The volcanics apparently flowed as sheets into the shallow water or onto playas without significant associated agglomerates or tuffs. An Oligocene-Miocene age is tentatively assigned to this group.

A 750 ft section that is largely volcanic agglomerate was penetrated between 3500 and 4250 ft. It is surprising that no ignimbrites — particularly the Needles Range ignimbrites — were found in this well. The volcanic grains in this section were primarily acidic-type volcanics typical of the late Eocene-early Oligocene ignimbrites, and it is assumed that this agglomerate section reflects erosion of this widespread ignimbrite sequence. The large grains are not all volcanic, and include an abundance of quartz, chert, quartzite, and limestone, with some intercalated grey, green, and tan claystones. The relative abundance of the volcanic and sedimentary grain types can be observed alternating vertically, and it would appear that there was nearby erosion of both Paleozoic and volcanic source areas. On the basis of the possible correlation of this unit with the widespread ignimbrite sequence, it was assigned an early Oligocene age, but could very well extend back into late Eocene time when acidic volcanics were also known in adjacent areas.

At 4250 ft there is a rather abrupt change to finer sand sizes and an increase in claystone. Close examination of the sands does not reveal a great difference in the composition of grains. There is still an abundance of quartz, quartzites, chert, some limestones, and finer-grained volcanics come and go, but the sizes are primarily fine- to medium-grained, subrounded to angular, and usually friable. The color of the claystones and siltstones is predominately grey, and in places the fissile grey shales and siltstones reminded the writer of upper Coloradan or early Montanan marine shales and silts from the craton. There are also zones of red, green, tan, and even lavender claystones, but unlike the variegated or mottled shales up the hole, these are deep drab colors. They are reminiscent of some of the variegated shales in the North Horn Formation. Some white to brown limestones were also observed.

It will be noted that the electric log appears considerably sandier than the sample log. The electric log was available when the samples were logged, and the writer was constantly aware of this problem. In part, it is probably caused by the great frequency of siltstones and arenaceous material in the clays; however, there are also some good claystone sections that just do not show up on the log. It was finally concluded that the SP curve was excessively nervous, and indicates more sand than is actually present. Changes between 4250 and 6250 ft are gradual and gradational, and no attempt was made to subdivide this 2000 ft section. Volcanic material in the sands becomes less frequent toward the base, but there is considerable clayey material that is attributed to alteration of tuffs. Volcanism was taking place, but apparently not in close proximity. The unit appears fluvial at the top, but grading downward into a primarily lacustrine sequence. A tentative middle to late Eocene age estimate has been made for this interval. For those familiar with the central Utah section, this would suggest correlating this 2000 ft interval with all or some part of the Green River, Crazy Hollow, Bald Knoll, and/or Grey Gulch formations. There is more volcanic material in the cuttings samples than contained in most of these tentative homologs, but the correlation is believed a good possibility.

The last 1800 ft, 6250 ft to total depth of 8064 ft, is primarily a red bed sequence, although there are fair amounts of grey argillaceous sandstones and siltstones in the upper part. The reason Heylum designated this sequence Triassic is immediately apparent. The red colors are mostly deep brick-red to brownish-red and are very similar to Triassic colors. On the basis of the sand fabric and constituent grain material, however, it is believed the sequence is definitely post-Sevier orogeny. This is a rather monotonous sequence, and descriptions of the four cores scattered through the interval will give a good overall picture of the detailed lithologies:

Core 6494-6518 ft. Several pieces of this core were in envelopes that did not designate various depths, but the pieces available were of the same lithology, described as follows:

Very fine- to medium-grained, grey, subrounded to angular, silty, very hard and tight sandstone with a brownish cast. Individual quartz grains are brown, and contain unidentifiable black accessory minerals.

Core 6914-6924 ft.—Several pieces in envelopes not assignable to specific depths, described as follows:

Brick red shale, arenaceous, with numerous quartz grains, subround to angular, milky to dark. Siltstone, light grey, very hard, calcareous. Sandstone, brown and calcareous. Grains from silt-size to coarse-grained, mostly very fine- to medium-grained. Brown quartz grains, tan and black unidentifiable minerals, rounded to angular; clear angular quartz shards, some small quartz crystals.

Core 7410-7430 ft.—In envelopes with notations of ¼ or ⅛, described as follows:

¼, Brick red siltstone, clayey and sandy, friable, some glauconite, abundant alteration kaolin.

¼, Brick red arenaceous claystone with abundant white specks of apparent alteration clay.

⅛, Arenaceous siltstone as in first ¼.

⅛, Sandstone, brick red, friable, rounded to angular grains, fine to coarse sizes, abundant white, altered kaolin?

¼, Above described sandstone and siltstone interbedded.

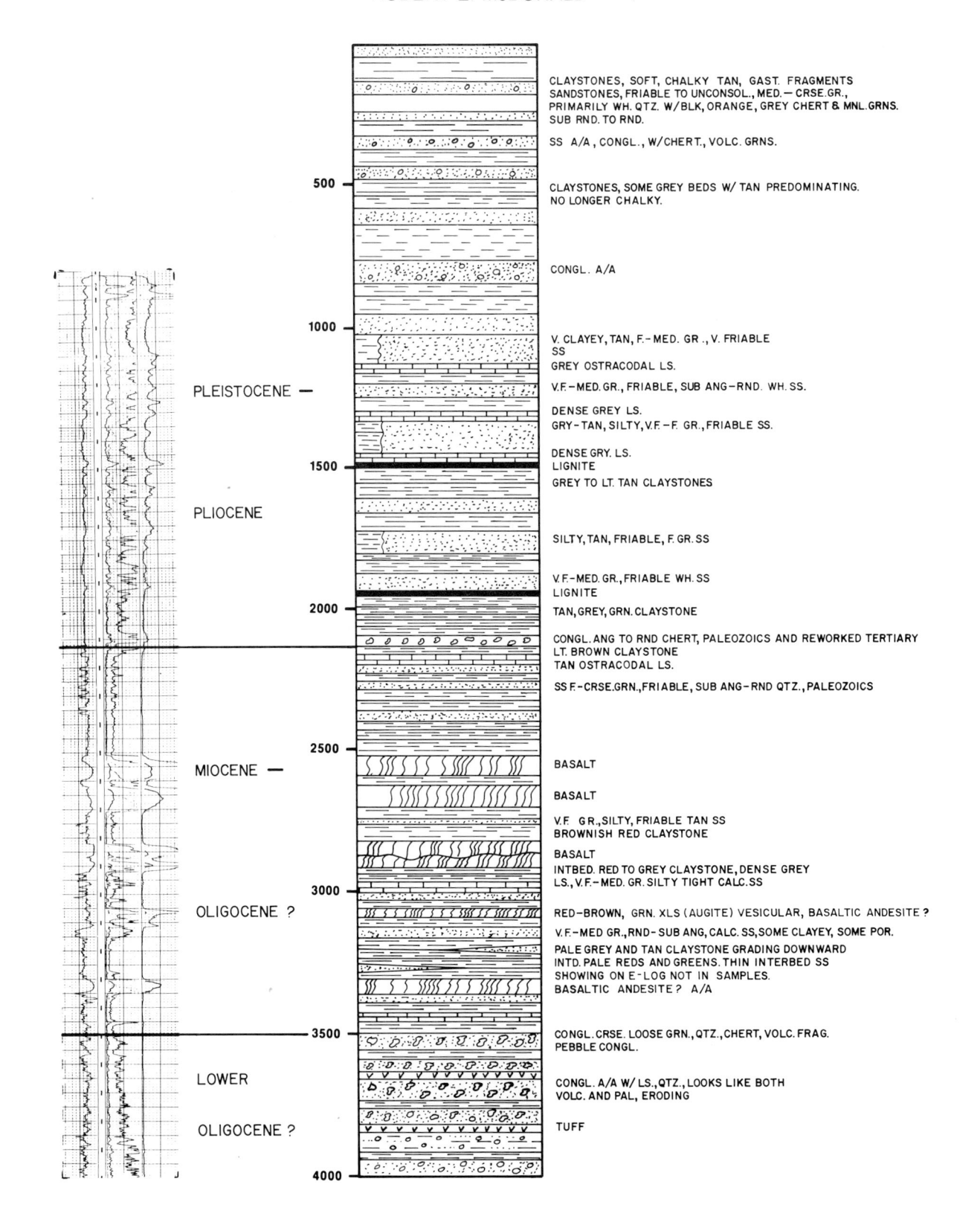

Depth	Age	Description
500	PLEISTOCENE	CLAYSTONES, SOFT, CHALKY TAN, GAST. FRAGMENTS
		SANDSTONES, FRIABLE TO UNCONSOL., MED.– CRSE. GR., PRIMARILY WH. QTZ. W/BLK, ORANGE, GREY CHERT & MNL. GRNS. SUB RND. TO RND.
		SS A/A, CONGL., W/CHERT., VOLC. GRNS.
		CLAYSTONES, SOME GREY BEDS W/ TAN PREDOMINATING. NO LONGER CHALKY.
		CONGL. A/A
1000		V. CLAYEY, TAN, F.– MED. GR., V. FRIABLE SS
		GREY OSTRACODAL LS.
		V.F.– MED. GR., FRIABLE, SUB ANG–RND. WH. SS.
		DENSE GREY LS.
		GRY–TAN, SILTY, V.F.– F. GR., FRIABLE SS.
1500		DENSE GRY. LS.
		LIGNITE
	PLIOCENE	GREY TO LT. TAN CLAYSTONES
		SILTY, TAN, FRIABLE, F. GR. SS
		V.F.– MED. GR., FRIABLE WH. SS
2000		LIGNITE
		TAN, GREY, GRN. CLAYSTONE
		CONGL. ANG TO RND CHERT, PALEOZOICS AND REWORKED TERTIARY
		LT. BROWN CLAYSTONE
		TAN OSTRACODAL LS.
		SS F.–CRSE.GRN., FRIABLE, SUB ANG–RND QTZ., PALEOZOICS
2500	MIOCENE	BASALT
		BASALT
		V.F. GR., SILTY, FRIABLE TAN SS
		BROWNISH RED CLAYSTONE
		BASALT
		INTBED. RED TO GREY CLAYSTONE, DENSE GREY LS., V.F.– MED. GR. SILTY TIGHT CALC. SS
3000	OLIGOCENE ?	RED–BROWN, GRN. XLS (AUGITE) VESICULAR, BASALTIC ANDESITE ?
		V.F.– MED GR., RND– SUB ANG, CALC. SS, SOME CLAYEY, SOME POR.
		PALE GREY AND TAN CLAYSTONE GRADING DOWNWARD INTD. PALE REDS AND GREENS. THIN INTERBED SS SHOWING ON E-LOG NOT IN SAMPLES.
		BASALTIC ANDESITE ? A/A
3500		CONGL. CRSE. LOOSE GRN., QTZ., CHERT, VOLC. FRAG. PEBBLE CONGL.
	LOWER	CONGL. A/A W/ LS., QTZ., LOOKS LIKE BOTH VOLC. AND PAL, ERODING
	OLIGOCENE ?	TUFF
4000		

Fig. 5 — Sample and Electric Logs, Gulf Oil Corporation

LATE

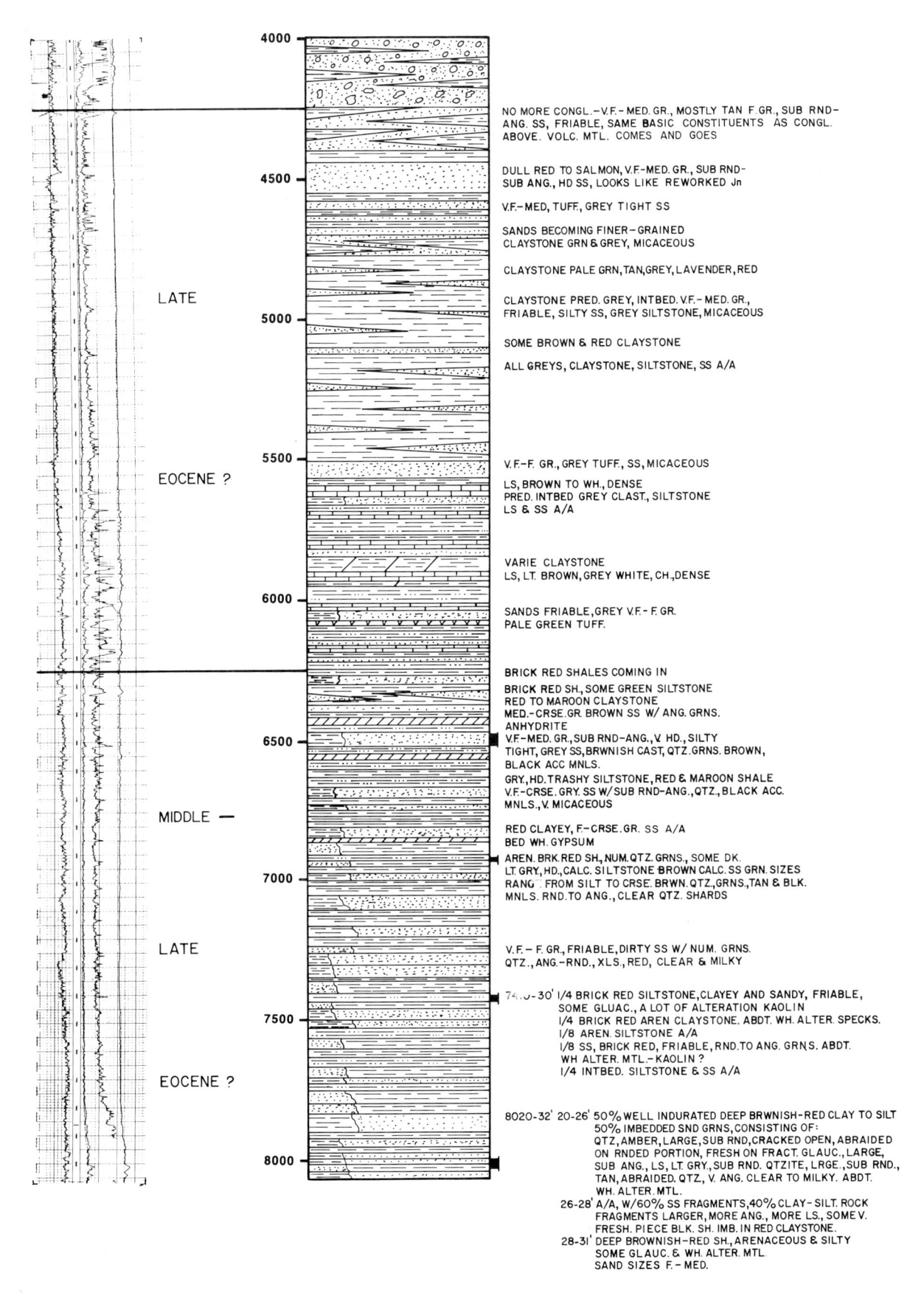

4000

NO MORE CONGL.-V.F.-MED. GR., MOSTLY TAN F.GR., SUB RND-ANG. SS, FRIABLE, SAME BASIC CONSTITUENTS AS CONGL. ABOVE. VOLC. MTL. COMES AND GOES

4500

DULL RED TO SALMON, V.F-MED. GR., SUB RND-SUB ANG., HD SS, LOOKS LIKE REWORKED Jn

V.F.-MED, TUFF., GREY TIGHT SS

SANDS BECOMING FINER-GRAINED
CLAYSTONE GRN & GREY, MICACEOUS

CLAYSTONE PALE GRN, TAN, GREY, LAVENDER, RED

CLAYSTONE PRED. GREY, INTBED. V.F.-MED. GR., FRIABLE, SILTY SS, GREY SILTSTONE, MICACEOUS

5000

SOME BROWN & RED CLAYSTONE

ALL GREYS, CLAYSTONE, SILTSTONE, SS A/A

EOCENE ?

5500

V.F-F. GR., GREY TUFF, SS, MICACEOUS
LS, BROWN TO WH., DENSE
PRED. INTBED GREY CLAST., SILTSTONE
LS & SS A/A

VARIE CLAYSTONE
LS, LT. BROWN, GREY WHITE, CH., DENSE

6000

SANDS FRIABLE, GREY V.F.-F. GR.
PALE GREEN TUFF.

BRICK RED SHALES COMING IN
BRICK RED SH., SOME GREEN SILTSTONE
RED TO MAROON CLAYSTONE
MED.-CRSE.GR. BROWN SS W/ ANG. GRNS.
ANHYDRITE

6500

V.F-MED. GR., SUB RND-ANG., V. HD., SILTY TIGHT, GREY SS, BRWNISH CAST, QTZ.GRNS. BROWN, BLACK ACC MNLS.

GRY, HD. TRASHY SILTSTONE, RED & MAROON SHALE
V.F-CRSE. GRY. SS W/ SUB RND-ANG., QTZ., BLACK ACC. MNLS., V. MICACEOUS

MIDDLE —

RED CLAYEY, F-CRSE.GR. SS A/A
BED WH. GYPSUM

AREN. BRK. RED SH., NUM. QTZ. GRNS., SOME DK.
LT. GRY, HD., CALC. SILTSTONE BROWN CALC. SS GRN. SIZES

7000

RANG. FROM SILT TO CRSE. BRWN. QTZ., GRNS., TAN & BLK. MNLS. RND. TO ANG., CLEAR QTZ. SHARDS

LATE

V.F.- F. GR., FRIABLE, DIRTY SS W/ NUM. GRNS.
QTZ., ANG.-RND., XLS., RED, CLEAR & MILKY

7500

74.0-30' 1/4 BRICK RED SILTSTONE, CLAYEY AND SANDY, FRIABLE, SOME GLUAC., A LOT OF ALTERATION KAOLIN
1/4 BRICK RED AREN CLAYSTONE. ABDT. WH. ALTER. SPECKS.
1/8 AREN. SILTSTONE A/A
1/8 SS, BRICK RED, FRIABLE, RND. TO ANG. GRNS. ABDT.
WH ALTER. MTL.-KAOLIN ?
1/4 INTBED. SILTSTONE & SS A/A

EOCENE ?

8020-32' 20-26' 50% WELL INDURATED DEEP BRWNISH-RED CLAY TO SILT
50% IMBEDDED SND GRNS, CONSISTING OF:
QTZ, AMBER, LARGE, SUB RND, CRACKED OPEN, ABRAIDED ON RNDED PORTION, FRESH ON FRACT. GLAUC., LARGE, SUB ANG., LS, LT. GRY, SUB RND. QTZITE, LRGE., SUB RND., TAN, ABRAIDED. QTZ, V. ANG. CLEAR TO MILKY. ABDT. WH. ALTER. MTL.

8000

26-28' A/A, W/ 60% SS FRAGMENTS, 40% CLAY-SILT. ROCK FRAGMENTS LARGER, MORE ANG., MORE LS., SOME V. FRESH. PIECE BLK. SH. IMB. IN RED CLAYSTONE.
28-31' DEEP BROWNISH-RED SH., ARENACEOUS & SILTY SOME GLAUC. & WH. ALTER. MTL.
SAND SIZES F.- MED.

#1 Gronning, Sec. 25, T. 16 S., R. 8 W., Millard County, Utah.

Core 8020-8032 ft.— Recovered 11 ft, described as follows:

8020-26 ft.—50 percent well indurated, deep brownish-red claystone and siltstone, layered and laminated. 50 percent imbedded sand grains in the above claystones and siltstones, consisting of:

> Quartz, amber, large subrounded grains, some cracked open, abraided on round portion, fresh on fractures.
> Quartz, very angular, clear to milky, large grains.
> Quartzite, large subrounded grains, tan, abraided.
> Glauconite, large and subangular.
> Limestone, light grey, subrounded.

8026-28 ft.—As above with 60 percent sand grains imbedded in 40 percent claystone and siltstone. Rock fragments larger, more angular. More limestone, some very fresh; piece black shale imbedded in red claystone.

8028-31 ft.— Deep brownish-red shale, arenaceous and silty, some glauconite, and specks of white altered clay. Sand grains in shale range from fine- to medium-grained, same material as above.

The only significant constituent of the red bed sequence not encountered in the above described cores are several thin beds of white, sucrosic anhydrite. The abundance of poorly sorted sands and sand grains in the shales, and the composition of the sand grains certainly leads to the conclusion that a Paleozoic highland was being eroded. The frequently abundant specks of white clayish alteration material in the sands and shales alike are believed to be altered tuffs, indicating volcanism, although not nearby. On the basis of stratigraphic relations — lying just beneath a unit with considerable tuffaceous material — and the presence of a limited amount of altered volcanic detritus in the red bed sequence, a middle to late Eocene age is believed possible. The presence of airfall tuff is not conclusive, however, since there are significant accumulations of pyroclastics, even in the Upper Cretaceous, that are quite remote from volcanic centers. The red bed sequence could be as old as Upper Cretaceous, or with only the two andesitic flows to offer a tentative and nebulous dating, the test could have bottomed in Miocene strata.

The coloration of the red bed sequence suggests a primarily oxidizing depositional environment, unless the source material was a red bed sequence. This is probably not the case, since the surrounding mountains are believed to have been cut into the Paleozoics by Eocene time. The abundance of sand grains indicates no shortage of source material, probably of Paleozoic origin, but does suggest a lack of energy to sort and clean the sands. The presence of some glauconite indicates at least occasional lacustrine depositional conditions. It is visualized that the area around the Gulf Gronning No. 1 was probably a broad floodplain, along which slow-moving streams meandered toward an area of primarily lacustrine depositional conditions. Much of the coarser grained detritus probably was deposited in these streams as point bars along the bends of meanders across the floodplain while streamloads were primarily fine-grained.

One of the interesting features of the Gulf well is that the entire sequence below 2100 ft is transitional in nature between fluvial and lacustrine depositional environments, with the exception of the 750 ft of volcanic agglomerate, which apparently had a nearby source. No significant thickness of section ever becomes a good, high-energy fluvial sequence, nor is there a good, thick series of lacustrine lake beds. It is believed that both probably exist in the Sevier Desert basin area, but the Gulf test was marginal to such depositional environments.

The nearest outcrops of Tertiary sediments (Fig. 6) are the Rush Valley section, 55 mi northeast of the Gronning No. 1, the Tintic Junction section located 35 mi northeast, the Canyon Range section about 27 mi to the east, and the Corn Creek section 50 mi southeast. The ages of Tertiary represented and the lithologies are quite varied. The Rush Valley and Tintic Junction sections are both Miocene and early Pliocene fluvio-lacustrine sediments. The Rush Valley section, (Fig. 7), measures over 8000 ft without being fully exposed, and is extremely significant since it is nearly all lacustrine and was located in a depositional setting that is believed to be similar to a portion of the Sevier Desert basin. As can be seen from the section, limestones and marlstones dominate, with interbeds of calcareous tuff, pumicite, and a few sandstones. The marlstones range from greyish-green to grey to brown in color, and the limestones are frequently algal and oolitic. The section dips westward at an average magnitude of about 15°, with a range of 3° to 40°

The Tintic Junction section outcrops between the East Tintic and West Tintic Mountains, at the extreme northeastern margin of the Sevier Desert basin. This exposure also dips west and has not been measured, but the writer has visited the area and these Miocene to early Pliocene strata (mapped as Salt Lake Formation on the geologic map of Utah) apparently occupy most of the valley between the two ranges, although vegetation covers most of the area. The section is quite thick and probably on the same order of magnitude as the Rush Valley section. The most prominent rock type on outcrop is chalky-white to grey limestone and greyish-green claystone and siltstone. There are many vitric tuff beds, and dark-grey, fetid, pellet and micritic limestone is also common.

The Canyon Range section (Fig. 6) possesses some problems in dating, but nevertheless is a tremendously thick section of Late Cretaceous and early Tertiary sediments that accumulated at least along the east side of the Canyon Range. There is also a middle to late Tertiary section of conglomerate that outcrops primarily on the west side of the Canyon Range, that has been variously dated from Oligocene to Pliocene. The problem for the entire sequence has been lack of faunal evidence of age. Without discussing the correlation problem, it may be said that along the east side of the Canyon Range there are 16,000 ft of Late Cretaceous and Paleocene conglomerate, sandstone, shale, and limestone; very coarse in the lower part, becoming less coarse in the upper portion. This is a very impressive section of fluvio-lacustrine strata, and whether any of this Late Cretaceous-early Tertiary sequence was deposited on the west side of the range is unknown, but will be discussed further under the following section on "Tectonic History".

There is up to 1200 ft of middle or late Tertiary Fool Creek Conglomerate exposed in the Canyon Range, before it disappears beneath Quaternary cover. This section was originally believed to be Oligocene by Christiansen (1951); however, the geologic map of Utah (1963) shows the outcrop as Pliocene. The evidence, however, in the opinion of the writer favors an earlier age than Pliocene, and it is believed Christiansen was more nearly correct. The angular unconformity between the Paleocene and the Fool Creek, where present, ranges from 15° to a maximum of 25° on the east flanks of the range. On the southwest flank the Fool Creek dips westward as much as 40° and appears to have been uplifted along the west side of older north-south trending normal faults — faults older than the present range-bounding normal faults. The Fool Creek is also exposed in many places high in the Canyon Range, extending up into the summit areas. These evidences point to an age earlier then Pliocene.

The Corn Creek section to the southeast lies along the southeast flank of the Pavant Range, which like the Canyon Range, contains one of the easternmost major thrusts of the Sevier orogeny. The lower post-Sevier unit is a boulder conglomerate consisting primarily of red conglomerate and containing large boulders of quartzite and limestone. It is 1900 ft thick and believed to grade northward and eastward into the latest Cretaceous Price River Formation as well as the Cretaceous-Paleocene North Horn Formation.

Overlying the boulder conglomerate is a distinctly separate unit that is best described as red and yellow, mottled, calcareous sandstone, pebble and cobble conglomerate, and impure limestone beds. It is 1100 ft thick and when traced northeastward appears to grade into both the Flagstaff Limestone and Crazy Hollow Formation, which would date it as Eocene.

Above this sand conglomerate is a lacustrine limestone and shale sequence that weathers an easily identifiable light tan, and contains a good fresh-water fauna identified as latest Eocene or early Oligocene. This formation is probably a western extension of the Bald Knoll Formation. Also believed to be part of the Bald Knoll equivalent is a conformably overlying sequence of sandstone, siltstone, shale, and argillaceous limestone that weathers to red, green, orange, and white bands. The lower limestone and shale formation is 400 ft thick and the upper sandstone, siltstone, shale, and limestone sequence is about 600 ft in thickness. This 1000 ft of primarily lacustrine strata is believed to be early Oligocene on the basis of stratigraphic position.

The youngest Tertiary strata, also probably Oligocene, is 400 ft of light grey sandstone, interbedded with red and greenish-grey siltstone and shale. The shales weather to orange bands, and contain gypsum veinlets. This may be the equivalent of the Grey Gulch Formation.

The latest Cretaceous through about middle Eocene portion of the section thickens rapidly in a northeasterly and northerly direction along the east side of the Pavant Range; however, the late Eocene and Oligocene portions of the section do not. In a westerly direction all units thin rapidly against the Pavant Range. The Clear Creek section is largely covered by alluvium and is not measured; however, it is doubtful if more than 500 ft of the 3500 ft Corn Creek section remains in a distance of about 10 miles.

Tectonic History of the Sevier Desert

Fig. 8 is a physiographic map of Utah on which several faults and probable faults have been noted. Along a northeasterly-trending zone of normal faults is a notation, "Las Vegas-Wasatch Line", which is an ancient lineament that has been a dominant factor throughout the geologic record in Utah. The inclusion of "Las Vegas" to the "Wasatch Line" is relatively recent, since it is only in the past 20 yr that the proof has accumulated that the Wasatch Line and the Las Vegas Line form a continuous fault zone. Actually, it has been convincingly demonstrated that this zone of faulting and shearing ties into the Garlock fault sytem in California. In the following discussion, this system will often be referred to as the "Wasatch Line" for the sake of brevity, but with recognition that it is a part of the total system.

Stokes has succinctly described the Wasatch Line". . . as a relatively narrow tectonic hinge zone between wide regions of strongly contrasting geologic history". It has been in existence since at least Late Precambrian, and has been accentuated many times during geologic history. Along it is a concentration of volcanic activity, fault zones, earthquake epicenters, and thermal springs. Kenneth Cook, of the geophysics department at the Unitversity of Utah, has been extremely active in studying and delineating the orientation, magnitude, and relative movements of the rifting associated with the Wasatch Line, as well as initiating a series of heat flow studies. He believes the Wasatch Line is situated over a deep-seated ascending thermal current. This thermally weakened and fractured zone has been the site of repeated disturbances, especially of hinge-like action between

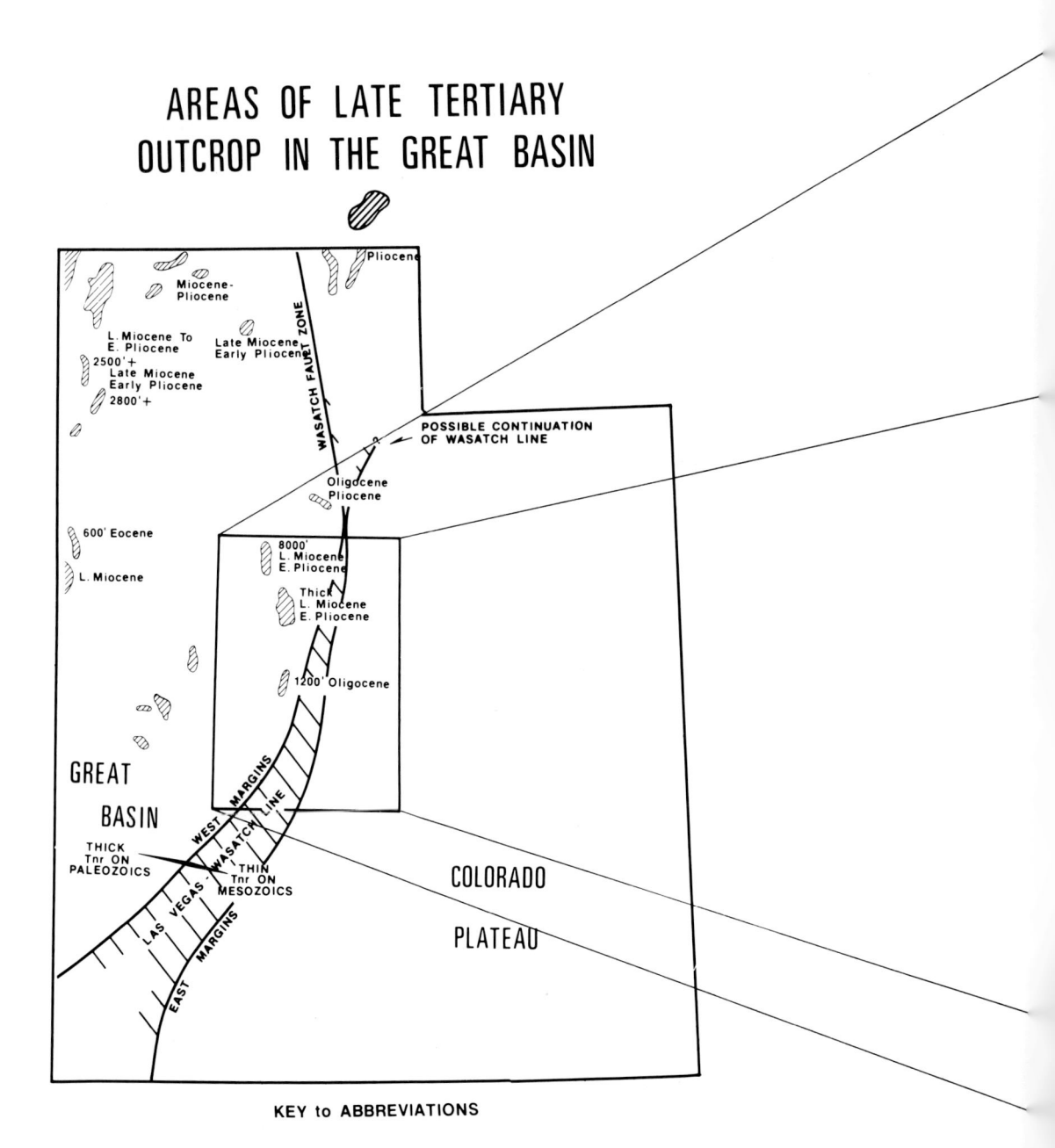

AREAS OF LATE TERTIARY OUTCROP IN THE GREAT BASIN

KEY to ABBREVIATIONS

LATE EOCENE-EARLY OLIGOCENE	
NEEDLES RANGE FORMATION	Tnr
EARLY EOCENE-MIDDLE EOCENE	
GREEN RIVER FORMATION	Tgr
COLTON FORMATION	Tc
LATE PALEOCENE-EARLY EOCENE	
FLAGSTAFF LIMESTONE	Tf
LATEST CRETACEOUS-EARLY PALEOCENE	
NORTH HORN FORMATION	TKn
LATE CRETACEOUS (MONTANAN)	
PRICE RIVER FORMATION	Kpn
MIDDLE CRETACEOUS (COLORADAN)	
INDIANOLA	Ki

FIG. 6

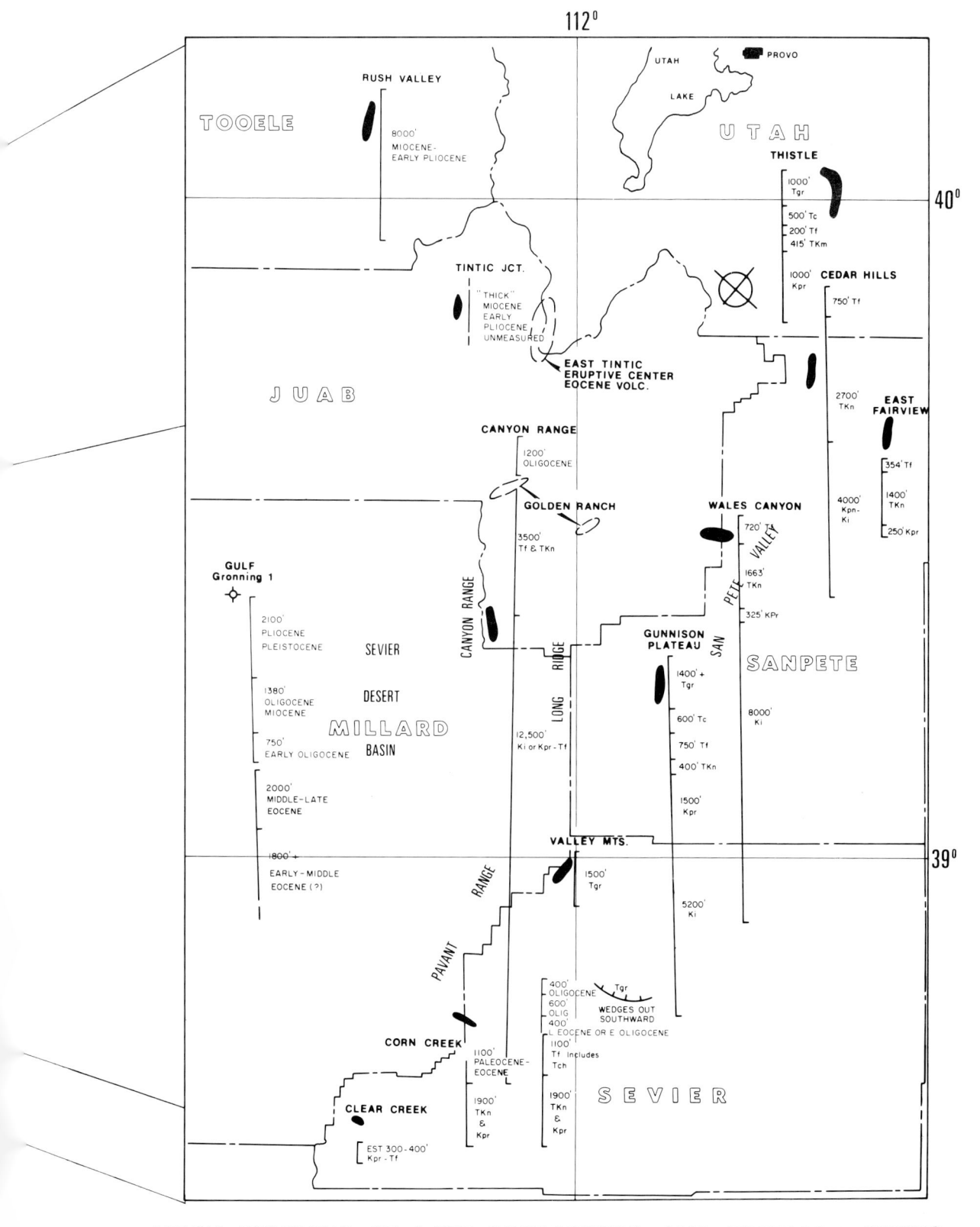

STICK SECTIONS OF LATE CRETACEOUS AND TERTIARY ROCKS SEVIER DESERT BASIN AREA

FIG. 6

STRATIGRAPHIC SECTION
MIOCENE – E. PLIOCENE ROCKS – RUSH VALLEY
T. 7S, R. 4 and 5E, TOOELE COUNTY, UTAH
MEASURED BY: EDGAR B. HEYLMAN

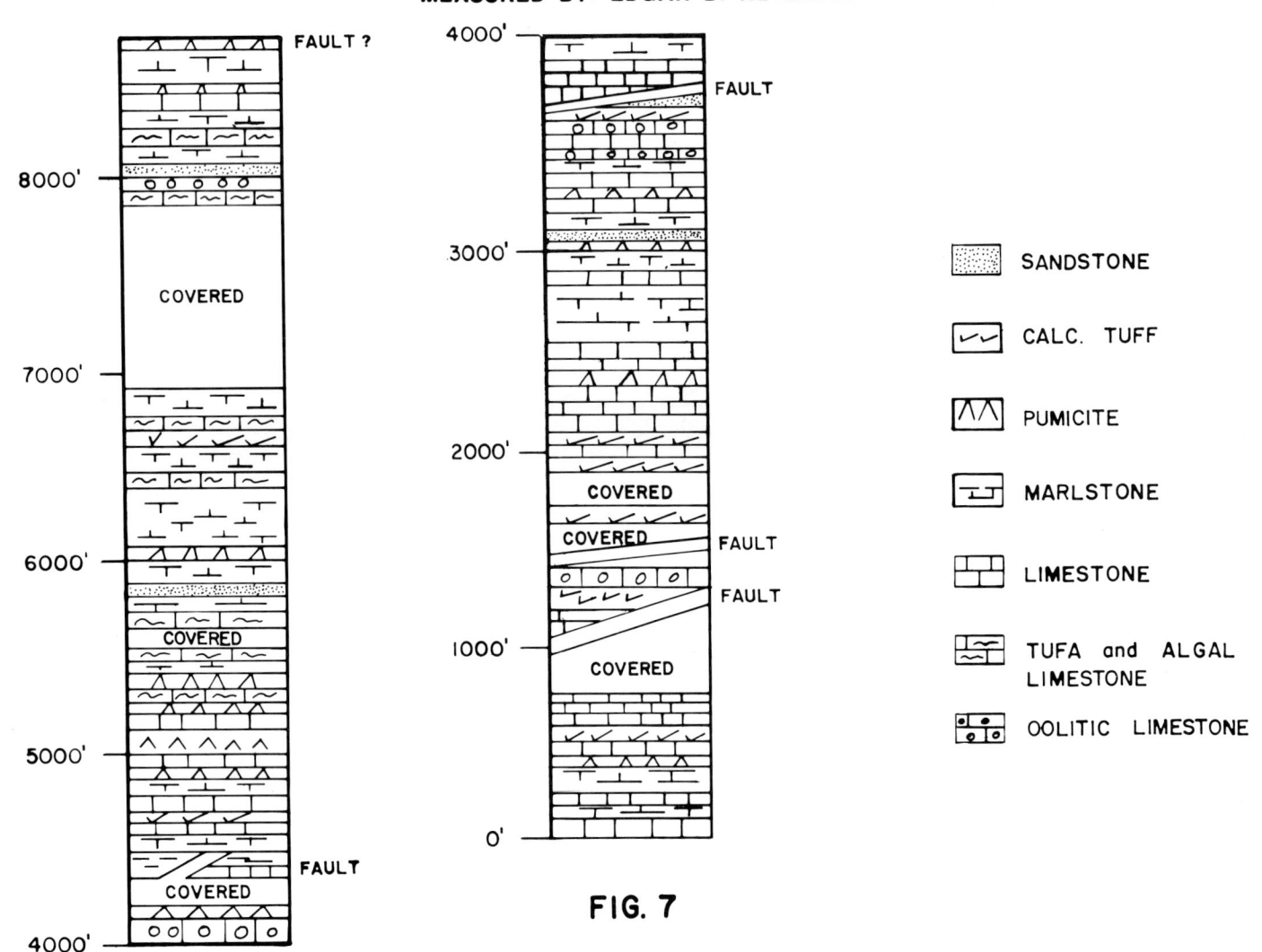

FIG. 7

Legend:
- SANDSTONE
- CALC. TUFF
- PUMICITE
- MARLSTONE
- LIMESTONE
- TUFA and ALGAL LIMESTONE
- OOLITIC LIMESTONE

the adjacent parallel belts. Throughout the Paleozoic, the Wasatch Line was the "hinge" between the persistently negative miogeosyncline on the west and the stable to moderately negative regions to the east. Stokes (1963) believes the first emergence of the area west of the Wasatch Line in Mesozoic time took place during the Triassic, and proposed the name Mesocordilleran highland. This is difficult to prove, since all direct evidence has been obscured by later subsidence, erosion, and volcanic cover. A complete reversal of configuration along the Wasatch Line definitely took place by Late Jurassic time, however, with elevation of all or nearly all of the region west of the Wasatch Line accompanied by extensive thrust faulting.

In detail, the structure of these Late Jurassic-Cretaceous thrust ranges is usually imbricate, but grossly much of the structure is relatively simple. Subsequent Cenozoic orogeny has made relationships at the close of the Sevier orogeny difficult to untangle; however, during the past 10 yr much has been published to clarify the picture. Fig. 8 includes a paleogeologic reconstruction of the major thrust faults as they might have appeared at the end of the Sevier orogeny, or the end of Cretaceous time. Parts of this interpretation has been adapted from others (Armstrong, 1968) and part is the writer's interpretation.

The Sevier orogenic belt was first named the Sevier arch, and a broad uplifted area was visualized commencing in Late Jurassic, with thrusting considered to be the climax of arching during Late Cretaceous. Considerable evidence has since accumulated, however, that does not support the concept of a

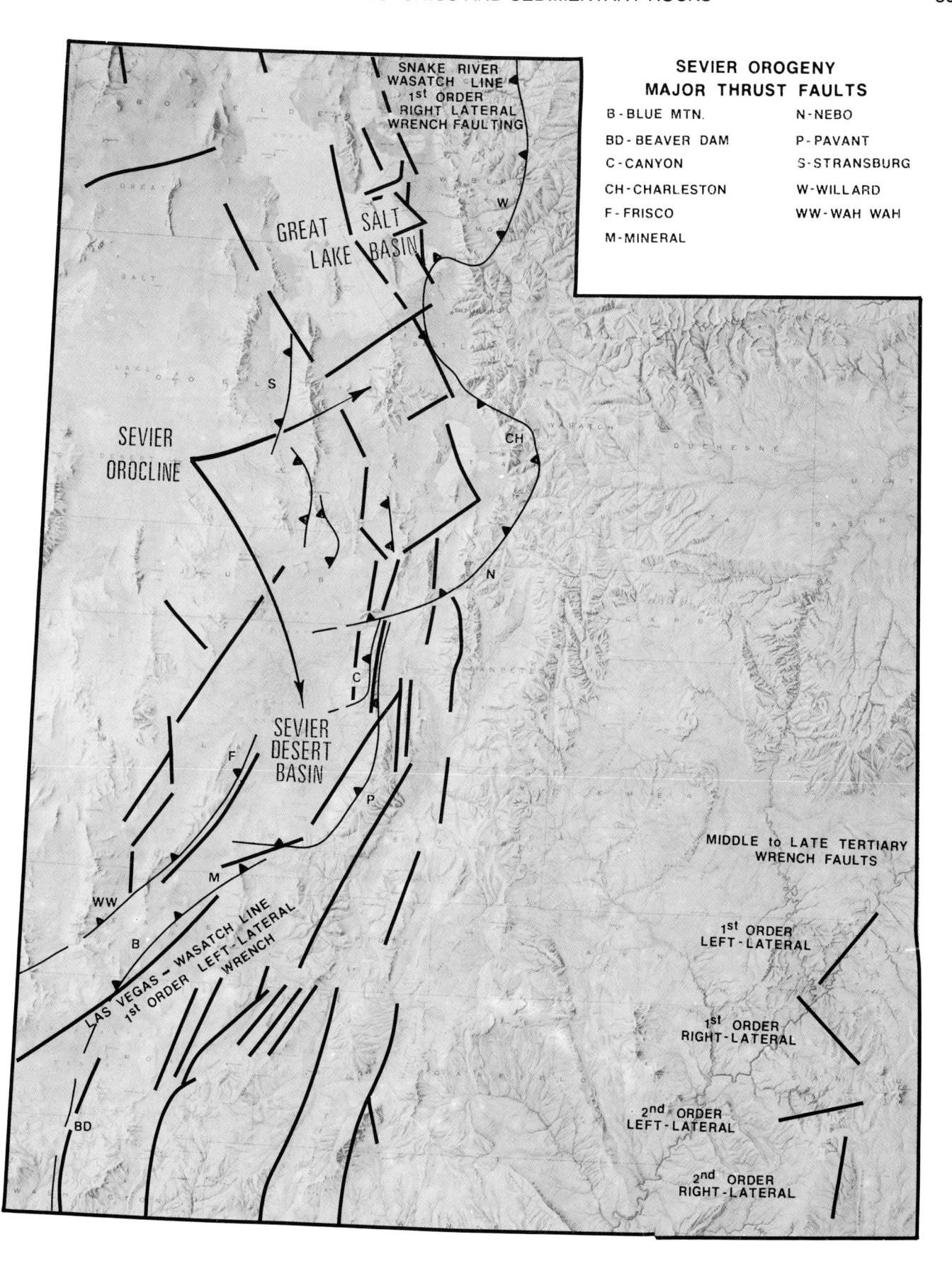

FIG. 8

simple arch and nowhere is an eastern limb evident. There probably was minor early arching as proposed by Stokes, and the pre-Triassic unconformity in southern Nevada indicates this was the case along the southwestern end of the Wasatch Line. The first conclusive evidence of orogenic deformation is the appearance of lower Paleozoic clasts in early Colorado age sediments in Utah and Nevada. The only possible source is the Sevier belt. As pointed out by Armstrong (1968), "The Eocambrian rocks of the belt occur almost exclusively in the sole of major thrusts. Their appearance at the end of Colorado time is evidence that thrust displacements of tens of miles existed by then". Evidence now strongly points to a cessation of movements along these major thrusts outlined on Figure 8 by the end of Cretaceous. There was further thrusting in front of this primary orogenic thrust front in early Tertiary to the northeast in Wyoming, and also to a lesser extent in central Utah, but this was minor by comparison. The Sevier orogeny is visualized by the writer to be primarily a southward continuation of the Wyoming overthrust belt, as opposed to uplift or arching being the main tectonic force with thrusting being secondary.

A brief reference to the Upper Cretaceous sections immediately east of the Canyon Range is pertinent, since it indicates a very negative area immediately contiguous with and east of that range. There is 12,500 ft of Indianola (Colorado) according to Christiansen (1951). There is some good argument that this section could be all Price River (Montanan) through Flagstaff (Paleocene-early Eocene) age, but if so it only adds strength to the argument that the area along the east side of the Canyon Range subsided very rapidly, relative to the upthrust range on the west. Using the conservative approach and Christiansen's age assignment for the Indianola, this is still the thickest section of post-Sevier orogeny strata known to the writer. The sections immediately east (Gunnison Plateau and Wales Canyon) are in the 6000 to 8000 ft range, and the section continues to thin eastward from there.

In latest Montanan and early Paleocene, the fluvio-lacustrine North Horn Formation was deposited, followed by the fluvial Colton Formation (not everywhere present) and the late Paleocene to early Eocene Flagstaff Formation which is primarily a lacustrine unit. The thickest section of combined North Horn and Flagstaff is again immediately east of the Canyon Range, with the single exception of a reported 3550 ft of North Horn and Flagstaff in the Cedar Hills section. Although no measured section is shown, the writer knows that as much as 1500 ft of Flagstaff alone is present south of the Canyon Range section along the eastern side of the Pavant Range.

The North Horn-Colton-Flagstaff facies relationships are difficult to determine regionally, because they were deposited on and around folds and positive features that were formed in the Mesozoic rocks in front of the Sevier orogenic belt. In an excellent paper, Malcom Weiss (1969) traced the old shorelines and positive features reflected by these formations from early Paleocene into early Eocene. Figure 9 illustrates the paleogeographic maps from this paper, which reflect much detailed field and paleoecologic data. The significance of these paleogeographic maps is that, if correct, they indicate the first downwarping west of a major Sevier orogenic thrust sheet was as early as late Paleocene, and was the embryonic beginning of Juab and Utah Valleys. Weiss did not extend his work west of the Canyon Range as there are no exposures to the west to study. There are Flagstaff outcrops on the Nebo-Charleston sheet between Nebo and Spanish Fork, however, so his projection of the Flagstaff Lake into Utah Valley has some basis other than speculation.

It is felt that the early to middle Eocene Green River Formation may possibly exist in the subsurface of the Sevier Desert basin. Both Eardley and Stokes have written repeatedly for many years that in the central Wasatch area, north-south folding over-rode the existing structures, accompanied by block faulting that began, in general, by dropping the synclinal areas and enhancing the elevation of the anticlinal fold axes. They have timed the commencement of this warping and faulting as middle to late Eocene. Work by the writer in the Wasatch Formation east of the Wasatch Range in northern Utah most certainly seems to substantiate their conclusions for this area. An accumulating body of evidence is beginning to point to a tendency for positive and negative areas developed during this period to retain this relationship along the Wasatch Line from at least the Farmington graben on the north to the Sevier Desert on the south.

There are several lines of indirect evidence immediately east of the Sevier Desert that tend to date the commencement of the basin and range-type structure in the Eocene. Weiss's conclusion that sagging occurred along the Wasatch Line west of a major Sevier orogenic thrust plate has been pointed out. Volcanism also commenced along the Wasatch Line in middle to late Eocene, earlier than most Great Basin volcanism. There is Eocene volcanism along the Wasatch Line in the East Tintic eruptive center, East Tintic Mountains, just northeast of the Sevier Desert basin. South and east of the East Tintic eruptive center an apron of volcanic detritus spread southward, which has been named the Golden Ranch Formation. In the vicinity of the northwestern area of the Golden Ranch outcrop on Figure 6, the volcanics interfinger with early Tertiary strata in the syncline between the Canyon Range and Long Ridge. At the southeastern outcrop area of the Golden Ranch, an ignimbrite closely similar to the dacitic unit observed to the northwest occurs near the base of the formation. It rests conformably on platy limestone and tuffaceous sediments regarded as Green River Formation. It is overlain several hundred feet higher by a limestone member carrying middle Eocene flora.

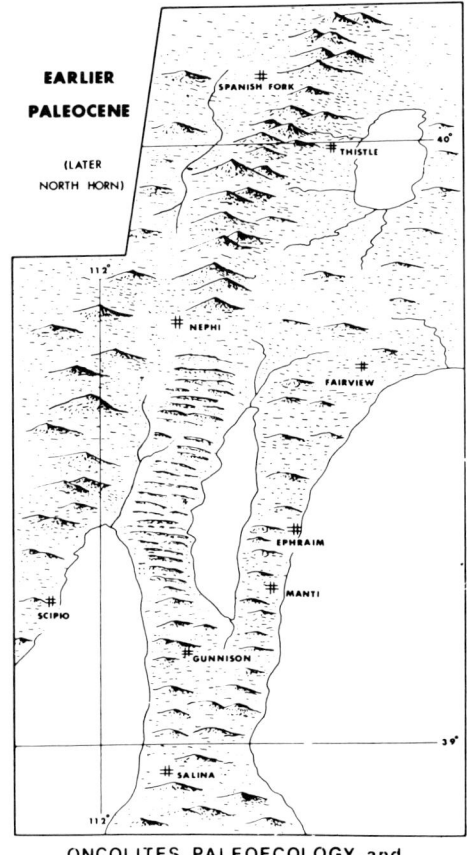

FIG. 9 — ONCOLITES, PALEOECOLOGY, and LARAMIDE TECTONICS, CENTRAL UTAH

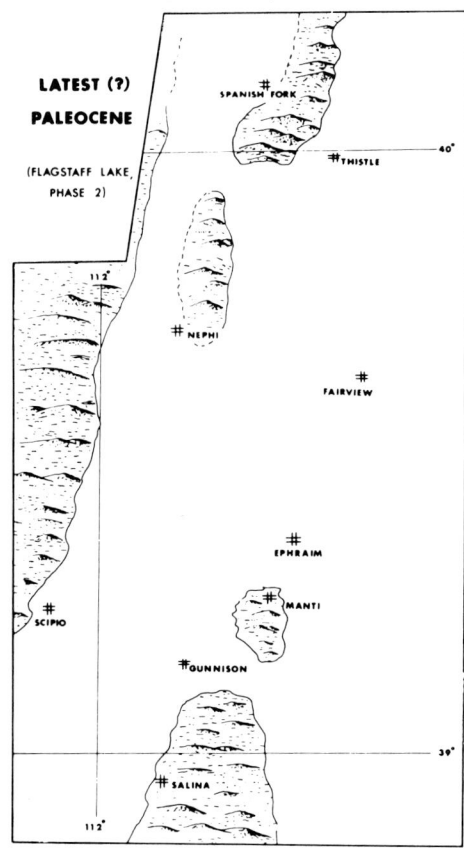

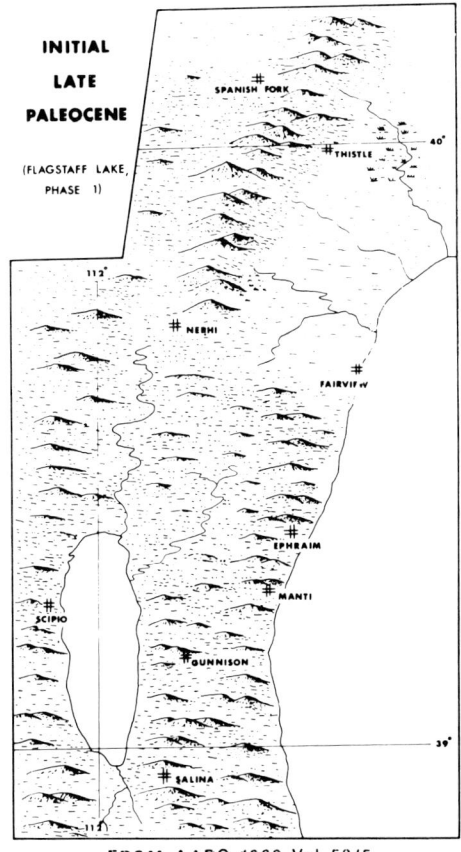

FROM AAPG 1969 Vol. 53/5
By: MALCOM P. WEISS

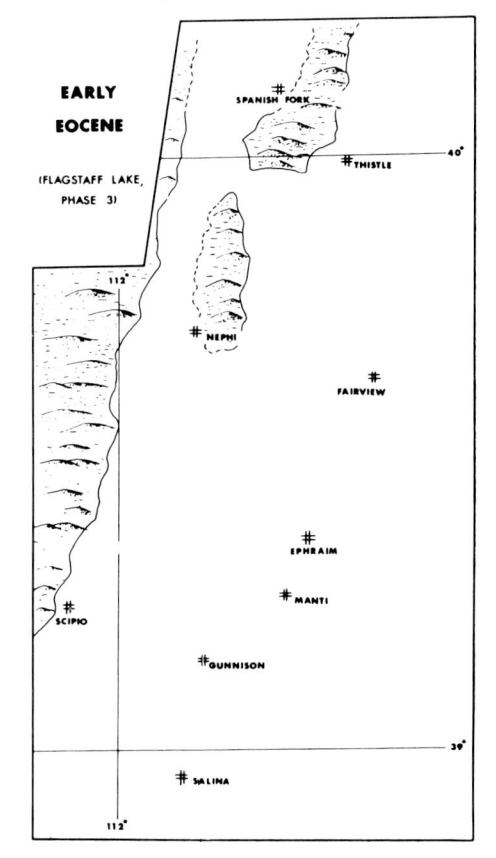

Maps shown:
- EARLIER PALEOCENE (LATER NORTH HORN)
- INITIAL LATE PALEOCENE (FLAGSTAFF LAKE, PHASE 1)
- LATEST (?) PALEOCENE (FLAGSTAFF LAKE, PHASE 2)
- EARLY EOCENE (FLAGSTAFF LAKE, PHASE 3)

Volcanic activity does not prove the formation of north-south folding and warping, but it does indicate the beginning of rift-faulting that typifies the present Wasatch Line and much of the basin and range structure. Many workers state that normal faulting is post-volcanic, and this is true of some faulting in the Great Basin. Observations along the Wasatch Line, however, suggest that faulting commenced with or prior to the volcanism in many places, although the magnitude of displacement may have greatly increased later. Mackin (1963) recognized this also. In working the Tushar volcanic pile southwest of the Sevier Desert and attempting to correlate what should be relatively simple units on a local scale, he wrote: "Several lines of evidence suggest that these and other stratigraphic anomalies are due to faulting during Needles Range time."

In addition to the above, there are several periods of early Cenozoic tectonic activity reflected in the well-exposed geologic record immediately adjacent to the Sevier thrust front that strongly indicate that warping and subsidence was widespread along the Wasatch Line. (1) There is documented Paleocene normal faulting during deposition of the North Horn and prior to deposition of the late Paleocene-early Eocene Flagstaff Limestone, it should be remembered that at least the major thrusting of the Sevier orogeny had concluded prior to Paleocene, and the events being enumerated are believed to be associated primarily with the tectonics that created the basin and range structure. (2) There are areas of pre-Flagstaff monoclinal arching as indicated by angular unconformities between the North Horn and the Flagstaff (not everywhere true). (3) There are cases of post-Colton, pre-Green River normal faulting on the west flank of the Gunnison Plateau, which places the deformation in early to middle Eocene time. (4) Documented instances of post-Green River, pre-Crazy Hollow normal faulting place additional normal faulting in middle to late Eocene time. (5) The flexing and faulting of the Wasatch monocline is an extremely important date, since this tectonism formed the Sanpete Valley, and much of its associated major drainage pattern. The tilting of the Wasatch monocline also marks a time of definite uplift and upwarping of the area east of the Wasatch Line in relation to the area on the west. This flexing commenced by late Eocene.

Geophysical Interpretation.—Figure 10 is a gravity map of the Sevier Desert and vicinity published by the Utah Geological and Mineral Survey. The Cricket Mountains are reflected by a north-south trending positive, with an adjacent trough paralleling this positive on the east. The positive feature just east of Delta, Utah actually lies along the eastern side of the Sevier Desert area and just west of the Pavant and Canyon Ranges. From this positive, gravity values decrease gradually eastward across the ranges and into the Sevier River Valley. The writer has access to a more detailed gravity survey which broadens the north-south trending gravity minimum along the western Sevier Desert and tends to narrow the positive feature east of Delta.

Plates I through IV are seismic sections across the Sevier Desert. The location of the seismic lines are indicated on Figure 10. In examining the lines, a prominent reflector can be noted dipping gently westward on every section. This dominant reflector is believed to be a detachment plane along which a major thrust fault has moved. It has been suggested that this may represent an erosion surface, in which case it would have been necessary to peneplain the surface and tilt it more or less uniformly westward. There are problems with the latter interpretation, but the greatest objection is the fact that the subject reflection continues westward beneath apparent overthrust rocks.

The Gulf Gronning No. 1 well is plotted on Line 1, Plate I, and at this solitary control point we do know 8,000+ ft of fluvio-lacustrine Tertiary sediments were penetrated. When tied into surface and gravity, it is believed that overthrust pre-Tertiary rocks, probably Paleozoics, are present a short distance west of the Gulf well. These overthrust rocks can be seen encroaching into the western portion of every seismic section, and would represent a northward continuation of the Cricket Range.

East of the overthrust and above the detachment plane is a filled sedimentary basin with anomalous seismic features. Several interpretations have been advanced and considered, but in the opinion of the writer only one so far fully explains all the phenomena manifested on these sections. The theory is not original and was suggested by Frank Royse. One of the main keys is the structure of the sediments indicated as Miocene-Pliocene on the seismic sections. These demonstrate structure typical of Gulf Coast Miocene-Pliocene, where basinward gravity creep of the sedimentary blanket controls fault patterns, and structure is typified by folds with reverse drag and antithetic faults. The reader not familiar with this pattern is referred to Cloos (1968). Fault and structure patterns produced by Cloos experimentally in graben structures appear to duplicate what is seen in the Sevier Desert, perhaps even more closely than in the Gulf Coast. This type of structure is clearly extensional in character, but the puzzler is the fact that the detachment plane is scarcely ever seen to be broken by faulting on the seismic sections. This is interpreted as an indication that as the eastern Great Basin made the transition from a region of compression to one of tension; overthrust rocks retreated westward along the same glide plane or planes over which they were originally thrust eastward. The writer has not had the opportunity to investigate all the mechanics and ramifications of such an interpretation, but it appears quite possible that this has transpired.

Deeper-rooted rifting in latest Eocene and early Oligocene initiated the great outpouring of Great Basin ignimbrites, and during Oligocene time the intervening valleys and basins of

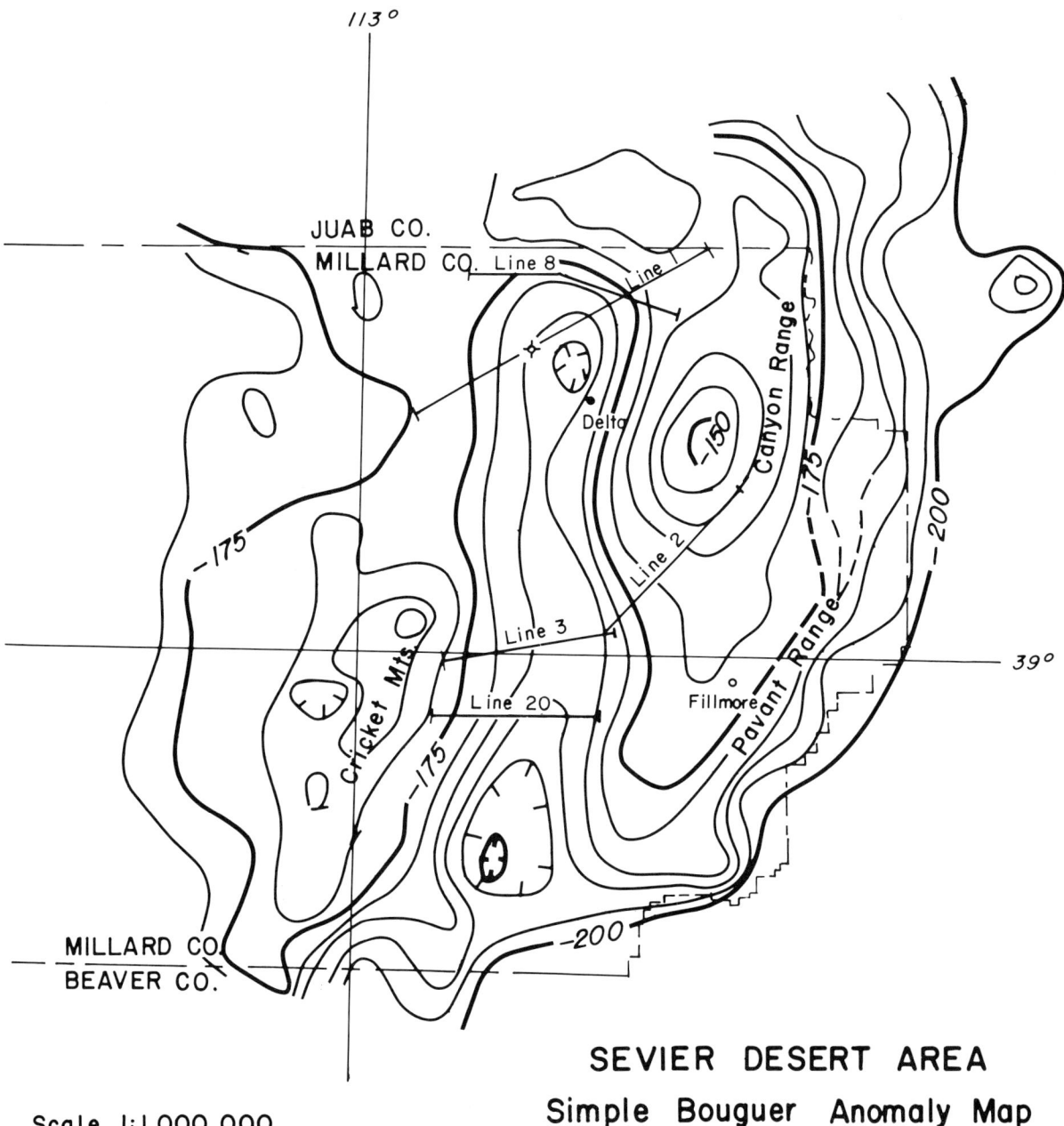

SEVIER DESERT AREA
Simple Bouguer Anomaly Map
from
Cook, Montgomery, Smith and Gray
Utah Geological and Mineral Survey
Fig. 10   Map 37

Scale 1:1,000,000
Contour Interval 5 Milligals
Seismic Lines Shot By
Pancanadian: Line 1

southwestern Utah were probably nearly filled with fluvial sediments, tuffs and ignimbrites. Lake formation was probably minimal during Oligocene. The seismic sections suggest another episode of westward retreat along detachment planes occurred after the accumulation of these earlier Tertiary rocks. It is not certain that the assignment of Miocene and Pliocene ages to these later rocks filling the new rift valleys is correct, but seems a logical assumption. Volcanism was still active, but not as prolific as during Oligocene. Extensive shallow lakes persisted, and continued minor episodic withdrawal of the thrust planes to the west kept the newly deposited sediments in tension. This resulted in the formation of typical Gulf Coast structure in the graben valleys during late Tertiary time.

The foregoing interpretation of tectonic and depositional framework is apparent on the seismic sections. The anticlinal structure in the middle of Line 8 (Plate II) is believed to be a remnant of Eocene-Oligocene strata that broke loose from its counterpart on the west as it moved westward along the glide plane. The anticlinal structure on the underlying detachment surface is believed to be real, at least in part, and not due entirely to velocity pull-up. If so, this might be partially responsible for providing the resistance necessary to detach the older Tertiary strata at this particular point. It has also been suggested that this feature may be salt. This would either require the preservation of substantial Jurassic Arapien in the Sevier Desert, or the accumulation of unusually thick mirabalite deposits in the Tertiary sediments. Neither seems too likely, but more significantly a salt feature of this size should react negatively on the gravity data, which it does not. An igneous feature has also been suggested, but it is doubtful if the beautiful reflections that continue beneath the feature would be observed beneath an igneous instrusive of this magnitude; further, a smooth gravity gradient exists across the area. One of the few obvious breaks in the detachment plane occurs on Line 2. It appears to be a small graben valley with late Tertiary sediments slumping into its interior. This may represent one of the deeply-rooted rifts that provide an avenue for episodic volcanism. There is a small volcanic cone surrounded by a Quaternary age flow on the surface a few miles northwest of this apparent rift.

TERTIARY SEDIMENTARY ROCKS: NORTHWESTERN UTAH

The earliest and also one of the most comprehensive periods of oil and gas exploration in northwestern Utah occurred in the 1890's. About 1891 or 1892, a farmer drilling for water in Sec. 36, T. 3 N., R. 1 W., just north of Salt Lake City, encountered a flow of natural gas. More wells were drilled and the gas was piped to Salt Lake City through a wooden pipeline in 1896 and 1897. There were about 20 producing gas wells in the Farmington gas field, and as much as 17 mmcf of gas per month was marketed. The largest flows were encountered at depths of 400 to 700 ft, and 9 gas-bearing horizons were encountered within 800 ft of the surface. The gas occurred in sand lenses enclosed in soft clays, the reservoirs were very porous and reportedly ranged from 3 to 20 ft in thickness. The gas was apparently free from water in the upper horizons but encountered hot water at 800 to 900 ft. Pressures varied between 200 and 250 psi, and the area produced an estimated 150,000 MCF before being abandoned. All wells drilled in the immediate vicinity were believed to have penetrated only Quaternary sediments, with the exception of the Carbon well in the SW¼, Sec. 26, T. 3 N., R. 1 W., which was drilled to a total depth of 3525 ft and is believed to have reached the upper Pliocene. Several gas shows were reported in this "deep" test, with one good show reported in possible Pliocene strata between 2700 and 2800 ft, but no attempt was made to complete the well.

Another interesting area is in Sec. 35, T. 9 N., R. 4 W., where the U.S. Biological Survey spudded a water well, on the Bear River Bird Refuge in November 1931. The well blew out at 322 ft, with sand, gas, mud, and water being blown 200 ft into the air creating a crater 30 ft in diameter around the top of the well. A second well was drilled 100 ft away, which recorded gas at 322 ft, 340 ft, and 357. Gas with artesian flows of water was reported at 520 ft, and 1000 ft. The well was completed as a water and gas well at 1028 ft, and supplied the headquarters of the Bird Refuge with natural gas for fuel and heating. The gas was analyzed as 87.9 percent methane, 5.5 percent nitrogen, 1.9 percent carbon dioxide, and 0.7 percent oxygen. In the fall of 1975 a water well that also made some natural gas at one of the duck clubs in the area experienced casing failure. The well blew out of control for a few weeks, the club apparently having neither the funds nor expertise to bring it under control. A strong flow of gas persisted, necessitating closure of roads in the area. It is the writer's understanding that the State finally moved in and killed the well with a special appropriation of funds for this purpose.

Other wells were drilled in the vicinity, but their exact locations are unknown. Within 1.5 mi, two other gas producing wells about 900 ft deep were drilled by farmers. About 3 mi to the east is another well that was a large producer of inflammable gas. The wells spud in recent river silts and sands, and at least the shallower gas is from the Quaternary. Whether or not the Tertiary is penetrated at total depth is unknown.

In and around Sec. 8, T. 8 N., R. 7 W., a large area of asphalt seeps on the north shore of the Great Salt Lake prompted the drilling of several wells at Rozel Point, commencing in 1904. Most of the drilling in the area is quite old and any dependable records are very difficult to come by. There probably have been somewhere between 30 and 40 wells drilled in the seep, mostly in the 200 to 300 ft depth range, and reported completions are in the 5 to 10 bbl oil/day category. It is doubtful if many produced at these rates on a sustained basis, however. At Rozel Point a series of seeps emerge from soft lake clays, and when the lake is high and the area is covered with water, small worm-like stringers of viscous oil rise to the surface. Surface rocks are a series of interbedded, conspicuously vesicular black basalts and irregular, thin-bedded to massive, greyish-tan limestones. These strata are part of the Salt Lake Group and are believed to be Miocene in age. They dip in an easterly direction toward the mountains.

The most significant drilling in the area was by Gulf Oil Company in 1963. The Gulf State Rozel No. 1, Sec. 17, T. 8 N., R. 7 W., was drilled to a total depth of 3505 ft, and topped the Paleozoics at 2356 ft. Three cores were taken in the Tertiary section, two of which between 221 ft and 243 ft recovered basalt, described as highly fractured, vesicular, and with pore spaces filled 25 to 40 percent with black, sticky asphalt and tar. Four cores in the Paleozoic rocks reported no shows and no tests were run. Vertical dips were reported in the last core taken at total depth. Gulf's other test in Sec. 9, T. 8 N., R. 7 W. crossed a fault and encountered the Paleozoics first at 1450 ft, and a second time at 2015 ft. No coring was done in the Tertiary section, however, four cores in the Paleozoics recovered marine shale and limestone, part of which was reported as vuggy. A drill stem test from 2020 to 2160 ft recovered 80 ft slightly gas-cut mud, 450 ft mud-cut saltwater, and 1370 ft saltwater. Shut-in pressures were 905 psig initial and 915 psig final.

Another significant test in the Salt Lake basin was Western Petroleum's Nebeker No. 1, Sec. 21, T. 1 N., R. 1 W., drilled to a total depth of 1985 ft just north of the Salt Lake airport. Asphaltic sand was found at 694 ft and a gas sand was reportedly encountered between 750 and 756 ft, both in the Quaternary. A water well nearby produced 875 Btu gas for several years for use in a nearby farmhouse. The Geotronic well in Sec. 30 of the same township and range reported five separate gas shows between 188 ft and 551 ft and an oil show at an unknown depth. During 1975 and 1976 Hiko Bell, Western United Mines, and Willard Pease have been drilling a series of tests in this area, however no information has been released to date.

The Morton Salt test in Sec. 24, T. 1 N., R. 3 W., was drilled to a total depth of 4231 ft. The Utah Geological & Mineral Survey reports the Quaternary-Tertiary contact to be at about 1432 ft and the base of the Tertiary at 3654 ft. Gravity work suggests this test was located on the Antelope Island horst block.

Two wells are reported in T. 9 N., R. 2 W., both of which recorded gas shows. The deepest test, in Sec. 30, was drilled by Rhine Petroleum in 1958 to a total depth of 2308 ft, penetrating Quaternary and Tertiary sediments, and questionably bottoming in Paleozoics or Precambrian. Gas shows were reported in several zones, but the well was abandoned. Gravity work suggests that the Tertiary is not thick in this particular area. Unpublished reports credit several wells immediately west and northwest of Brigham City with substantial flows of gas.

An interesting cluster of wells were drilled on the south end of the Great Salt Lake in T. 1 and 2 S., R. 5 W. An 1127 ft test was drilled into the Pennsylvanian in Sec. 31, T. 1 S., R. 5 W., reporting gas shows in the Tertiary. This well was obviously located on the Stansbury Mountains horst block. In 1955, a test well was drilled by Hickey in Sec. 13, T. 2 S., R. 5 W., to a total depth of 5486 ft. Quaternary and Tertiary clays, marls, limestones, and sands were reported to 4380 ft, where a Pennsylvanian limestone and sandstone sequence was penetrated. A Tertiary sand at 2925 ft reportedly had shows of gas, while zones at 5290 ft, 5400 ft, and 5460 ft in the Pennsylvanian were reported by the operator to have been saturated with high gravity oil and also to have contained gas. The well initially was temporarily abandoned, and the operator reportedly was going to attempt a completion, which never materialized. The following year the Walker-Wilson Drilling Company located a well less than a mi west in Sec. 14, and drilled to 7,979 ft. There was very little data released on this well, but it is known that it remained in the Tertiary Salt Lake Group to total depth. The well was promotional and whether any shows were encountered is unknown, but apparently none were encountered which the operator felt were commercial. Both of these latter two wells were in the Toole graben, and the Hickey well apparently reflects a buried horst block within the graben area, or he was incorrect in his report that Pennsylvanian limestones were penetrated.

Aside from the cluster of very shallow wells drilled in the Farmington gas field and at Rozel Point, there have been about 18 wildcat wells drilled in the area peripheral to the Great Salt Lake Tertiary basin, and most of these are in groups of two or three where shows were encountered. Of these, 12 reportedly had significant shows of oil and gas in Quaternary and Tertiary sediments. With the exception of two tests drilled by Gulf Oil Company, all were promotional and were drilled with little or no geological advice. For the area of roughly 4,500 sq mi covered by the basin, this amounts to an exploratory well density of 0.17 wells/township, with 65 percent of the wells reporting significant shows and only 20 percent exceeding 3000 ft in depth.

Tertiary Tectonic and Depositional History, Great Salt Lake Tertiary Basin

The basic geological and geographical setting of the Great Salt Lake basin and surrounding area is one of north-south trending mountain ranges, usually composed of Paleozoic rocks that reveal a complex structural history, separated by broad, flat north-south trending valleys or basins with Quaternary age sediments at the surface. The writer has attempted to piece together a probable Tertiary history for the Great Salt Lake basin, and some assumptions that now seem logical will undoubtedly be found in error, especially as subsurface data is developed. In order to place the Tertiary setting in its proper perspective, it is helpful to understand briefly the pre-Tertiary history of the area.

The Salt Lake City area has continuously been an area of "geologic conflict" between major east-west and north-south tectonic alignments. The oldest recognizable regional feature is an east-west structural trough that appears to have roughly paralleled the present Uinta Mountains axis, and is presumed to have been a branch of the extensive Precambrian geosyncline to the west. By Late Precambrian the outlines of the Cordilleran geosyncline, occupying western Utah and most of Nevada, had taken shape and the Middle Precambrian trough of the Uinta Mountains area had become emergent. Deposition commenced in the Cordilleran geosyncline to the west in Nevada, and gradually onlapped eastward, reaching the Wyoming-Utah shelf in Middle Cambrian time.

Early to middle Paleozoic data indicates that western Utah and eastern Nevada was the site of shelf-type carbonate deposition, intermediate between the Utah-Wyoming shelf on the east and the eugeosynclinal depositional environment prevailing in western Nevada. The edge of the continental shelf probably trended roughly north-northeast through central to west-central Nevada. In Early Ordovician, there was uplift along the Cortez-Uinta axis, which is an old positive element trending east-west through the present Uinta uplift and extending westward through Salt Lake City into east-central Nevada. In late Devonian, the Stansbury disturbance again arched the Cortez-Uinta axis, resulting in widespread erosion of earlier Paleozoic rocks that probably stripped away the early Paleozoic carbonates and eroded into the Precambrian in places. At about the same time, extensive orogeny commenced along the Antler orogenic belt in central Nevada, extending through central Nevada and into Idaho in a north-south to northeast-southwest configuration, and was the dominant tectonic element overriding the east-west Cortez-Uinta positive.

Paleozoic sedimentation patterns shifted after the Antler orogeny. Prior to that time the source for the miogeosynclinal facies of western Utah and eastern Nevada was from the shelf area to the east. There was now, for the first time since Precambrian, a western source.. A trough developed along the eastern side of the Antler orogenic belt, designated as the Chainman-Diamond Peak trough, which received great volumes of dark shale, chert, quartzite, and volcanic debris from the west. During this period shallow water carbonate deposition (Madison) still predominated in western Utah, but by late Mississippian time, Cambrian quartzites from the orogenic belt furnished quartz sands which were interbedded with bioclastic limestones of the Deseret Limestone in western Utah. Also in late Mississippian, depositional basins began to form in northwestern Utah to the north and south of the Cortez-Uinta axis.

The well known Oquirrh basin of western Utah and eastern Nevada accumulated great thicknesses of Pennsylvanian and Permian sediments, mostly sandstones and clastic or bioclastic limestones. The Oquirrh basin extended east-west from the Antler orogenic belt in central Nevada to the Utah-Wyoming shelf. Its north-south limits are not well defined, but it certainly extended from southern Nevada into south-central and southeastern Idaho. The deepest part of the basin was apparently in northwestern Utah, where it was divided into a northern and southern lobe by the Cortez-Uinta axis. An accumulation of 15,000 ft of Oquirrh is known in the northern lobe, and as much as 26,000 ft has been measured and described in the southern lobe.

Late Permian time was characterized by very widespread seas of shallow to moderate depths. The Phosphoria and Park City formations were deposited in a broad seaway extending through western Wyoming, southeastern Idaho, western Utah, and about the east half of Nevada. A northeast-southwest trending orogenic belt in northwestern Nevada contributed clastic debris and is reflected by a coarse clastic facies in central Nevada. This orogenic belt is sometimes considered an extension of the Antler orogeny, and is also called the Sonoma orogenic belt. In any event, it lies further to the northwest than the ancestral Antler orogenic area. Through central Wyoming and into northeastern Utah, the Permian carbonates grade into red silty shales and sandy shales, sometimes associated with evaporities. In Triassic time marine deposition became much more localized and apparently was restricted to southeastern Idaho, northeastern Nevada, western Utah, and probably southeastern Nevada.

West of the Wasatch Front in northwestern Utah there are no exposures of Mesozoic rocks. Some Mesozoics may be preserved beneath the Tertiary and Quaternary fill of the graben valleys, but in the mountain ranges the entire section has been removed by erosion. By projection of regional relationships, particularly from southeastern Idaho, it is apparent that during Triassic and into Lower Jurassic time, the geosyncline persisted in northwestern Utah, accumulating several thousand ft of largely marine carbonates, evaporites and shales. In the Wasatch Mountains immediately east of Salt Lake City, the Triassic-Jurassic interval is approximately 7500 ft thick, and represents a marginal facies between the geosyncline on the west and the largely continental shelf facies further east. The section at this locality indicates alternating continental deposition, typified by red shales, silts, and sands, with marine depositional environments which are represented by thick, fossiliferous limestones.

At about the end of Jurassic or in earliest Cretaceous, a new orogeny commenced in western Utah and eastern Nevada that was part of an orogenic belt that extended from Canada to southern California. Commonly called the Sevier orogenic belt, the disturbance occurred over a broad belt as much as 100 mi in width, and was accompanied by major thrusting in northwestern Utah, southeastern Idaho, and southwestern Wyoming. The thrusting in northwestern Utah resulted in the apparent superposition of thick Paleozoic sections from the Oquirrh basin in allochthonous blocks onto the thinner shelf section in large portions of the Wasatch Range.

The paleogeographic map (Fig. 11) is an idealized reconstruction of the northwestern Utah paleogeography and paleogeology in latest Cretaceous, about Maestrictian time, and at the time the basal portion of the Evanston Formation was being deposited. Immediately east of Salt Lake City in the Wasatch Range, and further eastward in the East Canyon and Coalville areas are Cretaceous sections that permit a fair reconstruction of Cretaceous depositional patterns. The lower-most Cretaceous unit, the Kelvin Formation, reflects the beginning of the Sevier orogeny. Immediately east of Salt Lake it is 1500 ft thick, consisting of essentially reddish-brown to purple siltstones, and thick, interbedded conglomerates. Overlying the Kelvin in this locality is about 8000 ft of Frontier Formation that is typically yellowish-brown sandstones interbedded with yellow or pale-red tuffaceous clays. Occasional conglomerates are interbedded in the lower portion, but become more frequent toward the top of the formation as the eroding highlands of the Sevier orogeny neared the present Wasatch Front.

Overlying the Frontier Formation is the Echo Canyon Conglomerate, probably Montanan in age. The Echo Canyon unconformably overlies all formations from Jurassic Twin Creek to Cretaceous Frontier with strong angular discordance, and is as much as 3000 ft thick 30 to 40 mi east of Salt Lake City. It certainly represents major movement and thrusting along the Bannock-Willard thrust system. Conglomerate clasts are 60 percent carbonate rock (Mullen, 1971), sandstone, and siltstone. Most of the remaining clasts are rounded cobbles and small boulders of tan and pink quartzite that resemble the Cambrian Tintic and Pennsylvanian Weber Quartzites of the autochthon of the Willard thrust. A few purple and green quartzite clasts are reported that may be derived from Cambrian and Precambrian sediments of the allochthon. On the basis of similarity of clasts, a conglomerate sequence believed to be Echo Canyon age is present in the Salt Lake salient just north of Salt Lake City, and may also be present in the Farmington graben beneath the area of interest. Either the major movement of the Willard thrust had not yet taken place or the sediments of the allochthon had not yet been eroded into the Precambrian during the major period of Echo Canyon deposition.

Unconformably overlying the Echo Canyon is the Late Cretaceous-Paleocene Evanston Formation. Like the Echo Canyon, the Evanston is exposed in the Salt Lake salient in the Wasatch Range immediately north of Salt Lake City, and in more extensive outcroppings east of the Wasatch Range, such as East Canyon, Echo Canyon, and the Coalville area. It was during deposition of the basal Evanston that good evidence supports major emplacement of the Bannock-Willard thrust and its probable continuation—or at least its contemporary to the south—the Charleston-Nebo thrust. Another latest Cretaceous period of orogeny and thrusting is indicated by the angular unconformity between the Echo Canyon and Evanston formations and the sudden appearance in the Evanston of a preponderance of clasts composed of quartzites from the Cambrian and Precambrian of the allochthon of the Willard thrust (Mullen, 1971). Thicknesses of measured Evanston sections in northern Utah range up to 1500 ft and its presence in the Salt Lake salient suggests that it may be present in places beneath the grabens of the Salt Lake Tertiary basin. The upper part of the Evanston is much finer-grained, and consists of as much as 1,000 ft of interbedded conglomeratic sandstone, grey micaceous sandstone, grey mudstone, and carbonaceous shale.

Overlying the probable Evanston sequence in the Salt Lake salient above Salt Lake City is about 800 ft of poorly consolidated red sandstone, sandy limestone, white to grey limestone, and water-laid tuff. Well preserved snails identified as *Oreoncus* have been recovered from this unit, which do not date it very precisely. It does give it a probable range of somewhere between upper Paleocene (Flagstaff) to upper Eocene, any of which are possible. Extensive laking commenced in Late Cretaceous and continued into at least middle Eocene along the eastern front of the Sevier orogenic belt slightly further south, and there could easily have been extensive areas of lacustrine deposition further north in the Salt Lake basin. As will be pointed out later, somewhere in middle to late Eocene, extensive north-south faulting and horst and graben formation commenced in northwestern Utah, and later Eocene laking and ponding could easily be associated with this period. The existence of considerable volcanic material suggests late Eocene, since volcanism became widespread at this time and the Norwood Tuff (latest Eocene) is attributed to the Park City volcanics. On the other hand, comparison of stratigraphic succession in the Salt Lake salient with similar sections a short distance eastward strongly suggests the correlation of this unit with the upper fine-grained unit of the Evanston formation.

Strongly supporting this latter correlation is the last Tertiary unit exposed north and east of Salt Lake City, which is a nearly flat-lying conglomerate that overlaps onto basement rocks of all ages. The conglomerate is more heterogeneous than any of the others and is not involved in any of the major folding that affected the older conglomerates or the Mesozoics. At Parleys Canyon and in the Salt Lake salient it overlaps prominent faults which cut all of the older rocks. This conglomerate should correlate with the Wasatch (usually Knight in the older literature) conglomerates of East Canyon, Lost Creek, Morgan Canyon, and several other areas to the east and northeast.

For those familiar with the Tertiary of northeastern Utah and southwestern Wyoming, but not up to date on developments of the past ten years, a word of explanation is necessary. Veatch originally subdivided the Wasatch (Eocene) into three units: (1) a basal member composed of reddish-yellow sandy clays, in many places containing pronounced conglomerate beds, which he named the Almy; (2) a great thickness of light-colored rhyolitic ash beds containing intercalated lenses of white limestones with fresh-water shells and leaves—the Fowkes Formation; and (3) a group of reddish-yellow sandy clays with irregular sandstone beds which was called the Knight Formation. For

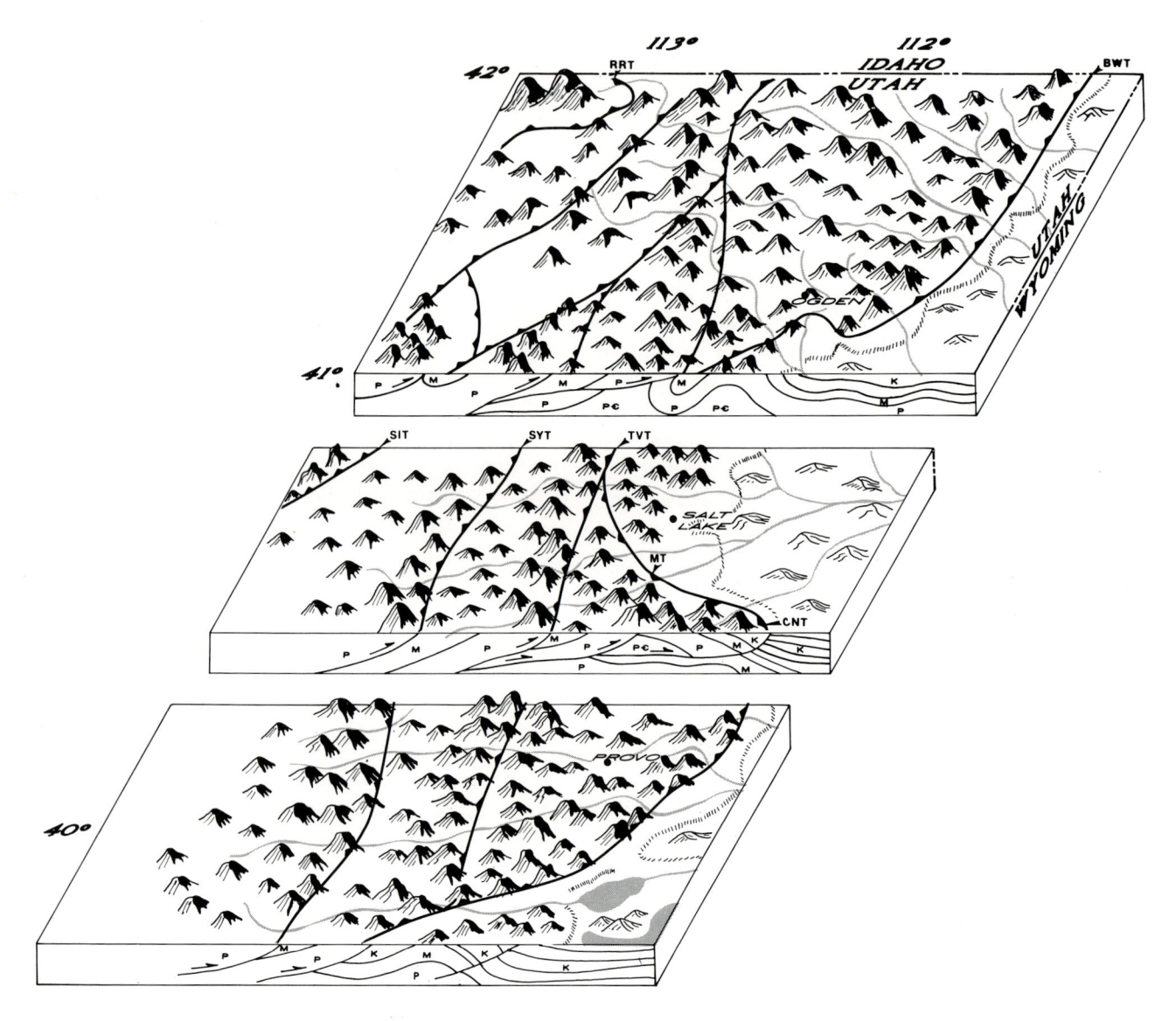

FIG. II

LATE CRETACEOUS BASAL EVANSTON TIME

PROBABLY MAESTRICTION

Idealized Paleogeographic Map
NORTH CENTRAL UTAH

FIG. 12

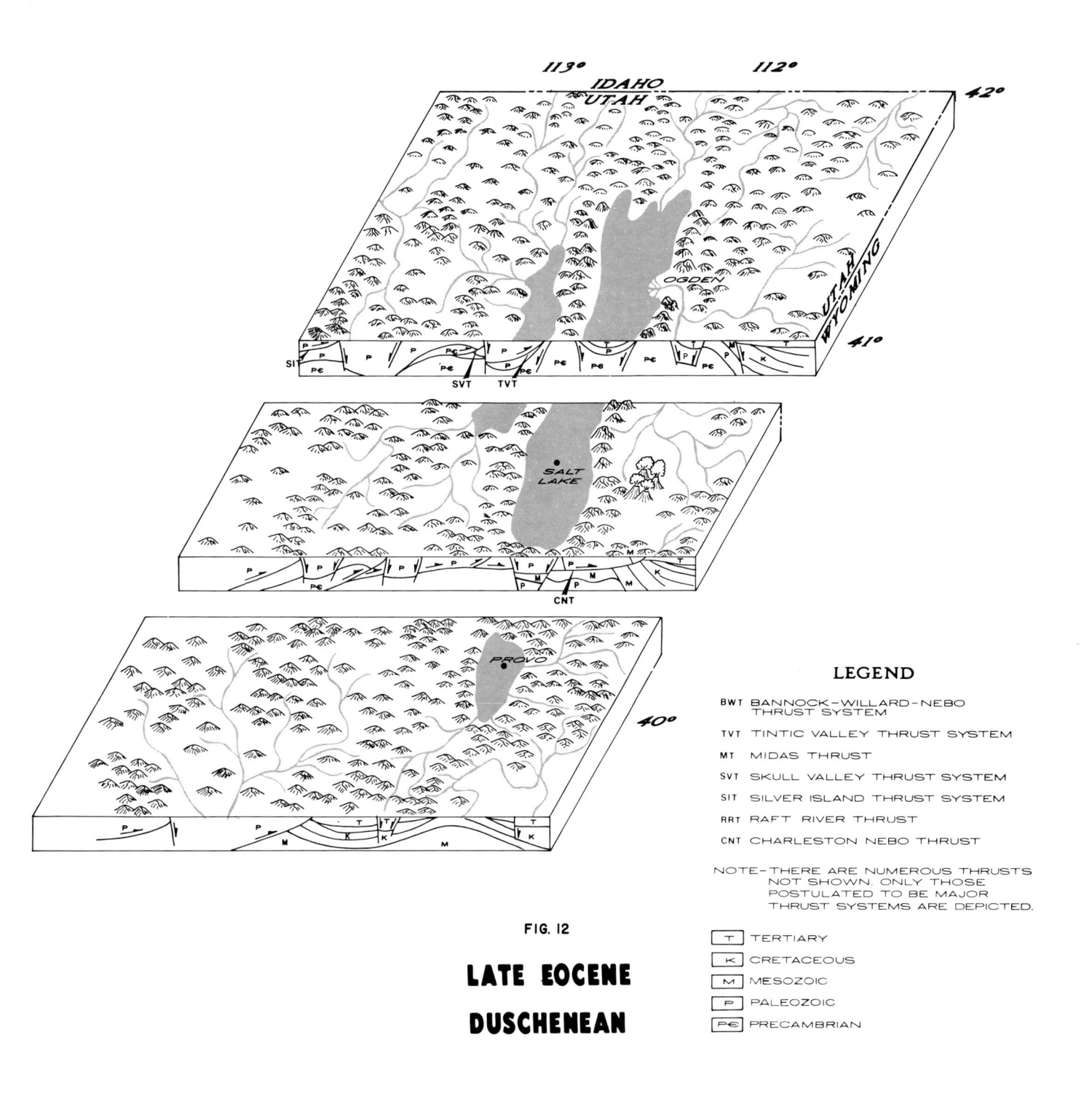

LATE EOCENE

DUSCHENEAN

Idealized Paleogeographic Map
NORTH CENTRAL UTAH

many years these divisions were pulled into the Wasatch Mountain areas and the names Almy, Fowkes, and Knight were applied to the various post-Sevier orogeny stratigraphic units. Gradually it was realized that many of the units were older than Eocene and other names were applied to the basal strata, such as Echo Canyon. Then it was discovered that because of faulting relationships unrecognized by Veatch in the type section, the Fowkes is actually younger than the Knight and overlies it. Further work indicated that the conglomeratic Almy Member of the Wasatch is a peripheral facies unit close to various sources, and grades laterally as well as vertically into the Knight Member, which is a basin facies of the Wasatch. There has not been sufficient work in the Wasatch east of the Wasatch Front to divide it into peripheral and basinal facies, and for the time being it is better to refer to the late Paleocene and Eocene facies as Wasatch.

The Wasatch is unconformable on all older rocks, but it overlies strata older than Echo Canyon and Evanston with only slight angular unconformity. There are sharp local exceptions to this, but where there is strong angular discordance between Evanston or Echo Canyon and Wasatch, it appears to be related to post-thrusting faulting and folding. As a matter of fact, the Wasatch appears to have been deposited well after emplacement of the Bannock-Willard thrust system. The Wasatch conglomerate facies appears more controlled by local topographic relationships and in places consists of diamictites and consolidated alluvial fans.

The evidence is strong that by latest Cretaceous the Bannock-Willard-Charleston thrust systems were emplaced and eroded into the Cambrian and Precambrian in the northern part of the system. North-south folding that overrode the existing structure, accompanied by block faulting that dropped the synclinal areas and enhanced the elevation of the anticlinal fold axes, set the stage for the present day topography. It has usually been suggested that this folding and faulting commenced in middle to late Eocene time. If in fact, the Wasatch does represent a time interval from late Paleocene to middle or late Eocene, it appears certain that the ancestral flexing and faulting that gave rise to the basin and range structure had certainly commenced by middle Eocene, and perhaps in late Paleocene to early Eocene time.

The important point insofar as the Great Salt Lake Tertiary basin is concerned is that by middle to late Eocene, such features as Morgan Valley may have been formed and the primary drainage patterns and directions of flow of the main river systems were set. For the first time since Jurassic this places the area west of the Wasatch Front lower than the hinterlands of northeastern Utah, southwestern Wyoming, and southeastern Idaho, and from this point on in geologic time the tremendous drainage area of the Weber River was pouring its debris into the Great Salt Lake Tertiary basin.

Figure 12 is an idealized paleogeographic map of the area in late Eocene. The lake in the Great Salt Lake Basin is speculative, but logical. The mountains are eroded down and will become even less prominent by the end of Eocene time. Sometime during latest Eocene and early Oligocene the great flood of Norwood Tuff will accumulate through the valley areas. There was local lake formation throughout the area, as evidenced by several local limestones east of the Wasatch in basinal facies, and if the interior drainage had commenced west of the Wasatch Front at this time, the formation of extensive shallow lakes is likely.

For discussion of probable Oligocene, Miocene and Pliocene facies, it becomes necessary to largely infer what is likely to be in the Great Salt Lake basin by a reconstruction of geologic history of the area from a few scattered middle to late Tertiary exposures, usually quite remote from the immediate area of interest. Since we are dealing with an intermontane basin in which depositional patterns are controlled by very localized conditions, there is little to be gained by going into detail on partially exposed sections located 30 mi to the south, 50 mi to the northeast, or 100 mi to the west. A summation of the middle to late Tertiary stratigraphy exposed as erosional remnants on the flanks of mountain ranges peripheral to the Great Basin area is helpful, however, in projecting the probable stratigraphic sequence within the deeper parts of the basin.

In the Jordan Narrows, about 30 mi south of Salt Lake City, there are exposures of Tertiary that are believed to range from Oligocene to Pliocene. They dip beneath the Quaternary cover and thick homologs probably exist basinward in the Jordan Valley and northward along the Wasatch Front into the Farmington graben area. Here, the lower Oligocene is represented by andesitic and latitic flow material and pyroclastics. Overlying this lower Oligocene volcanic sequence is a unit of lacustrine beds thought to range from middle Oligocene to middle Miocene in age, and described as oolitic, argillaceous, and cherty limestones, sandstones, and clays. The upper Miocene is essentially fluvial in character, consisting of mudstones, siltstones, and poorly consolidated sandstone. The Pliocene at this locality is primarily fanglomerates, which is not surprising since the section in the Jordan Narrows is exposed because of late Tertiary tectonism associated with the Traverse Mountains.

There are several exposures of Miocene and Pliocene lacustrine, fluvial, and volcanic rocks further to the southwest, west and northwest. Whether any Oligocene strata other than volcanics were deposited in these more peripheral areas is unknown. The presence of the Oligocene at only the south end of the Jordan Valley adds further evidence to the idea that the area immediately adjacent to the Wasatch Front has been the lowest part of the immense internal drainage basin of northwestern Utah for most of late Tertiary time.

TERTIARY TECTONICS AND SEDIMENTARY ROCKS 313

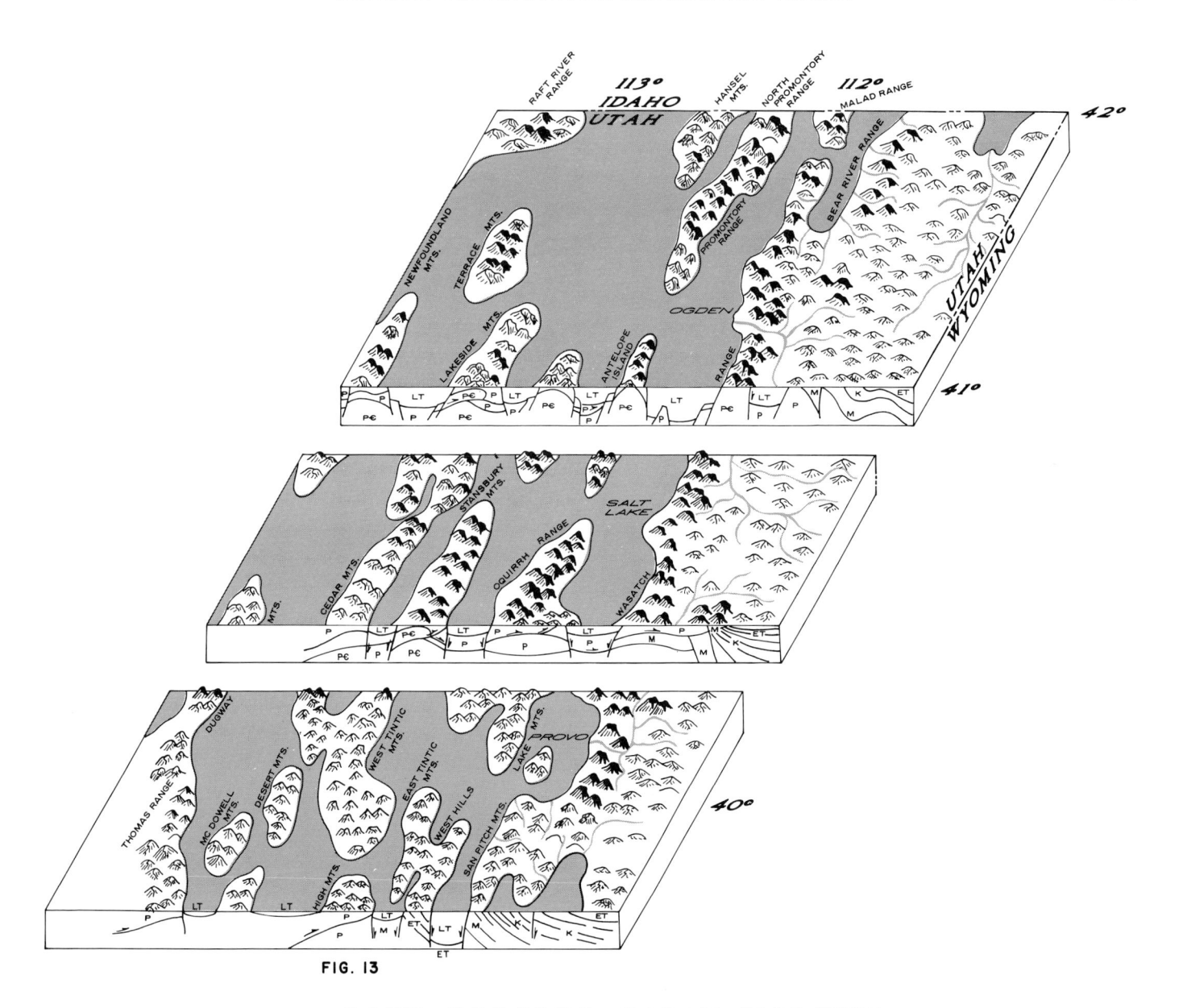

FIG. 13

LATE MIOCENE - EARLY PLIOCENE

ET	EARLY TERTIARY
LT	LATE TERTIARY
K	CRETACEOUS
M	MESOZOIC
P	PALEOZOIC
PЄ	PRECAMBRIAN

Idealized Paleogeographic Map

NORTH CENTRAL UTAH

Miocene and Pliocene age exposures occur extensively in the Grouse Creek District and Raft River Range in northwestern Utah, and to a lesser extent in the Pilot Range, Silver Island Range, Boulter Summit-Tintic Valley, and the southern end of Rush Valley (Fig. 6). These areas are located roughly 100 mi west to 100 mi southwest and south of the primary area of interest, and would be essentially peripheral basinal facies. The sediments here are characterized by yellowish to tan to white lacustrine limestone and silicified limestone, drab claystones, tuffs, lignites, green to brown sandstones, and Paleozoic pebble conglomerates. Petroliferous limestones and carbonaceous shales are also found in these units, and these strata correlate in part with the late Eocene to Pliocene Humboldt Formation of northeastern Nevada.

The middle to late Tertiary rocks have been consistently referred to by series designation rather than formational names. This is because geologists working in the Great Basin have usually called all Tertiary sedimentary units, and sometimes volcanic units, the Salt Lake Group. The Salt Lake Group has often been used to designate strata all the way from late Eocene through Pliocene age, and has become a nearly useless term. Figure 13 is a paleogeographic map depicting northwest Utah as it might have appeared in late Miocene to early Pliocene time. Moderate additional warping of the north-south trending structures, accompanied by further block faulting, had occurred since the late Eocene period, and relief between the basin and range areas was somewhat greater. The presence of significant thicknesses of late Miocene and early Pliocene lacustrine strata as far west as the Utah-Nevada line, and as far south as the Sevier desert, suggests that laking and distribution of lacustrine strata was very extensive.

Mention should be made of more than 7000 ft of apparent middle to late Pliocene lacustrine tuffs, limestones, sandstones, and conglomerates in the Cache Valley areas northeast of the Great Salt Lake basin. This area, which is part of the Bear River drainage system, was apparently emergent until about mid-Pliocene, when the graben faulting that forms the valley dropped the valley block relatively rapidly and accumulated this thick lacustrine sequence. Many of the limestones are reported as petroliferous, and gas shows have been reported in the few test wells drilled in the valley.

Although the lake level undoubtedly fluctuated and became playas and marshes for short periods of time, a major lake basically appears to have been extant for most of the period from Pliocene through Pleistocene and into Quaternary time. Localized pulsating uplift and faulting continously changed its outline around the edges and modified the configuration of basin interiors and projecting mountain ranges. These were changes of detail, however, and did not change the gross picture. R. B. Morrison, U.S.G.S. geologist who is probably the leading authority on Pleistocene laking in the Great Basin area, confirms this in his article (1966), wherein he states: "During middle and late Tertiary time, intermontane basins occupied much of northwestern Utah and adjacent parts of Idaho and Nevada . . . . , Lacustrine strata are common in these units (Oligocene, Miocene, and Pliocene) and testify to locally interrupted drainage and probably closed-basin conditions that in some areas obviously existed for long periods. The disrupted drainage probably resulted both from faulting and warping, and from intermittent volcanism. However, these Tertiary sediments have been so little studied, are so spotty in exposure, and commonly are so difficult to correlate from place to place, that very little is known about the configuration and fluctuation history of the Tertiary lakes. From the distribution of the younger part of the Salt Lake Formation and its lithofacies relations, *it is known that during the late Tertiary the intermontane basins within and west of the Wasatch Range appear to have had the same general outlines as the present basins,* but the relief between the basins and mountain ranges generally was less than it was during most of Quaternary." (italics mine)

Evidence is strong, therefore, that the Great Salt Lake Tertiary basin became a sag west of the Wasatch Range by mid-Eocene, and perhaps by late Paleocene, and since that time the basin has been an area of internal drainage for a vast mountainous area. Gravity interpretation suggests that as much as 10,000 ft of Tertiary sediments have accumulated in the graben valleys of the Great Salt Lake, most of which have accumulated in lacustrine, marsh, and playa-type depositional environments. Petroliferous limestones, lignites, and carbonaceous shales suggest that source beds may be present in the stratigraphic sequence. Periodic pulsating uplifts and block faulting in the surrounding mountainous areas have provided periods of rapid erosion and high clastic input, particularly around the mouth of the Weber River.

Geophysical Reconnaissance

Figure 14 is a gravity compilation map of the Great Salt Lake basin. It represents reconnaissance gravity surveys performed by the U.S. Geological Survey and the University of Utah (Cook *et al.*, 1961, 1963; Mabey *et al.*, 1967; and Stokes, 1966). These gravity surveys were performed at various times, utilizing existing roads and areas of easy access. They do not represent a gridded exploration-oriented survey and in places control points leave much room for error. Nevertheless, they do point out the areas of greatest graben subsidence and probable orientation of the deeper parts of the basin.

Interpretations across the Farmington trough between Salt Lake City and Ogden, Utah, indicate that approximately 10,000 ft of rock is present possessing densities anticipated in Tertiary sediments. Below this there is a density change, which probably indicates Paleozoic or Precambrian sediments.

TERTIARY TECTONICS AND SEDIMENTARY ROCKS

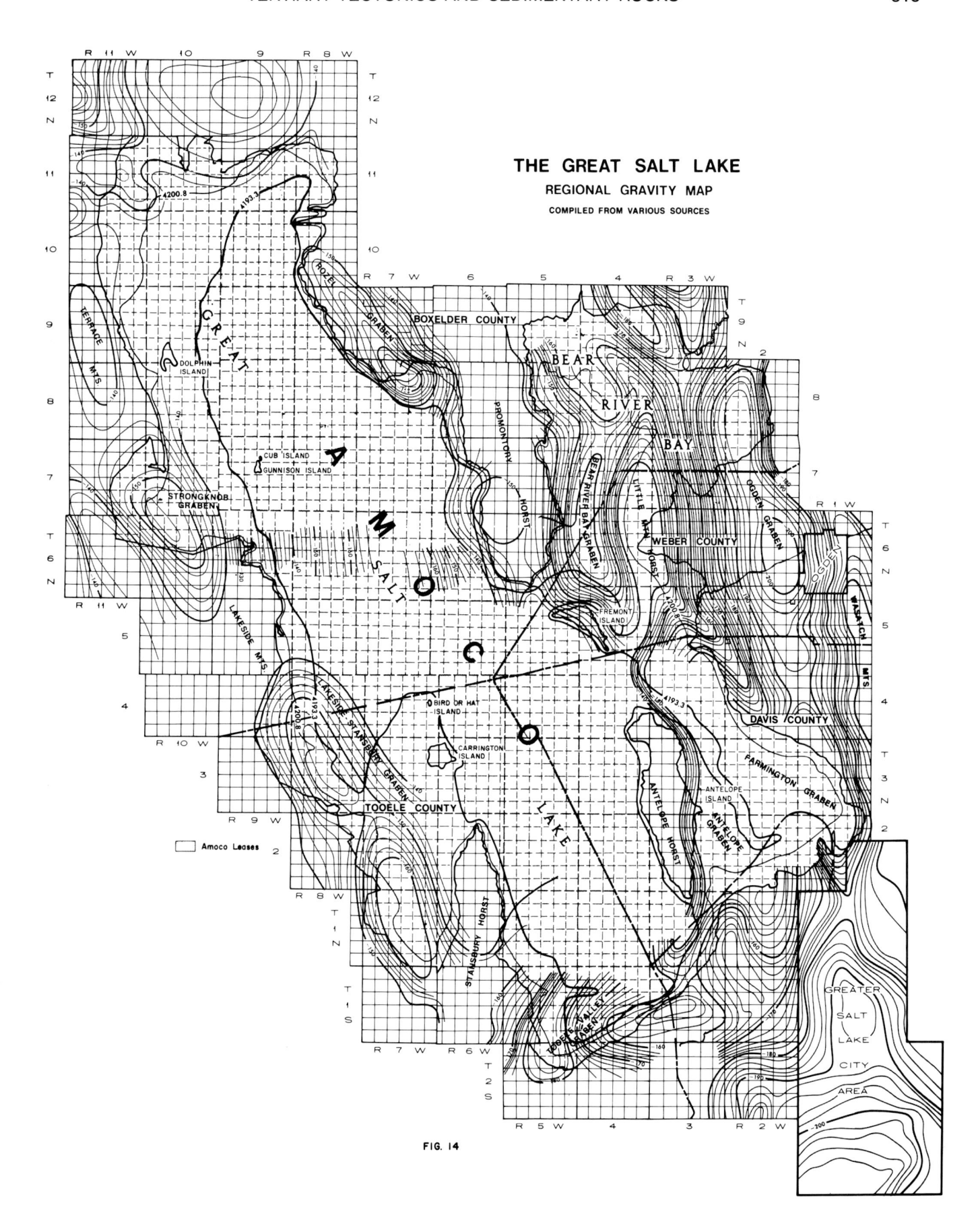

THE GREAT SALT LAKE
REGIONAL GRAVITY MAP
COMPILED FROM VARIOUS SOURCES

FIG. 14

DIAGRAMMATIC CROSS SECTION ACROSS THE SOUTHERN PORTION OF THE GREAT SALT LAKE BASIN

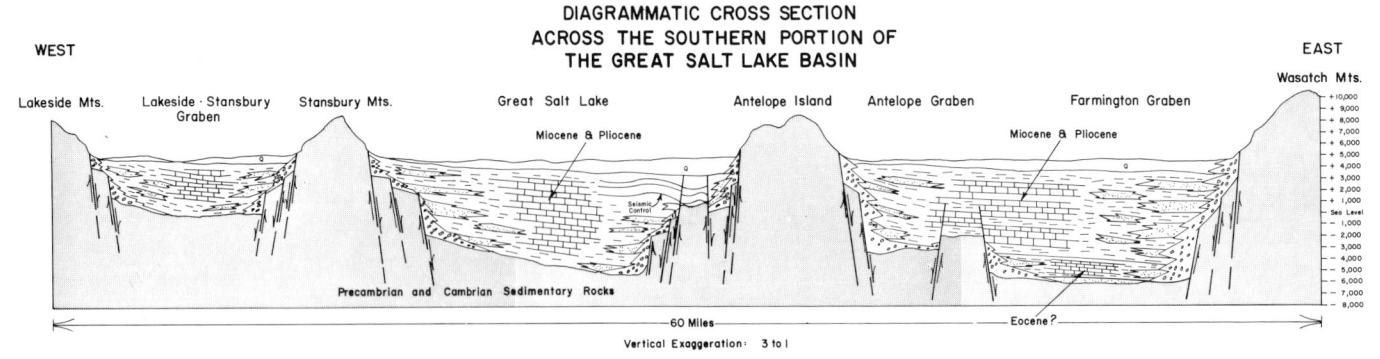

Fig. 15

Figure 15 is a diagrammatic cross section across the southern portion of the Great Salt Lake basin utilizing gravity, limited seismic, and aerial magnetometer surveys (Mikulich and Smith, 1974). These surveys also suggest about 10,000 ft of Tertiary and Quaternary sediments under the deepest portion of the Great Salt Lake, and show Tertiary and Quaternary sediments faulted into and resting on Paleozoic and Precambrian rocks. Along the eastern side of the lake, horst and graben faulting and folding are common in Tertiary sediments, while in the central and western portions of the lake homoclinal east dip prevails. Tertiary and Quaternary sediments apparently thin from about 10,000 ft beneath the south end of the lake to about 1600 ft beneath the northern portion. Amoco Production Company currently holds oil and gas leases on nearly all the Great Salt Lake and has announced two tentative locations, although no drilling plans have yet been finalized. The eastern portion of the Great Salt Lake Basin beneath the Farmington, Bear River, and Ogden grabens also warrants exploration. Unfortunately, however, Ogden, Utah and Hill Air Force Base overlie the Ogden graben, and the State of Utah has refused to issue leases beneath the Farmington and Bear River grabens because of potential visual pollution.

SELECTED REFERENCES

Armstrong, R. L., 1968, Sevier orogenic belt in Nevada and Utah: Geol. Soc. America Bull., Vol. 79, no. 4, p. 429-458.

Bjorklund, J. J., and Robinson, G. B., Jr., 1968, Ground-water resources of the Sevier River Basin between Yuba Dam and Leamington Canyon, Utah: U.S. Geol. Survey Water-supply Paper 1848, 79 p.

Bowen, C. F., 1911, Coal at Horseshoe Bend and Jerusalem Valley, Boise County, Idaho in Contributions to Economic Geology: U.S. Geol. Survey Bull. 531, p. 245-251.

_____ 1911, Lignite in the Goose Creek district, Cassia County, Idaho in Contributions to Economic Geology: U.S. Geol. Survey Bull. 531, p. 252-262.

Bullock, K. C., 1951, Geology of Lake Mountain, Utah: Utah Geol. and Mineral Survey Bull. 41, 46 pp.

Callaghan, Eugene, and Parker, R. L., 1962, Geology of the Delano Peak quadrangle: U.S. Geol. Survey Map GQ-153.

_____ 1962, Geology of the Sevier quadrangle, Utah: U.S. Geol. Survey Map GQ-158.

Chatfield, John, 1972, Case History of Red Wash Field, Uinta County, Utah in Stratigraphic Oil and Gas Fields: Am. Assoc. Petroleum Geologists Memoir 16, p. 342-353.

Christiansen, F. W., 1951, A summary of the structure and stratigraphy of the Canyon Range in Guidebook to the Geology of Utah, no. 6, Geology of the Canyon, House and Confusion Ranges, Millard County, Utah: Intermtn. Assoc. Petroleum Geologists, p. 5-18.

Cloos, Ernst, 1968, Experimental analysis of Gulf Coast fracture patterns: Am. Assoc. Petroleum Geologists Bull. vol. 52, no. 3, p. 420-444.

Cook, E. F., 1967, Geology of the Pine Valley Mountains, Utah: Utah Geol. and Mineral Survey Bull. 58, 111 p.

_____ 1960, Geologic Atlas of Utah, Washington County: Utah Geol. and Mineralogical Survey Bull. 70

_____ 1960, Great Basin ignimbrites in Guidebook to the Geology of East Central Nevada: Intermtn. Assoc. Petroleum Geologist, Eleventh Annual Field Conf., p. 134-141.

_____ 1965, Stratigraphy of Tertiary volcanic rocks in eastern Nevada: Nev. Bur. Mines Report 11, 61 p.

Cook, K. L., and Berg, J. W., Jr., 1961, Regional gravity survey along the central and southern Wasatch Front, Utah: U.S. geol. Survey Prof. Paper 316-E, 88 p.

_____, et al., 1964, Regional gravity survey of the northern Great Salt Lake Desert and adjacent areas of Utah, Nevada and Idaho: Geol. Soc. America Bull., vol. 75, p. 715-740.

_____ 1966, Regional gravity survey of Beaver and Millard Counties, Utah (abst.) in Utah Academy Science, Arts and Letters: vol. 43, pt. II, p. 62.

_____, and Iverson, R. M., and Strohmeier, M. T., 1967, Bottom gravity meter survey of the Great Salt Lake, Utah (abs.). EOS, Trans. Amer. Geophys. Union, 50, 321, 1967a.

_____, and Hardman, Elwood, 1967, Regional gravity survey of the Hurricane Fault and Iron Springs, District, Utah: Geol. Soc. America Bull., vol. 78, p. 1063-1076.

_____, and Iverson, R. M., and Strohmeier, M. T., 1969, Bottom gravity meter survey of the Great Salt Lake, Utah: Geol. Soc. America Abst., pt. 5, (Rocky Mt. Sec.) p. 16.

Crawford, A. L., and Buranek, Alfred M., 1948, A Reconnaissance of the geology and mineral deposits of the Lake Mountains, Utah County, Utah: Utah Geol. and Mineral Survey Circ. no. 35, 33 p.

_____ (Editor), 1963, Oil and gas possibilities of Utah, re-evaluated: Utah Geol. and Mineral Survey Bull. 54, 564 p.

Crittendon, M. D., Jr., 1961, Magnitude of the thrust-faulting in northern Utah: U.S. Geol. Survey Prof. Paper 424D, p. 128-132.

Crosson, R. S., 1964, Regional gravity survey of parts of Millard and Juab Counties, Utah: MS Thesis, Univ. Utah.

Doelling, H. H., 1972, Tertiary strata, Sanpete-Sevier region: Utah Geol. Assoc. Pub. No. 2, p. 41-54.

Eardley, A. J., (editor) 1955, Tertiary and Quaternary geology of the eastern Bonneville Basin: Guidebook to the Geology of Utah Number 10, Utah Geol. Soc., 132 p.

_____, and Hardy, C. T., (editors), 1956, Geology of parts of northwestern Utah: Guidebook to the Geology of Utah Number 11, Utah Geol. Soc., 97 p.

Feth, John H., 1964, Review and annotated bibliography of ancient lake deposits (Precambrian to Pleistocene) in the western United States: U.S. Geol. Survey Bull. 1080, 119 p.

Gilbert, G. K., 1928, Studies of basin-range structure: U.S. Geol. Survey Prof. Paper 153, 92 p.

Granger, Arthur E., Bell, M. M., Simmons, G. C., and Lee, Florence, 1957, Geology and mineral resources of Elko County, Nevada: Nev.Bur. Mines Bull. 54, 190 p.

Gregory, H. E., 1950, Geology and geography of the Zion Park Region, Utah and Arizona: U.S. Geol. Survey Prof. Paper 220, 200 p.

Hardy, C. T., 1952, Eastern Sevier Valley, Sevier and Sanpete Counties, Utah: Utah Geol. and Mineral Survey Bull. 43, 97 p.

_____, and Zeller, H. D., 1963, Geology of west-central part of the Gunnison Plateau, Utah: Geol. Soc. America Bull., vol. 64, p. 1261-1278.

Harris, H. D., 1959, Late Mesozoic positive area in western Utah: Am. Assoc. Petroleum Geologists Bull., vol. 43, no. 11, p. 2636-2652.

Heylmun, E. B., 1965, Reconnaissance of the Tertiary sedimentary rocks in western Utah: Utah Geol. and Mineral Survey Bull. 75.

Hill, D. P., Baldwin, H. L., and Pakiser, L. C., 1961, Gravity, volcanism, and crustal deformation in the Snake River Plain, Idaho; U.S. Geol. Survey Prof. Paper 424B, p. 246-428.

Hintze, L. F. (editor), 1962, Geology of the southern Wasatch Mountains and vicinity, Utah: Brigham Young Univ. Geology Studies, vol. 9, part I, 104 p.

Interior, Department of, 1971, Environmental impact statement for the geothermal leasing program: Distributed by U.S. Geol. Survey.

Isherwood, W. F., 1969, Regional gravity survey of parts of Millard, Juab and Sevier Counties, Utah: Geol. Soc. America Abs., pt. 5 (Rocky Mt. Sec.), p. 35.

Kirkham, V. R. D., 1935, Natural gas in Washington, Idaho, eastern Oregon, and northern Utah in Geology of Natural Gas: Am. Assoc. Petrol. Geol., p. 221-244.

LeFehr, T. R., and Pakiser, L. C., 1962, Gravity, volcanism and crustal deformation in the eastern Snake River Plain, Idaho: U.S. Geol. Survey Prof. Paper 450D, p. 76-79.

Lee, W. T., 1908, Water resources of Beaver Valley, Utah: U.S. Geol. Survey Water-supply Paper 217, 57 p.

Lindgren, Waldemar, 1898, Geologic folio of Boise, Idaho: U.S. Geol. Survey Geologic Folio no. 45, 7 p.

Lintz, Joseph, Jr., Nevada oil and gas drilling data, 1906-1953: Nev. Bur. Mines, Bull. 52, 80 p.

——————— (editor), 1969, Second basin and range geology field conference guidebook: Mackay School of Mines, Univ. of Nev.

Lovejoy, E. M. P., 1973, Major Early Cenozoic deformation along Hurricane fault zone, Utah and Arizona: Am. Assoc. Petroleum Geologists Bull., vol. 57, no. 3, p. 512.

Mabey, D. R., and Armstrong, F. C., 1962, Gravity and magnetic anomalies in Gem Valley, Caribou County, Idaho: U.S. Geol. Survey Prof. Paper 450D, p. 73-75.

———————, 1966, Relation between Bouguer gravity anomalies and regional topography in Nevada and the eastern Snake River Plain, Idaho: U.S. Geol. Survey Prof. Paper 550B, p. 108-110.

———————, and Morris, H. T., 1967, Geological interpretation of gravity and magnetometer surveys, Tintic Valley and adjacent areas: U.S. Geol. Survey Prof. Paper 516D.

Mackin, J. H., 1963, Reconnaissance stratigraphy of the Needles Range Formation in southwestern Utah in Guidebook to the Geology of Southwestern Utah: Intermtn. Assoc. Petroleum Geologists 12th Annual Field Conf., p. 71-78.

Mapel, W. J., and Hail, W. J., Jr., 1959, Tertiary geology of the Goose Creek district, Cassia County, Idaho, in Uranium in Coal in the Western United States: U.S. Geol. Survey Bull. 1055, p. 217-254.

Marsell, R. E. (editor), 1952, Geology of the central Wasatch Mountains, Utah: Guidebook to the Geology of Utah No. 8, Utah Geol. Soc., 71 p.

McGookey, D. P., 1960, Early Tertiary stratigraphy of part of central Utah: Amer. Assoc. Petroleum Geologists Bull., vol. 44, no. 5, p. 589-615.

Meinzer, O. E., 1911, Ground water in Juab, Millard, and Iron counties, Utah: U.S. geol. Survey Water-supply Paper 277, 162 p.

Mikulich, M. J., and Smith, R. B., 1974, Seismic reflection and aeromagnetic surveys of the Great Salt Lake, Utah: Geol. Soc. America Bull., vol. 85, no. 6, p. 991-1004.

Moody, J. D., and Hill, M. J., 1956, Wrench-fault tectonics: Geol. Soc. America Bull., vol. 87, p. 1207-1246.

Moore, J. G., 1969, Geology and mineral deposits of Lyon, Douglas and Ormsby Counties, Nevada: Nev. Bur. of Mines Bull. 75.

Morrison, R. B., 1964, Lake Lahotan: Geology of southern Carson Desert, Nevada: U.S. Geol. Survey Prof. Paper 401, 153 p.

———————, 1966, Predecessors of Great Salt Lake: Guidebook to the Geology of Utah, No. 20, Utah Geol Soc., p. 77-104.

Mudgett, P. M., 1964, Regional gravity survey of parts of Beaver and Millard Counties, Utah: MS Thesis, Univ. Utah.

Mullen, T. E., 1971, Reconnaissance study of the Wasatch, Evanston, and Echo Canyon formations in part of northern Utah: U.S. Geol. Survey Bull. 1311D.

Nevada Bureau of Mines, Nevada Mining Analytical Laboratory, Desert Research Institute and Atomic Energy Commission, 1964, Final report, geological, geophysical, chemical and hydrological investigations of the Sand Springs Range, Churchill County, Nevada for Shoal Event, Vela Uniform Program, Atomic Energy Commission: Clearinghouse for Federal Scientific and Technical Information, National Bureau of Standards, U.S. Dept. of Commerce, Springfield, Va.

Peck, D. L., 1964, Geologic reconnaissance of the Antelope-Ashwood area, north-central Oregon: U.S. Geol. Survey Bull. 1161-D.

Powers, H. A., and Malde, H. E., 1961, Volcanic ash beds as stratigraphic markers in basin deposits near Hagerman and Glenns Ferry, Idaho: U.S. Geol. Survey Prof. Paper 424B, p. 167-169.

Reed, R. D., and Hollister, J. S., 1951, Structural evolution of southern California: Am. Assoc. Petroleum Geologists, 146 p.

Roberts, R. J., Crittenden, M. D., Jr., Tooker, E. W., Morris, H. T., Hose, R. K., and Cheney, T. M., 1965, Pennsylvanian and Permian basins in northwestern Utah, northeastern Nevada and south-central Idaho: Am. Assoc. Petroleum Geologists Bull., vol. 49, no. 11, p. 1926-1956.

Rose, R. L., 1969, Geology of the Wadsworth and Churchill Butte Quadrangles, Nevada: Nev. Bur. Mines Bull. 71, 27 p.

Ross, C. P., 1961, A redefinition and restriction of the term Challis volcanics: U.S. Geol. Survey Prof. Paper 424C, p. 177-180.

———————, and Forrester, J. D., unpublished mapping for Idaho Bureau of Mines and Geology.

Ross, D. C., 1961, Geology and mineral deposits of Mineral County, Nevada: Nev. Bur. Mines, Bull. 58, 98 p.

Ruppel, E. T., 1964, Strike-slip faulting and broken basin-ranges in east-central Idaho and adjacent Montana: U.S. Geol. Survey Prof. Paper 501C, p. 14-17.

Rush, R. W., 1951, Stratigraphy of the Burbank Hills, Western Millard County, Utah: Utah Geol. and Mineral. Survey Bull 38, 23 p.

Russell, I. C., 1902, Geology and water resources of the Snake River plains of Idaho: U.S. Geol. Survey Bull. 199, 192 p.

———————, 1903, Preliminary report on artesian basins in southwestern Idaho and southeastern Oregon: U.S. Geol. Survey Water-supply Paper 78, 53 p.

———————, 1903, Notes on the geology of southwestern Idaho and southeastern Oregon: U.S. Geol. Survey Bull. 27, 83 p.

Schilling, J. H., and Garside, L. J., 1968, Oil and gas developments in Nevada, 1953-1967: Nev. Bur. Mines Report 18, 43 p.

Schneider, M. C., 1967, Early Tertiary continental sediments of central and south central Utah: Brigham Young Univ. Geol. Studies, v. 14, p. 143-194.

Spieker, E. M., 1949, The transition between the Colorado Plateaus and the Great Basin central Utah: Guidebook to the Geology of Utah No. 4, Utah Geol. Soc., 106 p.

Stearns, H. T., Crandall, Lynn, and Steward, W. G., 1939, Geology and ground-water resources of the Snake River Plain in southeastern Idaho: U.S. Geol. Survey Water-supply Paper 774, 268 p.

Stokes, W. L., and Heylmun, E. B., 1963, Tectonic history of southwestern Utah in Guidebook to the Geology of southwestern Utah: Intermtn. Assoc. Petroleum Geologists 12th Annual Field Conf., p. 19-25.

Stokes, W. L. (editor), 1966, The Great Salt Lake: Guidebook to the Geology of Utah No. 20, Utah Geol. Soc., 173 p.

Threet, R. L., 1952, Geology of the Red Hills area, Iron County, Utah: Ph. D. dissertation, Univ. Utah, 107 p.

Tschanz, C. M., 1960, Regional significance of some lacustrine limestones in Lincoln County, Nevada, recently dated as Miocene: U.S. Geol. Survey Prof. Paper 400B, p. 293-295.

———————, and Pampeyan, E. H., 1970, Geology and mineral deposits of Lincoln County, Nevada: Nev. Bur. Mines Bull 73, 187 p.

Van Houten, F. B., 1956, Reconnaissance of Cenozoic sedimentary rocks of Nevada: Am. Assoc. Petroleum Geologists Bull., vol. 40, no. 12, p. 2801-2825.

Wang, Y. F., 1970, Geology and geophysical studies of the Gilson Mountains and vicinity, Juab County, Utah: Ph. D. Thesis, Univ. Utah.

Waring, G. A., 1965, Thermal springs of the United States and other countries of the world — a summary: U.S. Geol. Survey Prof. Paper 492.

Washburne, Chester W., 1909, Gas and oil prospects near Vale, Oregon in Contributions to Economic Geology: U.S. Geol. Survey Bull. 431, p. 26-55.

Weaver, C. E., 1937, Tertiary stratigraphy of western Washington and northwestern Oregon: Univ. Washington Pub. in Geology, 264 p.

Weiss, M. P., 1969, Onocolites, paleoecology, and Laramide tectonics, central Utah: Am. Assoc. Petroleum Geologists Bull., vol. 53, no. 5, p. 1105-1120.

Willden, Ronald, 1964, Geology and mineral deposits of Humboldt County, Nevada: Nev. Bur. Mines Bull. 59, 154 p.

———————, and Speed, R. C., 1968, Geology and mineral deposits of Churchill County, Nevada: U.S. Geol. Survey Open File Report, 116 p.

Wiley, M. A., 1963, Stratigraphy and structure of the Jackson Mountain — Tobin Wash area, southwest Utah: M. A. thesis, Univ. Texas.

Winfrey, W. M., Jr. 1960, Stratigraphy, correlation and oil potential of the Sheep Pass Formation, east-central Nevada in Guidebook to the geology of east-central Nevada: Intermtn. Assoc. Petroleum Geologists Eleventh Ann. Field Conf., p. 126-133.

Young, J. C., 1955, Geology of the southern Lakeside Mountains, Utah: Utah Geol. and Mineral. Survey, Bull. 56, 110 p.

Youngquist, Walter, and Kiilsgaard, T. H., 1951, Recent test drilling, Snake River plains, southwestern Idaho: Am. Assoc. Petroleum Geologists Bull., vol. 35, no. I, p. 90-95.

OIL SHOWS IN SIGNIFICANT TEST WELLS OF THE CORDILLERAN HINGELINE

by

Harry K. Veal[1]

The following charts are constructed from AmStrat sample logs and well histories where available. Petroleum information scout cards and tickets are also utilized. American Stratigraphic Co. and Petro-Well Libraries of Denver allowed the examination of all their data and the cooperation is most appreciated. Interviews were carried out with various geologists familiar with certain of the listed wells and in Wyoming and Idaho faults and some of the shows were taken from Monley's excellent article (1971). The list of oil shows is by no means complete as not all AmStrat logs and/or well-site geologic reports were available to me. Test recovery and completion data is very brief due to space limitations, but of course is available on PI scout cards. The quality of oil shows is the same as that shown on AmStrat sample logs whether or not the show data came from AmStrat. The purpose of the paper is to perhaps aid in the exploration of the area by pointing out where oil shows occur in the geologic column along the Cordilleran Hingeline.

A brief inspection of the table indicates the following;

1. Shows in Nevada, Arizona, and central and southern Utah south of the Uinta Range occur primarily in rocks of Permian and Triassic age with lesser amounts of oil shows in the Pennsylvanian, Mississippian, Devonian, and Cambrian. Jurassic, Cretaceous and Tertiary oil shows also occur. One unusual oil show north of Salt Lake City occurs in Tertiary volcanics (Well 26).

2. Oil shows in northern Utah, Wyoming, and Idaho north of the Uinta Range are most numerous in the Jurassic and Cretaceous with lesser numbers occurring in rocks of Permo-Pennsylvanian, Mississippian, Ordovician, and Tertiary age.

3. Only minor production had been found along the hingeline south of the Uinta Range in rocks of Cretaceous, Permian and Pennsylvanian age. However, the high incidence of oil shows combined with numerous large untested deep structures indicate a favorable future for this area. North of the Uinta Range major production has been recently established in the Twin Creek and Nugget formations of Jurassic age at Pineview, Yellow Creek, and Ryckman Creek fields. Rocks of Cretaceous, Permo-Pennsylvanian, Mississippian, and Ordovician age also produce and may hold potentially large reserves within the Wyoming-Idaho overthrust belt. Although relatively low in the number of reported oil shows the Devonian should not be overlooked.

REFERENCES CITED

Monley, L. E., 1971, Petroleum potential of the Idaho-Wyoming overthrust belt: Am. Assoc. Petroleum Geologists Memoir 15, p. 509-529.

---

[1] Independent, Denver, Colorado

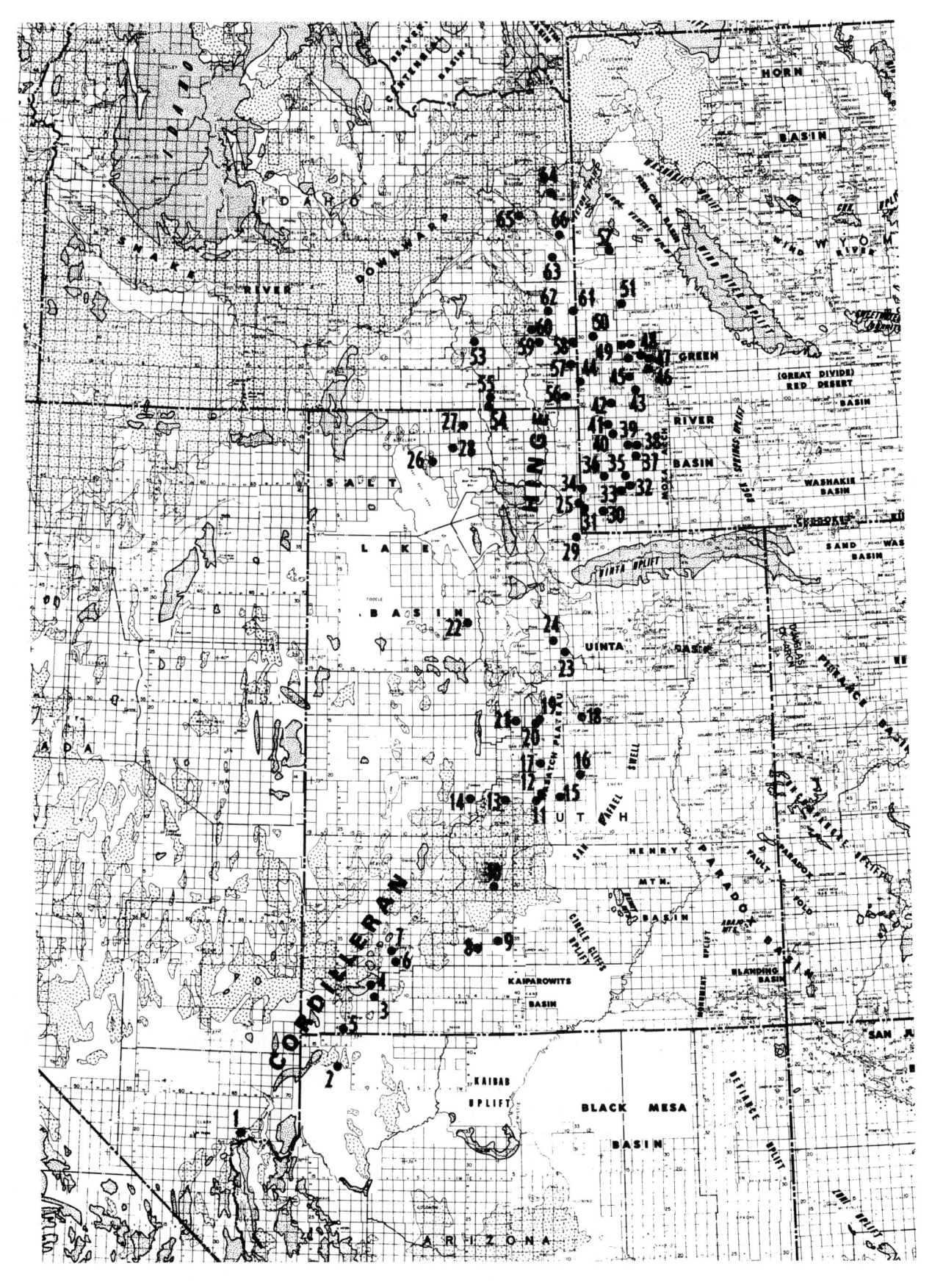

Fig. 1. — Location of Significant Test Wells, Cordilleran Hingeline.

OIL SHOWS — CORDILLERAN HINGELINE

FORMATION ABBREVIATIONS

Qal - Quaternary Alluvium		Tr - Triassic Undifferentiated		Mu - Miss. Undifferentiated	
Tu - Tertiary Undifferentiated		Tra	Ankareh	UM	Upper Miss.
Tsl	Salt Lake	Trth	Thaynes	LM	Lower Miss.
Tv	Volcanics	Trw	Woodside	Mrs	Rogers Springs
Tgr	Green River	Trm	Moenkopi	Md	Deseret
Tw	Wasatch	Trmm	Mid. Moenkopi	M"d"	Younger Miss.
Twt	Wasatch Tongue	Trmv	Moenkopi Virgin Mbr.	M"a"	Oldest Miss.
Tb	Black Shale Facies	Trms	Moenkopi Sinbad Mbr.	Mga	Gardison
Tal	Almy	Trs	Shinarump	Mhm	Horseshoe Mesa
Tf	Flagstaff	Trmt & Trt	Timpoweap	Mmf	Mooney Falls
		Trsk	Shnabkaib Mbr.	Mh	Humbug
Kpr - Cretaceous Price River		Trd	Dinwoody	Mb	Brazer
Kws	Wahweap - Straight Cliffs			Mf	Fitchville
		Pk - Permian	Kaibab	Mm	Madison
Ksp	Star Point	Pt	Toroweap	Mrw	Redwall
Kmv	Mesaverde	Ptu	Toroweap Upper		
Kbh	Blackhawk	Ptm	Toroweap Middle	D & Du-Devonian Undifferentiated	
Kav	Adaville	Pta	Toroweap Lower	Dmp	Muddy Peak
Kba	Baxter	Phe	Hermit	Dj	Jefferson
Ka	Aspen	Pq	Queantoweap	Dp	Pilot Shale
Kmb	Bluegate	Pec	Elephant Canyon	Dpp	Pinyon Peak
Kf	Frontier	Pk & Ppk	Pakoon	Dbb	Bluebell
Kmf	Ferron	Pwr	White Rim	UD	Upper Devonian
Kmt	Tununk	Ppc	Park City		
Km	Muddy	Pco	Coconino	Ob-Ordovician Big Horn	
Kbr	Bear River	Pcm	Cedar Mesa		
Kd	Dakota			Є & Єu-Cambrian Undifferentiated	
Kk	Kelvin	Pu & Pp	Phosphoria	Єly	Lynch
				Єmx	Maxfield
KJr - Cret.-Jur. Beckwith		Pca - Penn.	Callville	Єt	Tintic
KJg - Gannett		Po	Oquirrh	Єg	Gallatin
		Ph	Hermosa	Єgv	Gros Ventre
		Psl	L. Supai Tongue	Єdc	Death Canyon
Jrm - Jurassic Morrison		Pa	Amsden		
Jrs	Stump	Pm	Morrow	OIL SHOW AND FAULT LEGEND	
Jp	Preuss	Pi	Illipah	"O"	Weak Oil Show
Jra	Arapien	Pw	Weber - Wells	"◊"	Spotty Oil Show
Jca	Carmel	Pm	Morgan	"●"	Good Oil Show
Jrtc	Twin Creek			"D"	Dead Oil Show
Jrn	Nugget - Navajo			nf	normal fault
				tf	thrust-fault

Significant Test Wells Within the Cordilleran Hingeline

Map No.	Well Name	Location	Date Comp.	Spud Fm.	T.D. Fm.	Old. Fm. Tested	T.D.	Status	Remarks and Shows
1.	Shell #1 Bowl of Fire Unit	**NEVADA** 5-20S-66E	'59	Tr	Єu	Єu	8967	P&A	"◊" Trm, Pk, Pt, Mrs; "D" Pk, Pq, Ppk, Pca, Mrs, Dmp; "O" Pt, Ppk, Pca, Mrs, Dmp.
2.	Tennessee Gas #1 Schreiber	**ARIZONA** 35-39N-13W	'60	Pk	D-Є?	Є?	4015	P&A	Hole Bridged, not fully logged; "O-◊" Pt, "◊" Pi, Mrw; "D" Pk, Mrw.
3.	Buttes #1 Pease Federal	**UTAH** 25-40S-13W	'67	Trm	Mrw	Mrw	5236	IP 2BO 18BWPD Pca	"D" Pt; "●" Pca; "O"? Pca; Pca 2300 feet fill up with oil in 72 hours; Perf. - rapid decline
4.	Pan American #1 Pintura Unit	33-39S-13W	'62	Jrn	Du	Du	9501	P&A	"D" Trmt, Jrn, Ppk, Pca, M"d", Du; "O" Trmv, Ptm; Trsk test Mud; Du test Mud, MW & W.
5.	Calco #1 St. George Unit	19-43S-15W	'51	Pk	Du	Du	6347	P&A	tf@5340, 6215, 6250; "D" M"d"; Tests all recovered fresh wtr.
6.	Mountain Fuel #1 Shurtz Ck.	9-37S-11W	'73	Trm	Pq?	Pq?	5996	P&A	"O"-"●" Pk, Ptu, Pta; tf@3410; nf@5070.
7.	Odessa-Southland #1 Cedar City	18-36S-11W	'75	Qal	Du	Du	11700	P&A	"●"-"◊" Pk, "O" Pt, Mmf; Pk test MCW.

Map No.	Well Name	Location	Date Comp.	Spud Fm.	T.D. Fm.	Old. Fm. Tested	T.D.	Status	Remarks and Shows
8.	Lion #1 Bryce	10-36S-4W	'57	Tw	Pcm	Pcm	11221	P&A	"◇" Trmt, Pk, Pt; Jr test W; Pt test Mud; Pt test OCW, PtPcm test W; Pcm test W. f@3230 & 5070.
9.	Tenneco #1 Johns Valley	35-35S-2W	'69	Kws	Mmf	Mmf	11180	P&A	"●" Trs, Trt, Pk, Pt, Phe, Pcm, Ph, Mhm, Mmf; "D" Trs; Set pipe Mmf, swb 1% oil; Mmf test O&OCW; Pcm test OCW.
10.	Tenneco #1 Antimony Can.	30-30S-2W	'65	Pwr	€t	€t	7926	P&A	"○"-"◇" Pt, "○"-"◇"-"●"-"D" Pec; "○"-"◇"-"D" Md, "○"-"D" Mga; "D"-"○" Cly; "D" Cmx; Cly test sulf.w.
11.	Pan-American #1 Porcupine Unit	30-22S-3E	'63	Kpr	Kmt	Kmt	6308	P&A	"○"-"◇"-"D" Kbh, Ksp, Kmf; Kmf test SWCM. ?f@5510
12.	Phillips #1 USA "D"	20-22S-3E	'73	Kbh	Pt	Pt	14140	P&A	"○" Kbh, Jca; "○"-"◇" Trmv; "◇" Pk; "D" Pwr; Jca?test FW.
13.	Standard Cal., #1 Sigurd U.	32-22S-1W	'57	Jra	Jrn	Jrn	9640	P&A	"○" Jra, Jrn; ? 3MCF on test w/wtr @ 9309-9477 from Jrn?
14.	Shell #1 Sunset Canyon	21-22S-4W	'60	?Trs Dbb-Sbb			8962	P&A	"○"-"◇"-"D" Pk, Pt; "D" Mh; "●"-"◇" Md; "○"-"◇" Mga; "○" Mf, Dbb-Sbb; "D" Dpp; "○"-"●" Dbb-Sbb.
15.	Skelly #1 Emery	34-22S-5E	'62	Kmb	Du	Du	10740	P&A	"○" Kmf, Pk, M"d"; "D" Pc, Pec, M"a"; Pci & M"d"-M"c" Test CO^2 & wtr.
16.	Pan-American #3 Ferron Unit	21-20S-7E	'64	Kmb	€t	€t	10022	IP 43BO 34BWPD	"●" Kmf; "○" Jca, Pk; "●"-"○" Pcm, Pk "○" Pec.
17.	Phillips USA "E"	27-19S-3E	'75	Tf	€t	€t	20450	P&A	"○"-"D" Ksp; "●"-"◇"-"○" Kmb; "○"-"◇" Kmf; "◇"-"D" Kmt; "◇"-"D" Jca; "D" Trmm, Trms, Pt, Pec; "○"-"◇" Du; "D"-"○" €mx: Kmf test WCM; Attp'd test Pt failed; perfed Trms - recovered show oil and gas.
18.	Atlantic-Richfield #1 Hiawatha	13-15S-7E	'66	Kbh	Md	Md	15882	P&A	"D"-"○" Pk; "D" Pco, Md; Md test W.; Set pipe Attmp'd comp. Md.
19.	Tennessee Gas #1 Irons	16-15S-3E	'58	Tv	Jrm	Jrm	9995	P&A	"●"-"○" Tgr. "○" Kbh, Ksp, Kmf, Kd; Gas show Kmf tests rec. GIP, Mud w/tr oil; Tests of Kd rec GCW.
20.	Phillips #1 Price "N"	29-15S-3E	'72	Tgr	Jra	Jra	12332	P&A	"○" Tgr; "◇" Tf; "○" Kmf: Set pipe perf. Ferron, later plugged.
21.	Std. Oil of Calif. #1 Levan	17-15S-1E	'60	Jra	Trth	Trth	7526	P&A	"D" Jrn.
22.	Gulf #1 Alkali Canyon	32-6S-4W	'66	Tgr	Tw	Tw	5370	P&A	"◇" Tgr, Twt.
23.	Mountain Fuel #1 Thistle	7-9S-6E	'67	Jrtc	Trw	Trw	8207	P&A	"○" Jrtc; 50° dip @ 6341 fault?
24.	Feldman Oil #1 Diamond State	16-8S-5E	'55	Tra	Ppc	Ppc	5639	P&A	"○" Tra; "○" U. Ppc; "D" Mu, Ppc; tf @ 2954; Ppc test M & SW.
25.	Utah So. & Phillips #1 Hatch	28-6N-8E	'52	Tal	Jrn	Jrn	8837	P&A	KJr (Beckwith) test SW & WCM.
26.	Utah So. #1 Bar B	17-10N-7W	'52	Po	LM/UD	LM/UD	7918	P&A	"○"-"●" Tsl (basalt) "○"-Pm test rec OCM & OCMW.
27.	Promotory Oil #1 Jensen	7-13N-4W	'51	Po	Po	Mu?	7320	P&A	"○"-"◇"-"D" Po; "◇" Mu; Po Set pipe rec. wtr.
28.	Gulf #1 Adams	18-11N-5W	'63	Tsl	Dj	Dj	8967	P&A	"●"-"◇"-"○" Dj; Dj test GCW
29.	American Quasar-Energetics #1 Newton Sheep Company	4-2N-7E	'75	Tk	Kbr?	Trth	14440	IPF 550BO 250MCF 160BW-Jrn	"◇"-"●" Jrn-pay; Jrtc "○"-"◇"; Jrs "◇"; tf@ 12,990. Pineview field Nugget Disc.
	American Quasar, et al. #3-3 Union Pacific	3-2N-7E	'76	Kk	Jrn	Jrn	10400	IPF Jrtc-2392 BOPD 3, 153MCFG/D IPF Jrn-2360 BOPD 2,798MCFG/D	"●" Jrtc "●" Jrn
		Wyoming							
30.	Big Piney #3 UPRR	27-14N-119W	'54	Kf	Kd	Kd	5594	P&A	"●" Kf; Bailed 15-20 gals. oil-Kf.
31.	Amoco #1 Amoco Gulf Yellow Creek	2-14N-121W	'76	Tw	Jrn	Jrn	9063	IPF 2,750 MCFG/D Jrtc	"◇"-"●" Jrtc; "○"-"◇" Jrn. Yellow Crk. field Twin Creek Discovery

OIL SHOWS — CORDILLERAN HINGELINE

Map No.	Well Name	Location	Date Comp.	Spud Fm.	T.D. Fm.	Old. Fm. Tested	T.D.	Status	Remarks and Shows
32.	Doheney #1 Govt	9-16N-116W	'50	Tw	Kmv?	Pw	6289	P&A	"◇" Pw; Pw test GCM. tf@4465 & 5860.
33.	Shell #3 Leroy Unit	33-16N-117W	'51	Tw	Pw	Pw	4450	P&A	"◇" Trd; Trd test mud with tr oil
34.	Slosson #1 Cole	2-16N-121W	'55	Tw	Jp	Jp	5012	P&A	"O" Kbr; Kbr test show gas
	Wasatch #1 Govt.	26-16N-121W	'54	Tp	KJb	KJb	5010	P&A	"O" Ka; Ka test GCM
35.	Calif Co. #1UPRR	7-17N-116W	'30	KJb	Trt	Trt	4815	P&A	"O" Tra; (Cable tools/Sli. Show G in Tra)
36.	Amoco #1 Champlin 224 Amoco A	19-17N-118W	'76	Tw	Kf-Kbr?		14600	IPF 522 BOPD 530 MCFG/D Jrn	"◇" Jrn; Jrn test gas & cond @ 3-5,000 MCF/D: tf@ 8456, 14226. "O"? Kf-Kbr? Ryckman Crk. field Nugget Discovery.
37.	Belco #30-2 Hams Fork	30-19N-115W	'60	Tgr	Kf	D	14970	P&A	"O" Trth; "D" Trw, Pp, **Pw**, **Pa**; tf@ 7410 & 9182
38.	Union-Carter #1 Govt.	30-20N-115W	'48	Tw	Pw	Pw	6155	P&A	"O"-"◇" Trth; "O" Trw, Pp; "●" Pp, **Pw**; tf@ 4395 & 5350
39.	Conoco #23-1 Hams Fork	23-20N-116W	'59	Tw	Pw	Pw	5990	P&A	"◇" Trw, Tra, **Pw**, "D" Trw; Tra test Mud with tr. oil
40.	Amerada #1 Fossil Unit	23-21N-117W	'59	Tw	Kf	Є	9980	P&A	Ob text GCW; tf@ 3410
	Palisade #1 Govt	23-21N-117W	'48	Tw	Kav	Mb	6005	P&A	tf@ 5720
41.	Amerada #1 Chicken Creek	30-22N-117W	'59	Tw	Mm	Mm	8000	P&A	"◇" Trd, Pp, Pa, Mm; "●" **Pa**, **Mm**; "D"-"◇" Pwb; tf @ 1328, 3005, 5190
42.	National Coop #1 Larson	33-24N-117W	'59	Tw	Jrp	Mm	5267	P&A	"D" Pa, tf @ 4335, 4684
43.	Phillips #A-1 Fort	18-25N-114W	'63	Jrtc	Jrn	Єp	17345	P&A	tf@ 5125, 7432, 10700, 13740.
44.	Pan American #1 Etcheverry	13-26N-120W	'62	Qal	D	D	8935	P&A	"O" Mm; "D" **Pm**?; tf@4634
45.	Carter #1 Meridian Ridge	10-26N-115W	'59	Jrtc	Ka	Єdc	14397	P&A	"◇" Jrn, Mm, Ob; "●" Mm; "D" Mm, Pw; tf@ 4623, 8650, 9845, 12,952.
46.	Mobil 42-31G Hogsback	31-27N-113W	'63	OЄu	Jrn	OЄu	10953	Oil	"◇" Kf, Km, Kd; "O" Jrn, KJg; IPF 73 BOPD & 213 BW Kd & Jrn; tf @ 960, 5583. Hogsback Field
47.	Mobil #22-19 Tip Top Unit	19-28-113W	'62	Ob	Єgv	Єgv	15435	Gas	"O" Kh, Kf, Kbr, Kd, KJg, Jrn, Mm; "◇" (CO$_2$) Pp, Pw, Mm & Ob; IPF 931 MCFG/D Kf; 1,085 MCFG/D Km; Tip Top Field.
48.	General Pet #71-11-G	11-28N-114W	'53	Tp	Jrn	Jrn	10037	Oil	"O" Kf, Km; "●" Jrn: IPF 450 BOPD Jrn
49.	Calco #1 Deadline Ridge	13-28N-115W	'58	Tp	KJb	Єdc	9574	P&A	"●" Mm; "◇" Mm, Du, Ob, Kav, Kf, Kbr. tf @ 3620; Kbr test GCM & W.
	Calco-Humble #1 Pinegrove	14-28N-115W	'68	Mm	Tra	Єdc	12420	P&A	"O" Kf; test show gas Kf; tf @ 3530.
	Mountain Fuel #1 Deadline	22-28N-115W	'54	Pw	Jrn	Єdc	12605	P&A	tf @ 2255, 4550.
	General Pet. #43-19 Govt.	19-29N-114W	'52	Tp	Pw	Pw	14720	P&A	"●" Jrn, Pp, Pw; "O" Km & Kd; Test show gas Kd, Jrn, Pp, Pw.
	Calco #1 Tierney	15-29N-115W	'48	Ka	Jrtc	Trth	9544	P&A	"◇" Kbr, KJr, Jrn, Jra, Trth, Kf, Ka, Kbr; tf @ 5458.
50.	Shell #1 Smith Fork Unit	16-30N-118W	'49	Qal	Pw	Pw	7630	P&A	"◇" Tra, Trth, Trw, Pp, **Pw**, **Pa**.
51.	Phillips #A-1 Hoback	13-33N-115W	'60	Qal	KJg	Tra	13267	P&A	"O"-"◇" Kbr; "O" Jrn, Kmv, Kh. tf @ 5200.
52.	Mobil #F-22-3G Camp Davis	3-38N-115W	'66	Dd	KJg	Єg	13336	P&A	"O" Kmv; test show gas Kmv; tf @ 1163, 2723, 7823.
	IDAHO								
53.	Norton #2 Arimo Valley	19-10S-37E	'50	Tsl	?	?	2700	P&A	"◇" at 1928, ? fm.
54.	Utah-Idaho #1 Jensen	15-16S-38E	'57	Tsl	Pw	Pw	5233	P&A	
55.	Willett Flying #1 Willett	36-15S-38E	'66	Tsl	?	?	4473	P&A	T.D. possibly in Miss.
56.	Calco #1 Sheep Creek Unit	34-15S-45E	'52	Tra	Pw	Pw	6768	P&A	
57.	Rocky Mtn., #1 Bear Lake	30-12S-46E	'54	Jrtc	Jrtc	Jrtc	5017	P&A	"◇"-"●" Jrtc
58.	Amerada #1 Wild Tract	28-10S-46E	'63	Tsl	Jrn	Jrn	4115	P&A	"O"? Jrtc (possible gas show)
59.	East Idaho Dev., #1 State	21-10S-43E	'56	Tsl	Tsl	Tsl	3920	P&A	"O" Tsl-gas show
60.	Frazier #1 Ellis	27-9S-42E	'56	Qal	Tu	Tu	3540	P&A	"O" Tu-gas show

Map No.	Well Name	Location	Date Comp.	Spud Fm.	T.D. Fm.	Old. Fm. Tested	T.D.	Status	Remarks and Shows
61.	Great Western #1 Tygee	34-7S-46E	'27	Ju	?	?	2641	P&A	
62.	Calco #1 Dry Valley Unit	32-7S-44E	'52	Mb	Mm	Mm	7868	P&A	
63.	Sun #1 Big Elk Unit	23-2S-44E	'50	Trth	Pw	Pw	5611	P&A	"◇" Pp, Pw; Pw test non-flam. gas.
	Pan American #2-A USA Weber	24-2S-44E	'64	Tr	Mb	Mb	9720	P&A	"O" Mb, test non-flam. gas; "●" Trd.
64.	Phillips #1 Dewey-Horseshoe	28-5N-44E	'53	Kf	Mu	Mu	12720	P&A	"◇" Jrtc, Pp, Pw; "O" Ku, Kf, Ka, Kbr, KJg, Jrtc, Jrn, Tra, Trth, Trd, Pp, Pw. tf @ 4616, 4842, 6230.
65.	Calco #1 Sorenson	33-3N-41E	'30	Jrn	Pw	Pw	3780	P&A	
66.	Allday #1 Govt	24-1N-44E	'66	Єu	Єu	Єu	5760	P&A	tf@1650, 4120.

BRIGHAM YOUNG'S OIL SPRING

Brigham Young's oil spring (or oil well, as it was called after it had been deepened to 57 feet total depth) is located two miles west of Hilliard in the NE NE quarter of Section 4, Township 13 North, Range 119 West, Uinta County, Wyoming.

It was known to Jim Bridger, who probably learned of it from the Indians. The first published account of the spring was in The Latter-Day Saints Emigrant's Guide, published in 1848, where it is described as

.... a "tar" or "oil spring" covering a surface of several (square) rods of ground. There is a wagon trail running within a short distance of it. It is situated in a small hollow, on the left of the wagon trail, at a point where the trail rises to a higher bench of land. When the oil can be obtained free from sand, it is useful to oil wagons. It gives a nice polish to gunstocks, and has been proved to be highly beneficial when applied to sores on horses, cattle, etc.

Capt. Howard Stansbury mentioned the oil spring in his report on the exploration of the Salt Lake region (1852), and it was noted by the geologist Henry Engelmann (1859). Brigham Young is said to have had a well dug here. Oil was skimmed off the water in the well and sold to emigrants or taken to Salt Lake City. A. C. Veatch (1907) says the production was "a few gallons a day."

A mile north of the Brigham Young oil spring in the SE NW quarter of Section 33, Township 14 North, Range 119 West, a second oil seep became known as the Judge C. M. White oil spring. Judge White dug a large pit here in 1867. Oil skimmed off the surface of water in the pit was sold to Salt Lake City tanners. Judge White drilled a 480-foot dry hole here in 1868.

The Carter oil spring is located eight miles to the northeast in the SW NE quarter of Section 31, Township 15 North, Range 118 West. It was mentioned by Fielding B. Meek of the Hayden Survey in 1871, and by Samuel F. Emmons of the King Survey in 1877. Shallow wells and pits dug here in 1868 yielded eight to ten gallons of heavy black oil per day. Three tests, ranging from 225 to 300 feet in depth, were drilled about 1886. Two made six barrels of oil a day, and the third was dry.

Another oil spring known from early exploration was one near Twin Creek three and one-half miles southeast of Fossil, located in the SE quarter of Section 14, Township 21 North, Range 117 West, Lincoln County, Wyoming. It was referred to, in vague terms, in an 1859 report by Col. F. W. Lander (for whom the town of Lander was named), who was checking possible wagon routes west of South Pass. The oil seep was responsible for early drilling in the Fossil area.

The generalization can be made that Brigham Young's oil spring, and others in southwestern Wyoming are related, in one way or another, to faults of the overthrust belt. This was known by A. C. Veatch, and reported by him in 1907.

— W. Lyle Dockery

RESERVOIR VARIATIONS AT UPPER VALLEY FIELD GARFIELD COUNTY, UTAH

by

George C. Sharp[1]

ABSTRACT

The Upper Valley field, Garfield County, Utah, is the only significant production found to date in the Kaiparowits basin of south-central Utah. Cumulative production to January, 1976 is 14.9 million bbls of oil from four distinct zones within the Triassic Timpoweap and Permian Kaibab formations. All of the reservoirs are carbonates which have been dolomitized to varying degrees and were deposited in environments ranging from supratidal to shallow marine. Significant facies variations and diagenetic alterations are present in all four reservoir zones and have affected distribution of porosity, permeability and fracturing. Production has been offset along the western flank and down the southern plunge of the Upper Valley anticline by a hydrodynamic drive that appears to have created a curvilinear oil-water contact due to variation in the densities of the crude within the field. The accumulation has been further complicated by an apparent lack of hydrocarbon charge along certain portions of the structure in the main pay zone.

INTRODUCTION

The Upper Valley oil field is located in Townships 36 and 37 South, Ranges 1 and 2 East, Garfield County, Utah. It is situated in the western portion of the Colorado Plateau structural province and is the only significant production found to date in the Kaiparowits basin of south-central Utah (Fig. 1). The field was discovered in 1964 by Tenneco Oil Company and produces from dolomitized carbonates in the Triassic Timpoweap and Permian Kaibab formations at depths between 6350 ft and 7630 ft. Significant variations in porosity, permeability, and fracturing occur within the reservoirs and have affected productivity. The accumulation is localized on the Upper Valley anticline, but has been offset along the western flank and down the southern plunge of the structure by a hydrodynamic drive.

The objectives of this study are: 1 — describe the Timpoweap and Kaibab rock sequences and interpret their depositional environments, 2 — identify individual reservoir rock types and describe the distribution of porosity, permeability, and fracturing, and 3 — hypothesize on additional influences to production, namely hydrodynamics and limited oil charge. The study included detailed core examination of the Timpoweap and Kaibab formations from eleven wells (representing approximately 800 ft) within or in close proximity to the field. Each pay zone was cored in at least two wells and the distribution of control is shown in Figure 2. It is appropriate to mention here that communication of oil and water between zones has been indicated behind casing in a number of producing and injecting wells. This communication made it difficult in some cases to compare geologic observations in individual zones with production history or injection response.

DEVELOPMENT HISTORY

The first well on the Upper Valley structure was drilled in 1947 and 1948 by the California Company (Calco, No. 1, NW/4, Sec. 12, T36S, R1E) to 8857 ft in the Mississippian. This location was on the crest of the anticline and encountered water and mud with oil and gas shows from numerous tests of

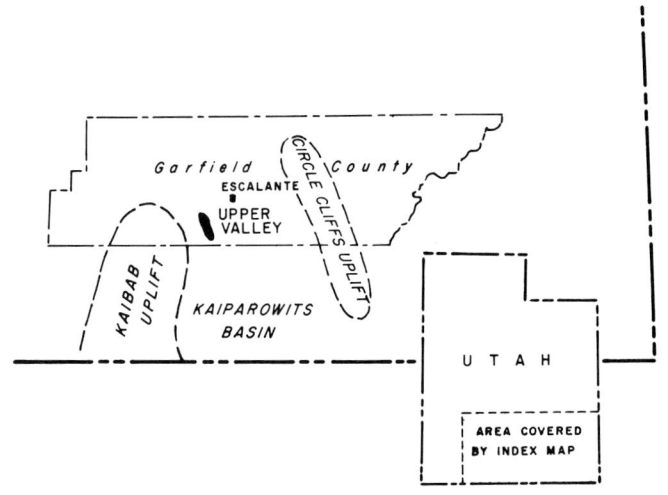

Fig. 1 — Index map of Southeastern Utah showing location of Upper Valley field.

---

[1] Tenneco Oil Company, Denver, Colorado.

The writer wishes to express his appreciation to Tenneco Oil Company for permission to publish this paper. Grateful acknowledgement and recognition is given to a number of staff in the Production and Exploration Departments who contributed significantly to the knowledge of the field and who derived most of the engineering data used in this paper. Many thanks are also owed to Richard Louden for critiquing the manuscript and to the Tenneco Drafting Department for preparing the illustrations.

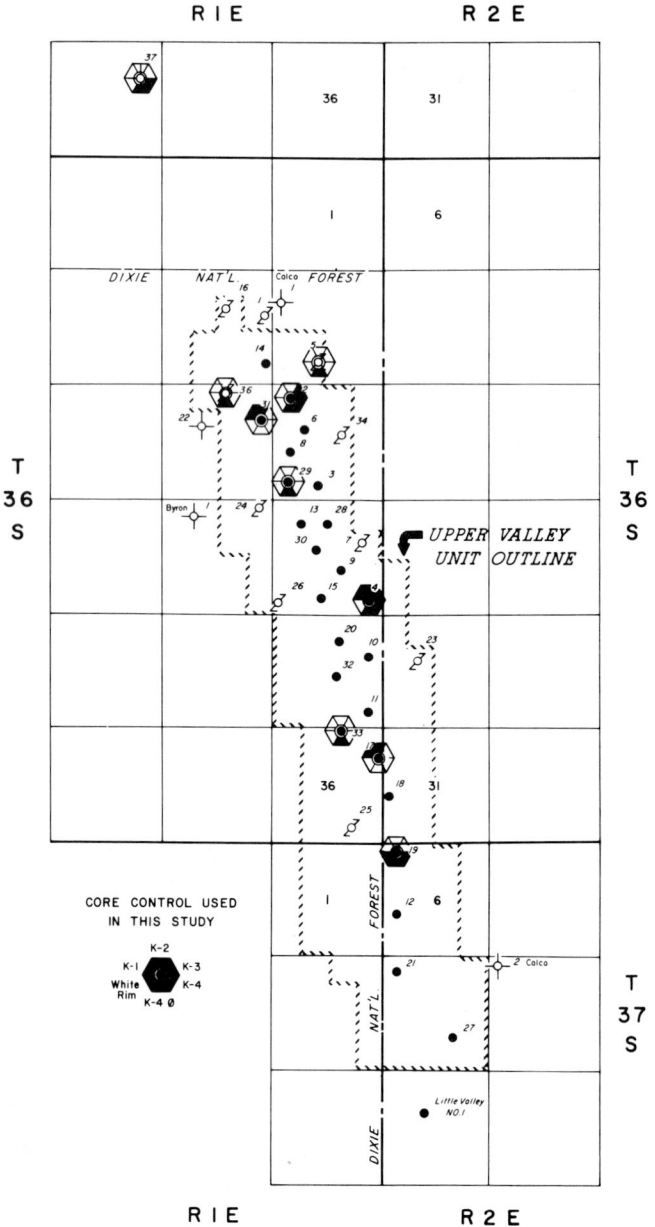

Fig. 2 — Core control in Upper Valley field used in this study.

the Permian Kaibab, White Rim, Toroweap, and Elephant Canyon formations, and the Mississippian Redwall Formation. The well was eventually completed in the Mississippian and produced approximately 17,000 bbl oil (17° API) with a high water cut before being abandoned as noncommercial. Calco deepened the well in 1951 to 10,120 ft in the Cambrian, but was unsuccessful in establishing production. A second test was drilled by the California Company in 1952 (Calco, No. 2, NW/4, Sec. 8, T37S, R2E) six mi downdip of their first well and slightly east of the axial trace of the anticline. This dry hole was drilled to a total depth of 7114 ft in the White Rim Formation and tested gas-cut mud and water from the Timpoweap, Kaibab and White Rim formations.

The third well to test the structure was Tenneco Oil Company Upper Valley No. 1 (NE/4 Sec. 11, T36S, R1E) drilled in 1962 on a farm-out from Sun Oil Company. This location was less than 1000 ft southwest of Calco No. 1 and was proposed as a Mississippian test until a prolonged fishing job terminated operations at 7931 ft in the Permian Elephant Canyon Formation. Casing was subsequently set through the Cedar Mesa Formation and completion was attempted in the Cedar Mesa and Kaibab formations. Although oil was recovered from swab tests of the Kaibab, the well was temporarily abandoned because of high water cuts.

Following a detailed geologic and engineering analysis of the first three dry holes, Tenneco persisted in their belief that a substantial accumulation existed on Upper Valley anticline and drilled the fourth well on the structure in 1963 (TOC, Upper Valley No. 2, NW/4, Sec. 13, T36S, R1E) about 3/4 mi south of the TOC, No. 1 test. This well reached a total depth of 9424 ft in the Mississippian and was plugged back to 6775 ft in the Kaibab. Cores and tests of the Timpoweap/Kaibab section indicated the presence of commercial hydrocarbons, and the well was completed in 1964 as the discovery well for Upper Valley field with an IPP of 300 bbl oil/day.

Tenneco, as operator of the Upper Valley Unit, has completed 33 wells within the unit and 4 wells outside the present unit boundaries. Of these wells, 25 have been completed in the Timpoweap and Kaibab formations as oil wells, 10 as water injection wells, and 2 as dry holes. A development program is continuing in the southern end of the field and the TOC Trap Canyon No. 1 (NE/4, Sec. 19, T37S, R2E) is undergoing completion attempts at the time of this writing. Cumulative production to 2 Jan. 1976 is 14.9 million bbl oil and current daily production is 3,027 bbl oil/day. Pressure maintenance through peripheral water injection in the northern portion of the field was initiated in March, 1969.

STRUCTURE

The Upper Valley anticline is a doubly-plunging structure whose axis can be traced at the surface as far as 50 mi (Campbell, 1969). The anticline is one of several major northwest-trending structural folds in the Kaiparowits basin between the Kaibab uplift to the west and the Circle Cliffs uplift to the east (Fig. 1). In the vicinity of the field, the structure is strongly asymmetric (Fig. 3) with surface dips on the western flank ranging up to 40° and dips on the eastern flank generally less than 10°. It is bounded to the west by the Table Cliff syncline, and to the east by the Alvey syncline. The steep western flank of the anticline, which locally may be faulted at depth, has been termed the Dutton monocline by Zeller (1973a, 1973b) and Bowers (1973) — (Upper Valley monocline of Kelly, 1955). Subsurface control and seismic data suggest the structural axis at Kaibab level coincides closely with the trace of the axis in

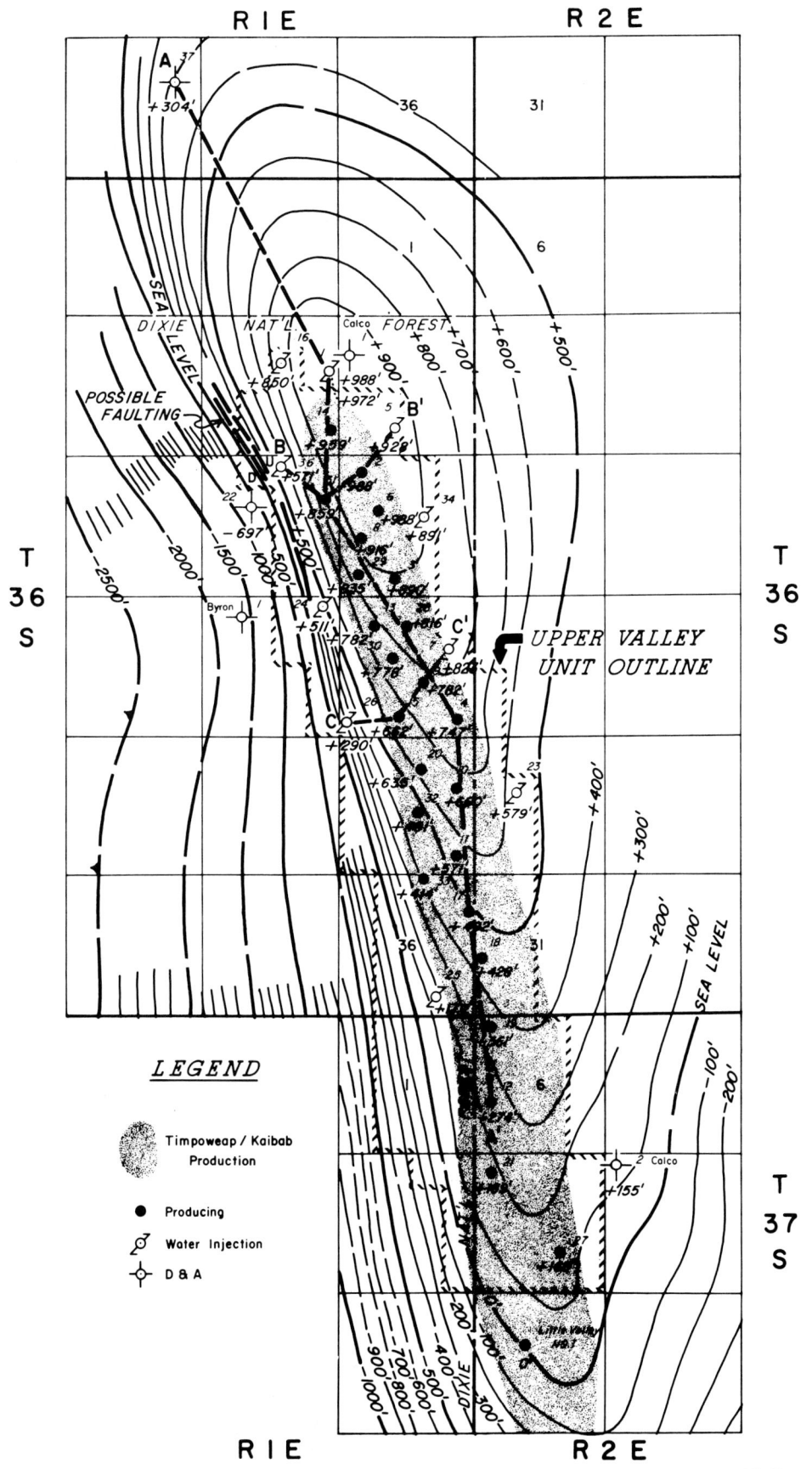

Fig. 3 — Structure map contoured on top of K-4ϕ zone (Kaibab Formation), Upper Valley field. Contour interval — 100 ft.

Upper Cretaceous sediments exposed at the surface. The amount of structural closure on the Kaibab is unknown due to lack of control on the eastern flank, but is probably equivalent to the 850 ft of closure expressed in the Cretaceous at the surface. Production is offset along the western flank and down the southern plunge of the structure and is presently seven miles in length. Structural relief between the highest and lowest producers to date is approximately 1000 ft. Age of the structure is postulated to be Laramide based on field work by Kelly (1955), Zeller (1973a, 1973b), and Bowers (1973).

STRATIGRAPHY

Upper Valley field produces from four separate and distinct reservoirs that are correlatable within the limits of production. These pay zones are informally numbered K-1, K-2, K-3, and K-4 in descending order on the basis of gamma ray characteristics. The main pay in the field is a porous unit within the K-4 zone and is described separately as the K-4ϕ zone. Although precise age determinations of the producing units was not the purpose of this project, regional correlations suggest that the K-1 and K-2 zones are Triassic in age and correlative to the Timpoweap Member of the Moenkopi Formation, whereas the K-3, K-4 and K-4ϕ zones are Permian in age and correlative to the Kaibab Formation. McKee (1938) summarized and subdivided the Kaibab Limestone regionally into the Alpha, Beta and Gamma Members. Campbell (1969) concluded that the principal Kaibab production at Upper Valley was from the Beta Member and the K-3 zone of this report appears to be correlative in part with his "Beta Marker".

All of the reservoirs at Upper Valley are carbonates and have been dolomitized to varying degrees. The nature and distribution of porosity and permeability in each reservoir was of prime importance in this investigation and in many cases appeared to be strongly influenced by secondary diagenesis. In order to emphasize the causes and timing of these post-depositional diagenetic events, the zones have been described in descending sequence from youngest to oldest. Porosity cutoffs and averages for each zone were derived from relationships between cross-plots (computer-calculated) of all available porosity logs and core data (porosity, permeability and capillary pressures). Cross-plots utilizing all porosity logs are necessary because calculations from individual porosity logs were sometimes inconsistent in the same well. Upper Valley No. 4 had the most continuous core through all the zones and is considered for correlation purposes as the type log for the field (Fig. 4). A chart summarizing observations and data for individual Timpoweap and Kaibab zones is included as Figure 5. Facies relationships across the field are illustrated by cross section A-A' (Fig. 6).

K-1 Zone

The K-1 zone is the upper portion of the Triassic Timpoweap Formation and has a gradational contact with the overlying Lower Red Member of the Moenkopi Formation and a sharp unconformable contact with the underlying K-2 zone. The unit has a relatively uniform thickness over the field area that ranges from 45 ft to 62 ft. Core control was provided by wells No. 4, 17 and 31.

Lithology and stratigraphic relations. — Three dominant rock types exist in the K-1 unit: laminated dolomite wackestone, oolitic dolomite grainstone, and bioturbated dolomite wackestone.

The laminated dolomite wackestone facies is characterized by thin, even, horizontal laminations, and low-angle cross laminations which are produced by concentrations of silt and fine-size quartz grains. Anhydrite nodules and rosettes up to 1 cm in diameter are abundant and the facies is occasionally argillaceous. A very-fine crystalline dolomite matrix obscures most original carbonate grains, although faint pellets, rarely glauconitic, and occasional shell fragments were observed. Desiccation features and signs of bioturbation are absent. This facies is generally light colored except when argillaceous or oil stained.

Oolitic dolomite grainstones and packstones are interbedded with the laminated wackestones. These grain-supported beds range from one to 15 ft in thickness with an average of five ft and generally have sharp upper and lower contacts. Bedding within this facies is generally indistinct, although horizontal and low- to medium-angle laminations were observed. Oolites are the dominant constituent and are spherical, well-sorted, fine- to medium-grained, and rarely glauconitic. Most of the oolites have a single outer coating and an interior that is structureless or contains opaque organic debris; a few display faint shadows of concentric rings. This facies is usually dark brown in color due to heavy oil staining.

Bioturbated dolomite wackestones are a persistent rock type at the base of the K-1 and have a gradational contact with the overlying wackestones and grainstones. This facies is characterized by its high degree of bioturbation with distinct vertical and horizontal burrows up to 1.5 cm in diameter. It is very argillaceous and silty and is associated with a high radioactive gamma-ray response. Fossil fragments and small anhydrite nodules are common. The bioturbated facies exhibits a relatively consistent thickness of approximately 20 ft throughout the field area.

Depositional environment. — The bioturbated dolomite wackestones at the base of the K-1 were deposited as lime muds in a restricted subtidal (lagoonal?) environment. This lagoonal (?) environment grades upward into a shallow, nearshore marine to intertidal environment for the interbedded laminated wackestones and oolitic grainstones. A current-dominated, saline environment for the laminated wackestone facies is suggested by the horizontal and low-angle laminations, abundance of anhydrite, scarcity of fauna, and lack of

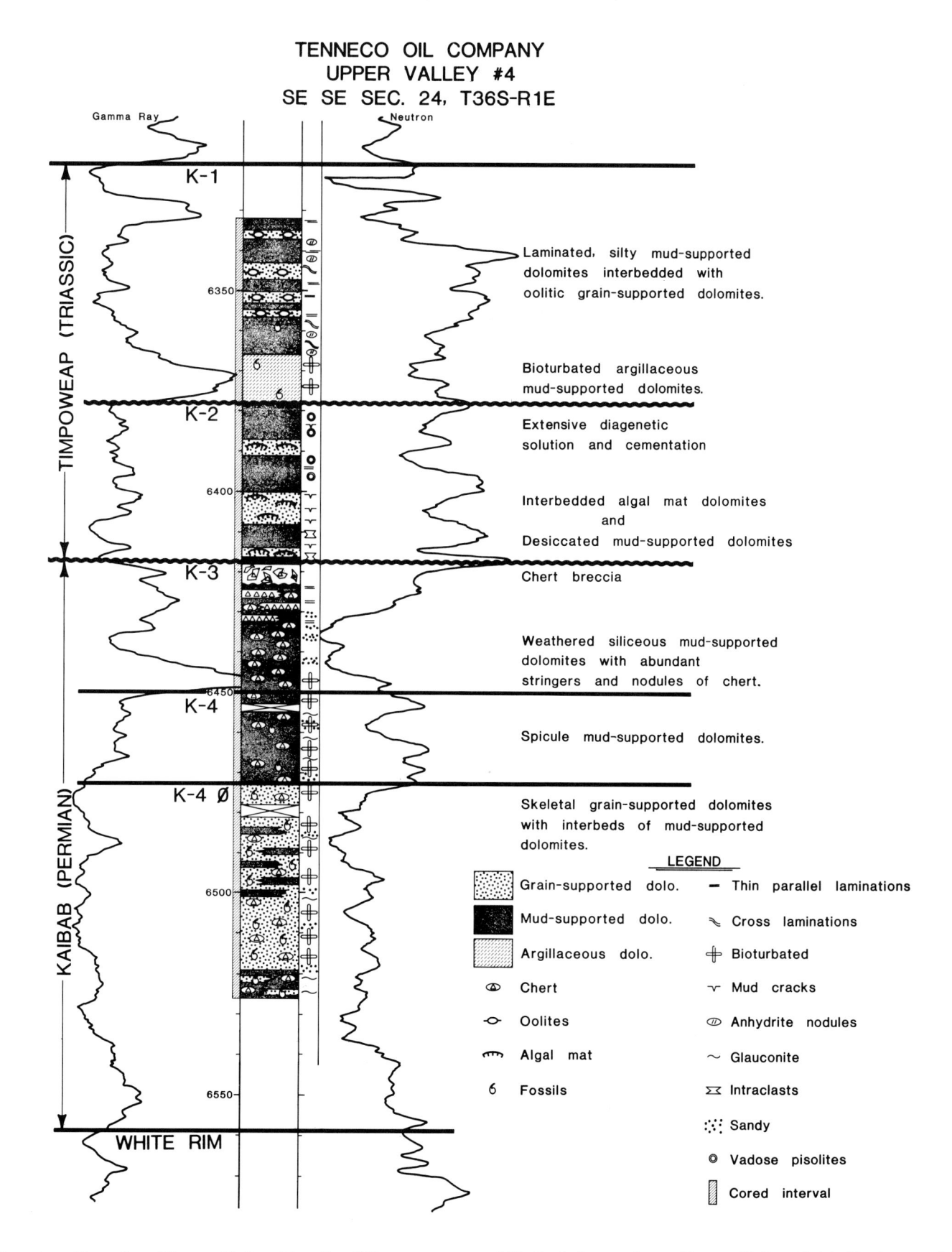

Fig. 4 — Type log (Upper Valley No. 4) of Timpoweap and Kaibab formations, Upper Valley field. Facies and sedimentary features observed from slabbed core.

330 GEORGE C. SHARP

ZONES	WELLS WITH CORE CONTROL	FACIES	DEPOSITIONAL ENVIRONMENT	RESERVOIR ROCK TYPE	POROSITY TYPE	RESERVOIR DATA	RESERVOIR PROBLEMS	FRACTURES	CONTINUITY OF RESERVOIR
K-1 (TIMPOWEAP, TRIASSIC)	UPPER VALLEY NO'S. 4,17,31	Laminated, silty mud-supported dolomites interbedded with oolitic grain-supported dolomites. Grades to bioturbated argillaceous mud-supported dolomites at the base	Nearshore Marine or Intertidal ↔ Restricted Marine (lagoon)	Oolitic grain-supported dolomites	Interoolitic porosity. Minor fractures	Weighted porosity average: 6.5% Log cutoff for gross pay determinations: 4.8% ∅ Core Data Max ∅: 7.5% Max K: 2.5 md.	Porosity reduction within reservoir due to: a) blocky dolomite spar cement b) anhydrite plugging	Moderately fractured (tectonic) Basal 10-15' (bioturbated argillaceous wackestone) generally not fractured	Lateral Continuity: Poor Vertical Continuity: Poor Basal 10-15' may act as a vertical barrier to oil
		EXPOSURE—SURFACE							
K-2	UPPER VALLEY NO'S. 4,17,31	1) Interbedded sequence of algal mat dolomites and desiccated mud-supported dolomites 2) Skeletal grain-supported dolomites to west 3) Local argillaceous mud-supported dolomites. Diagenetic solution and cementation superimposed over all primary facies	West ← → East Open Restricted Supra-tidal and Shallow Shallow inter-Marine Marine tidal	1) Algal mat dolomites 2) Skeletal grain-supported dolomites 3) Calichefied diagenetic dolomites resulting from subaerial exposure	1) Fenestral and vuggy in algal mats 2) Intercrystalline, vuggy and leached fossil-moldic in skeletal dolomites 3) Solution vugs, channel and "earthy" in diagenetically altered dolomites 4) Fracture	Weighted porosity average: 7.2% Log cutoff for gross pay determinations: 4% ∅ Core Data Algal Mats Max ∅: 8.9% Max K: 60 md. Skeletal Dolomites Max ∅: 13.2% Max K: 141 md.	Porosity in algal mats reduced by: a) dolomite druse cement b) later blocky dolomite spar cement c) anhydrite plugging Vadose cementation destroyed some solution porosity	Extensively fractured (Tectonic, solution-collapse brecciation and desiccation)	Lateral Continuity: Poor in algal mat dolomites Fair in skeletal dolomites Vertical Continuity: Fair-Good due to fractures
		EXPOSURE—SURFACE							
K-3	UPPER VALLEY NO'S. 2,4	Siliceous mud-supported dolomites with abundant stringers and nodules of chert. Reworked chert breccia at top. Diagenetic weathering and silicification superimposed over primary facies	Restricted Marine	Silicified, earthy, diagenetic dolomite resulting from subaerial exposure	Solution voids Intercrystalline "Earthy" (highly altered) Fracture	Weighted porosity average: 16.3% Log cutoff for gross pay determinations: 13% ∅ Core Data Max ∅: 19.9% Max K: 162 md.	Extensive silicification destroyed some secondary porosity	Extensive fracturing primarily in cherts and siliceous dolomites (tectonic)	Lateral Continuity: Fair-Good Vertical Continuity: Fair-Good due to fractures
K-4 (KAIBAB, PERMIAN)	UPPER VALLEY NO'S. 2,4,19,37	Spicule mud-supported dolomites	Restricted Marine Open Subtidal Marine	No Reservoir	Minor intercrystalline porosity	No Reservoir	Unfavorable facies	Minor fractures (tectonic)	Poor Continuity May act as a vertical barrier to oil
K-4 ∅	UPPER VALLEY NO'S. 2,4,5,19,33,36,37 & SOUTH UPPER VALLEY NO. 1	Skeletal grain-supported dolomites. Becomes mud-supported and sandy at base. Extensive oil staining and dolomitization commonly obscures primary facies	Open Shallow Marine	Sucrosic skeletal grain-supported dolomites	Intercrystalline Vuggy (pin point vugs to 2 cm.) Leached fossil-moldic	Weighted porosity average: 18.2% Log cutoff for gross pay determinations: 13.8% ∅ Core Data Max ∅: 27.3% Max K: 281 md.	Reservoir quality reduced by a) Variations in amount of secondary leaching b) facies change to mud-supported dolomites c) anhydrite plugging	Minor fractures (tectonic)	Lateral Continuity: Good Vertical Continuity: Good

Fig. 5 — Summary chart of lithologic, environmental and reservoir characteristics in Timpoweap and Kaibab zones at Upper Valley field.

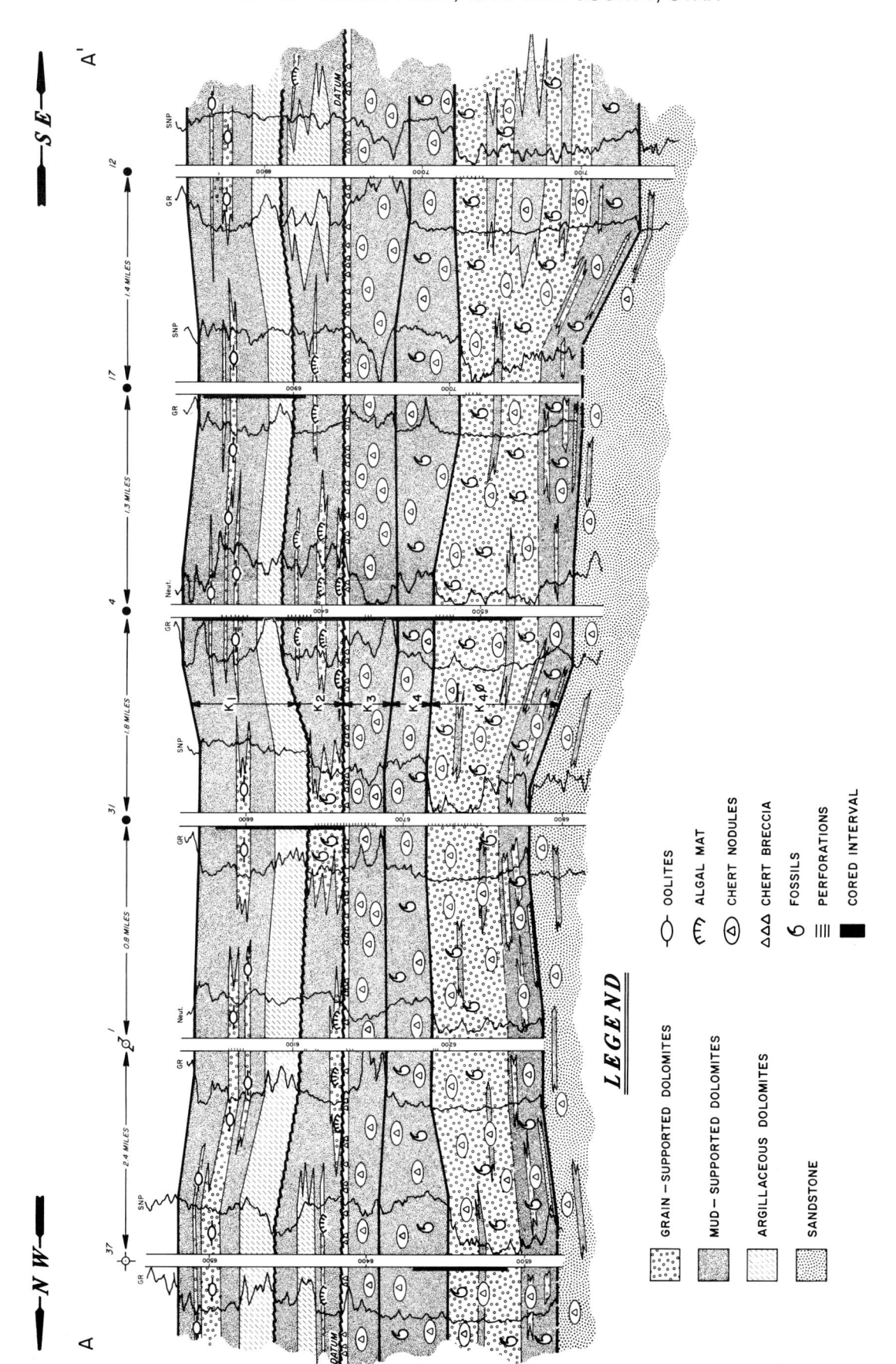

Fig. 6 — Stratigraphic cross section A-A' showing distribution of facies within Timpoweap and Kaibab reservoir zones, Upper Valley field. Datum is top K-3 zone (top Kaibab).

TOC, Upper Valley No. 4, depth 6344 ft, Timpoweap Fm. (K-1 zone). Photomicrograph of oolitic dolomite grainstone facies. Porosity (white spaces) is inter-oolitic. Note sharp upper contact with laminated dolomite wackestone facies.

bioturbation. The oolitic grainstones were most likely deposited as thin, discontinuous banks or bars on shoal areas or in near-shore marine channels.

Reservoir description. — The K-1 is the least important reservoir at Upper Valley in terms of productivity but has been perforated in 12 producing wells and 6 injection wells. The pay zone in this unit is associated with the oolitic dolomite grainstone facies. Porosity is inter-oolitic and has been enhanced by a moderate amount of fracturing. Maximum porosity and permeability measurements from core data is 7.5 percent and 0.3 md. Weighted porosity average for the pay section from calculated log data is 6.5 percent and a porosity cutoff for gross pay calculations from logs was determined to be 4.8 percent. A gross pay isopach using these cutoffs shows discontinuous porosity over the field, and large areas with no pay (Fig. 7).

Lateral and vertical continuity of the oolitic grainstone reservoirs within the field is relatively poor. This is attributed to: 1 — the discontinuous nature of the oolite facies (correlation of individual beds from well to well is difficult), and 2 — porosity destruction within the facies due to blocky dolomite spar cementation and/or anhydrite infill. Continuity within the reservoir has been enhanced somewhat by a moderate amount of fracturing. The bioturbated wackestone facies at the base of the K-1 may act as a vertical barrier between the K-1 and K-2, since it is widespread, tight, and relatively unfractured.

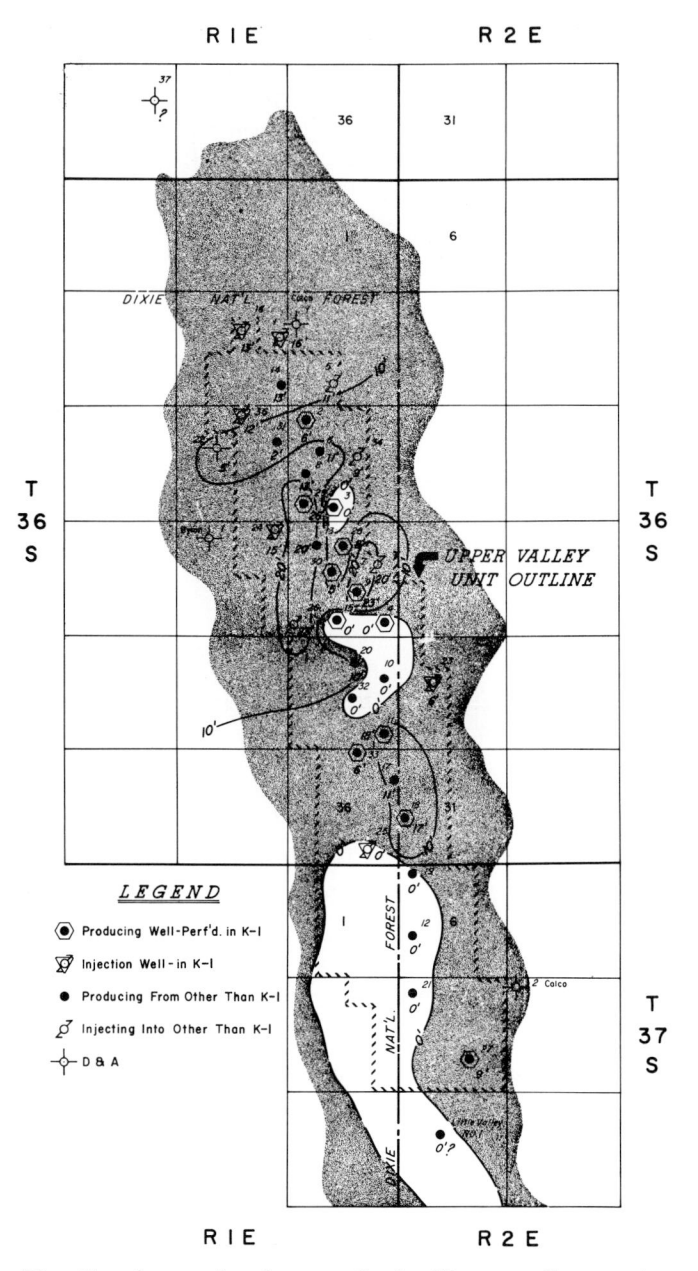

Fig. 7 — Isopach of gross feet with porosity greater than 4.8 percent in the K-1 zone (Timpoweap Formation), Upper Valley field. Contour interval is 10 ft.

K-2 Zone

The K-2 zone comprises the lower portion of the Triassic Timpoweap Formation. It has a sharp erosional contact with the overlying K-1 and an abrupt contact with the chert breccias of the underlying K-3 zone. The unit displays a generalized eastward thickening along the length of the field, and ranges from a minimum of 14 ft in No. 22 to a maximum of 52 ft in No. 27. The upper portions of the unit and its contact with the K-1 was cored by Nos. 17 and 31; the entire K-2 zone was cored by No. 4.

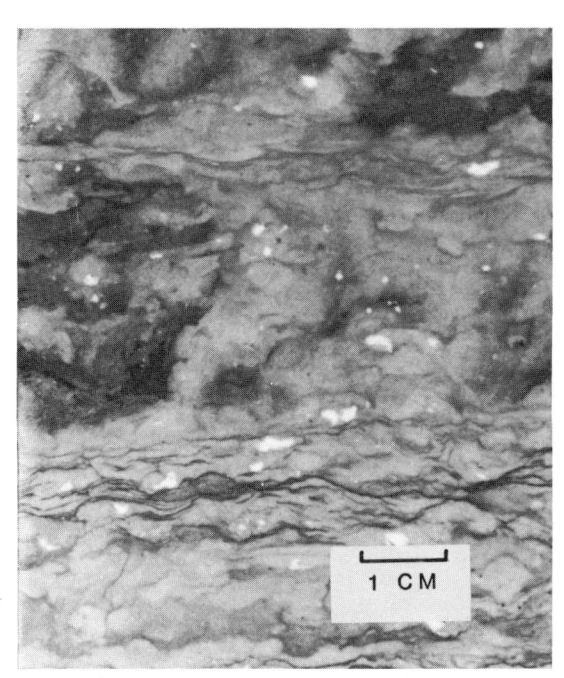

TOC, Upper Valley No. 4, depth 6367 ft, Timpoweap Fm. (K-1 zone). Polished core slab of bioturbated dolomite wackestone facies at base of K-1 zone.

Lithologies and stratigraphic relations. — A variety of rock types were observed within the K-2 unit. These include: a — skeletal dolomite packstones and grainstones, b — an interbedded sequence of dolomites ranging from laminated and desiccated mudstones to stromatolitic (algal mat) packstones, and c — an argillaceous mud-supported dolomite interpreted from gamma-ray log response.

The skeletal dolomite packstone and grainstone facies was cored in No. 31. These grain-supported dolomites are characterized by medium- to coarse-grain skeletal debris which include crinoid stems, algal plates, gastropods, and other unidentified fragments. Dolomitization is extensive and occasionally obscures original carbonate grains. Fine- to medium-size quartz grains are common and bedding is indistinct.

The skeletal dolomites grade eastward into a sequence of stromatolitic (algal mat) grain-supported dolomites interbedded with laminated and desiccated mud-supported dolomites. Irregular laminations, mud cracks and lithoclasts are common throughout the sequence. Fossil debris is rare, although gastropods were occasionally observed. The algal mat dolomites exhibit "birdseye" or fenestral fabric and are composed primarily of very fine- to coarse-grained, poorly sorted, spherical pellets.

Gamma-ray logs indicate the local presence of anomalously high radioactive portions within the K-2 zone. These radioactive sections were not cored, but on the basis of their stratigraphic position are interpreted to be argillaceous mud-supported dolomites or dolomitic shales.

All rock types in the K-2 have undergone extensive diagenetic alteration related to an unconformity at the top of the unit. Original textures and structures are often totally obscured by vadose solution and cementation. Dense, laminated, caliche-like cementation, solution-collapse brecciation features, and cavities filled with detrital carbonates, vadose silt, and caliche crusts are common throughout the interval. Spherical, poorly-sorted vadose pisolites up to 2.5 cm in diameter are present primarily in the upper portions of the unit. These pisolites are concentrically laminated and have either shell fragments or composite grains as nuclei.

Depositional environment. — The skeletal grain-supported dolomites were deposited as skeletal lime sands in an open, shallow marine environment. These skeletal sands could have accumulated in a number of marine subenvironments such as banks, bars, or channels and may not be as widespread as depicted in Figure 8. The skeletal facies grades eastward into the sequence of stromatolitic dolomites interbedded with laminated and desiccated mud-supported dolomites. This sequence was deposited in a high-intertidal to low-supratidal environment as suggested by desiccation features, irregular laminations, lithoclasts, lack of fossils, and presence of algal mats. The radioactive facies of the K-2 zone generally occurs between the supratidal-intertidal sequence and the shallow marine facies. The "shaley" appearance on logs is interpreted to represent an argillaceous mud-supported dolomite deposited in a restricted, perhaps lagoonal, marine environment. Its position is seaward of the supratidal-intertidal sequence and could have been created by local buildups of marine bars or banks of the skeletal grain-supported facies in front.

The interface between supratidal-intertidal deposition and marine deposition trends northwest-southeast through the field and is close to the present-day steep flank of the Upper Valley structure. It is not known whether this relationship is fortuitous, or whether the structure had an early expression which influenced deposition during K-2 time.

As previously mentioned, all of the rock types in the K-2 have undergone extensive diagenetic alteration, primarily from vadose solution and cementation. This is indicative of subaerial exposure and is related to an exposure surface recognized in cores at the top of the K-2 zone. This unconformity probably represents only a moderate hiatus between the K-2 and overlying bioturbated lagoonal facies of the lower K-1, but was of long enough duration to superimpose important diagenetic changes over the original sedimentary fabric.

Reservoir description. — The K-2 is the third most important producing zone at Upper Valley and has been perforated in 19 producing and 10 injection wells. Three rock types produce within the zone and include: a — skeletal grain-supported dolomites, b — stromatolitic (algal mat) dolomites,

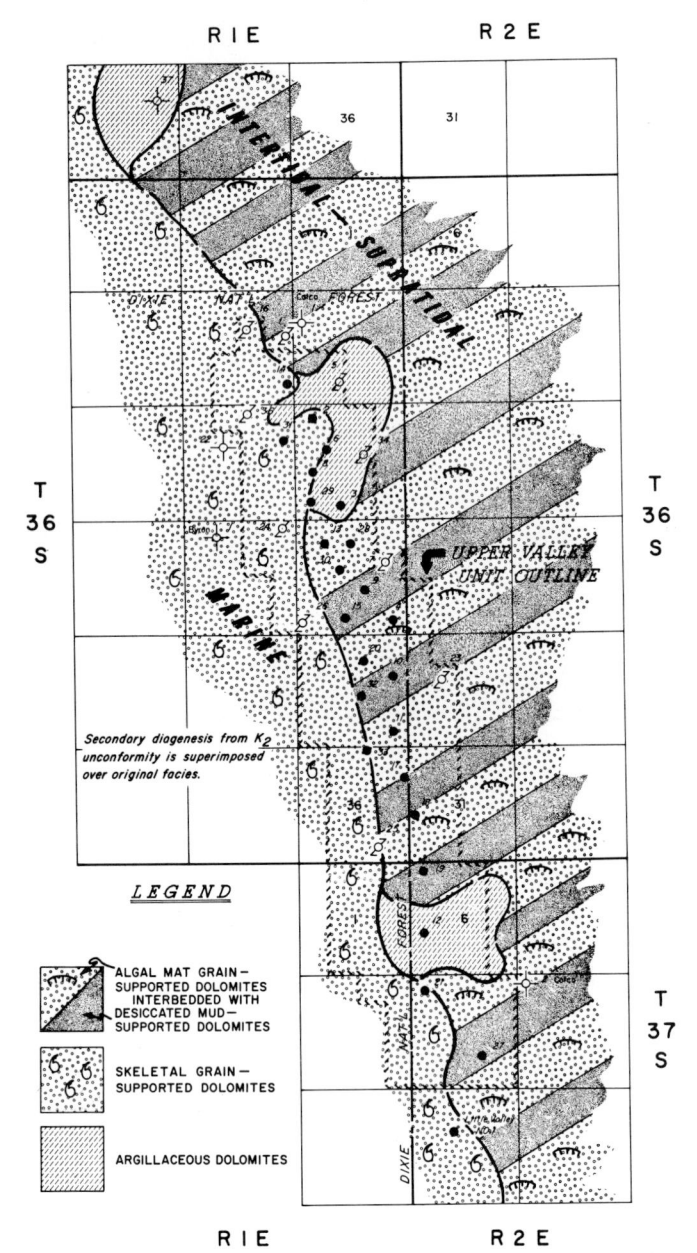

Fig. 8 — Facies distribution within K-2 zone (Timpoweap Formation), Upper Valley field.

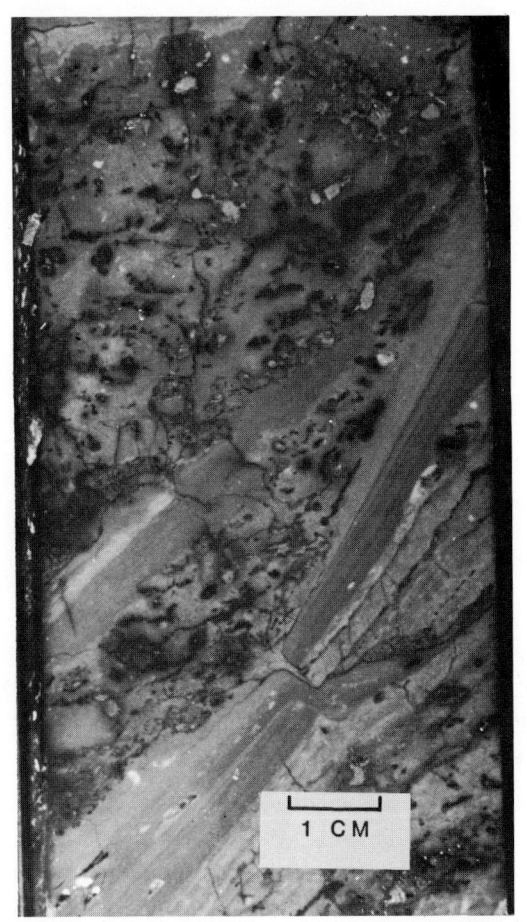

TOC, Upper Valley No. 4, depth 6383 ft, Timpoweap Fm. (K-2 zone). Polished core slab of diagenetic solution-collapse brecciation, upper K-2 zone. Solution cavity has been filled with detrital carbonates and vadose silt.

and c — calichefied, diagenetic dolomites. Porosity and permeability are variable and are dependent upon reservoir rock type and amount of superimposed diagenesis. Porosity type ranges from intercrystalline, vuggy, and leached fossil-moldic porosity in the skeletal grain-supported dolomites to fenestral and vuggy porosity in the algal mat dolomites to solution vugs, channel, and "earthy" porosity in the diagenetically altered dolomites.

Fracturing was extensive throughout the K-2 and contributes to the productivity of each rock type. Three varieties of fractures were observed in cores: 1 — desiccation fractures related to syndepositional exposure and drying, 2 — early diagenetic non-tectonic fractures related to solution-collapse brecciation, and 3 — later tectonic fractures related to structural bending of the Upper Valley structure. These later tectonic fractures appear to be more abundant in the K-2 than in the overlying K-1 and may reflect the more brittle nature of the calichefied dolomites.

Weighted porosity average from calculated log data for all reservoirs in the K-2 is 7.26 percent and porosity cutoff for gross pay calculations from logs was determined to be 4 percent. Maximum porosity and permeability measurements from core data are 13.2 percent and 141 md for the skeletal dolomites and 8.9 percent and 60 md for the algal mat dolomites. A gross pay isopach (Fig. 9) constructed from the above cutoffs shows pay thickness ranging from 0 to 41 ft. Comparison of this gross pay isopach with facies distribution (Fig. 8) suggests porosity is random and discontinuous in the intertidal-supratidal and lagoonal facies, whereas it is relatively continuous in the marine facies.

Examination of thin sections and slabbed core indicate that the diagenesis related to subaerial exposure at the top of the

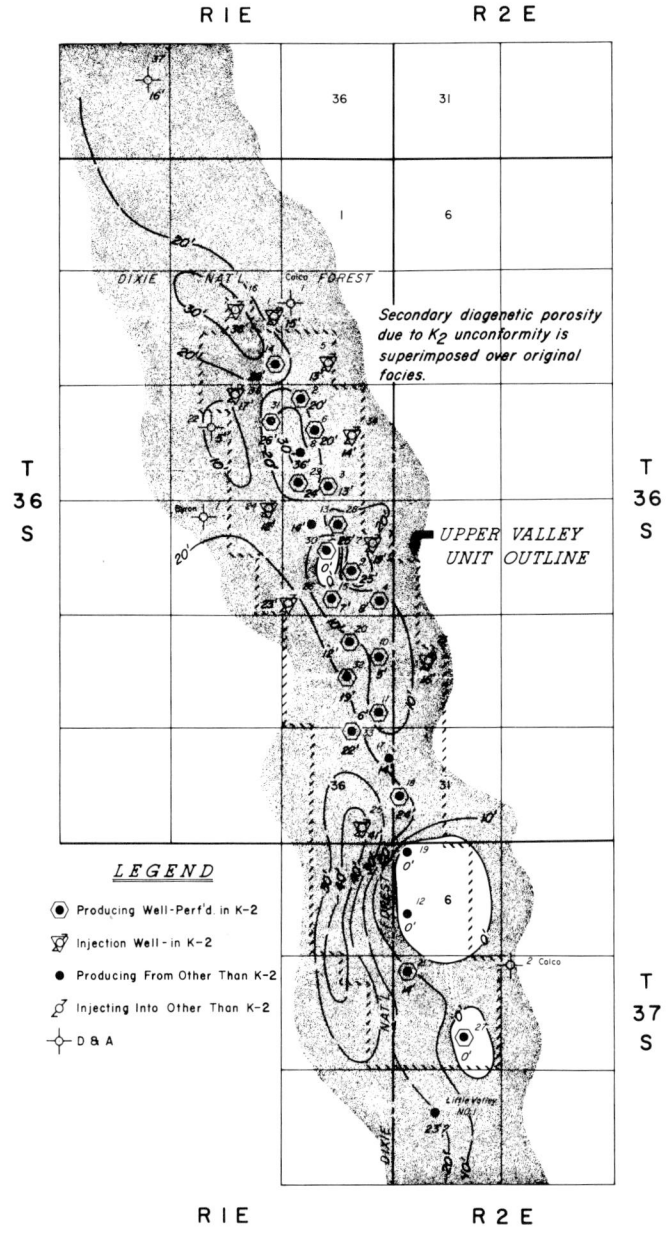

Fig. 9 — Isopach of gross feet with porosity greater than 4 percent in the K-2 zone (Timpoweap Formation), Upper Valley field. Contour interval is 10 ft.

K-2 not only enhanced and created porosity, but also destroyed it. This porosity destruction was accomplished primarily by vadose, caliche-like cementation which was superimposed to some degree over all reservoir rock types. Porosity was further reduced, particularly in the algal-mat facies, by dolomite druse cement and a later phase of dolomite "dogs tooth" spar cementation and anhydrite plugging.

Lateral continuity of the stromatolitic reservoirs is poor due to the discontinuous development of the algal-mats and variations in the amount of porosity plugging by dolomite spar or anhydrite. Lateral continuity of the skeletal grain-supported

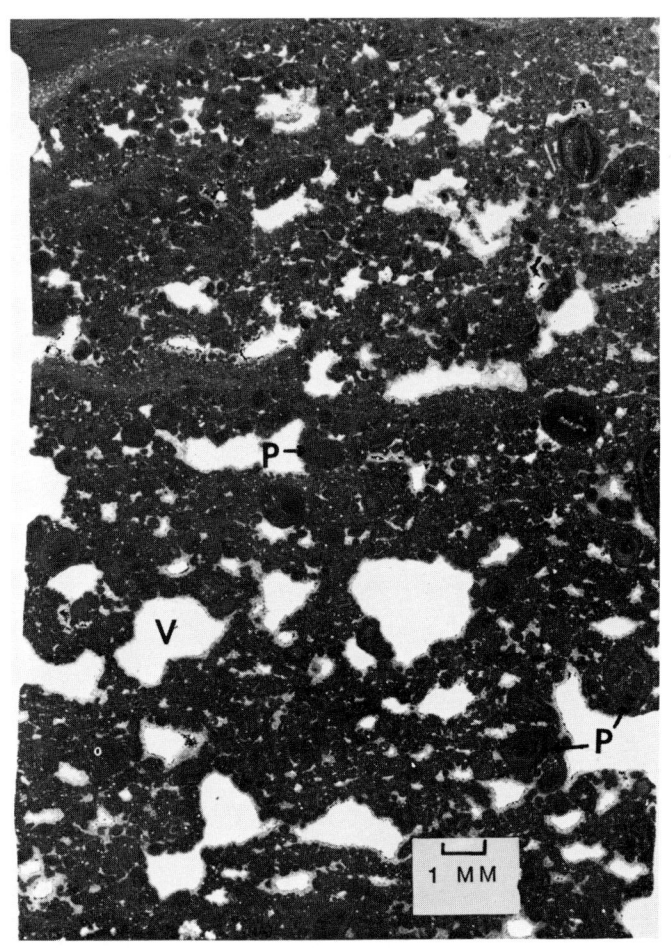

TOC, Upper Valley No. 4, depth 6392 ft, Timpoweap Fm. (K-2 zone). Photomicrograph of diagenetic dolomite with abundant vadose pisolites (P). Vugs (V) lined with dolomite druse cement.

dolomites, on the other hand, appears to be better than that of the algal mat because of the more widespread nature of the marine facies. Good vertical continuity exists within the K-2 reservoirs because of extensive fracturing.

K-3 Zone

The K-3 zone is the upper portion of the Permian Kaibab Formation in the Upper Valley area and is separated from the overlying K-2 zone of the Timpoweap Formation by a major unconformity. This unconformity is regional in extent and defines the boundary between the Permian and Triassic systems. The K-3 is distinctive on gamma-ray logs in the mapped area because of its high radioactive response. A general increase in thickness to the south from 23 ft in No. 7 to 47 ft in Little Valley No. 1 (Sec. 18, T37S, R2E) occurs within the field area. Core control is provided by wells No. 2 and No. 4.

Lithologies and stratigraphic relations. — The K-3 consists primarily of siliceous wackestones and mudstones with abundant stringers and nodules of chert. A thin, reworked chert breccia overlies these mud-supported dolomites and cherts.

TOC, Upper Valley No. 4, depth 6410 ft, Timpoweap Fm. (K-2 zone). Polished core slab of desiccated supratidal dolomite. Note the abundance of fractures.

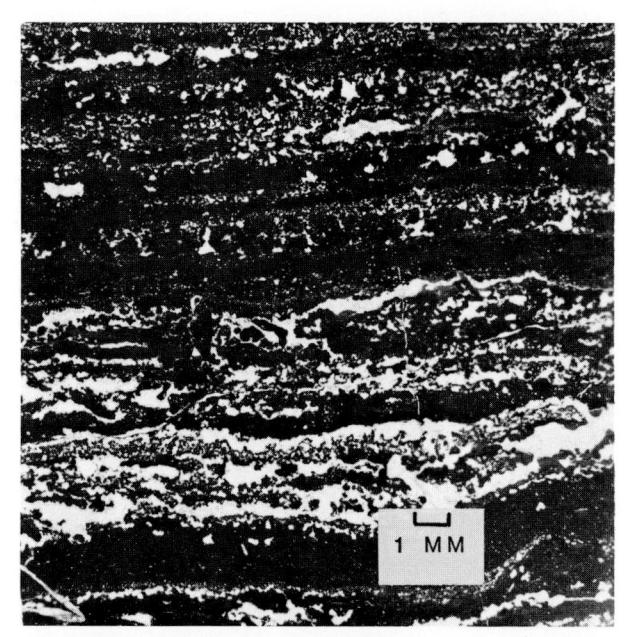

TOC, Upper Valley No. 4, depth 6415 ft, Timpoweap Fm. (K-2 zone). Photomicrograph of laminated stromatolitic (algal mat) dolomite. Fenestral porosity (white) partially filled with dolomite druse cement and anhydrite.

The chert breccia at the top of the unit is a reworked rubble of chert and dolomite from the underlying Kaibab Formation and was deposited on the Triassic-Permian unconformity surface. The breccia is most likely Triassic in age, but was included in the K-3 zone for isopach purposes because of difficulties in separating it on logs from the underlying lithologies. It was cored in well No. 4, where it was 4 ft thick and composed of unsorted, angular to sub-angular clasts of white and gray chert and a few clasts of mud-supported dolomite. The chert and dolomite clasts range up to 2.5 cm and show a general increase in size downward. Bedding is generally indistinct, although occasional vague bedding planes dip up to 15°. Matrix of the breccia is an earthy, pale yellow-brown dolomite mud.

Siliceous dolomite mudstones and wackestones are the dominant rock types in the K-3. These dolomites are very-fine crystalline, tan to brown in color, sandy, and occasionally argillaceous. Sedimentary structures include both thin laminations and burrows. Dolomitization and silicification of this facies is extensive, but shadows of pellets and shell debris, including sponge spicules, were observed. Although the dolomites appear to be predominantly mud-supported, these pellets and shell fragments occasionally form grain-supported lenses or patches.

Chert is abundant throughout the zone and is present in the form of layers, stringers, and nodules which are oriented parallel to bedding and are up to 3 cm in thickness. The chert ranges from light grey to bluish-grey in color, and usually has a thin outer white rim which is somewhat less siliceous. Shell fragments and relic laminae similar to those in adjacent dolomites are preserved in many nodules or layers and are suggestive of a replacement origin for the chert. Some laminated dolomites show minor draping around chert, but this is interpreted to represent later differential compaction over the hard nodules.

The dolomites of the K-3 have undergone extensive weathering and silicification as a result of the unconformity at the top of the unit. In some dolomites this diagenesis is so extreme that all primary depositional structures and textures are obscured. These highly altered zones have an "earthy" appearance and, although patchy and erratic in distribution, are usually in close proximity to chert nodules or layers.

Depositional environment. — The siliceous mud-supported dolomites of the K-3 are interpreted to have been deposited as lime muds in a restricted marine environment as suggested by the presence of pellets, skeletal fragments, argillaceous material, and burrowing. The thinly laminated beds indicate lack of organic activity and were probably deposited during periods of unfavorable water conditions. The K-3 was deposited during the final regressive phase of the Kaibab and

TOC, Upper Valley No. 4, depth 6436 ft, Kaibab Fm. (K-3 zone). Photomicrograph of siliceous mud-supported dolomite with intercrystalline and vuggy (V) porosity (19.9 percent). Siliceous patches are light colored; stained dolomite is dark colored.

was eventually exposed to subaerial conditions following total withdrawal of the Kaibab seas.

Origin of chert nodules in the K-3 and underlying zones appears to be related to subaerial exposure during the Kaibab unconformity. Chert was not observed above the Kaibab unconformity in the K-1 or K-2 zones, but is present immediately below the exposure surface and extends down through the K-3, K-4, K-4ϕ zones and into the White Rim sandstones. The abundance of chert generally decreases with depth beneath the unconformity. As noted above, chert nodules in the K-3 as well as the underlying zones appear to have a replacement origin as evidenced by the preservation of sedimentary structures, shell fragments and accessory minerals. These data suggest that the chert was formed by percolating ground waters (surface runoff or artesian upwelling) during subaerial exposure at the end of Kaibab time. The percolating ground waters and associated leaching and weathering may also be responsible in part for the high radioactivity on gamma-ray logs which is so characteristic of the K-3. Possible sources of silica for the chert include sponge spicules and/or detrital quartz in the K-3 or underlying K-4 dolomites.

Reservoir description. — The K-3 is the second most important producing reservoir at Upper Valley and has been perforated in 21 producing and 8 injection wells. The pay zone is associated with highly altered, "earthy" diagenetic dolomites. As noted above, these porous zones are usually in close proximity to chert and commonly form "halos" around chert nodules and beds. The porous "halos" are easily recognized in cores because they are so heavily stained. It is interpreted that while the percolating ground waters from the unconformity were forming the replacement chert, they were also leaching the matrix carbonate in the immediate vicinity of the nodules and beds. These highly altered leached zones were preserved as the porous "halos" of dolomite and are very productive. Porosity types in the dolomite include solution voids, intercrystalline porosity, and "earthy" porosity. Fracturing is extensive throughout the unit, although most fractures were observed to be associated with chert and/or highly siliceous dolomites. The fracturing appears to be related to tectonic folding of the Upper Valley structure, and was selective in that it affected the brittle cherts and siliceous dolomites.

The weighted porosity average for this reservoir from calculated log data is 16.3 percent and porosity cutoff from logs for gross pay determinations is 13 percent. Maximum porosity and permeability measurements from core data are 19.9 percent and 162 md. An isopach of gross pay using this porosity cutoff shows erratic distribution of pay ranging in thickness from 0 to 24 ft (Fig. 10). These local variations of pay thicknesses may be related to differences in amount of weathering along the unconformity or to areas in which extensive silicification actually reduced or destroyed porosity. Overall lateral and vertical continuity within K-3 is fair to good because of the widespread effects of the unconformity and high degree of fracturing.

K-4 Zone

The K-4 zone is a subdivision of the Kaibab Formation which is not productive at Upper Valley field. It has gradational contacts with the overlying K-3 and underlying K-4ϕ zone. Thickness within the field area is erratic, and ranges from 17 to 46 ft. Portions of the unit were cored by wells No. 2, 4, 19 and 37.

Lithologies and stratigraphic relations. — Spicule dolomite wackestone is the dominant rock type in the K-4. This mud-supported facies is well burrowed, light greenish-grey to cream in color, and has a matrix of microcrystalline to very-fine crystalline dolomite. Very fine-grain quartz sand is common and the unit is locally argillaceous. Fine- to coarse-grained glauconite pellets are locally abundant and give the rock a "grainy" look with a greenish coloration. Sponge spicules are common, and other identifiable skeletal debris includes fragments of crinoids, bryzoans, and brachiopod shells. The amount of fossil debris, particularly crinoids, generally increases toward the base of the unit.

Large ovoid chert nodules up to 10 cm in diameter are abundant throughout the K-4. These nodules are light grey to bluish-grey in color, and have a less siliceous, chalky white rim. Fossil fragments and glauconite pellets are preserved in the

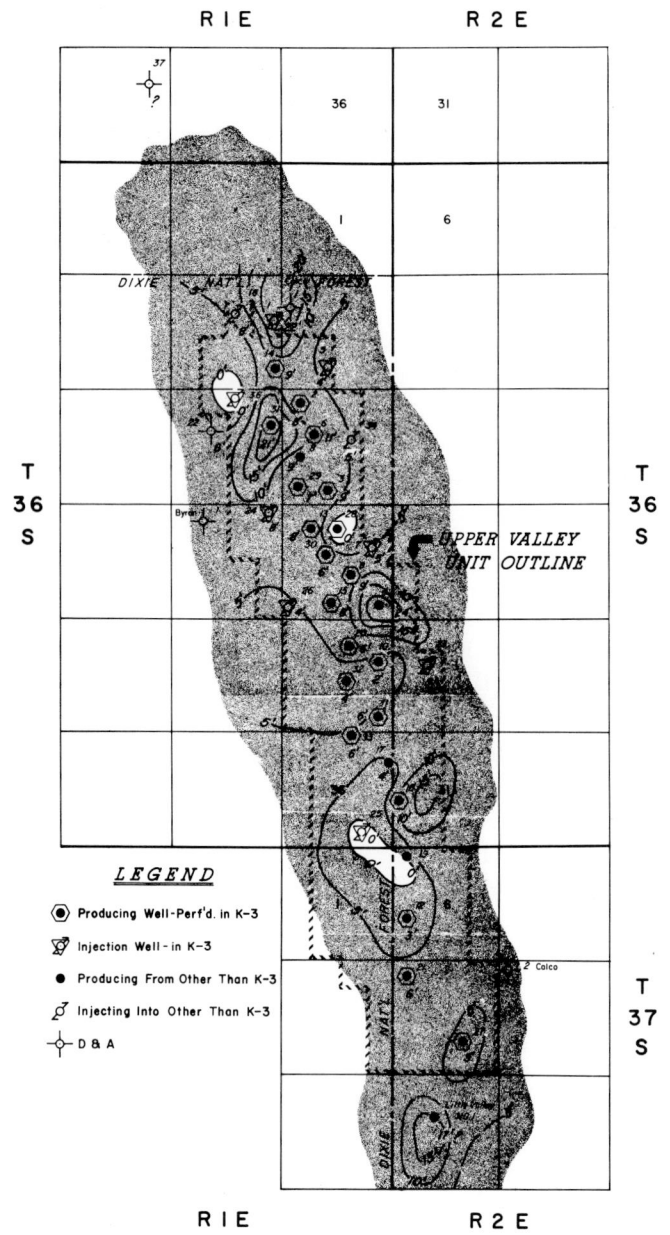

Fig. 10 — Isopach of gross feet with porosity greater than 13 percent in the K-3 zone (Kaibab Formation), Upper Valley field. Contour interval is 5 ft.

nodules. The cherts are similar to those described in the K-3, and are replacement nodules whose origin is related to subaerial exposure at the end of Kaibab deposition.

Depositional environment. — Fossil content, bioturbation, presence of glauconite, and stratigraphic relationships with adjacent facies suggest that the spicule dolomite wackestones were deposited as lime muds in an open marine, subtidal environment.

Reservoir description. — The K-4 has an unfavorable reservoir facies, and is not productive in Upper Valley field. The dolomite wackestones generally have low porosities and permeabilities with the exception of small local weathered patches surrounding chert nodules. Fractures are relatively uncommon in the K-4. The widespread nature of the unit coupled with its low porosities and permeabilities and general absence of fractures suggest that it may act as a barrier to oil between the K-3 and K-4ϕ reservoirs.

K-4ϕ Zone

The K-4ϕ zone is the lower portion of the Kaibab Formation in the Upper Valley area and is the most prolific reservoir in the field. It has gradational contacts with the nonproductive K-4 dolomites above and the porous sandstones of the White Rim Formation below. Its upper contact in the field is based on the first occurrence of significant porosity below the top of the K-4. The unit increases in thickness to the east and southeast and ranges from 62 ft to 116 ft. Portions of the K-4ϕ were cored by 9 wells and 8 were examined in this study.

Lithologies and stratigraphic relations. — The dominant rock types in the K-4ϕ are skeletal dolomite grainstones and packstones. The dolomites are sucrosic, fine- to coarse-grained, bioturbated, and locally glauconitic. Very fine- to fine-size quartz grains are common and increase in abundance downward toward the gradational contact with the White Rim sandstones. Porous zones, particularly in the upper portions of the unit, are heavily stained and range in color from black to medium brown. Unstained portions lower down in the unit range in color from light brown to cream with pinkish and greenish tinges. Dolomitization and staining is so intense over much of the upper portions that original composition and texture is obscured, even in thin-section. Vague outlines of poorly sorted, fine- to coarse-grained fossil fragments, however, suggest that the dolomite for the most part was originally a skeletal sand.

Interpretations regarding the nature of original facies in heavily dolomitized and stained portions is aided by observing composition of chert nodules. Ovoid replacement nodules up to 10 cm in diameter are abundant throughout the zone, and, as in the overlying K-3 and K-4 zones, have preserved the texture and composition of the original limestone prior to dolomitization. The nodules in the K-4ϕ are commonly composed of silicified fossil hash including bryzoans, crinoids, brachiopod shells, spines or spicules, gastropods, cup corals, and unidentified skeletal debris. Skeletal fragments that are well preserved in chert and extend beyond edges of nodules are usually unrecognizable in the sucrosic dolomite. The nodules appear to be less siliceous than those in the overlying K-3 and K-4, and have a chalky white color which stands out against the heavily stained, brown dolomite.

Although the bulk of the K-4ϕ zone appears to have been originally grain-supported, interbeds of mud-supported dolomites are common. These mud-supported beds are difficult to recognize in heavily stained and extensively dolomitized areas,

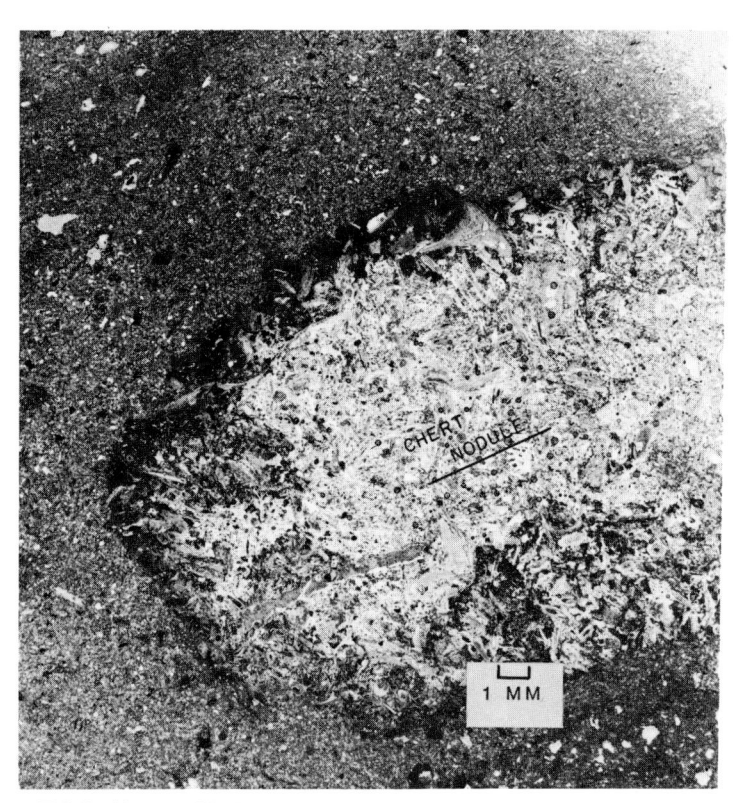

TOC, Upper Valley No. 4, depth 6506 ft, Kaibab Fm. (K-4ϕ zone). Photomicrograph of fossiliferous chert nodule in sucrosic, porous dolomite. Chert has preserved the original constituents (fossil hash) prior to dolomitization. Porosity in the dolomite (15.9 percent) is intercrystalline, vuggy and fossil-moldic.

but are characterized by smaller crystal size of dolomites, denser texture, greater amount of mud content preserved in adjacent chert nodules, and increased difficulty in recognizing even vague shadows or "ghosts" of fossil debris. These mud-supported beds are generally associated with a slight increase in radioactivity on gamma-ray logs, and exhibit lower porosities and/or permeabilities than the adjacent grain-supported rock. The percentage of mud-supported dolomite increases towards the base of the K-4ϕ.

A generalized facies map of the K-4ϕ (Fig. 11) shows skeletal grain-supported dolomites with lesser amounts of mud-supported interbeds over the length of the field. An increase in mud-supported dolomites, however, is interpreted in the vicinity of wells No. 12, 21, and Calco No. 2. Core was not available for study in these three wells, but this subtle facies change is suggested by a loss of porosity, decrease in the quality of production, and an increase in thickness of the K-4ϕ zone in this vicinity.

Depositional environment. — The skeletal grain-supported dolomites were deposited as widespread skeletal lime sands in an open, shallow-marine environment. This is suggested by the abundance and diversity of marine skeletal debris, presence of glauconite, high degree of bioturbation, and widespread, continuous nature of the facies. Interbeds of mud-supported dolomites were most likely deposited as lime mud in somewhat deeper or more restricted marine conditions. The increase in thickness of the K-4ϕ unit to the southeast could represent slightly deeper-water deposition.

Reservoir description. — The K-4ϕ is the main reservoir in the Upper Valley field and has been perforated in all producing and injection wells. Pay is associated with the sucrosic, skeletal grain-supported dolomites. Porosity is intercrystalline, vuggy, (pin-point vugs to 2 cm), and leached fossil-moldic. Leaching of the reservoir was extensive and is responsible for the vuggy and fossil-moldic porosity. This leaching probably occurred during the period of subaerial exposure at the end of Kaibab time and was selective in that it affected primarily the grain-supported facies of the K-4ϕ and not the mud-supported beds within the K-4ϕ or overlying K-4. The effects of leaching appear to decrease downward within the reservoir, a fact that may be related to either the increase in mud-supported interbeds in the lower part of the unit or a decrease in amount of leaching fluids at greater depths below the unconformity surface. Dolomitization of the original skeletal lime

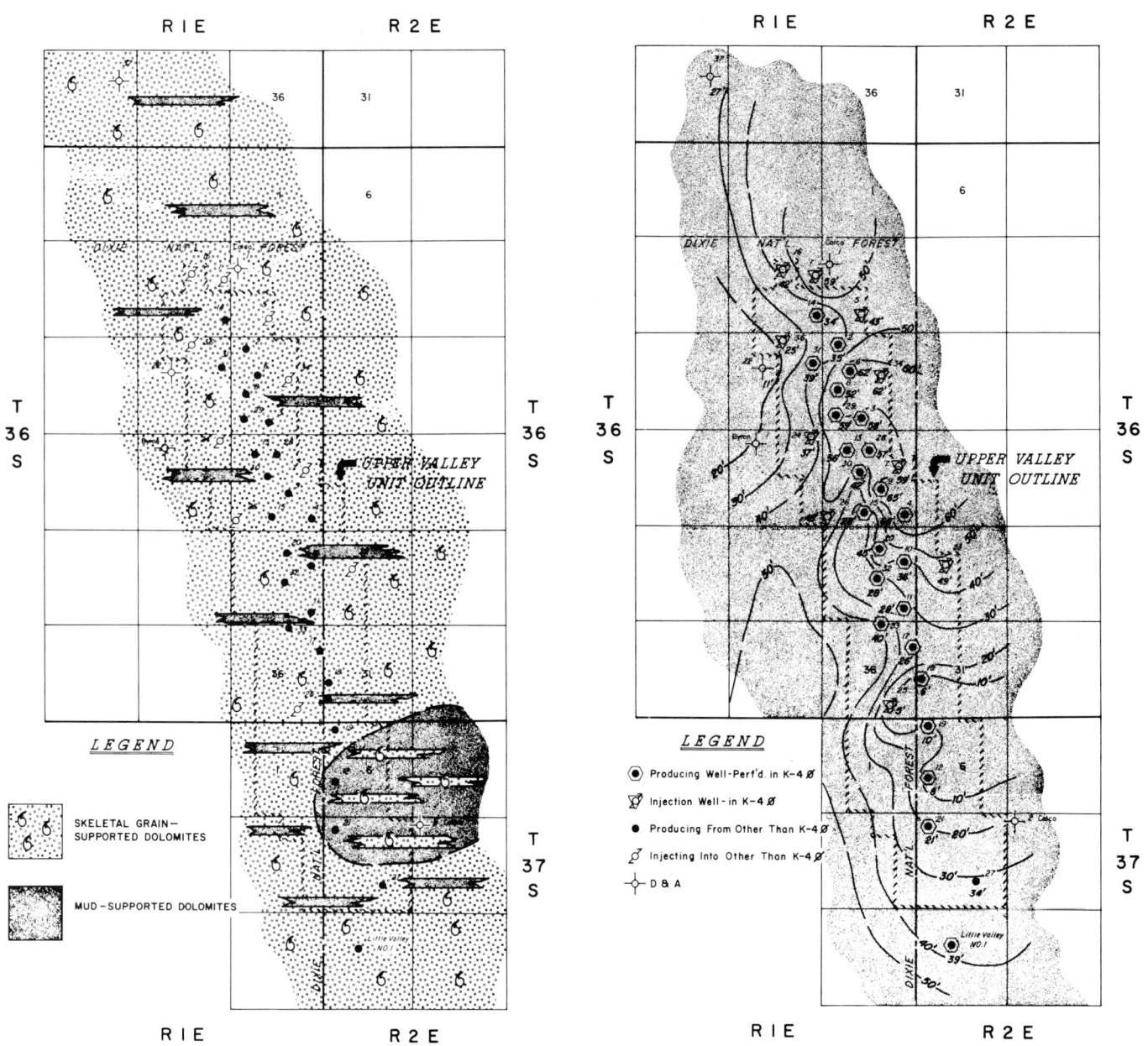

Fig. 11 — Generalized facies distribution within the K-4ϕ zone (Kaibab Formation), Upper Valley field.

Fig. 12 — Isopach of gross feet with porosity greater than 13.8 percent in the K-4ϕ zone (Kaibab Formation), Upper Valley field.

sands occurred after leaching and formation of the replacement chert nodules. This dolomitization created sucrosic intercrystalline porosity which provides excellent communication between vugs and leached fossil molds.

Calculated weighted porosity average for the K-4ϕ is 18.2 percent and a porosity cutoff from logs for gross pay determinations was calculated to be 13.8 percent. Porosity and permeability measurements from core data range up to 27.3 percent and 281 md. An isopach of gross pay using the 13.8 percent porosity cutoff demonstrates widespread porosity over much of the field (Fig. 12). A general decrease in reservoir quality, however, exists from the northern to the southeastern portion of the field and pay thicknesses range from 66 ft in No. 4 to 6 ft in No. 18.

Local variations of porosity and permeability as well as the overall loss of reservoir quality in the southeastern part of the field appear to be related in part to variations in amount of secondary leaching and/or subtle facies changes to more mud-supported dolomites. Several wells cored skeletal grain-supported dolomites in the lower portion of the K-4ϕ zone which were not as leached as similar rocks in the upper part of the zone. This may reflect a decrease in the influence of percolating leaching ground waters at greater depths below the Kaibab unconformity. Permeability is also affected by an increase in mud-supported dolomites as observed in No. 4

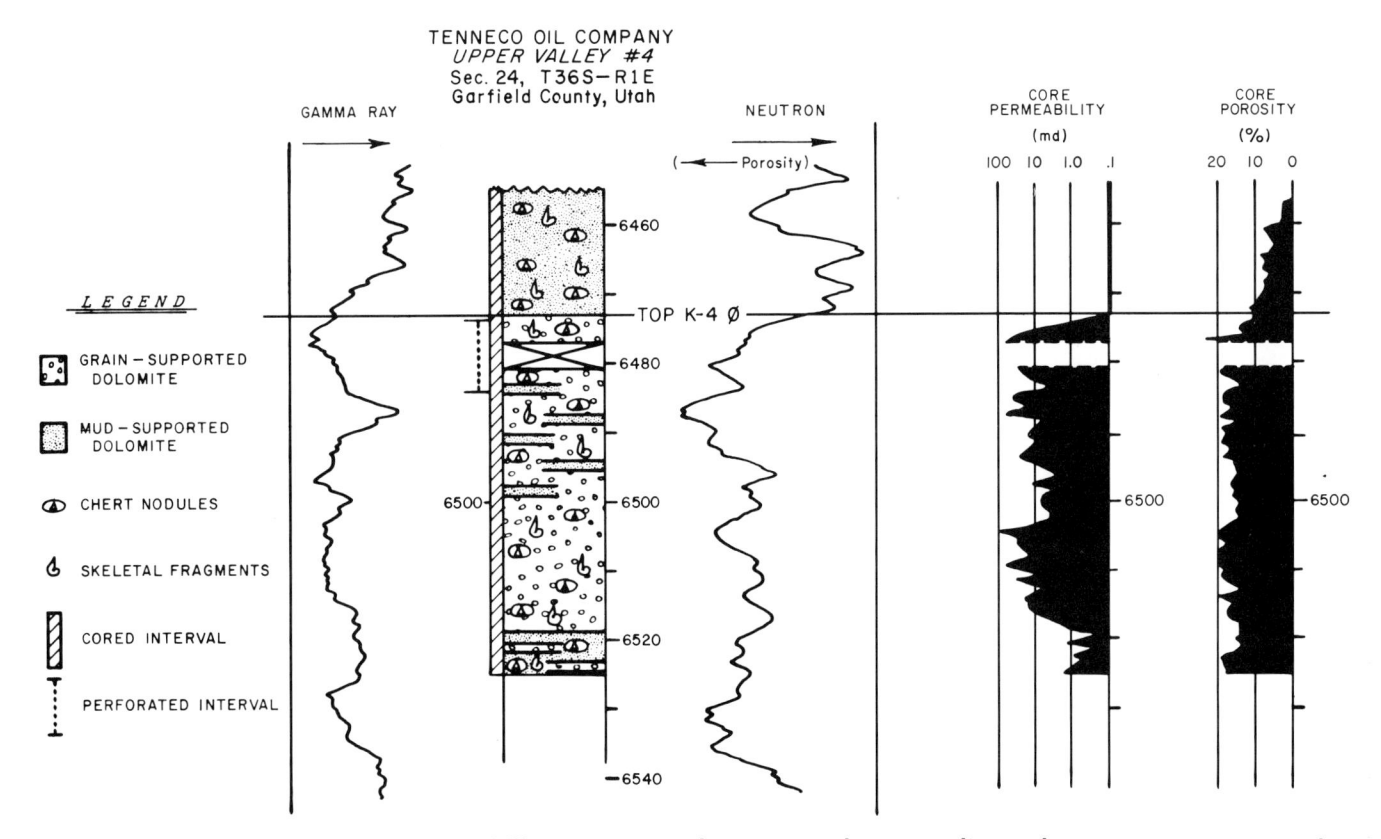

Fig. 13 — Relationship of core permeability, core porosity, neutron log porosity, and gamma-ray response to facies in the K-4ϕ zone (Kaibab Formation), Upper Valley field. Note the abrupt loss of core permeability at 6518 ft which corresponds to a facies change from grain-supported to mud-supported dolomites. This loss of permeability is not reflected in core porosity measurements or neutron log porosity.

(Fig. 13). In this well an increase in mud content corresponds generally to a loss of permeability, but not necessarily to a loss of porosity. This is probably related to smaller crystal size and correspondingly smaller pore throat size of the mud-supported dolomites. Anhydrite plugging may also locally reduce reservoir quality and was observed as the primary factor for porosity reduction in No. 19.

Fracturing in the K-4ϕ does not appear to be as extensive as that observed in the K-2 or K-3 reservoirs. Occasional fractures were observed in the K-4ϕ, but their overall intensity and frequency is low, and their influence as a major aid to productivity is questionable. A possible explanation for the differences in fracture intensity between the K-4ϕ and the overlying K-2 and K-3 reservoirs may be related to the fact that the K-4ϕ zone is not as brittle as either the calichefied K-2 or silicified K-3.

Overall lateral continuity within the K-4ϕ is good due to the widespread nature of the producing facies and the excellent communication between vugs and leached fossil molds provided by intercrystalline porosity of the sucrosic dolomite. Local areas may have poor lateral continuity due to increases in mud content, lack of leaching, or anhydrite plugging. Vertical communication, particularly in the upper portions of the unit, appears to be good because of the absence of bedding planes or other consistent permeability barriers. Lower portions of the K-4ϕ, however, may have poor vertical continuity because of the increase in mud-supported dolomites toward the base.

HYDRODYNAMICS

Oil accumulation at Upper Valley field is displaced along the western flank and down the southern plunge of the anticline as illustrated by a number of crestal wells which tested wet, yet are structurally higher than offset producing wells (Figs. 14 and 15). This southwestward tilt of the oil-water contact is attributed to a hydrodynamic drive which is regionally present in the Kaibab of the Kaiparowits basin. The average direction of water flow in the Timpoweap-Kaibab reservoirs within the field area was determined to be approximately S45°W. This is based on regional hydrodynamic studies and pressure data within the field. The amount of tilt for the oil-water contact in each reservoir was determined by first estimating oil-water contact in individual reservoirs from production data, test results, or core data and then correlating these contacts from well to well. Most control for this analysis exists in the northern part of the field where an average gradient of 270 ft/mi was determined for oil-

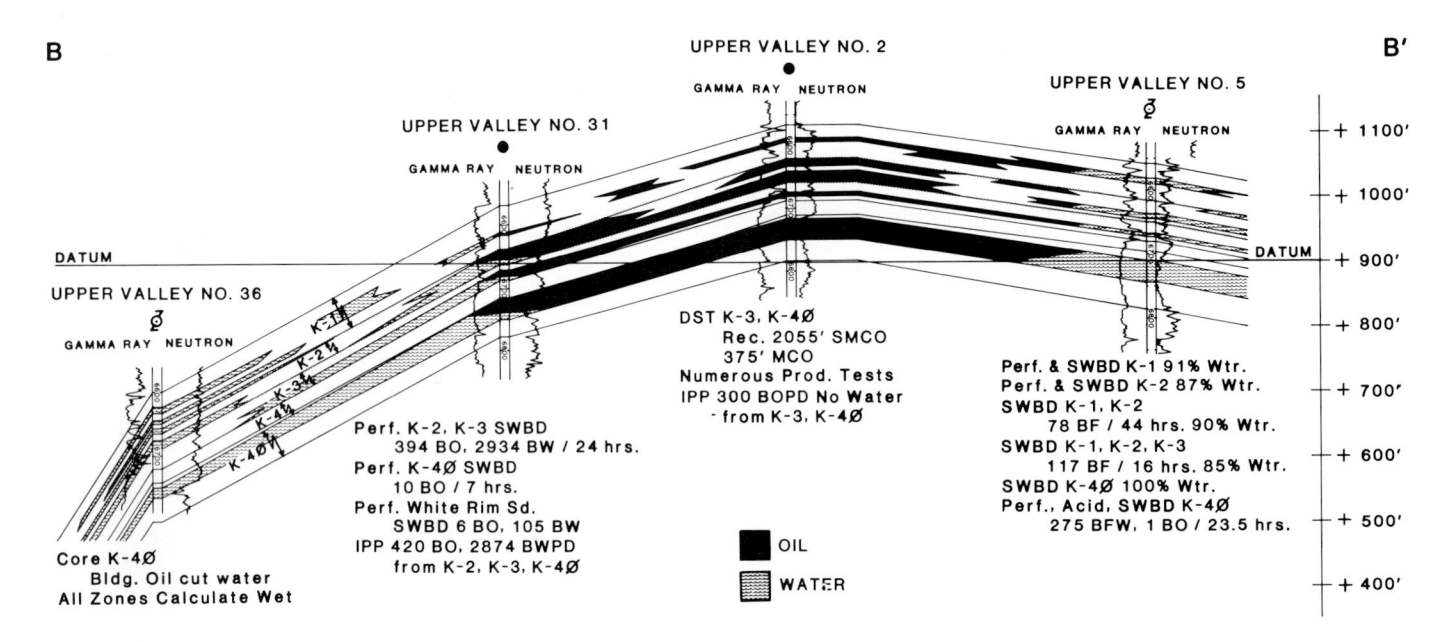

Fig. 14 — Structural cross section B-B' showing tilted oil-water contacts in Timpoweap and Kaibab reservoirs, Upper Valley field. Note that U.V. No. 5 tested water in all zones, yet is structurally higher than the producing U.V. No. 31. See Fig. 3 for location of cross section.

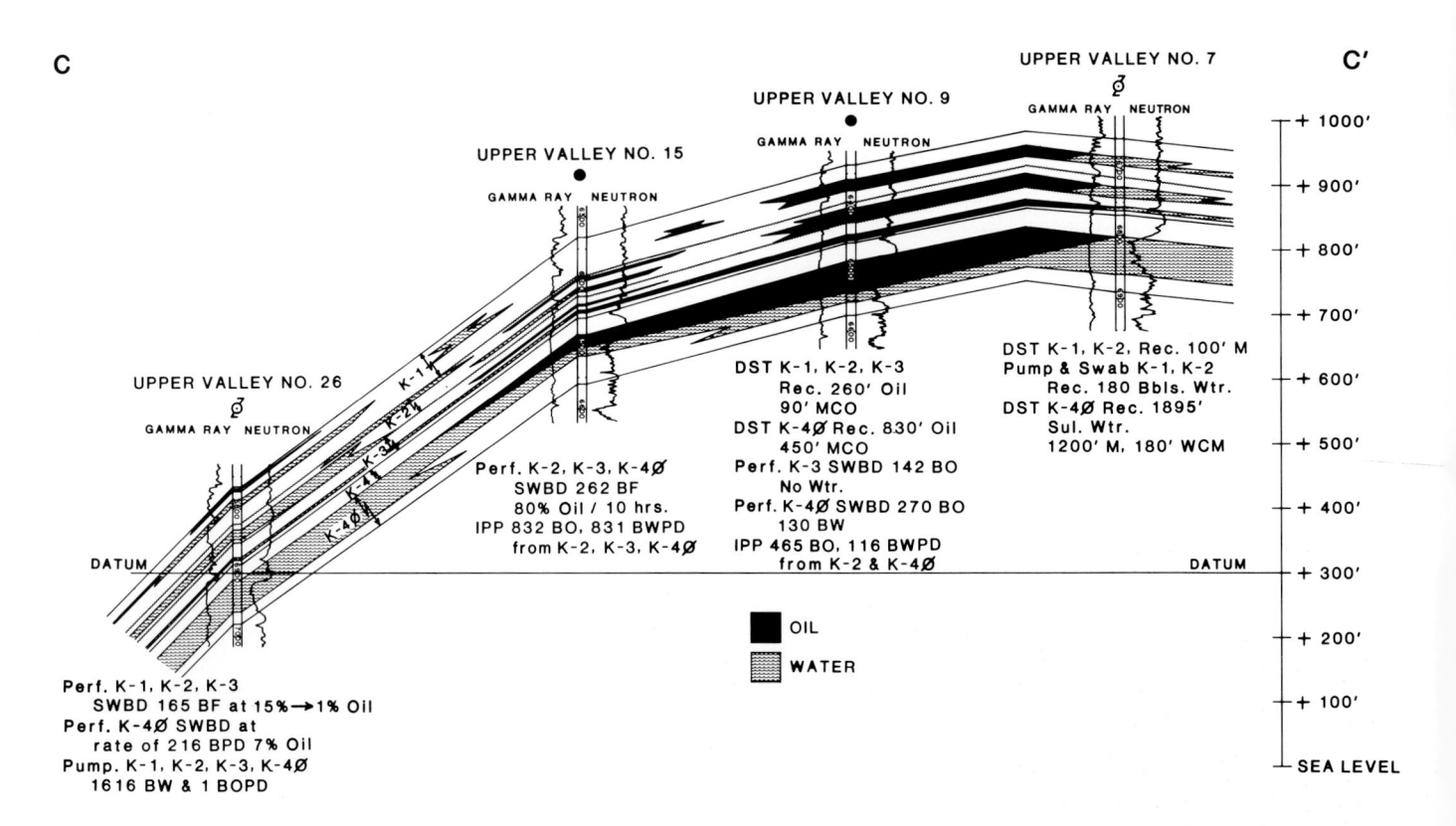

Fig. 15 — Structural cross section C-C' showing tilted oil-water contacts in Timpoweap and Kaibab reservoirs, Upper Valley field. Note that U.V. No. 7 tested wet, yet is structurally higher than producing wells U.V. Nos. 9 and 15. See Fig. 3 for location of cross section.

water contacts in the K-2, K-3, and K-4φ reservoirs (the K-1 reservoir was too discontinuous for this analysis).

Many local factors such as permeability changes, faulting, or changes in fluid density can affect the slope of an oil-water contact. At Upper Valley field, API gravities of the produced crude display a gradational decrease from about 27° near the apex of the structure in well No. 2 to approximately 19° at the southern end of the field in well No. 27. This gradational decrease in oil gravities down the field most likely has an effect on the slope of the oil-water contact since oils with lower gravities (higher densities) can be tilted a greater amount than oils with higher gravities (low densities). Tilt factor calculations indicate the 19° gravity crude at the south end of Upper Valley field can be tilted approximately 1½ times greater than the 27° oil at the north end of the field. This suggests that the tilted oil-water contact may not be a planar surface, but may actually be a curvilinear surface, and that the gradient changes from 270 ft/mi in the north to about 380 ft/mi at the southern end of the field. The intersection of this curvilinear surface with structure maps on top of each reservoir defines limits of production for each zone and is consistent with most production data and test results.

LIMITS OF OIL CHARGE IN THE K-4φ

The extent of production at Upper Valley field is not only influenced by hydrodynamics, but is also complicated by the fact that the main reservoir (K-4φ) apparently did not receive an oil charge along certain portions of the structure. This lack of oil charge is suggested by core data from well No. 37 (Sec. 34, T35S, R1E), Calco No. 2 (Sec. 7, T37S, R2E), and TOC, No. 1 South Upper Valley (Sec. 16, T38S, R2E). The upper K-4φ was cored in these dry holes and, with the exception of 3 ft in No. 37, none of these cores had any analyzed residual oil saturations. The movement of oil through a reservoir rock, particularly a low-gravity, heavy, asphaltic crude like that produced at Upper Valley, will almost certainly leave a percentage of residual, non-recoverable oil trapped in the pore system. The absolute lack of residual oil saturations from analyzed core data suggests the K-4φ zone in Calco No. 2, No. 1 South Upper Valley and all but 3 ft in No. 37 never received an oil charge.

This local absence of an oil charge in the K-4φ zone is puzzling and the author was unable to fully resolve the problem with the data at hand. An organic-rich source rock facies was not observed in cores of the Timpoweap or Kaibab, and it is assumed that oil at Upper Valley migrated laterally and/or vertically to the present site of accumulation. Perhaps the position of structural closure at the time of migration was shifted somewhat south and west of its present position. If this was the case, oil initially accumulated along the crest of the interpreted paleostructure and never spread far enough south to reach No. 1 South Upper Valley, far enough east to reach Calco No. 2, and barely spread far enough north to stain only 3 ft in No. 37. A later structural readjustment and southwestern hydrodynamic drive redistributed the oil to its present site, but left these three wells uncharged.

Several other interpretations and hypotheses are no doubt possible to explain this phenomenon. Additional information regarding regional identification of source rocks, maturity, time of expulsion, and migration paths may aid these interpretations. The overall observation that wells with no apparent oil charge can be situated close to an accumulation may be significant when evaluating other structures in the region.

CONCLUSIONS

1. Production at Upper Valley is from four separate and distinct zones within the Timpoweap and Kaibab formations. All of the reservoirs are carbonates which have been dolomitized to varying degrees and were deposited in environments ranging from shallow marine to supratidal.

2. Facies variations and diagenetic alteration are present in all four reservoir zones and affect porosity and permeability distribution.

3. Two subaerial exposure surfaces were recognized in the producing sequence: one exists within the Timpoweap at the top of the K-2 and the other exists at the boundry between the Permian and Triassic systems at the top of the Kaibab (top K-3). Both unconformities were important in reservoir enhancement and destruction.

4. Fractures occur to some degree in all of the reservoirs, but are more intense in the K-2 and K-3. This selective fracturing may be due to the more brittle nature of the calichefied K-2 and silicified K-3.

5. Local communication of fluids behind pipe have complicated the integration of geologic observations with reservoir response in individual zones.

6. A hydrodynamic gradient produces a tilted oil-water contact in all zones in the field. This oil-water contact may be curvilinear (concave downward) due to variations in densities of the crude within the field.

7. The northern and southern ends and at least one portion along the eastern flank of the Upper Valley structure never received an oil charge in the K-4φ.

REFERENCES CITED

Bowers, W. E., 1973, Geologic map and coal resources of the Upper Valley Quadrangle, Garfield County, Utah, U.S. Geol. Survey Coal Investigations Map C-60.

Campbell, J. A., 1969, The Upper Valley oil field, Garfield County, Utah *in* Geology and natural history of the Grand Canyon region: Four Corners Geol. Soc., 5th Field Conf. Guidebook, p. 195-200.

Kelley, V. C., 1955, Monoclines of the Colorado Plateau: Geol. Soc. America Bull., v. 66, no. 7, p. 789-804.

McKee, E. D., 1938, The environment and history of the Toroweap and Kaibab Formations of northern Arizona and southern Utah: Carnegie Institute of Washington, Publication no. 492, 268 p.

Zeller, H. D., 1973a, Geologic map and coal resources of the Canaan Creek Quadrangle, Garfield County, Utah, U.S. Geol. Survey Coal Investigation Map C-57.

Zeller, H. D., 1973b, Geologic map and coal resources of the Death Ridge Quadrangle, Garfield County, Utah, U.S. Geol. Survey Coal Investigation Map C-58.

THE GEOLOGY OF THE PINEVIEW FIELD AREA SUMMIT COUNTY, UTAH

by

P. D. Maher[1]

INTRODUCTION

The Pineview field is located approximately 7 mi from the southwest corner of the state of Wyoming. It is in T. 2 N., R. 7 E., Summit County, Utah. From a geologic standpoint, the field is located just north of the Uinta Mountains, and well back within the Overthrust Belt (Fig. 1).

The recent Nugget discovery by Amoco and Chevron, at their Ryckman Creek prospect, is located approximately 40 mi northeast of the Pineview field. The Nugget production from the Tiptop, Dry Piney, and Hogsback fields on the greater LaBarge uplift is located approximately 100 mi northeast of Pineview.

The Pineview field discovery well, the No. 1 Newton Sheep Company, was drilled by American Quasar Petroleum Company, Energetics, Inc., and North Central Oil Company on a farmout from Amoco, Occidental and Sun Oil Company. It is located in the NE/4SE/4, Sec. 4, T. 2 N., R. 7 E., Summit County, Utah. The well was drilled to a total depth of 14,500 ft.

STRUCTURE

There is some expression of the Pineview anticline on the surface. The Kelvin Formation is exposed on the east flank of the structure and dips to the east and northeast. The Tertiary sediments, which unconformably overlie the Cretaceous in this

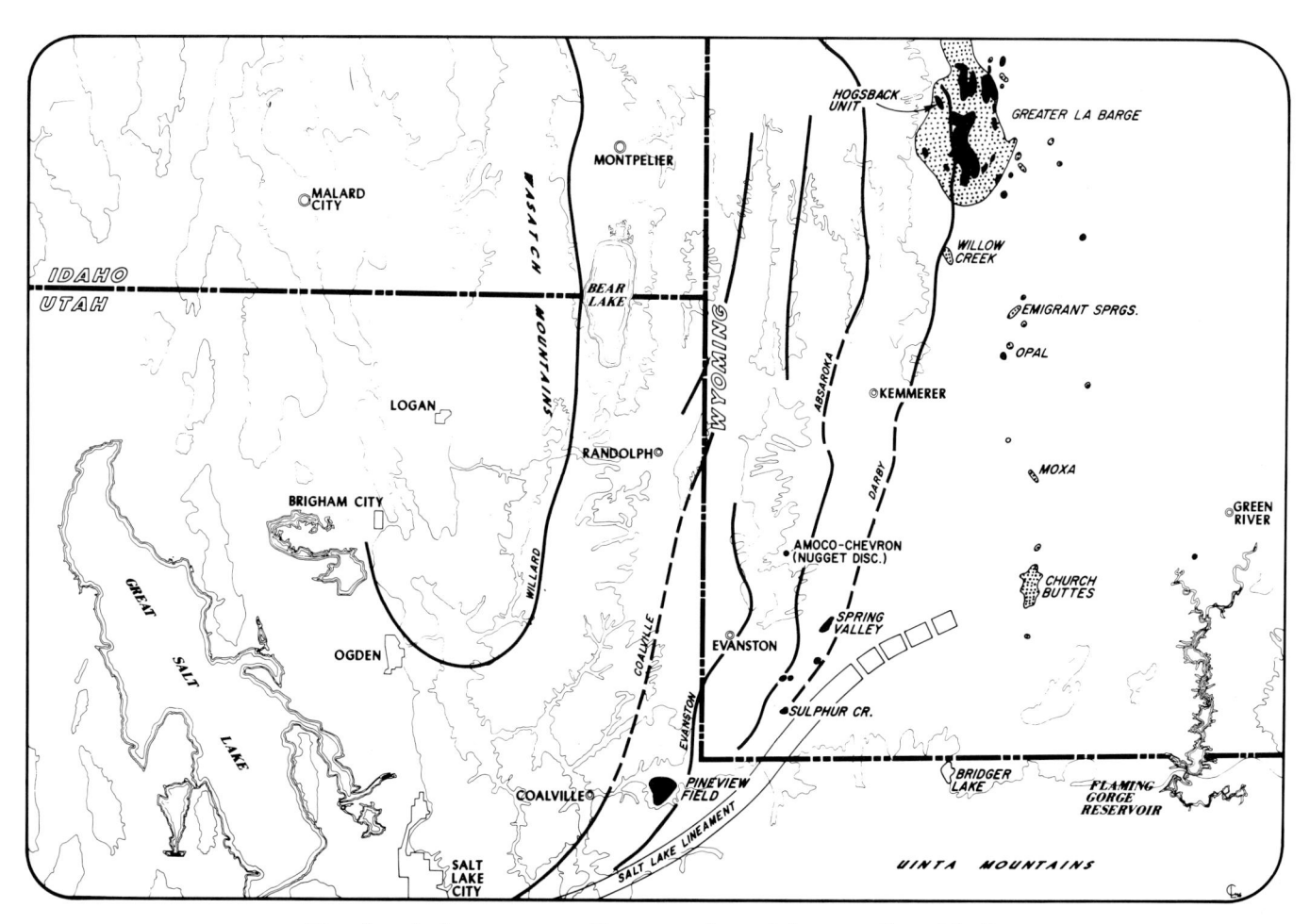

Fig. 1 — Index map, southern portion of the overthrust belt.

[1] Executive Vice President, Energetics, Inc., Denver, Colorado.

region, mask most of the structure in the Pineview area. The Pineview structure was originally mapped by Occidental Petroleum Corporation using seismic data. Seismic profiles show very clearly a complicated overthrust structural closure, a major thrust fault at approximately 13,000 ft and a very large, broad structure beneath the major thrust fault. Occidental Petroleum Corporation drilled the initial test well in the area, the No. 1 Pineview, to a total depth of 10,530 ft during 1971 and 1972. This well apparently was located in a position to test the sub-thrust structural high. It had excellent oil shows in the Nugget from a depth of 10,220 ft to total depth of 10,530 ft. A drill stem test of the upper portion of the Nugget recovered 360 ft of mud and 8,767 ft of slightly gas-cut salt water. The well was subsequently abandoned. The only structural complication in the well is a 705 ft thrust fault displacement encountered at 7,210 ft, which repeats parts of the Stump and Preuss formations.

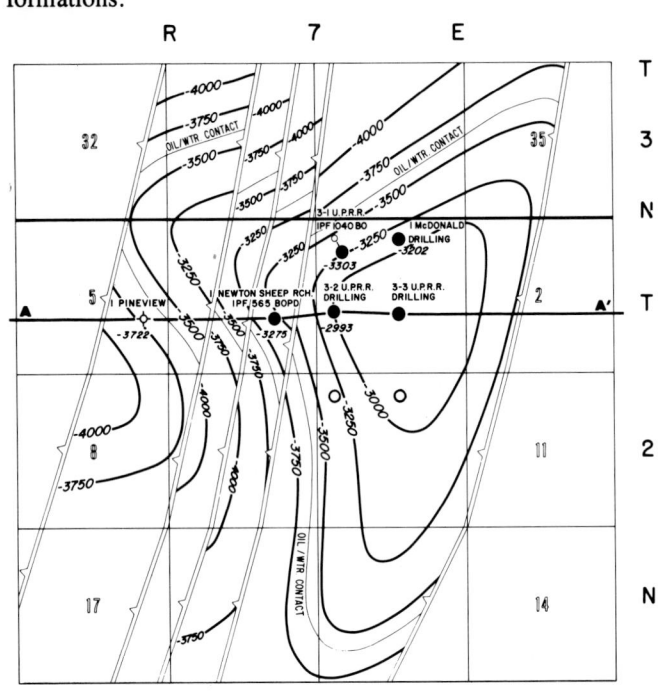

Fig. 2 — Structural contour map on top Nugget Sandstone, Pineview field.

The No. 1 Newton Sheep Company, the second well drilled, encountered several thrust faults in the interval from 5,500 ft to 6,224 ft. These faults repeat 1,300 ft of the Stump and portions of the Preuss formations. The dipmeter suggests the presence of a fault at a depth of approximately 8,300 ft in the Preuss salt zone. The divergence of dips above and below the fault zone may be due to a high-angle reverse fault, or it could reflect bedding plane movement along the salt layer. On the cross section (Fig. 3), the bulk of the faulting occurs in the Stump and upper part of the Preuss formations, which lie above the Preuss salt zone. There is a strong possibility that some lateral movements occurred along the salt zone resulting in shallow thrusts. This thrusting serves to complicate the shallow structure. Seismic profiles indicate a number of these faults do penetrate the salt zone and continue on down to the main Absaroka fault. A small fault is probably present in the No. 1 Newton Sheep Company at a depth of 11,730 ft, which repeats approximately 350 ft of the Ankareh Formation. The Absaroka thrust was cut at a depth of 12,974 ft where the well went from the Triassic Thaynes Formation into the dark grey-black Cretaceous shale sequence of the Aspen and Bear River formations.

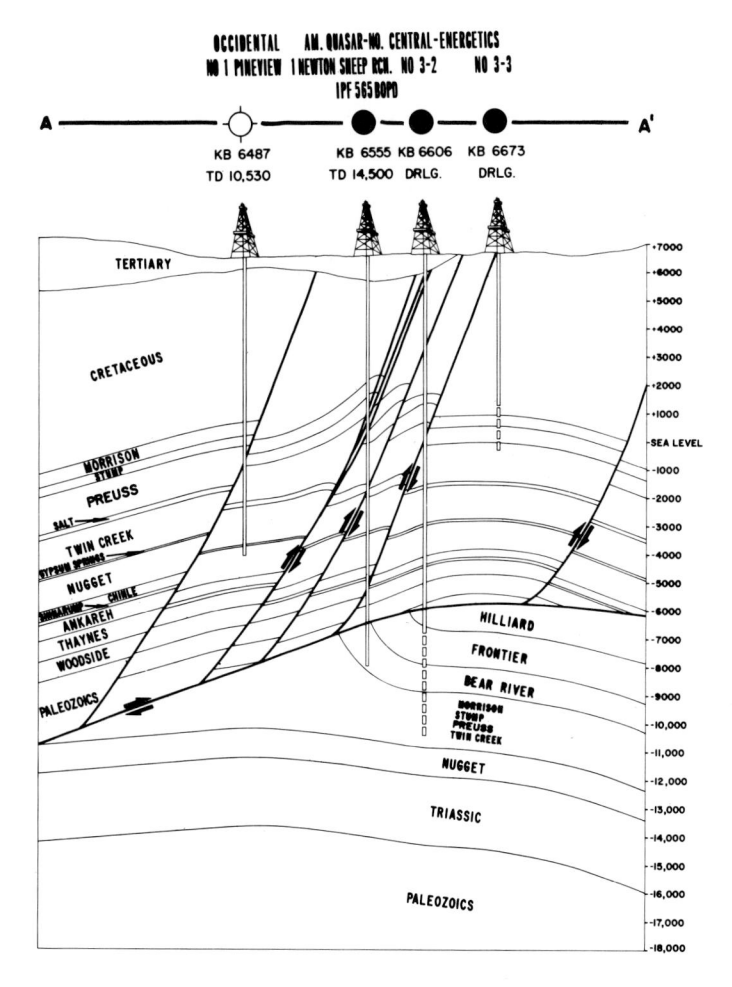

Fig. 3 — Structural cross section A-A', Pineview field.

The No. 3-2 UPRR, located in the NW/4SW/4, Sec. 3, T. 2 N., R. 7 E., is currently drilling below 15,000 ft. Logs have been run and intermediate casing set through the Nugget Sandstone. The logs show the presence of an 820 ft thrust fault displacement at 6,480 ft. This well encountered the Absaroka thrust at 12,540 ft, where again the Lower Triassic is thrust over Cretaceous shales. It appears that the well may have penetrated the lower part of the Hilliard Formation and then into the Frontier. This is a very interesting well, in that it will give us our first look at a normal stratigraphic section of Lower Cretaceous and Jurassic rocks in the sub-thrust at Pineview.

Six miles to the southwest the No. 35-1 UPRR in Sec. 35, T. 2 N., R. 6 E., cut the Absaroka thrust at 15,990 ft. Here again the well went from Lower Triassic into the Cretaceous Aspen and then into a sandy section of the Bear River.

The No. 3-1 UPRR, located in the NW/4NW/4, Sec. 3, had a 370 ft displacement thrust fault at 5,703 ft.

STRATIGRAPHY

A summarized description of the stratigraphic section in the Pineview area is as follows:

Tertiary

The Tertiary consists of a series of varicolored shales, silts, sandstones, and conglomerates, which overlie the Cretaceous sediments unconformably.

Cretaceous

The Frontier consists of several thousand ft of paludal sediments consisting of light to dark grey and brown bentonitic shales, siltstones, and very fine-grained, grey, glauconitic sandstones, interbedded with coals throughout most of the sequence.

The Aspen is 200 to 300 ft thick, consisting of medium to dark grey-brown bentonitic marine shales.

The Kelvin Formation is about 3,000 ft thick. It consists of a continental series of varicolored grey, green, reddish-brown, and tan shales with interbedded grey, tan, and red sandstones which grade from very fine- and fine-grained to coarse and conglomeratic. At the base of the Kelvin Formation is a sandstone which is fine- to medium-grained, becoming coarse and conglomeratic with abundant loose chert grains.

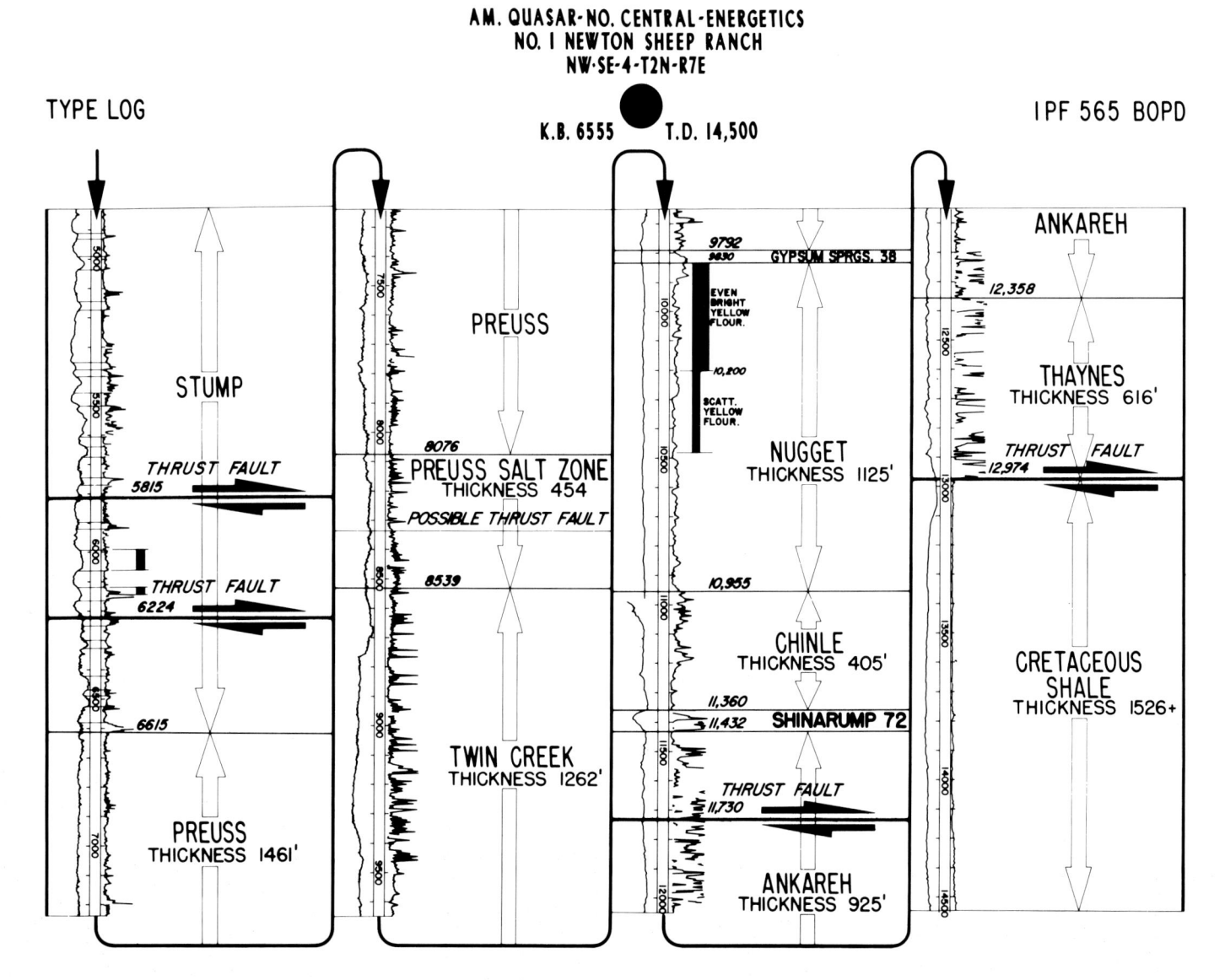

Fig. 4 — Type log, Pineview field discovery well.

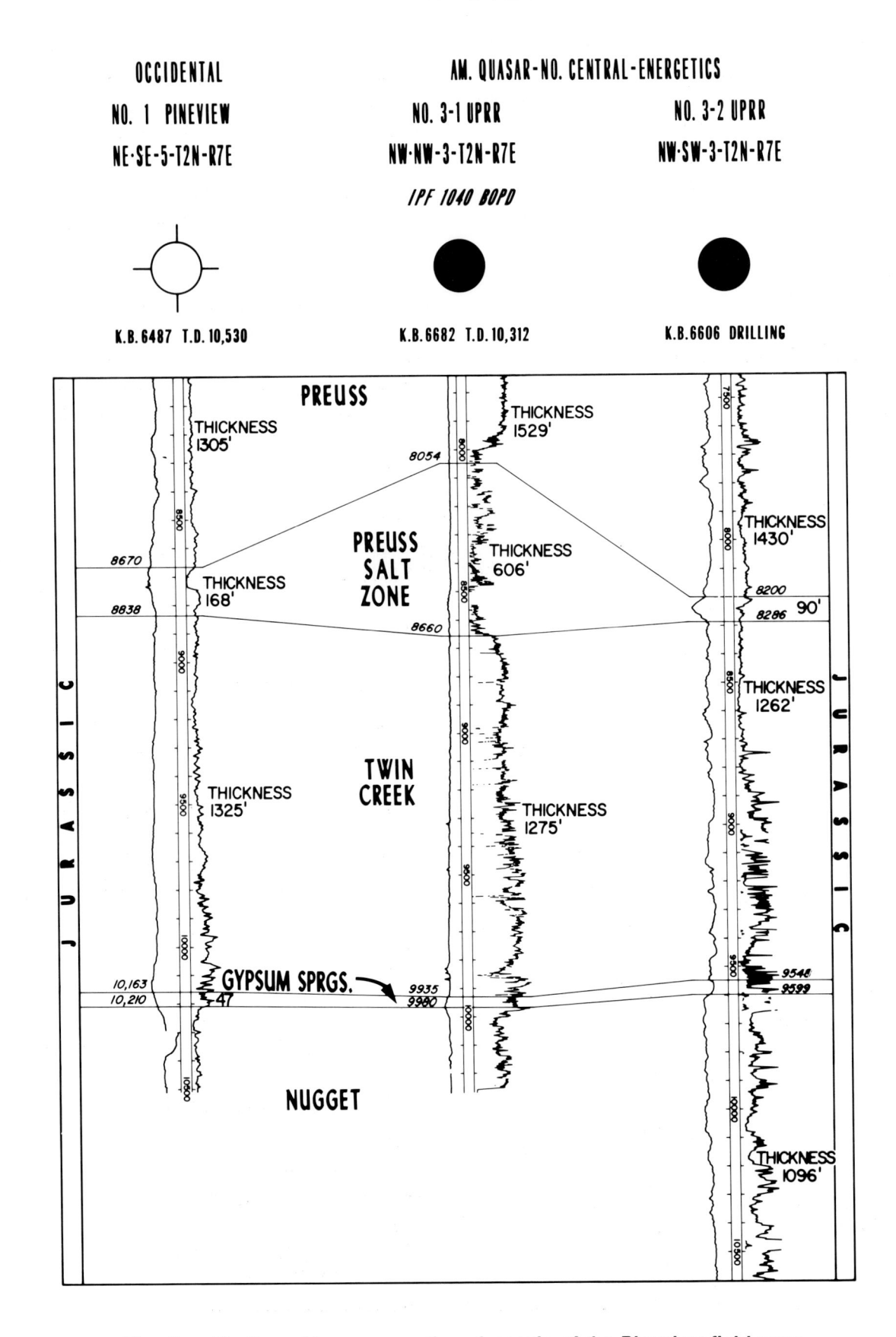

Fig. 5 — Stratigraphic cross section, Jurassic of the Pineview field area.

Jurassic

The Morrison Formation is 400 to 500 ft thick, consisting predominately of grey, grey-green, reddish-brown, maroon, and some purple shales with a few stringers of grey to pink limestone and some thin sands.

The Stump Formation is 300 to 400 ft thick and is a sandy series which is characterized by glauconite and cherty conglomeratic sands interbedded with varicolored shales.

The Preuss is 1,300 to 1,500 ft thick and is predominately a brown, silty shale and siltstone series with a sandstone interval developing about 500 ft above the base. There are occasional stringers of buff to grey, dense limestone.

The Preuss salt zone ranges from 86 to 1,100 ft thick in the Pineview area. It consists primarily of salt and anhydrite with light grey to grey, moderately soft, micaceous shale.

The Twin Creek Formation ranges from 1,262 to 1,325 ft in thickness. It is composed of interbedded light to dark limestone and hard grey calcareous shales (Fig. 5).

The Gypsum Springs is 40 to 50 ft thick and consists of reddish-brown, non-calcareous shale.

The Nugget Formation is approximately 1,100 ft thick and is a white to tan, fine- to coarse-grained sandstone, grading downward into a red and salmon-colored sandstone.

Triassic

The Chinle is 300 to 400 ft thick and consists of red, maroon, and orange-red, silty, hard shale.

The Shinarump is a 70 ft thick sandstone; grey, very fine-grained, hard and tight.

The Ankareh is approximately 600 ft thick and consists of red to dark red, silty, hard shale with some lavender, ocher, brown, and light grey shale. It also has interbedded grey to white, very fine-grained, hard and tight sandstone.

The Thaynes is 700 to 800 ft thick and consists of light to dark grey, tan, hard, dense, micro-crystalline limestone with interbedded red, brown, and purple shale.

Subthrust Cretaceous

The youngest formation beneath the thrust fault in the No. 1 Newton Sheep Company is the Aspen Formation, which is approximately 500 ft thick with very steep dip. It consists of black, soft, non-calcareous shale with a silky luster. Below the Aspen, the Bear River Formation is encountered. The No. 1 Newton Sheep Company penetrated approximately 1,000 ft of light to medium brown and tan shale with some grey, red, and pink calcareous siltstone and was in the Bear River at total depth.

PRODUCTION

The Pineview field is currently producing from both the Nugget Sandstone and the Twin Creek Limestone. There is a tight interval at the top of the Nugget which varies from 10 to 60 ft thick, but does have some fracture porosity and permeability. Beneath this tight zone is a 90 to 100 ft interval with core porosities of 8 to 14% and permeabilities ranging from 1 to 116 md. Fracturing is undoubtedly an important factor in the effectiveness of the reservoirs at Pineview. The balance of the Nugget Sandstone ranges from 4 to 6% porosity with streaks going to 8 and 10%. The net pay thickness in the No. 3-2 UPRR, based on log and core data is as follows: 75 ft of greater than 10% porosity; 205 ft of 8 to 10% porosity; 222 ft of 6 to 8% porosity; for a total of 502 ft of net pay with greater than 6% porosity. The Twin Creek Limestone is also an excellent pay in the Pineview field. The reservoir in the Twin Creek consists of fractured limestones that are located in the basal part of the formation.

There are currently two producing wells in the Pineview field. The No. 1 Newton Sheep Company produced an average of 250 bbl of oil per day during January, 1976 from the Nugget Formation. The No. 3-1 UPRR is dually completed. It produced an average of 547 bbl of oil per day from the Twin Creek and 314 bbl of oil per day from the Nugget during January of 1976. To date, none of the producing zones at Pineview have been stimulated in any manner.

CONCLUSION

Let us now take another look at the overthrust belt and try to understand the oil potential of this vast region. As all of us are well aware, the three essential features in looking for oil fields are traps, reservoir rocks, and oil source rocks.

At this stage of exploration, the overthrust belt must be explored for anticlinal-type closures. Both surface geology and the seismograph are essential in the exploration for this type of trap. The work that has been done to date indicates that there are literally hundreds of closed structures within the overthrust belt, both on the shallow thrust sheets and within the lower thrust sheets. Reservoir rocks are also abundant with sandstones developed in the Frontier, Dakota, Kelvin, Stump, Preuss, Nugget, the Triassic section, and the Weber; as well as carbonates of the Twin Creek and Thaynes formations. It has also been established by the recent discoveries of Pineview and Ryckman Creek that parts of the overthrust belt are rich in source rocks. The Cretaceous shales, which underlie the Pineview and Ryckman Creek fields, are probably the source of oil which is being found in the Nugget and Twin Creek formations. Certainly the Absaroka thrust sheet offers tremendous potential for additional oil discoveries to be made in this overthrust sheet. It would appear, anywhere that Cretaceous marine source rocks can be found in the sub-thrust sheets, that there will be excellent oil potential in the reservoirs above them. In addition to the Cretaceous shales, the dark grey to black shales of the Preuss salt section in the Pineview area rapidly thicken westward toward Coalville, where they reach a thickness of 1,000 ft and are located immediately below 1,100 ft of evaporites. These Jurassic shales are the equivalent of the Arapien section in central Utah. Geochemical work done on the Arapien rocks indicate that they are excellent oil source shales. Another potential source is the shales and limestones of the Park City Formation.

All of the ingredients for one of the nation's largest oil producing provinces are present in the overthrust belt. With the recent discoveries of Pineview and Ryckman Creek fields, there is little doubt that the overthrust belt will be one of the nation's significant oil and gas producing regions.

SELECTED REFERENCES

Cockran, K., 1959, Results of pre-Cretaceous exploration in the overthrust belt of western Wyoming: Intermtn. Assoc. Petroleum Geologists Guidebook, 10th Ann. Field Conf., p. 200-203.

Eardly, A. J., 1944, Geology of the north-central Wasatch Mountains, Utah: Geol. Soc. America Bull., v. 55, p. 819-894.

Eyer, J. A., 1969, Gannett Group of western Wyoming and southeastern Idaho: Am. Assoc. Petroleum Geologists Bull., v. 53, no. 7, p. 1368-1390.

Gregg, Clare C., 1976, Pineview field opens new Rockies oil province: World Oil, June 1976, p. 75-77.

Imlay, R. W., 1952, Marine origin of Preuss Sandstone of Idaho, Wyoming, and Utah: Am. Assoc. Petroleum Geologists Bull., v. 39, no. 9, p. 1735-1753.

_____, 1967, Twin Creek Limestone (Jurassic) in the western interior of the United States: U.S. Geol. Survey Prof. Paper 540.

Kummel, B., 1954, Triassic stratigraphy of southeastern Idaho and adjacent areas: U.S. Geol. Survey Prof. Paper 254-H, p. 165-194.

Loucks, G. G., 1975, The search for Pineview field, Summit County, Utah, in Rocky Mtn. Assoc. Geologists Symposium on deep drilling frontiers in the central Rocky Mountains, p. 255-264.

Monley, L. F., 1971, Petroleum potential of Idaho-Wyoming overthrust belt, in Future petroleum provinces of the United States — their geology and potential: Am. Assoc. Petroleum Geologists Mem. 15, v. 1, p. 509-529.

Royse, F., Jr., Warner, M. A., and Reese, D. L., 1975, Thrust belt structural geometry and related stratigraphic problems, Wyoming-Idaho-northern Utah, in Rocky Mtn. Assoc. Geologists Symposium on deep drilling frontiers in the central Rocky Mountains, p. 41-54.

GEOTHERMAL ENERGY CORDILLERAN HINGELINE — WEST

by
B. Greider[1]

PRINCIPLES

Before describing the typical geothermal projects that are present from the Cordilleran "hingeline" of central Utah westward, a review of geothermal principles may be helpful. Geothermal energy is simply the natural heat occurring below the surface of the earth. It is believed that the majority of the heat generated in the earth is due to radioactive decay of uranium, thorium and potassium in the mantle and crust. The heat generated in the interior is transferred to the earth's surface by conduction through the rock framework of the crust and by convection of water in a liquid or gaseous phase where rocks are porous and permeable. Heat is the energy contained in a body whose molecules are in motion. When heat is transferred from one substance to another, energy is transferred to that substance. Heat flow is a measure of the amount of heat (energy) being transferred from a substance of higher temperature to a substance of lower temperature. Heat flow (Q) is equal to the temperature difference between points one and two ($T_1 - T_2$) divided by the distance between the two points and multiplied by the thermal conductivity K. $Q = \frac{(T_1 - T_2) K}{L}$ is the familiar heat flow equation where L is the distance between the two temperature measurements. The earth's heat flow at any given location is a product of the geothermal gradient and the thermal conductivity of the rocks. Normal heat flow at the surface of the earth is 1.5 microcalories per cm²/sec. Figure 1 summarizes these principles.

The geothermal gradient to a given depth is the difference of the temperature of the earth at that depth and the surface temperature divided by the distance between the points of measurement. The normal geothermal gradient is usually 1.0 to 2.0°F per 100 ft. An area with a heat flow or geothermal gradient two to five times the normal for that area is considered to be a geothermally anomalous prospect.

Mantle heat flow into continental crustal provinces is consistently low over large areas. Heat is transferred into the sedimentary cover primarily by conduction through the crystalline basement rocks. These impermeable igneous and metamorphic rocks do not allow rapid flow of heat into specific areas.

Transfer of heat by conduction through the crustal rocks is inefficient, as the conductivity of rock is not high enough to allow rapid transfer of heat to provide local accumulations that are useful.

GEOTHERMAL ENERGY
○ NATURAL HEAT OF THE EARTH ○
○ PRIMARY SOURCE IS RADIOACTIVE DECAY ○

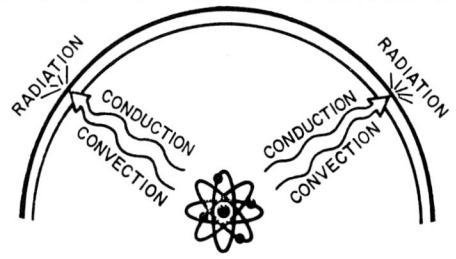

NORMAL CONDITIONS
○ HEAT FLOW = 1.5 μCAL/CM² - SEC.
○ GRADIENT = 1°F. TO 2°F. PER 100'
○ 500°F. = 22,250' TO 44,500'

Fig. 1 — Schematic Diagram of Thermal Energy within the Earth.

The high temperatures recorded in near-surface, high intensity geothermal anomalies appear to be concentrated by heat being transported by water from hot, deeper sources to reservoirs at lesser depths. Water contained within the near-surface rocks can store a large quantity of heat. As the amount of heat contained in the water increases, the density of the water decreases. This establishes a density contrast (buoyancy) with the normal temperature water. If vertical permeability exists over several thousand ft, the less dense (hot) water can rise toward the surface and carry the heat rapidly toward economical depths for development. Faults and associated fracture zones provide sufficient vertical permeability to allow strong movement of the heat-carrying waters. The heat flows from the water upward and outward to the cooler rocks. In this manner, the thermal energy is mined from deeper rocks and stored in the more shallow rock framework and in water-saturated pores and fractures.

Most of the heat available in the outer crust of the earth (3×10^{26} cal.) is not concentrated into useable volumes. Areas near large molten magma bodies intruded into anomalously

---

[1] Member Executive Committee, Geothermal Resources Council, Larkspur, California.

shallow depths may have heat concentrations sufficient for commercial power production. These anomalies must have sufficient heat concentrated to result in contained rocks and fluids having a temperature above 400°F and be located above 10,000 ft in depth. The volume of hot reservoir should be between 7 and 8 cu mi.

If a well is drilled into a fluid-saturated system, the heat is transported from the rocks to the well bore by either vapor (steam) or liquid. To have an effective transfer of heat to a well bore there must be sufficient horizontal and vertical permeability. A 6,000 ft to 8,000 ft well must sustain flow rates of more than 100,000 lbs of steam per hr, or 500,000 lbs of water (above 325°F) per hr for 20 to 25 yrs to be considered commercial for electricity generation.

Geothermal anomalies occur around the world in association with structural trends of the active mid-ocean ridges, major subduction zones and major plate boundaries. These features are areas where material from the mantle is being added to the crust (spreading centers) or where crustal material is being dragged into the mantle, melted and then buoyantly moves upward into the crust (Cenozoic volcanism). These conditions form anomalous concentrations of heat.

The geologic model that is generally accepted by geothermal explorers and developers is shown in Figure 2 which is a simplification of D. E. White's (1973) classical diagrams. There are three basic requirements for this model to function:

1. A heat source (presumed to be an intrusive body) that is above 1200°C and within 16 Km of the surface.

2. Meteoric waters circulating to depths of 10,000 ft - 20,000 ft where heat is transferred from the conducting impermeable rocks above the heat source.

3. Water expanding upon being heated and moving buoyantly upward in a hot concentrated plume. Cold waters move downward and inward from the basin's margins to continue the heat transfer process. Heat is transported by convection in this part of the model.

Figure 3 depicts the solution to numerical modeling of a geothermal reservoir by P. Cheng, et al. (1970) of the University of Hawaii. Though these are for an island aquifer, the results are similar to the conditions created by a thick, predominantly sandstone aquifer surrounding a cylindrical geothermal anomaly. As the rate of flow and Ra (Raleigh Number) is increased, the pluming effect increases, creating a large volume of hot water at shallow depths below the surface; Θ represents zones with the same temperature. If conductance becomes the primary heat transfer mechanism the mushroom shape of the isotherms disappears and the values for Θ describe a conical shaped center within a domal shaped outer isotherm. Moderate depth observation holes can be used to obtain detailed temperature logs. These data can be plotted

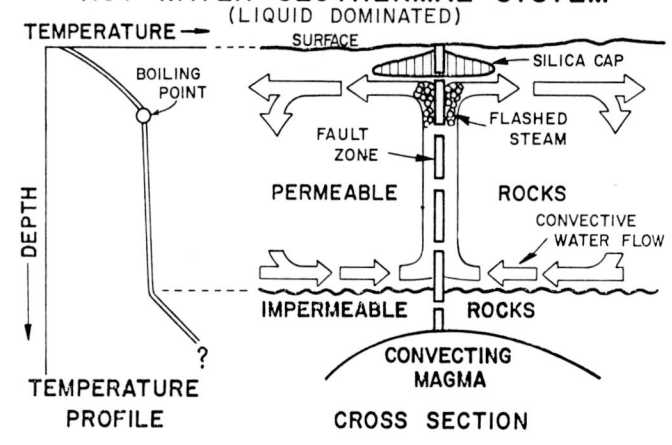

Fig. 2 — Geological Model of a Hot Water Geothermal System (after White, 1973).

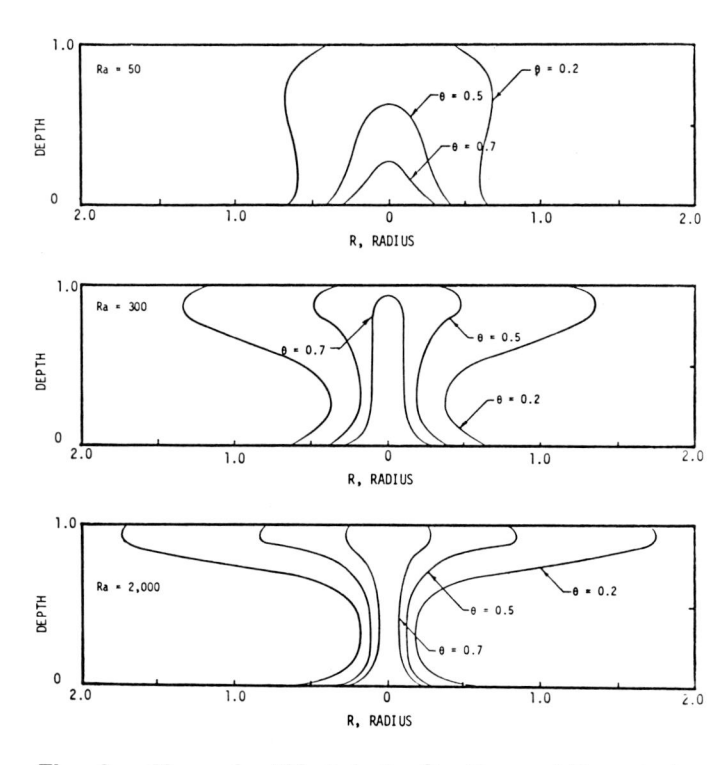

Fig. 3 — Numerical Model of a Geothermal Reservoir (after Cheng, et al., 1975)
R_a is the flow rate relationship (Raleigh number), Θ are isothermal lines, R is radius of cylindrical shape at certain depths.

in section to reveal the nature of the system present and will have the appearance of these theoretical models.

EXPLORATION METHODS

Figure 4 is a schematic showing the exploration methods useful in defining a commercial field. The first steps in geologic reconnaissance must be planned to locate areas with a local heat source and the presence of vertical conduits that will provide plumbing sufficient for the water movement that is required to mine the heat from deeper sources. Geologic history, tectonic setting, and a thorough surface examination of thermal occurrences will define areas for geophysical surveys.

Fig. 4 — Schematic of Methods Useful in Defining a Commercial Geothermal Field (after Greider, 1975).
Land is not shown in this figure; however, it is an integral part of the planning strategy.

Geochemistry

U.S. Geological Survey Professional Paper 492, by G. A. Waring (1965), is the most complete reference on thermal springs of the world. A study of the chemistry of waters produced from springs or shallow well bores may determine their recent thermal environments if the waters are in chemical equilibrium with their reservoir and have moved rapidly to the surface. There is a close association of warm springs with areas of recent volcanic activity. The importance of these waters as geothermometers is dependent upon whether chemical equilibrium has been maintained between the reservoir rock and the water. The temperature, discharge rate, composition, and elemental ratios of their solutes are analyzed to estimate the temperatures of the reservoir rocks within which they were resident.

The solubility of silica in water is usually a function of temperature and the type of silica being dissolved. Silica solubility decreases rapidly as temperatures decrease to 350°F but changes slowly below 350°F. Therefore, the usefulness of dissolved silica as a thermometer above that temperature is suspect. A system of using the molar concentrations of Na, K, and Ca occurring in thermal waters was devised by Fournier and Truesdell (1973). This is more reliable than the older Na to K ratio method. As salinity and temperature increase, the Na-K-Ca method becomes more reliable than the silica method. Work by Harmon Craig, Donald White, Robert Fournier, Patrick Muffler and Alfred H. Truesdell of the U.S. Geological Survey established the background for the modern chemical evaluation of temperature in the reservoir accepted by exploration geochemists today. Deposits of sinter or travertine around extinct springs can indicate their thermal history. Waters produced from a vapor-dominated reservoir differ chemically from those rising from a liquid-dominated reservoir.

Geophysics

Geophysical surveys are useful in predicting the general area and depth of high-temperature rocks and water. Rocks at depth are better conductors of electricity (natural and induced currents) when there is an increase in temperature, an increase in porosity, an increase in clay minerals, or an increase in salinity. Hydrothermal alteration of rocks may result in the formation of minerals such as zeolites and clays with a high ion-exchange capacity. The increase in conductivity is limited as temperatures increase above 350°F. However, salinity may increase sharply above this temperature and provide an additional decrease in resistivity. The contrast between a heat anomaly and normal background is most definitive in volcanic or igneous rocks. Porous sedimentary sections have a natural low resistivity, so the effects of temperature increases are less definitive in these rocks.

A number of electrical and electromagnetic methods can measure these changes in conductivity at depth and are useful in establishing the expected size and boundary shapes of the geothermal anomaly. The basic configuration of the common dipole profiling system is shown on Figure 5. There are several variations of this basic configuration.

Magnetic surveys are helpful in determining thickness of sedimentary cover and location of near-surface intrusive

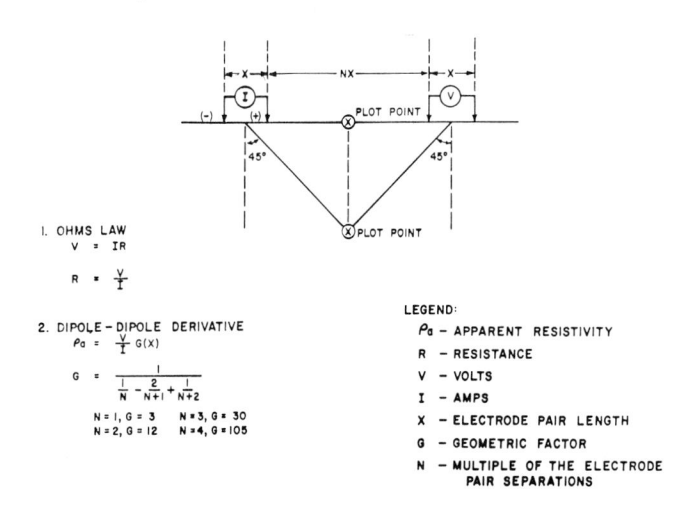

Fig. 5 — Dipole-Dipole Profiling with Basic Equations

masses. The contact between country rock and intrusive bodies frequently has local magnetic deposits. This results in a series of low-amplitude magnetic anomalies ringing the intrusive that may be low in magnetic response if the high temperatures exceed the Curie point.

Gravity surveys are useful in determining faults and basement structural trends.

In a sedimentary section where waters of greater than 400°F are rising, a zone of silicification may develop in the area above the thermal system as the waters cool below the solubility temperature for silica. This cap of silica should result in a positive gravity contrast with the surrounding sedimentary rocks. Near the water table, a boil off of a rising hot water column will also precipitate a silica cap. $CaSO_4$ and $CaCO_3$ solubility decrease as the temperature increases. These may precipitate around the flanks of the anomaly thereby increasing the local density of the warmer rocks. Figure 6 illustrates this principle (White, 1971).

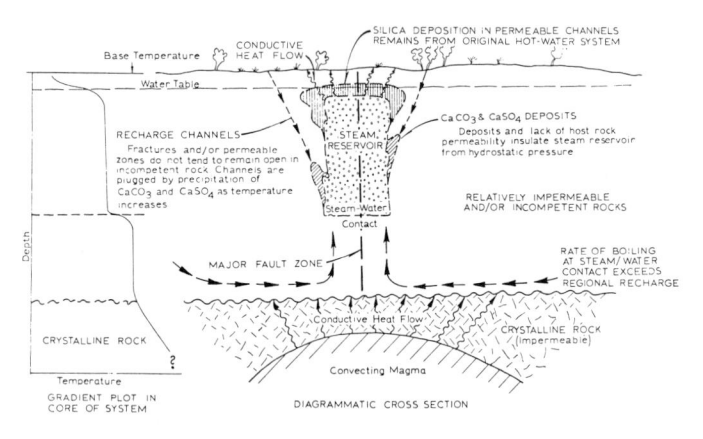

Fig. 6 — Generalized Model of High Intensity Vapor-Dominated (Dry Steam) System (after White, 1971).

Heat flow and temperature gradients measured in the upper 100 to 500 ft-depth are useful in describing the area where the heat transfer is most intense. When mapped, these do give a qualitative analysis as to the location and shape of the hottest near-surface heat accumulation. Linear projection of temperatures obtained near the surface cannot be used to predict the temperatures that will be encountered 2000-3000 ft below the surface, even if the section below has a uniform lithology and the geothermal gradient is a straight slope. The temperature for a fluid-saturated system cannot be projected to a maximum above that for boiling water at the pressure calcuated for the depth of projection. At any point along the boiling point curve, the temperature of the system may become isothermal and the rocks and fluids will have the same temperature for many hundreds of feet deeper. The rock temperature may decrease as a hole is drilled deeper if the hole is on the descending edge of a plume of hot water or merely below the spreading top of a plume. Heat flows from a hot body to a cooler body. This is not a function of being above or below a reference point of depth.

The slope of the temperature log may steepen for highly permeable beds and flatten for impermeable beds. In a temperature vs. depth plot, the more porous sections will have a low conductivity. As porosity decreases, the slope of the plot will increase; the more dense the rock becomes, the higher the conductivity. However, if a convection cell is encountered, the slope steepens to near vertical because the rapid movement of heat by convecting water produces an isothermal condition.

Test Wells

To effectively predict the performance of the geothermal cell, deep tests must be drilled. These holes must be of sufficient size to adequately determine the ability of the reservoir to produce fluids above 365°F at rates of more than 100,000 lbs of steam per hr or 500,000 lbs of liquid per hr. Although it is desirable that these fluids have less than 32,000 ppm dissolved solids and less than 1 percent non-condensable gases in solution, they may be extremely corrosive and dangerous to test.

The geochemistry of the produced fluids should confirm that the temperature of the reservoir from which they were produced was as high or higher than their measured temperatures in the bore hole.

The following series of temperature graphs are from "A Cooperative Investigation of the Geothermal Resources in the Imperial Valley of California." (Combs, 1971). Figure 7 depicts a rapid rise in temperature down to 125 ft, the curve becomes isothermal for 50 ft, then the temperature decreases gradually with depth increase to below 500 ft. Figure 8 illustrates an isothermal porous permeable section down to 150 ft beneath which there is a low permeability section with a strong temperature gradient. Below this interval from 350 ft to total depth of 542 ft the temperature plot is isothermal. These plots can be used to predict zones of conductive flow of heat and zones that are part of a convective system. Changes in porosity and effective permeability also can be estimated from these plots. Figure 9 is an isothermal plot indicating very small temperature gradient below near surface. Figure 10 indicates pluming with a flow of heat away from a rising source. The lower temperature interval from 300 ft to total depth is considerably better than normal and is within the geothermal anomaly.

TYPICAL PROJECTS

In the Western United States, there are abundant areas with temperatures between 300°F and 400°F. The Energy

Fig. 7 — (Combs, 1971)

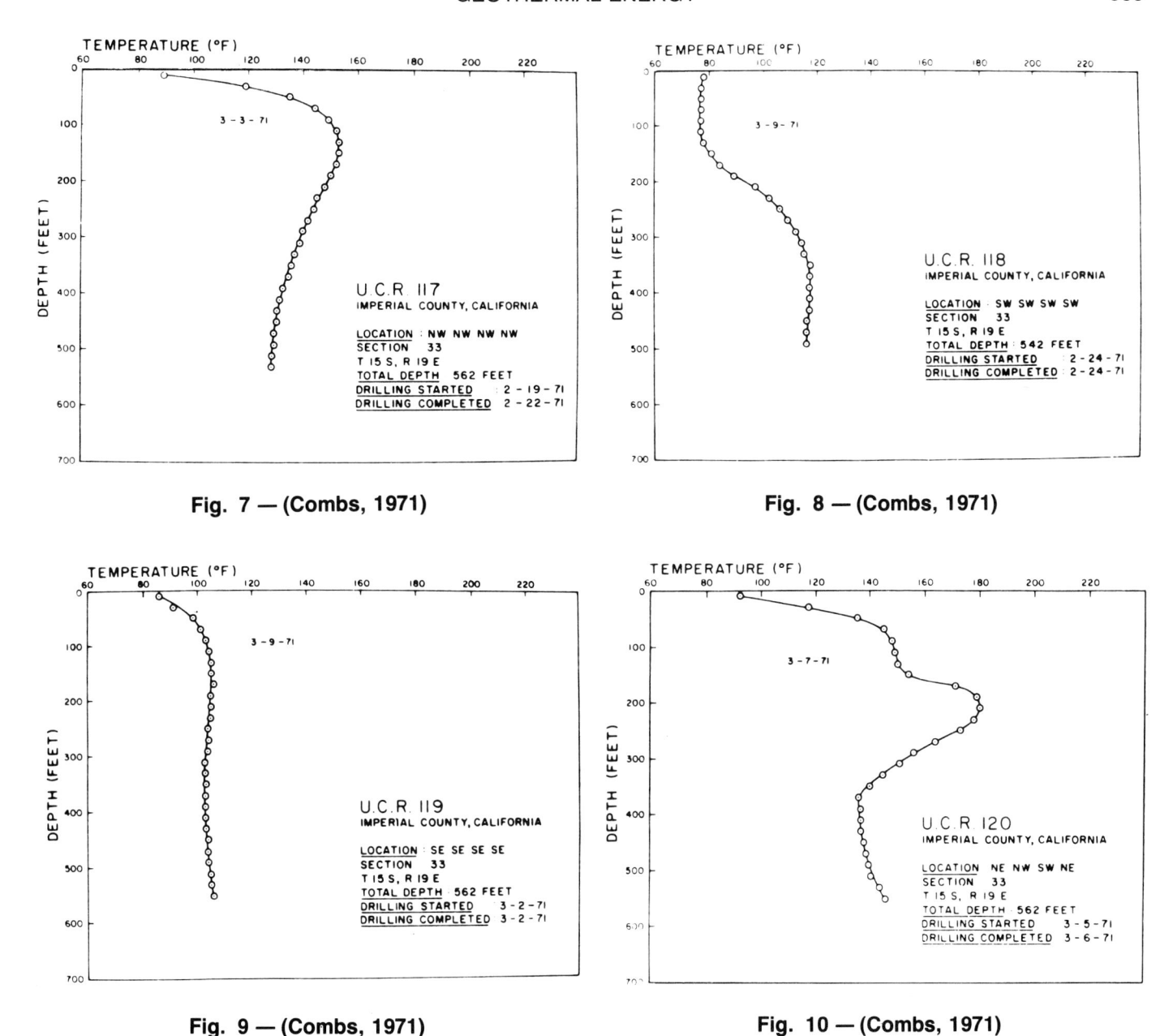

Fig. 8 — (Combs, 1971)

Fig. 9 — (Combs, 1971)

Fig. 10 — (Combs, 1971)

Research and Development Administration (ERDA), directly or with the National Science Foundation, has sponsored programs to locate, assess, and initially develop prospects in the lower half of that temperature range (300°F-350°F) because industry programs had objectives with more intense heat. A second phase has been to learn how to explore for the higher intensity anomalies. Geothermal energy will become more important as an energy source when the abundant mid-range temperature resource areas can be put to work. Geothermal areas with temperatures above 450°F are clearly commercial if the waters carrying the heat are low in salinity and can be produced at 50,000 bbl/day as at Cerro Prieto, Mexico.

The Basin and Range Province extending westward from the Colorado Plateau of eastern Utah contains more than a third of the strong surface indications of geothermal anomalies in the United States. This region includes the western half of Utah, southern Idaho, Nevada, the western half of New Mexico, the northern half of Arizona and the extreme eastern border area of California. The Sierra Nevada-Cascade Ranges form the western limit of this unusually hot geothermal province.

There are three easily accessible areas in this province that demonstrate the type of work being done by industry and federally funded projects. These are: (1) the Raft River project for low-temperature energy conversion to electricity; (2) the Battle Mountain high, where intermediate temperatures from 350°F to 400°F are being catalogued in an effort to develop sophisticated exploration techniques that may provide better resource assessment methods; (3) the development of geophysical techniques that can define the high-temperature areas of western Utah, such as the Roosevelt Known Geothermal Resource Area.

Raft River Low-Energy Project

Work sponsored by the Energy Research and Development Administration for low-energy geothermal resources is being managed by the Idaho Operations Office (formerly AEC). In the Raft River Valley of Cassia County, Idaho, a 300°F reservoir is being developed to support an electrical generation system being developed by Aerojet Nuclear Company. The U.S. Geological Survey has conducted geological, geophysical, and geothermal research to select the site for this development. Boiling water springs in the Raft River south of Malta, Idaho, were confirmed as coming from a moderate temperature (300°F) reservoir by extensive geochemical tests. The large amounts of stored heat known to exist in these low-energy reservoirs offered an attractive objective for developing a system that might make such a resource useful. To visit this area from Salt Lake City, take U.S. Highway 50 north to U.S. 80 north to Snowville, Utah, then turn northwest on U.S. 305. In about 50 mi you will cross the Raft River. Drive one-half mi, turn to the southwest, and continue about 4 mi to Sec. 23, where the low-energy geothermal project is located.

In 1975, two wells were drilled in Sec. 23, T15S, R26E. Each has encountered artesian flows at 295°F. The basic data (Table 1) obtained from extensive flow tests of both wells confirms the low-salinity (2,000 ppm), high-permeability reservoir extends at least 0.5 mi and will provide a large volume of fluid for power plant design tests. A third well has been completed this year and is similar to the first two.

Table 1. Results of initial test wells, Raft River low-energy project

	RRGE #1	RRGE #2
Location	NENESE Sec. 23 T15S, R26E	NENENE Sec. 23 T15S, R26E
Depth	4,989 ft	5,988 ft
Producing Interval	4,350-4,900 ft	4,300-5,000 ft
Max. Temp.	294°F	297°F
Flow Rate	650 gpm	800 gpm

Figure 11 is a schematic diagram of the method used to complete the wells in the Raft River project. The section above the major producing zone is isolated by a cemented 13-3/8 in. casing. A 12-1/4 in. hole is uncased as the producing interval.

The economics of using the moderate temperature geothermal type of fuel in this area for electrical generating plants is not favorable at this time. Additional engineering is needed to design a plant that is economically competitive, and reduced drilling costs must be effected.

It will be very difficult to lower costs for well drilling and operation of the field, or to lower plant capital costs to provide electricity as cheaply as Utah or Wyoming coal fuels. Agricultural applications will provide the greatest immediate benefit from these geothermal resources. If there is a local demand for the direct heat of the hot water, this would be economical if it can be produced for about $1.30 per million Btu.

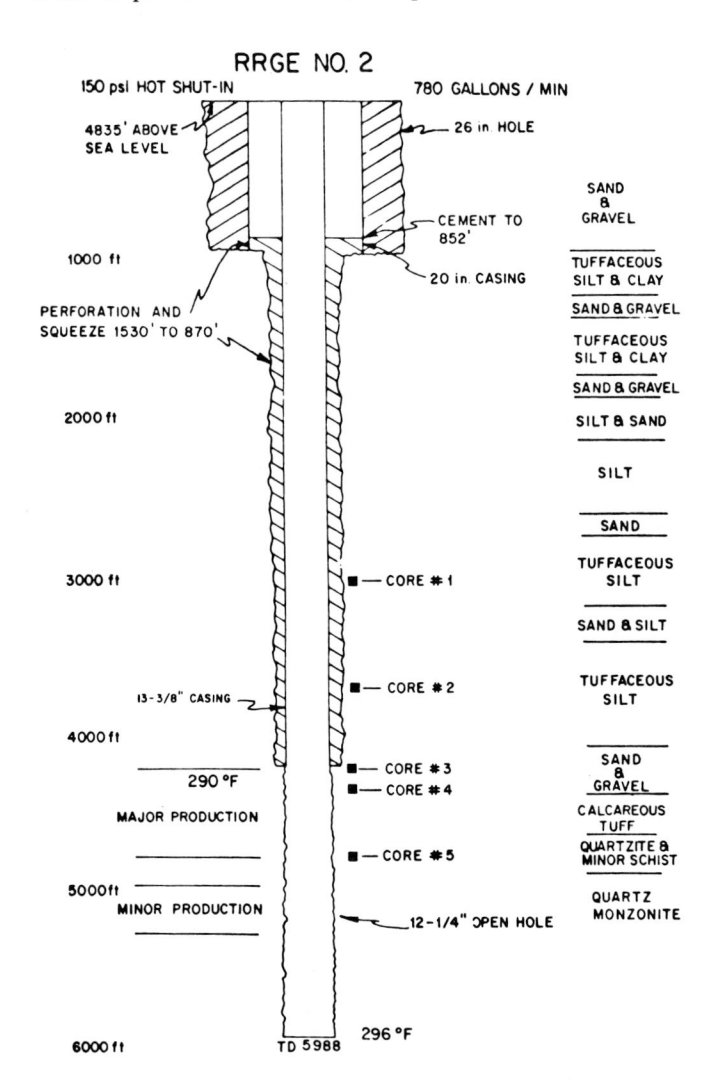

Fig. 11 — Schematic diagram of completion methods, Raft River Exploratory Well #2 (Kunze and Miller, 1975).

Battle Mountain High

The Battle Mountain High is an area of unusually high heat flow measurements in north central Nevada. Figure 12 shows the location of important thermal springs within this area in relation to other well known thermal areas in northwestern Nevada. The individual prospective areas can be reached from U.S. Interstate Highway 80 since this route circumscribes the western and northern perimeter of the area.

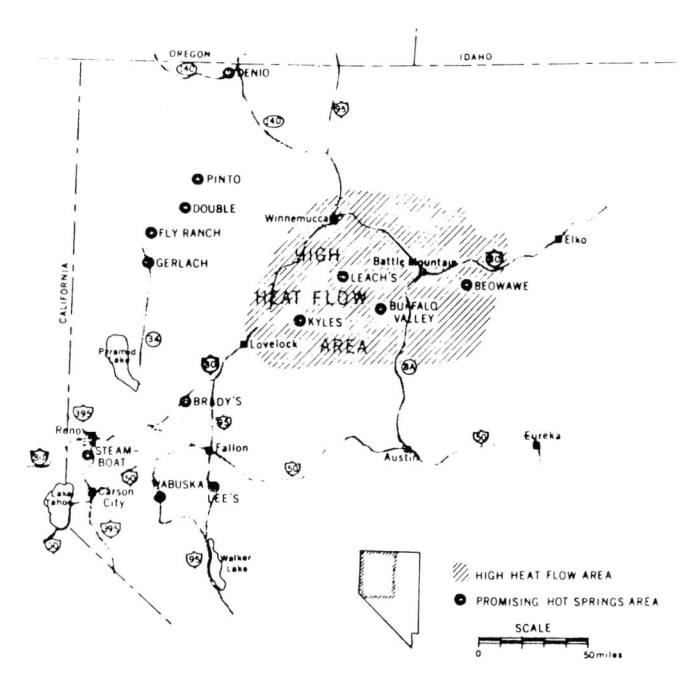

Fig. 12 — Hot Springs in Northwestern Nevada (Wollenberg, et al., 1975).

The federal government withdrew almost 100,000 ac from leasing within this area. The objective was to learn how to explore for geothermal resources with the hope that if all methods of geophysics and geochemistry were brought to bear on the problem a commercial discovery would result. A demonstration plant would then be justified to explore the optimum utilization scheme. A comprehensive report was published on the geology, geophysics and geochemistry conducted during this program (Wollenberg, et al., 1975).

Three areas within the Battle Mountain High are well known by industry and thoroughly surveyed by the recent federal assessment programs. These are shown in Figure 13.

Area A is Grass Valley, 20 mi south of Winnemucca. This is the locale of the Leach's Hot Springs. Area B is Buffalo Valley. Area C is Whirlwind Valley. This valley typifies the Basin and Range province with the steeply dipping ENE trending Malpais fault zone intersected by a NNW trending fault zone creating a locale for the prominent Beowawe Hot Springs and Geysers.

Each of these areas has active hot springs associated with warm ground temperatures. The springs are associated with steeply dipping faults that serve as plumbing zones to provide conduits for rapidly rising hot waters. Fault zones along the edges of the basins provide permeable fracture systems for the cold meteoric waters to descend deep enough to be heated to around 400°F. The heat is then carried upward to concentrate in the valley-fill alluvium, Tertiary sediments, or fractured volcanic rocks beneath the alluvium. Published geochemical data have not indicated that temperatures above 400°F can be expected at reasonable depths.

The occurrence of extensive lacustrine deposits in the valley alluvium creates resistivity lows. This creates a difficult condition to properly interpret the geophysical surveys in these areas. Lawrence Berkeley's geophysical interpretations demonstrate heavy emphasis must be given to obtaining accurate geologic data on the stratigraphy of the valleys to develop proper models of the prospective area. Heat flow or thermal gradient holes provide this important dimension.

Upthrown blocks of Tertiary volcanics form the northern and southern borders of Whirlwind Valley. The Malpais fault escarpment forms the southern edge of the Valley. The escarpment has extensive alteration and siliceous sinter deposits which define an area of near-surface hydrothermal activity.

Magma Power Company discovered the near-surface geothermal reservoir associated with this hydrothermal alteration. The first two Magma wells failed to find the permeable reservoir. The next three wells drilled into a permeable fault zone. At depths between 590 and 660 ft, high flow rates were encountered with water temperatures of 410°F. The discovery well encountered one of the strongest flows ever measured from a liquid-dominated geothermal reservoir. Almost 4,700 barrels an hour were measured with 116.5 p.s.i.g. and a well-head temperature of 342°F. Figure 14 is a schematic section across Whirlwind Valley showing the hot springs location of these wells (Wollenberg, et al., 1975). A more extensive reservoir must be defined before commercial development can be expected.

In 1973, an electrical resistivity bipole-dipole survey was funded by the predecessors of ERDA. This survey measured a low-resistivity zone (30-20 ohm meters) that circumscribed the geysers-hot springs area and the Magma wells. To the east, a prominent north-south elongated low (20 ohm meters) is believed to be a fault intersection with the Malpais fault zone. Figure 15 illustrates the values recorded by this survey (Wollenberg, et al., 1975). The location of the schematic section (Fig. 14) is shown on this map for comparison.

Roosevelt Area, Utah

This area is located 8 mi northeast of Milford in Beaver County, Utah. The area can be reached by turning west from U.S. Interstate Highway 15 at Beaver and driving west northwest on State Highway 21 about 30 mi to Milford, then north on State Highway 257 about 5 mi, where a county road leads to the northeast into the geothermal area.

This is the area of the most recent geothermal discovery in the region. Phillips Petroleum Company drilled two failures in the Roosevelt KGRA, then followed these with three successful wells in an area between the Domal fault and the Mineral Range.

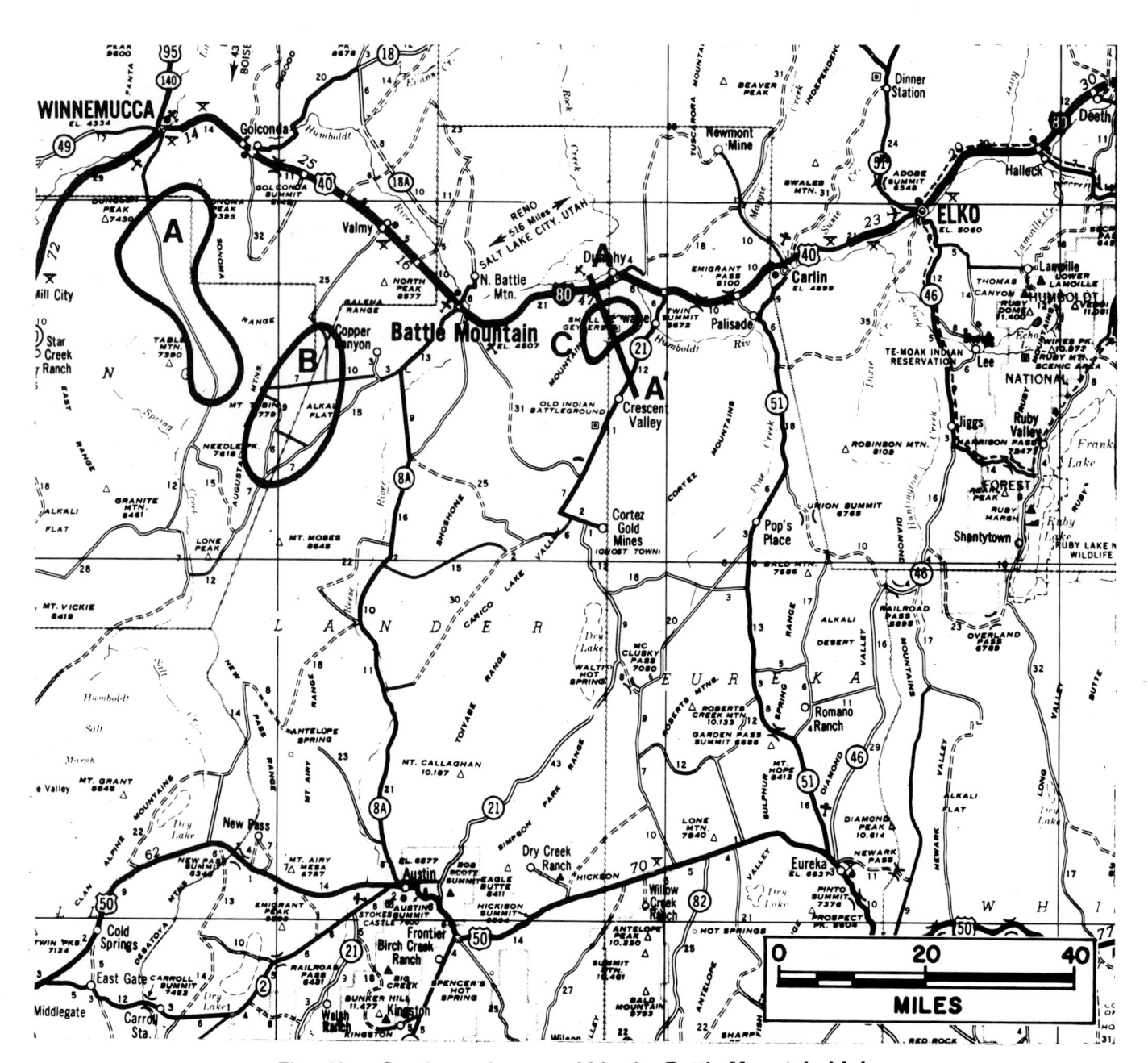

Fig. 13 — Geothermal areas within the Battle Mountain high.

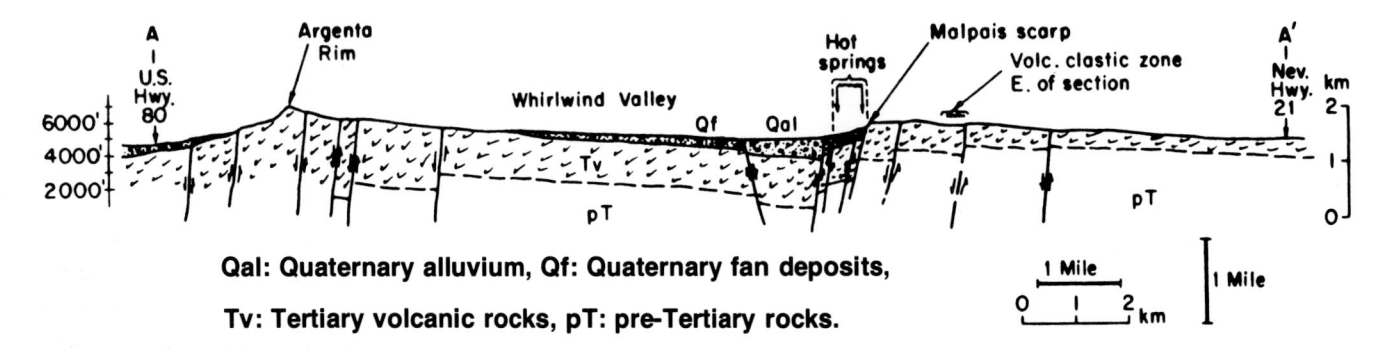

Qal: Quaternary alluvium, Qf: Quaternary fan deposits,
Tv: Tertiary volcanic rocks, pT: pre-Tertiary rocks.

Fig. 14 — Schematic Geologic Cross Section A-A', Whirlwind Valley area (Wollenberg, et al., 1975).

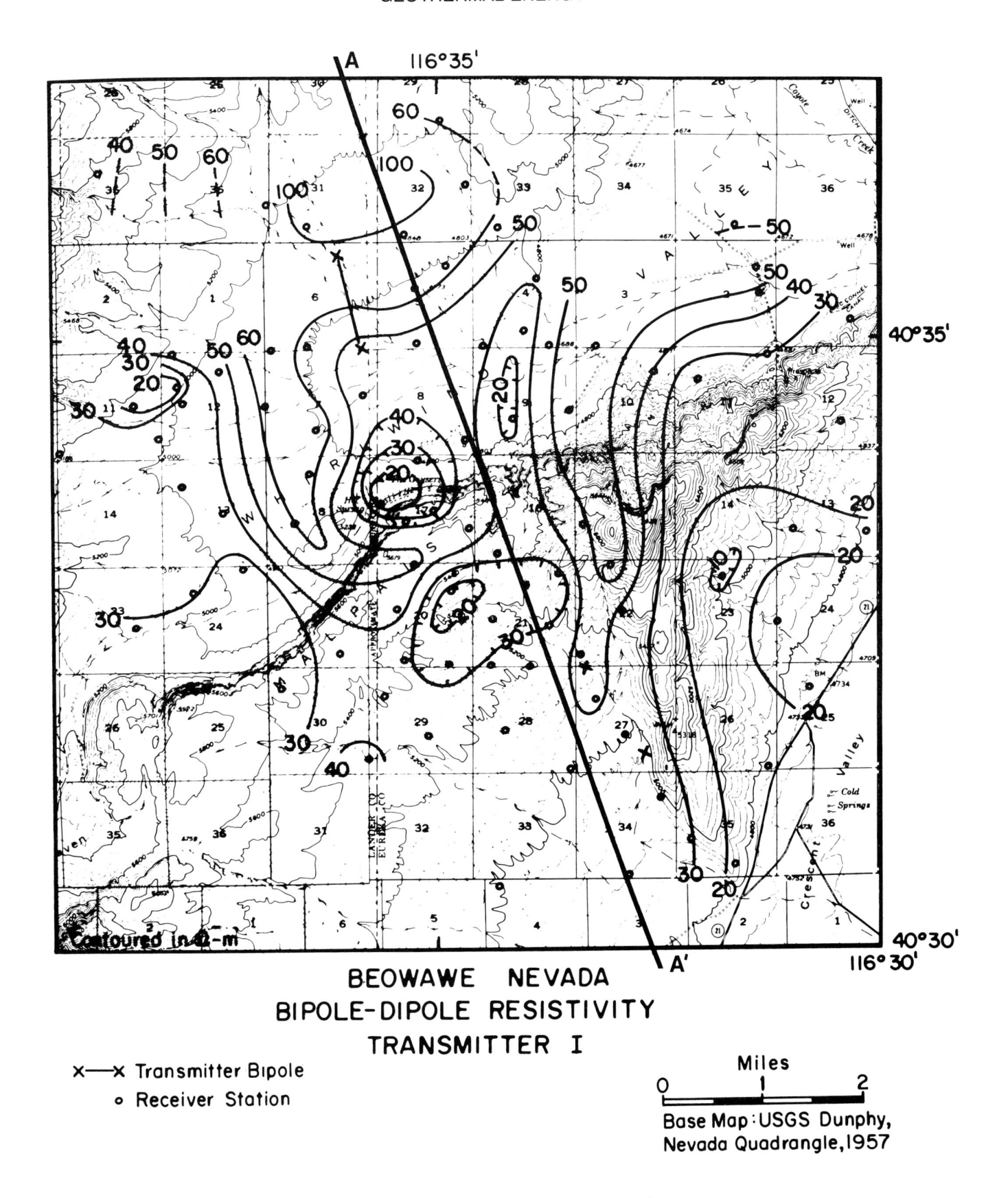

BEOWAWE NEVADA
BIPOLE-DIPOLE RESISTIVITY
TRANSMITTER I

x—x Transmitter Bipole
o Receiver Station

Base Map: USGS Dunphy,
Nevada Quadrangle, 1957

Fig. 15 — (Wollenberg, et al., 1975)

The west side of the Mineral Range is bounded by several parallel north-south trending high-angle normal faults. Where these are intersected by faults striking east-west, centers of increased vertical permeability are created. To the west of this block fault system, the valley is filled with 4,000-6,000 ft of Quaternary or Tertiary alluvium.

The Domal fault is one of the north-south fault features and has probably been the main conduit for hydrothermal fluids rising to the surface. A siliceous-cemented alluvial surface unit with dimensions about one-third of a mi east-west and one-half mi north-south is located at the south end of the Domal fault. This appears to be the result of silica-rich waters reaching the surface and forming the cap of a self-sealed geothermal system (Petrick, 1974). The Domal horst lies west of the Domal fault and forms the western edge of the Roosevelt geothermal prospect.

The University of Utah's geothermal team, under the direction of S. H. Ward, has conducted a comprehensive geological and geophysical evaluation of the western Utah seismic belt from St. George northward through Cove Fort. A.S.F. Grant GI43741 provided funds to concentrate on the Roosevelt Hot Springs KGRA.

Figure 16, from the Energy Resource Map of Utah, outlines areas thought to be prospectively valuable for geothermal resource exploration. The Cove Fort and Roosevelt Hot Springs areas are receiving the greatest industry activity in Utah. The large area between Lund and Roosevelt Hot Springs is the locale chosen for the assessment of geothermal/geophysical techniques by the University of Utah. Originally designated Known Geothermal Resource Areas are shown in hachured patterns. Lands involved in competitive interest are shown in black.

The conceptual model developing from their work is similar to that reported by Phillips Petroleum Company team (Birge, et al., 1976). The uplift of the Tertiary granite in the Mineral Range induced intense fracturing. This was followed by a silicic intrusion-extrusion of Quarternary age. A convective hot water system was formed with these young silicic rocks being the source of the heat and silica being carried by the thermal waters. The heat carried in these silica-rich waters provided the energy to form rock alteration minerals of opal, alunite, quartz, kaolinite, and montmorillonite.

Temperatures of the geothermal system calculated by fluid geochemistry match the temperatures in the wells being developed by Phillips Petroleum Company. These appear to be 500 to 510°F (260°C to 265°C). The Na-K-Ca method and the SiO_2 method give the same temperatures. However, analysis of waters obtained from seeps calculate to be 465°F (241°C) by the Na-K-Ca method and 257°F (125°C) by the SiO_2 method. The system appears to be leaking and mixing with near-surface waters which then seep to the surface.

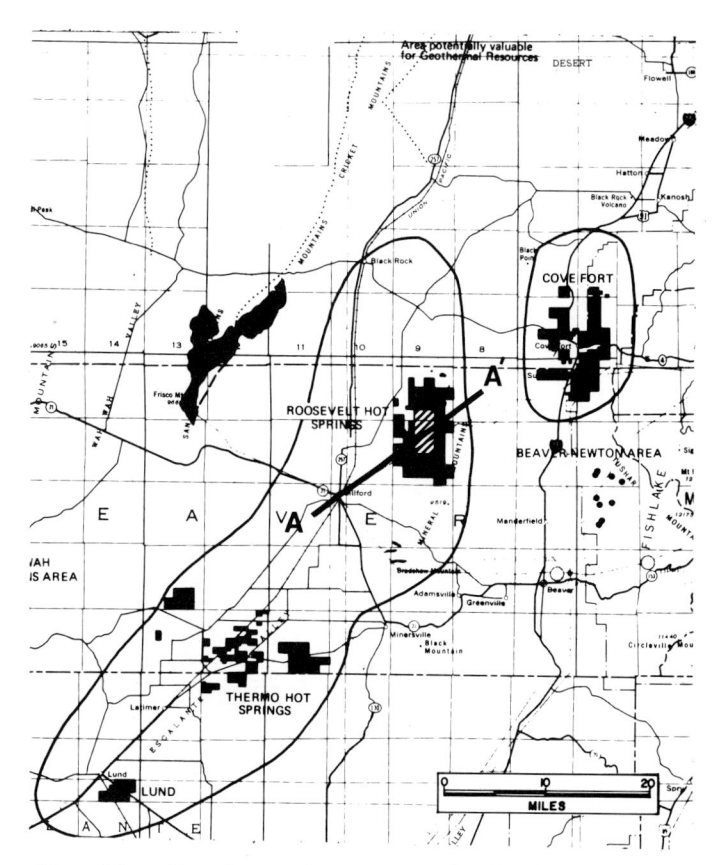

Fig. 16 — Geothermal prospects in the Roosevelt KGRA area (Utah Geological and Mineral Survey, 1975).

Dipole resistivity, magneto-tellurics, and shallow temperature surveys outlined an anomalous area that coincides with the location of the hot water reservoir. The low-resistivity zone is thought to be the result of a hot-water filled fracture system between 3,000 ft and 4,000 ft in depth. Figure 17 summarizes the structural concepts presented by Berge, et al. (1976).

ECONOMICS

In the last five years, it has become evident that established sources of energy are becoming more expensive. People planning natural resource exploration and development have recognized that an increased effort must be made to locate and develop energy sources that can provide economical space heat, transportation, and industrial power. Because electricity is easy to transport from point of generation to places of consumption, there has been a continuing growth (3-4 percent/yr) in electricity demand at the expense of other forms of energy. Interest in using the heat of the earth to provide an indigenous source of energy has begun to increase almost as rapidly as energy bills. Government agencies are funding projects with the objective of increasing investors' desires to develop this widespread source of energy. Natural resource development com-

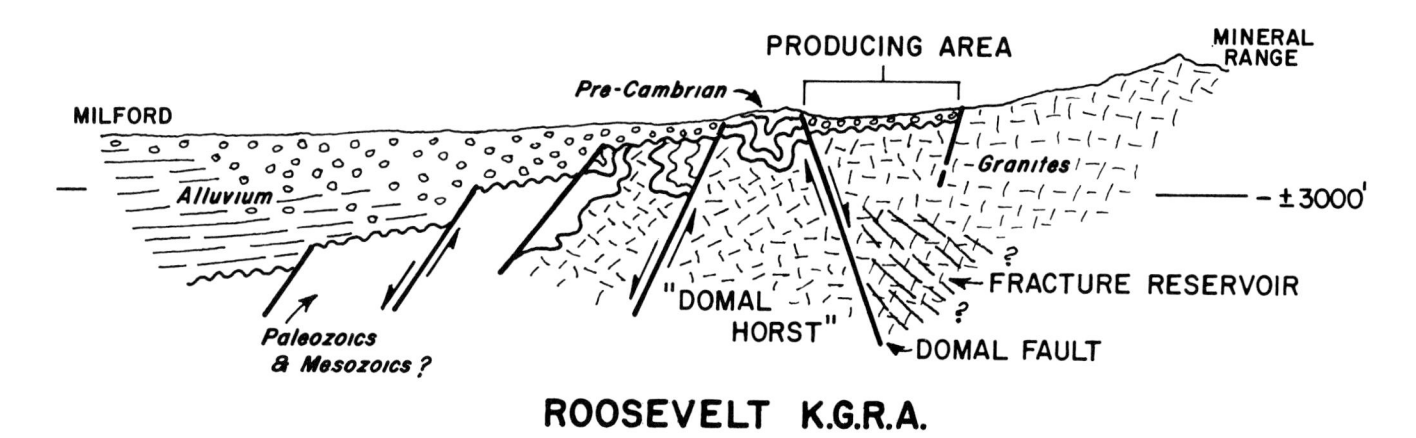

Fig. 17 — Generalized structural interpretation Roosevelt KGRA (Berge, et al., 1976).

panies and groups of investors are increasing their exploration for accumulations of heat that can be used in electrical generation, space heating and cooling, agriculture, and industrial process heating.

Developers expect these natural sources of heat in the Western United States to produce electricity at prices competitive with low sulfur coals shipped from the Powder River basin of Wyoming to the electricity generating centers supplying western Nevada and California. If a 300-Megawatt plant produces electricity at the price of coal per kilowatt hour, this should be 6 mills less than oil-fueled plants. It could provide an annual savings for customers of 15.75 million dollars. Of equal importance, more than 100 million bbls of fuel oil will be saved for other uses during a 30-year plant lifetime. Water in the low-energy 150°F temperature range can provide processing heat. Savings in fueling a 1,500-2,000 ton per day uranium mill, for instance, would be between $350,000 and $450,000 per yr.

It is expected that sulfur limits of fuel oil will be set similar to coal; 0.3 percent sulfur content is the maximum amount allowed in California. It is expected that other areas will adopt similar standards. To meet such standards, additional investment and costs will be required to prepare acceptable fuel. With such increases in cost, new uses for geothermal heat (energy) will become practical. When that happens, more people will become interested in joining the exploration search to find new deposits of heat for production of energy.

The development of a geothermal reservoir is capital-intensive, requires expert planning, and long times from initial expenditure until positive income is achieved. The utilization of a developed project requires extensive engineering, approximately two yrs in negotiation with governmental agencies, and a lot of money.

The costs of maintaining and operating the producing fields is about four to five times greater than the capital investment. An important portion of this cost is associated with the injection system that collects the water after the heat is removed and then returns it to the subsurface reservoirs. Reducing these costs is an essential objective if geothermal is to be competitive with other fuels.

REFERENCES CITED

Berge, C. W., Crosby, G. W., and Lenzer, R. C., 1976, Geothermal exploration of Roosevelt KGRA, Utah [abs.], Rocky Mtn. Sec. Amer. Assoc. Petroleum Geologists and Soc. Econ. Paleontologists & Mineralogists, 25th Ann. Mtg., Billings, Mont.

Combs, Jim, 1971, Heat flow and geothermal resource estimates for the Imperial Valley, Cooperative Geological-Geophysical-Geochemical Investigations of Geothermal Resources in the Imperial Valley of California: University of California, Riverside, p. 5-27.

Cheng, P., Lau, K. H., and Lau, L. S., 1975, Numerical modelling of geothermal reservoirs: The Hawaiian Geothermal Project, summary report for phase 1, p. 83-110.

Diment, W. H., Urban, T. C., Sess, J. H., Marshall, B. V., Munroe, R. J. and Lacherbruch, A. H., 1975; Temperatures and heat contents based on conductive transport of heat: U.S. Geol. Survey Circ. 726, p. 84-121.

Energy Resources Map of Utah, 1975, Utah Geol. and Mineral Survey.

Fournier, R. O., and Rowe, J. J., 1966, Estimation of underground temperatures from the silica content of water from hot springs and wet stream wells: Am. Jour. Science, v. 264, p. 685-697.

Fournier, R. O., and Truesdell, A. H., 1973, An emperical Na-K-Ca geothermometer for natural waters: Geochim. et Cosmochem. Acta, v. 37, p. 1255-1275.

Fournier, R. O., White, D. E. and Truesdell, A. H., 1974, Geochemical indicators of subsurface temperatures, part 1, basic assumptions: U.S. Geol. Survey Jour. Research, v. 2, no. 3, p. 259-262.

Grieder, B., 1975, Status of economics and financing geothermal power production, in 2nd U.N. Symposium on the Development of Geothermal Resources (in press).

Hose, R. K., and Taylor, B. F., 1974, Geothermal systems of northern Nevada: U.S. Geol. Survey Open-File Report 74-271, 27 p.

Kunze, J. F., and Miller, L. G., 1975, Geothermal R&D project report for period July 1, 1975, to September 30, 1975: ANCR 1281 Aerojet Nuclear Co., 54 p.

Oesterling, W. A., 1964-1965, Reports and lithologic well logs of Beowawe wells to Sierra Pacific Power Co.

Olmsted, F. H., Glancy, P. A., Harrill, J. R., Rush, F. E. and Van Denburgh, A. S., 1975, Preliminary hydrogeologic appraisal of selected hydrothermal systems in northern and central Nevada: U.S. Geol. Survey Open-File Report 75-56, 267 p.

Petrick, W. R., 1974, Test electromagnetic soundings, Roosevelt Hot Springs KGRA: University of Utah — NSF Technical Report 74-1, 17 p.

Renner, J. L., White, D. E. and Williams, D. L., 1975, Hydrothermal convection systems: U.S. Geol. Survey Circular 726, p. 5-57.

Truesdell, A. H., and White, D. E., 1973, Production of superheated steam from vapor-dominated reservoirs: Geothermics, v. 2, p. 145-164.

Waring, G. A., 1965, Thermal springs of the United States and other countries of the world — a summary: U.S. Geol. Survey Prof. Paper 492, 383 p.

White, D. E., 1973, Characteristics of geothermal resources: Geothermal Energy, Stanford Univ. Press, p. 69-93.

──────────, Muffler, L. J. P., and Truesdell, A. H., 1971, Vapor-dominated hydrothermal systems compared with hot-water systems: Econ. Geol., v. 66, p. 75-97.

Wollenberg, H. A., Asaro, F., Bowman, H., McEvilly, T., Morrison, R., and Witherspoon, P., 1975, Geothermal energy resource assessment: Energy and Environment Division, Lawrence Berkeley Laboratory, University of California, UCID 3762, 92 p.

Young, H. C., and Mitchell, J. C., 1973, Geothermal investigations in Idaho, part 1, Idaho Dept. of Water Administration, Water Information Bulletin No. 30, 39 p.

STRATIGRAPHIC AND STRUCTURAL SETTING OF THE COTTONWOOD AREA, UTAH

by

Max D. Crittenden, Jr.[1]

ABSTRACT

The site of the Cottonwood area, Utah, lay astride the hingeline of the Cordilleran miogeocline from at least early Precambrian Z time (800 my) until the end of the Paleozoic. Full understanding of the stratigraphic relations was long delayed, however, because basin deposits have been thrust eastward over those deposited on the shelf. Stratigraphic thickening that marked the westward transition to the basins can be seen only locally beneath the leading edges of the thrusts.

Thickness contrasts across the hingeline amount to 15 or 20 to 1 in rocks of Precambrian Z age, 20 or 30 to 1 in rocks of early Paleozoic age (in large part owing to uplift and erosion during a Late Devonian disturbance), and 10 or 12 to 1 during late Paleozoic. Stratigraphic contrasts appear to be negligible during early Mesozoic, but so few rocks of this age are preserved within the western basins that little is known of their continuity. Still less can be said of late Mesozoic—for by the beginning of Cretaceous, reversal of the drainage initiated during the Jurassic had become permanent, and uplift and erosion were yielding large volumes of coarse clastic debris to foredeep basins which lay just east of the Cottonwood area.

The Cottonwood area is uniquely situated also in that it lies at the intersection of the north-trending Sevier orogenic belt, and the east-trending Uinta arch, and the record of their interaction is preserved in the adjoining Parleys Canyon syncline. It is evident that thrust sheets were being emplaced from the west at the same time that uplift and folding were taking place on the Cottonwood portion of the Uinta arch. The earliest thrusts (Alta and Mt. Raymond) are folded with the arch, the latest one (Charleston-Nebo) overrode the Uinta arch entirely. As a consequence, it appears that west to east thrusting alternated with folding on this and adjoining east-trending fold axis, implying, if interpreted in the simplest fashion, a spasmodic 90° reversal of stress directions that is difficult to believe, much less to understand.

The anomalous contrast in timing between the Idaho-Wyoming thrust belt in which thrusts are progressively younger eastward, and the Cottonwood area where they are younger to the west is inferred to result from thrust sheets moving unimpeded into the growing foredeeps in Wyoming at the same time they were piling up against the end of the growing Uinta Mountains to the south. If true, these relations imply complex geometric and temporal relations including tear faults along the margins of the western Uinta Mountains.

INTRODUCTION

Selection of a site at Alta in Little Cottonwood Canyon (Fig. 1 and 2) for a symposium on the stratigraphy and structural features of the Cordilleran hingeline is particularly appropriate because it lies exactly on the Wasatch Line used by Kay (1951, p. 14) as the boundary between the stable platform and the geosyncline (Fig. 1). Moreover, this area is surrounded by a wealth and variety of geologic features excelled by few other areas in the world. Within an hour's drive it is possible to visit extensive exposures of Precambrian igneous and metamorphic rocks at least 1.5 billion years old, sections of essentially unmetamorphosed sedimentary rocks ranging in age from more than one billion years to deposits formed in Lake Bonneville only ten to twenty thousand years ago. These sedimentary rocks include Precambrian deposits formed in glacial or glaciomarine environments, redbeds typical of the Triassic and Jurassic section in many parts of the world, and coarse orogenic deposits of Cretaceous and Tertiary age which provide a record of tectonic events of maximum interest to the symposium.

---

[1] U.S. Geological Survey, Menlo Park, California

I am indebted to H. T. Morris and N. K. Huber for thoughtful and rapid review of the manuscript, and to R. W. Kopf for painstaking attention to geologic names.

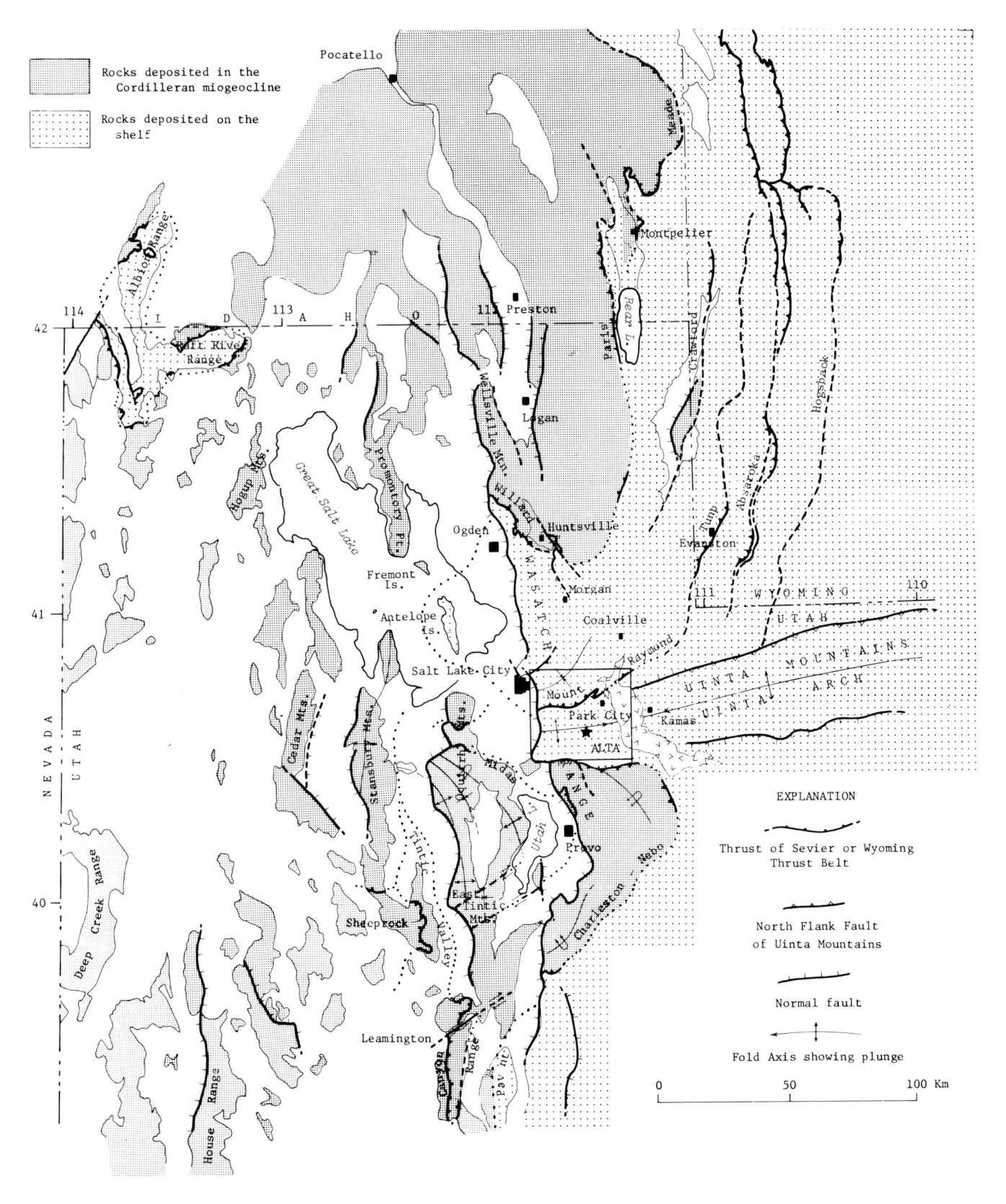

Fig. 1 — Index map showing generalized sedimentary facies in northern Utah, and structural setting of Cottonwood area (outlined) and of conference site at Alta (star). Keetley Volcanics shown by "V" pattern. Geology adapted from Stokes, *et al.* (1961-1964), and Royse, *et al.* (1975).

In another sense, the Cottonwood area is uniquely situated with respect to features of the hingeline inasmuch as it lies within a re-entrant of the thrust belt surrounded on the west and south by allochthons which bring into close view sedimentary facies originally deposited far to the west, providing easy access, and emphasizing the contrast in depositional history. The purpose of the present paper is to provide a generalized view of the resulting stratigraphic contrasts, of the structural setting of the Little Cottonwood area, and particularly of the enigmatic relations in northeastern Utah between the Sevier and Idaho-Wyoming thrust belts and the Uinta Mountains.

Stratigraphic descriptions will emphasize those parts of the section that illustrate contrasts across the hingeline or that demonstrate specific phases of the structural history; intervening parts of the section have been described repeatedly elsewhere and will only be summarized here.

STRATIGRAPHY

Rocks of Precambrian X age (1600 to 2500 m.y.)

Farmington Canyon Complex. — The Precambrian rocks most extensively exposed in the Salt Lake area are part of the unit called the Farmington Canyon Complex by Eardley and Hatch (1940). These rocks are exposed in the Wasatch Range from a point about five mi northeast of Salt Lake City where they form the rugged summit of the Wasatch Range, continuing north to a point about fifteen mi north of Ogden. They are also exposed on Antelope Island in Great Salt Lake. The Farmington Canyon Complex consists largely of gneissic rocks that are the product of moderate- to high-grade regional metamorphism. They are commonly strongly banded at all scales and

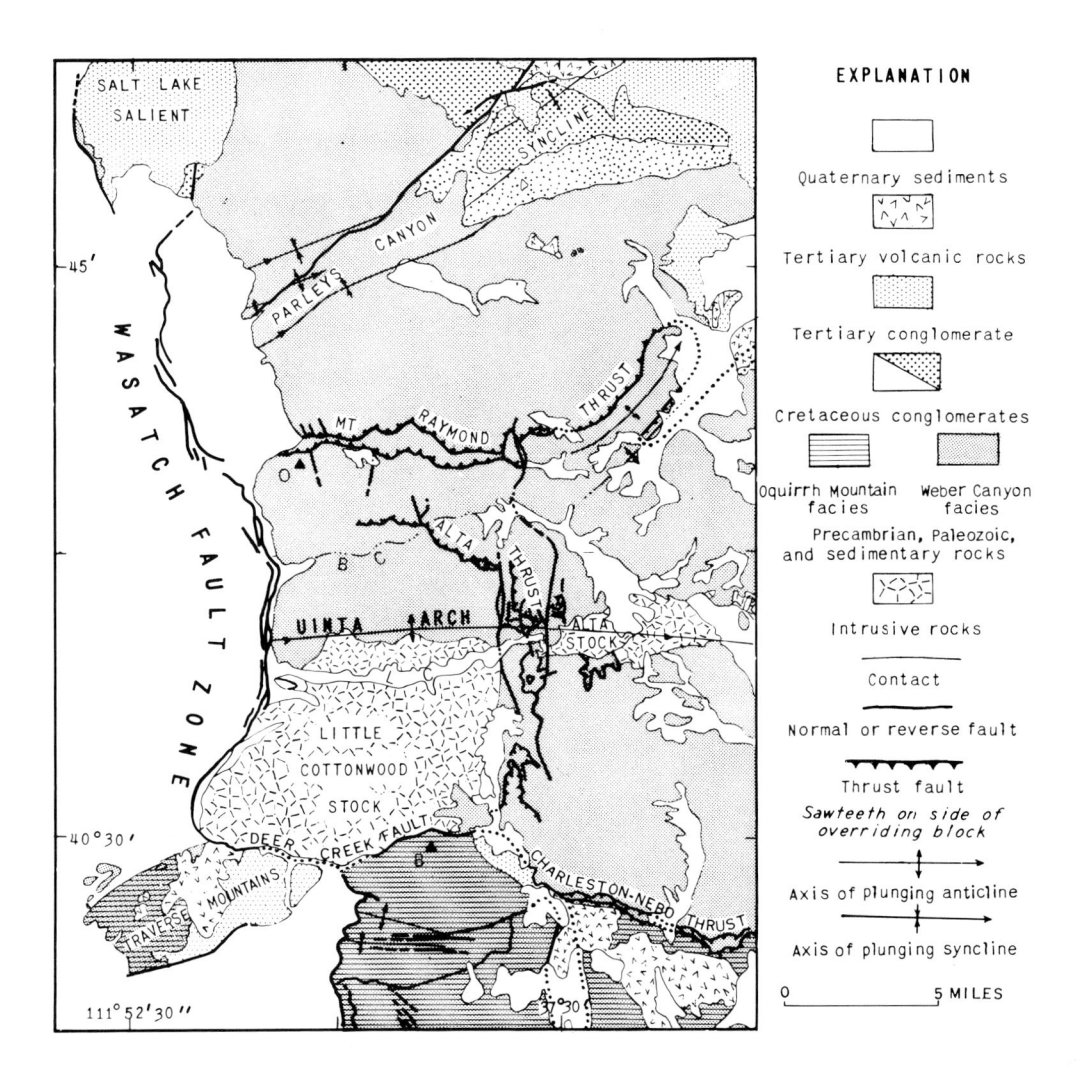

Fig. 2 — Generalized structure of central Wasatch Range east of Salt Lake City, northern part after Granger and Sharp (1952). O-Mt. Olympus, B-Boxelder Peak, BC-Big Cottonwood Canyon, LC-Little Cottonwood Canyon.

range widely in composition from mafic rocks containing hypersthene, biotite, and garnet, which may have been derived from mafic igneous rocks, to gneisses consisting largely of quartz, feldspar, and muscovite which may have been derived from quartzite, aluminous sediments, or tuffs.

The dominant characteristic of the complex as a whole is the pervasive development of pegmatite in dikes and lenses at all scales. The most abundant pegmatites consist of quartz and salmon pink microcline along with smaller amounts of biotite and muscovite. Gray plagioclase-rich pegmatites are present locally.

The age of the Farmington Canyon Complex is known to be at least 1.6 to 1.8 billion years on the basis of both potassium-argon and rubidium-strontium dating largely of muscovite, biotite, and hornblende. These dates represent a major episode of metamorphism and orogeny, but until further studies are carried out on zircons and other minerals using uranium-lead techniques it remains uncertain whether the rocks may have been deposited and initially deformed much earlier. These crystalline Precambrian rocks represent the western margin of the North American craton. Condie (1969) regards them as a part of the Churchill Province, which extends east-northeast across extensive ares of central North America. As a consequence they are representative of the crystalline basement beneath the thick prisms of sedimentary rock throughout the area of northeastern Utah, southeastern Idaho, and western Wyoming.

Little Willow Formation. — A sequence of strongly folded gneissic quartzites, mica schists, and cobble or pebble schists, intruded by mafic igneous rocks now altered to amphibolite, is exposed in an area a few miles square at the mouth of Little Cottonwood Canyon (Fig. 2) and in its type locality in Little Willow Creek immmediately to the north. These rocks are known to be Precambrian because they are overlain unconformably by the Big Cottonwood Formation now regarded as an approximate equivalent of the Belt Supergroup of Montana and Idaho. The Little Willow is lithologically distinct from the Farmington Canyon Complex because it is dominated by schistose rocks predominently of sedimentary origin, but more significantly because it contains almost none of the pegmatites that are so characteristic of the Farmington Canyon Complex.

The exact age of the Little Willow Formation is unknown because it is so strongly metamorphosed by the Little Cottonwood Stock that rubidium-strontium and potassium-argon systems have been almost completely reset. In general character it most resembles the Red Creek Quartzite, a Precambrian unit, exposed at the east end of the Uinta Mountains which has been determined to be approximately 2.3 billion years old (Hansen 1965, p. 31). Nevertheless, it is strikingly lower in metamorphic grade than either the Red Creek or the Farmington, and may possibly be younger than both.

Rocks of Precambrian Y age (800 to 1600 m.y.)

Big Cottonwood Formation. — The Big Cottonwood Formation is a 16,000 ft thick unit consisting of mostly quartzite and siltstone that forms the core of the Uinta arch where that structure reaches the front of the Wasatch Range. Except where affected by the Little Cottonwood stock, it is essentially unmetamorphosed. The type locality is in the lower part of Big Cottonwood Canyon, about 5 mi west north-west of Alta (Fig. 3). The base of the formation is exposed in Little Willow Canyon where the basal beds rest unconformably on the type Little Willow Formation. At its top, the Big Cottonwood Formation is overlain unconformably by the Mineral Fork Tillite, or where that is missing, the Mutual Formation. The lower third of the Big Cottonwood consists of bluish-purple thin-bedded shale or siltstone interbedded with texturally mature pinkish-gray orthoquartzite. The middle third is dominated by greenish-gray or gray shale interbedded with gray and greenish orthoquartzite. The upper third is characterized by variegated green and red shale and siltstone intercalated with buff and white quartzite. Pelitic units are thickest and slightly more abundant in the lower half. Overall the ratio of shale and siltstone to quartzite is almost one to one. Cross bedding, ripple marks, mud cracks, and rain-drop prints are found throughout and suggest that the entire unit was deposited in shallow water.

The Big Cottonwood Formation is now tentatively believed to be correlative with the Uinta Mountain Group exposed in northeastern Utah and both are tentatively regarded as an equivalent of the Belt Supergroup in Montana and Idaho (Crittenden and Peterman, 1975).

The Big Cottonwood Formation and Uinta Mountain Group are presumed to have been deposited at the same time in an east-trending basin whose northern margin lay a short distance north of Salt Lake City, and extended almost due east along the present Utah-Wyoming state line. The possible westward extent of this basin is unknown as is its possible connection with the area of Belt deposition further north. These strata are absent in northwestern Utah in the Raft River-Albion area (Compton, *et al.*, 1977) indicating that they were either eroded from the western edge of the craton in that area or were never present.

Rocks of Precambrian Z (570 to 800 m.y.) and Early Cambrian Age

Thick sequences of Precambrian strata younger than the Big Cottonwood Formation and generally presumed to be equivalent in age to parts of the Windermere Group in Washington and southern British Columbia have been recognized from Pocatello, Idaho, to southeastern California (Crittenden, *et al.*, 1971; Crittenden, *et al.*, 1972; Stewart, 1972). Sections deposited along the strike of the Cordilleran geosyncline extending

Fig. 3 — View east across Big Cottonwood Canyon showing type section of Big Cottonwood Formation (pЄbc) in foreground, Mineral Fork Tillite (pЄmf), and Mutual Formation (pЄm). Unconformity at base of Tintic Quartzite (Єtq) is visible in cliffs above Mineral Fork.

from Pocatello, Idaho, to the Wah Wah and Beaver Mountains of west-central Utah can be correlated on the basis of a distinctive and consistent sequence of lithologic units. Sections at right angles to the depositional strike are much harder to correlate, owing to east-west facies changes complicated by later overthrusting and extensive sedimentary cover. The thickest known section (25,000 ft) is at Pocatello (D. E. Trimble, unpub. data). The same units are recognized near Huntsville, Utah, where the section is estimated to be about 15,000 ft thick. This section, described briefly below (Fig. 4), begins above the Willard thrust, along the Wasatch front and extends eastward across the Wasatch Range, terminating just north of Huntsville. These rocks are described here because they constitute the oldest rocks extensively exposed in the upper plate of the Willard-Paris thrust (the Cache allochton), hence are relevant to reconstructing the geometry and emplacement of these and related thrust sheets. Moreover, they provide the key to an understanding of the attenuated and intermittently deposited section that extended eastward across the hingeline, toward southeastern Wyoming and into the Cottonwood area in Utah.

Sequence at Huntsville, Utah.—Throughout much of northeastern Utah and southeastern Idaho, only the upper, dominantly quartzitic parts of this sequence are exposed. In such areas these rocks are commonly lumped as the Brigham Quartzite (Oriel and Armstrong, 1971). But from approximately Preston, Idaho, west to Pocatello, and south into Utah, the older parts of the sequence also appear in the thrust plates, and the upper part can be separated into well-defined formational units; there the Brigham is given group status.

The lowest unit of the Huntsville sequence (Fig. 4), informally designated the formation of Perry Canyon, is a laterally intertonguing complex that consists of massive black diamictite, black carbonaceous and pyritic boulder mudstone, and black laminated argillite, all grading laterally into olive-drab-weathering siltstone and graywacke. Diamictite ranges from massive featureless bodies 60 to 600 ft thick to rudely bedded units 3 to 30 ft thick that show evidence of slumping and subaqueous mass flow. At greater distances from their source, the diamictites are winnowed and sorted, yielding proximal turbidites, exposed in the Wasatch, and distal turbidites, now exposed to the west in the islands of Great Salt Lake (Condie, 1967). Dropstones, up to 6 ft in diameter showing penetrative deformation of the enclosing laminite, and sand clots enclosed in fine-grained laminated mudstone, give clear evidence of ice rafting. In several places the diamictites contain lenses of greenstone, locally showing well-developed pillow structure. In a gross sense, the diamictites of the Perry Canyon are believed to be correlative with the Mineral Fork Tillite of the Cottonwood area.

Diamictite-bearing rocks grade upward into the Maple Canyon Formation, the lowest unit of the Huntsville sequence exposed east of the Wasatch Range (Crittenden, et al., 1971). This unit consists of olive-drab argillite (900 to 1200 ft), greenish fine-grained well-sorted arkose, and at the top, thin

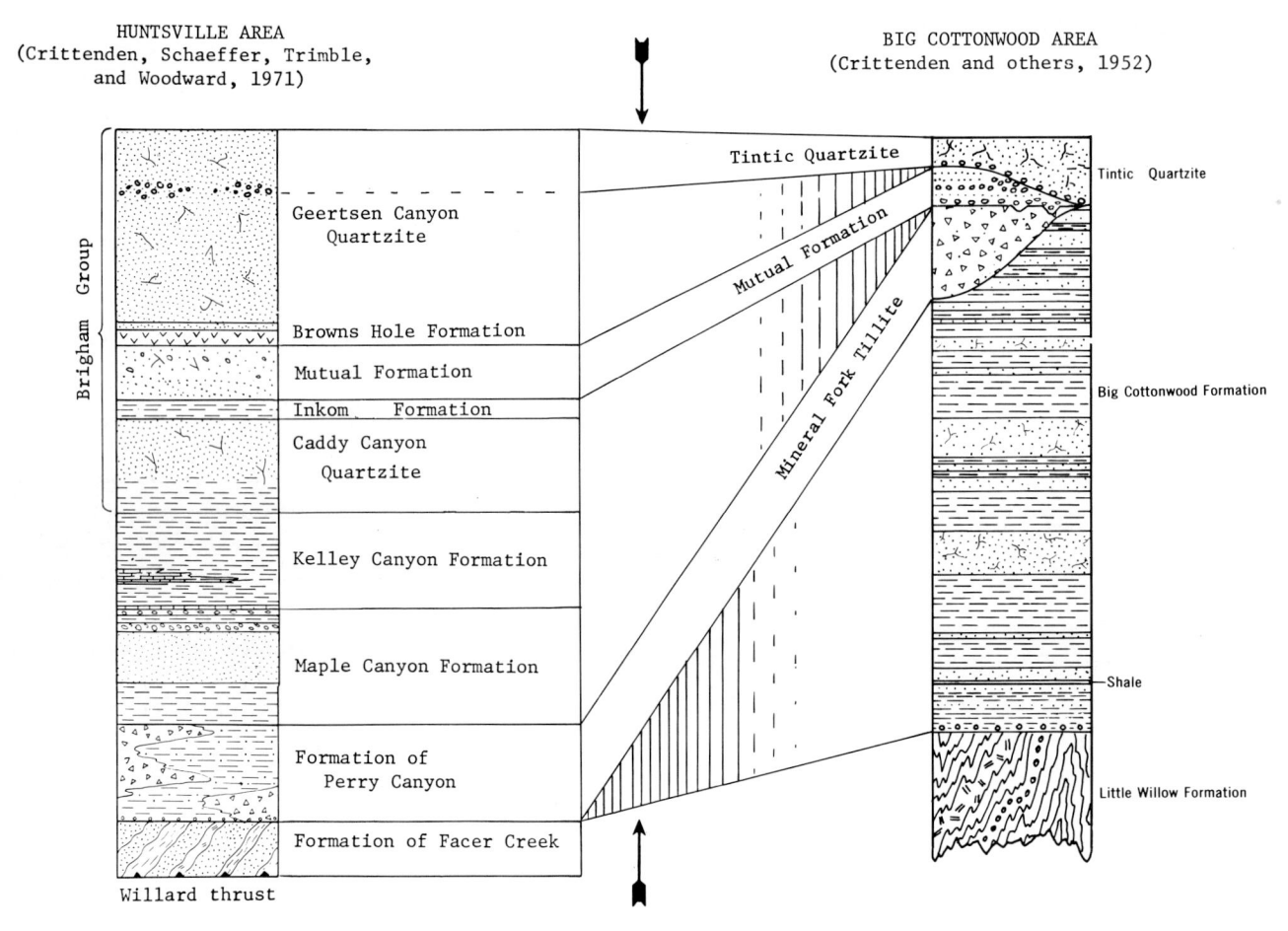

Fig. 4 — West to east correlation chart showing change in thickness in rocks of Precambrian Z age across the Cordilleran hingeline from Huntsville area (allochthonous) to the Big Cottonwood area (autochthonous). Vertical ruling indicates nondeposition. Bold arrows show location of Willard-Paris-Charleston-Nebo thrusts.

units (180 ft) of conglomerate and quartzite. These rocks may be equivalent to a thin upper layer of diamictite recognized by Trimble in the Pocatello area (Crittenden, *et al.*, 1971, Fig. 7).

Overlying the Maple Canyon Formation is a dominantly argillaceous unit, the Kelley Canyon Formation (1800 ft), consisting of lutite and siltstone that is dominantly lavender gray at the base, grading up to olive drab near the top. At the base is a distinctive 10 ft (3 m) tan-weathering laminated (locally stromatolitic?) dolomite. A thicker (100 ft) lenticular body of thin-bedded silty limestone occurs near the top of the lower third of the formation, and is suspected to be approximately equivalent to the Blackrock Canyon limestone of the Pocatello area (Crittenden, *et al.*, 1971, p. 585).

Near the east end of its area of exposure, north and east of Huntsville, the Kelley Canyon grades imperceptibly into the overlying Caddy Canyon Quartzite. To the northwest, however, near Brigham City, the Kelley Canyon grades up into a distinctive 450 ft thick unit of olive-drab siltstone and fine-grained quartzite containing crumpled shrinkage cracks that charcterize

the Papoose Creek Formation of the Pocatello area (D. E. Trimble, unpub. data). This unit in turn grades upward into resistant quartzites of the Caddy Canyon.

The Caddy Canyon Quartzite (1500 to 2500 ft), the stratigraphically lowest formation of the Brigham Group, consists of medium-grained thick-bedded cliff-forming quartzite. To the southeast, near Huntsville, the color is greenish or pinkish gray, bedding is thinner, and small-scale crossbedding is abundant.

The overlying Inkom Formation (300 to 450 ft) is a bench-forming unit that is largely siltstone. In exposures east of Preston, Idaho and of Huntsville, Utah, it is dominantly red or grayish red. To the west, red hues intertongue with olive drab, and in the Promontory Range, Utah, it is entirely green. Locally it contains reworked vitric tuff, intimately interlaminated with fine-grained siltstone. The Inkom is overlain by the cliff-forming Mutual Formation, (1000 to 3500 ft), a key unit in the correlation of these sequences. The Mutual consists of grayish-red to red-purple, crossbedded, pebbly feldspathic quartzite with local lenses of red siltstone and basalt or other volcanic rocks. It can be recognized readily in the central part of the depositional trough from Pocatello to central Utah but not in more easterly exposures, as in the Portneuf Range (Bright, 1960) where both the adjoining units are red instead of white or pale pinkish gray (Oriel and Armstrong, 1971).

Above the Mutual Formation is a thin (150 to 450 ft) terrestrial unit, the Browns Hole Formation, which consists of basaltic flows and reworked volcanic rocks ranging from basalt to alkali trachyte, plus an overlying fine-grained, well-sorted, pale-tan quartzite whose rounded and frosted grains suggest eolian transport.

The stratigraphically highest unit in the Brigham Group, the Geertsen Canyon Quartzite, is a thick (3500 ft) quartzite, generally equivalent to the Tintic Quartzite of central Utah and the Prospect Mountain Quartzite (restricted) of the Great Basin. At the basal part of the Geersten Canyon is coarse-grained arkosic (quartzite, commonly streaked with purple or green hues. The bulk of the unit is typically tan, pale gray, or pale greenish-gray non-feldspathic quartzite. It contains a medial zone 30 to 100 ft thick of pebble and cobble conglomerate that resembles the basal conglomerate of the much thinner Tintic Quartzite. The upper 300 to 450 ft locally contains abundant *Scolithus*, and bedding surfaces commonly are marked by tracks, trails, and phosphatic pellets. Olenellid trilobites were found in southeastern Idaho in similar rocks at the top of the Brigham (Oriel and Armstrong, 1971, p. 10), but because of the great thickness of quartzites below, the unit was there assigned a Precambrian to Cambrian age. More recently a date of 570 m.y. was obtained on the underlying unit (see next paragraph) which suggests that the entire Geertsen may be Cambrian.

The age of the Huntsville sequence is still poorly defined. Fresh hornblende obtained from volcanic breccia fragments in the Browns Hole Formation near the top of the sequence has been dated by M. A. Lanphere at 570 m.y. (Crittenden and Wallace, 1973). This supports the inference that deposition of this sequence continued into Early Cambrian without major lapses of time. Attempts to date rhyolitic tuffs in the Inkom Formation and the widespread basaltic rocks associated with the diamictites have so far been frustrated by alteration and low-grade metamorphism. The diamictites and volcanics at the base closely resemble rocks in the Metalline district of northwestern Washington on which dates of 825 to 900 m.y. are reported by Miller, *et al.*, (1973) suggesting that the base of the Huntsville sequence also may be of very late Precambrian Y age. Because it is bracketed by the dates noted above, the bulk of the sequence is presumed to be of Precambrian Z age.

The Huntsville sequence is inferred by Stewart, (1972) to represent the earliest deposits of the Cordilleran geosyncline and to have been deposited without tectonic interruption on a newly formed continental slope. The black pyritic shales and associated diamictites suggest moderate water depth and a locally euxinic environment that may have been caused by water density stratification. Ice tongues are inferred to have extended into this seaway from the east. Some diamictite was reworked and redeposited in deeper parts of the basin by turbidity currents as suggested by Condie (1976). Shale, sandstone, minor conglomerate, and thin-bedded carbonate of the overlying Kelley Canyon Formation were deposited in a low- to moderate-energy marine environment. Sands of the Caddy Canyon Quartzite were deposited in a nearshore area, which implies shallowing that may have resulted from decrease in subsidence rate or an increase in the sedimentation rate. The Mutual Formation probably represents a fluvial environment in which thin lava flows occurred sporadically. Later, during deposition of the Browns Hole Formation, basalt flows and volcanic breccias accumulated to greater thicknesses than in the Mutual Formation. Much of the resulting volcanic detritus was reworked into fluvial debris. Shallow-water marine conditions returned during deposition of the Geertsen Canyon Quartzite, a product of the widespread marine transgression that characterized the latest Precambrian and Early Cambrian throughout the Cordillera.

Sequence in Big Cottonwood Area. — In striking contrast to the thick and continuously deposited sequences present in western Utah and in the Huntsville area east of Ogden, only three equivalent units are present in the Big Cottonwood area, all separated by marked unconformities (Fig. 3). The Mineral Fork Tillite lies in shallow troughs cut into the underlying Big Cottonwood Formation (Fig. 4). The tillite is inferred to be temporally equivalent to diamictite wedges intercalated with the base of the Huntsville sequence. Resting unconformably on the tillite is the 1,200 m (3940 ft) thick Mutual Formation. It is similar in color, lithology, internal stratigraphy, and primary bedding structures to correlative rocks in the Huntsville area, but is thinner and contains boulders and large blocks derived from the underlying tillite. The Mutual Formation in the Big Cottonwood area is separated by another unconformity from the Cambrian Tintic Quartzite, which lithologically resembles the upper beds of the Geertsen Canyon Quartzite. These relations indicate that the latest Precambrian and Early Cambrian rocks of the Big Cottonwood area accumulated in an intermittently inundated platform environment very different from the continuously subsiding basin to the west. Altogether as much as 10,000 to 12,000 ft of sediments may have been deposited during the intervals represented by these unconformities. Moreover, local derivation of debris in the Mutual suggests that the lesser thicknesses of these strata in the Big Cottonwood area represent the feather edges of westward-thickening wedges that are thin because of nondeposition, rather than subsequent erosion.

Rocks of Early Paleozoic Age

Stratigraphic contrasts in rocks ranging from Middle Cambrian through Devonian age in the eastern part of the Cordilleran miogeocline have been summarized by Stewart and Poole (1974). Although rocks of this stratigraphic interval (Fig. 5) show thickness changes from the deep basins of the miogeocline to the edge of the craton in the Cottonwood or western Uinta areas on the order of 20 or 30 : 1, a major part of this thickness change results from extensive erosion associated with a regional unconformity of Late Devonian to earliest Mississippian age. Middle Cambrian rocks in the Cottonwood area total only about 1,000 ft of the shaly Ophir Formation at the base and the Maxfield Limestone at the top. Rocks of approximately the same age in the Tintic District are about 3,000 ft thick and in western Utah and eastern Nevada are 6,000 to 8,000 ft thick.

Rocks of Ordovician and Silurian age are entirely absent in the Cottonwood area of the Wasatch Range. In contrast, rocks of Ordovician and Silurian age in western Utah and easternmost Nevada total more than 6,000 ft thick. Similarly, rocks of Devonian age are either absent, as in the Cottonwood area, or at most about 150 ft thick just northeast of Salt Lake City, whereas in the basins to the west along the Utah-Nevada line correlative rocks total from 4,000 to 6,000 ft thick (Stewart and Poole, 1974).

Regional Unconformity of Late Devonian Age

It is evident from an east-west stratigraphic reconstruction (Fig. 5) that a large part of these drastic changes of thickness are the result of erosion and removal of Lower Devonian, Ordovician, Silurian, and Middle Cambrian rocks by progressive eastward downcutting beneath uppermost Devonian rocks generally included with the Pinyon Peak Limestone. Rigby (1958) and Morris and Lovering (1961) show that an area of uplift and erosion extended from approximately the site of the Uinta Mountains westward through the Wasatch, Oquirrh, and Stansbury Ranges defining a nearly west-trending anticlinal arch whose eastern portion was nearly coincident with the axis of the present day Uinta arch. Uplift was greatest in the Uinta Mountains where uppermost Devonian and Lower Mississippian rocks rest in places directly on Precambrian rocks of the Uinta Mountain Group. Another area of sharp uplift on the site of the Stansbury Mountains resulted in a local high where uppermost Devonian rocks were deposited on the Cambrian Ophir Formation and consist in part of a coarse clastic wedge, the Stansbury Formation (Stokes and Arnold, 1958). A thin wedge of this unit is present in City Creek, immediately northeast of Salt Lake City, where it was tentatively called Swan Peak Quartzite by Granger (1953). To the south, west, and north of the uplift, younger and younger rocks appear, until in the area along the Utah-Nevada line, in the East Tintic Mountains, Utah, and in the area near Logan, Utah, essentially complete stratigraphic sections are present.

In the area of Little Cottonwood Canyon the thickness of Cambrian rocks exposed above and below the Alta thrust, the most striking structural feature of the northern wall of Little Cottonwood Canyon (Fig. 6), is largely controlled by the position of this unconformity. In the block below the Alta overthrust in exposures at the foot of the blue and white banded cliffs immediately north of Hellgate Spring, the lower Mississippian rocks rest on the Maxfield Limestone within a few feet of its base. To the east, in the upper plate of the Alta thrust, the basal Mississippian rocks rest on the thickest sections of Maxfield, exposed in this area (more than 500 ft).

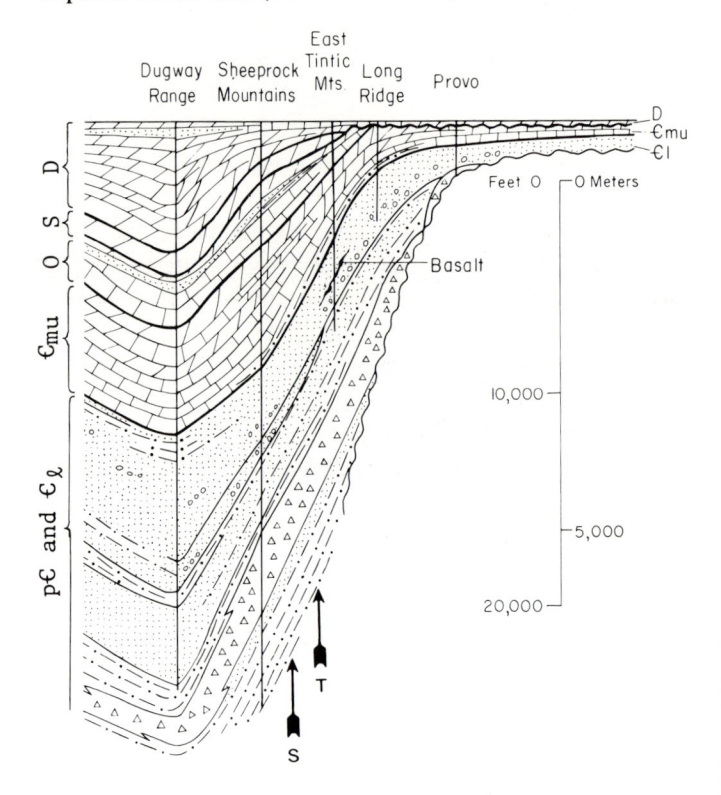

Fig. 5 — West-east stratigraphic diagram showing thickness changes in rocks of late Precambrian and early Paleozoic age across the Cordilleran hingeline in central Utah. After Stewart and Poole (1974). Local effects of Tintic Valley (T) and Sheeprock (S) thrusts not shown.

Rocks of Mississippian Age

Throughout much of northern and central Utah Early Mississippian age rocks are relatively consistent in both lithology and thickness across the Wasatch Line in both the Oquirrh Mountain and Weber Canyon facies (Table 1). The lowest formation, the Fitchville, is largely dolomite and ranges from 120 ft in the Cottonwood area to 265 ft in the Provo area. Although it has a marked unconformity at the base, the lowest beds contain only

Fig. 6 — View of north wall of Little Cottonwood Canyon northwest of Alta showing trace of Alta, Columbus, and Reed and Benson thrust zones. Alta thrust is cut by high angle faults of Superior fault system. Big Cottonwood Formation (pЄbc), Mineral Fork Tillite (pЄmf), Tintic Quartzite (Єtq), Ophir Shale (Єo), Maxfield Limestone (Єm), Fitchville Formation (Mf), Deseret and Gardison Limestones (Mdg).

rounded grains of sand instead of a basal conglomerate.

Above the Fitchville is the 450 to 500 ft thick Gardison Limestone (Morris and Lovering, 1961). This unit, entirely Lower Mississippian, is the most fossiliferous in the area. In early reports it was termed the Madison Limestone in the Cottonwood area, and the Gardner Dolomite in the East Tintic Mountains.

Table 1. Average Thickness (in feet) of rocks of Mississippian age in Northern Utah

Oquirrh Mountain Facies (Allochthon)		Weber Canyon Facies (Autochthon)	
Manning Canyon Shale	1100	Doughnut Formation	400
Great Blue Formation	3000	Humbug Formation	800
Humbug Formation	650	Deseret Limestone	900
Deseret Limestone	900	Gardison Limestone	450
Gardison Limestone	500	Fitchville Formation	120
Fitchville Formation	300		2600
Average total thickness	6300		

The Gardison is overlain conformably in most areas by about 900 ft of dark-gray cherty dolomite termed the Deseret Limestone. At its base is a thin but persistent unit of carbonaceous shale commonly containing pelletal phosphorite and anomalous amounts of uranium, vanadium, chromium, zinc, and rare earth elements.

The carbonate rocks of the Deseret grade upward through an interval of several tens of feet into the overlying Humbug Formation, which consists of alternating quartzitic sandstones and sandy or pure limestones and dolomites. The Humbug ranges from 600 to 800 ft thick. Similar lithologies are present in the Humbug further north, though in Ogden Canyon sandstone and quartzite apparently equivalent in age to the Deseret Limestone were deposited.

Rocks of Late Paleozoic Age

The greatest disparity in thickness between rocks of the miogeocline and those deposited on the craton are in rocks of late Paleozoic age (Fig. 7). In the Oquirrh Mountains, for example, rocks between the top of the Humbug Formation and the base of the Park City Formation are nearly fifteen times as thick as the rocks in that same interval in the Cottonwood area immediately to the east. In recognition of these differences the thicker sequences to the west are informally called the Oquirrh Mountain facies. The thinner rocks to the east are informally designated the Weber Canyon facies. The extent of the Oquirrh Mountain facies is approximately shown in fine-stippled pattern on Figure 1.

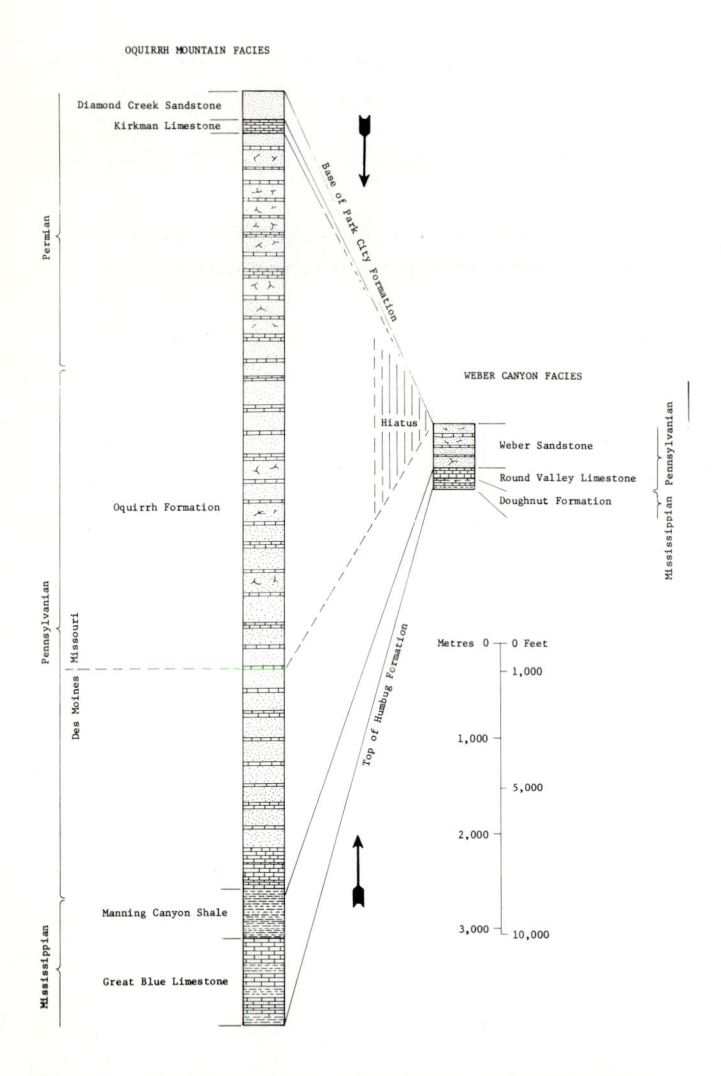

Fig. 7 — Stratigraphic contrast across the Cordilleran hingeline in rocks of late Paleozoic age. Oquirrh Mountain facies after Baker, et al. (1949), Weber Canyon facies after Calkins and Butler (1943), and Crittenden (unpublished data). Location of Charleston-Nebo thrusts shown by bold arrows.

Weber Canyon Facies. — In the central part of the Wasatch Mountains east of Salt Lake City, between the trace of the Charleton thrust on the south and the Willard thrust on the north, the interval between the Humbug and Park City Formations is represented by about 2,150 ft of rocks. Beginning at the base, these are: the Doughnut Formation (Late Mississippian) consisting of dark shale and limestone; the Round Valley Limestone (Early Pennsylvanian) consisting of about 400 ft of pale gray limestone with characteristic salmon-pink chert and fossils; and the Weber Quartzite (Middle Pennsylvanian), about 1,500 ft thick, consisting of intermittently cherty limestone and limy quartzite in the lower part and vitreous white quartzite in the upper part. These rock units can be traced north and northeast to the type locality of the Weber near Morgan, Utah, and eastward into the western Uinta Mountains. To the west it is inferred that these rocks originally extended across the site of Antelope Island because it is clear from stratigraphic relations in late Precambrian rocks that the trace of the thrust fault which connects the Willard and the Charleston-Nebo thrusts must be concealed beneath the Great Salt Lake in the space between the Oquirrh Mountains and Antelope Island on the south and between Antelope Island and Fremont Island on the north.

Oquirrh Mountain Facies. — In much of northwestern Utah, surrounding the area of the Great Salt Lake, the same sequence of formations can be recognized in rocks of late Paleozoic age, again bracketed between the Humbug Formation at the base and the Park City Formation at the top. These include the Great Blue Limestone, Manning Canyon Shale, Oquirrh Formation, Kirkman Limestone, and Diamond Creek Sandstone. This facies can be recognized northward from a major east-northeast trending tear fault in the vicinity of Leamington, Utah, at the south edge of the East Tintic Mountains, throughout the Charleston-Nebo allochthon east of Provo, Utah, and from a point west of Antelope Island in the Great Salt Lake as far north as Pocatello in southern Idaho. To the west the units appear to thin rapidly between the Cedar Mountains southwest of the Great Salt Lake and in the vicinity of the Hogup Mountains north and west of the Great Salt Lake. To the northeast the facies can be recognized in the Wellsville Mountain area southwest of Logan but thins rapidly to the east and does not extend too far east of Logan. During Pennsylvanian time the deepest part of this remarkable basin was centered in what is now the Oquirrh and Charleston-Nebo areas but during Permian time it shifted northwestward into the area just west and north of the Great Salt Lake, possible connecting to the Sublett basin of southern Idaho.

The Kirkman Limestone and Diamond Creek Sandstone, which together total 2300 ft, are recognized mainly in the Charleston-Nebo part of the basin.

Park City Formation. — The disparate rocks of the Weber Canyon and Oquirrh Mountain facies are overlain by rocks of the Permian Park City Formation. In the autochthon this unit is particularly well exposed near the mouth of Mill Creek southeast of Salt Lake City where it is about 850 ft thick. Limestones in the lower part contained some of the rich bodies of silver ore mined at Park City in the early days of that camp. The medial part of the formation is largely shale containing beds of phosphorite and is regarded as a tongue of the Meade Peak Phosphatic Shale Member of the Phosphoria Formation. The upper part is largely limestone and has been designated the Grandeur Member of the Park City, of which this is the type locality. Similar lithologies are exposed on the Weber River east of Morgan, Utah.

The Park City Formation is also recognized above the Oquirrh Mountain facies where it is generally thicker, richer in phosphate, and contains tongues of diverse lithologies, particularly chert (Roberts, *et al.*, 1965).

Rocks of Triassic and Jurassic Age

The lithology, thickness, and nomenclature of rocks of Triassic age in the Cottonwood area are summarized in Figure 8. They are relatively constant in character across the hingeline and will be described only briefly. The formations shown here are those first described in the Park City District (Boutwell, 1907) and were extended to the Cottonwood area by Calkins and Butler (1943). Correlation with the units commonly used in the Colorado Plateau is also shown diagrammatically. It should be noted that the detailed stratigraphic relations between the Gartra Grit Member of the Ankareh Formation in the Wasatch Range, the Gartra Member of the Chinle Formation in the Uinta Mountains and the Shinarump Member of the Chinle Formation remain uncertain, hence the two members continue to be distinguished in recent studies (Stewart, *et al.*, 1972).

Red beds of typical Triassic aspect are overlain in the Cottonwood area by pale orange to pale pinkish-tan, fine-grained sandstone assigned to the Nugget Sandstone. This unit, originally named by Veatch (1907) in southwestern Wyoming is well exposed near the mouth of Parleys Canyon and along the highway to the east. In many areas it shows characteristic large scale cross beds, in other areas, even, 2 to 4 in. thick beds are extensively quarried for ornamental and structural stone. In this area the Nugget is commonly regarded as Early Jurassic in age.

The Nugget Sandstone is overlain with apparent conformity by about 2800 ft of pale gray limestone interbedded with greenish-gray limy shale or siltstone assigned to the Twin Creek Limestone. The entire unit is soft and forms low rounded slopes. Although it covers large areas in Parleys and Emmigration Canyons it is well exposed only in artificial road cuts near the summit of Parleys Canyon and along the highway a short distance east. Limestone of the Twin Creek has been extensively mined for cement rock in a quarry immediately north of the highway mid-way up Parleys Canyon. The Twin Creek is exceedingly incompetent and is intensely folded and crumpled,

Wasatch Mountains		Thickness in feet	Lithology	Eastern Utah and NW Colorado (Stewart & Others, 1972)
Ankareh Formation	Upper Member	700	Reddish brown to bright red shale, siltstone and sandstone	Chinle Formation
	Gartra Grit Member	75	White coarse grained pebbly quartzite (sandstone to east)	Gartra Member / Shinarump Member
	Mahogany Member	800	Reddish brown to bright red shale, siltstone and sandstone	Mahagony Formation
Thaynes Formation		1,000	Gray limestone, sandstone and greenish shale	Thaynes Formation / Moenkopi Formation
Woodside Shale		1,000	Deep maroon shale, siltstone and thin-bedded sandstone	

Fig. 8 — Diagram showing relations between Triassic unit names used in the Wasatch and in eastern Utah and northwestern Colorado.

even where the adjoining formations are comparatively undisturbed. The Twin Creek in this area is regarded by Imlay (1950) as a lateral extension of that at the type locality, which contains an abundant Middle and Late Jurassic fauna.

The Twin Creek Limestone is overlain conformably by about 1000 ft of pale red shale and sandstone, termed Preuss Sandstone, of Late Jurassic age. The thickness and lithology are similar to those in southeastern Idaho.

It should be noted that the Morrison Formation, an Upper Jurassic continental unit widely exposed in eastern Utah and partially equivalent to the Stump Sandstone exposed in southeastern Idaho and southwestern Wyoming, is believed to be absent from the Cottonwood area of the Wasatch Range. The nearest exposures are near the town of Peoa, Utah, on the Weber River. It is believed that these units are cut out to the west by an unconformity that develops below the Early Cretaceous Kelvin Formation.

Rocks of Cretaceous Age

The red beds of the Preuss Sandstone are overlain in the Parleys Canyon area by a strongly contrasting white or pale lavendar unit about 150 ft thick that is readily distinguished both on the ground and in aerial photographs. This unit was termed Morrison(?) by Granger and Sharp (1953) but was later redefined as the Parleys Member of the Kelvin Formation and assigned an Early Cretaceous age (Crittenden, 1963). These beds can be traced eastward to Peoa, Utah, where they were shown to overlie a unit containing charophytes characteristic of the Morrison Formation. As now defined, the Parleys Member is presumed to be equivalent to the Cedar Mountain Formation of eastern Utah.

The upper member of the Kelvin Formation consists of pale red to dark grayish-red sandstone, siltstone, and conglomerate whose color contrasts with the underlying nearly white Parleys Member. In western exposures, as on the ridge between Emmigration and Parleys Canyon near Salt Lake City, the basal conglomerate is thick and contains both pebbles and large cobbles. The thickness of the Kelvin in this area was reported to be 1565 ft (Granger, 1953). Although the unit thickens eastward to about 2700 ft at Peoa, Morris (1953) reports only pebble conglomerate in the lower part.

The Kelvin Formation is overlain conformably by another contrasting unit consisting of pale gray to black thin-bedded siliceous shale assigned to the Aspen Shale. Although the Aspen is incompetent and seldom well exposed it can be traced readily by the abundance of splintery fissile shale fragments containing fish scales. The Aspen Shale is about 525 ft thick in the section along the Weber River near Peoa, but is only about 350 ft thick in exposures near the north end of Rockport Reservoir. The Aspen Shale is present immediately below the basal Longwall Sandstone Member of the Frontier Formation (Crittenden, 1974, Fig. 2). Rocks to the south of the marker but above the Dry Canyon fault, underlying the area labeled "Rockport", are therefore part of the Kelvin Formation. I am indebted to T. A. Ryer, of Yale University, for pointing out that these exposures were omitted on a reconnaissance map of this area (Crittenden, 1974, Fig. 2).

The Aspen Shale of this area appears to be a lithologic and temporal equivalent of part of the much thicker Mowry Shale of northwestern Colorado and is assigned an uppermost Lower Cretaceous age.

The Aspen Shale is overlain with probable slight hiatus, but without marked discordance, by the Frontier Formation, a nearly 5000 ft thick unit consisting of alternating thick sandstone and shale units with local beds of coal. The unit has been studied extensively by Cobban and Reeside (1952), Trexler (1955, 1966), Hale (1960, 1962), and Hale and VanDeGraaff (1964).

The Frontier Formation is widely recognized as representing a marginal facies of a sea that lay to the east and in which was deposited the still thicker Mancos Shale and part of the Hilliard Formation. The succession of transgressive and regressive units recognized in the Coalville area has been summarized by Trexler (1955, 1966) and Hale (1960). Of major significance are westward thickening wedges of conglomerate and sandstone that mark pulses of uplift and erosion in the Sevier orogenic belt to the west. An intraformational unconformity of great strucural significance is present locally at the base of the Dry Hollow Member of Hale (1960), which can be followed along the strike to the south and west from Coalville, Utah, into the area near Rockport Reservoir where the Dry Hollow and overlying portions of the classic Frontier Formation were designated the Wanship Formation by Williams and Madsen (1959). Although appearing conformable near Coalville, the basal conglomerate of this unit rests on underlying parts of the Frontier and on still older rocks with angular discordance of more than 90 degrees near Rockport Reservoir. The structural significance of this unconformity lies in the fact that it truncates thrust faults that appear to represent an extension of the Mt. Raymond thrust of the central Wasatch Range (Crittenden, 1974).

Following the sharp local uplift and erosion represented by this unconformity, marginal marine and shallow marine conditions continued to alternate until the end of Frontier deposition marked by the 450 ft thick, light-colored ridge-forming Upton Sandstone Member of Hale (1964). Following this are approximately 2500 ft of shale, sandstone and conglomerate assigned to the Henefer Formation of Eardley (1944). These rocks also were included in the upper part of the Wanship Formation by Williams and Madsen (1959).

Another major unconformity of latest Cretaceous age separates the Henefer Formation from the overlying coarse boulder conglomerate designated the Echo Canyon Conglomerate by Williams and Madsen (1959) and the Almy Conglomerate by Eardley (1944). This same unconformity, farther west in the core of the Parleys Canyon syncline, separates the Frontier Formation from coarse conglomerate originally referred to simply as Cretaceous(?) conglomerate by Granger and Sharp (1953).

Conglomerates of Latest Creaceous and Early Teriary Age

Detailed stratigraphic relations between conglomeratic units well known in southwestern Wyoming and those in the vicinity of Salt Lake City have long remained enigmatic, in large part, because workers have approached from diverse directions using different sets of nomenclature (Table 2). For present purposes the Late Cretaceous and Tertiary rocks of southwestern Wyoming are grouped with those of the Coalville area and are designated as Echo Canyon Conglomerate, Evanston Formation, and Wasatch Formation, following Mullens (1971). These appear to correspond, so far as can be determined, with the Almy, Fowkes, and Knight formations of Eardley (1944). In the Wasatch Range, just east of Salt Lake City, these same units appear to correspond with Cretaceous(?) conglomerates, Tertiary conglomerate No. 1, and Tertiary conglomerate No. 2, mapped and described briefly by Granger and Sharp (1953). The limestones, tuffs, and andesites that separate the two Tertiary conglomerates appear to represent equivalents of the Evanston Formation of Mullens.

The major structural significance of these orogenic deposits, and the intervening unconformities, is that they represent pulses of mountain building activity, either taking place locally by folding along the Uinta arch and the Parleys Canyon syncline, or by emplacement of thrust sheets derived from the Sevier orogenic belt to the west.

Rocks of Tertiary Age

Coarse-grained intrusive igneous rocks of the Little Cottonwood, Alta, and Clayton Peak stocks form some of the most impressive scenic features of the Cottonwood area. These three stocks (Fig. 2) are approximately aligned along the Uinta arch, and are both younger in age and progressively more felsic in composition from west to east. The easternmost, the Clayton Peak stock (not labelled on Fig. 2) is dominantly diorite, the Alta stock is dominantly granodiorite, and the Little Cottonwood stock is dominantly quartz monzonite. It is the massive sparsely jointed phase of this rock that forms the nearly vertical glaciated canyon walls of the lower part of Little Cottonwood Canyon.

The age of the three stocks has been determined by radiometric dating as 37-41 m.y. (Clayton Peak), 32-33 m.y. (Alta), and 24-31 m.y. (Little Cottonwood). The uncertainty in age, particularly for the Little Cottonwood stock, appears to result from its deep seated emplacement and slow cooling.

Two volcanic units of early Tertiary age will be noted for completeness. The oldest is the Norwood Tuff (Eardley, 1944), which lies in a gentle syncline behind the frontal horst of the Wasatch Range from the vicinity of Morgan, Utah, to well north of Huntsville, Utah, and perhaps into Cache Valley. It consists of white to pale greenish reworked tuff and tuffaceous sediment, locally extensively zeolitized. Vertebrate fossils collected from it by Eardley were dated as latest Eocene and earliest Oligocene by Gazin (1959); it also yielded potassium-argon dates of 37.5 m.y. (Evernden, *et al.*, 1964).

A slightly younger sequence of andesitic breccias, lahars, and local flows, termed the Keetley Volcanics, extends from the area of Park City north nearly to Coalville (shown in "V" pattern on Fig. 1). These rocks, which appear to be an extrusive

Table 2. Nomenclature of conglomerates of latest Cretaceous and early Tertiary age in northeastern Utah

Parleys Canyon and City Creek	Coalville area	
(Granger and Sharp, 1953)	(Mullens, 1971)	(Eardley, 1944, and Utah State Geologic map)
Tertiary conglomerate No. 2	Wasatch Conglomerate	Knight Conglomerate
Limestone, tuff, andesite	Evanston Formation	Fowkes (?) Formation
Tertiary conglomerate No. 1		
Cretaceous(?) conglomerate	Echo Canyon Conglomerate	Almy Conglomerate
(Absent)	Henefer Formation	Henefer Formation
(Absent)	Wanship Formation of Williams and Madsen, 1959	Frontier Formation
Frontier Formation	Frontier Formation	

equivalent of stocks in the Park City District, have yielded potassium-argon dates of 33 to 35 m.y. (Crittenden, *et al.*, 1973).

STRUCTURE

Regional Setting

In addition to its fortuitous location with regard to stratigraphic changes across the Wasatch line, the Cottonwood area lies at the intersection of two major tectonic elements of western North America (Fig. 1). The first is the zone of north-trending thrusts and folds generally known within Utah and Nevada as the Sevier orogenic belt (Armstrong, 1968), but which is widely recognized to be a part of a much larger zone of thrusting and folding which extends essentially the full length of the North American continent from southern Nevada to the northwest corner of Alaska. This belt has been called, appropriately, the Cordilleran foldbelt by King (1969).

The second major tectonic element is the east-trending Uinta arch. This structure, long recognized as an anomaly in the western United States, can be traced from the northwest corner of Colorado westward through the length of the Uinta Mountains. Near the west end of that range the anticlinal axis plunges gently beneath the Keetley Volcanics between Kamas and Park City, Utah, and emerges in the central Wasatch. Here the anticlinal form is outlined by the contact between Precambrian and Paleozoic rocks which strikes east near Mount Olympus, bends gradually to a southeasterly strike in the vicinity of Alta and finally swings southwest again in roof pendants barely preserved within the Little Cottonwood stock. The axis of the fold reaches the Wasatch Front in Little Willow Canyon where erosion has exposed rocks of probable Precambrian age. Because rocks of Cretaceous and Tertiary age have been eroded from the anticline, temporal relations between these east-trending structures and those of the Sevier orogenic belt are best demonstrated in the adjoining Parleys Canyon syncline which extends in an east-northeasterly direction from the mountain front near Parleys Canyon to the vicinity of Coalville in northeastern Utah.

Structures of the Sevier Orogenic Belt

The earliest thrusts in the Wasatch were recognized independently in the Alta area by F. F. Hintze and G. F. Loughlin about 1913 and were described in detail by Calkins and Butler (1943). The structurally lowest and apparently oldest of these features is the complexly braided fault zone to which I shall refer simply as the Alta thrust. This structure, beautifully displayed on the ridge north of Little Cottonwood Canyon (Fig. 6), largely involved Precambrian and Paleozoic rocks and locally controlled mineralization in the Alta area. Although the Alta thrust now dips northward or eastward, Calkins and Butler (1943, p. 54) concluded on the basis of eastward-leaning overturned drag folds that this attitude is a result of folding, and that the thrust actually moved from west to east. This tectonic transport also is confirmed by stratigraphic relations between the thrust sheets, in which each successively higher thrust plate contains a thicker section of Cambrian carbonate rocks than the one below. The age of the Alta thrust is difficult to ascertain because the youngest rocks involved are Paleozoic. It is clearly penetrated by the Alta stock and judging from metamorphic effects also pre-dates the Little Cottonwood stock, but this indicates merely that it is pre-mid-Tertiary.

A structurally higher and probably later thrust in the Cottonwood area is the Mount Raymond thrust (Fig. 2). This structure begins at the mountain front just north of Mount Olympus and extends eastward across the range to a point north of Park City where it is bent into a double hairpin-like loop by a northeast-trending anticline and syncline that developed on the flank of the Uinta arch. Beyond this it disappears beneath the Keetley Volcanics. I have inferred (Crittenden, 1974) that it reappears east of the volcanics and crosses the Rockport Reservoir where it consists of two strands called the Dry Canyon and Crandal Canyon faults, by Williams and Madsen (1959). This inferred connection is important because movement on this fault system can be more closely dated than any other in the area. Both faults cut rocks of the Frontier Formation of Early to Late Cretaceous age, but are themselves truncated by an unconformity and overlapped by a conglomerate that Williams and Madsen called the Wanship(?) Formation. The age of thrusting can therefore be confined to the age span of the unconformity, which Hale (1960, p. 141) has estimated as spanning middle and late Carlile time. That the Dry Canyon and Crandal Canyon faults are strands of a thrust fault rather than high-angle normal faults is demonstrated by overturning of rocks in the Frontier Formation below the lowest strand, by overturning that has resulted in the development of antiforms and synforms in the middle plate, and by the fact that the Parleys Member of the Kelvin Formation in the middle plate rests directly on the Preuss Formation, whereas in the lower plate the Parleys Member rests on the westernmost exposures of the Morrison and Stump formations.

The latest and most extensive episode of thrusting in the Sevier orogenic belt is the one that produced the Charleston-Nebo thrust south of the Cottonwood area and its presumed counterpart, the Willard-Paris thrust of northeastern Utah and western Wyoming. The Nebo thrust was first recognized by Eardley (1933) but was not then connected with faults farther north. The existence of a major structure in the Charleston area east of Provo was recognized by Baker (1947) who pointed out that it was this fault that brings the 26,000 ft thick Oquirrh Mountain facies underlying Mount Timpanogos northeast of Provo into justaposition with the 2,500 ft thick Weber Canyon facies in the Cottonwood area. This thrust crosses the Wasatch Range almost at right angles in the area between Little Cottonwood Canyon and Boxelder Peak (Fig. 2). Where it reaches the mountain front between the Little Cottonwood stock and the

Traverse Mountains it has been reactivated as the Deer Creek fault. This is a mid-Tertiary normal fault which, like the thrust, has a moderate to gentle southwest dip. I inferred (Crittenden, 1959) that the thrust is downthrown by the Wasatch fault, that its traces beneath Salt Lake Valley are shifted north some 6 to 10 miles, and that it passes northwestward separating the allochthonous Paleozoic section in the Oquirrh Mountains from autochthonous older Precambrian crystalline rocks exposed on Antelope Island. From there it must bend northward and northeastward in order to pass between the autochthonous rocks of Antelope Island and the thick late Precambrian sections exposed on Fremont Island and Promontory Point which match those above the Willard thrust in the area east of Ogden. It is uncertain where the fault reaches the Wasatch front, but it presumably is 3 to 4 mi north of Ogden. Here also the thrust must be displaced by the Wasatch fault and shifted northward to appear in the upthrown block of the Wasatch Range some ten miles to the north in the vicinity of Willard. There it strikes north-northwest, dips gently northeast, and can be traced southeastward across the entire range to a point southwest of Huntsville. A small window is present east of Huntsville, but beyond that it disappears beneath a cover of Cretaceous(?) and Tertiary conglomerate beneath which it apparently curves east and north to connect with the exposures of the Paris thrust west of Bear Lake.

Although the Willard thrust was first recognized by Blackwelder (1910) because a thick section of late Precambrian rocks is present in the upper plate, but entirely absent in the lower plate, the structural continuity between the Willard thrust and the Charleston-Nebo thrust to the south is based on the presence of thick sections of the Oquirrh Formation in both. Such sections are present above the Willard thrust in Wellsville Mountain and in extensive exposures across northern Utah west from Promontory Point to the Raft River Mountains.

The vast extent of movement on the Charleston-Nebo and Willard-Paris thrust systems is indicated not only by the stratigraphic contrasts between the rocks above and below the thrust, but by the fact that the Uinta arch and Parleys Canyon syncline have no counterparts in the Oquirrh Mountains (Gilluly, 1932). Their absence cannot be explained by normal faulting alone because together the two folds have an amplitude of more than 30,000 ft and simple normal displacement on the Wasatch fault of even as much as 20,000 ft would not suffice to eliminate folds of this amplitude. The structural discontiuity across the Salt Lake Valley is evident in Fig. 1 which shows the striking discordance between the eastward convex arc of folds in the central Oquirrh Mountains and the relatively simple west-trending Uinta arch, which is continuous as far as the front of the Wasatch Range. It must be inferred that the arch is simply depressed and overridden by the thrust plate of the Charleston-Nebo thrust which brought with it not only the thick Pennsylvanian-Permian sections from the basin which lay to the west but its own structural features resulting from the eastward transport. It is hard to escape the impression that the allochthon of the Charleston-Nebo thrust bulged eastward across the present site of the Wasatch Range in the fashion of a lobe of glacial ice, probably impelled in large part by gravity.

The age of the Charleston-Nebo thrust cannot be determined in the Cottonwood area but in the Uinta basin this fault displaces the Late Cretaceous Mesa Verde Formation and is overlapped by conglomerates of either the Late Cretaeous Price River or the Late Cretaceous and Paleocene North Horn Formation (Spieker, 1946). The thrust is thus clearly of very late Cretaceous age (i.e., at least post-Santonian and perhaps as young as Maestrichtian) (Schoff, 1937, 1951). Mullens (1969) has shown that the Willard segment of this fault system east of Ogden, is overlapped by conglomerates containing pollen of very late Cretaceous age and therefore presumably continuous with the Cretaceous-Tertiary Evanston Formation to the east. In any case, inasmuch as the Charleston-Nebo fault truncates not only the east-trending Uinta arch but the Alta and Mount Raymond thrusts that are folded with it, it is clear that the Charleston-Nebo thrust must postdate all of the earlier structures, although the time available for these events is very short.

Other structurally higher imbricate thrusts have been recognized in the upper plate of the Charleston-Nebo allochthon, on the basis of contrasting sedimentary facies or structural discontinuities. These include the Midas, Tintic Valley, and Sheeprock thrusts (Fig. 1), all of which trend northerly and appear to have resulted in generally eastward tectonic transport. The North Oquirrh thrust (unlabelled on Fig. 1) is anomalous in that it trends easterly and its upper plate appears to have moved south, or to the south-southeast. The age sequence within this series of faults is unknown on any except structural grounds. The combined Sheeprock and Tintic Valley thrust, east of the Stansbury Mountains, for example, appears to truncate east-trending arcuate folds carried in the upper plate of the Midas thrust. But, without stratigraphic analysis to provide a basis for actual dating of individual thrusts, even the sequence of these structures must remain in doubt.

Structural History of the Uinta Arch

Although our primary interest is in the post-Paleozoic history of the Uinta arch, the sedimentary record which records its structural history, particularly during the Cretaceous, is best preserved on the flanks of the anticline and in the adjoining Parleys Canyon syncline. The first significant unconformity east of Salt Lake City is that between the Preuss Sandstone of Late Jurassic age and the overlying Parleys Member of the Kelvin Formation believed to be of Early Cretaceous age. This hiatus apparently records gentle uplift that resulted in erosion and removal of the Morrison and Stump formations from the area west of approximately Peoa on the Weber River. Coarse conglomerates in the Kelvin Formation nearer to Salt Lake City suggest that uplift was much more pronounced further west. The absence of strata of this age west of the Wasatch makes it impossible to be certain what structural events are involved but I have inferred that this may be an interval during which the movement on the Alta overthrust took place.

The next structural event that can be recognized in this area is the movement on the Mount Raymond thrust, the age of which is bracketed between the age of the Frontier Formation (late Carlile) and that of the coarse conglomerates of the Wanship Formation of Williams (1959) (early Niobrara). Although the deformation accompanying this episode appears to have been most severe to the east in the area near the present Weber River, conglomerates in the upper part of the Frontier Formation near Salt Lake City indicate that uplift also occurred farther west.

Emplacement of the Mount Raymond thrust was followed by an episode of strong folding which, in the area nearer Salt Lake City, separates the Frontier Formation—apparently including equivalents of the Wanship Formation of Williams—from the overlying Echo Canyon Conglomerate. I believe most of the folding of the Mount Raymond thrust occurred at this time. As this folding preceded the movement on the Charleston-Nebo thrust, it too clearly took place during latest Cretaceous and probably before deposition of the Evanston Formation, which is partly Cretaceous and partly Paleocene. The final episode of folding postdates the Evanston, which is unconformably overlain in the area immediately east of Salt Lake City by an extensive sheet of conglomerate here correlated with the Wasatch Formation. Although the Wasatch is strongly folded elsewhere it is nearly flat for a considerable distance along this portion of the crest of the Wasatch Range.

The age relations between the structures of the Sevier orogenic belt and those of the Uinta arch are combined and summarized in Fig. 9. In terms of the underlying processes, this diagram suggests that thrusting and folding alternated in this area as follows: (1) emplacement of the Alta thrust from the west, (2) gentle north-south compression, (3) emplacement of the Mount Raymond thrust from the west, (4) strong north-south compression, (5) emplacement of the Charleston-Nebo thrust from the west (6) strong north-south compression. If thrust emplacement is presumed to result from east-west compression, this sequence of events presents a picture of alternating east-west and north-south compression that is hard to accept. On the other hand, if thrust emplacement is regarded as a gravitational response to uplift taking place far to the west, following the model suggested by Price and Mountjoy (1970), perhaps the episodes of periodic thrust emplacement can be related to pulses of uplift in the hinterland of the thrust belt and thereby sufficiently removed from the local scene to be more compatible with the episodes of north-south compression evident within this portion of the craton.

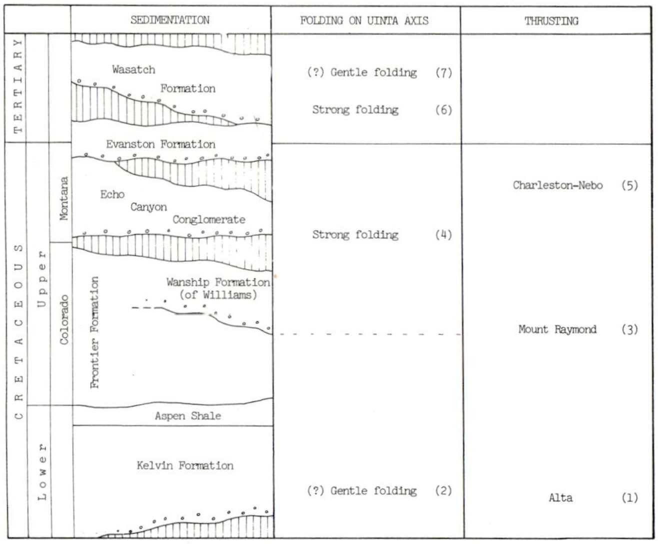

Fig. 9 — Diagram showing inferred sequence of sedimentation and structural events resulting from interaction of Sevier orogenic belt and Uinta arch.

On a much broader scale this relationship between the Sevier orogenic belt and the adjoining Laramide block uplifts has been addressed specifically by Armstrong (1974), who ascribes the change from Sevier to Laramide style of deformation to an abrupt cutoff of magmatic activity in the arc complexes of the Sierra far to the west. This suggestion of a direct cause and effect relationship between an active volcanic arc of Andean type and the orogenic response within the craton far inland is provocative, but appears to ignore the details of interaction of the kind described here.

REFERENCES CITED

Armstrong, R. L., 1968, Sevier orogenic belt in Nevada and Utah: Geol Soc. American Bull., v. 79, p. 429-458.

_____, 1974, Magmatism, orogenic timing, diachronism in the Cordillera from Mexico to Canada: Nature, v. 247, no. 5540, p. 348-351.

Baker, A. A., 1947, Stratigraphy of the Wasatch Mountains in the vicinity of Provo, Utah: U.S. Geol. Survey Oil and Gas Inv. Prelim. Chart no. 30.

_____, Huddle, J. W., and Kinney, D. M., 1949, Paleozoic geology of north and west sides of Uinta Basin, Utah: Am. Assoc. Petroleum Geologists Bull., v. 33, no. 7, p. 1161-1197.

Blackwelder, Eliot, 1910, New light on the geology of the Wasatch Mountains, Utah: Geol. Soc. American Bull., v. 21, p. 517-542.

Boutwell, J. M., 1907, Stratigraphy and structure of the Park City mining district, Utah: Jour. Geology, v. 15, p. 433-458.

Bright, R. C., 1960, Geology of the Cleveland area, southeastern Idaho: unpublished M.S. thesis, Univ. of Utah, 262 p.

Calkins, F. C., and Butler, B. S., 1943, Geology and ore deposits of the Cottonwood-American Fork area, Utah: U.S. Geol. Survey Prof. Paper 201, 152 p.

Cobban, W. A., and Reeside, J. B., Jr., 1952, Frontier Formation, Wyoming and adjacent areas: Am. Assoc. Petroleum Geologists Bull., v. 36, no. 10, p. 1913-1916.

Crompton, R. R., Todd, V. R., Zartman, C. A., and Naeser, C. W., 1977, Oligocene and Miocene metamorphism, folding, and low angle faulting in northwestern Utah: Geol. Soc. American Bull., 41 p. [in press].

Condie, K. C., 1967, Petrology of the late Precambrian tillite(?) association in northern Utah: Geol. Soc. America Bull., v. 78, no. 11, p. 1317-1343.

―――, 1969, Geologic evolution of the Precambrian rocks in northern Utah and adjacent areas, in Guidebook of northern Utah: Utah Geol. and Mineral Survey Bull. 82, p. 71-95.

Crittenden, Max D., Jr., 1959, Mississippian stratigraphy of the central Wasatch and western Uinta Mountains, Utah, in Intermtn. Assoc. Petroleum Geologists Guidebook, 10th Ann. Field Conf., p. 63-74.

―――, 1963, Emendation of the Kelvin Formation and Morrison(?) Formation near Salt Lake City, Utah, in Geological Survey Research 1963: U.S. Geol. Survey Prof. Paper 475-B, p. B95-B98.

―――, 1974, Regional extent and age of thrusts near Rockport Reservoir and relation to possible exploration targets in northern Utah: Am. Assoc. Petroleum Geologists Bull., v. 58, no. 12, p. 2428-2435.

―――, Sharp, B. J., and Calkins, F. C., 1952, Geology of the Wasatch Mountains east of Salt Lake City,—Parleys Canyon to the Traverse Range: Utah Geol. Soc., Guidebook to the Geology of Utah, no. 8, p. 1-37.

―――, Stuckless, J. S., Kistler, R. W. and Stern, T. W., 1973, Radiometric dating of intrusive rocks in the Cottonwood area, Utah: U.S. Geol. Survey Jour. Research, v. 1, no. 2, p. 173-178..

―――, Shaeffer, F. E., Trimble, D. E., and Woodward, L. A., 1971, Nomenclature and correlation of some upper Precambrian and basal Cambrian sequences in western Utah and southeastern Idaho: Geol. Soc. American Bull., v. 82, p. 581-602.

―――, Steward, J. H, and Wallace, C. A., 1972, Regional correlation of upper Precambrian strata in western North America: 24th Internat. Geol. Cong. Rept., sec. 1, p. 334-341.

―――, Wallace, R. E., 1973, Possible equivalents of the Belt Supergroup in Utah, in Belt Symposium, University of Idaho, v. 1: Moscow, Idaho, University of Idaho, p. 116-138.

―――, and Peterman, Z. E., 1975, Provisional Rb/Sr age of the Precambrian Uinta Mountain Group, northeastern Utah: Utah Geology, v. 2, no. 1, p. 75-77.

Eardley, A. J., 1944, Geology of the north-central Wasatch Mountains, Utah: Geol. Soc. America Bull., v. 55, p. 819-894.

―――, and Hatch, R. A., 1940, Proterozoic(?) rocks in Utah: Geol. Soc. America Bull., v. 51, no. 6, p. 795-843.

Evernden, J. F., Savage, D. E., Curtis, G. H., and James, G. T., 1964, Potassium-argon dates and the Cenozoic mammalian chronology of North America: Am. Jour. Sci., v. 262, no. 2, p. 145-198.

Gazin, C. L., 1959, Paleontological exploration and dating of the early Tertiary deposits in basins adjacent on the Uinta Mountains [Utah-Wyo.-Colo.], in Intermountain Assoc. Petroleum Geologists Guidebook, 10th Ann. Field Conf., p. 131-138.

Gilluly, James, 1932, Geology and ore deposits of the Stockton and Fairfield quadrangles, Utah: U.S. Geol. Survey Prof. Paper 173, 171 p.

Granger, A. E., 1953, Stratigraphy of the Wasatch Range near Salt Lake City, Utah: U.S. Geol. Survey Circ. 296, 14 p.

―――, and Sharp, B. J., 1952, Geology of the Wasatch Mountains East of Salt Lake City (City Creek to Parleys Canyon): Utah Geol. Soc. Guidebook, no. 8, p. 1-37.

Hale, L. A., 1960, Frontier Formation—Coalville, Utah, and nearby areas of Wyoming and Colorado: Wyoming Geol. Assoc. Guidebook, 15th Ann. Field Conf., p. 136-146.

―――, 1962, Frontier Formation—Coalville, Utah, and nearby areas of Wyoming and Colorado (revised): Wyoming Geol. Assoc. Guidebook, 17th Ann. Field Conf., p. 211-220.

―――, and Van DeGraff, F. R., 1964, Cretaceous stratigraphy and facies pattern—northeastern Utah and adjacent areas: Intermtn. Assoc. Petroleum Geologists Guidebook, 13th Ann. Field Conf., p. 115-138.

Hansen, W. R., 1965, Geology of the Flaming Gorge area, Utah-Colorado-Wyoming: U.S. Geol. Survey Prof. Paper 490, 196 p.

Imlay, R. W., 1950, Jurassic rocks in the mountains along the west side of the Green River Basin: Wyoming Geol. Assoc. Guidebook, 5th Ann. Field Conf., 1950, p. 37-48.

Kay, Marshall, 1951, North American geosynclines: Geol. Soc. America Mem. 48, 143 p.

King, P. B., 1969, The tectonics of North America—A discussion to accompany the Tectonic Map of North America, scale 1:5,000,000: U.S. Geol. Survey Prof. Paper 628, 95 p.

Miller, F. K., McKee, E. H., and Yates, R. G., 1973, Age and correlation of the Windermere Group in northeastern Washington: Geol. Soc. America Bull., v. 84, p. 3723-3730.

Morris, E. C., 1953, Geology of the Big Piney area, Summit County, Utah: unpublished M.S. thesis, Univ. of Utah, 66 p.

Morris, H.T., and Lovering, T.S., 1961, Stratigraphy of the East Tintic Mountains, Utah: U.S. Geol. Survey Prof. Paper 361, 145 p.

Mullens, T. E., 1969, Geologic map of the Causey Dam quadrangle, Weber County, Utah: U.S. Geol. Survey Geol. Quad. Map GQ-790.

―――, 1971, Reconnaissance study of the Wasatch, Evanston, and Echo Canyon Formation in part of northern Utah: U.S. Geol. Survey Bull. 1311-D, 31 p.

Oriel, S. S., and Armstrong, F. C., 1971, Uppermost Precambrian and lowest Cambrian rocks in southeastern Idaho: U.S. Geol. Survey Prof. Paper 394, 52 p.

Price, R. A. and Mountjoy, E. W., 1970, Geologic structure of the Canadian Rocky Mountains between Bow and Athabasca Rivers—A progress report: Geol. Assoc. Canada, Spec. Pap. no. 6, p. 7-25.

Rigby, J. K., 1958, Geology of the Stansbury Mountains, eastern Tooele County, Utah: Utah Geol. Soc. Guidebook, no. 13, p. 1-134.

Roberts, R. J., Crittenden, M. D., Jr., Tooker, E. W., Morris, H. T., Hose, R. K., and Cheney, T. M., 1965, Pennsylvania and Permian basins in northwestern Utah, northeastern Nevada and south-central Idaho: Am. Assoc. Petroleum Geologists Bull., v. 49. p. 1926-1956.

Royse, F., Jr., Warner, M. A., and Reese, D. L., 1975, Thrust belt structural geometry and related stratigraphic problems, Wyoming-Idaho-northern Utah, in D. W. Bolyard, ed., Deep drilling frontiers of the central Rocky Mountains: Rocky Mtn. Assoc. Geologists Symposium, p. 41-54.

Schoff, S. L., 1951, Geology of the Cedar Hills, Utah: Geol. Soc. America Bull., v. 62, no. 6, p. 619-645.

Spieker, E. M., 1946, Late Mesozoic and early Cenozoic history of central Utah: U.S. Geol. Survey Prof. Paper 2;5D, p. 117-161.

Stewart, J. H., 1972, Initial deposits in the Cordilleran geosyncline: evidence of a late Precambrian (<850 m.y.) continental separation: Geol. Soc. America Bull., v. 83, no. 5, p. 1345-1360.

Stewart, J. H. Poole, F. G., and Wilson, R. F., 1972, Stratigraphy and origin of the Triassic Moenkopi Formation and related strata in the Colorado Plateau region: U.S. Geol. Survey Prof. Paper 691, 195 p.

―――, and Poole, F. G., 1974, Lower Paleozoic and uppermost Precambrian Cordilleran Miogeocline, Great Basin, Western United States: Soc. of Econ. Paleontologists and Mineralogists, Spec. Pub. no. 22, p. 1-57.

Stokes, W. L. and Arnold, D. E., 1958, Northern Stansbury Range and the Stansbury Formation: Utah Geol. Soc. Guidebook no. 13, p. 135-149.

―――, et al., 1961-1964, Geologic map of Utah: [Salt Lake City] Utah State Land Board.

Trexler, D. W., 1955, Stratigraphy and structure of the Coalville area, northeastern Utah: Unpublished Ph.D. dissertation, Johns Hopkins University.

―――, 1966, Stratigraphy and structure of the Coalville area, northeastern Utah: Colo. School of Mines, Prof. Contr. 2, 69 p.

Veatch, A. C., 1907, Coalfields of east central Carbon County, Wyoming: U.S. Geol. Survey Bull. 316, p. 244-260.

Williams, N. C. and Madsen, J. H., Jr., 1959, Late Cretaceous stratigraphy of the Coalville area, Utah: Intermtn. Assoc. Petroleum Geologists Guidebook, 10th Ann. Field Conf., p. 122-125.

GEOLOGY OF THE COALVILLE ANTICLINE, SUMMIT COUNTY, UTAH

by

Lyle A. Hale[1]

INTRODUCTION

The Coalville anticline is located 33 mi northeast of Salt Lake City in west-central Summit County, Utah. Geomorphically, the region is a foothill topography marginal to the adjacent highlands of the Wasatch and Uinta Ranges. The relief is about 2,000 ft in the map area ranging from about 5,400 ft in the Weber Valley to 7,500-plus ft in adjacent ridges. In the vicinity of Coalville and Echo Junction, there are several prominent escarpments composed of Cretaceous strata. These are overlooked by plateaus consisting of Tertiary conglomerates. The principal stream of the area is the Weber River which drains the northwest slopes of the Uinta Mountains and environs. It flows north past the town of Coalville where it is joined by Chalk Creek from the east. Two other main tributaries of the Weber, Grass Creek and Echo Canyon, enter just downstream from Echo Reservoir. The main line of the Union Pacifc Railroad runs through Echo Canyon from Wyoming thence northwest down the Weber Valley to Ogden, Utah. The newly constructed Interstate Highway 80 connects Coalville with Salt Lake City and Evanston, Wyoming.

The Mountain Fuel Supply Company's main natural gas transmission lines enter the east side of the map area from gas fields in Wyoming and follow the course of Chalk Creek to the Coalville compressor station. From this point it branches with a 16 in. line leading northwest down the Weber River Valley to Ogden and two parallel lines (20 in. and 24 in.) extending south-southwest to Salt Lake City. Two gas storage areas have been developed along Chalk Creek in the map area by Mountain Fuel.

With the discovery in December 1974 of prolific Jurassic oil production at Pineview some 12 mi east of Coalville, new attention was focused on the oil and gas possibilities of the Coalville anticline. This motivated the Guidebook editor to invite me to update, insofar as possible, earlier papers on the general geology of the area.

PREVIOUS INVESTIGATIONS

In 1859 Capt. J. H. Simpson reported on the Coalville area during an early reconnaissance survey published by the Engineer Department of the U.S. Army. In the report of the exploration of the Fortieth Parallel, Clarence King described the general geology of the area, its coal resources and early mining operations, as did F. V. Hayden in his Fourth Annual Report.

In 1893 the stratigraphy of the area was described in some detail by Stanton. A. C. Veatch (1907) briefly compares the Coalville section with an equivalent coal-bearing section in Wyoming.

Carrol H. Wegeman published his report of the Coalville coal field in 1915. Much of his data outside the local vicinity of the gas storage fields have been combined with mapping by the writer in preparing the accompanying map (Plate I, in pocket).

More recent investigations have been published by A. J. Eardley and Norman C. Williams of the University of Utah. Several theses by University students have been published under the direction of these two men. In addition, D. W. Trexler published a doctoral thesis from Johns Hopkins University in 1955. Recently, the writer published his observation on both the structures as related to gas storage and details of the Frontier Formation (Hale 1960, 1963).

REGIONAL SETTING

The Coalville anticline is situated near the junction of three major tectonic elements, the Wasatch Range on the west, the Uinta Mountain uplift to the southeast, and the Idaho-Wyoming overthrust belt to the north and east. The structural style of the Coalville area clearly implies a genetic relationship to the overthrust belt. Eardley (1944) describes at least eight episodes of deformation that shaped the area during the Laramide orogeny. Locally, the area is flanked on the east by the recumbent Clark Canyon syncline that separates the Coalville anticline from the now-productive Pineview-Evanston lineament, and on the west by the Stevenson Canyon syncline (Trexler, 1955).

The anticline has a northeasterly trend typical of structures in this region of the overthrust belt. The southerly plunge is obscured south of Chalk Creek, thus its regional continuity and relations are uncertain. Its northeast plunge is likewise poorly exposed but can be traced to Echo Canyon where it disappears beneath the Tertiary cover. The fold reappears 32 mi to the north as the Crawford Mountain uplift whose core is composed of rocks of Mississippian Age.

---

[1] Consultant, Salt Lake City, Utah

The Cretaceous rocks exposed on the Coalville anticline are believed to have been displaced several miles eastward along underlying thrust planes from their original depositional site. Supporting factors for this tentative conclusion are: 1) the anticline is known to be undercut by one or more thrust faults that have been penetrated to date by at least two wells, 2) comparing the lithology and thickness of the Kelvin and Frontier Formations on the Coalville anticline with equivalent strata penetrated by wells in the Pineview field strongly suggests crustal shortening, and 3) the red-bed sequence of the lower Frontier, indicative of a foreland depositional environment, appears to be missing at Pineview.

STRATIGRAPHY

The surface stratigraphy of the Coalville anticline has previously been described (Hale, 1960). Therefore, only a general summary of the stratigraphic section is shown on Table 1. Chart 1 shows the nomenclatural history of the Cretaceous rocks in the Coalville area.

On the structure map (Plate I) and accompanying cross-section, only the Kelvin Formation and the lower Frontier Formation are shown in some detail. Those members of the Frontier lying above the "Oyster Ridge" are shown as Frontier undivided (Kfu), which roughly corresponds to the Wanship Formation of Williams and Madson (1959). Detailed mapping did not include the Echo Canyon Conglomerate or the Tertiary Knight Formation, although the general distribution of both is roughly indicated on the map. Key mapping horizons used by the writer in gas storage investigations were the sandstone above the "Wasatch" coal (Coalville Member), the Oyster Ridge Sandstone, and the Longwall Sandstone. A generalized surface-subsurface measured section of the lower Frontier Formation on Coalville anticline is illustrated in Figure 1.

Kelvin Formation

The name Kelvin Formation was first applied by Eardley (1944) in the Peoa-Wanship area southeast of Coalville. It includes a lower sequence of conglomerates and shales up to 350 ft thick and an upper variegated shale and sandstone sequence 2,000 ft or more in thickness. About 4,200 ft of variegated shales and coarse-grained to conglomeratic sandstones were penetrated by the Ohio Oil Company No. 1 Wilde in Sec. 9, T2N, R5E. This section between 2,650 ft and 6,840 ft is nearly twice the thickness of the type-section. Wells later drilled on the structural crest (Texota #F-1) encountered 2,300 ft of Kelvin above the Chalk Creek thrust. Adding this to the estimated 850 ft of section exposed in the escarpment above the well gives a thickness of 3,150 ft. This compares with the 3,000 ft reported by Maher (1976) from wells drilled in the Pineview field.

The Kelvin is dated Early Cretaceous age and equivalent to the upper sequence of the Beckwith or Gannett farther north in the thrust belt and the Bear River Formation near Evanston, Wyoming.

LOCAL STRUCTURE

The Coalville anticline is a complex, asymmetrical, doubly plunging arch with an axial trend that varies from N15°E to N40°E. The northwest flank dips an average 20° northwest into the Stevensen Canyon syncline, and is interrupted by several normal faults. The structural rise to the southest is disrupted by a central graben and culminates in a sharp reversal in the northwest corner of T2N, R6E. The east flank is overturned and broken by west-dipping thrust faults into a complex imbricate structure. The magnitude of shortening is unknown at this time. Details of the southwest plunge are completely obscured by the unconformably overlapping Tertiary conglomerates. The west flank, as mentioned, is disrupted by two large normal faults. One of these, named the "Great" fault by Wegeman (1915), has caused a two-mile offset in the strata but cannot be traced south of Chalk Creek Valley. This fault, along with the Spring Hollow fault, have formed what was initially believed to be a simple graben some two mi wide. This graben has produced a separate area of fault closure against the Spring Hollow fault east of Coalville. Complex structure observed in a hill one-half mi east of the Coalville station was initially interpreted as evidence of an east-west trending tear fault zone. The outcrop in question consists of vertical beds of maroon shales, sandstone, and conglomeratic sandstones tentatively assigned to the Kelvin Formation. The vertical dips and abrupt strike changes markedly oppose the normal structure of the Frontier Formation across the valley. Maher (1976) reports 86 to 1,100 ft of evaporites in the Preuss Formation at Pineview consisting primarily of salt and anhydrite and moderately soft micaceous shale. Therefore, on the basis of new subsurface data, the anomalous structure is tentatively interpreted as a collapsed Jurassic diapir. Inferred structure is diagrammatically shown in the structure cross-section accompanying Plate I. Deep oil and gas tests scheduled for this area should help solve this question.

West of the graben (or collapsed diapir) the Coalville anticline continues to plunge to the southwest. This fault closure was drilled first by Mountain Fuel Supply Company in 1938 to a depth of 5,538 ft. In 1959 Ohio Oil Company deepened this well to 11,434 ft. The objective formation, the Nugget Sandstone, was never reached, due to mechanical problems. No oil and gas shows were reported nor were sufficient details available to accurately date or diagram the section penetrated.

STRUCTURAL HISTORY

The structural features of the Coalville anticline evolved from two principal orogenies. Initially, compressive forces acting from the west began to fold the area in early Montanan time and culminated at the close of Late Cretaceous. The

Table 1. Cretaceous and Tertiary Stratigraphic Section, Coalville Area.

Age	Formation/Member	Thickness in Feet
EOCENE	Knight Formation—redbeds, conglomerate, sandstone, claystone	2000
	—Unconformity—	
UPPER CRETACEOUS (?)	Echo Canyon Conglomerate—conglomerate, massive, nonmarine	3000
UPPER CRETACEOUS	Henefer Formation—sandstone, claystone, carbonaceous at base, nonmarine	2450-2500
UPPER CRETACEOUS	*Frontier Formation:* Upton Sandstone Member—regressive marine sandstone (early Niobrara)	450
	Judd Shale Member—marine shale tongue of the Hilliard (early Niobrara)	350(W)-760(E)
	Grass Creek Member—mixed marine and nonmarine sandstone and shale	875(E)-1025(W)
	Dry Hollow Member*—basal conglomerate, middle nonmarine sandstone and shale and conglomerate, Dry Hollow coal zone and upper white sandstone hogback (early Niobrara)	1000-1220 (varies locally)
	—Unconformity— (Hiatus spans middle and late Carlile)	
	Oyster Ridge Member—basal regressive marine "Oyster Ridge" sandstone; upper part nonmarine shale and sandstone (early Carlile)	200-280 (varies locally)
	Allan Hollow Shale Member—marine shale (early Carlile)	780
	Coalville member—"Wasatch" coal overlying and underlying marine sandstone (late Greenhorn)	175(E)-223(W)
	Chalk Creek Member—nonmarine redbed facies; pink to red claystone, coarse sandstone and conglomeratic sandstone (possibly Belle Fourche to late Greenhorn)	3150
LOWER CRETACEOUS (?)	Spring Canyon Member*—carbonaceous shale, sandstone and thin coals (possibly Mowry or Belle Fourche)	350
	Longwall Sandstone Member—regressive marine sandstone (possibly Mowry age)	70(W)-100(E)
LOWER CRETACEOUS	Aspen (?) Shale—dark gray shale and tan sandstone with interbedded light gray shale containing Teleost fish scales	210
	Kelvin Formation—nonmarine redbeds: shale and sandstone, lenses of conglomerate	3150

*Member names not cleared by U.S. Geological Survey Geologic Names Committee.

resulting anticlinal folding and imbricate thrust faulting observable on the east flank may have triggered diapiric movement of Jurassic evaporites. With the waning of compressional stresses, the structure was further complicated by tensional faulting in early Tertiary time. There ensued a period of extensive erosion that breached the anticline to the Kelvin Formation and produced an extensive drainage pattern. Further uplift of the early Wasatch Range triggered rapid sedimentation that buried the old drainage system with coarse material of the Knight Formation. A subsequent orogeny of probable Miocene age further accentuated the structure of the area. A more detailed account of the geologic history of the area is outlined by Eardley (1944, 1959).

GAS STORAGE PROSPECTS

In 1959 the writer examined in detail two areas of structural closure for gas storage along the Mountain Fuel Supply Company natural gas transmission line. Although, two areas were examined specifically, initial emphasis was placed on the crest of the Coalville anticline centered in Sec. 6, T2N, R5E.

Following an analysis of surface data, along with meager subsurface data from old wells, Mountain Fuel began experiments in 1960 to check the feasibility of gas storage in an unproven structural closure. Initially, the company drilled its Government No. 1 for stratigraphic information to a depth of 2,349 ft. The Chalk Creek thrust was encountered at 2,299 ft, and the well penetrated 50 ft of west-dipping (probably overturned) Frontier Formation. A sandstone in the Kelvin Formation between 1,783 and 1,870 ft was selected as the gas storage reservoir, and the well was completed with facilities for both gas injection and withdrawal. The old Ohio Oil Company Government No. 1 and Texota Oil Company Federal "L" No. 1 were reentered and likewise completed with storage-withdrawal facilities. Two additional wells were completed as observation wells to monitor pressures and possible vertical migration of the gas out of the reservoir sandstone.

Upon completion of five wells, permanent surface facilities were installed. These consist of a six in. spur line from the main transmission line, a 550-HP compressor and a four in. high-pressure line. A dehydration unit with a capacity of 6,000 MCF gas day was installed initially. Testing operations began on November 23, 1960, with the injection of about 4,000 MCF gas day into the storage sand. As injection continued, it became apparent there was no escape for connate water in the reservoir. Thus, to increase storage volume, a water withdrawal well was drilled down-structure in the southeast corner of Sec. 6. During the initial stage of operation, nearly 7 BCF gas was stored and withdrawals of up to 30,000 MCF gas per day were obtained to meet peak demands for natural gas in the Utah Valley area during cold winter days.

Increasing demands for gas storage in Salt Lake Valley motivated further research of the Coalville anticline with interest focused on the fault closure east of Coalville. The largest sandstone in this area was the Longwall Sandstone described by the author in a previous paper (Hale, 1960). It is a littoral marine sandstone with a porosity of about 17 percent and is associated with the lower Frontier coal member.

FIG. 1

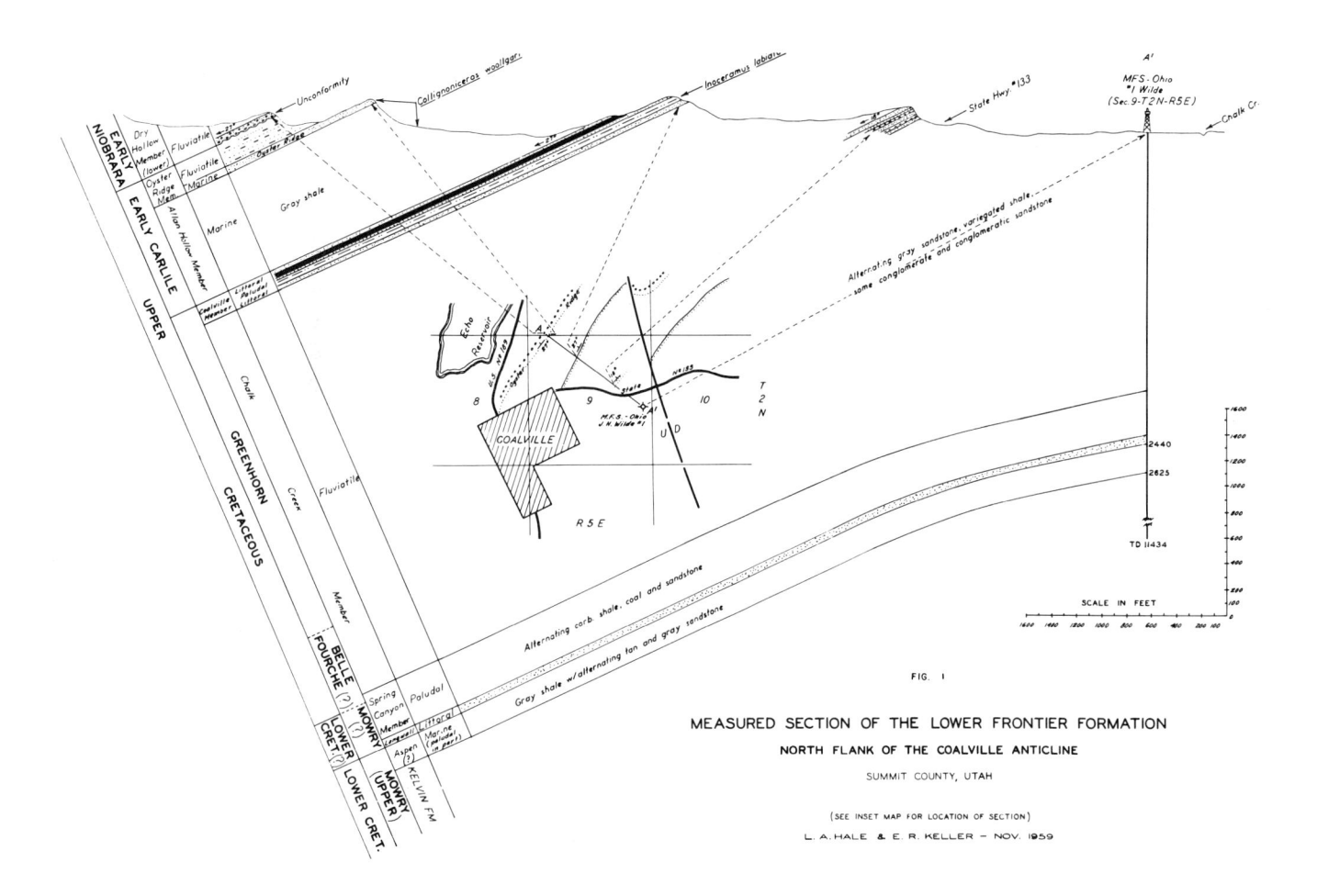

MEASURED SECTION OF THE LOWER FRONTIER FORMATION
NORTH FLANK OF THE COALVILLE ANTICLINE
SUMMIT COUNTY, UTAH

(SEE INSET MAP FOR LOCATION OF SECTION)
L. A. HALE & E. R. KELLER — NOV. 1959

Mountain Fuel reentered the old Ohio-Wilde well in 1973 and after logging through the lower Frontier Formation, completed the well for storage experiments in the Longwall Sandstone. To date, at least five wells have been completed, including recompletion of the old Western Empire well on the southwest plunge of the structure. Of special interest, these wells revealed that the Spring Hollow fault plane previously thought to be nearly vertical, dips eastward at about 65-70°. Some of the wells were located east of the fault trace, intersect the fault plane, and encounter the Longwall Sandstone at a higher structural position than earlier wells.

From the standpoint of oil and gas possibilities, the foregoing research provides an insight to the surficial geology. Undoubtedly, structure will change markedly at depth due to the influence of faulting and plastic flowage of the Jurassic Preuss evaporites.

OIL AND GAS POSSIBILITIES

The Coalville anticline has been an occasional target for exploratory drilling since Wegeman's (1915) pioneering work. Since 1921, seven oil and gas tests have been drilled on various structural segments of the anticline. Minor oil shows were reported from only one, the Ohio Oil Company well drilled in 1921 near the apex of the structure in Sec. 6, T2N, R6E. Encouraging oil shows found in 1973 in the Occidental well on the Pineview structure to the east led to an extensive seismic survey of the Coalville area by Colorado Energetics, Inc. of Denver.

Following the American Quasar, *et al.* oil discovery at Pineview, a deep test was drilled on the Coalville anticline by Colorado Energetics and others. Located one mi west of the gas storage field, the well was drilled to a depth of 13,124 ft in repeated Cretaceous strata. This well was probably located too far east on the thrust plate to encounter primary Jurassic objectives. At the present time a 16,500 ft test of Pennsylvanian Weber is being drilled by Colorado Energetics, *et al.* one mi northeast of Coalville. This well should find older primary objectives on the Chalk Creek thrust plate plus subthrust objectives below the Preuss. Therefore, the productive potential of the Coalville anticline area remains to be determined.

Despite the structural complexities, the writer is optimistic that commercial quantities of oil and gas will be discovered. The results at Pineview have proven that excellent reservoir and source rocks exist in the area. Specifically, these are to date the Jurassic Nugget Sandstone and overlying Twin Creek

Limestone. Excellent possibilities still not tested or evaluated are sub-thrust Cretaceous rocks, intertongued Triassic carbonate and sandstone facies and carbonate reservoirs in the Paleozoics. These rocks all underlie the Coalville map area.

The region constitutes a fascinating challenge to the explorationist; and the key to future discoveries here, as in other areas, will be an accurate understanding of the stratigraphic section, and an intimate knowledge of tectonic events combined with modern geophysical exploration methods.

REFERENCES

Baker, A. A., 1959, Faults in the Wasatch Range near Provo, Utah: Intermtn. Assoc. Petroleum Geologists Guidebook, 10th Ann. Field Conf., p. 153-158.

Eardley, A. J., 1939, Structure of the Wasatch-Great Basin Region: Geol. Soc. America Bull., v. 50, p. 1277-1310.

—————————, 1944, Geology of the northcentral Wasatch Mountains, Utah: Geol. Soc. America Bull., v. 55, p. 821-893.

—————————, 1959, Review of geology of northeastern Utah and southwestern Wyoming: Intermtn. Assoc. Petroleum Geologists Guidebook, 10th Ann. Field Conf., p. 166-171.

—————————, 1961, Relation of uplifts to thrusts in Rocky Mountains: Am. Assoc. Petroleum Geologists Bull., v. 45, no. 3, p. 407.

Hale, L. A., 1960, Frontier Formation, Coalville, Utah, and nearby areas of Wyoming and Colorado: Wyoming Geol. Assoc. Guidebook, 15th Ann. Field Conf., p. 137-146.

—————————, 1963, Northern Utah in Oil and gas possibilities of Utah, re-evaluated: Utah Geol. and Mineral Survey Bull. 54, p. 183-192.

Maher, P. D., 1976, Pineview field, Utah: Oil and Gas Journal, v. 74, no. 24, p. 96-100.

Mansfield, G. R., 1927, Geography, geology, and mineral resources of part of southeastern Idaho: U.S. Geol. Survey Prof. Paper 152, 409 p.

Richardson, G. B., 1941, Geology and mineral resources of the Randolph Quadrangle, Utah-Wyoming: U.S. Geol. Survey Bull. 923, 55 p.

Ritzma, H. R., 1959, Geologic atlas of Utah—Daggett County: Utah Geol. and Mineral Survey Bull. 66, 111 p.

Stanton, T. W., 1893, The Colorado formation and its invertebrate fauna: U.S. Geol. Survey Bull. 106, 288 p.

Trexler, D. W., 1955, Stratigraphy and Structure of the Coalville area, northeastern Utah: Unpublished Ph.D. dissertation, Johns Hopkins University.

—————————, 1966, Stratigraphy and structure of the Coalville area, northeastern Utah: Colo. School of Mines Prof. Contr. 2, 69 p.

Wegemann, C. H., 1915, The Coalville coal field, Utah: U.S. Geol. Survey Bull. 581-E, p. 161-184.

Williams, N. C., and Madsen, J. H., 1959, Late Cretaceous stratigraphy of the Coalville area, Utah: Intermtn. Assoc. Petroleum Geologists Guidebook, 10th Ann. Field Conf., p. 122-125.

Vetch, A. C., 1907, Geography and geology of a portion of southwestern Wyoming: U.S. Geol. Survey Prof. Paper 56, 178 p.

HINGELINE SEDIMENTS OF THE OVERTHRUST BELT

Field Trip Committee

Chairman
Howard H. "Tom" Odiorne
Crystal Exploration and Production Co.

Assistant Chairman
Michael W. Boyle
Crystal Exploration and Production Co.

Field Trip Leaders

Willson W. Bell	Robert E. Near
Odessa Natural Corp.	Energy Reserves Group, Inc.
Denver, Colorado	Denver, Colorado

Bus Speakers

James Baer	Jock A. Campbell	Howard R. Ritzma	Floyd C. Moulton
Brigham Young University	Utah Geological and Mineral Survey		Phillips Petroleum Company
Provo, Utah	Salt Lake City, Utah		Denver, Colorado

Field Trip Stop Speakers

Max D. Crittenden, Jr.	Lyle A. Hale	James H. Madsen, Jr.	Patrick D. Maher
U.S. Geological Survey	Geological Consultant	University of Utah	Energetics, Inc.
Menlo Park, California	Salt Lake City, Utah	Salt Lake City, Utah	Englewood, Colorado

Richard W. Moyle	Milan S. Papulak	Peter R. Rose
Weber State College	Quantex Corporation	Energy Reserves Group, Inc.
Ogden, Utah	Salt Lake City, Utah	Houston, Texas

Photo by Jack Rathbone

Pineview Oil Field. The discovery well, American Quasar, Energetics Inc., North Central Oil, #1 Newton Sheep Company well-head in the foreground. Drilling well, (Parker rig #118), is the American Quasar, Union Pacific 3-2, drilling below 15,800 in mid-June.

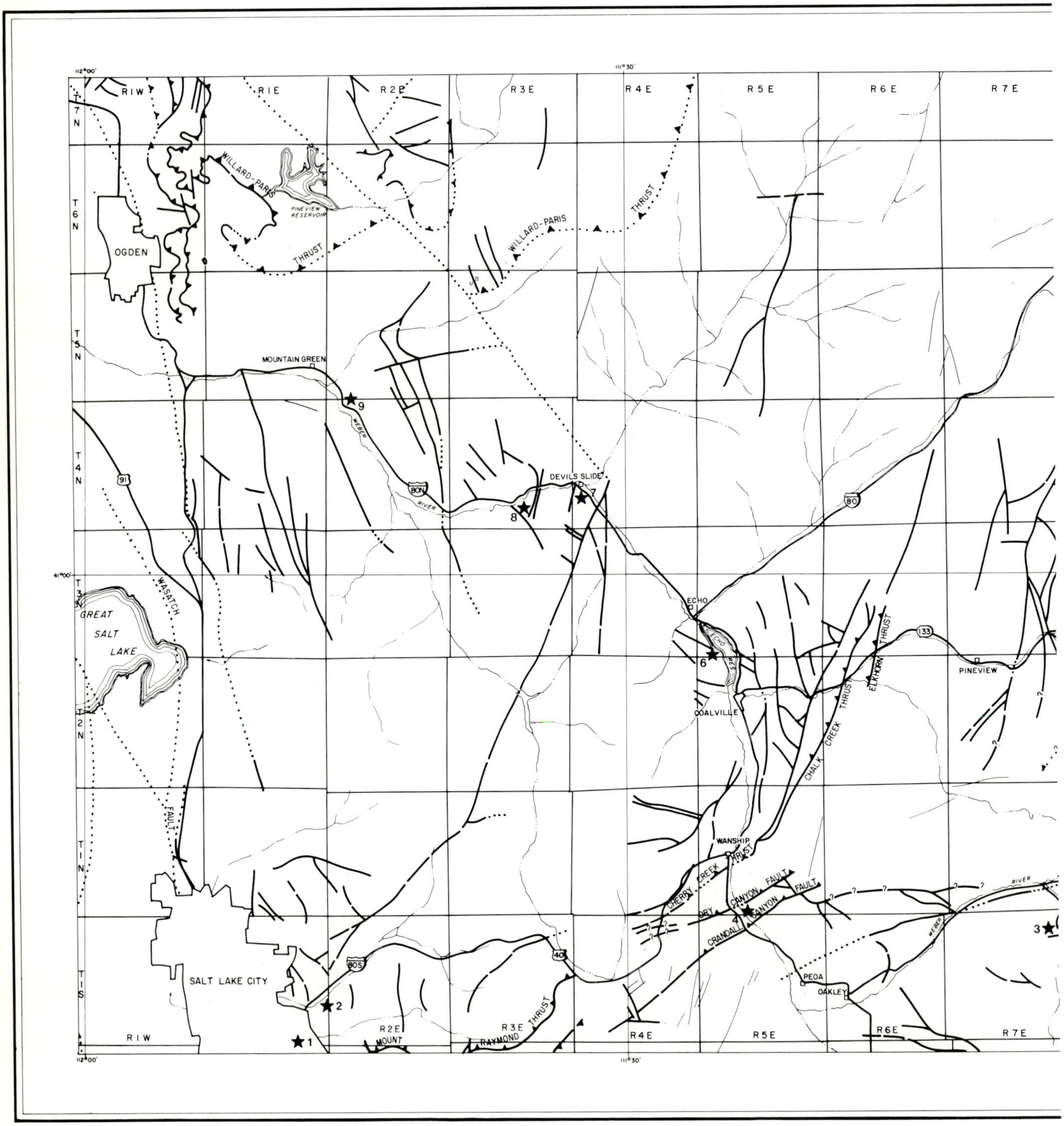

112°00'
111°30'
R1W
R1E
R2E
R3E
R4E
R5E
R6E
R7E
T7N
T6N
T5N
T4N
T3N
T2N
T1N
T1S
41°00'
OGDEN
WILLARD-PARIS
THRUST
PINEVIEW RESERVOIR
MOUNTAIN GREEN
WEBER
RIVER
DEVILS SLIDE
ECHO
ECHO RES.
COALVILLE
GREAT SALT LAKE
WASATCH
FAULT
WANSHIP
CHERRY CREEK
THRUST
DRY CANYON FAULT
CRANDALL CANYON FAULT
CHALK CREEK THRUST
ELKHORN THRUST
PINEVIEW
PEOA
OAKLEY
SALT LAKE CITY
MOUNT RAYMOND THRUST
RIVER
91
80N
80
80S
40
133
1
2
3
4
6
7
8
9

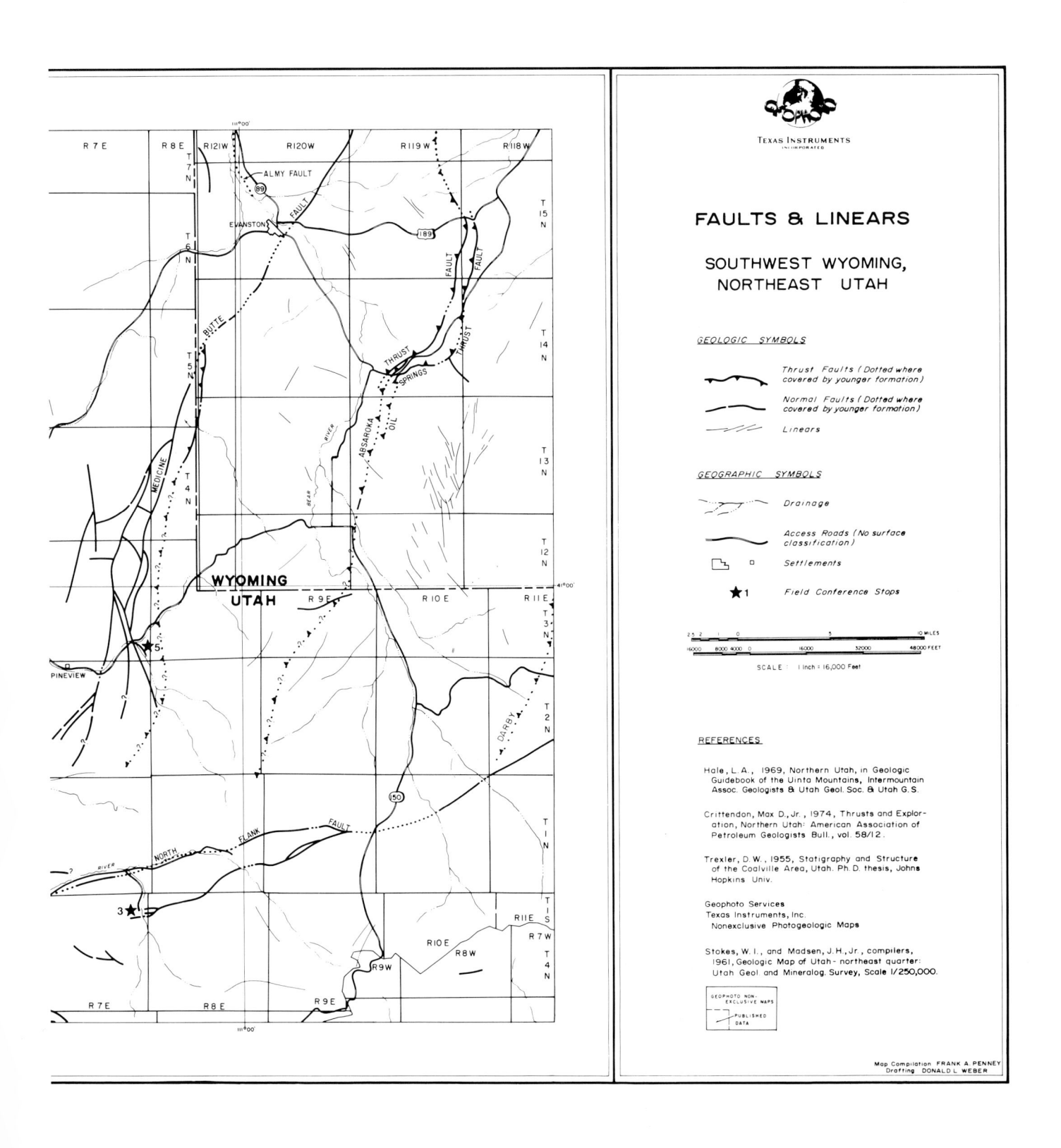

ROAD LOG — HINGELINE SEDIMENTS OF THE OVERTHRUST BELT

FAULTS & LINEARS

SOUTHWEST WYOMING, NORTHEAST UTAH

GEOLOGIC SYMBOLS

- Thrust Faults (Dotted where covered by younger formation)
- Normal Faults (Dotted where covered by younger formation)
- Linears

GEOGRAPHIC SYMBOLS

- Drainage
- Access Roads (No surface classification)
- Settlements
- ★1 Field Conference Stops

SCALE: 1 Inch = 16,000 Feet

REFERENCES

Hale, L. A., 1969, Northern Utah, in Geologic Guidebook of the Uinta Mountains, Intermountain Assoc. Geologists & Utah Geol. Soc. & Utah G.S.

Crittendon, Max D., Jr., 1974, Thrusts and Exploration, Northern Utah: American Association of Petroleum Geologists Bull., vol. 58/12.

Trexler, D. W., 1955, Statigraphy and Structure of the Coalville Area, Utah. Ph.D. thesis, Johns Hopkins Univ.

Geophoto Services
Texas Instruments, Inc.
Nonexclusive Photogeologic Maps

Stokes, W. I., and Madsen, J. H., Jr., compilers, 1961, Geologic Map of Utah - northeast quarter: Utah Geol and Mineralog. Survey, Scale 1/250,000.

Map Compilation FRANK A. PENNEY
Drafting DONALD L. WEBER

FIELD TRIP ROUTE—SEGMENT NO. 1
Mouth of Parleys Canyon to Wanship, Utah
Driving Distance: 28.7 Miles

Introduction

This segment of the field trip will cover geologic highlights along Interstate Highway 80, starting at the mouth of Parleys Canyon located in the foothills of the Wasatch Range and culminating at the town of Wanship, approximately 35 miles east of downtown Salt Lake City. Surface exposures of sedimentary rocks from Paleozoic through Quaternary age are present along the route of this segment of the field trip and stops are scheduled to permit observation and discussion of outstanding geologic features. Parleys Canyon roughly parallels a prominent east-west trending structural syncline. Outcrops of Triassic and Jurassic rocks are exposed at the mouth of the canyon. Proceeding eastward for a distance of approximately eleven miles to Parleys Summit, Jurassic and Cretaceous stratigraphy can be observed. The route of the field trip from this point will descend through the stratigraphic section back into the Triassic. In the vicinity of Kimball Junction, Eocene sediments and Tertiary pyroclastics are represented, and Cretaceous age formations are again exposed before reaching Wanship.

Credits

The R.M.A.G. Field Trip Committee is deeply indebted to Milan S. "Mel" Papulak, Quantex Corporation, and James H. Madsen, Jr., Department of Geology and Geophysics, University of Utah, Salt Lake City, for compiling geologic data for this segment of the field trip route. Sincere appreciation is extended by the Committee to Max D. Crittenden, U.S.G.S., Menlo Park, California, for lecturing at selected stops along the entire route of the field trip and for graciously offering professional advice which contributed to the field trip.

Note: Start of the field trip will be at the Olympus Hills Shopping Center just south of the mouth of Mill Creek Canyon (Wasatch Blvd. at 39th). From Snowbird Lodge to the Olympus Hills Shopping Center is 17 miles.

Dist.	Cum Miles	
0.	.0	**STOP NO. 1.** Lecture by Max D. Crittenden on Paleozoic and Mesozoic stratigraphy including structural complexities associated with the Wasatch Front north of Mt. Olympus. Proceed north to traffic light, turn left and merge onto U.S. Interstate Highway 215 North.
1.0	1.0	Cheyenne turnoff, proceed right onto U.S. Interstate Highway 80 East.
.7	1.7	"Suicide Rock" at bottom of canyon painted by local high school students represents the Triassic Gartra Grit (Shinarump equivalent) underlying the Chinle Formation. Ahead, on the north side of the canyon are exposures of salmon to orange colored beds of the Jurassic Nugget Sandstone overlain by typical gray-weathering, highly fractured and splintered shales and limestones of the Twin Creek Limestone. The caravan will enter the mouth of Parleys Canyon along the steeply dipping south limb of a breached syncline. To the right, on the south side of the canyon, red beds of the Triassic Ankareh Formation are exposed and are cut locally by a dike.
.9	2.6	**STOP NO. 2.** Pull off freeway to right at outcrops of fractured Twin Creek limestones and shales. Discussion of outcrops exposed at mouth of Parleys Canyon by James H. Madsen, Jr.
.1	2.7	Traveling along base of Parleys Canyon syncline. Talus slides are developed on both sides of canyon due to highly fractured nature of the Twin Creek and Nugget formations.
.9	3.6	Utah Portland Cement quarry in the Twin Creek Limestone. Note outcrops displaying such structural features as fractures, joint patterns, and local folding, and primary features such as ripple marks.
3.4	7.0	Mountain Dell Dam site. On the right are outcrops of the Twin Creek Limestone and red beds of the Preuss Sandstone. Observe from this point to the crest of Parleys Summit numerous slumps and landslides along the south side of the canyon road created by highway construction of questionable design. Removing the toe of these slides is an exercise in futility, since it simply rejuvenates the slide. Slide debris overlies northward plunging dip-slopes along much of the Interstate 80 route between Salt Lake City and Wanship, and to the north in Weber Canyon between Ogden, Utah and Evanston, Wyoming. A little geology in action requiring continuing attention by the highway crews.
1.0	8.0	Ahead and to the east, outcrops of the Kelvin

Mount Olympus from Stop #1 (Olympus Hills Shopping Center). Steeply dipping Precambrian quartzites form the light-colored cliffs near the top of the mountain with flatirons of Paleozoic rocks propped against the intermediate slopes.

Formation (Lower Cretaceous) are represented by reddish beds in the bottom of the canyon which are overlain by the Frontier Sandstone, Wanship Formation, and Echo Canyon Conglomerate (Upper Cretaceous). The Aspen Shale found between the Kelvin and overlying beds of the Frontier east of Parleys Summit is apparently absent in this area. North, to the left, are exposed the orange colored beds of the Nugget Sandstone.

.9   8.9 Observe slumping of landslide debris along south side of canyon.

.9   9.8 Mouth of Lamb's Canyon—more slumping of unstable rock and soil overlying northward-dipping outcrops.

1.7 11.5 Twin Creek Limestone exposed on the right or south side of the canyon. On the left or north side, red beds and white nodular limestones are presently assigned to the Lower Cretaceous Kelvin Formation. Formerly, these beds were regarded as being in the Jurassic Morrison Formation. An open discussin of Kelvin and Morrison stratigraphy is scheduled at Stop No. 3, Smith and Morehouse Reservoir.

1.2 12.7 Parleys Summit. Jurassic Preuss Sandstone outcrops in left or north side of canyon followed by exposures in road cut of the underlying Twin Creek Limestone. East of the summit the caravan will be going stratigraphically down section.

2.4 15.1 Gorgosa Ski Area. The Jurassic Nugget Sandstone is exposed in small pits and rock quarries along the right hand side of the road. Tracks of amphibians and reptiles have been found in the Nugget at this location and at others as far south as Heber. The rock is popular for use as flagstone in construction. Before reaching the valley flatlands exposures of the underlying Thaynes Formation and Woodside Shale can be observed on the left or north side of the highway.

2.2 17.3 Kimball's Junction. Although almost completely obscured by vegetation, beds of Triassic and Jurassic age have been mapped nearby in the upper plate of the folded Mt. Raymond thrust.

1.2 18.5 Park City West ski runs on flank of high hills to the south.

.3   18.8 Mountain Meadows Ranch on left side of highway.

1.5 20.3 Silver Creek Junction, approaching Silver Creek

Photo by Jack Rathbone

Twin Creek Limestone overlying Nugget in the vicinity of Stop #2 in Parley's Canyon. View down canyon.

Canyon. This valley is filled with reworked Tertiary age sediments and volcanics. Approximately ten years ago, a building and development project known as Silver Creek Estates required drainage improvements in the vicinity of the ranch houses to our left on the north side of the valley floor. Ditching machines unearthed a large mammoth tooth. Salvage funds were appropriated to explore this new paleontological site in light of eminent highway construction plans to complete the Silver Creek Interchange resulting in a prolific find of Pleistocene vertebrate fauna. The hills immediately north of the valley are capped by Tertiary conglomerates.

1.4   21.7   Entering Silver Creek Canyon. The weathered gray andesites, breccias, and agglomerates are of Tertiary age and are locally known as the Silver Creek Volcanics. These volcanics are distributed throughout many of the valleys between this area and the town of Kamas (located approximately 12 miles southeast of Silver Creek Junction). Flows related to Silver Creek-age volcanics have been mapped to the south in a band that wraps around the nose of the Uinta Mountains.

2.2   23.9   Occasional exposures of the Frontier Sandstone and reddish beds of the Kelvin Formation and underlying Preuss Sandstone.

1.0   24.9   Outcrops of Frontier overlying Kelvin, the strike of which roughly parallels the highway.

2.6   27.5   Volcanics along both sides of the highway approaching the town of Wanship. Gray beds of the Frontier at 1:00. At 12:00, the Wanship Formation overlies Frontier. The Tertiary conglomerate forms prominent benches below the crests of high hills present on the horizon.

1.2   28.7   Wanship exit. The caravan will proceed south on U.S. Highway 189 Alt. to Rockport Reservoir.

FIELD TRIP ROUTE—SEGMENT NO. 2

Wanship, Utah to Smith and Morehouse Reservoir to Coalville, Utah

Driving Distance: 65 Miles

Introduction

This segment of the field trip will permit observation of geology between Wanship, Peoa, Oakley, and Smith and Morehouse Reservoir. The caravan will return to Wanship and proceed to the town of Coalville after an intervening stop at the Rockport State Park area. Surface exposures of rocks ranging in age from Precambrian through Quarternary are recognized along this route.

Credits

Compilation of this portion of the field trip route was assembled by Michael W. Boyle, District Geologist, Crystal Exploration and Production Co., Denver, Colorado, in cooperation with Milan S. "Mel" Papulak, Quantex Corporation, Salt Lake City, Utah.

Dist.	Cum Miles	
1.2	29.9	On right, red beds of Kelvin Formation unconformably resting upon Jurassic Preuss Sandstone in fault contact with Frontier Formation.
1.5	31.4	Wanship Dam and Rockport Lake. To left, on east side of reservoir, outcrops of Frontier Formation are folded into the north-plunging, asymmetrical Dry Creek anticline which is truncated, and overlain by Cretaceous Wanship Formation. The southeastern limb of the anticline is terminated by the Dry Canyon fault. Rockport Lake is a popular recreation facility, offering excellent sailing and fishing.
.4	31.8	Road to Utelite plant on right. Swelling clay minerals are open-pit mined from Frontier shales and marketed locally as lightweight aggregate in the manufacture of prestressed concrete beams and similar construction materials.
1.2	33.0	On right, outcrops in roadcut are Preuss and Kelvin formations in the fault slice between the Dry Canyon and Crandall Canyon faults. Crandall Canyon fault zone immediately ahead.
.6	33.6	Right, outcrops exposed in road-cut identified as Oyster Ridge Member of Frontier Formation in IAPG 1959 Field Conference road log.
1.1	34.7	Entrance to Rockport State Park on left. Outcrops on immediate right are Frontier Formation. As we cross the Weber River, we are moving down through the Lower Cretaceous and underlying Jurassic sections which are steeply dipping towards us.
.2	34.9	To left in valley, outcrop of Aspen Formation with the typical fish scales. Immediately below the Aspen, reddish beds of the Kelvin are exposed. Ahead at the curve in road, a resistant sandstone member of the Kelvin is cut by the road.
.2	35.1	To left, brilliant orange-red shales and nodular limestones, formally mapped as Jurassic Morrison Formation (Critteden, 1974, and Stokes and Madsen, 1961) are now included in the lower part of the Kelvin Formation (Madsen-personal communication). Ahead, Jurassic Stump, Preuss, and Twin Creek formations are present—all the contacts are visable. To right, at 4:00 on the horizon, the Mountain Meadow surface is formed by Eocene-Oligocene volcanics which overlie truncated Mesozoic strata.
.4	35.5	On left, contact between reddish-brown Preuss Sandstones and underlying gray Twin Creek Limestone. Unconformably overlying Jurassic strata on the crest of the hill are tuffaceous beds of the Fowkes Formation—recently dated by Mike Nelson at the University of Utah as early to middle Eocene. The tuffs which contain vertebrate fragments are related to the Silver Creek Volcanics. Cross road at 3:00 is Brown's Canyon Road.
.3	35.8	On left, Twin Creek—Nugget contact. (Jurassic)
.2	36.0	On left, Peoa Cemetery, reddish-maroon sandstones of the Nugget are found in fault contact with Twin Creek Limestone up small gulley.

Fault trends northeast-southwest.

.5 36.5 Town of Peoa, turn left in center of town, continue to town of Oakley. Nugget Sandstone exposed on left.

.4 36.9 Highly cross-bedded Nugget Sandstone in outcrop immediately on left side of road near buildings.

1.1 38.0 Traversing terraces of Pleistocene age boulders. To left at 9:00 is Big Piney Mountain, composed of Upper Jurassic and Cretaceous strata. Ahead lies Hoyt Mountain at 12:00.

1.3 39.3 Junction at Oakley, U.S. Highway 189 Alt. and Utah Highway 213. Intersection unmarked, turn left at gas station. On skyline from 11:00 to 12:00 north-plunging dip slopes are in Weber Formation. Canyon to right of Hoyt Mountain at 1:00 is cut along the strike of the Park City Formation. The high point on gable-shaped ridge is formed by a lower member of the Thaynes Limestone. The bright red outcrop to the left of the erosional surface on the skyline is in the Ankareh Formation. Shinarump, Chinle, and Nugget follow in sequence to the north.

1.3 40.6 On left, outcrops of Thaynes Limestone of Lower Triassic age.

1.1 41.7 High on left, ridge known as Mahogany Hills is formed by outcrops of the Triassic Thaynes and Ankareh (Moenkopi Group), Shinarump (Gartra Grit), and Chinle (Popo Agie equivalent) formations and the Jurassic Nugget and Twin Creek formations. As we proceed eastward up the canyon, we will be going up section across all these formation boundaries, although most of them are poorly exposed.

2.5 44.2 On left, exposures of Twin Creek Limestone overlying Nugget Sandstone in roadside quarry.

.5 44.7 We are now following the North Flank fault zone which forms the northern edge of the Uinta Mountain uplift. On the right, Triassic and late Paleozoic rocks form the south canyon wall. On the left, the north canyon wall is composed of Cretaceous rocks capped unconformably by Tertiary Knight Conglomerate.

.5 45.2 Crossing Weber River at Upper Weber River Country Store.

.5 45.7 On left, Windy Ridge is topped with Tertiary conglomerates which are correlative with the Wasatch beds of the Bridger Basin.

2.6 48.3 Note three river terraces on left, cabin once property of famous Beefeater, Harry Veal.

1.3 49.6 On right is mouth of Smith and Moorehouse Canyon. Weber Quartzite outcrops on east canyon wall. Note the glacial morain topography on right hand side of road. At 12:00 on skyline is Moffit Peak, composed of vertical Weber Quartzite strata.

.4 50.0 Junction, turn right up Smith and Morehouse Canyon.

.5 50.5 Smith and Morehouse Creek.

.1 50.6 On left side of canyon, thin-bedded Round Valley Limestone of Lower Pennsylvanian age.

.5 51.1 On right, outcrop of Upper Pennsylvanian Weber Quartzite.

1.1 52.2 Ahead on skyline, Bear Trap Hollow—the densely wooded area rests upon a dipslope of the Humbug Formation.

.8 53.0 On left is Smith and Morehouse Reservoir. Note terminal moraines at dam site. Rolling topography east of dam is ancient landslide mass.

1.1 54.1 Keep left, cross bridge and continue into public campsite.

.1 54.2 **STOP NO. 3.** Discussion of Wasatch, Echo Canyon, and Kelvin formations as interpreted in field trip area, James H. Madsen, Jr., University of Utah, Salt Lake City, Utah. Regional discussion of Lower Paleozoics, Peter R. Rose, Energy Reserves Group, Inc., Houston, Texas, and Max Crittenden, U.S.G.S., Menlo Park, California.

Return to Rockport Reservoir.

20.9 75.1 Rockport Reservoir State Park turnoff—turn right.

2.1 77.2 **STOP NO. 4.** Campground on left. Discussion of stratigraphy and faulting and possible connection of thrust faulting observed in this area with thrusting observed northeastward in Wyoming and westward near Mt. Raymond by Milan S. Papulak and Max D. Crittenden, Jr. Return to U.S. Highway 189 Alt.

2.1 79.3 Rockport Reservoir State Park entrance. Turn right and return to Wanship.

6.0 85.3 Wanship, Utah. Underpass, proceed straight ahead.

Lower Mississippian limestone cliffs above the Precambrian Red Pine Shale. View west from campground south of Smith and Morehouse Reservoir.

.4	85.7	Fork in road, turn right onto Utah Highway 333, proceed north.
.1	85.8	Ahead on skyline, Tertiary Knight Conglomerate.
1.0	86.8	At 11:00, volcanics of Eocene-Oligocene age. Hills at 3:00 have Lower Cretaceous Kelvin exposed along the base.
.8	87.6	On right, terrace deposits.
2.7	90.3	Hoytsville, Utah. At 10:00 on skyline are hogbacks formed by resistant sandstone of the Wanship Formation.
1.7	92.0	On both sides of valley, Tertiary Knight Conglomerate unconformably overlies Cretaceous formations.
.8	92.8	Entering Coalville.
.5	93.3	Ahead, across valley, note two hogbacks of the Wanship Formation.
.4	93.7	Intersection with Utah Highway 181, turn right.

Photo by Jack Rathbone

Resistant sandstone hogbacks of the Wanship Formation from Hoytsville, south of Coalville.

FIELD TRIP ROUTE—SEGMENT NO. 3

Coalville, Utah to Pineview Oil Field to Echo Reservoir and Echo Junction

Driving Distance: 35.6 Miles

Introduction

Complex folding and faulting associated with Coalville anticline is revealed on the surface between Echo Reservoir and Pineview field. The route of the field trip will permit observation of the Cretaceous and Tertiary sediments east and north of Echo Reservoir.

Credits

Compilation of this segment of the field trip was made by Milan S. "Mel" Papulak, Quantex Corporation, Salt Lake City, Utah, and Michael W. Boyle, District Geologist, Crystal Exploration and Production Co., Denver, Colorado.

The R.M.A.G. Field Trip Committee is indebted to Lyle Hale, Geological Consultant, Salt Lake City, Utah, and Patrick D. Maher, Executive Vice-President, Energetics, Inc., speakers at stops on this portion of the field trip route.

Dist.	Cum. Miles		Dist.	Cum. Miles	
	93.7	Coalville, proceed east on Utah Highway 181. WARNING: Stay on right side of road and be on lookout for crude transport trucks.	1.8	97.0	On left, structural deformation in Frontier beds associated with the Great fault.
.5	94.2	Ridge on left formed by Oyster Ridge Member of Frontier Formation.	.2	97.2	Crossing the Great fault, a north-south trending normal fault downdropped to the west.
1.0	95.2	North-south trending faults in this vicinity.	.3	97.5	On left, Lower Cretaceous Kelvin exposed in core of Coalville anticline.

Angular unconformity between horizontal Knight conglomerates (Eocene) above the dipping Echo Canyon Conglomerate.

1.1	98.6	Mountain Fuel Supply Chalk Creek Station compressors.
.4	99.0	Approximate position of north by northwest trending axis of Coalville anticline in upper plate of Chalk Creek thrust.
.3	99.3	On left, near vertical Kelvin beds on the asymetrical east flank of the Coalville anticline. Continuing through zone of thrusting and up section stratigraphically for several miles.
2.8	102.1	On left, outcrops of Frontier Formation.
1.8	103.9	Upton, Utah. Proceed straight, following Chalk Creek. **Watch out for chuckholes in road.**
.9	104.8	To left, Tertiary Knight Conglomerate on crest of hill.
.8	105.6	To left, outcrops of Wanship Formation (Upper Frontier—Hale, 1969) unconformably overlain by flat-lying Knight Conglomerate.
.5	106.1	To right, Knight (?) Conglomerate.
3.1	109.2	**STOP NO. 5.** Pineview Field. Discussion of geology of the Pineview field by Patrick D. Maher, Energetics, Inc., Denver, Colorado. Return to Coalville.
15.5	124.7	Intersection with Utah Highway 333, turn left.
.3	125.0	Right turn.
.2	125.2	Right turn to U.S. Interstate Highway 80.
.9	126.1	On left, Wanship Formation (Upper Frontier - Hale, 1969). On right, Echo Reservoir.
.3	126.4	On left, several secondary northwest-southeast trending normal faults cut the Wanship beds.
1.2	127.6	**STOP NO. 6.** Scenic view area. Turn off interstate to right. Discussion of the geology of Coalville anticline by Lyle Hale, Geological Consultant, Salt Lake City, Utah. After lecture proceed north to Echo Junction on U.S. Interstate Highway 80.
.8	128.4	On left, contact between the Cretaceous Wanship Formation and the red conglomerates of the Echo Canyon Formation. To the west and north, Upper Cretaceous Wanship and Echo Canyon formations unconformably overlain by Tertiary Knight Conglomerate.
.9	129.3	Echo Junction, take U.S. Interstate Highway 80N (Ogden) and proceed northwest. To right, Echo Canyon conglomerates form canyon walls.

Echo Canyon in 1878. The Cliffs are composed of Echo Canyon conglomerate of Late Cretaceous Age. From King, 1878.

FIELD TRIP ROUTE—SEGMENT NO. 4

Echo Junction to mouth of Weber Canyon and south to Olympus Hills Shopping Center

Driving Distance: 71.9 Miles

Introduction

This segment of the field trip contains more geologic and scenic attractions than can be described within the limitations of this road log. The area on the field trip route between Echo Junction and the mouth of Weber Canyon near Ogden, Utah is included in what Eardley (1952) called the Wasatch hinterland. Structural and stratigraphic records preserved in rocks in this Wasatch hinterland offer classic examples of the type of geology encountered to the northeast of the overthrust belt area. A panoramic view of the Wasatch Front between Ogden and Salt Lake City will add "frosting to the cake" and the 1976 field trip of R.M.A.G. will terminate at the original starting point in the Olympus Hills shopping center. There is disagreement about the age and nomenclature of certain units along the field trip route. We have chosen to follow the Utah State Geologic map where there is disagreement.

Credits

The R.M.A.G. gratefully acknowledges Richard W. Moyle and Sidney R. Ash, Weber State College, Ogden, Utah for compiling the road log and geologic discussions for this segment of the field trip.

The Witches in 1878. From King, 1878.

Dist.	Cum. Miles	
	129.3	Echo Canyon Junction. Proceed right (northwest) on U.S. Interstate Highway 80N. The field trip route will follow the Weber River from here to Uintah Junction at the mouth of Weber Canyon. Echo Canyon Conglomerate of Late Cretaceous age (Pulpit Conglomerate of Eardley, 1944), exposed on both sides of the highway for several miles. In this area it consists mainly of red-brown conglomerate and is about 3100 ft thick (Williams and Madsen, 1969). The Echo Canyon Conglomerate is thought to be approximately equivalent to the Price River or upper Indianola of central Utah.
.5	129.8	Echo, Utah. Just over the hill to right.
1.0	130.8	Crossing approximate trace of axial plane of the north-trending Stevenson Canyon syncline.
.4	131.2	Hoodoos. Unique erosional remnants on right at about 3 o'clock on hillside. Locally these are called "The Witches". They are eroded from the Echo Canyon Conglomerate.
.6	131.8	Contact between the Echo Canyon Conglomerate and the underlying Wanship Formation of Late Cretaceous age at base of cliff to the right.
.2	132.0	Anticlines and synclines exposed on both sides of highway ahead are in the Wanship Formation. These local structural features probably formed during the Laramide orogeny.
1.3	133.3	Steeply dipping beds of the Wanship Formation in the hills to the right.
.2	133.5	Henefer to left. Continue straight ahead. The Mormon pioneers turned left near here and entered Salt Lake Valley through Main, East, and Emigration Canyons. They did not continue down Weber Canyon because Devil's Gate made the canyon impassable for wagon trains. South end of Beaver River Range to right. Durst

Mountain uplift of Eardley (1952) ahead. The escarpment is composed principally of Knight Conglomerate of Eocene age (Echo Canyon Conglomerate of Mullens and Laraway, 1964). The unit rests with angular unconformity on Triassic and Jurassic rocks.

.4   133.9   Clay pits on the right from 2-3 o'clock are in flat-lying beds of the Wanship Formation of Late Cretaceous age. The pits were worked by early settlers to make bricks used in the construction of many homes and buildings in Salt Lake City, Ogden, and other communities in the area (Stringham and Cahoon, 1959). Some of the pits are still being operated on a limited basis.

1.4   135.3   Cross Weber River.

1.2   136.5   Cross approximate trace of Croyden (East Canyon) fault. It is a normal fault with the downthrown block on the east and is probably of Laramide age. To the left, the fault brings the Knight Conglomerate into contact with the Kelvin Formation of Early Cretaceous age. The road cuts ahead on both sides of the road are in the Knight Conglomerate.

.7   137.2   Cross approximate position of angular unconformity (covered) between southeast-dipping beds of the Knight Conglomerate and the nearly vertical beds of the Twin Creek Limestone.

.3   137.5   Nearly vertical beds of upper and middle Twin Creek Limestone to left at 11 o'clock. The angular unconformity separating the Twin Creek Limestone from the Knight Conglomerate is exposed high on the cliff to the left. Note the ripple marks on the bedding planes in the Twin Creek to the left near the base of the cliff at about 9 o'clock.

.2   137.7   Mouth of Lost Creek Canyon to right. The Ideal Cement Company has a plant on the floor of the canyon which uses limestone from the middle part of the Twin Creek.

.2   137.9   Prepare to exit to right at Devil's Slide viewpoint exit.

.3   138.2   **STOP NO. 7.** Devil's Slide viewpoint area. Vertical outcrop of two resistant limestone beds in the lower Twin Creek Limestone forms the chute-like Devil's Slide on the west side of the highway. The Devil's Slide nearly blocked the canyon at this point and prevented the Mormon pioneers from proceeding further down this canyon. Since that time man has modified the canyon so that there is now room for both the railroad and Interstate Highway 80N.

Devil's Slide in 1878. From King, 1878, at stop No. 7.

.1   138.3   Return to Interstate Highway 80N. Contact between Twin Creek Limestone and Nugget Sandstone of Jurassic age on right at 2 o'clock where the brown shale of the Twin Creek is in contact with the orange sandstone of the Nugget. The rocks in this area are coated with a thin crust of gray cement which obscures their natural color. This coating came form the dust produced by the nearby Ideal Cement Company plant. Recently a $2,000,000 precipatator has been installed at the plant to reduce air pollution in the area. The precipitator now collects about 210 tons of dust each day.

.1   138.4   Approximate contact between Nugget Sandstone and Ankareh Formation of Late Triassic age exposed on right.

.2 138.6 The contact between the siltstone and claystone member and the Mahogany Member of Ankareh Formation is at the mouth of the side canyon at right.

Photo by Jack Rathbone
Devils Slide—Twin Creek Limestone.

.2 138.8 Approximate contact of the Mahogany Member of the Ankareh Formation (reddish-brown siltstones and sandstones) with the underlying gray limestones of the Thaynes Limestone of Late Triassic age. As we move westward we will continue to go stratigraphically down section and we will see from a distance the Thaynes, Woodside, Dinwoody, Park City, and Weber formations. Shortly, however, we will back track and it will be possible to examine closely each one of these units.

.2 139.0 Contact between the Thaynes Limestone and Woodside Formation at the mouth of the side canyon to the right.

.9 139.9 Old open-cut phosphate mine in the Permian Park City Formation on canyon walls to left. Park City Formation to right. Notice folding to the north.

.2 140.1 Prepare to exit Interstate to right.

.4 140.5 Exit at Taggarts. Turn left at stop sign and pass under Interstate Highway 80N. Follow old highway. The following units are exposed in the left side of the old highway:

Weber Quartzite of Pennsylvanian and Permian age. The Weber is about 1200-1500 ft thick, but only the upper part is exposed. The Weber consists of fine-grained quartzite interbedded with limestone.

Park City Formation of Permian age, consisting of about 600 ft of limestone, shale, and sandstone.

Dinwoody Shale of Early Triassic age, consisting of about 100 ft of red shale and sandstone.

Thaynes Formation of Early Triassic age, consisting of about 2400 ft of gray limestone, siltstone, and sandstone.

STOP NO. 8. Buses will progress up the canyon to the east on the old highway and will stop at an outcrop of the Woodside Formation where those who wish can walk up the canyon through the Woodside and Thaynes formations. In the meantime the buses will turn around below an underpass and will pick up passengers on the way back to Interstate Highway 80N. Two miles are allowed in this log for the side trip.

2.0 142.5 Enter Interstate Highway 80N. Continue westward. A series of anticlinal and synclinal folds in Weber Quartzite are exposed in the walls of the canyon. A large asymmetrical anticline with the steep dips on the east flank exposed in the canyon wall at 12 o'clock.

.5 143.0 Flat-lying beds of cherty limestone in the Weber Quartzite exposed in the road cuts ahead.

1.1 144.1 Note minor fault to right in Weber Quartzite.

.4 144.5 Contact between the gray Weber Quartzite and red sandstones of the Morgan Formation of Middle Pennsylvanian age in the valley to the right. This is the type area of the Morgan Formation which was named by Blackwelder (1910). Here it consists of about 200 ft of red sandstone and shale. The hills on the south side of the valley consist of Wasatch Formation.

.8 145.3 The contact between the Morgan Formation and Round Valley Formation of Early Pennsylvanian age is in the strike valley to the right at about 2 o'clock. The contact occurs where there is a change from the reddish rocks of the Morgan to the gray color of the Round Valley Formation.

Photo by Jack Rathbone

Thaynes Formation standing at a high angle at Stop #8, Taggert Exit.

.2	145.5	Contact between Round Valley Formation and Brazer Formation of Mississippian age. Outcrops ahead are in the Brazer. Note channels and caverns exposed in the road cuts.
1.0	146.5	Crossing trace of the northward-trending Como fault. The Como fault is exposed on the canyon wall to the left and is a normal fault with the downthrown block to the west. It brings the Mississippian Madison Formation (on the east) in contact with the Eocene Wasatch Formation (on the west). Drag folds are prominent in the Mississippian rocks in the upthrown block to the left at 7 o'clock. Springs to the left at about 11 o'clock in the recreation area are fed by hot mineralized waters moving upward along the Como fault. The Como fault is on the east end of Morgan Valley and forms the west boundary of the Durst Mountain uplift. Entering Morgan Valley syncline.
.1	146.6	Exit Interstate Highway 80N and follow old highway northward. Note pioneer limestone kilns at base of hill to right. The quarry behind the kilns is in the Madison Limestone of Mississippian age. Hill at right is scarp of Como fault. Entering North Morgan. Hitching Post Lounge on right.
.4	147.0	Red siltstones and sandstones of Wasatch Formation to the right in the downthrown block of Como fault.
.1	147.1	Wasatch Formation in the road cut to right.
.2	147.3	Stop sign. Proceed straight ahead (northwest) through the outskirts of North Morgan. The city of Morgan is to the left. It is now the site of the Browning Arms Corporation which moved here in 1964 from Ogden. This is the corporation that developed the Browning Automatic Rifle (BAR) and many other firearms.
.3	147.6	The highway has just passed onto sands and gravels deposited by Pleistocene Lake Bonneville when it was at its highest or Bonneville level (Gilbert, 1890). When it was at its highest level, Lake Bonneville formed an estuary into Morgan Valley. These outcrops of sands and gravels mark the farthest eastern extent of the lake. Since Lake Bonneville subsided, streams have greatly modified the lake terraces and delta in Morgan Valley, and in some cases redeposited the sands and gravels.
2.5	150.1	Conoco Station at Stoddard Inn. Lake Bonneville sediments form the hills on either side of valley for the next several miles.
1.1	151.2	Morgan County sanitary (?) land fill at right. They say that rat hunting is outstanding at this land fill. Wasatch Range to the left ahead from about 7 to 12 o'clock. The prominent peak at 12 o'clock is Strawberry Peak, elevation 9572 ft. Weber Canyon at 11:30. Francis Peak at 10 o'clock with microwave facility on top. Note the cirques on Francis Peak.
1.6	152.8	The patches of Norwood Tuff of Tertiary age to right from 1-3 o'clock have been redeposited or otherwise modified by Lake Bonneville. Bonneville level of the lake present on the southwestern side of Durst Mountain to the right at 3-5 o'clock. In this area the level is about 80 ft above the road.
1.9	154.7	Norwood Tuff exposed in canal-cut to the left across the Weber River at 9 o'clock. Red beds of Mesozoic and Tertiary age are exposed high on the west-facing flanks of the Wasatch Range

.2	154.9	to the left.
.2	154.9	Norwood Tuff exposed to the right in hill.
1.3	156.2	**STOP NO. 9.** Mountain Green — Peterson Church. The Church and the adjacent area are on the Provo level of Lake Bonneville. The Provo level is about 300 ft below the uppermost (Bonneville) level of the lake.
.4	156.6	Descend off Provo level of Lake Bonneville.
.2	156.8	Cross Cottonwood Creek.
.3	157.1	The top of the terraces on the right and ahead from 1 to 4 o'clock are at the Provo level of Lake Bonneville.
.2	157.3	Mountain Green.
1.1	158.4	Numerous geologic problems such as landsliding have been encountered in the new housing developments to the right and ahead. The problems are principally due to the development being built on deposits of the Norwood Tuff which swells and becomes unstable when wet (Kaliser, 1972).
.6	159.0	Entrance to Gateway Tunnel of the Weber Basin Project to the left at about 10 o'clock at the foot of the Wasatch Range. The Gateway Tunnel is used to transport irrigation water from reservoirs in the Wasatch hinterland to the valley of the Great Salt Lake. The tunnel is about 3.3 miles long and extends through metamorphic rocks of the Farmington Complex of Precambrian age in the Wasatch Range.
.1	159.1	Intersection to left. Continue straight ahead.
.1	159.2	Return to Interstate and Highway 80N and continue westward.
.1	159.3	Approximate contact between Precambrian rocks of the Wasatch Mountains (to the west) and the Quaternary sediments in Morgan Valley (to the east).
.1	159.4	Entering Weber Canyon. The canyon walls are composed of quartzites, schists, and gneisses of the Farmington Complex of early Precambrian age. Patches of Lake Bonneville sediments are preserved at places along the lower walls of the canyon.
1.0	160.4	Crossing Weber River. Devil's Gate ahead at 1 o'clock.
.2	160.6	Devil's Gate to the right and old landslide to the left. The Devil's Gate was formed by the Weber River when a landslide blocked the canyon and the river was diverted to the north around the nose of the slide where it cut a meander-like channel into Precambrian strata of the

Weber Canyon looking west at Devil's Gate (D). The old highway and Weber River go through Devil's Gate. A portion of the slide that formerly blocked most of the canyon is visible just to the right of the railroad.

Farmington Complex. In this area there are potholes and abandoned river channel debris about 50 ft above the present river level. The Devil's Gate was a pronounced barrier to emigrants over which they used windlasses and ropes to lower their wagons. The Devil's Gate was also a barrier for the railroad and a large trestle was built across the structure. Recently the toe of the slide has been removed to make room for the Interstate highway—any predictions?

1.3	161.9	Entrance to Weber Generating Station of Utah Power and Light Company to left.
.2	162.1	Precambrian rocks to left and right.
.7	162.8	Mouth of Weber Canyon.
.1	162.9	Crossing approximate trace of Wasatch fault, mouth of Weber Canyon. The Wasatch fault extends from near Nephi, Utah to Brigham City, Utah, a distance of about 150 mi. It is a normal fault, downthrown to the west. The dip of the fault plane varies from 20° to 70° W but dips only about 30° in the vicinity of Ogden. The total amount of throw on the fault in the Weber Canyon area is unknown but it has been estimated to be possibly on the order of about 10,000 ft (Feth, et al., 1966). In some places the Wasatch fault consists of a series of individual faults that branch and join. Near the mouth of Weber Canyon it seems to consist of only one major fault (Cluff, et al., 1970). Some authors suggest that the Wasatch fault is actually a zone of shattering, movement, and slippage that may be a mile or more in width (Feth, et al., 1966). Scarps of the Wasatch fault and its branches are visible at many places along the remainder of the field trip route. The scarps range up to 100 ft or more in height. Springs and seeps occur at a number of places along the strike of the fault. The actual fault plane has been exhumed in North Salt Lake City gravel pits. The Wasatch fault is only one of a series of active faults that extends northward from southern Utah into Idaho and Montana. It is considered to be active on the basis of geological and seismological evidence (Cluff, et al., 1970).
.2	163.1	The highway traverses the floodplain of the Weber River. In this area the river has excavated a broad valley into the Weber Delta which was built by the ancestral Weber River when Lake Bonneville was at a high level. The Weber Delta is the largest delta formed in Lake Bonneville. It is about 12 miles wide along the front of the Wasatch Range and extends about 7 mi westward from the mouth of Weber Canyon. The floodplain of the Weber River is about 300 to 400 ft below the top of the Weber Delta. The Weber Delta in this area is composed principally of alternating beds of pink and gray silt and fine grained sand (Feth, et al., 1966).
.1	163.2	Approaching Uintah Junction and the settlement of Uintah.
.1	163.3	Prepare to turn off Interstate at the second or Hill Field Air Force Base and Salt Lake exit.
.4	163.7	Exit Interstate, enter U.S. Highway 89, and proceed south under Interstate 80N.
.5	164.2	Crossing Weber River. Gravel pits on both sides of highway are in Weber River Delta deposits. The floodplain of the Weber River within a mile and a half of the Wasatch Mountains is the principal recharge area for the important artesian aquifers that underlie the Weber Delta district.
.3	164.5	Debris from the Gateway Tunnel above the Weber Basin Job Corps Center to the left at about about 9 o'clock. The Bonneville and Provo levels of Lake Bonneville are visable on the mountain front to the left.
1.3	165.8	The highway begins to climb upward toward the top of the Weber Delta through a depression which is thought to be a small north-south-trending graben. Road cuts ahead expose fine-grained sediments of the Alpine Formation which were deposited by Lake Bonneville. Large west-facing scarp of the Wasatch fault to the left from about 8-11 o'clock. The irregular hummocky ground near the top of the north-facing slopes to the right from 1 o'clock to about 3 o'clock are the results of recent landslides.
.3	166.1	Intersection with Utah Highway 193 to Hill Field of the U.S. Air Force. Continue straight ahead. Provo level of Lake Bonneville.
.6	166.7	Crossing peat bog at the Provo level.
1.3	168.0	Panoramic view of the Great Salt Lake to the right from about 1-4 o'clock. Antelope Island at 2-3 o'clock; north end of the Oquirrh Mountains between 1 and 2 o'clock. The Bingham Canyon open pit is further south in the Oquirrh

.3	168.3	Mountains and is not visible from here.
		Excavations to the left at about 10 o'clock are the source of sand and gravel used in road building in this area. The gravel pits are in the Lake Bonneville Group between the Bonneville and Provo levels. The surface of the delta is beginning to dip downward as we near its southern edge.
.5	168.8	Intersection with road to Layton. Continue straight ahead. The Bonneville and Provo levels to the left.
.1	168.9	The stone walls on the left at about 3 o'clock were built during the 1930's to control mudflows and floods in this area.
.9	176.8	Town of Bountiful directly ahead in the amphitheater on this side of the Salt Lake salient of the Wasatch Range.
.8	170.6	Foundations of a uranium mill to the left at 9
1.2	171.8	Sea cliffs at left formed by Lake Bonneville. Panoramic view of the Great Salt Lake to the right.
.8	172.6	Intersection with Utah State Highway 106 to Kaysville and Cherry Camp Ground to right. Continue around curve on U.S. 89. Utah State University Agricultural Experimental Station to the left. The highway has just passed onto what is said to be the "largest failure by lateral spreading landslide" in the United States (Van Horn, 1975). It covers about 8 km² and is partially overlain by a younger slide of the same type which covers about 9 km². The younger landslide is less than 2000 years old and the older slide is 2,000-5,000 years old. According to Van Horn the slides "started to move when the pore water pressure of ground water saturating the deposit became great enough to overcome the internal resistance to movement of the deposit.
.4	173.0	Francis Peak (with radar domes) to left at 9 o'clock.
.8	173.8	Mouth of Farmington Canyon to the left at about 9 o'clock. Type locality of the Farmington Canyon Complex of Early Precambrian age as defined by Eardley (1939).
.3	174.1	Prepare to enter Interstate Highway 15 going south.
.6	174.7	Enter Interstate.

The front of the Wasatch Mountains near Farmington showing the Bonneville (B) and Provo (P) levels of Lake Bonneville. From Gilbert, 1890.

.1	174.8	Settlement of Farmington and Lagoon Amusement Park to left. Numerous mud-rock debris flows came out of major canyons in this area during the 1930's due to heavy spring rains and overgrazing. Since then the watersheds have been managed by the U.S. Bureau of Reclamation and U.S. Forest Service and there haven't been any further floods. The U.S. Bureau of Reclamation tried recharge experiments in this area which reportedly were successful.
3.4	178.2	All three major levels of Lake Bonneville are clearly visible to the left from about 9 to 12 o'clock. The uppermost level, the Bonneville level, is below the "V" on the mountainside at 11 o'clock, the Provo level is about halfway down the hillside and the lowest, the Stansbury level is in the vicinity of the housing developments.
2.2	180.4	Interchange. Bountiful to the left. Stay on Interstate Highway 15. Many of the towns in this area contain beautiful old pioneer homes built from bedrock.
2.0	182.4	In this area the housing developments go right up to the Bonneville level on the left. Note the brown condominiums in the Bonneville level itself at 9 o'clock.
.5	182.9	The large building with orange and brown stripes is Woods Cross High School on left. Wasatch Mountains to left from 7 to 9 o'clock. Salt Lake salient of the Wasatch Mountains from 9 to 12 o'clock. The Salt Lake salient is composed of Eocene Knight which is in fault contact with

2.4	185.3	Prepare to leave Interstate Highway 15 to right at Beck Street.
.2	185.5	Exit at Beck Street exit on to U.S. Highway 89. Lake Bonneville gravels to left along the nose of the Salt Lake salient.
1.1	186.6	Mississippian limestones in escarpment to left have been exposed in gravel pits. Pioneer limestone kiln above white rock at 9 o'clock.
.2	186.8	Go under Interstate Highway 15.
.4	187.2	Exhumed fault plane of Warm Springs segment of Wasatch fault at left in gravel pits. The fault plane dips about 70° to the west along here.
.6	187.8	Ensign Peak to left with microwave relay tower on skyline. Tuff and gravel deposits to the left. Also additional exposures of fault plane.
.7	188.5	Stay right at intersection and continue on U.S. Highway 89. Lake Bonneville levels well exposed to left. Pioneer limestone kiln at left.
.4	188.9	Wasatch Springs Plunge (now closed). The water in this spring apparently moves upward along the Warm Springs segment of the Wasatch fault. Keep right and prepare to return to Interstate Highway 15 at 6th North.
.5	189.4	Turn right (west) and return to Interstate Highway 15 on 6th North. The Bonneville and Provo levels of Lake Bonneville are visable to the right on the south side of the Salt Lake salient. Enter Interstate Highway 15 and proceed south.
5.1	194.5	Stay in left lanes. Prepare to exit Interstate Highway 15 and enter Interstate Highway 80N.
.2	194.7	Enter Interstate Highway 80N, and proceed east toward Wasatch Range.
4.1	198.8	Prepare to turn right on to Interstate Highway 215 (Belt Route).
.5	199.3	Exit from Interstate Highway 80N and enter Interstate Highway 215 (Belt Route). Proceed south on Interstate Highway 215 to 39th South.
1.6	200.9	Exit at 39th South Exit.
.1	201.0	Turn left on 39th South and go under Interstate.
.1	201.1	Intersection of 39th South and Wasatch Blvd. Olympus Hills Shopping Center to right on east side of Wasatch Blvd.

Precambrian and Paleozoic rocks in the Wasatch Mountains to the east. The extreme western end of the salient consists of folded Paleozoic, Cretaceous, and Cenozoic rocks. The Warm Springs segment of the Wasatch fault bounds the salient on the west and forms the steep westward inclined escarpment that is visible at 11 o'clock. Industrial area to right in North Salt Lake City. The complex includes several oil refineries.

End of Road Log.

REFERENCES CITED

Blackwelder, E., 1910, New light on the geology of the Wasatch Mountains, Utah: Geol. Soc. America Bull., v. 21, p. 517-542.
Cluff, L. S., Brogan, G. E., and Glass, C. E., 1970, Wasatch fault, northern portion: Woodward-Clyde and Associates, Oakland, California.
Eardley, A., 1939, Structure of the Wasatch-Great Basin region: Geol. Soc. America Bull., v. 50, p. 1277-1310.
_____ 1944, Geology of the north-central Wasatch Mountains: Geol. Soc. America Bull., v. 55, p. 819-894.
_____ 1952, Wasatch hinterland: Utah Geol. Soc., Guidebook to the geology of Utah, No. 8, Geology of the central Wasatch Mountains, p. 52-60.
Feth, J. H., Baker, D. A., Morre, L. G., Brown, R. J., and Veirs, C. E., 1966, Lake Bonneville: geology and hydrology of the Weber Delta district, Including Ogden, Utah: U.S. Geol. Survey Prof. Paper, 153, 92 p.
Gilbert, G. K., 1890, Lake Bonneville: U.S. Geol. Survey Monograph 1, 438 p.
_____, 1928, Studies of Basin-Range structure: U.S. Geol. Survey Prof. Paper 153, 92 p.
Kaliser, B. N., 1972, Geologic hazards in Morgan County with applications to planning: Utah Geol. and Mineral Survey Bull. 93, 56 p.
King, C., 1878, Systematic Geology: United States geological explorations of the fortieth parallel, v. 1, 803 p.
Mullens, T. E., and Laraway, W. H., 1964, Geology of the Devil's Slide quadrangle, Morgan and Summit Counties, Utah: U.S Geol. Survey Min. Inv. Map MF-290.
Stringham, B., and Cahoon, H. P., 1959, Ceramic red clay near Henefer Utah: Intermtn. Assoc. Petroleum Geologists Guidebook, 10th Ann. Field Conf., p. 204-206.
Van Horn, R., 1975, Largest known landslide of its type in the United States; a failure by lateral spreading in Davis County, Utah: Utah Geol., v. 2, p. 82-87.
Williams, N. C., and Madsen, J. H., Jr., 1959, Late Cretaceous stratigraphy of the Coalville Area, Utah: Intermtn. Assoc. Petroleum Geologists Guidebook, 10th Ann. Field Conf., p. 122-125.

ALTERNATE FIELD TRIP

Morgan Section, Morgan, Utah

Credits

This field trip was prepared by Peter B. Rose, Energy Reserves Group, Inc., Houston, Texas.

Going southeast on U.S. Interstate Highway 80N:

— 0.0 miles	Morgan railroad station
0.6 — 0.9	Lodgepole Limestone well exposed on left at road level as medium- to thin-bedded, cherty, black micritic limestone beds.
0.9 — 1.2	Deseret Limestone crops out to left about halfway up the hillside as prominent, thick, gray limestone ledges protruding above talus rubble.
1.2 — 1.5	Lower Humbug sandstones are in covered slope to left, but crop out to the right across the canyon as the less resistant ledges

1.5 — 1.6	below the resistant brownish upper Humbug cliffs.
	Upper Humbug limestones and sandstones in road cut at left; these beds crop out to the right across the canyon as the dipping resistant ledges high above the creek.
— 1.6	Lower Doughnut silty shales in covered gully to left.
1.6 — 1.7	Upper Doughnut limestone and black shale in road cut at left.
1.9 — 2.2	Round Valley Limestone crops out to left as low ledges in reddish soil.
2.2 — 2.6	Morgan Formation to left, exposed as prominent reddish ledges.
2.6 — 3.5	Cliffs of Weber Sandstone to left.

PEPPARD-SOUDERS
& ASSOCIATES *Petroleum Consultants*

DALLAS • DENVER • HOUSTON • MIDLAND

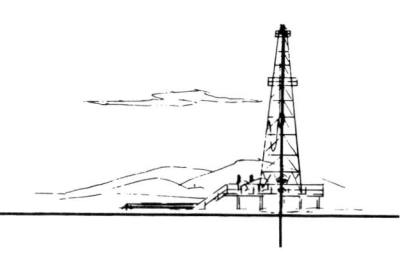

PROVIDING PROPRIETARY AND NON-EXCLUSIVE GEOLOGIC, GEOPHYSICAL, ENGINEERING AND PROPERTY MANAGEMENT SERVICES TO THE OIL AND GAS INDUSTRY WORLDWIDE

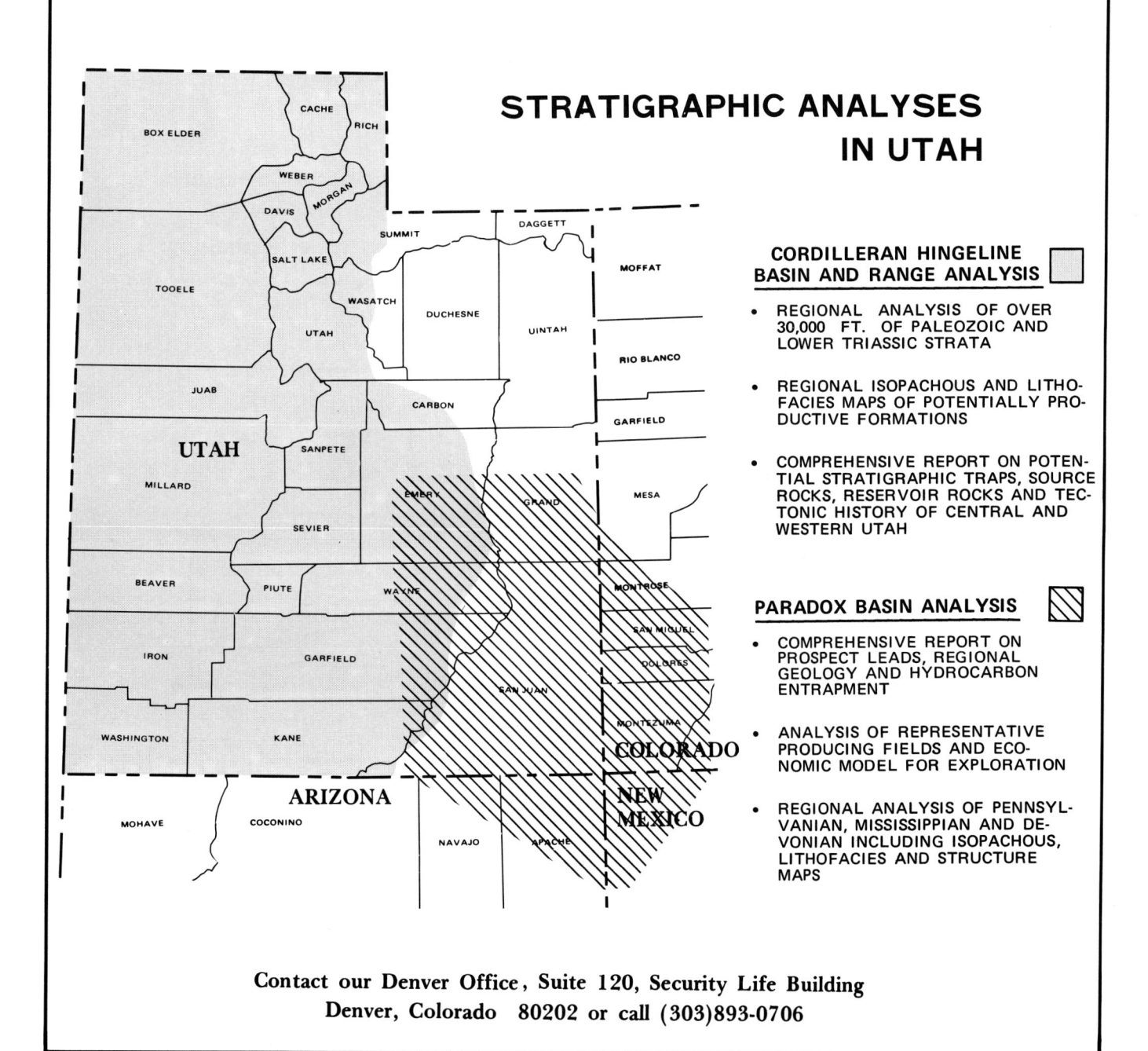

STRATIGRAPHIC ANALYSES IN UTAH

CORDILLERAN HINGELINE BASIN AND RANGE ANALYSIS

- REGIONAL ANALYSIS OF OVER 30,000 FT. OF PALEOZOIC AND LOWER TRIASSIC STRATA

- REGIONAL ISOPACHOUS AND LITHOFACIES MAPS OF POTENTIALLY PRODUCTIVE FORMATIONS

- COMPREHENSIVE REPORT ON POTENTIAL STRATIGRAPHIC TRAPS, SOURCE ROCKS, RESERVOIR ROCKS AND TECTONIC HISTORY OF CENTRAL AND WESTERN UTAH

PARADOX BASIN ANALYSIS

- COMPREHENSIVE REPORT ON PROSPECT LEADS, REGIONAL GEOLOGY AND HYDROCARBON ENTRAPMENT

- ANALYSIS OF REPRESENTATIVE PRODUCING FIELDS AND ECONOMIC MODEL FOR EXPLORATION

- REGIONAL ANALYSIS OF PENNSYLVANIAN, MISSISSIPPIAN AND DEVONIAN INCLUDING ISOPACHOUS, LITHOFACIES AND STRUCTURE MAPS

Contact our Denver Office, Suite 120, Security Life Building
Denver, Colorado 80202 or call (303)893-0706

Long Co
TECHNICAL SERVICE

Post Office Box 1028 ● Casper, Wyoming 82601
Telephone AC 307 237-2256

●

MUD LOGGING
WELL SITTING
WELL CONTROL

---

jensen/mark

822 LINCOLN TOWER BUILDING

DENVER, COLORADO 80203

(303) 861-1813

WILLARD OWENS ASSOCIATES INC

Geotechnical Consultants
Engineering Geology
Ground Water Hydrology

7391 W. 38th Ave., Wheatridge, Colo. 80033   (303) 424-5564

VEEZAY GEOSERVICE INC.
CONSULTING GEOLOGISTS

24 D BROOKS TOWERS
1020 15th STREET
DENVER, COLORADO 80202
PHONE: (303) 263-2102

NOW AVAILABLE:
1975/1976 PROJECT OVERTHRUST
  A. Comprehensive Field-Stratigraphic Study
  B. Detailed Structural Analysis

NOW IN PROGRESS:
1976 UTAH HINGE LINE PROGRAM
  Comprehensive Field-Stratigraphic Analysis

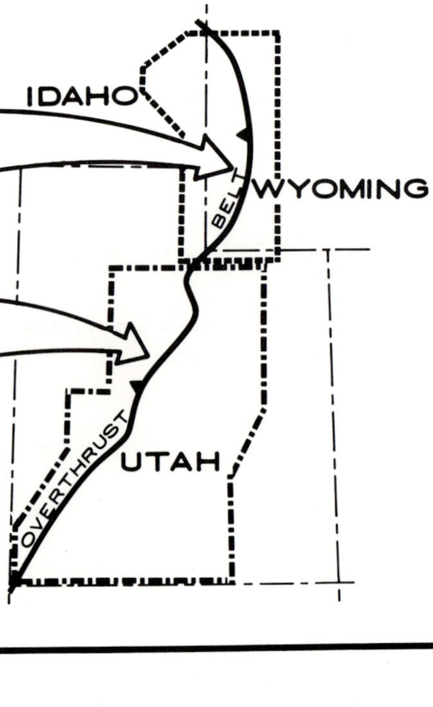

Affiliated with: V. ZAY SMITH ASSOCIATES LTD.
and
VEEZAY GEODATA LTD.

1143-17th Ave. S.W. CALGARY, ALBERTA.
T2T 0B6   PHONE: (403) 244-5551.

Utah Hingeline seismic data available

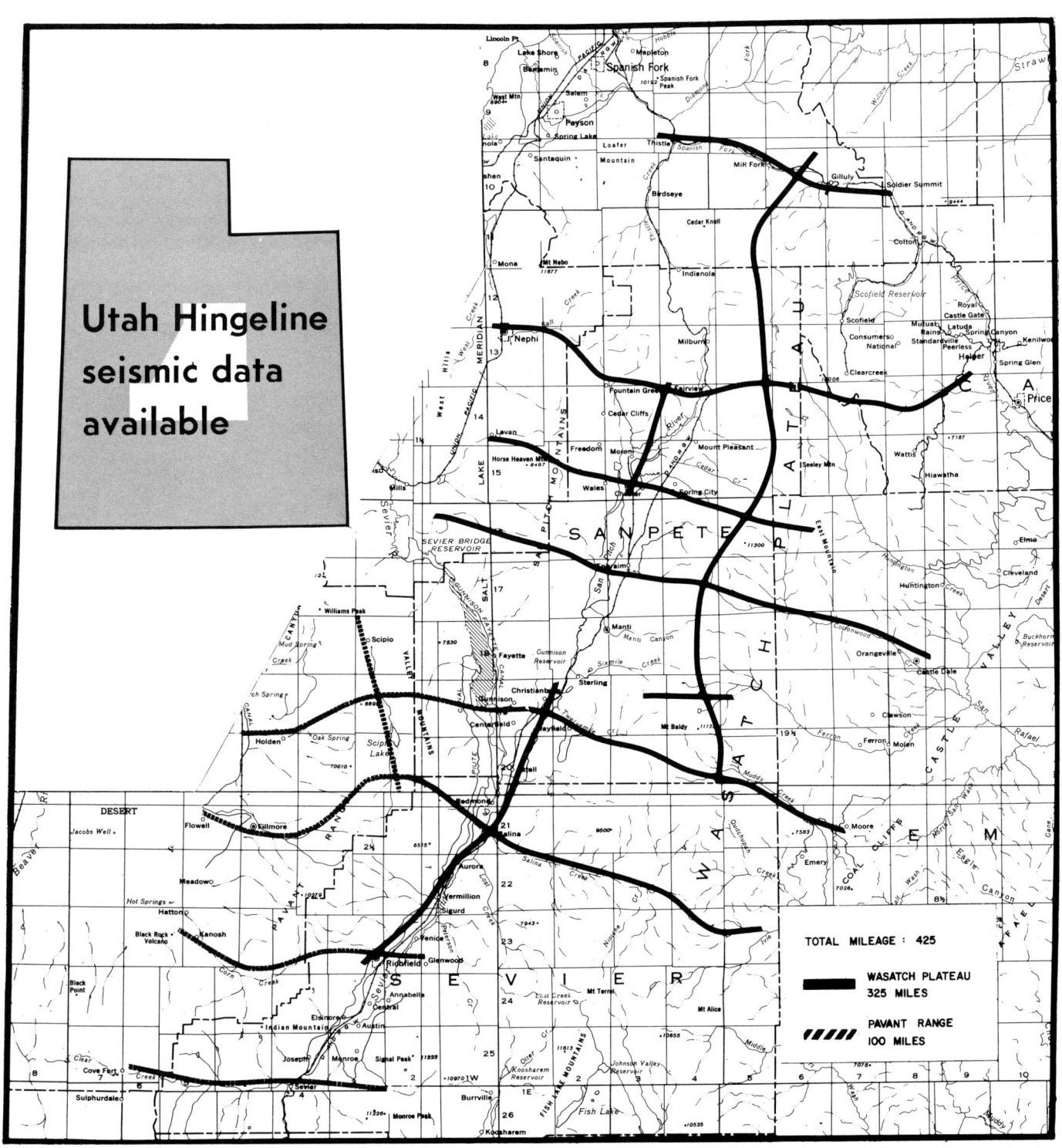

Seismic Coverage — Hingeline Area — Utah   Wasatch Plateau Survey   Pavant Range Survey

TOTAL MILEAGE : 425
WASATCH PLATEAU 325 MILES
PAVANT RANGE 100 MILES

COMPAGNIE GENERALE de GEOPHYSIQUE
One Park Central
Suite 1255
1515 Arapahoe Street
Denver, Colorado 80202
(303) 571-1143

I. Information
   A. Source-Vibroseis
   B. Fold 12 and 24
   C. All data gathered in 1976
   D. Data is available **NOW**

II. Contact
   A. Floyd Wilson or Pierre Benichou
      Denver (303) 571-1143
   B. Charlie Smith
      Houston (713) 781-9790

RAINBOW RESOURCES INC.

PH. (303) 458-5663
10 LAKESIDE LANE, SUITE 308
DENVER, COLORADO 80212

PH. (307) 265-1090
305 GOODSTEIN BUILDING
CASPER, WYOMING 82601

Tooke Engineering

Complete Hydrocarbon Logging Service
Automatic Chromatography Manufacturing

P.O. BOX 3200
CASPER, WYOMING 82601

307-265-1630 CASPER
303-573-5824 DENVER

Signal Drilling Company Inc.

F. M. STEVENSON
CHAIRMAN OF THE BOARD

PHONE 629-1050
1200 SECURITY LIFE BUILDING
DENVER, COLORADO 80202

EDWARDS OIL PROPERTIES, INC.

STAN EDWARDS
PRESIDENT
BUS. PHONE 303-761-2201
RES. PHONE 303-794-8428

SUITE 300
770 W. HAMPDEN
ENGLEWOOD, COLO. 80110
P. O. BOX 1214
ENGLEWOOD, COLO. 80110

MOUNTAIN FUEL Supply Company

An investor-owned, fully integrated natural gas utility. Engaged in exploration, production, transmission and distribution.

Exploration Headquarters:
180 East 1st South, Salt Lake City, Utah 84139
Telephone (801) 534-5555

Division Exploration Office:
Suite 820 Republic Bldg., Denver, Colorado 80202
Telephone (303) 573-0212

Division Exploration Office:
P.O. Box 2329, Farmington, New Mexico 87401
Telephone (505) 327-4024

W.A. WAHLER & ASSOCIATES

Founded in 1960

P.O. Box 10023
PALO ALTO, CALIFORNIA 94303
(415) 968 6250

ENGINEERS & CONSULTANTS
Specialists in

EARTH & ROCK FILL DAMS
TAILING & REFUSE EMBANKMENTS
UNDERGROUND EXCAVATIONS
OPEN PIT SLOPES
GROUNDWATER
INSTRUMENTATION

Impulsive seismic data...

Wherever the prospect, you can depend on SSC for the seismic method best suited to your data acquisition needs. When impulsive techniques are called for, we offer a wide selection of energy sources . . . plus equipment to fit the terrain . . . trucks, tractors, marsh buggies and portable equipment, to name a few.

To couple the right equipment with the right people and field techniques, as well as complete data processing capabilities, contact the geophysical problem solvers at our office nearest you.

Seismograph Service Corporation, P.O. Box 1590, Tulsa, Okla. 74102.
(918) 627-3330, Telex: 497428.

Seismograph Service Corporation
A SUBSIDIARY OF RAYTHEON COMPANY

Seismograph Service Limited Compagnie Française de Prospection Sismique

...we can get it.

NORTH AMERICAN EXPLORATION CO., INC.

Geophysical Consultants

Geophysical Interpretations

Agents for seismic data sales
Western U.S. and Alaska

936 East 18th Ave., Denver, Colo. 80218
(303) 534-4191   Tx 45612

1625 So. Boston Ave., Tulsa, Okla. 74119
(918) 587-0069   Tx 497585

SPEEDY COPY SERVICE, Inc.

FAST, DEPENDABLE, CONFIDENTIAL SERVICE

—: PHONE: 222-9746 or 222-5378 :—

2014 Welton Street   Denver, Colo. 80205

LADD

LADD PETROLEUM CORPORATION
Suite 830 • Denver Club Building
Denver, Colorado 80202
(303) 292-3080

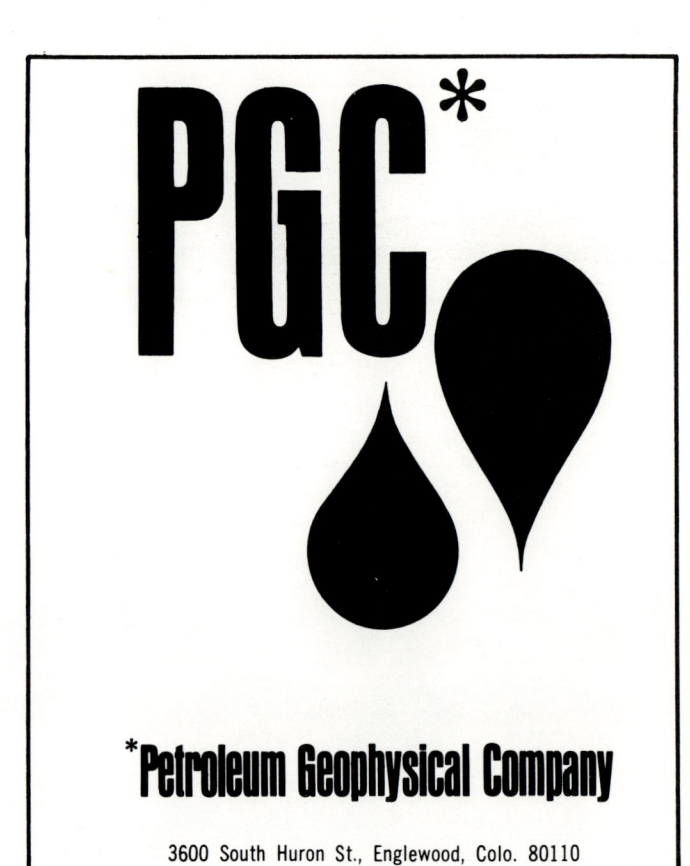

A Subsidiary of Utah International Inc.

Walter Duncan OIL PROPERTIES

1300 WRITERS CENTER IV
1720 SO. BELLAIRE ST.
DENVER, COLORADO 80222
PHONE 759-3303

BILLIE J. McALPINE
PETROLEUM GEOLOGIST

224 PATTERSON BUILDING   OFF. 303: 572-8916
DENVER, COLORADO 80202

PGC*

*Petroleum Geophysical Company

3600 South Huron St., Englewood, Colo. 80110
Area Code 303-761-3541

414

Φ Energy Reserves Group

We have changed the name of Clinton Oil Company to more accurately reflect our new direction and management.

Energy Reserves Group represents our expansion into coal and uranium from our traditional base of oil and gas.

Energy Reserves Group, Inc. Northern Division Exploration and Reservoir Sections	First of Denver Plaza - Suite 3200 Denver, Colorado 80202 303 - 572-3323

American Mud Co.
Div. of PETRO CHEM. INC.

GENERAL OFFICES
744 Metro Bank Bldg., Denver, CO
PHONE: 303/893-0728

SERVING THE ROCKY MOUNTAIN AREA
WITH 24-HOUR DRILLING MUD SERVICE

NORTHERN ROCKY MOUNTAIN AREA
Duane Peterson, Area Manager
Williston, North Dakota
Phone: (701) 572-8000 (24 hr.)

WEST CENTRAL ROCKY MOUNTAIN AREA
Keith Stetson, Area Manager
Casper, Wyoming
Phone: (307) 237-7856 (24 hr.)

SOUTHERN ROCKY MOUNTAIN AREA
Ron Mahan, Area Manager
Farmington, New Mexico
Phone: (505) 327-2525 (24 hr.)

MINERALS DIVISION
Harold Bittle, Manager
Grand Junction, Colorado
Phone: (303) 242-3072 (24 hr.)

Energetics Inc.

333 WEST HAMPDEN AVE. • SUITE 1010
ENGLEWOOD, COLORADO 80110
PHONE 303 • 761 • 4541

Diamond Shamrock

J. R. BOSHARD
CONSULTING GEOLOGIST

OFFICE
(303) 534-2984
HOME 771-1463

THE DENVER CENTER BUILDING
1776 LINCOLN STREET, SUITE 918
DENVER, COLORADO 80203

DEPCO, Inc.
1025 Petroleum Club Bldg.
DENVER, COLORADO 80202
303-292-0980

Production   Exploration

Birdwell serves
THE OIL & GAS INDUSTRY
and the Mining and Construction Industries

with wireline services including:

- Resistivity, nuclear, acoustic and auxiliary logging
- Seismic velocity surveys
- Perforating services
- Pipe recovery services

Contact us in Denver at 1600 Broadway, Tel. (303) 861-9340; in Grand Junction at 2466 Industrial Blvd., Tel. (303) 245-1610; and in Lafayette at 106 S. Public Rd., Tel. (303) 665-3320.

BIRDWELL DIVISION
Seismograph Service Corporation
A SUBSIDIARY OF RAYTHEON COMPANY

Arapaho Petroleum Incorporated

Arapaho's Aim: Exploration

Arapaho Petroleum, Inc.
Rocky Mtn. Exploration Division
1700 Security Life Building
Denver, Colorado 80202

Gear Drilling Company

623-4422
470 Denver Club Bldg.
Denver

Ashland

Ashland Exploration Company
DIVISION OF ASHLAND OIL, INC.

2500, THE FIRST OF DENVER PLAZA, 633 SEVENTEENTH ST.
DENVER, COLORADO 80202•(303) 534-2338

ROCKY MOUNTAIN REGION

CASPER   BRIGHTON   VERNAL   GLENDIVE

oil well perforators
inc.

KIMBALL   ROCK SPRINGS   GILLETTE

RES: 2275 QUAIL DRIVE
LAKEWOOD, COLORADO 80215
(303) 237-1424

Earl S. Griffith
GEOLOGIST
INTERNATIONAL PETROLEUM EXPLORATION

438 GUARANTY BANK BLDG. 817 SEVENTEENTH ST. DENVER, COLO. 80202
(303) 893-5134

Toltek Drilling Company

340 DENVER CLUB BUILDING
DENVER, COLORADO 80202

R. D. "DICK" GASCH

OFFICE 255-5255
RESIDENCE 771-9630

Amoco Production Company

AMOCO

Subsidiary of Standard Oil (Indiana)
Sixth Largest Oil Company

An Equal Opportunity Employer

Monaco engineering, inc.

PROFESSIONAL SERVICES

WELLSITE
GEOLOGICAL SUPERVISION

HYDROCARBON
WELL LOGGING

DRILLING AND
COMPLETION ENGINEERING

BILL PUGH
DENVER, COLORADO
(303) 892-1887

MARSHALL TILLEY
LAMAR, COLORADO
(303) 336-5740

BILL SMALL
SALT LAKE CITY, UTAH
(801) 772-5165

KIMBARK
SUITE 808 LINCOLN TOWER BUILDING
1860 LINCOLN STREET
DENVER, COLORADO 80203
303-534-7161

WALTER K. ARBUCKLE
GEORGE WALLACE BAYNE
WILLIAM R. THURSTON

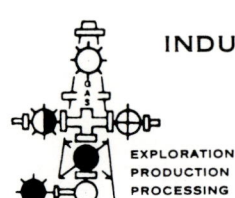

INDUSTRIAL GAS SERVICES, INC.

EXPLORATION
PRODUCTION
PROCESSING
TRANSMISSION

4501 WADSWORTH BOULEVARD
DENVER, COLORADO 80033

TELEPHONE (303) 422-3400

C.I.G. EXPLORATION, INC.
GAS PRODUCING ENTERPRISES, INC.
COLORADO OIL COMPANY, INC.
(Units of Coastal States Gas Corporation)

2100 PRUDENTIAL PLAZA
BOX 749   DENVER, COLORADO 80201
(303) 572-1121

GEOPHOTO SERVICES
TEXAS INSTRUMENTS
INCORPORATED

9725 E. Hampden Ave., Suite 301
Denver, Colorado 80231
303/751-1780

THE MOST COMPLETE ROCKY MOUNTAIN COVERAGE

- PHOTOGEOLOGY
- GEOMORPHOLOGY
- PLATE TECTONICS

JERRY CHAMBERS
OIL PRODUCER

1660 Lincoln Street, Denver, Colorado 80203   303/892-0464

"When competence counts"™

Milchem®

Milchem Incorporated
Houston, Texas
A Baker International Corporation company.

TGA INC.

SPECIALIZING IN COMPREHENSIVE, DETAILED
PHOTOGEOLOGIC-GEOMORPHIC ANALYSIS

NEW NON-EXCLUSIVE STUDIES AVAILABLE or in progress in.......
- MONTANA - WYOMING - N. DAKOTA - S. DAKOTA - IDAHO - UTAH - COLORADO - ARIZONA - NEW MEXICO - KANSAS - OKLAHOMA - TEXAS - MISSOURI - ARKANSAS - LOUISIANA - ILLINOIS - INDIANA - MICHIGAN - OHIO - WEST VIRGINIA - KENTUCKY - TENNESSEE - MISSISSIPPI - ALABAMA & FLORIDA

INCLUDING..... NEW DETAILED RC-9 MAPPING, COMPLETED OR IN PROGRESS, IN THE........
- BIG HORN BASIN
- RED DESERT BASIN
- WASHAKIE BASIN
- WIND RIVER BASIN
- POWDER RIVER BASIN
- IDAHO-WYOMING THRUST BELT

TROLLINGER GEOLOGICAL ASSOCIATES, INC.

TGA Bldg. / 2150 S. Bellaire St.
Denver, Colo. 80222 / Phone (303) 757-7141

PETROLEUM, INC.

500 COLORADO STATE BANK BUILDING
DENVER, COLORADO 80202

PHONE 893-9921
AREA CODE 303

Chaparral Resources, Inc.

444 Seventeenth Street   Denver, Colorado 80202
Telephone (303) 825-2505

573-3885

EDWARD J. ACKMAN
PETROLEUM GEOLOGIST
SUITE 516 DENVER CENTER BUILDING
1776 LINCOLN STREET
DENVER, COLORADO 80203

Petroleum Information

INFORMATION SERVICES

Drilling and production reports

Statistical and geological publications

Map and log services

Completion cards

Petroleum Information Exchange Ltd.

Well Record Service

Well History Control System

ENERGY SERVICES

Petroleum consulting

Technical exploration services

Engineering

Custom production reports

Digital Log & Map Service

Petroleum Information
CORPORATION
A Subsidiary of A.C. Nielsen Company

Box 5040 TA, Denver, CO 80217

Box 1702
Houston, TX 77001
2600 Southwest Freeway
713/526-1381

7200 Stemmons Freeway
Suite 1315
Dallas, TX 75247
214/630-6011

DISTRIBUTORS
In the Rocky Mountain Area
Dupont Explosives - Blasting Supplies

DUPONT

BUCKLEY POWDER CO.
4701 JACKSON STREET
DENVER, COLORADO 80216
TELEPHONE 303-333-4235

---

filon
Exploration Corporation

1700 BROADWAY — SUITE 2216
DENVER, COLORADO 80202

TWX: 9109312282 FILON DVR (303) 892-7055

EXPLORATION

EQUITY OIL COMPANY

EXECUTIVE OFFICE
10 West Third South, Suite 806 / Salt Lake City, Utah 84101

EXPLORATION OFFICE
2220 Western Federal Savings Building / Denver, Colorado 80202

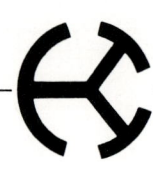

---

webb

webb resources, inc.
first of denver plaza
suite 2200 / 633 17th street
denver, colorado 80202
(303) 892-5504

Basic Core Analysis
Advanced Core Analysis Studies
Hydrocarbon Well Logging
Reservoir Fluid Analysis
Source-Bed Evaluation
Research and Development
Engineering and Consulting
Petroleum Software and Services
Environmental Services

CORE LAB

CORE LABORATORIES, INC.
Serving the Petroleum Industry Since 1936

The
GRAYROCK
Corporation

1600 Broadway
Suite 2300
Denver, Colorado 80202
(303) 534-3267

EDCON

PHONE (303) 989-1550
TWX 910-320-2912

EXPLORATION DATA CONSULTANTS, INCORPORATED
345 S. UNION BLVD. · SUITE 212 · DENVER, COLORADO 80228

Integrated Geophysical Interpretation
and Gravity Surveys in the Overthrust Belt Region

DOUGLAS J. GUION, Geophysicist

American Stratigraphic Company

DENVER, COLORADO
6280 E. 39TH AVE.

CASPER, WYOMING
524 EAST YELLOWSTONE

BILLINGS, MONTANA
17 NORTH 31ST STREET

ANCHORAGE, ALASKA
BOX 2127

CANADIAN STRATIGRAPHIC SERVICE, LTD.
705 11TH AVENUE S.W.
CALGARY, ALBERTA

STRATIGRAPHIC LOGS   DIGITIZED LITHOLOGY

SUBSURFACE STUDIES   GEOLOGICAL CONSULTING

WELL-SITE CONSULTING   SAMPLE LIBRARIES

SERVING THE OIL AND MINING INDUSTRY — WORLDWIDE

TransOcean Oil, Inc.

1700 First City East Building
1111 Fannin, Houston, Texas 77002.

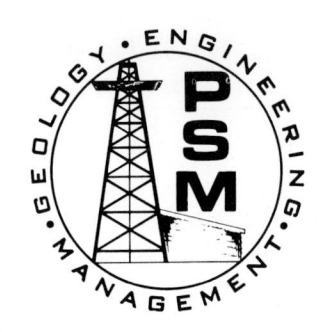

GEOLOGY · ENGINEERING · MANAGEMENT
PSM

GEOLOGISTS;
R. L. Wagner
J. J. Haverfield

ENGINEERS;
E. G. Toombs
J. O. Cox
J. E. Vaughn

PETROLEUM SUPERVISION AND MANAGEMENT, INC.

SUITE 230, WESTERN RESOURCES BLDG.
CASPER, WYOMING 82601
PH: 307-266-1276

PENNZOIL COMPANY

ROCKY MOUNTAIN DIVISION

COLORADO STATE BANK BUILDING
P. O. DRAWER 1139 • DENVER, COLORADO 80201
PHONE 303/892-7070

MOUNTAIN PETROLEUM LTD.

712 DENVER CENTER
1776 LINCOLN STREET
DENVER, COLORADO 80203

GENERAL PARTNERS
John P. Lockridge
Clyde E. Thompson

---

richard b. ross
consulting petroleum geologist

2880 south locust, apt. 204 - north
denver, colorado   80222

303 - 758 - 8979

Dresser Atlas
WIRELINE SERVICES

DRESSER ATLAS DIVISION
DRESSER INDUSTRIES, INC.

475 - 17TH ST.
SUITE 1532
DENVER, COLO. 80202
629-0294

Pacific West Exploration Co.

DENVER, COLORADO

TELE-(303) 758-3712

Seismic Surveys

For

Oil & Gas

Minerals

Engineering Projects

♦ ♦ ♦ ♦

Utilizing Modern Digital
Recording Instrumentation
With Specialized Field Techniques
And Data Processing

Hydrocarbon Well Logging

Manned and Portable Logging Units -
Formation evaluation, penetration
rate, lithology, mud gas and
cuttings gas analysis, plus other mud
monitoring and rig monitoring services.
H_2S, CO_2 and Helium detection available.
Quality, professional service.

Offices Located At:

1744 Mullowney Lane
Billings, Montana 59102
(406) 252-8424

P.O. Box 1270
Mills, Wyoming 82644
(307) 234-1314

4610 - 12th Street N.E.
Calgary, Alberta T2E 4R4
(403) 276-7966

220 Patterson Building
Denver, Colorado 80202
(303) 892-0202

Continental Laboratories

Haskins · Pfeiffer · Owings, Inc.

International Geophysical Consultants

1449 Denver Club Building
Denver, Colorado 80202
(303) 573-8135

BASS ENTERPRISES PRODUCTION CO.

AND

PERRY R. BASS

✦ ✦ ✦

Oil and Gas Producing
and Exploration Cos.
Domestic and Foreign

✦ ✦ ✦

3100 Fort Worth National Bank Bldg.

Fort Worth, TX 76102

Anderson Drilling Co.

A SUBSIDIARY OF ANDERSON OIL CO.

DENVER — 303-623-5600

CASPER — 307-265-5102

Woods Petroleum Corporation

Howard E. Marlow
Manager, Rocky Mountain Division
3525 South Tamarac Drive, Suite 130
Denver, Colorado 80237
303-771-3732

Robert F. Dewey
Senior Exploration Geologist
3555 N.W. 58, Suite 500
Oklahoma City, Oklahoma 73112
405-947-7811

Ronald E. Hando
Geologist
3525 South Tamarac Drive, Suite 130
Denver, Colorado 80237
303-771-3732

TERRA RESOURCES, INC.

ROCKY MOUNTAIN EXPLORATION DISTRICT OFFICE
900 SECURITY LIFE BUILDING
DENVER, COLORADO 80202

ENERGY CONSULTING ASSOCIATES

ENHANCED OIL RECOVERY
RESOURCE EVALUATION & MANAGEMENT
FEASIBILITY STUDIES
WASTE DISPOSAL
PROFIT ANALYSIS

DWAYNE A. CHESNUT, PRESIDENT
817-17th STREET, SUITE 602
DENVER, COLORADO 80202
(303) 623-8706

PHONE: OFFICE (307) 234-1953
RESIDENCE (307) 234-1686

WILLIAM H. DUNLAP
Consulting Geologist

209 C.B.C. Building
254 North Center   Casper, Wyoming 82601

A. THOMAS GRAHAM, JR.
Consulting Geologist

930 Hillside Drive   Off: 307-875-5151
Green River, Wyo. 82935   Mobile J.S. 382-0452

Give your meetings a big lift.

If you're planning a business meeting, party or convention, call us.

Besides fresh air, pines, rushing water and aerial tram rides, Snowbird has complete convention facilities.

We accommodate groups of four to 450 for your most refreshing group meeting ever.

Call 521-6040 snowbird

BAYLES LABORATORIES
CASPER, WYOMING

Complete Oil Well Logging Services
32 Valley Drive
Casper, Wyoming 82601
307-235-5356   307-265-1978

ED BAYLES

BEREN CORPORATION

Oil and Gas Producers

Western Division
2160 First of Denver Plaza
633 17th Street
Denver, Colorado 80202
(303) 892-6541

Mid-Continent Division
970 Fourth Financial Center
Wichita, Kansas 67202
(316) 265-3311

CODE 303
893-5057

ROBERT ZINKE
PETROLEUM EXPLORATION

SUITE 732
REPUBLIC BUILDING
1612 TREMONT
DENVER, COLORADO 80202

Burdette A. Ogle
A. Saterdal
E. G. Griffith

ARGONAUT OIL AND GAS COMPANY
438 guaranty bank bldg.
817 seventeenth street
denver · colorado · 80202
telephone (303) 893·5134

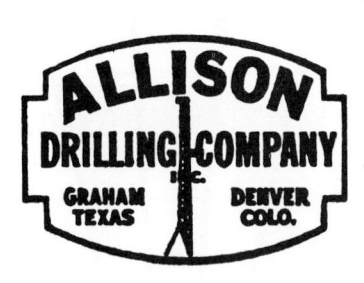

ALLISON DRILLING COMPANY
GRAHAM, TEXAS — DENVER, COLO.

BILL ALLISON — PRESIDENT

202 ALLISON OIL BLDG.
1275 Sherman
Denver, Colo.

Phone 825-2355

SOUTHLAND ROYALTY COMPANY
ESTABLISHED 1924

1010 Denver Club Building
518 17th Street
Denver, Colorado 80202
Telephone: (303) 534-0286

Corporate Office:
1600 First National Building
Fort Worth, Texas 76102
Telephone: (817) 336-9801

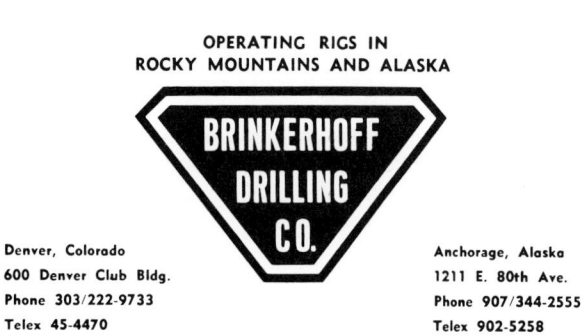

OPERATING RIGS IN
ROCKY MOUNTAINS AND ALASKA

BRINKERHOFF DRILLING CO.

Denver, Colorado
600 Denver Club Bldg.
Phone 303/222-9733
Telex 45-4470

Anchorage, Alaska
1211 E. 80th Ave.
Phone 907/344-2555
Telex 902-5258

Helton Engineering & Geological Services, Inc.
CONSULTING PETROLEUM ENGINEERS AND GEOLOGISTS

Drilling, Completion, Workover Programs, & Supervision
Well Site Geology
Sub-Surface & Surface Studies
Prospect & Property Evaluation
Reserve Appraisals
Unitization Studies & Representation
Equipment Design & Installation
Secondary Recovery Programs & Facility Design

301 Fratt Building
Billings, Montana 59101
Tel. 406-248-3101

616 Metropolitan Bldg.
1612 Court Place
Denver, Colo. 80202
Tel. 303-571-1026
Telex Interdrill 4-5648

625 Fourth Ave. SW
Calgary, Alberta T2P OK2
Tel. 403-263-0894
Telex Interdril 24827

Affiliated with Heisler-Helton International Ltd.

SKELLY

SKELLY OIL COMPANY
DENVER DISTRICT
EXPLORATION & PRODUCTION DEPT

1860 LINCOLN STREET
DENVER, COLORADO 80203
(303) 292-3660

Chevron Oil Company

Denver, Colorado

EXETER EXPLORATION COMPANY

EXPLORATION

&

PRODUCTION

Phone (303) 623-5141
2300 Lincoln Center Building
Denver, Colorado
80203

DAVIS OIL COMPANY

1100 METROBANK BLDG.
475 17TH STREET
DENVER, COLORADO 80202
TELEPHONE 623-1000

NEW YORK
NEW ORLEANS
HOUSTON
TULSA

THE LOUISIANA LAND AND EXPLORATION COMPANY

DENVER DIVISION

ROCKY MOUNTAINS—WEST TEXAS—ALASKA
SUITE 1500 • DENVER CLUB BUILDING
518 17TH STREET
DENVER, COLORADO 80202
PHONE 303-623-5759

Four of Denver's finest Restaurants

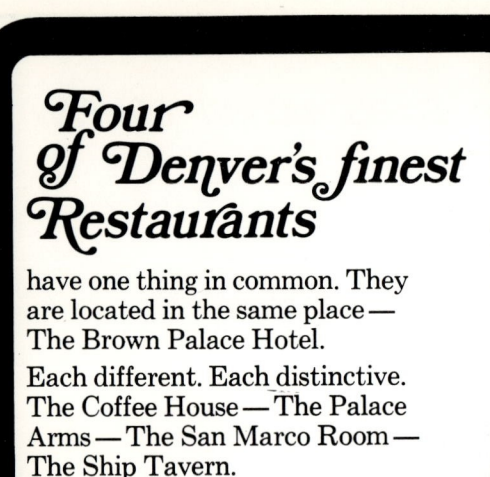

have one thing in common. They are located in the same place — The Brown Palace Hotel.

Each different. Each distinctive. The Coffee House — The Palace Arms — The San Marco Room — The Ship Tavern.

The Brown Palace Hotel
Denver, Colorado 80202
Karl W. Mehlmann, General Manager
Reservations: Mrs. Iazzetta
(303) 825-3111

Denver's *Preferred* Hotel
Major Credit Cards Accepted

PHILLIPS 66

The Performance Company

PHILLIPS PETROLEUM COMPANY
EXPLORATION AND PRODUCTION
WESTERN REGION
SECURITY LIFE BUILDING
1616 GLENARM PLACE
DENVER, COLORADO

TELEPHONE: 303-573-6611

Exploring For Gas And Oil In The Rocky Mountains

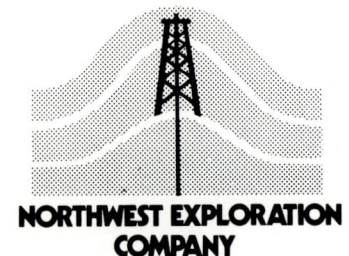

NORTHWEST EXPLORATION COMPANY

Operating Headquarters:
One Park Central, Denver, Colorado 80202
Tel: (303) 623-9303

WESTERN WELL LOGGING, INC.

Exploration borehole geophysical surveys for coal, uranium and other minerals.

- Gamma
- Density
- SP
- Resistance
- Drift
- Caliper
- Temperature

For more information call:

(303) 423-7823

7391 W. 38th AVE. • WHEAT RIDGE, COLORADO 80033

qp

825-3368

PEERLESS PRINTING

1989 BROADWAY DENVER, COLORADO 80202

CITIES SERVICE COMPANY
WESTERN REGION

EXPLORATION

AND

PRODUCTION

SUITE 900 · 1600 BROADWAY

DENVER, COLORADO · 80202

(303) 892-0263